FROM NEURON TO BRAIN

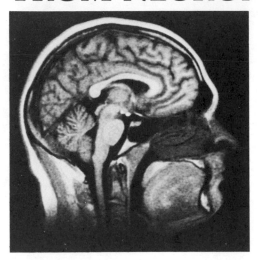

A
Cellular Approach
to the Function of
the Nervous System

SECOND EDITION

STEPHEN W. KUFFLER
Late,
Harvard Medical School

JOHN G. NICHOLLS
Biocenter,
Basel University

A. ROBERT MARTIN
University of Colorado
School of Medicine

Sinauer Associates Inc.
Publishers
Sunderland, Massachusetts

We wish to thank our colleagues who kindly provided original illustrations from published and unpublished work. We also thank the editors of the Journal of Physiology and the Journal of Neurophysiology, from which many of the illustrations were taken.

Cover design by Laszlo Meszoly

FROM NEURON TO BRAIN:
A Cellular Approach to the
Function of the Nervous System

Second Edition

Library of Congress Cataloging in Publication Data

Kuffler, Stephen W.
    From neuron to brain.

    Bibliography: p.
    Includes index.
    1. Neurophysiology.   2. Brain.   3. Neurons.
I. Nicholls, John G.   II. Martin, A. Robert, 1928-
III. Title.   [DNLM: 1. Nervous system—Physiology.
WL 102 K945f]
QP355.2.K83   1984   591.1'88   83-27138
ISBN 0-87893-444-8

Manufactured in U.S.A.

10   9   8   7   6   5   4   3

*This book is dedicated to
our friend and colleague,
Steve Kuffler.*

CONTENTS

PART THREE
THE SPECIAL ENVIRONMENT OF NERVE CELLS IN THE BRAIN FOR SIGNALING   321

PART FOUR HOW NERVE CELLS TRANSFORM INFORMATION 377

PREFACE TO
THE SECOND
EDITION

The aims of this new edition remain similar to those outlined in the original preface: "to describe how nerve cells go about their business of transmitting signals, how these signals are put together, and how out of this integration higher functions emerge. This book is directed to the reader who is curious about the workings of the nervous system but does not necessarily have a specialized background in biological sciences." Again, as in the first edition, we have chosen to present examples that lend themselves to a narrative description and with which we have some first-hand experience. The scope of the book has, however, been broadened. One entirely new chapter on the control of movement, somatic sensation, and pain has been added. In addition, to help readers unfamiliar with the structure of the mammalian brain, Laszlo Meszoly has drawn for us a new section on neuroanatomy; this appears in the form of a short appendix. Another new feature is the greater use of "Boxes." These provide short, self-contained descriptions of important topics that could detract from the flow of the argument if included in the text. Into this category fall derivations of equations, descriptions of techniques, and points of interest related to but outside the main thrust of the chapter.

In addition to bringing all the chapters up to date, we have drastically rewritten most of them. An appreciation of how much new material is now available and how many new concepts have developed in the last few years has been vividly impressed on us as we set out to face each chapter. This is apparent from a cursory glance at the table of contents. Consider just a few examples: the microcircuitry and the laminar structure of the visual cortex, patch clamp analysis of single channel currents, peptide transmitters and neuromodulators, demyelination and remyelination, the development and application of monoclonal antibodies, descending control of pain, the role of the basal lamina in regeneration, long-term changes in the *Aplysia* nervous system, retraction of geniculate axons in the neonatal cortex—all these represent just a few of the problems in which major experimental progress has been made since the last edition. Inevitably the book is longer than it was. What we have tried to do, however, is to retain the flavor of the original.

The pleasure and satisfaction that we might hope to feel in recreating a book that has seemed to fill a need has been diminished by the death of our friend and colleague, Steve Kuffler. We have tried to produce a book he would not have minded keeping his name on.

J. G. N.
A. R. M.

ACKNOWLEDGMENTS

We are grateful to the numerous colleagues who have encouraged us and influenced our thinking. We particularly thank Drs. Marla Luskin, Ken Muller, and Wesley Thompson, who critically read the whole work for us, and Drs. D. Baylor, W. Knox Chandler, R. Davis, I. Dietzel, P. Drapeau, R. Hoy, D. Kuffler, R. Levinson, U. J. McMahan, L. Mendell, I. and H. Parnas, J. P. Quilliam, R. Rahamimoff, B. Salzberg, C. Shatz, E. Shooter, S. Feinstein, H. Vánegas, B. Wallace, and T. Wiesel, who discussed portions of the book with us.

Original plates for this edition were kindly provided by Drs. A. Aguayo, D. Baylor, A. Grinvald, D. Hubel, A. Kaneko, S. LeVay, U. J. McMahan, K. Muller, S. Palay, C. Shatz, E. Weber, D. Weisblat, and T. Woolsey.

We also wish to thank Ms. Diane Honnecke, Ms. Dorothy Scally, and Ms. Sandy Lewis for their unfailing help with the manuscript and Ms. Jane Choi for her dedicated work and essential contribution at a critical stage.

To Laszlo Meszoly, who has done the artwork, Joseph Vesely, who has handled production matters, and Andy Sinauer, our editor, we owe special thanks not only for their skill, insight, and taste but for making the collaboration such a pleasure.

PREFACE TO
THE FIRST EDITION

Our aim is to describe how nerve cells go about their business of transmitting signals, how these signals are put together, and how out of this integration higher functions emerge. This book is directed to the reader who is curious about the workings of the nervous system but does not necessarily have a specialized background in biological sciences. We illustrate the main points by selected examples, preferably from work in which we have first-hand experience. This approach introduces an obvious personal bias and certain omissions.

We do not attempt a comprehensive treatment of the nervous system, complete with references and background material. Rather, we prefer a personal and therefore restricted point of view, presenting some of the advances of the past few decades by following the thread of development as it has unraveled in the hands of a relatively small number of workers. For example, in Part One (Neural Organization for Perception) we emphasize the approach used by Hubel and Wiesel, which we were fortunate to witness step by step in laboratories next to our own. Similarly, Part Two (Mechanisms for Neuronal Signaling) leans heavily on the work of Hodgkin, Huxley, Katz, Miledi, and their colleagues, and omits comprehensive treatment of many other aspects. A survey of the table of contents reveals that many essential and fascinating fields have been left out: subjects like the cerebellum, the auditory system, eye movements, motor systems, and the corpus callosum, to name a few. Our only excuse is that it seems preferable to provide a coherent picture by selecting a few related topics to illustrate the usefulness of a cellular approach.

We describe the more complex functions first, because the visual systems of the cat and the monkey lend themselves well to an initial presentation of the neuronal events that are clearly correlated with such higher functions as perception. This approach puts in perspective the subsequent discussion in Parts Two and Three of the cellular machinery that is used to bring about the brain's more complex activity. Throughout, we describe experiments on single cells or analyses of simple assemblies of neurons in a wide range of species. In several instances the analysis has now reached the molecular level, an advance that enables one to discuss some of the functional properties of nerve and muscle membranes in terms of specific molecules.

Fortunately, in the brains of all animals that have been studied there is apparent a uniformity of principles for neurological signaling. Therefore, with luck, examples from a lobster or a leech will have relevance for our own nervous systems. As physiologists we must pursue that luck, because we are convinced that behind each problem that appears extraordinarily complex and insoluble there lies a simplifying principle that will lead to an unraveling of the events. For example, the human brain consists of over 10,000 million cells and many more connections that in their detail appear to defy comprehension. Such complexity is at times mistaken for randomness; yet this is not so, and we can show that the brain is constructed according to a highly ordered design, made up of relatively simple components. To perform all its functions it uses only a few signals and a stereotyped repeating pattern of activity. Therefore, a relatively small sampling of nerve cells can sometimes reveal much of the plan of the organization of connections, as in the visual system.

In Part Three and especially in Part Six, we discuss "open-ended business," areas that are developing and whose direction is therefore uncertain. As one might expect, the topics cannot at present be fitted into a neat scheme. We hope, however, that they convey some of the flavor that makes research a series of adventures.

From Neuron to Brain expresses our approach as well as our aims. We work mostly on the machinery that enables neurons to function. Students who become interested in the nervous system almost always tell us that their curiosity stems from a desire to understand perception, consciousness, behavior, or other higher functions of the brain. Knowing of our preoccupation with the workings of isolated nerve cells or simple cell systems, they are frequently surprised that we ourselves started with similar motivations, and they are even more surprised that we have retained those interests. In fact, we believe we are working toward that goal (and in that respect probably do not differ from most of our colleagues and predecessors). Our book aims to substantiate this claim and, we hope, to show that we are pointed in the right direction.

S. W. K.
J. G. N.
Woods Hole
August 1975

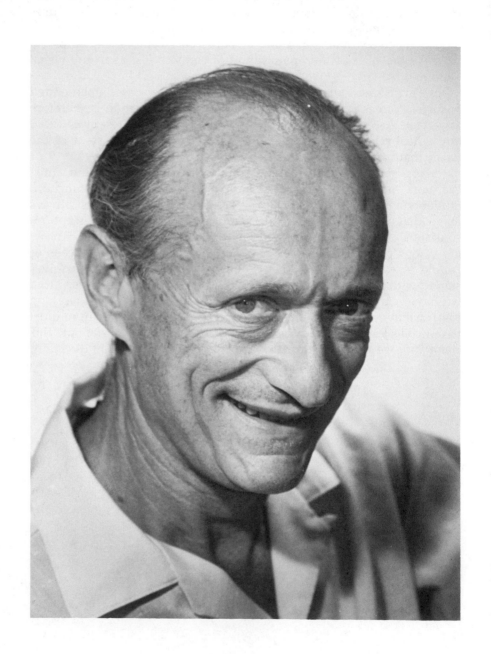

THE AUTHORS

STEPHEN KUFFLER (1913–1980)

Time after time, in a career that spanned 40 years, Stephen Kuffler made experiments on fresh topics, hitherto confused or ignored, in which he revealed fundamental mechanisms and laid paths for future research to follow. In each instance a striking feature of his work is the way in which the right problem was tackled at the right time, using the right preparation. Examples that spring to mind are his studies on denervation supersensitivity, stretch receptors and muscle spindles, efferent control, presynaptic and postsynaptic inhibition, GABA and peptides as transmitters, integration in the retina, the properties of glial cells, and the detailed analysis of synaptic transmission. In books on neurobiology, Kuffler's papers form a sizeable fraction of the reading list and one is struck by the clear-cut answers that were provided to well-defined problems that continue to be important.

What was it that gave each new paper by Stephen Kuffler that special quality which made it such a pleasure to read? Partly it was the unremittingly high standards of evidence, partly the elegance of the approach and the beautiful figures, and partly the underlying excitement of wondering—what would he tackle next? In addition, most of the experiments combined high technical virtuosity with directness of approach and clarity of thought matched by the style of the writing. Moreover, one knew that, right up to the end, he himself had done *every* experiment he described.

A striking feature of Stephen Kuffler's work is the multidisciplinary approach. To this end, he, more than anyone else, gave meaning to the idea of "Neurobiology"—a discipline in which the nervous system is studied in terms of cell biology, using biochemical, physiological, immunological, and anatomical approaches. At Harvard he created for the first time a department of Neurobiology in which he brought together people from widely different disciplines who actively collaborated, and thereby allowed new ways of thinking to evolve. He helped also to create interdisciplinary courses for young scientists at Woods Hole.

The list of his personal attributes is difficult to describe adequately. Those who knew him remember that unique combination of tolerance and firmness, kindness without sentimentality, good sense with enduring humor, with jokes and puns that often had an end but no beginning but still made one laugh. Long walks, long talks, relaxed meals, and quiet silences with friends were among his pleasures and contributed to the indelible memories he gave his friends.

He was the John Franklin Enders University Professor at Harvard and was closely associated with the Marine Biological Laboratory at Woods Hole. Among his many honors and distinctions was his election as a foreign member of the Royal Society.

JOHN G. NICHOLLS is Professor of Pharmacology at the Biocenter, Basel University, and until recently was Professor of Neurobiology at Stanford University School of Medicine. He was born in 1929 in London, where he graduated in medicine from Charing Cross Hospital, London University, and received a doctorate in physiology from the Department of Biophysics at University College. He has taught at University College, London, at Oxford, and at Yale, Harvard, and Stanford Medical Schools. During the summers he has given courses at the Marine Biological Laboratory in Woods Hole and at the Cold Spring Harbor Laboratory. His research has contributed to sensory and nerve–muscle physiology and to the physiology of neuroglial cells, an area in which he and Stephen Kuffler collaborated. For some years he has used the relatively simple nervous system of the leech to study synaptic transmission and the regeneration of synaptic connections.

A. ROBERT MARTIN is Professor and Chairman of the Department of Physiology at the University of Colorado School of Medicine. He was born in Saskatchewan in 1928 and majored in mathematics and physics at the University of Manitoba. He received a doctorate in biophysics from University College London in 1955, where he and John Nicholls studied together, with Sir Bernard Katz as their advisor. He has taught at McGill University, the University of Utah, Yale University and the University of Colorado Medical Schools, and has been a visiting professor at Monash and Edinburgh Universities. His major research interests are synaptic transmission in the central and peripheral nervous systems, processes of release of neurotransmitter from presynaptic nerve terminals, and membrane channels activated by neurotransmitters. Recently, he has turned his attention to synaptic mechanisms in the central nervous system of the lamprey.

PART ONE
NEURAL ORGANIZATION
FOR PERCEPTION

ANALYSIS OF SIGNALS IN THE CENTRAL NERVOUS SYSTEM

CHAPTER ONE

The brain uses stereotyped electrical signals to process all the information it receives and analyzes. The signals are virtually identical in all nerve cells. They are symbols that do not resemble in any way the external world they represent, and it is therefore an essential task to decode these signals. There is good evidence that the origins of nerve fibers and their destinations within the brain determine the content of the information they transmit. Thus, fibers in the optic nerve carry visual information exclusively, while similar signals in another type of sensory nerve—for example, one arising in the skin—convey a quite different meaning. Individual neurons can encode complex information and concepts into simple electrical signals; the meaning behind these signals is derived from the specific interconnections of neurons.

The brain is an unresting assembly of cells that continually receives information, elaborates and perceives it, and makes decisions. At the same time, the central nervous system can also take the initiative and act upon various sense organs to regulate their performance.

From neuronal signals to perception

To carry out its tasks of determining the many aspects of behavior and of controlling directly or indirectly the rest of the body, the nervous system possesses an immense number of lines of communication provided by the nerve cells (NEURONS). These cells are the fundamental units or building blocks of the brain, and it is our task to find out the meaning behind their signaling. A start has already been made in studying behavior in terms of the meaning of signals in nerve cells. It is not enough, however, simply to register impulses—the sign that a neuron is conveying information. When, for example, a picture is presented to the eye, an understanding is required of the specific, unique relation of the signals carried by the neurons to higher visual functions, such as the perception of color, form, or depth. For this, neuronal signals must be traced, from their origin in sense organs, through a succession of relays that includes the cerebral cortex.

That neurobiologists who study single nerve cells should be in a position to discuss higher functions of the brain, such as perception, is a recent, somewhat unexpected development. It has long been realized, of course, that knowledge of the cellular properties of neurons is essential for any detailed study of the brain. Nevertheless, there were no clear indications of how an understanding of membrane properties, signaling, or connections could help to explain intricate psychological

3

phenomena such as depth perception and pattern recognition. It seemed quite possible that the workings of the cerebral cortex would still remain a mystery even when a great deal was known about signaling in individual nerve cells. We hope to show that such a pessimistic view has now lost much of its force.

A comparison of the brain with other organs of the body throws the problem into sharper focus. The brain as an "organ" is much more diversified than, for example, the kidney or the liver. If the performance of relatively few liver cells is known in detail, there is a good chance of defining the role of the whole organ. In the brain, different cells perform different, specific tasks. Frequently, such specificity is linked with a distinct chemistry for individual cells; a good example is provided by comparing neurons that excite or start signals with neurons that inhibit or suppress signaling. In addition to their different chemistry, inhibitory and excitatory neurons obey different plans of connections. Only rarely can aggregates of neurons be treated as though they were homogeneous. Above all, the cells in the brain are connected with one another according to a complicated but specific design that is of far greater complexity than the connections between cells in other organs.

Fortunately, there are many simplifying features in the nervous system. First, it has only two basic types of electrical signals, one for short and the other for long distances. Second, these signals are virtually identical in all nerve cells of the body, whether they carry messages to or from centers, or are the result of painful stimuli or touch, or simply interconnect various portions of the brain. Better still, signals are so similar in different animals that even a sophisticated investigator is unable to tell with certainty whether a photographic record of a nerve impulse is derived from the nerve fiber of whale, mouse, monkey, worm, tarantula, or professor. In this sense, nerve impulses can be considered to be stereotyped units. They are the universal coins for the exchange of communication in all nervous systems that have been investigated. Similarly, they secrete various chemical substances (the TRANSMITTERS, used for conveying signals from one cell to the next), which are often identical in different species of animals.

The neurophysiologist has learned to deal reasonably well with the initial stages of sensory signaling that eventually lead to perception and also with certain "control" or "executive" tasks of the nervous system, particularly in relation to movement of skeletal muscle; but problems of different scope and magnitude arise in connection with questions concerning the neural basis of perception. The complications introduced by "higher functions" can no longer be avoided. For example, the functions of the cerebral cortex cannot be considered without reference to consciousness, sensation, perception, and recognition. What tools are now available for making a meaningful analysis of cortical mechanisms?

What type of information does an individual neuron convey?

First, is it useful to study complex problems by recording the activity of individual cells or small groups of cells? To account for the neural events involved in the perception of touch, we can start by recording signals from a neuron that terminates in the skin and can be generally satisfied about the meaning of its signals. These signals consist of brief electrical pulses, about 0.1 V in amplitude, that last for about 0.001 second (1 msec). They move along the nerve at a speed up to 270 miles/ hr (120 meters/sec). Although the impulses in a cell may appear identical with those in other nerve cells, the significance and meaning are quite specific for that cell; for example, they indicate that a particular part of the skin has been pressed. A most important generalization, first made by Adrian,[1] is that the frequency of firing in a nerve cell is a measure of the intensity of the stimulus. In the above example, the stronger the pressure applied to the skin, the higher the frequency and the better maintained the firing of the cell. As Hodgkin has written, Adrian "made what in the jargon of today (which Adrian detested) would be called a break-through".[2] Adrian himself described the circumstances of the experiment:

E. D. Adrian

> "That particular day's work, I think, had all the elements that one could wish for. The new apparatus seemed to be misbehaving very badly indeed, and I suddenly found that it was behaving so well that it was opening up an entire new range of data. I'd been bogged down in a series of very unprofitable experiments and here suddenly was the prospect of getting direct evidence instead of indirect, and direct evidence about all sorts of problems which I had set aside as outside the range of the techniques that one could use . . . it didn't involve any particular hard work, or any particular intelligence on my part. It was one of those things which sometimes just happens in a laboratory if you stick apparatus together and see what results you get."[2]

As Hodgkin says, "The comment that one wants to make about the last sentence is that when most people stick apparatus together and look around they do not make discoveries of the same importance as those of Adrian."

So far, then, there seems little difficulty in interpretation, and we can go a step further and discuss a simple reflex involving two or three sequential steps (Chapter 4). The whole cycle of events in the pathway from the sensory stimulus to the motor response can be traced because the role of the individual nerve cells is known. However, much less is known about the meaning of signals generated by a neuron deep within the brain, a neuron that receives its input from many cells and in turn supplies many others. Before the analysis can be started, a great deal of information is needed. Does the neuron under study handle information derived from the skin, the eye, the ear, or all three? If it is influenced by the eye, does it in turn regulate the size of the pupil, does it move the eye, or is it involved in perception of form? Or does it perhaps secrete a transmitter or hormone that profoundly influences the emotional state of the animal?

Analyzing the situation within the brain is somewhat similar to

---

[1]Adrian, E. D. 1946. *The Physical Background of Perception.* Clarendon Press, Oxford.
[2]Hodgkin, A. L. 1977. *Nature 269*: 543–544.

examining a small portion of a circuit in a large, complex computer. This analogy is overworked but nevertheless still appropriate. The properties of some of the basic circuit elements are known, and the electrical signals can be recorded at various stages; but unless the design of the instrument is known, there is no clue to the role the circuit serves. The meaning of the measurements may therefore be very limited, akin to recognizing letters in a foreign language without understanding the words. And yet, the remarkable lesson of the last two decades is that, in spite of such difficulties, considerable progress can be made in understanding higher functions of the brain by correlating the activity of individual nerve cells with complex behavioral or perceptual activities.

Pattern of neuronal connections determines meaning of electrical signals

At first it may seem surprising that the nervous system uses only two types of electrical messages. The signals themselves cannot be endowed with special properties because they are stereotyped and much the same in all nerves. The mechanisms by which signals are generated are also similar, though with interesting variants. The brain deals only with symbols of external events, symbols that do not resemble the real objects any more than the letters *d o g*, taken together, resemble a spotted Dalmatian. Rather, a particular set of signals must have a precise and special relation to an event.

Theoretically, there is no reason why a great deal of information could not be conveyed by any agreed-upon symbol, including a code made up of different frequencies. In the nervous system, however, the frequency or pattern of discharges will not do on its own as a code, for the following reason: Even though impulses and frequencies are the same in different cells responding to light, touch, or sound, the content of information is quite different. The quality or meaning of a signal depends on the origins and destinations of the nerve fibers, that is, on their connections. Various types of sensory modalities (light, sound, touch) are linked to different parts of the brain; even within each modality and in each area of the cortex, specific stimuli (such as lines or rectangles in the visual system) act selectively on specific populations of neurons. This organization is brought about by strictly determined connections. Frequency coding is used by the nervous system to convey information about the intensity of a stimulus rather than about its quality. Occasional exceptions to this rule do occur, for example, in vibration sense or in the pathways for deep-pitched sounds, where the frequency of firing may indicate the frequency of the source.

Many lines of evidence, some gross but nevertheless convincing, bring home the essential point of the importance of connections. A blow to the eye or a current passed across the eyeball produces the sensation of a light flash. The phantom limb phenomenon is another, familiar example. Amputees frequently report sensations in a missing member, vividly referring to a specific region, such as the toe or the knee. The sensations usually arise in the nerve stump when severed sensory axons are irritated; the irritation is caused by scar formation or by the development of a swelling produced by a disorderly growth of nerve fibers at the cut end of the stump. Judicious electrical stimulation of sensory axons along their course in the body can also produce sensations that vary in modality according to the origin of the sensory fibers. At present it is not possible to reproduce with electrodes, or by other artificial means, the natural discharge pattern in a nerve composed of many fibers. This is what would be required to evoke complex sensations artificially while bypassing sense organs. In the same way, a complex message cannot be sent through a bundle of wires making up a telegraph cable by simply passing currents through it somewhere along its course.

An understanding of the importance of the pattern of connections is enhanced by looking at examples that show how the information about external events is inherent in connections. A radio, a computer, and a TV set use commonplace components and stereotyped signals, yet perform a variety of tasks. The specialization resides in the design of the wiring. The diversity of connections, not the types of signals, increases the complexity of the tasks that can be undertaken. In much the same way the nerve cells in the brain are made up of "standard," commonplace chemical materials. What endows them with their diverse capabilities and gives meaning to their signals is the manner in which they are linked to each other.

A further requirement that goes with complex, satisfactory computer performance is an adequate number of components; this condition is also met by the nervous system. The numbers of cells in the cortex are so great (probably more than 20,000 cells/mm^3) that they do not as yet present a limitation to speculation. The brain, then, is an instrument, made of 10^{10} to 10^{12} components of rather uniform materials, that uses a few stereotyped signals. What seems so puzzling is how the proper assembly of the parts endows the instrument with the extraordinary properties that reside in the brain.

It is worth pointing out that the conclusions presented here were expressed in 1868 by the German physicist–biologist Helmholtz. Starting from first principles, long before the facts as we know them were available, he reasoned:[3]

> The nerve fibers have often been compared with telegraphic wires traversing a country, and the comparison is well fitted to illustrate the striking and important peculiarity of their mode of action. In the network of telegraphs we find everywhere the same copper or iron wires carrying the same kind of movement, a stream of electricity, but producing the most different results in the various stations according to the auxiliary apparatus with which they are connected. At one station the effect is the ringing of a bell, at another a signal is moved, at a third a recording instrument is set to work. . . . In short, every one of the hundred different actions which electricity is capable of producing may be called forth by a telegraphic wire laid to whatever spot we please, and it is always the same process in the wire itself which leads to these diverse consequences. . . . All the difference which is seen in the excitation of different nerves depends only upon the

[3]Helmholtz, H. 1889. *Popular Scientific Lectures.* Longmans, London.

difference of the organs to which the nerve is united and to which it transmits the state of excitation.

The preceding sections point out some of the difficulties that enter into considerations of conscious perception. These difficulties are much reduced in the visual system, particularly when responses in cortical cells are analyzed. This is illustrated in Chapter 2, which deals mainly with recent experiments made by recording from single cells in the visual pathways. Although much information is available about the auditory and other sensory systems in the body, the mammalian visual system has several advantages, owing particularly to the relative technical simplicity of many of the experiments and their direct relevance to perception. It offers clues for an understanding of the code used by neurons to transmit not just simple information about light and darkness but also sophisticated concepts. For example, reasonable hypotheses can now be formulated, in terms of neural signals and organization, relating to the following questions: What neural mechanisms can explain the recognition of shapes, such as light edges or corners of certain dimensions, positioned at one angle rather than another in the visual field? How can triangles or squares be recognized independent of their position on the retina or of their brightness and size? How can we perceive with both eyes one image rather than two, even though it is known that each eye really sees a somewhat different part of the world? With regard to the problems just mentioned, the visual system is different from sensory systems dealing with input from the skin; these systems are discussed in Chapter 17. For example, mechanical stimuli, like a gentle touch, applied to the skin of two corresponding areas on both sides of the body do not usually give rise to a single, fused sensation in the midline. Similarly, the visual system is different from a photographic plate in that it takes account of contrast or differences rather than of the absolute level of brightness; visual perception ignores information about absolute levels and can detect subtle differences even if the background illumination changes over a range of many orders of magnitude. An apparent paradox pointed out by David Hubel is that black newsprint seen in sunshine on the beach reflects more light than the white part of a page seen with an ordinary electric lamp; yet the print appears black and the paper white in both situations. The visual system also provides one of the most favorable systems for studying fundamental questions relating to the development and maturation of the nervous system. Are the neural circuits used for perception already present at birth or are they formed as a result of visual experience? What sort of stimuli must be present in the environment to prevent sensory systems from becoming atrophied and useless? These questions are considered in Chapter 20.

The advances that make it possible to discuss these problems in physiological terms have resulted from the discovery that a highly specific set of cell connections exists in the visual system and in other sensory systems. The arrangement of these connections accounts for the almost infinite wealth of information that reaches us as we look at the world around us. This is not to deny that we still remain profoundly ignorant about higher functions such as perception. But it seems reasonable to expect that the processes that lead to perception can be analyzed by the use of the same principles that govern the other functions of the nervous system.

BACKGROUND INFORMATION

For an easier understanding of the material in Chapters 2 and 3, we first present a few basic facts about the structure of neurons, their interconnections, and the methods of recording from them. Some key terms and definitions appear at the end of this chapter. (A fuller description of signaling is given in Part Two, and other terms used are defined in the glossary.)

Shapes and connections of neurons

The shape of a neuron, as well as information about its position, origin, and destination in the neural network, supplies valuable clues to its function. For example, the arborization of a neuron provides a notion of how many connections a cell can accommodate and to how many sites it sends its own processes.

In practice it is difficult to find out about the configuration of neurons because they are so densely packed. Early anatomists had to tease nervous tissue apart to see individual neurons. Figure 1 shows a spinal motoneuron dissected and drawn by Deiters more than 100 years ago. Staining methods that impregnate all neurons are virtually useless for investigating cell shapes and connections because a structure like the cortex appears as a dark blur of intertwined cells and processes. Many of the pictures in Figures 1 and 2 were made with the Golgi staining method, which has become an essential tool because by some unknown mechanism it stains just a few random neurons out of the whole population. Furthermore, the technique tends to stain individual cells in their entirety.

Many of the illustrations in this chapter are based on the work of Ramón y Cajal, done before the turn of the century. Ramón y Cajal was one of the greatest students of the nervous system, selecting samples from a wide range of the animal kingdom with an almost unfailing instinct for the essential. The illustrations show several distinct cell types, some relatively simple, such as the bipolar cell, others with a highly complex arborization.

Ramón y Cajal, about 1914

In recent years greater selectivity has been obtained by a number of new techniques that have enabled single cells or cells with common properties to be stained or marked. For example, cells can be injected with a fluorescent dye such as Lucifer Yellow, or with a metal such as cobalt, or with the enzyme horseradish peroxidase. These methods enable the investigator to obtain the entire outline and geometry of cells from which he has recorded and whose physiological performance he

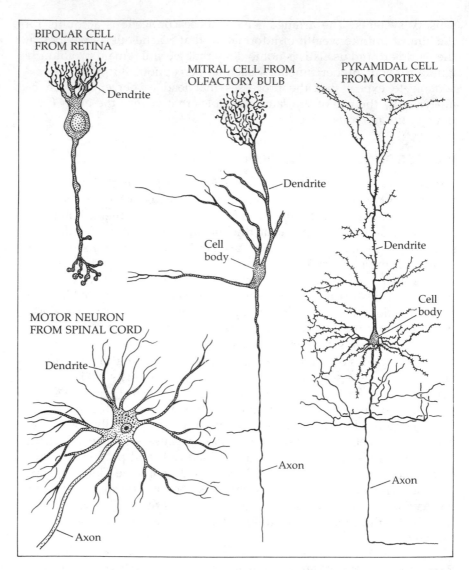

BIPOLAR CELL
FROM RETINA

Dendrite

MITRAL CELL FROM
OLFACTORY BULB

PYRAMIDAL CELL
FROM CORTEX

Dendrite

Cell
body

Dendrite

Cell
body

MOTOR NEURON
FROM SPINAL CORD

Dendrite

Axon

Axon

Axon

1 SHAPES AND SIZES OF NEURONS. The cells have processes, the dendrites, upon which other neurons form synapses. Each cell in turn makes connections with other neurons. The motor neuron, drawn by Deiters in 1869, was dissected from a mammalian spinal cord. The other cells, stained by the Golgi method, were drawn by Ramón y Cajal. The bipolar cell is from the retina of a dog, the pyramidal cell from the cortex of a mouse, and the mitral cell from the olfactory bulb (a relay station in the pathway concerned with smell) of a cat.

knows. Moreover, after horseradish peroxidase injection, the detailed morphology of the cell and its contacts can be seen by electron microscopy.[4]

Methods for tracing the course taken by axons from their cell bodies

[4]Muller, K. J. and McMahan, U. J. 1976. *Proc. R. Soc. Lond. B 194*: 481–499.

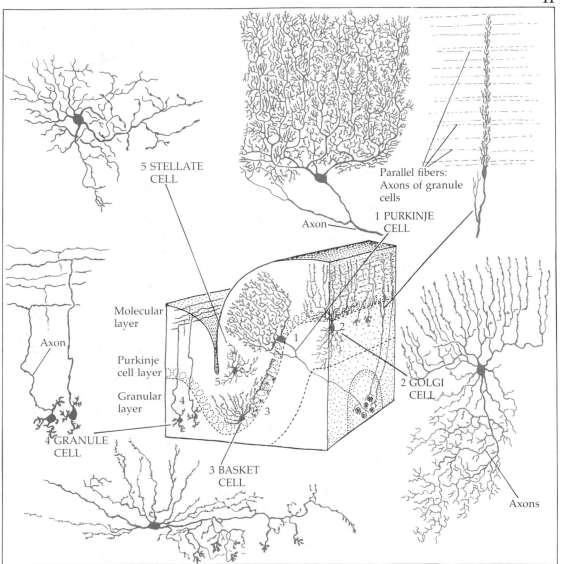

2 CEREBELLAR NEURONS. The human cerebellum has over 10^{10} cells but only five neuronal types, the cell bodies of which are confined to distinct layers; they are shown in drawings made from Golgi-stained preparations. The Purkinje cell (1), whose axons constitute the only output from the cerebellum, has its processes aligned in one plane. The axons of granule cells (4) traverse and make connections with the Purkinje cell process in the molecular layer. The Golgi cell (2), the basket cell (3), and the stellate cell (5) have characteristic positions, shapes, branching patterns, and synaptic connections. See Figure 1 in Chapter 2 for the position of the cerebellum in the brain. (1 and 2 after Ramón y Cajal, 1955; 3–5 after Palay and Chan-Palay, 1974.)

to their final destinations are also now available. For example, the terminals of axons in the central nervous system or in a muscle take up horseradish peroxidase that has been injected extracellularly in their vicinity. Such "retrograde" transport of the enzyme along the axon carries it to the cell body, which can then be stained and identified in another, perhaps distant, region of the central nervous system. A neuron that has axons distributed to two separate regions of the brain can be doubly labeled by retrograde transport. Two markers with different colors can be injected into the areas containing the terminals. This technique enables one to distinguish which individual nerve cells in the nucleus suppy both areas.[5] "Anterograde" transport from the cell body to the terminals can also be used to define cells and their pathways. Amino acids that are taken up by the cell body, converted into protein, and shipped down the axon have proved particularly valuable. In the visual system, transynaptic transfer occurs from one cell to the next along the pathway. When biochemical techniques that involve the uptake of a radioactive analogue of glucose (2-deoxyglucose) are used, those cells in a region of the nervous system that respond to a particular stimulus can be distinguished from those that do not. (In principle, the active cells can take up the radioactive glucose but cannot metabolize it; the radioactivity is then detected in those active cells. Examples are shown in Chapters 3 and 20.)

Cytology has demonstrated that what appears at first sight to be a staggering array of shapes and processes can be divided into meaningful groupings; thus, cells can be recognized and classified in much the same way as trees can. Although differences within a group can be considerable, one can distinguish a spinal motor neuron from a pyramidal cell, just as one can always tell a birch tree from a palm tree.

Design of connections as exemplified by the cerebellum

The cerebellum stands as one of the best examples of the orderliness of the design of the neuronal connections in the nervous system; the retina is a close competitor. Although the cerebellum has only five distinct kinds of cells, the total number of its neurons is truly staggering. The cell population for the entire human brain is frequently given as 10,000 million. This number is not likely to include the cells of the cerebellum, in which one cell type alone, the granule cell (Figure 2), is supposed to number around 10^{10} to 10^{11}.[6]

Cerebellar cytology came to life with the work of Ramón y Cajal[7] almost a century ago and received a new lease on life in the past two decades with advances that arose from developments in electron microscopy and from new studies by physiologists in the laboratories of Eccles, Szentágothai, Palay, Ito, Llinás, and others.[8–10]

The input to the cerebellum comes from the various sensory structures in muscles, skin, and joints; from the visual and auditory cortex,

[5]Luskin, M. B. and Price, J. L. 1982. *J. Comp. Neurol. 209*: 249–263.

[6]Braitenberg, V. and Atwood, R. P. 1958. *J. Comp. Neurol. 109*: 1–33.

[7]Ramón y Cajal, S. 1955. *Histologie du Système Nerveux.* II., C.S.I.C., Madrid.

[8]Eccles, J. C., Ito, M. and Szentágothai, J. 1967. *The Cerebellum as a Neuronal Machine.* Springer-Verlag, Berlin.

[9]Palay, S. L. and Chan-Palay, V. 1974. *Cerebellar Cortex.* Springer-Verlag, Berlin.

[10]Llinás, R. R. 1981. Chapter 17, pp. 831–976, *Handbook of Physiology*, V. Brooks, ed., Amer. Physiol. Soc., Bethesda, MD.

and so on. All the information is handled by repeated groupings of a few cell types—granule cells, basket cells, stellate cells, and Golgi cells—all of which directly or indirectly act upon the Purkinje neurons. The latter provide the only output from the cerebellar cortex.

The schematic presentation in Figure 2 gives an idea of the orderly arrangement of neurons. Within the cortex of the cerebellum the cell bodies of the various neuron types lie in distinct layers, sending their processes to other specific regions and to particular portions of target cells. The Purkinje cell dendrites (processes receiving synaptic inputs) are aligned in one plane, like a many-pronged candelabrum, extending close to the cerebellar surface. The granule cells send their axons from their own layer, near the white matter, straight toward the surface, where they bifurcate and run parallel to the cerebellar surface. They traverse the arborization of the Purkinje neurons at right angles and make synapses with their processes. It has been estimated that one Purkinje cell in the monkey receives synapses from about 80,000 granule cells, all of them restricted to small spiny protrusions. Other synaptic contacts on Purkinje cells are made by basket and stellate cells, mainly in the region of the cell body. The Golgi cells do not make contact directly but do so indirectly by acting on granule cells. A single neuron may accommodate as many as 200,000 synapses.

The cerebellum is concerned with the regulation of movement. Disease or lesions that damage the cerebellum result in disorders of coordination. For example, after damaging or removing part or all of the cerebellum in monkeys, the animals can still move their eyes in the correct direction when the head is rotated, but the eye movements are no longer accurate.[11] In man, disorders of gait and of eye movements result from damage to the cerebellum. Detailed theories about the types of analyses performed by the cerebellum have been proposed.[12,13] In principle, with information about the inputs, outputs, and circuitry and about the effects of lesions, one might hope for a practical and comprehensive solution to the problem of its role in controlling movement—a goal that still seems elusive in spite of so much progress.

At the level of the electron microscope, the nervous system seems even more complex and, at first glance, insoluble. Yet orderliness becomes apparent once again by following processes through successive sections, by selectively staining cells, and by correlating the appearance with that seen in light micrographs. Figure 3 is a section through the region of the cerebellum and shows a climbing fiber ending on the processes of a Purkinje cell. The synapses, sites at which information is transferred from cell to cell, appear as well-defined structures. The presynaptic terminals contain numerous vesicles close to the cell membrane. Separating the membranes of the two cells is a cleft that is filled with extracellular fluid and is somewhat wider than elsewhere. Dense

Fine structure of synapses

[11]Optican, L. M. and Robinson, D. A. 1980. *J. Neurophysiol. 44*: 1058–1076.
[12]Marr, D. 1969. *J. Physiol. 202*: 437–470.
[13]Llinás, R. and Simpson, J. I. 1981. *Handbook of Behavioral Neurobiology 5*: 231–302.

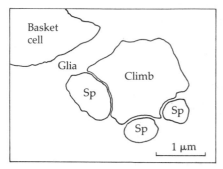

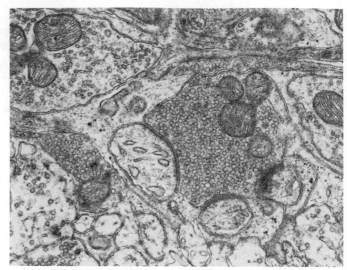

3 SYNAPSES IN CEREBELLAR CORTEX OF RAT, molecular layer; climbing fiber terminal making synaptic contact with three spines of Purkinje cell dendrite. The climbing fiber terminal (Climb) is packed with round vesicles and it makes contact with the thorns or spikes (Sp). The complex is partially surrounded by neuroglial processes. A large basket axon varicosity is nearby (flattened vesicles), but it does not have a synaptic junction in the field. (Photograph courtesy of S. L. Palay, unpublished.)

material is often seen in the cleft and on the two membranes. Many aspects of synaptic transmission are now well enough understood for the morphology to be correlated with the functional behavior observed by recording electrically.

Recording techniques   The electrical activity of a neuron can be monitored by an electrode placed outside of the cell membrane (extracellular recording) or by a microelectrode that actually penetrates the cell (intracellular recording). Most of the recordings of electrical activity presented in the next two chapters were made with extracellular electrodes. The electrode itself can be either a fine wire insulated almost to its tip or a capillary tube filled with salt solution. Figure 4 depicts diagramatically the arrangement for extracellular recording. This technique supplies information about whether a cell is firing or is quiescent and about whether the rate of firing is increasing or decreasing. With care, the signals from a single neuron can be identified.

Intracellular recording is used to obtain information about the processes of excitation and inhibition and the mechanisms that initiate nerve impulses. The tip of a microelectrode is inserted into the cell, as shown in Figure 5. The electrode is a fine glass capillary with a tip 0.1 μm in diameter or smaller and is completely filled with a salt solution such as 3 M potassium chloride or 4 M potassium acetate. It measures the potential difference between the inside and the outside of the cell. The intracellular electrode can also be used for passing electrical currents

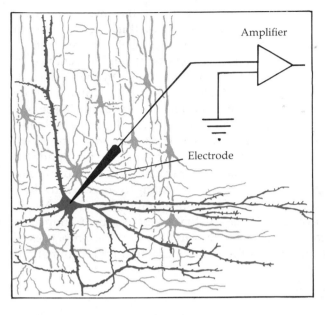

Amplifier

Electrode

4

EXTRACELLULAR RECORDING with a fine wire electrode. The electrode tip has been drawn close to a nerve cell in the cortex.

into or out of the cell and for staining it in its entirety with a dye or with horseradish peroxidase.

A serious difficulty is encountered when recording from the mammalian central nervous system with external or intracellular microelectrodes, namely, ascertaining the exact position of the tip, which cannot be seen. Another difficulty arises from the pulsations caused by the

The slice technique

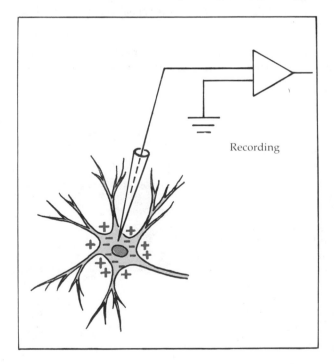

Recording

5

INTRACELLULAR RECORDING. The tip of a microelectrode has been inserted into a nerve cell. In a neuron that is at rest there is a potential difference of about 70 mV; the inside is negative with respect to the outside.

circulation of the blood, which tend to dislodge the microelectrodes. Both problems are eliminated by working with thin slices of neural tissue in vitro. Well-organized areas, such as cerebellar cortex, retina, cerebral cortex, or hippocampus (see Appendix B) can be sectioned in one plane to make a slice about 0.5 mm thick; this can then be maintained in a dish for several hours or days. Each slice can contain all the elements of the circuit. Under these conditions, it is possible to see one particular cell under the microscope (for example, a Purkinje cell) and to record from it without movement while stimulating incoming fibers or interneurons. With precise localization of the electrode tip, dendrites as well as cell bodies can be impaled. In addition, the composition of the fluid surrounding the cells can be rapidly changed so as to introduce pharmacological agents or to vary the concentrations of ions. An attractive feature of the slice technique is that tissue can be examined in detail, first in vitro to establish the microcircuitry and the properties of the neural elements and then in the natural situation in the animal.[14,15]

SUGGESTED READING

Carpenter, M. B. 1978. *Core Text of Neuroanatomy*, 2nd Ed., Williams & Wilkins, Baltimore.

Ottoson, D. 1983. *Physiology of the Nervous System*, Oxford University Press, New York.

Palay, S. L. and Chan-Palay, V., eds. 1982. *The Cerebellum—New Vistas.* Springer-Verlag, New York.

Peters, A., Palay, S. L. and Webster, H. de F. 1976. *The Fine Structure of the Nervous System.* Saunders, Philadelphia.

Williams, P. L. and Warwick, R. 1975. *Functional Neuroanatomy of Man.* Saunders, Philadelphia.

[14]Llinás, R. and Sugimori, M. 1980. *J. Physiol. 305*: 171–195; 197–213.

[15]Andersen, P., Silfrenius, H., Sundong, S. H. and Sveen, O. 1980. J. Physiol. 307: 273–299.

BOX 1   REVIEW OF KEY CONCEPTS AND TERMS

The ACTION POTENTIAL in a nerve fiber is fixed in size. It is a brief, stereotyped electrical event lasting for about 1 msec. It moves rapidly along the nerve from one end to the other. IMPULSE is another term for action potential.

Impulses on a fast time scale

Impulses on a slow time scale

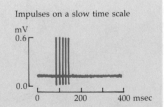

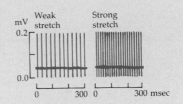

Stronger stimuli produce HIGH-ER FREQUENCIES of impulse firing. For example, a sensory nerve responding to stretch of a muscle fires at a rate proportional to the stretch.

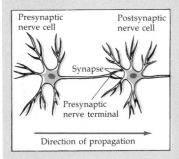

The junctions between nerve cells are called SYNAPSES. These are the sites at which cells transfer signals.

Nerve cells influence each other by (1) EXCITATION (shown in blue), that is, they tend to produce impulses in another cell; and by (2) INHIBITION, that is, they tend to prevent impulses from arising in another cell.

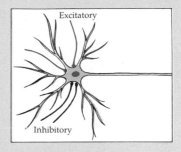

A cell receives many excitatory and inhibitory inputs from other cells (called CONVERGENCE) and in turn supplies many others (DIVERGENCE).

The process whereby a cell adds together all the incoming signals that excite and inhibit it is known as INTEGRATION.

DEPOLARIZATION is a reduction in magnitude of the membrane potential toward 0 mV, the inside becoming less negative with respect to the exterior; HYPERPOLARIZATION is an increase in the magnitude of the potential, the inside becoming more negative. Depolarization to a critical potential level, the THRESHOLD, causes the initiation of an impulse. At its peak the inside of the cell becomes positive with respect to the outside.

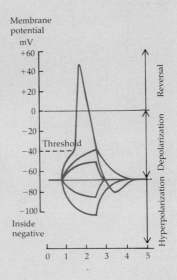

In a typical sensory nerve the effective or ADEQUATE STIMULUS (for example, stretch) depolarizes and sets up impulses.

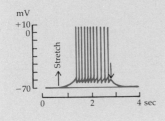

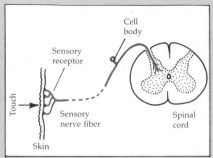

The specialized ending of a sensory nerve that responds to an external physical stimulus (touch, light, heat) is called a sensory RE-CEPTOR.

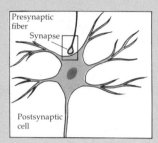

At chemical synapses the presynaptic terminal liberates a chemical TRANSMITTER SUBSTANCE in response to a depolarization.

Characteristically, at the site of a CHEMICAL SYNAPSE, the two nerve cell membranes are thickened and are farther apart than elsewhere. Vesicles containing transmitter are aggregated close to the presynaptic membrane. Dense material is often present in the synaptic cleft and under the membrane of the post-synaptic cell.

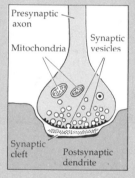

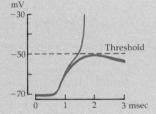

At an EXCITATORY SYNAPSE the chemical liberated by the presynaptic ending depolarizes the postsynaptic cell, driving its membrane potential toward threshold.

At an INHIBITORY SYNAPSE the transmitter tends to keep the membrane potential of the postsynaptic cell below threshold.

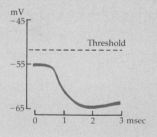

THE VISUAL WORLD: CELLULAR ORGANIZATION AND ITS ANALYSIS

Much progress has been made in determining how the activity of an individual neuron in the visual system is related to specific features of visual perception. Visual information is processed in successive stages: The neural responses to light start at the receptors, progress through the sequential layers of cells in the retina, ascend through a relay station (the lateral geniculate nucleus), and continue through a series of stages in the cerebral cortex. The transformation or integration that occurs at each level is best analyzed in terms of the receptive fields of neurons. This term refers to the restricted area on the retinal surface that influences, upon illumination, the signaling of an individual neuron in the visual system. Receptive fields are the building blocks for the synthesis and perception of the complex visual world. They demonstrate a general rule in vision: The system is designed to perceive differences in intensity (that is, contrast) rather than absolute intensities of light.

To reveal the functional organization of connections, one must find for each neuron the characteristic light stimulus to which it will give the best response. This has been done most completely for the various cell types in the retina. For example, a ganglion cell, which represents the output of a small fraction of the retina, responds best to a small spot of light in the center of its circular receptive field or to a line or edge that passes through the receptive field. Some ganglion cells respond better to light spots, others to dark ones; in addition, moving spots or edges and brisk changes in intensity are more effective stimuli for the larger ganglion cells with faster conducting axons. Nevertheless, all the ganglion cells share the common property of responding best to small spots and relatively poorly to diffuse illumination.

The requirement for contrast is emphasized still more in cortical neurons, which practically ignore uniform illumination. Their activation requires highly specific shapes or forms—in particular, lines or edges with a certain orientation and position on the retina. Some categories of neurons are specialized to respond to angles or corners or to movements in one direction but not in another. According to the type of information they carry, neurons in the cortex have been classified as simple and complex.

A scheme that assumes a hierarchically ordered series of ascending connections can be used to explain many features of how neurons respond selectively to specific stimuli, for example, to a bar of light, a corner, or a square. At each stage, cells with relatively simple properties combine to form fields of progressively greater complexity and visual content. So far, transformation of visual information has been analyzed only for the first few stages between peripheral receptors and nerve cells within the visual cortex.

RETINA AND LATERAL GENICULATE NUCLEUS

To convey the relevance of the cellular approach to higher functions, this chapter deals principally with the performance of nerve cells at successive stages or relays of the visual systems of the cat and the monkey. However, because the physiological analysis of processing in the retina leading to perception has not been so fully studied in mammals, this aspect is illustrated by examples from work in a variety of species.

A comprehensive critical treatment is not possible within the scope of this book; the past few years have provided an overwhelming body of work on the structure and function of the visual system alone. In addition, psychophysics, color vision, dark adaptation, retinal pigments, transduction, and the organization of the retina could each form the basis of a self-contained monograph (see references at the end of the chapter). The same applies to comparative aspects of the visual system in invertebrates (such as the horseshoe crab, *Limulus*, or the barnacle) and in lower vertebrates (fish, frog, mud puppy, and turtle), as well as in mammals (rabbits and squirrels).

Nevertheless, a brief description of studies on cat and monkey, chiefly along the lines pioneered by Hubel and Weisel in their work, provides a clear, continuous thread extending from signaling to perception. These studies also help to relate the structure of the nervous system to questions of genetics and development (discussed in Part Five).

A crucial factor in the physiological analysis is the use of stimuli that mimic those occurring under natural conditions. For example, edges, contours, and simple patterns presented to the eye reveal features of the organization that could never be detected by using bright flashes without form.

Another key to the success of Hubel and Wiesel's approach lies in asking not simply what stimulus evokes a response in a particular neuron, but what is the most effective stimulus. Since all the impulses in any one cell have the same amplitude and time course, the "optimal" stimulus is defined as the one that produces the highest frequency discharge. Pursuit of this question through the various stages of the visual system has elicited many surprising and remarkable results.

Finally, the orderly, layered arrangement of neurons in the retina and within the brain itself suggests on its own that information processing is carried out in hierarchically arranged levels, proceeding from

21
THE VISUAL
WORLD:
CELLULAR
ORGANIZATION
AND ITS
ANALYSIS

one functionally related group of cells to the next. Thus, it is reasonable to suppose that the behavior of one group of neurons can be explained by understanding the effects produced by converging fibers which supply it.

This section describes first the principal anatomical features of the visual pathway and then the signals recorded at the successive stages. Chapter 3 deals with the cellular organization and anatomy of the visual cortex in greater detail.

The eye acts as a self-contained outpost of the brain. It collects information, analyzes it, and hands it on for further processing by the brain through a well-defined pathway, the optic nerve.

Anatomical pathways in the visual system

The pathways from the eye to the cerebral cortex are illustrated in Figure 1A; Figure 1B to D depicts some of the major landmarks of the human brain that are useful in the context of the following discussion. The optic nerve fibers arise from ganglion cells in the retina and end on nests of cells in a relay station (the lateral geniculate nucleus), whose axons in turn project through the optic radiation to the cerebral cortex. From here on the progression becomes ever more complex, with no end station in sight.

Figure 1A also shows how the output from each retina divides in two at the optic chiasm to supply the lateral geniculate nucleus and cortex on both sides of the animal. As a result, the right side of each retina projects to the right cerebral hemisphere. Figure 1 also shows that the right side of each retina receives the image of the visual world on the left side of the animal. Each cerebral hemisphere, therefore, "sees" the visual field of the opposite side of the world. Accordingly, patients with damage to the right side of the cortex caused by trauma or disease become blind to the left side of the world and vice versa.

Other pathways that branch off to the midbrain are not described here. In higher vertebrates they are primarily concerned with regulating eye movements and pupillary responses and are not directly relevant for the types of pattern recognition to be considered.

By merely examining the cellular anatomy of the various structures in the visual pathway, one can exclude the possibility that information is handed on unchanged from level to level. The neurons CONVERGE and DIVERGE extensively at any stage; that is, each cell makes and receives connections with a number of other cells. For example, the human eye contains over 100 million primary receptors, the rods and cones, but only about 1 million optic nerve fibers are sent by ganglion cells into the brain. In the monkey and the cat the same principle holds: a stepdown in neuronal numbers from receptors to ganglion cells. Therefore, within the eye as a whole there occurs a funneling of information. As a result, an individual neuron that receives impulses from several incoming nerve fibers cannot reflect separately the signals of any one of them. Instead, converging impulses of different origin are combined at each stage into an entirely new message that takes account of all the inputs. This process is called INTEGRATION (Chapters 1 and 16).

What makes the retina so specially inviting for physiological research

22

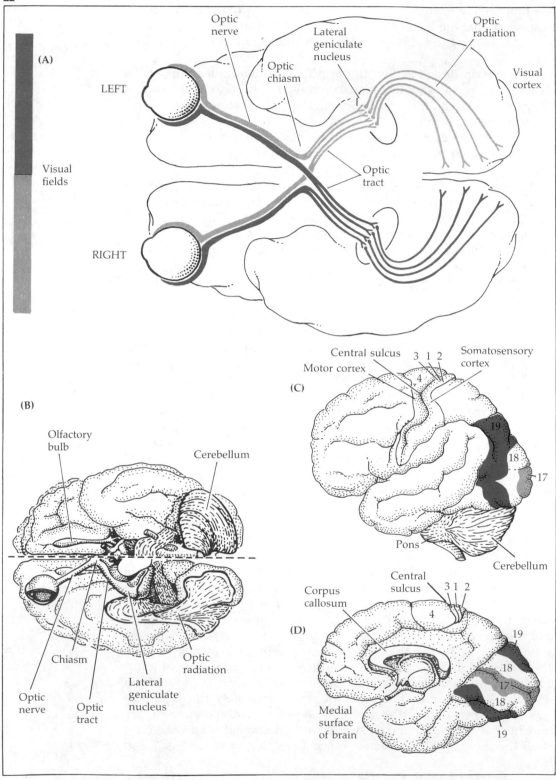

(A)

LEFT

RIGHT

Optic
nerve

Optic
chiasm

Lateral
geniculate
nucleus

Optic
radiation

Visual
cortex

Optic
tract

Visual
fields

(B)

Olfactory
bulb

Cerebellum

Chiasm

Optic
nerve

Optic
tract

Lateral
geniculate
nucleus

Optic
radiation

(C)

Central sulcus

Motor cortex

3 1 2

4

Somatosensory
cortex

19

18

17

Pons

Cerebellum

(D)

Corpus
callosum

Central
sulcus

3 1 2

4

19

18

17

18

19

Medial
surface
of brain

23
THE VISUAL
WORLD:
CELLULAR
ORGANIZATION
AND ITS
ANALYSIS

is the neat layering and orderly repetition of the relatively few cell types—there are only five main classes. The arrangement and the typical positions of various cells are illustrated in Figure 2, which shows a cross section of a human retina. On the deep surface, farthest from the incoming light, lie the RODS and CONES, which are concerned with night and daytime vision, respectively. They are connected to the BIPOLAR CELLS, which in turn connect to the GANGLION CELLS and so to the optic nerve fibers. Apart from this through-line, there are other cells that make predominantly lateral or side-to-side connections. These are the HORIZONTAL CELLS and the AMACRINE CELLS. Within each of these major classes, there are subgroups exhibiting important differences in structure and function; the role played by these cells and their interconnections will be discussed later. We shall see that the aggregation of comparable cells into layers is also a feature of the lateral geniculate nucleus and the visual cortex, although the arrangement is not so homogeneous and the boundaries not so sharp.

Retinal ganglion cells and the concept of receptive fields

A number of developments set the stage for the single-neuron analysis of the mammalian visual system. Among these was the elegant and lucid work of Adrian (see Chapter 1). Hartline's[1] pioneering work on the horseshoe crab, *Limulus,* also foreshadowed intriguing and exciting developments. If so much could be gleaned from a relatively simple invertebrate eye, would the vertebrate visual apparatus not yield corresponding insights, if similar methods could be used? Hartline himself advanced his studies into the retina of the frog;[2] these experiments, in which he recorded from single neurons and analyzed their responses to illumination, set the stage for many of the approaches to be used later. Recordings from single neurons of the mammalian eye were made by Granit and his colleagues.[3]

These investigations were the main points of departure for the exploration of the mammalian retina in cellular terms by Kuffler.[4] The principal new approach was not so much a matter of technique; rather, it consisted of formulating the following question: What is the best way to stimulate individual ganglion cells whose axons carry information to the higher centers through the optic nerve? This question led logically to the use of discrete circumscribed spots and patterns of light for

[1]Hartline, H. K. 1940. *J. Opt. Soc. Am. 30*: 239–247.
[2]Hartline, H. K. 1940. *Am. J. Physiol. 130*: 690–699.
[3]Granit, R. 1947. *Sensory Mechanisms of the Retina*. Oxford University Press, London.
[4]Kuffler, S. W. 1953. *J. Neurophysiol. 16*: 37–68.

1 VISUAL PATHWAYS. (A) Outline of the visual pathways seen from below (base of the brain) in primates. The right side of each retina (shown in color) projects to the right lateral geniculate nucleus and the right visual cortex receives information exclusively from the left half of the visual field. (B) Visual pathways in a partially dissected human brain seen from below. (C, D) Lateral and medial views of the cortical surface. Area 17 is also known as the striate cortex or visual area I; areas 18 and 19 are visual areas II and III. In addition, area 4 (the motor cortex) and areas 3, 1, and 2 of the sensory cortex are labeled.

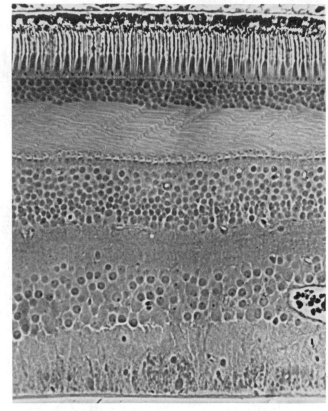

Nuclei of rods
and cones

Horizontal,
bipolar, and
amacrine cells

Ganglion cells

Optic nerve
fibers

2 SECTION THROUGH A HUMAN RETINA, showing the five principal cell types arranged in layers. Light enters the retina at the ganglion cell layer and reaches the photoreceptors (rods and cones), where it is absorbed, starting excitation in the outer segments. (From Boycott and Dowling, 1969.)

stimulation of selected areas of the mammalian retina, a procedure used on the frog at about the same time by Barlow.[5]

A methodological feature of Kuffler's experiments that became essential was the use of the practically intact, undissected eye, the normal refracting channels of which served as pathways for stimulation. Not only was it technically easier to record the signaling of ganglion cells, but the analysis was also simplified by beginning with the end result of activity in the eye. This presentation, therefore, discusses first the impulse patterns emerging from the eye and then the preceding steps that have occurred within the retina.[4]

Illumination of selected areas of the retina introduced the important concept of the receptive field, a concept which provided the key to understanding the significance of the signals. The term *receptive field* was coined originally by Sherrington in relation to reflex actions and

[5]Barlow, H. B. 1953. *J. Physiol. 119*: 69–88.

25
THE VISUAL
WORLD:
CELLULAR
ORGANIZATION
AND ITS
ANALYSIS

was reintroduced by Hartline.[2] Applied to the visual system, the RE-CEPTIVE FIELD OF A NEURON can be defined as THE AREA ON THE RETINA FROM WHICH THE DISCHARGES OF THAT NEURON CAN BE INFLUENCED BY LIGHT. For example, a record of the activity of one particular fiber in the optic nerve of a cat (Figure 4) shows that that fiber increases or decreases its rate of firing only when a defined area of retina is illuminated. This area is its receptive field. By definition, illumination outside the field produces no effect at all. The area itself can be subdivided into distinct regions, some of which act to produce firing and others to suppress impulses in the cell.

The best way of illuminating particular portions of the retina is to anesthetize the animal lightly and place it facing a screen at a distance for which its eyes are properly refracted. When one then shines spots or patterns of light onto the screen, these will be well focused on the retinal surface (Figure 3). A convenient way to supply finely controlled visual images is to use a TV screen with a computer-generated display. The experiments described below were made on immobilized eyes in the light-adapted state, but neuronal discharges have also been recorded from unrestrained, waking cats with chronically implanted microelectrodes.

When one is recording from a particular nerve cell, the first task is to find the location of its receptive field. Characteristically, most cells in the eye, and throughout the visual system, show continued discharges at rest even in the absence of illumination. Appropriate stimuli do not necessarily initiate activity but may modulate the background firing; responses of the cell can therefore consist of either an increase or a decrease of ongoing discharges.

At first it may seem puzzling and unexpected that uniform illumi-

Receptive fields of ganglion cells and optic nerve fibers

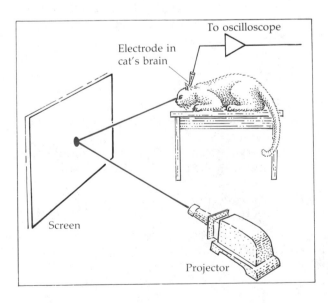

3

STIMULATION OF RETINA with patterns of light. The eyes of an anesthetized, light-adapted cat (or monkey) focus on a screen onto which various patterns of light are projected. Alternatively, a TV screen is used, with patterns generated by a computer. An electrode records the responses from a single cell in the visual pathway. Light (or shadow) falling onto a restricted area of the screen may accelerate (excite) or slow (inhibit) the signals given by a neuron. By determining the areas on the screen from which a neuron's firing is influenced, one can delineate the receptive field of the cell. The positions of cells in the brain and the tracks of electrode penetrations can be reconstructed histologically after the experiment.

nation of the eye by flashes of light is not the best way to influence discharges of ganglion cells. Uniform illumination, in fact, produces a bewildering array of responses, usually a transient burst of signals when the light is flashed on or turned off. A small spot of light, 0.2 mm in diameter, shone onto a receptive field is generally far more effective. Furthermore, the same spot of light can have opposite effects, depending on the exact position of the stimulus within the receptive field. For example, in one area the spot of light excites a ganglion cell for the duration of illumination. Such an "on" response can be converted into an inhibitory "off" response by simply shifting the spot by 1 mm or less across the retinal surface. The same spot of light, therefore, suppresses the firing of the same ganglion cell. When small spots are used to map large numbers of receptive fields, a constant and simplifying feature of the neural organization of the cat's retina emerges. There are two basic receptive field types: the ON-CENTER and the OFF-CENTER. The receptive fields are all roughly concentric, with the ganglion cell in the geometrical central region of any field.[4] Figure 4 shows these features and some of the responses that can be obtained. In an on-center receptive field, light produces the most vigorous response if it completely fills the center, whereas for most effective inhibition of firing it must cover the entire ring-shaped surround (annular illumination in Figure 4). Inhibition is always followed, when the light is turned off, by an "off" discharge. Another uniform feature is that the spotlike center and its surround are antagonistic; therefore, if they are illuminated simultaneously, they tend to cancel each other's contribution (diffuse illumination in Figure 4). There then occurs merely a relatively weak "on" component and a similar "off" discharge when the light is turned off. The off-center field has a converse organization, with inhibition arising in the circular center.

All these properties of receptive fields explain the initially puzzling finding that illumination covering an entire field has a much weaker action in arousing a ganglion cell than does a well-placed small spot or a line or an edge passing through the center.

A remarkable and simplifying consideration therefore emerges about the performance of the retina with its 100 million neurons. Their STEREOTYPED ARRANGEMENT with repeating units GIVES ONLY A FEW BASIC TYPES OF SIGNALS. Several variations in the types of concentric receptive fields are noted below. Of particular importance is the X-Y classification,[6-8] which describes the characteristics of two main groups of ganglion cells responding to stationary patterns and changes of illumination (Box 2). These characteristics appear to be preserved by successive cells at relay after relay.

The initial choice of the cat for receptive field analyses was a lucky

[6]Enroth-Cugell, C. and Robson, J. G. 1966. *J. Physiol.* 187: 517–552.
[7]Cleland, B. G., Dubin, M. W. and Levick, W. R. 1971. *J. Physiol.* 217: 473–496.
[8]Rodieck, R. W. 1979. *Annu. Rev. Neurosci.* 2: 193–225.

27
THE VISUAL
WORLD:
CELLULAR
ORGANIZATION
AND ITS
ANALYSIS

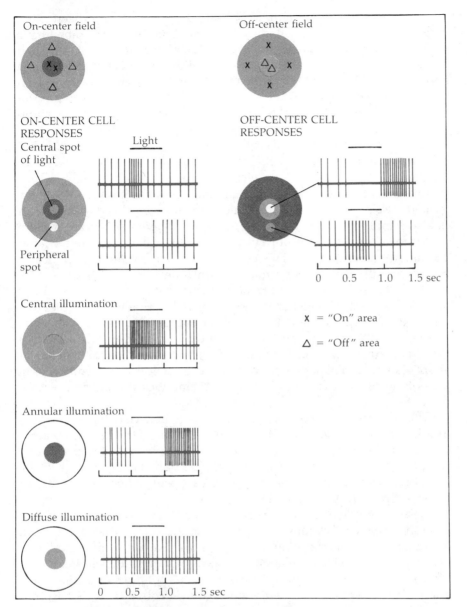

4 RECEPTIVE FIELDS OF GANGLION CELLS in the retinas of cats and monkeys are grouped into two main classes: on-center and off-center fields. On-center cells respond best to a spot of light shone onto the central part of their receptive fields. Illumination (indicated by bar above records) of the surrounding area with a spot or a ring of light reduces or suppresses the discharges and causes responses when the light is turned off. Illumination of the entire receptive field elicits relatively weak discharges because center and surround oppose each other's effects. Off-center cells slow down or stop signaling when the central area of their field is illuminated and accelerate when the light is turned off. Light shone onto the surround of an off-center receptive field area excites the corresponding cell. (After Kuffler, 1953.)

one; in the rabbit, for example, the situation would have been more complicated. As shown by Barlow,[9] the ganglion cells in the rabbit have more elaborate receptive fields and can respond specifically to such complex features as edges or to movement in one direction rather than another. Equally complex are lower vertebrates, such as frogs, which were investigated by Hartline[2] and Barlow,[5] and later by Lettvin, Maturana, Michael, and their colleagues.[10,11] In monkeys and other animals, with color vision, color-coded ganglion cells are found.[12,13]

Neighboring ganglion cells collect information from very similar, but not quite identical, areas of the retina. Even a small (0.1 mm) spot of light on the retina covers the receptive fields of many ganglion cells that have diverse responses, some ganglion cells being inhibited, others excited. (In practice it is inconvenient to describe the dimensions of fields as millimeters on the retina; instead, degrees of arc are a more useful measure. Thus, in our eyes 1 mm corresponds to about 4°, and the moon subtends 0.5°.) This characteristic organization, with neighboring groups of receptors projecting onto neighboring ganglion cells in the retina, is retained at all levels in visual pathways. The systematic analysis of receptive fields demonstrates the general principle that NEURONS PROCESSING RELATED INFORMATION ARE CLUSTERED TOGETHER. In sensory systems this means that the central neurons dealing with a particular area of the surface can communicate with each other over short distances. This appears to be an economical arrangement, as it saves long lines of communication and simplifies the making of connections.

The size of the receptive field of a ganglion cell depends on its location in the retina.[8] The receptive fields situated in the central areas of the retina have much smaller centers than those at the periphery; receptive fields are smallest in the area centralis (a region that in the cat corresponds to the fovea in the human eye), where the acuity, or resolving power, of vision is highest.

As an example of the generalization of the functional importance of receptive fields of various sizes, consider sensory areas in the skin, activated by touch or pressure. There is a strikingly similar gradation of receptive field size in relation to fine resolution or discrimination. A higher order sensory neuron in the brain responding to a fine touch applied to the skin of the fingertip has a receptive field that is very small compared to that of a neuron having a field on the skin of the upper arm. To discern the form of an object, we use our fingertips and foveas, not the less discriminating regions on the receptor surfaces with poorer resolution.

[9]Barlow, H. B., Hill, R. M. and Levick, W. R. 1964. *J. Physiol. 173*: 377–407.

[10]Maturana, H. R., Lettvin, J. Y., McCulloch, W. S. and Pitts, W. H. 1960. *J. Gen. Physiol. 43*: 129–175.

[11]Michael, C. R. 1973. *N. Engl. J. Med. 288*: 724–725.

[12]Rodieck, R. W. 1973. *The Vertebrate Retina: Principles of Structure and Function.* Freeman, San Francisco.

[13]Schiller, P. H. and Malpeli, J. G. 1977. *J. Neurophysiol. 40*: 428–445.

The receptive field studies on ganglion cells have led logically into the exploration of higher centers. They also have stimulated an examination of the neural machinery that synthesizes the ganglionic receptive fields within the retina. The fields come about by synaptic interaction in the maze of retinal connections. One would like to know, therefore, both the wiring diagram and the mechanisms of synaptic action within the network.

How are retinal neurons connected to form receptive fields of ganglion cells?

Owing to methodological difficulties, a number of years passed before serious inroads were made on these problems. Among the essential improvements of methods was the development of finer capillary electrodes that would penetrate the various cells. However, identification of neurons from which intracellular recordings had been made necessitated marking them with dyes or the enzyme horseradish peroxidase (see later, Figure 6). The introduction of improved methods through the pioneering work of Svaetichin,[14] Tomita,[15] and their colleagues gave the impetus for a quiet, but real, revolution in retinal physiology. This development promises to gain further momentum by the use of cellular neurochemistry, a methodology that has contributed greatly to the steady progress in understanding chemical communication between nerve cells. One has the impression, however, that more remains to be done before the diverse pieces fall into place for a coherent understanding of the entire retinal system.

To find out how the receptive fields of ganglion cells are constructed, it is plainly necessary to examine the performance of the other cell types within the retina: These are the photoreceptor, bipolar, horizontal, and amacrine cells—the neurons that occupy the bulk of the retina and form the input to the ganglion cells. But even in the absence of detailed information, on the basis of the physiological results discussed above one could predict some general features of intraretinal organization. To form concentric receptive fields, a set of specially arranged inhibitory and excitatory lines of communication must run between the photoreceptors and the ganglion cells. The crucial problem, therefore, is to find out what is happening in the two intraretinal synaptic stations, in the OUTER and INNER PLEXIFORM LAYERS.

A general guide for this discussion is supplied by Dowling and Boycott's[16] diagrammatic representation of a primate retina (Figure 5). The scheme is based on structural and physiological observations. The pathway from receptors through bipolar cells to ganglion cells is far more complex than a simple through-line. There exists, in addition, a variety of interconnections among cells in the various layers. For example, the horizontal cells shown in Figure 5 receive synapses from many receptors and in turn feed back onto them as well as onto bipolar cells. Similarly, amacrine cells, which receive their inputs from bipolar cells, send synapses back to them as well as to the ganglion cells. From

[14]Svaetichin, G. 1953. *Acta Physiol. Scand. 29*: 565–600.
[15]Tomita, T. 1965. *Cold Spring Harbor Symp. Quant. Biol. 30*: 559–566.
[16]Dowling, J. E. and Boycott, B. B. 1966. *Proc. R. Soc. Lond. B 166*: 80–111.

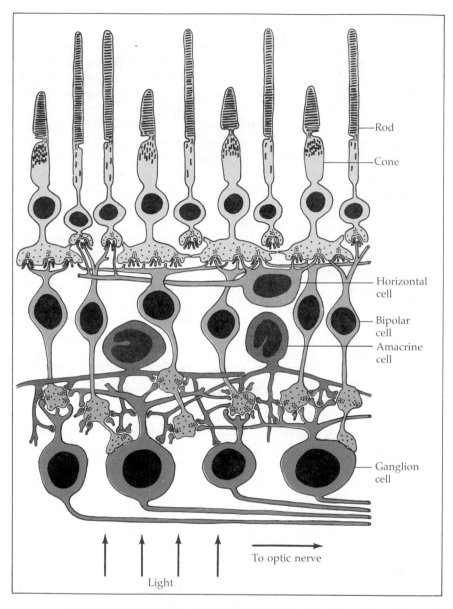

Rod

Cone

Horizontal
cell

Bipolar
cell

Amacrine
cell

Ganglion
cell

To optic nerve

Light

5 **ORGANIZATION OF PRIMATE RETINA. (After Dowling and Boycott, 1966.)**

such structural and physiological considerations, one can conclude that horizontal cells and amacrine cells modify or influence the transfer of information through the retina. The receptive field of a ganglion cell, therefore, is a composite constructed by the receptive fields of the cells along the lines leading to it.

Each of the major classes of neurons shown in Figure 5 is known to have various morphological subtypes, often with clear implications for

31
THE VISUAL
WORLD:
CELLULAR
ORGANIZATION
AND ITS
ANALYSIS

function (see Box 2). For example, in monkeys, ganglion cells receiving input from the fovea are small and receive inputs from only a few bipolar cells, or even from only a single one.

For technical reasons, lower vertebrates are especially favorable for studying the properties and receptive fields of the various types of cells within the retina. In these animals the cell types appear similar to those of the cat and are often easier to impale with microelectrodes. Accordingly, the brief account that follows is taken largely from the work of Tomita, Kaneko, Dowling, Baylor, Fuortes, and their colleagues, who used a variety of fish, amphibian, and turtle retinas.[15,17-19] Once the general outlines were known, it became possible to correlate structure and function in greater detail and with modern techniques to make a similar analysis of the cat retina.[20]

The techniques of intracellular recording and dye injection are described in detail later. In brief, the procedure is to record from a cell with an intracellular microelectrode that measures the potential across the surface membrane. By appropriate illumination one can determine the membrane properties, the responses to light, and the receptive field organization of a particular cell. After recording, a small amount of dye or enzyme is injected into the cell—often a difficult procedure. Subsequently, the tissue is examined histologically and the cell that had been impaled by the microelectrode is identified. Examples of various types of injected cells are shown in Figure 6. The different classes of cells in the retina give highly distinct responses, which are sufficient for reliable identification once the initial correlation between structure and function has been made. However, within each class, further injections are required for the recognition of the cell types that correspond to subgroups,[21] for example, α ganglion cells with Y cells.[22]

From the outset, the intracellular records revealed that cells in the vertebrate retina gave different signals from what might have been predicted on the basis of knowledge of impulses in invertebrates and in the central nervous system of vertebrates. Only ganglion cells and amacrine cells gave impulses of the type usually observed in typical neurons (Chapter 4). The other cell types responded to illumination or darkness with relatively slow, graded potentials that in most cells are found only in synaptic regions. Particularly surprising and somewhat confusing, when first observed, was the electrical response given by receptors to illumination: HYPERPOLARIZATION (an increase in internal negativity). The sign of this potential is the opposite of what is seen in other sensory receptors of the body. Direct measurements have shown that a depolarizing current flows continually through the receptor "at

Photoreceptors

[17]Kaneko, A. 1970. *J. Physiol. 207:* 623–633.
[18]Dowling, J. E. and Werblin, F. S. 1971. *Vision Res. 3:* 1–15.
[19]Baylor, D. A., Fuortes, M. G. F. and O'Bryan, P. M. 1971. *J. Physiol. 214:* 265–294.
[20]Nelson, R., Famiglietti, E. V. Jr. and Kolb, H. 1978. *J. Neurophysiol. 41:* 472–483.
[21]Sterling, P. 1983. *Annu. Rev. Neurosci. 6:* 149–185.
[22]Cleland, B. G., Levick, W. R. and Wässle, H. 1975. *J. Physiol. 248:* 151–171.

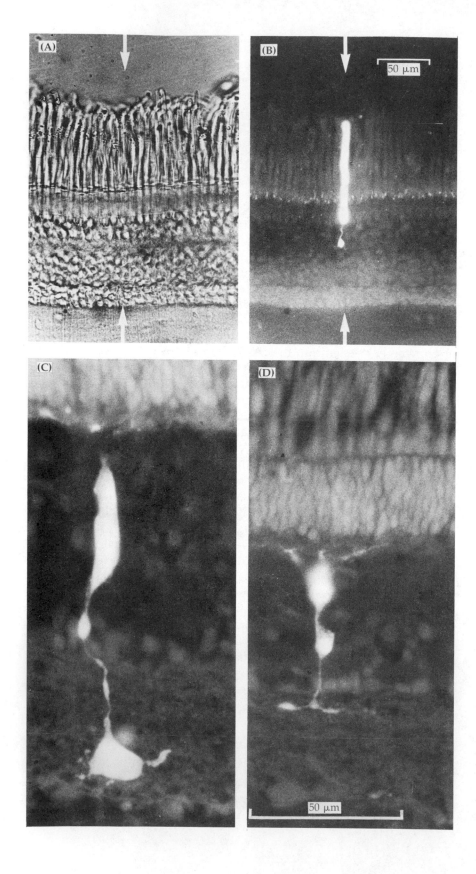

33
THE VISUAL
WORLD:
CELLULAR
ORGANIZATION
AND ITS
ANALYSIS

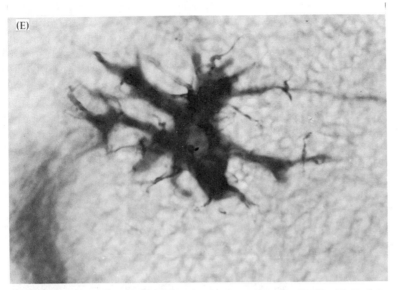

(E)

6 MARKING AND IDENTIFICATION OF NEURONS. (A, B) Rod in retina of toad injected with fluorescent dye, Lucifer Yellow, as seen in normal and ultraviolet light. Arrows mark identical points on retina. Scale, 50 μm. (C, D) Bipolar cells of goldfish injected with procion yellow. Note that the bipolar cell in (C) which is depolarizing (on-center) has a different morphology from that in (D) which is hyperpolarizing (off-center); the two types of bipolar cells end in different sublayers of the inner plexiform layer. Scale, 50 μm. (E) Horizontal cell in dogfish retina injected with horseradish peroxidase. (A and B courtesy of B. Nunn, unpublished. C, D, and E courtesy of A. Kaneko, unpublished.)

rest" in the dark and that this "dark current" is turned off by light[23] (see Box 1). Such responses have been observed not only in lower vertebrates but also in photoreceptors of monkeys.[24]

What is the interpretation of this result and how does it influence signaling? Chapter 15 describes in detail how sensory cells outside the eye respond to stimulation and how one cell hands on information to the next by secreting a chemical transmitter. For the moment it is convenient to illustrate critical points about the difference between signaling in photoreceptors in vertebrate and invertebrate eyes. Readers who are unfamiliar with this kind of material may prefer to read Chapter 4 before continuing.

Figure 7A shows the response of a photoreceptor cell in the eye of the horseshoe crab (*Limulus*).[25] The sequence of signals is similar to that in all other sensory cells that have so far been studied. They all become DEPOLARIZED by the appropriate stimulus. Heating, cooling, pressing, and so on always lead to the generation of a current and depolarization: The inside of the cell becomes more positive than when at rest (see definitions in Chapter 1). This change in potential is excitatory and

[23]Hagins, W. A., Penn, R. D. and Yoshikami, S. 1970. *Biophys. J. 10*: 380–412.
[24]Nunn, B. J. and Baylor, D. A. 1982. *Nature 299*: 726–728.
[25]Fuortes, M. G. F. and Poggio, G. F. 1963. *J. Gen. Physiol. 46*: 435–452.

causes impulses to be initiated in the nerve fibers. When these impulses reach the end of an axon, they in turn cause the liberation of a chemical transmitter from the endings, and this substance influences the next cell in line. Much is now known about these processes; in particular, it is clear that the release of transmitter is triggered by depolarization of the nerve endings and suppressed by hyperpolarization (increased internal negativity). Depending on the chemistry of the transmitter substance and the membrane of the second-order cell, its response may be depolarization or hyperpolarization; in other words, excitation or inhibition. In contrast, recording from a vertebrate photoreceptor (Figure 7B) shows that illumination causes a hyperpolarization, the size of which is graded with the intensity of the light flash.

This is why the hyperpolarization response of rods and cones to light was so surprising. It represents the turning off of a steady current. By analogy with other receptors, vertebrate photoreceptors behaved as though dark was the stimulus, since that was the condition in which they were depolarized. Light apparently signaled the absence of a stimulus. Again by analogy, this would imply that illumination turned off the liberation of transmitter, a release that was proceeding continuously in the dark. It is now clear that the explanation of this apparent paradox is that photoreceptors do continually release transmitter in the dark and stop releasing it on illumination.

An elegant demonstration of the through-line from photoreceptor to ganglion cell was provided by Baylor and Fettiplace.[26] They changed the membrane potential of a single photoreceptor by passing current through an intracellular microelectrode while recording from a ganglion cell to which it was connected via bipolar cells. Figure 8 shows the

[26]Baylor, D. A. and Fettiplace, R. 1977a. *J. Physiol. 271*: 391–424.

7

RESPONSES OF PHOTORECEPTORS. (A) Photoreceptors of an invertebrate (horseshoe crab) respond to light (indicated by bar above record) by a depolarization that gives rise to impulses. This is the usual type of response elicited from vertebrate sensory receptors activated by various stimuli such as touch, pressure, or stretch. **(B)** Photoreceptors of a vertebrate (turtle) respond by a hyperpolarization that is graded according to the intensity of the light flash. (A after Fuortes and Poggio, 1963; B after Baylor, Fuortes, and O'Bryan, 1971.)

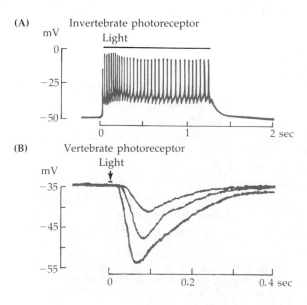

35
THE VISUAL
WORLD:
CELLULAR
ORGANIZATION
AND ITS
ANALYSIS

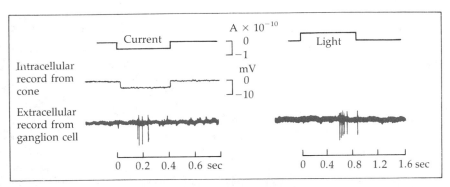

8 PHOTORECEPTOR–GANGLION CELL INTERACTION in turtle retina. Impulses are evoked in an on-center ganglion cell by current that hyperpolarizes a red-sensitive cone and by light. An intracellular electrode is used to measure the membrane potential of the cone and to hyperpolarize it. The impulses that follow in the ganglion cell are recorded with an extracellular electrode. A small spot of light shone onto the cone produces a similar effect. (After Baylor and Fettiplace, 1977a.)

firing initiated in the ganglion cell by hyperpolarization of the single receptor, in this case a cone sensitive to red light. Other off-center ganglion cells were activated by depolarization of single photoreceptors. These experiments showed further that light and membrane potential changes produced by artificially applied currents had identical effects on the firing of ganglion cells. As expected, appropriate currents could quantitatively counteract or enhance the effects of light. Hence, the hyperpolarization or depolarization per se of a receptor is the event that signals information to other cells in the brain about light or darkness in the outside world.

Although photoreceptors are influenced predominantly by light falling directly upon their outer segments, they can also activate each other. They do this in two ways: First, direct electrical interactions have been observed physiologically between photoreceptors, a result that is reinforced by the finding of anatomical junctions characteristic of electrical coupling (Chapter 9). Second, Baylor, Fuortes, and O'Bryan[19] have shown that in the turtle, activation of cones by light may also affect neighboring receptors through the intermediary of horizontal cells that, in turn, can act back onto cones. This feedback action can create a surround effect in a receptor. In this way diffuse illumination serves to reduce the excitability of the receptor. An unusual effect can be produced experimentally through the feedback pathway: Sudden removal of diffuse light from the surround can induce an action potential in the receptor under certain circumstances.[27]

HORIZONTAL CELLS, which receive synaptic inputs from many receptors, give a variety of maintained responses to illumination. In the carp or goldfish, for example, one type of horizontal cell responds with either

Horizontal cells, bipolar cells, and amacrine cells

[27]Piccolino, M. and Gerschenfeld, H. M. 1980. *Proc. R. Soc. Lond. B 206*: 439–463.

BOX 1 SENSITIVITY OF PHOTORECEPTORS

Psychophysical experiments have revealed that the eye is capable of perceiving dim flashes of light—so dim that only a few single photons are trapped by individual photoreceptors. Somehow a single photon can give rise to an electrical signal; and this signal must be sufficiently large and long to activate cell after cell in the relay leading to consciousness. The signal must also stand out as the symbol of a dim flash, to be separated from the "noise" or false signals occurring spontaneously at rest in the absence of light, signals caused by thermal agitation and other processes at the molecular level. The end result is that a subject can say, "Yes, the light was flashed."[28]

Much information is now available about how light is trapped by the pigment molecule rhodopsin within photoreceptors and about the chemical steps that result in changes in the structure of the molecule.[29] There is also evidence that the changes in configuration lead to liberation of chemical messengers inside the photoreceptor (probably calcium ions) that diffuse to the cell membrane.[30] It is the interaction of messenger with specific ion channels in the membrane that gives rise to an electrical signal. But what sort of electrical event could one single quantum pro-

(A)

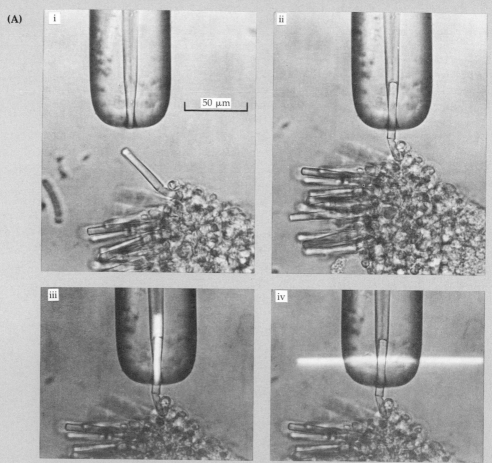

50 μm

MEMBRANE CURRENT OF ROD OUTER SEGMENT. (A) i, ii, iii, iv. A suction electrode with a fine tip is used to suck up the outer segment of a rod that protrudes from a piece of toad retina. Slits of light illuminate the receptor with precision. Since the electrode fits tightly around the photoreceptor, current flowing into or out of it is recorded. (B) Responses of toad rod outer segment to 40 consecutive dim flashes, marked by dots. The responses

duce and how could such a signal be measured?

To approach these problems, recordings have been made by Baylor and his colleagues using a novel technique.[31] The experimental arrangement for measuring the currents generated by an individual rod (the sensitive photoreceptor used for night vision in a toad or a monkey) is shown in the figure. A piece of retina is isolated from the animal and maintained in a chamber. Part of the outer segment of a rod—the region where light is trapped by rhodopsin—is sucked into a fine pipette (a snug fit is necessary). At rest in the dark (as mentioned above, page 33), a current flows continuously into the outer segment of the photoreceptors. This is called the "dark current." The effect of light is to close channels in the outer segment, causing a decrease in the steady dark current. It is this sort of change in the current that causes the electrical potentials recorded with intracellular microelectrodes such as those shown in Figure 7. With the system shown in the figure, the currents are recorded directly and with high resolution from particular areas of membrane. Panel B shows the results of very dim flashes corresponding to one or two quanta of light. As one might expect, the currents are small and quantal in nature. That is to say, sometimes a dim flash evokes a unitary response, sometimes a doublet, and sometimes nothing at all. In monkey rods, the unit current caused by a single photon is about 0.5×10^{-12} A. Typical rod responses to more intense, brief flashes are shown in panel C. Studies of the kinetics of transfer through the retina indicate that the properties of the synapses are exquisitely tuned to enable signals of the amplitude and waveform produced by single quanta to be carried from cell to cell without being lost in the process.[32]

These experiments provide a rare example of the way in which a process as complex as seeing the dimmest possible flashes of light can be correlated with the events that occur in single molecules.

[28]Hecht, S., Shlaer, S. and Pirenne, M. H. 1942. *J. Gen. Physiol. 25*: 819–840.

[29]Knowles, A. 1982. In H. B. Barlow and J. D. Mollon (eds.), *The Senses*, Cambridge University Press, New York, pp. 82–101.

[30]Yoshikami, S., George, J. S. and Hagins, W. A. 1980. *Nature 286*: 395–398.

[31]Baylor, D. A., Lamb, T. D. and Yau, K.-W. 1979. *J. Physiol. 288*: 589–611.

[32]Baylor, D. A. and Fettiplace, R. 1977b. *J. Physiol. 271*: 425–448.

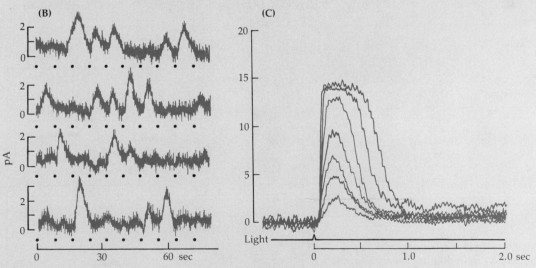

fluctuate in a quantal manner. The small deflections are the currents generated by single photons interacting with visual pigment. Often photoisomerizations fail to occur. (C) Records from rod in monkey retina with flashes of increasing intensity. These currents are the counterpart of voltage traces of Figure 7B (A, B from Baylor, Lamb, and Yau, 1979; C from Nunn and Baylor, 1982.)

depolarization or hyperpolarization, depending on the wavelength of light, while another type just hyperpolarizes in a graded manner according to light intensity;[33,34] neither kind of cell generates conducted impulses. Since the horizontal cells are supplied by receptors over a large area of retina, their fields are far larger than those of bipolar or ganglion cells. In addition, neighboring horizontal cells are coupled to each other with junctions that permit current to flow between them, and this coupling further increases the size of their receptive fields. In the cat, as in lower vertebrates, two main types of horizontal cell with large receptive fields have been described; they differ in their morphology and in the contacts they make with rods and cones.[21]

BIPOLAR CELLS respond to light with a sustained depolarization or hyperpolarization and, like horizontal cells, they do not generate impulses. The bipolar cells in the goldfish and in the mud puppy have a concentric receptive field organization, with an antagonistic center-surround arrangement. An example is shown in Figure 9: A spot of light in the center hyperpolarizes and an annular light stimulus depolarizes. For this type of cell a small spot of light surrounded by darkness is the most effective stimulus. Other bipolar cells are depolarized by light at the center. Diffuse illumination of the whole field is relatively ineffective.[17] Hence, already at this level some of the key features of the receptive fields of ganglion cells are evident (Figure 4).

The center of the receptive field of the bipolar cell is supplied directly by the receptors themselves and the extensive surround by the horizontal cells, the receptive fields of which spread laterally over a wide area.[35] The horizontal cells can drive bipolar cells and through them

[33]Kaneko, A. 1971. *J. Physiol. 213*: 95–105.
[34]Drujan, B. D. and Laufer, M. (eds.) 1982. *The S-Potential.* Alan R. Liss, New York.
[35]Kaneko, A. 1979. *Annu. Rev. Neurosci. 2*: 169–191.

9

RECEPTIVE FIELD OF BIPOLAR CELL in a goldfish retina responding by hyperpolarization to illumination of its center and by depolarization to a ring of light. Other bipolar cells respond in the opposite way (depolarization with central illumination); neither type generates impulses. (From Kaneko, 1970.)

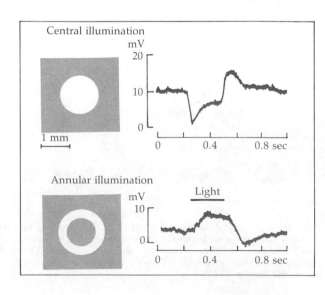

39
THE VISUAL
WORLD:
CELLULAR
ORGANIZATION
AND ITS
ANALYSIS

elaborate the surround part of receptive fields of ganglion cells. The output of the bipolar cells reports mainly the difference of the illumination between periphery and center of the receptive field, thereby providing a clear contrast mechanism. Among the various anatomical types of bipolar cells, some receive inputs only from cones, others only from rods. In the carp and turtle, both of which have color vision, center-surround bipolar cells have been found in which the two areas of the receptive field have different spectral sensitivities.[36,37] Moreover on- and off-center bipolar cells differ from one another in their morphology, projecting to different sublaminae of the inner plexiform layer, where they connect to different populations of ganglion cells and amacrine cells.[38] Similar classes of bipolar cells with comparable projections have also been found in the cat retina.[21] In this entire scheme the graded responses in receptors, bipolar cells, and horizontal cells are reflected in a graded output of transmitter; all the connections of these cells with each other are made in one layer (the outer plexiform layer).

In the second retinal layer, the AMACRINE CELLS receive no connection from the receptors, but only from bipolar cells and other amacrine cells. The amacrine cells also feed back onto the bipolar cells, besides making connections with ganglion cells. Unlike horizontal and bipolar cells, amacrine cells in *Necturus* (the mud puppy) and the turtle give impulses.[18] Again, in the mammalian retina, several distinct types of amacrine cell can be distinguished by their morphology, chemistry, arborization, and responses.[21] The precise roles played by amacrine cells are not known; they may be concerned with the ability of ganglion cells to detect moving stimuli and with directional sensitivity.

In birds, another important level of control in the retina has been established. EFFERENT FIBERS from the brain terminate upon amacrine cells and modulate their receptive field properties. This, in addition to the observed feedback control between cells, reminds one that although the main flow of information is from receptors to ganglion cells, there can also exist signaling in the opposite (centrifugal) direction (Chapter 15).[39,40] The ganglion cell discharges presumably reflect the balance between the bipolar and amacrine inputs, but the actual role of amacrine cells in producing the typical on- or off-center field is not clear.

In conclusion, the activity of ganglion cells is made up by the total contribution of four cell types. In the outer retinal layer, the receptor and horizontal cells interact with each other and with the bipolar cells. These in turn, together with the amacrine cells, determine the signals that arise in the ganglion cells. The rich system of retinal interconnections is not yet understood in detail, and there remain wide gaps in understanding other aspects of retinal function. For example, little is known of synaptic mechanisms that link neurons. In chemical studies

[36]Kaneko, A. and Tachibana, M. 1983. *Vision Res. 23*: 381–388.
[37]Marchiafava, P. L. and Weiler, R. 1982. *Proc. R. Soc. Lond. B 214*: 403–415.
[38]Famiglietti, E. V. Jr., Kaneko, A. and Tachibana, M. 1977. *Science 198*: 1267–1269.
[39]Cowan, W. M. and Powell, T. P. S. 1963. *Proc. R. Soc. Lond. B 158*: 232–252.
[40]Pearlman, A. L. and Hughes, C. P. 1976. *J. Comp. Neurol. 166*: 123–132.

on retina and on isolated neurons, information about transmitters is becoming available.[41] Thus, known transmitters such as γ-aminobutyric acid, acetylcholine, dopamine, and various peptides (the actions of which are discussed in Chapter 12) have been found in retina and located within specific identified cell types.[21] γ-aminobutyric acid, for example, is associated with horizontal and amacrine cells.[42-44] In comparison with what was known ten to fifteen years ago, the advances seem dramatic, and a definitive outline of the structure, chemistry, and function of the various cells in the organization of the retina has emerged.

What information do ganglion cells convey?

The most striking feature of ganglion cell signals is that they tell a different story from that of primary sensory receptors. They do not convey information about absolute levels of illumination because they behave in a similar fashion at different background levels of light. They ignore much of the information of the photoreceptors, which work more like a photographic plate or a light meter. Rather, they measure differences within their receptive fields by comparing the degree of illumination between the center and the surround. Apparently they are designed to notice simultaneous contrast, the transition from more to less or no light, whereas they are relatively insensitive to gradual changes in overall illumination. They are exquisitely tuned to detect such contrast as the edge of an image or a bar crossing the opposing regions of a receptive field.

What image of the world is presented to the brain by the retina? By comparison with our daily experience, it is a rather drab environment. The world is represented as a series of closely spaced dots and contours depending on DIFFERENCES in the level of illumination. Position and movement on the retinal surface are indicated and the picture is livened by color. On the other hand, by comparison with information obtained from primary receptors alone (Chapter 15), the information provided by the entire retina is relatively lively. It is apparent, therefore, that the three retinal layers extract and analyze a great deal of information about the outside world. By choosing some aspects of the information collected by the primary receptors, and not others, a start has been made in selecting features that are important for form vision while jettisonsing the level of background illumination.

Lateral geniculate nucleus

The optic nerve fibers running to the cortex from each eye terminate on cells of the right and left lateral geniculate nucleus, a distinctively layered structure (GENICULATE means "bent like a knee"). In the lateral geniculate nucleus of the cat there are three obvious, well-defined layers of cells (A, A₁, C), one of which (C) has a complex structure that has been further subdivided.[47] In the monkey the lateral geniculate nucleus

[41]Drujan, B. D. and Svaetichin, G. 1972. *Vision Res. 12*: 1777–1784.

[42]Dowling, J. E., Lasater, E. M., Van Buskirk, R. and Watling, K. J. 1983. *Vision Res.* 23: 421–432.

[43]Lam, D.M.-K. and Ayoub, G. S. 1983. *Vision Res.* 23: 433–444.

[44]Miller, R. F. and Dacheux, R. F. 1983. *Vision Res.* 23: 399–412.

[47]Guillery, R. W. 1970. *J. Comp. Neurol.* 138: 339–368.

BOX 2   X AND Y CELLS

In recent years the classification of concentric receptive fields of ganglion cells has been extended and refined, and almost certainly additional features of receptive field performance will be uncovered in the future. In the cat two main types of on-center and off-center fields have been distinguished on the basis of their responses to stationary and moving patterns of light.[6–8] One class—X cells—responds to stationary spots or gratings in a predictable manner. The responses show spatial summation that is approximately linear. For example, as a spot is made larger to encroach more on "off" areas, the firing is reduced progressively. The other main class—Y cells—responds briskly to changes in illumination or to moving stimuli without clear spatial summation. In practice the specific characteristics of the ganglion cells seem designed to enable X cells to distinguish the fine grain of stationary patterns (or higher frequencies of gratings) and to enable Y cells to detect objects moving across the visual field or to detect changes in intensity. The axons of the X ganglion cells conduct more slowly than those of the Y cells. In addition to the typical on- and off-center cells, other cells with more complex receptive field properties exist; for example, several laboratories have reported ganglion cells with centers that can be aroused by either light *or* dark spots. These neurons, called W cells,[8] have the slowest conduction velocities so far observed in any axons arising from ganglion cells. In the monkey, as in the cat, some ganglion cells behave like X cells—with spatial summation, sustained discharges, and slower conduction velocities; others resemble Y cells—with brisk discharges to changes in illumination and rapid conduction velocities. As one

might expect, ganglion cells whose receptive fields are specifically color-coded have been noted in various animals, including the monkey, the ground squirrel, and some fishes.[8] These animals, in contradistinction to the cat, possess excellent color vision and an intricate neural mechanism for processing color. In the monkey, cells with X-like properties are the ones that respond selectively to specific wavelengths of light shone onto the center or surround regions of their receptive fields; Y-like cells show no such spectral sensitivity.[12]

Characteristic, structural features of ganglion cells accompany their X- or Y-type properties. Thus, ganglion cells can be classified into two groups—α and β—by their sizes and their arborization.[45] The larger cells, known as α, have widely spread dendritic arborizations and correspond to Y-type ganglion cells (brisk responses to changes in illumination); the smaller β cells correspond to X cells (linear spatial summation, sustained responses, and, in the monkey, color sensitivity). Within the β group of ganglion cells there are two subgroups having dendritic processes that end characteristically in different layers of the inner plexiform layer. These correspond to on-center and off-center X ganglion cells. At the next relay station, in the lateral geniculate nucleus, the distinction between X and Y characteristics is preserved. Moreover, the X- and Y-type cells are situated in different layers (Figure 10) and have different morphologies, the Y cells again being larger.[46] X and Y geniculate axons maintain their segregation and end in different layers of the cortex, the Y fibers ending in the more superficial sublayers of layer IV (see later).

[45]Wässle, H., Peichl, L. and Boycott, B. B. 1981. *Proc. R. Soc. Lond. B* 212: 157–175.
[46]Friedlander, M. J., Lin, C.-S., Stanford, L. R. and Sherman, S. M. 1981. *J. Neurophysiol. 46*: 80–129.

has six layers of cells (Figure 10). The cells in the deeper layers 1 and 2 are larger than those in layers 3, 4, 5, and 6, giving rise to the terms *magnocellular* and *parvocellular* layers. With these layers are correlated different functional properties (see later). In both cat and monkey each layer is predominantly supplied by one or the other eye. In the cat the

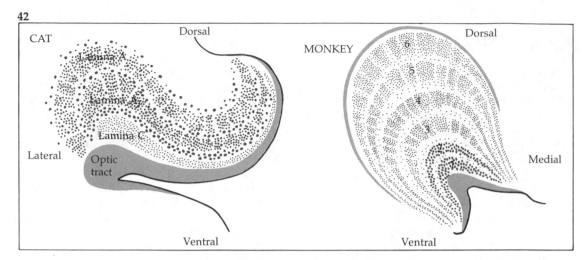

10 LATERAL GENICULATE NUCLEUS. (A) In the lateral geniculate nucleus of the cat, there are three layers of cells—A, A_1, and C. (B) in the monkey, the lateral geniculate nucleus has six layers. In the four dorsal layers (3, 4, 5, 6; parvocellular), the cells are smaller than in layers 1 and 2 (magnocellular). In both animals each layer is supplied by only one eye and contains cells with specialized properties. (From Szentágothai, 1973.)

inputs to layers A and C originate in ganglion cells in the eye on the other side of the animal, the fibers having crossed at the chiasm; A_1 is supplied with fibers that come from the eye on the same side and do not cross. In the monkey, layers 6, 4, and 1 are supplied by the opposite eye; layers 5, 3, and 2 by the eye on the same side. The segregation of endings from each eye into separate layers has been shown by recording electrically and also by a variety of anatomical techniques. Particularly striking is the arborization of a single optic nerve fiber that has been injected with the enzyme horseradish peroxidase (Figure 11): The ter-

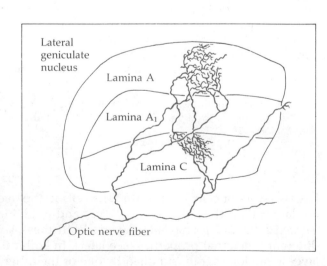

11

TERMINATION OF OPTIC NERVE FIBER in lateral geniculate nucleus of cat. A single on-center Y axon from the contralateral eye was injected with horseradish peroxidase. Branches end in layers A and C but not in A_1. (Modified from Bowling and Michael, 1980.)

43
THE VISUAL
WORLD:
CELLULAR
ORGANIZATION
AND ITS
ANALYSIS

minals are all confined to the layers supplied by that eye, with no spillover across the border.[48] Because of the orderly and systematic separation of fibers at the chiasm, the receptive fields of the cells in the lateral geniculate nucleus are all situated in the VISUAL FIELD on the opposite side of the animal.

A striking topographical feature is the highly ordered arrangement of receptive fields within individual layers of the geniculate. Neighboring regions of the retina make connections with neighboring geniculate cells, so that the receptive fields of adjacent neurons overlap over most of their area. The area centralis in the cat (the region of the cat retina with small receptive field centers) and the fovea in the monkey project onto the greater portion of each geniculate layer. There are relatively few cells devoted to the peripheral retina. This heavy representation (the same as in the cortex) presumably reflects the use of these regions for high-acuity vision and the need for fine-grained resolution. Although there are probably equal numbers of optic nerve fibers and geniculate cells, each geniculate cell receives its input from several optic fibers, which in turn split up to make synapses with several geniculate neurons.

The topographically ordered sequence of connections is not confined to one individual layer; even cells in the different layers are in register. Thus, as a microelectrode passes from one layer of the geniculate to the next, which is supplied by optic nerve fibers from the other eye, one records from successive cells driven by first one eye and then the other; the positions of the receptive fields remain in corresponding positions on the two retinas representing the same region in the visual field.[49] For the relay cells in the lateral geniculate nucleus, no extensive mixing of information or interaction between the eyes occurs, and binocularly excited cells (neurons with receptive fields in both eyes) are rare. However, interneurons that mediate inhibition and do receive inputs from both eyes have been recorded from.[50]

Surprisingly, perhaps, the responses from geniculate cells do not differ drastically from those of retinal ganglion cells (Figure 12). Geniculate neurons also have concentrically arranged antagonistic receptive fields, of either the off-center or the on-center type, but the contrast mechanism is more finely tuned by more equal matching of the inhibitory and excitatory areas. The geniculate neurons, like retinal ganglion cells, require contrast for optimal stimulation but give weaker responses to diffuse illumination. Both X and Y types are seen (Box 2).[7]

Studies of receptive fields of the lateral geniculate nucleus are still incomplete. For example, there are interneurons whose contribution has not been established and pathways of unknown function that descend from the cortex to end in the lateral geniculate nucleus. Knowledge of the synaptic organization is likely to be increased through

[48]Bowling, D. B. and Michael, C. R. 1980. *Nature 286*: 899–902.
[49]Hubel, D. H. and Wiesel, T. N. 1961. *J. Physiol. 155*: 385–398.
[50]Dubin, M. W. and Cleland, B. G. 1977. *J. Neurophysiol. 40*: 410–427.

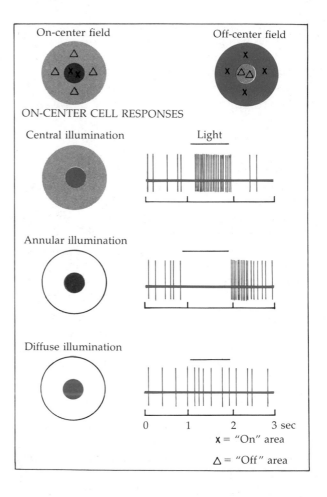

12

**RECEPTIVE FIELDS OF LATERAL GENICU-
LATE NUCLEUS CELLS.** The concentric receptive fields of cells in the lateral geniculate nucleus resemble those of ganglion cells in the
retina, consisting of on-center and off-center
types. The illustration is from an on-center cell
in the lateral geniculate nucleus of a cat. Bar
above records indicates illumination. The central
and surround areas antagonize each other's effects, so that diffuse illumination of the entire
receptive field gives only weak responses (bottom record), less pronounced than in retinal ganglion cells. (After Hubel and Wiesel, 1961.)

analyses made by intracellular electrodes. Such a technique seems essential for sorting out the excitatory and inhibitory synaptic contributions of various optic nerve fibers.

Functional significance
of layering

Why is there more than one layer devoted to each eye? It is now
becoming apparent that different functional properties are represented
in the different layers. For example, cells in the four dorsal layers of
the monkey lateral geniculate nucleus (the parvocellular layers) have X-
like properties and respond to steady illumination and to lights of
different colors (Box 2). In contrast, layers 1 and 2 (the magnocellular
layers) contain cells that give Y-like brisk responses to moving stimuli
and are not selective for wavelength.[51] Other types of neurons, the
classification of which is not yet clear, have also been found in layers 1
and 2.[52] In the cat, X and Y fibers end in different sublayers of A and
A_1. In some animals, such as the mink and the ferret, on- and off-center
cells are segregated into different layers.[53]

[51]Schiller, P. H. and Malpeli, J. G. 1978. *J. Neurophysiol. 41*: 788–797.
[52]Kaplan, E. and Shapley, R. M. 1982. *J. Physiol. 330*: 125–143.
[53]LeVay, S. and McConnell, S. K. 1982. *Nature 300*: 350–351.

45
THE VISUAL
WORLD:
CELLULAR
ORGANIZATION
AND ITS
ANALYSIS

In Chapters 3 and 20, the lateral geniculate nucleus is again discussed in connection (1) with the projection of the cortex and (2) with the changes that occur in Siamese cats and albino animals of various species in which fibers cross abnormally at the chiasm as a result of a genetic defect. Also discussed are the structural and physiological effects of use and disuse in immature animals.

THE VISUAL CORTEX

Proceeding from the retina and lateral geniculate nucleus to the cerebral cortex raises questions that go beyond simple matters of technique. It has long been acknowledged that understanding the workings of any part of the nervous system requires knowledge of the cellular properties of its elementary units, the neurons—how they conduct and carry information and how they transmit that information from one cell to the next at synapses or specialized points of contact. Yet thirty years ago it was not obvious that monitoring activity in single neurons could reveal many of the processes underlying higher functions. The argument usually took (and still does, at times) the following form. The brain contains some 10^{10} or more cells. Even the simplest task or event, like a movement or looking at a line or a square, engages hundreds of thousands of nerve cells in various parts of the nervous system. What chance does a physiologist have of gaining insight into complex actions within the brain when he samples only one or a few of those units, a hopelessly small fraction of the total number? This pessimistic view was in vogue until fairly recently.

General problems and question of numbers

On closer scrutiny, the logic of the argument about basic difficulties introduced by large numbers and complex higher functions is perhaps not so impeccable as it seems. As so frequently happens, some SIMPLIFYING PRINCIPLE turns up, opening a new and clarifying view. In the case of the retina, after all, well over 100 million cells deal with the infinite variety of the world. What simplifies the situation are the FEW CELL TYPES that are laid out in an apparently well-ordered manner, as REPEATING UNITS. Receptive field studies now indicate that this maze of over 100 million cells and many hundred million cross connections results in only a few stereotyped kinds of responses. What now holds back the investigator of the retina is not the question of numbers but rather the technical problems associated with determining the patterns of connections and their integrative mechanisms at the synaptic sites. In some respects the situation is similar in the cerebellum, which contains many more neurons than the retina, but again, apart from neuroglia, has only five principal cell types (Chapter 1, Figure 2).

The retina analyzes visual events in a stepwise manner. Hubel and Wiesel acted on the assumption that the visual centers would perform their processing according to similar principles, but at a more advanced level. At the outset, however, they faced completely unanswered questions; for example, how signals are processed after they arrive by way of the optic radiation at various cortical cells. As a start, they took their

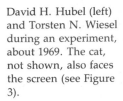

David H. Hubel (left) and Torsten N. Wiesel during an experiment, about 1969. The cat, not shown, also faces the screen (see Figure 3).

clue from the simple anatomical observation that neurons in the visual system tend to be arranged in ordered layers. This layering is also present in the cortex, but is less striking than in the retina or the lateral geniculate nucleus. Therefore, it seemed reasonable to assume, as a working hypothesis, that information was handed on according to hierarchically ordered levels, progressing from one layer to the next. Hubel has written, "I think that the most important advance was the strategy of making long microelectrode penetrations through the cortex, recording from cell after cell, comparing responses to natural stimuli, with the object not only of finding the optimal stimulus for particular cells, but also of learning what the cells had in common and how they were grouped."[54]

The experimental procedure consisted of recording from neurons and finding for each one the appropriate stimuli and the configuration of the receptive field. The search resulted in the discovery of a marked increase in specialization of neurons: They respond selectively to visual patterns of progressively greater complexity at ascending levels in the hierarchy of the cortex. From the transformation of the signal patterns at different stages, one can derive much information about the general wiring diagram of connections within the visual cortex.

Structure of the cortex

Visual information passes to the cortex from the lateral geniculate nucleus through the optic radiation. In the cat the part of the visual cortex where the optic radiation ends consists of a folded plate of cells about 1 to 2 mm thick (Figure 13). This region of the brain, area 17 (also called the striate cortex or visual area I), lies posteriorly in the occipital lobe and can be recognized in cross section by its characteristic appearance. Incoming bundles of fibers form in this area a clear stripe that can be seen by the naked eye—hence the name "striate." The regions of cortex surrounding the striate cortex are also concerned with vision. Two adjacent areas that can be recognized are called 18 and 19 (visual areas II and III) and receive inputs from area 17 (Figure 13). Their exact boundaries cannot be defined by simple inspection of the surface of the

[54]Hubel, D. H. 1982a. *Annu. Rev. Neurosci.* 5: 363–370.

47
THE VISUAL
WORLD:
CELLULAR
ORGANIZATION
AND ITS
ANALYSIS

13

VISUAL AREAS ON CAT CORTEX. As in primates (Figure 1), three distinct areas can be recognized by several criteria. Area 17 (striate cortex, visual area I) projects to areas 18 and 19 (visual areas II and III). Areas 18 and 19 are further subdivided into separate discrete areas. Parts of the areas are relatively inaccessible, lying on the medial surface of the hemisphere or within the folds of the cortex. (A) View from above. (B) Coronal section. (C) Medial view.

brain, but a number of structural criteria exist. For example, in area 18 the striate appearance of layer IV is lost; large characteristic cells are found in the third layer; and coarse, obliquely running, myelinated fibers are seen in the deeper layers. Over the past decade it has been shown that many more than these three areas exist in the occipital cortex of rats, guinea pigs, and monkeys. All of these areas are concerned with vision, and each contains its own representation of the visual field projected in an orderly manner.[55] From area to area there is much variation in the types of cells seen and the relative thickness of the different layers, and this provides additional criteria for demarcating boundaries. In addition, physiological experiments reveal functional

[55]Van Essen, D. C. 1979. *Annu. Rev. Neurosci.* 2: 227–263

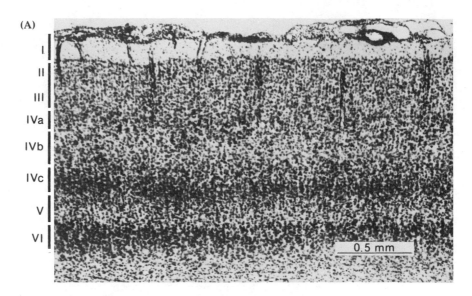

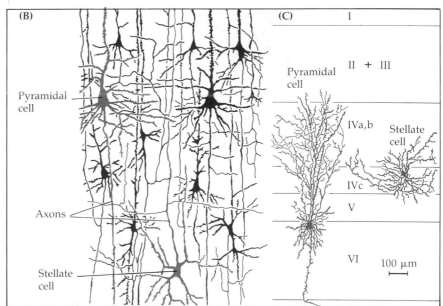

14 ARCHITECTURE OF VISUAL CORTEX. (A) Distinct layering of cells in a section of striate cortex of the macaque monkey, stained to show cell bodies (Nissl stain). Fibers arriving from the lateral geniculate nucleus end in layers IVa, IVb, IVc. (B) Drawing of pyramidal and stellate cells (Golgi stain) in the visual cortex of the cat. The connections for the most part run radially through the thickness of the cortex and extend for relatively short distances laterally. (C) Drawing from photographs of a pyramidal cell and a spiny stellate cell in the cat cortex which had been injected with horseradish peroxidase after their activity had been recorded. Both were simple cells. (A from Hubel and Wiesel, 1972; B after Ramón y Cajal; C from Gilbert and Wiesel, 1979.)

49
THE VISUAL
WORLD:
CELLULAR
ORGANIZATION
AND ITS
ANALYSIS

differences in the types of processing carried out by the structurally different areas.

The wealth of cell types and connections in the cortex immediately suggests that, in contrast to the geniculate, one can expect much greater integrating activity and transformation of signals to occur here. In monkey and man, the number of cells is even greater than in the cat, and histological examination shows the various layers and zones to be more clearly demarcated. A general feature of the mammalian cortex is that the cells are arranged in layers within the gray matter (Figure 14). Characteristically, the processes of the cells run for the most part in a radial direction, up and down through the thickness of the cortex (at right angles to the surface). In contrast, the majority (but not all) of their lateral processes are short.[56] The main lateral connections between areas are made by axons that dip down and run in bundles through the white matter to surface again elsewhere.

Cortical cells are classified on the basis of structural criteria, in the same way as are the cerebellar cells mentioned in Chapter 1. The two principal groups of neurons are STELLATE CELLS and PYRAMIDAL CELLS. Examples of these cells are shown in Figure 14B. The main differences are in the lengths of the axons and the shapes of the cell bodies. The axons of pyramidal cells are longer, dip down into white matter, and leave the cortex; those of stellate cells tend to terminate locally. These two groups of cortical cells exhibit variations, such as the presence or absence of spines on the dendrites, that bear on their functional properties (Chapter 3). Pyramidal cells are spiny; smooth as well as spiny stellate cells are found. In addition, there are other fancifully named neurons (double bouquet cells, chandelier cells, basket cells, and crescent cells) and the neuroglial cells (Chapter 13).[57] The shapes, connections, properties, and functions in the different layers are described more fully in Chapter 3. For the present discussion, which deals with receptive field properties, it is important to note that layer IV is the region in which many of the geniculate fibers end: In the monkey, layer IV is further subdivided into A, B, Cα, and Cβ; and in the cat into ab and c (according to a terminology not yet standardized). One might therefore expect processing of visual information to begin to a large extent within layer IV of the cortex.

Layering and morphological features of cortical neurons

As in the geniculate layers, visual fields from the eyes are represented in a strict topographical projection in each cortex. This was shown before the era of single-cell analyses by shining light onto small parts of the retina and recording with gross electrodes the potentials evoked on the surface of the cortex. Such potentials represent the summed activity of large numbers of neurons. Projection maps made in this manner by Talbot and Marshall[58] and by Daniel and Whitteridge[59]

Mapping visual fields in the striate cortex

[56]Gilbert, C. D. and Wiesel, T. N. 1983. *J. Neurosci.* 3: 1116–1133.
[57]Ramón y Cajal, S. 1955. *Histologie du Système Nerveux* II., C.S.I.C., Madrid.
[58]Talbot, S. A. and Marshall, W. H. 1941. *Am. J. Ophthalmol.* 24: 1255–1264.
[59]Daniel, P. M. and Whitteridge, D. 1961. *J. Physiol.* 159: 203–221.

demonstrate that much more cortical area is devoted to representation of the fovea than to representation of the rest of the retina—as expected, since form vision is principally confined to foveal and parafoveal areas. Clinical tests involving destruction of these areas in the visual cortex and ablation studies in animals support these conclusions. But for an explanation of the cellular mechanisms on which perception or form vision is based, an analysis with higher resolution is needed; this is provided by recording from single cells. (Chapter 3 discusses the finer grain of the architecture of the cortex and how cells are organized in columns of functionally related groups for the analysis of visual information.)

Cortical receptive fields

Responses of cortical neurons, like those of the retinal ganglion and geniculate cells, tend to occur on a background of maintained activity. One of the most consistent observations is that discharges of CORTICAL NEURONS ARE NOT SIGNIFICANTLY INFLUENCED BY DIFFUSE ILLUMINATION of the retina. Bright flashes of light onto the entire retina, presented in the dark or at various levels of background illumination, do not noticeably change the irregular background discharge rate of neurons in the visual cortex. The almost complete insensitivity to diffuse light is an intensification of the process already noted in the retina and the lateral geniculate nucleus; it results from an equally matched antagonistic action of the inhibitory and excitatory regions in the receptive fields of cortical cells. Neuronal firing rate is altered only under special conditions, that is, only when certain demands about the position and form of the stimulus on the retina are met. In other words, the receptive fields of most cortical neurons have configurations that differ from those of retinal or geniculate cells, so that spots of light often have little or no effect. In his Nobel address, Hubel[60] described the experiment in which Torsten Wiesel and he first recognized this essential property.

> Our first real discovery came about as a surprise. . . . for three or four hours we got absolutely nowhere. Then gradually we began to elicit some vague and inconsistent responses by stimulating somewhere in the midperiphery of the retina. We were inserting the glass slide with its black spot into the slot of the opthalmoscope when suddenly, over the audio monitor, the cell went off like a machine gun. After some fussing and fiddling, we found out what was happening. The response had nothing to do with the black dot. As the glass slide was inserted its edge was casting onto the retina a faint but sharp shadow, a straight dark line on a light background. That was what the cell wanted, and it wanted it, moreover, in just one narrow range of orientations. This was unheard of. It is hard now to think back and realize just how free we were from any idea of what cortical cells might be doing in an animal's daily life.

By following a progression of clues, Hubel and Wiesel worked out the appropriate light stimuli for various cortical cells; they named the receptive fields on the basis of relative simplicity of construction. In

[60]Hubel, D. H. 1982b. *Nature 299*: 515–524.

51
THE VISUAL
WORLD:
CELLULAR
ORGANIZATION
AND ITS
ANALYSIS

those halcyon days, for want of more descriptive terms the various field types were known as simple, complex, hypercomplex, and higher order hypercomplex. Now, with more detailed information available about the properties of cortical cells, about their connections, and about the layers in which they are located, this classification has been modified to two principal types of receptive fields: SIMPLE and COMPLEX.[61–63] Each of these categories includes a number of subgroups and important variables that bear on perceptual mechanisms.

The receptive fields of simple cells, like those of ganglion cells, are restricted in extent and exhibit several varieties. In the monkey, one class of cortical cells with properties more elementary than those of simple cells has a concentric center–surround receptive field, much like that of a geniculate cell or a ganglion cell. Such neurons are found exclusively in the lower part of layer IV, the region where most geniculate fibers end. Little or no obvious transformation has occurred in the properties or the meaning of the signal.

Simple cells, however, have quite different receptive fields. They are found in layers IV and VI and deep in layer III. All of these regions receive direct inputs from the geniculate. One type of simple cell has a receptive field that consists of an extended narrow central portion flanked by two antagonistic areas. The center may be either excitatory or inhibitory. Figure 15B shows a receptive field of a simple cell in the striate cortex mapped out with spots of light that excited weakly in the center [because the spots covered only a small part of the central area (marked with crosses)] and inhibited in the surround (triangles).

The requirements of such a cell with a simple field are exacting. For optimal activation it needs a bar of light that is not more than a certain width, that entirely fills the central area, and that is oriented at a certain angle. This is illustrated in Figure 15C by the records taken from the same cell. As one might expect for this cell, a vertically oriented slit of light (fifth record from top) is most effective. With a small deviation from this requirement, the response deteriorates.

Different cells have receptive fields appropriate for a wide range of different orientations and positions. A new population of simple cells is therefore brought in by rotating the stimulus or by shifting its position in the visual field. The distribution of inhibitory–excitatory flanks in various simple receptive fields may not be symmetrical, or the field may consist of two longitudinal regions facing each other—one excitatory, the other inhibitory. Figure 16 shows examples of four receptive fields, all with a common axis of orientation but with differences in the distribution of areas within the field. For receptive field A, a narrow light slit with appropriate orientation on the central rectangle elicits the best response; a dark bar in the same place with light flanks suppresses

Simple cells

[61]Hubel, D. H. and Wiesel, T. N. 1962. *J. Physiol. 160*: 106–154.
[62]Hubel, D. H. and Wiesel, T. N. 1977. *Proc. R. Soc. Lond. B 198*: 1–59.
[63]Gilbert, C. D. 1977. *J. Physiol. 268*: 391–421.

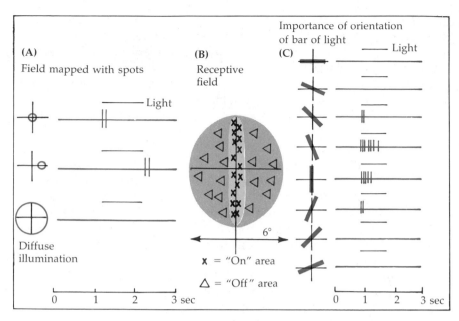

15 RESPONSES OF A SIMPLE CELL in cat striate cortex to spots of light (A) and bars (C). The receptive field (B) has a narrow central "on" area flanked by symmetrical antagonistic "off" areas. The best stimulus for this cell is a vertically oriented light bar (1° × 8°) in the center of its receptive field (fifth record from top in C). Other orientations are less effective or ineffective. Diffuse light (third record from top in A) does not stimulate. Illumination indicated by bar. (After Hubel and Wiesel, 1959.)

the background discharge. Cells with field shapes B and C produce the opposite responses. For field D, an edge with light on the left and darkness on the right is most effective for "on" responses, whereas reversing the darklight areas is best for "off" discharges.

Since these original descriptions, another type of simple cell has been found. Once again the orientation and the position of the stimulus are critical and the field is made of antagonistic "on" and "off" areas. But in addition the length of the bar or edge is important: stretching the bar beyond an optimal length reduces its effectiveness as a stimulus. It is as though there were an additional "off" area that exists at the top or the bottom of the fields shown in Figure 16 and that tends to suppress firing. Originally this region, which is not a simple "off" area, was unmasked by computer-generated gratings;[64] later it became clear that it is encroachment by a prolonged edge or slit beyond the conventional field that suppresses the response. Hence, for such simple cells, the best stimulus is an appropriately oriented bar or edge that *stops* in a particular place ("end-inhibition" or "end-stopping").[63]

In practice, one observes any number of asymmetrical receptive

[64]Bishop, P. O., Coombs, J. S. and Henry, G. H. 1971. *J. Physiol. 219*: 625–657.

53
THE VISUAL
WORLD:
CELLULAR
ORGANIZATION
AND ITS
ANALYSIS

fields in which the portions flanking the central strip are of unequal strength. The common properties of all simple cells are (a) that they respond best to a properly oriented stimulus positioned at the border between the antagonistic zones and (b) that stationary slits or spots can be used to define "on" and "off" areas. Another constant and remarkable feature is that in spite of all the different proportions of inhibitory and excitatory area relations, the two contributions match exactly and cancel each other's effectiveness, so that diffuse illumination of the entire receptive field produces a feeble response at best. The "off" areas in cortical fields are not always able to initiate impulses in response to dark bars. Frequently (particularly in end-inhibition and in the more elaborate fields to be described shortly) illumination of the "off" area can only be detected as reduction in an ongoing discharge that had been evoked from the "on" area.[63]

In these examples the width of the narrow light or dark bar is comparable to the various diameters of the on- or off-center regions in the doughnut-shaped receptive field of ganglion or lateral geniculate body cells (Figures 4 and 12). Once again there is a specialization for detecting differences, but the spotlike contrast representation of ganglion cells has been transformed and extended into a line or an edge. Resolution has not been lost but has been incorporated into a more complex pattern. In agreement with the preceding assumption, cortical cells that have fields derived from the fovea are best excited by narrower bars than those cells with fields in the peripheral area of the eye.

In initial studies on simple cells, it was noted that moving edges or bars of the appropriate orientation were highly effective in initiating impulses. Indeed, many of the higher order cells to be discussed shortly respond only to moving stimuli (see Figure 18). As might be predicted from the diverse arrangement of "on" and "off" areas, different cells may be aroused by different types of movement. Some, like the cell with receptive field D of Figure 16, might respond preferentially to a

Responses of simple cells to moving stimuli

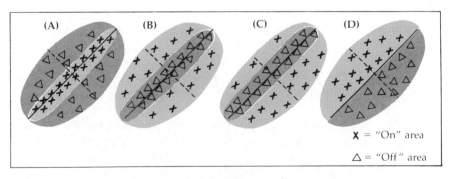

X = "On" area

$\triangle$ = "Off" area

16 RECEPTIVE FIELDS OF SIMPLE CELLS in cat striate cortex. In practice all possible orientations are observed for each type of field. The optimal stimuli are for (A), a narrow slit (or bar) of light in the center; for (B) and (C), a dark bar; and for (D), an edge with dark on the right. Considerable asymmetry can be present, as in (C). (After Hubel and Wiesel, 1962.)

movement of an edge in one direction, from right to left, when light moves from an "off" area onto an "on" area; for others with symmetrical flanks (cells with receptive fields A and B), movements in either direction may be equally effective. One would predict that cells with asymmetrical fields would be best designed for detecting movement in one direction and when light moves from an "off" area onto an "on" area. In practice, however, it is often not possible to explain the directional sensitivity of a cortical simple cell from the design of its field. Previously, Barlow and Levick had proposed cellular mechanisms to explain the directional sensitivity of certain ganglion cells in the retina of the rabbit.[65] In the cat cortex, X and Y neurons of the lateral geniculate nucleus, with their sustained and transient responses, presumably contribute to the movement sensitivity. A fuller explanation will probably have to await information about the synaptic mechanisms involved in the formation of cortical receptive fields.

Complex cells   There is an interesting parallel between movement sensitivity in the visual system and movement sensitivity in the somatosensory system. In the somatosensory cortex of monkey, cells have been found that respond selectively to stroking of the skin in one direction but not another. Again, this direction sensitivity is a property of higher order neurons and not of primary receptors in the skin.[66]

In recordings made from individual neurons in the visual cortex, one finds, in addition to simple cells, other neurons that behave quite differently. These complex cells have two important properties in common with simple cells: They require specific field-axis orientation of a dark–light boundary, and illumination of the entire field is ineffective. The demand, however, for precise positioning of the stimulus, observed in simple cells, is relaxed in complex cells. In addition, there are no longer distinct "on" and "off" areas. As long as a properly oriented stimulus falls within the boundary of the receptive field, most complex cells will respond, as in the examples illustrated in Figure 17. The meaning of the signals arising from complex cells, therefore, differs significantly from that of simple cells. The simple cell localizes the orientation of a bar of light (Figure 15) to a particular position within the receptive field while the complex cell signals the abstract concept of ORIENTATION WITHOUT STRICT REFERENCE TO POSITION.

Two main classes of complex cells can be distinguished; both respond to moving edges or slits of fixed width and precise orientation. The type of cell giving the responses shown in Figure 17 corresponds to what can be called the "standard" class of complex cells. For such cells the response improves as the edge or slit becomes longer, up to a point; prolonging the stimulus further gives rise to no additional effect. Other complex cells, like end-stopped simple cells, require slits or edges that stop. In the previous nomenclature, these would have been called

[65]Barlow, H. B. and Levick, W. R. 1965. *J. Physiol. 178*: 477–504.
[66]Hyvärinen, J. and Poranen, A. 1978. *J. Physiol. 283*: 523–537.

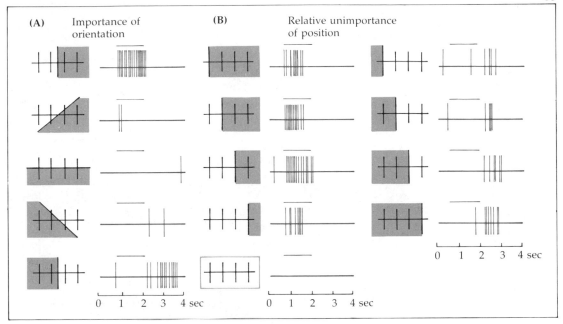

(A) Importance of orientation   **(B)** Relative unimportance of position

17   **RESPONSES OF A COMPLEX CELL** in the striate cortex of the cat. Cell responds best to a vertical edge. **(A)** With light on the left and dark on the right (first record), there is an "on" response. With light on the right (fifth record) there is an "off" response. Orientation other than vertical is less effective. **(B)** Position of border within field is not important. Illumination of entire receptive field (bottom record) gives no response. (After Hubel and Wiesel, 1962.)

hypercomplex cells, a term that is now obsolete. The best stimulus for such cells, therefore, requires not only a certain orientation but in addition some discontinuity, such as a line that stops, an angle, or a corner, as for the end-inhibition of simple cells. Figure 18 shows the responses of a cell that responds best to an edge with dark below and light above at an angle of about 45°. Diffuse illumination, other axis orientations, or spots are without effect. At first glance one might classify this as a complex cell similar to that giving the responses shown in Figure 17. The difference, however, becomes clear in the fourth and fifth records from the top in Figure 18, when the dark edge is extended; this elongation depresses the response of this complex cell. Interestingly, diffuse illumination of the right-hand field does not diminish the response (last record). It is therefore not a simple "off" area. One description of the best stimulus for this cell is a corner; moreover, the stimulus must move in one direction.

Other variants of complex cells may be mentioned. For example, the cell giving the responses shown in Figure 19 also exhibits end-inhibition. It responds to a narrow bar or slit; and, unlike the cell just discussed, it behaves as though an extra end-inhibitory effect were added on the left side. The field can be considered as made up of three components,

18

END INHIBITION OF A COMPLEX CELL in area 18 of the cat cortex. The best stimulus for this cell is a moving (arrows), oriented edge (a corner) that does not encroach on the antagonistic right-hand portion of the receptive field (third record from the top). The records also show the selective sensitivity of the cell to upward movement. (After Hubel and Wiesel, 1965a.)

two inhibitory and one excitatory. The best stimulus for this special cell is a moving bar or slit that covers the middle region. The stimulus, however, must not extend into *either* of the two inhibitory lateral portions of the receptive field. When widened in either direction, the stimulus is weakened or ineffective. This cell, therefore, is even more specific in its requirements. It signals that a narrow dark line is stopping in this part of the retina in a subdivision of its receptive field. It does not tell exactly where the line is in terms of position of the plane of movement (up or down), but it does indicate that (a) the line is not wider than the amount specified; (b) it stops within the central part of the field.

Still other complex cells, called "special" complex cells, respond best to narrow slits of the appropriate orientation moving in one direction anywhere within the receptive field. But the length of the slit must be

57
THE VISUAL
WORLD:
CELLULAR
ORGANIZATION
AND ITS
ANALYSIS

19

RESPONSES OF A COMPLEX CELL with end inhibition in area 19 of cat cortex. The best stimulus is a narrow dark tongue in the middle moving downward. Shifting the dark edge sideways or widening it encroaches on antagonistic flanking areas and diminishes the response. (After Hubel and Wiesel, 1965a.)

considerably shorter than the dimensions of the receptive field.[67] Lengthening it within the field reduces the response. There is, therefore, no summation for increased slit length. Specificity for orientation, length, and direction of movement are maintained, but the exact position of the stimulus along the orientation axis has become less important. Special complex cells are found largely in layers III and V (see Chapter 3).

Often it takes several hours of trying out different patterns on the screen in front of the animal before the most effective stimulus can be identified. Certain advantages are obtained by using computer-generated slits or sinusoidal gratings (light and dark bars alternating in a graded manner), which have proved effective for defining certain aspects of receptive field properties.[68]

At this stage the following question is frequently asked: How can you be sure that the best stimuli for the simple and complex cells are in fact straight lines rather than curves, the letter E, or perhaps the complex shape of a mouse or a hand? It is in practice not possible, of

[67]Palmer, L. A. and Rosenquist, A. C. 1974. *Brain Res. 67*: 27–42.
[68]Movshon, J. A., Thompson, I. D. and Tolhurst, D. J. 1978a. *J. Physiol. 283*: 53–77.

course, to try all combinations on each cell. In general, however, it has been found in many laboratories that curves, circles, and complex patterns do not stimulate cortical cells as effectively as straight bars or edges. Naturally, "straight" can only be considered in relation to the size of receptive fields, so that a segment of a large circle on the retina is virtually a straight line. Just as a line can be constructed out of closely spaced dots, any desired shape can be formed out of straight lines of the proper length.

Possible role of complex cells in perceptual disorders

The various specialized complex neurons convey information about a discontinuity, such as when a line stops or changes its direction. This provides a clue to the explanation of a clinical observation in man—the COMPLETION PHENOMENON, which occurs when small retinal or cortical lesions cause blind areas (scotomas). In this condition, forms or shapes projected onto the retina appear to contain empty areas in the visual field corresponding to the site of the lesion. Yet, if the patient looks at striped wallpaper, a zebra, or a simple straight-line pattern, he sees the pattern continue through the blind area.

An interesting self-observation made by Lashley, a perceptive reporter of psychophysical phenomena, during a migraine attack is worth quoting in this context:[69]

> Talking with a friend I glanced just to the right of his face whereon his head disappeared. His shoulders and necktie were still visible but the vertical stripes on the wall paper behind him seemed to extend right down to the necktie. Quick mapping revealed an area of total blindness covering about 30° just off the macula. It was quite impossible to see this as a blank area when projected on the striped wall paper of uniformly patterned surface although any intervening object failed to be seen.

The events described by Lashley can be interpreted in terms of complex cells that signal information only about the terminations of lines (Figure 20). With an array of such cells, silence of members in the area of temporary (in Lashley's case) blindness produces little change in signaling since they are normally silent anyway, unless the line stops or changes direction. This type of reasoning provides a conceptual scheme that uses known properties of cells to explain perceptual phenomena. Some of the weaknesses and strengths of this approach are discussed later. It is ironic, incidentally, that this interpretation rests on cortical cells that have specific connections. This is contrary to the principles enunciated with clarity by Lashley himself, who stressed absence of specificity and localization in the cortex.

Receptive fields from both eyes converging on cortical neurons

Binocular interaction is introduced here because it provides another example of part of the design the brain uses to end up with perception of form. When we look at an object with one or both eyes, we see only one image, even if the size and the position of the object's projection are slightly different on the two retinas. Sherrington posed the problem:[70]

[69]Lashley, K. S. 1941. *Arch. Neurol. Psychiat.* 46: 331–339.
[70]Sherrington, C. S. 1947. *The Integrative Action of the Nervous System.* Yale University Press, New Haven.

59
THE VISUAL
WORLD:
CELLULAR
ORGANIZATION
AND ITS
ANALYSIS

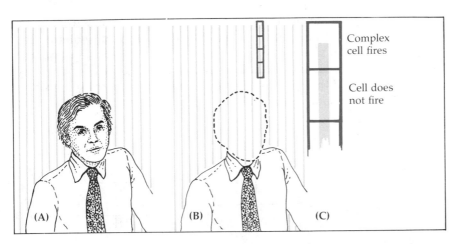

20 **THE COMPLETION PHENOMENON.** During a migraine attack (B), a small area of complete blindness occurs in the visual field occupied by the head of Lashley's friend (A). Yet the stripes on the wallpaper continue through the area. A possible explanation is that complex cells of the type shown in Figures 16 and 17 respond best to corners or lines that stop; hence, a stripe on the wallpaper acts as a good stimulus for a complex cell only at its end (C). A complex cell with its receptive field within the blind area would no longer be activated by the lines of the face; however, its silence could be interpreted by higher centers as an indication that the gray stripe had not stopped.

How habitually and unwittingly the self regards itself as one is instanced by binocular vision. Our binocular visual field is shown by analysis to presuppose outlook from the body by a single eye centered at a point in the midvertical of the forehead at the level of the root of the nose. It, unconsciously, takes for granted that its seeing is done by a cyclopean eye having a centre of rotation at the point of intersection just mentioned. In this visual field it obtains visual depth by unknowingly combining . . . crossed images of not too great lateral disparation. . . . Oneness is obtained by compromise between differences, it not too great, offered to the perceiving "self." There are other perceptual instances. The brightness of a binocular field differs hardly sensibly from that of either of two equally illuminated uniocular fields composing it. But the quantity of stimulus received by the eyes is roughly double in the binocular observation than that which it is in the uniocular.

Interestingly, well over 100 years ago Johannes Müller suggested that individual nerve fibers from the two eyes might fuse or become connected to the same cells in the brain. Thereby, he almost exactly anticipated Hubel and Wiesel's results. They found that about 80 percent of all cortical neurons in the visual areas of the cat can be driven from both eyes. Since the neurons in the various layers in the lateral geniculate nucleus are predominantly innervated from one eye or the other, the first opportunity for significant interaction between the eyes must occur in the cortex. In the cat and monkey, the separation is maintained in the fourth layer of area 17. Each simple cell is driven by only one eye, the other being without effect, a point referred to again in Chapters 3 and 20 in relation to the columnar organization of the cortex. Mixing between the two eyes occurs in the subsequent relay stations, that is, in layers deeper toward the white matter and in more superficial cortical layers (toward the roots of the hair).

Examination of the receptive fields of a binocularly driven cell shows that (1) they are usually in exactly corresponding positions in the visual

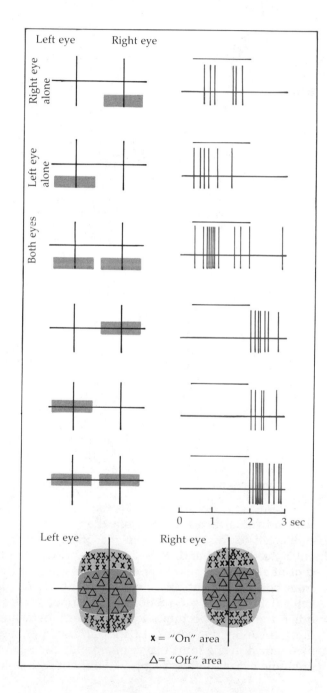

21

BINOCULAR ACTIVATION of a simple cortical neuron that has identical receptive fields in both eyes. Simultaneous illumination of corresponding "on" areas of right and left receptive fields is more effective than stimulation of one alone (upper three records). In the same way, stimulation of "off" areas in the two eyes reinforces each other's "off" discharges (lower records). (After Hubel and Wiesel, 1959.)

61
THE VISUAL
WORLD:
CELLULAR
ORGANIZATION
AND ITS
ANALYSIS

field of the two eyes, (2) their preferred orientation is the same, and (3) the corresponding areas in the receptive fields add to each other's effect. An example of synergistic action between the two eyes is shown for a simple cell in Figure 21. Shining light onto an "on" region in the left eye sums with illumination onto the "on" area of the right eye. Simultaneous illumination of antagonistic areas in the two eyes reduces or suppresses an ongoing response and increases "off" discharges. Such cells would be useful for unifying images from the two eyes.

For DEPTH PERCEPTION there exists another binocular specialization of receptive fields.[71,72] Such fields are not in exactly corresponding points in the two retinas. Figure 22 shows how objects out of the plane of focus cast images on disparate parts of the retina. Neurons with a performance that fits them for depth perception have been found in cat and monkey, in areas 17 and 18. In many respects they resemble the

[71]Fischer, B. and Poggio, G. F. 1979. *Proc. R. Soc. Lond. B 204*: 409–414.
[72]Ferster, D. 1981. *J. Physiol. 311*: 623–655.

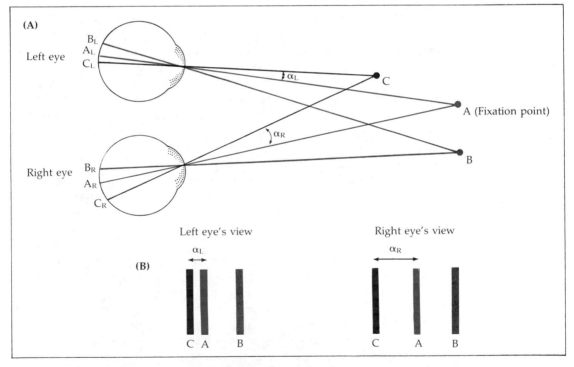

22 DISPARITY AND DEPTH PERCEPTION. For both eyes the object at A is fixated; its image falls on the fovea. B is in the same plane. Its image falls in each eye at the same distance from that of A, on corresponding points of the retina. For C, in front of the plane of focus, the angle is different for the left (α_L) and right (α_R) eyes. As a result the image is closer to A in one eye than in the other. If the bars below were displayed in a stereoscope, C would appear in front of B or A.

simple and complex cells described earlier. They fire best, however, when the bar, edge, or slit of the correct orientation is presented to the two eyes in front of or behind the plane of focus. This seems to be a good example of a high degree of specialization, in that a whole specific series of conditions must be met to make a stimulus suitable.

Connections for combining right and left visual fields

Each hemicortex is wired to perceive one half of the external world but not the other. This is equally true for sensations of touch and position and constitutes a general situation in relation to perception. It is natural to wonder what happens in the midline. How are the two cortices knitted together to produce a single image of the body and the world?

The obvious way to preserve continuity is for the two sides to be fused in register at the midline. This would allow a complete picture to be formed with a minimum number of connections between the two hemispheres. On the other hand, there would be little purpose in linking fields seen out of the corners of the two eyes that look on quite different parts of the world and observe, for example, a cathedral and a dolphin.

Corpus callosum

The general question of transfer of information between the hemispheres has been studied in a most rewarding manner in humans and in monkeys by Sperry, Myers, and Gazzaniga, and their colleagues.[73,74] Concentrating on the coordinating role of the corpus callosum, a bundle of fibers that runs between the two hemispheres, they have shown that the fibers are actually involved in the transfer of information and learning from one hemisphere to the other. To cite one example, a normal right-handed person can name an object, such as a coin or a key, when it is placed in either hand (stereognosis). After section of the corpus callosum, however, the object can be named only when it is placed in the right hand; the information from the right hand crosses before reaching the cortex and projects to the left hemisphere. It is in the LEFT hemisphere that the main area responsible for language lies. What happens when the object is placed in the left hand, which projects to the right hemisphere? Information still reaches consciousness when the key is placed in the left hand. There is, however, no way in which the concept "key" can be verbally expressed because the center for language, which is situated on the left side of the brain, cannot be reached without the corpus callosum. Thus, a man without a corpus callosum can recognize an object with his left hand and may be able to use it, but he cannot say the word *key*. Figure 23 expresses some of these ideas. Other experiments on the corpus callosum provide surprising insight into higher functions. For example, when deprived of cross connections, the two hemispheres can lead virtually separate existences.

One small aspect of this important area of work is noted here because it bears directly on visual perception. It suggests that the fusing, or

[73]Sperry, R. W. 1970. *Proc. Res. Assoc. Nerv. Ment. Dis. 48*: 123–138.
[74]Gazzaniga, M. S. 1967. *Sci. Am. 217*: 24–29.

63
THE VISUAL
WORLD:
CELLULAR
ORGANIZATION
AND ITS
ANALYSIS

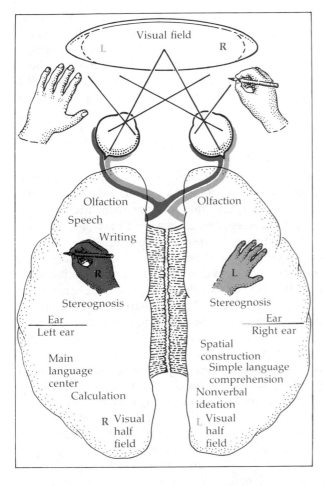

Visual field

Olfaction

Olfaction

Speech

Writing

Stereognosis

Stereognosis

Ear

Ear

Left ear

Right ear

Spatial
construction

Main
language
center

Simple language
comprehension

Calculation

Nonverbal
ideation

R Visual
half
field

L Visual
half
field

23

THE CORPUS CALLOSUM. Transection of the corpus callosum interrupts connections between the cerebral hemispheres, including those of receptive fields that straddle the midline. (After Sperry, 1970.)

knitting together, of the two fields of vision is mediated by fibers in the corpus callosum. Cells with receptive fields that straddle the midline provide information about both sides of the visual world. There is good evidence that such cells combine inputs from both hemispheres and that the corpus callosum bundle provides the fibers responsible for extending the receptive field across the midline. Thus, the projection from nerve cells in one hemisphere to the other has been shown anatomically by injecting horseradish peroxidase into the cortex at the boundary of areas 17 and 18. This site in the cortex is where the midline is represented. Enzyme taken up by terminals is transported back to and fills neuronal cell bodies that are situated at exactly corresponding sites in opposite hemispheres.[75] Interestingly, the cells lie in layer III, which (as will be seen in Chapter 3) is a layer containing neurons that send their axons to other regions of cortex. Cutting or cooling (which blocks conduction) the corpus callosum causes the receptive field to

[75]Shatz, C. J. 1977. *J. Comp. Neurol. 173*: 497–518.

shrink and become confined to just one side of the midline (the usual arrangement with cortical cells). Furthermore, recording from single fibers in the callosum that are concerned with vision shows that their fields lie close to the midline, not in the periphery.[76] The role of callosal fibers is clearly demonstrated in an experiment of Berlucchi and Rizzolatti.[77] They made a longitudinal cut through the optic chiasm, thereby severing all direct connections to the cortex from the contralateral eye. Yet, provided the corpus callosum was intact, some cells in the cortex with fields close to the midline could still respond to appropriate illumination of the contralateral eye.

The general conclusion to be drawn from these demonstrations of binocular interactions on individual neurons is that relatively simple neuroanatomical concepts can explain how images from the two eyes are fused and how depth is assessed, an achievement that has simplified thinking about two complex problems related to perception.

In the retina, with its layering and sequence of connections, it is obvious that the performance of ganglion cells, and of their receptive fields, is determined by the neurons in the two preceding strata. Yet, attempts at synthesis are still only partially successful; the problems are due largely to incomplete knowledge of synaptic connections and of their performance and mechanisms. Even less is known about the cortex. Some initial but important approaches will be discussed here. The experimental evidence will be considered in greater detail in Chapter 3.

Schemes for elaborating receptive fields

A scheme of organization was proposed early on by Hubel and Wiesel. This scheme had the advantage of using known mechanisms for understanding how a nerve cell can respond so selectively to a visual pattern—say, only to a horizontal line of a certain length that is moving upward through a small part of the visual field. The key idea is that more elaborate receptive fields are constructed by an ordered synthesis of simpler ones originating at a lower level. In the cortex, fields of simple cells behave as if they were built up of large numbers of geniculate fields. This idea is illustrated in Figure 24A, where the centers of concentric fields of geniculate neurons are lined up in such a way that a bar of light through their centers would excite them strongly and a parallel shift of the bar into the inhibitory surround would reduce or stop the excitatory output of the cells. This scheme results in a central excitatory area and its opposing surround. The value of such a scheme is not diminished by several obvious shortcomings. For example, a simple excitatory input to cortical cells does not account for their silence to diffuse stimulation. Geniculate cells tend to respond, although not vigorously, with "on" as well as "off" discharges when diffuse light is flashed onto them or when it is turned off. The diagram of connections could be expanded to include inhibitory pathways. Nevertheless, the scheme of Figure 24A provides a working model for

---

[76]Hubel, D. H. and Wiesel, T. N. 1967. *J. Neurophysiol. 30*: 1561–1573.
[77]Berlucchi, G. and Rizzolatti, G. 1968. *Science 159*: 308–310.

65
THE VISUAL
WORLD:
CELLULAR
ORGANIZATION
AND ITS
ANALYSIS

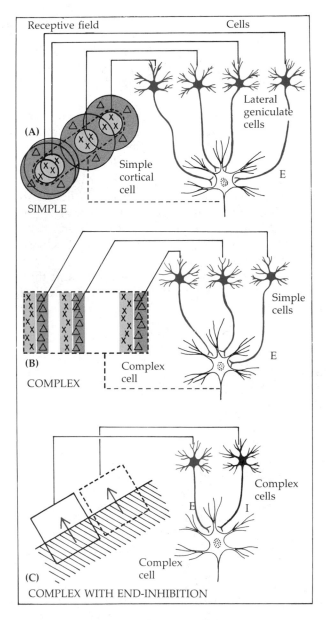

Receptive field Cells

Lateral
geniculate
cells

(A)

Simple
cortical
cell

SIMPLE

E

Simple
cells

(B)

Complex
cell

COMPLEX

E

Complex
cells

E   I

Complex
cell

(C)

COMPLEX WITH END-INHIBITION

24

SYNTHESIS OF RECEPTIVE FIELDS. Hypothesis devised by Hubel and Wiesel to explain the synthesis of simple and complex receptive fields. In each case lower order cells converge to form receptive fields of higher order neurons. (A) Fields of simple cells are elaborated by the convergence of many geniculate neurons with concentric fields (only four appear in the sketch). They must be arranged in a straight line on the retina according to the axis orientation of simple receptive fields. (B) Simple cells responding best to a vertically oriented edge at slightly different positions could bring about the behavior of a complex cell which responds well to a vertically oriented edge situated anywhere within its field. (C) Each of the two complex cells responds best to an obliquely oriented edge. But one cell is excitatory and the other is inhibitory to the next cell, producing end-inhibition. Hence an edge that covers both fields, as in the sketch, is ineffective, whereas a corner restricted to the left field would excite. (After Hubel and Wiesel, 1962, 1965a.)

further experimental approaches. In the same way, one can tentatively construct complex receptive fields by lining up appropriate rows of simple fields.

The scheme illustrated in Figure 24B, which again presents just three sample components out of a great number that would be needed, satisfies the requirements of a complex cell that is excited by a vertical edge stimulus that falls anywhere within the area of the receptive field. This is so because wherever the edge falls, one of the simple fields is traversed at its vertical inhibitory–excitatory boundary. None of the other components respond because they are uniformly covered by light or darkness. Diffuse illumination of the entire field covers all component fields equally and therefore none fires. One can postulate that only one or a few of the simple cells fire at any one position of the stimulus to evoke a response in a complex cell. Similarly, one can devise a way of combining complex cell fields to form fields with end-inhibition, as shown in Figure 24C.

Other ways for elaborating the fields of cortical cells from geniculate and intracortical connections have been proposed recently. The principal difference is that receptive fields would be synthesized by "parallel" instead of hierarchical processing. Thus, one set of geniculate fibers (Y-type) would provide the behavioral inputs for complex cells, another (X-type) for simple cells. Now that cortical cells can be studied with intracellular electrodes so that the synaptic potentials that excite or inhibit them can be seen, the relative strength and distribution of the excitatory and inhibitory influences can be better estimated. Such experiments (discussed in Chapter 3), together with knowledge of the circuitry leading into the cortex, through the various layers, and then out again, do provide support for the general idea of a hierarchical organization—albeit in a modified form. Clearly, parallel inputs occur

TABLE 1
CHARACTERISTICS OF RECEPTIVE FIELDS
AT SUCCESSIVE LEVELS OF THE VISUAL SYSTEM

Type of cell	Shape of field	What is best stimulus?	How good is diffuse light as a stimulus?
Receptor	⊗	Light	Good
Ganglion		Small spot or narrow bar over center	Moderate
Geniculate		Small spot or narrow bar over center	Poor
Simple (layers IV and VI only)		Narrow bar or edge (some end-inhibited)	Ineffective
Complex (not layer IV)		Bar or edge	Ineffective
End-inhibited complex (not layer IV)		Line or edge that stops; corner or angle	Ineffective

67
THE VISUAL
WORLD:
CELLULAR
ORGANIZATION
AND ITS
ANALYSIS

to cells outside layer IV, and segregation of X and Y inputs has to be taken into account. Nevertheless, the main point to be emphasized here is that the various schemes serve as a conceptual framework, the details of which still need to be worked out. It is also worth emphasizing that although the original scheme was proposed by Hubel and Wiesel in 1962 as a working hypothesis, it still seems clear, elegant, and reasonable.

RECEPTIVE FIELDS: UNITS FOR FORM PERCEPTION

Table 1 lists some of the characteristics of receptive fields at successive levels of the visual system. Each eye conveys to the brain information collected from areas of various sizes on the retinal surface. The emphasis is not on diffuse illumination or the energy absorbed by the photoreceptors but on contrast. The ganglion cells prefer spots that fill the centers of receptive fields, but they also respond well to narrow bars of light or darkness, provided the bars traverse the center; POSITION is important. A small movement from the central position greatly alters the discharge by crossing over the border into the surround. Therefore,

TABLE 1 (*continued*)

Is orientation of stimulus important?	Are there distinct "on" and "off" areas within receptor fields?	Are cells driven by both eyes?	Can cells respond selectively to movement in one direction?
No	No	No	No
No	Yes	No	No
No	Yes	No	No
Yes	Yes	Yes (except in layer IV)	Some can
Yes	No	Yes	Some can
Yes	No	Yes	Some can

one would expect receptive fields with small centers to be best for RESOLUTION as well as for CONTRAST. This property is exactly what is seen as receptive fields become progressively smaller toward the foveal region. The process of contrast becomes further emphasized in the geniculate, where the antagonism between center and surround is stronger.

As the cortex is approached, the demands made on a stimulus for activation of neurons rise greatly. To the importance of position on the retina is added ORIENTATION of the stimulus (light or dark), which must be a line or a bar rather than a spot. Directional sensitivities for a moving stimulus appear first in simple cortical cells; and at this level, arises also the first opportunity for interaction between the two eyes. In complex cells, the requirement of position is relaxed and orientation becomes generalized over a considerable area (up to 2 mm^2 in the retina of the cat).

What types of signals are generated if a square patch of light such as that in Figure 25 (or a shadow of the same shape) is presented to the retina? The receptors in the retina seeing the bright central area

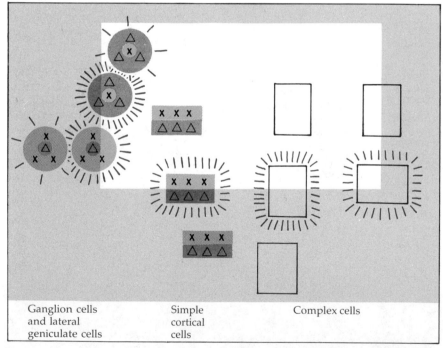

Ganglion cells
and lateral
geniculate cells

Simple
cortical
cells

Complex cells

25 **RESPONSES OF NEURONS TO A PATTERN. When a square patch of light is presented to the retina, signals arise predominantly from receptive fields close to the borders of the square. The ganglion cells and lateral geniculate cells whose receptive fields are situated close to the border fire better than those subjected to uniform light or darkness. The only simple and complex cells that fire are those with appropriate receptive fields situated along a border or corner with the correct orientation.**

69
THE VISUAL
WORLD:
CELLULAR
ORGANIZATION
AND ITS
ANALYSIS

absorb more light than those in the surround. Ignoring the receptors and bipolar cells and starting with the optic nerve fibers, the following types of signaling occur: the on-center ganglion cells within the square increase their discharge (at least for the initial period) while the off-center ganglion cells are suppressed. The best-stimulated ganglion cells are those subjected to the maximum of contrast, that is, those having centers lying immediately adjacent to the boundary between the light and the dark areas. At the geniculate this contrast is even more pronounced because the cells respond only poorly to diffuse light; hence those having receptive fields lying entirely within the dark or the light areas fire weakly, whether they are on-center or off-center cells. As is true of retinal ganglion cells, the geniculate cells that fire best have their centers close to the border. This process is still further enhanced in the cortex. Cortical cells having receptive fields lying completely either within the square or outside it send no signals (except for their usual maintained discharge) because diffuse illumination is not an effective stimulus. Only those simple cells with receptive fields having axis orientations coinciding with the horizontal or vertical boundaries of the square can be stimulated. Similar considerations apply to the stimulation of complex cells, which also require properly oriented bars or edges. There is an important difference, however, which depends upon the fact that, at rest, small rapid saccadic movements are made continually by the eyes. They are essential for vision to be preserved while the eyes fixate, but they are not perceived as motion. Each movement causes a new population of simple cells with exactly the same orientation to be thrown into action.

For those complex cells that "see" the square, however, a boundary of appropriate orientation anywhere within the field is the only requirement. Many of the same complex cells therefore continue to fire even during eye movements, as long as the displacement is small and does not pass outside the receptive field. For such cells the position of the square on the retina does not appear to change. In the lower right corner of the square is inserted a complex cell with end inhibition that responds best to a right angle—a corner.

If the preceding considerations are valid, the surprising conclusion is that the primary visual cortex receives little information about the absolute level of uniform illumination within the square. Signals arrive only from the cells with receptive fields situated close to the border. This hypothesis is supported by a well-known psychophysical experiment. A square that appears white when surrounded by a black border can be made to appear dark merely by increasing the brightness of the surround. In other words, we perceive the difference or contrast at the boundary and it is by that standard that the brightness in the uniformly illuminated central area is judged.

The eye does have an index of the maintained level of brightness, expressed in the constancy of the pupillary size according to the absolute strength of ambient light over a wide range. Pupillary size is adjusted by a feedback mechanism (Chapter 15), the incoming loop of

Hierarchical synthesis
of receptive fields

which leaves the eye through the optic nerve.[8] Outside area 17, in a neighboring visual area, cells have been found that do respond to alterations in the level of illumination, but their function is not known.[78]

The orderliness, repetition, and progression of receptive field organization constitute convincing evidence that the connections between neurons are specific and laid out according to a functionally related ground plan. Anatomical evidence, presented in Chapter 3, supports this basic idea. There is much less certainty about the detailed proposals invoked to explain the actual synthesis of receptive fields.

As information ascends from the eye, some of it is also used (or perhaps diverted) at each stage for various purposes, such as eye movements, regulation of the pupil, depth perception, and other functions. Additional types of cells and input from other parts of the brain intervene in the synthesis of fields, and some of the projections may miss a step. For example, certain geniculate axons in the cat project to layers of area 17 containing complex cells and directly to area 18. In connection with Figure 24, examples have been cited in which the known properties of visual neurons could not satisfactorily be explained. These include the absence of inhibition in the scheme for simple cells, the directional sensitivity of cortical cells to moving stimuli, and the progressively increasing demands for movement in complex cells. The scheme also fails to take into account the color coding of the cells in the monkey cortex, which has been studied by Zeki, Michael, Hubel, and others.

The role of visual cortical areas outside 17 is also not clear. As mentioned earlier, numerous visual areas have now been found in the cat and monkey cortex.[55] Cells there may be complex, color-coded, disparity detectors, responsive to movement in one direction, and may encode other variables. The idea that each area codes for one property is probably an oversimplification. As Hubel has said, "The idea . . . may turn out to be correct, but it nevertheless to me has a kind of naivete, like the notion that stripped of our cortex—a kind of double scalping, we become alligators, or the notion that our right hemisphere is for art and music and other nice things and our left is rational and analytic and propositional, in short a bore."[54]

Again, these considerations indicate that the hierarchical scheme is not complete, correct in detail, or the only possible explanation. Rather it represents a useful hypothesis and an effective way of describing how the complex behavior of cells in the visual system could be brought about. Although huge gaps in knowledge remain, the description of receptive fields has supplied for the first time an inkling of how such concepts as verticality, length, thickness, depth, and squareness can be derived from properties and connections of individual cortical neurons.

Where do we go from
here?

How much progress has been made in the last dozen years in explaining perception in neurophysiological terms? Earlier, the topographical representation of the retina on the cortex and ablation of

[78]Zeki, S. M. 1978. *J. Physiol.* 277: 245–272.

71
THE VISUAL
WORLD:
CELLULAR
ORGANIZATION
AND ITS
ANALYSIS

stimulation experiments showed that certain specific areas are active during visual perception. But it was difficult to put forward clear-cut hypotheses in cellular terms. For example, in 1943, Craik[79] discussed the kind of device that might be employed in the brain to recognize the three angles of a triangle in the visual field. He suggested that there might be an "electronic scanning device in which the scanning beam could be made to move in a straight line and would then continue to move in that direction until the line suddenly changed its course at an angle." Craik pointed out that "the essential thing is that the scanning beam should be made to acquire the habit, so to speak, of moving along a line so the system shall be disturbed by an abrupt change of direction." This explanation showed great insight and was formally correct. However, it is hard to see how such a postulate could predict actual neural mechanisms that might be involved in detecting a line that stops at a corner and then changes its direction. These are, however, just the properties ascribed to complex cells by virtue of their connections. The single-cell analysis, therefore, provides an idea of how the brain puts together and collates information used for the early steps in perception.

The responses certain neurons will give when they "see" various configurations of stimuli can be predicted. Yet, this is only the beginning, since even the most complex receptive fields, which provide sophisticated abstract information, still cover only a small part of what goes on in the large visual field each eye surveys. An extension of the hypothesis of hierarchical organization predicts that in the future cells should be discovered that bring together larger and larger parts of the information that appears in the field of vision. But how far can this concentration of information go? Will there be a small group of cells or, ideally, one pontifical cell that synthesizes and combines into a whole picture all the features perceived? The neurophysiologist obviously faces a dilemma. On the one hand, we know that visual information is synthesized into diverse meaningful units by various receptive fields. These are distributed among an incalculable number of cells; in this sense, perception is scattered and diffuse. On the other hand, there should be cells that see the "big picture", because without them we end up stating that each group of cells looks at the next group and vice versa. The most encouraging development is that perception can be thought of in terms of known cell properties after the signals have traversed only six or seven synapses, and there are many more synapses to come before the cells run out.

As one might perhaps expect, it becomes harder and harder to find out the roles of individual nerve cells with each succeeding step and with each discovery of a new area. An understanding of perception still seems a long way off. An approach based on quite different principles than ever more elaborate receptive fields has been proposed by Marr, Poggio, and their colleagues. Their analysis of the steps that lead to

[79]Craik, K. 1943. *The Nature of Explanation*. Cambridge University Press. London.

perception includes simple and complex cells as starting points. It then uses psychophysical observations and information theory (1) to break down visual images into spatial extents of gradations of light intensity, (2) to extract contours, and (3) to establish the contextual relationships of objects. Although this analysis does not indicate what roles are played by specific areas of the brain or by groups of neurons, it has provided a theoretical approach for thinking about what types of computation the brain itself may be making.[80,81]

The main lesson of the discussion so far is recognition of the importance of the layout of connections. It seems that the machinery of the brain will perform properly and provide the required information if its parts are correctly assembled, just as a radio will emit music or noise, depending on how its circuits are wired. This chapter can appropriately close with the following quotation from Sherrington, written long before visual fields were mapped for single cells:[82]

> The chief wonder of all we have not touched on yet. Wonder of wonders, though familiar even to boredom. So much with us that we forget it all our time. The eye sends, as we saw, into the cell-and-fibre forest of the brain throughout the waking day continual rhythmic streams of tiny, individual evanescent, electrical potentials. This throbbing streaming crowd of electrified shifting points in the spongework of the brain bears no obvious semblance in space-pattern, and even in temporal relation resembles but a little remotely the tiny two-dimensional upside-down picture of the outside world which the eyeball paints on the beginnings of its nerve-fibres to electrical storm. And that electrical storm so set up is one which affects a whole population of brain-cells. Electrical charges having in themselves not the faintest elements of the visual—having, for instance, nothing of "distance," "right-side-upness," nor "vertical," nor "horizontal," nor "colour," nor "brightness," nor "shadow," nor "roundness," nor "squareness," nor "contour," nor "transparency," nor "opacity," nor "near," nor "far," nor visual anything—yet conjour up all these. A shower of little electrical leaks conjures up for me, when I look, the landscape; the castle on the height, or, when I look at him, my friend's face and how distant he is from me they tell me. Taking their word for it, I go forward and my other senses confirm that he is there.

SUGGESTED READING

General reviews

Gazzaniga, M. S. 1970. *The Bisected Brain*. Appleton, New York.

Hubbell, W. L. and Bownds, M. D. 1979. Visual transduction in vertebrate photoreceptors. *Annu. Rev. Neurosci.* 2: 17–34.

Hubel, D. H. and Wiesel, T. N. 1977. Ferrier Lecture. Functional architecture of macaque monkey visual cortex. *Proc. R. Soc. Lond. B 198*: 1–59.

[80]Marr, D. 1982. *Vision*. Freeman, San Francisco.
[81]Stent, G. S. 1981a. *J. Sociol. Biol. Struct. 4*: 107–124.
[82]Sherrington, C. S. 1951. *Man on His Nature*. Cambridge University Press, London.

73
THE VISUAL
WORLD:
CELLULAR
ORGANIZATION
AND ITS
ANALYSIS

Hubel, D. H. and Wiesel, T. N. 1979. Brain mechanisms of vision. *Sci. Am. 241 (3)*: 150–162.

Kaneko, A. 1979. Physiology of the retina. *Annu. Rev. Neurosci. 2*: 169–191.

Knowles, A. 1982. *Biochemical aspects of vision*. In H. B. Barlow and J. D. Mollon, *The Senses*. Cambridge University Press, New York, pp. 82–100.

Mollon, J. D. 1982. Colour vision and colour blindness. In H. B. Barlow and J. D. Mollon, *The Senses*. Cambridge University Press, New York, pp. 165–191.

Sterling, P. 1983. Microcircuitry of the cat retina. *Annu. Rev. Neurosci. 6*: 149–185.

Van Essen, D. C. 1979. Visual areas of the mammalian cerebral cortex. *Annu. Rev. Neurosci. 2*: 227–263.

Retina and lateral geniculate nucleus

Baylor, D. A., Fuortes, M. G. F. and O'Bryan, P. M. 1971. Receptive fields of cones in the retina of the turtle. *J. Physiol. 214*: 265–294.

Baylor, D. A. and Hodgkin, A. L. 1973. Detection and resolution of visual stimuli by turtle photoreceptors. *J. Physiol. 234*: 163–198.

Baylor, D. A., Lamb, T. D. and Yau, K.-W. 1979. The membrane current of single rod outer segments. *J. Physiol. 288*: 589–611.

Dowling, J. E. and Boycott, B. B. 1966. Organization of the primate retina: Electron microscopy. *Proc. R. Soc. Lond. B 166*: 80–111.

Hubel, D. A. and Wiesel, T. N. 1961. Integrative action in the cat's lateral geniculate body. *J. Physiol. 155*: 385–398.

Kaneko, A. 1970. Physiological and morphological identification of horizontal, bipolar and amacrine cells in goldfish. *J. Physiol. 207*: 623–633.

Kuffler, S. W. 1953. Discharge patterns and functional organization of the mammalian retina. *J. Neurophysiol. 16*: 37–68.

Kuffler, S. W. 1973. The single-cell approach in the visual system and the study of receptive fields. *Invest. Ophthalmol. 12*: 794–813.

Nunn, B. J. and Baylor, D. A. 1982. Visual transduction in retinal rods of the monkey *Macaca fascicularis*. *Nature 299*: 726–728.

Schiller, P. H. and Malpeli, J. G. 1978. Functional specificity of lateral geniculate nucleus laminae of the rhesus monkey. *J. Neurophysiol. 41*: 788–797.

Wässle, H., Peichl, L. and Boycott, B. B. 1981. Morphology and topography of on- and off-alpha cells in the cat retina. *Proc. R. Soc. Lond. B 212*: 157–175.

Visual cortex

Gilbert, C. D. 1977. Laminar differences in receptive field properties of cells in cat primary visual cortex. *J. Physiol. 268*: 391–421.

Hubel, D. H. and Wiesel, T. N. 1962. Receptive fields, binocular interaction and functional architecture in the cat's visual cortex. *J. Physiol. 160*: 106–154.

Hubel, D. H. and Wiesel, T. N. 1965. Receptive fields and functional architecture in two non-striate visual areas (18 and 19) of the cat. *J. Neurophysiol. 28*: 229–289.

Hubel, D. H. and Wiesel, T. N. 1968. Receptive fields and functional architecture of monkey striate cortex. *J. Physiol. 195*: 215–243.

Stent, G. S. 1980. Thinking about seeing: A new approach to visual perception. In *The Sciences* (May/June Issue), The New York Academy of Sciences, pp. 6–11.

COLUMNAR ORGANIZATION AND LAYERING OF THE CORTEX

CHAPTER THREE

Throughout the visual system, neighboring groups of neurons respond to information from neighboring areas on the retinal surface. This principle has been well established at the cellular level in other sensory systems as well. Neighboring neurons in the visual cortex share other common functional properties; such cells are stacked in the form of columns or slabs which run at right angles to the cortical surface. For example, some columns of cells are best stimulated by vertical bars shone onto a small region of a particular area of the retina. Other columns of neurons in nearby areas of the cortex respond preferentially to a horizontal orientation, and so on for different angles. A comparable segregation into columns has been shown by recording from cells that are influenced preferentially by one or the other eye. These ocular dominance columns have also been demonstrated by anatomical techniques. In layer IV, geniculate fibers are seen to end in discrete, nonoverlapping territories. Narrow, alternating bands of cells are supplied by one eye or the other exclusively. Cells above and below layer IV show the corresponding eye preference. The aggregation of cortical neurons with related receptive field positions and functions makes it easier for them to interconnect so that they can perform the type of analysis required of them.

Studies of the laminar distribution of the various cell types found in layers I to VI and of their inputs, outputs, and interconnections have shed new light on how information is processed within the cortex. For example, in the cat, X and Y inputs from the geniculate nucleus end on cells in different sublayers; simple cells are restricted to layers IV and VI; and particular cell types in distinct layers give rise to fibers that dip down to supply lower centers and other regions of cortex. Intracellular recordings have begun to provide clues about how inhibitory and excitatory components of receptive fields are synthesized.

The retina is not simply represented as a map on the visual cortex; instead, each retinal area is analyzed over and over again in column after column, and again in neighboring cortical regions, with respect to a number of different variables such as position, orientation, and color. This type of cortical architecture provides some insight into the reason why so many components are needed for visual analysis. Cortical structure and functional organization go hand in hand. This emphasizes the great need for identifying the basic design behind the precise assembly of connections in the nervous system.

This chapter presents evidence to show how functionally related nerve cells are aggregated and connected to each other at successive stages in the central nervous system. The basic principle of organization was first elaborated by Mountcastle and his colleagues,[1,2] who studied the somatosensory cortex. They discovered that sensory neurons, such as those serving light touch of skin, are grouped together, segregated from those neurons that respond to other stimuli—for example, deep pressure or rotation of joints. The grouping occurs in columns that run radially from the cortical surface to the white matter, and connections between cells are principally up and down along the columnar axis. The columnar organization has now been studied in detail in the visual cortex of cats and monkeys. Information is also now available about the connections and properties of cells at different depths, in the various laminae of the visual cortex. Together, these findings form the basis of the following discussion.

In the visual system the pattern starts with the retina, where the functional unit consists of a group of photoreceptor cells connecting by way of bipolar cells to ganglion cells, with horizontal and amacrine cells feeding into the system. This unit, through its connections, elaborates the first integrated stage of visual information, which is handed on to the higher centers by the optic nerve fibers. As the impulses combine in the various cell stations in the cortex, the abstract significance of their message becomes transformed. It was suggested earlier, for example, that the field of a specific complex cell, which recognizes a corner, can be thought of as the convergence of fibers from the chain of lower order cells—complex, simple, geniculate, ganglion, bipolar, and primary receptor cells. However, as mentioned in Chapter 2, this is not a simple through-line; connections jump steps and also descend to influence antecedent levels.

Idealized schemes for building up receptive fields in a hierarchical fashion from layer to layer do not convey a structural picture of how cells are actually laid out to elaborate all the intricate receptive fields. Fortunately, analyses of receptive fields have clearly revealed the helpful, simplifying principle that neurons performing similar tasks tend to be grouped together, and this principle has been confirmed histologically. Thus, a circumscribed area in the striate cortex receives its principal input from a group of geniculate axons with receptive fields in the same region of the retina. Within this cortical area are contained the

[1]Mountcastle, V. B. 1957. *J. Neurophysiol.* 20: 408–434.
[2]Powell, T. P. S. and Mountcastle, V. B. 1959. *Bull. Johns Hopkins Hosp.* 105: 133–162.

very cells required for the synthesis of simple and complex fields with the same axis orientation, position, and ocular dominance. The outgoing information is then sent on to another area of cortex or to lower centers (such as the lateral geniculate nucleus or the superior colliculus). There another group of cells performs further processing, and so on.

The concept of grouping according to function adds new physiological insight to anatomical concepts. There is not simply a complete map of the retinal surface projecting onto the visual cortex, but rather a series of cell clusters, each of which performs its own special analysis and synthesis of the information coming from the retina. Further, circumscribed areas of the retina are represented over and over again in the various cortical regions.[3]

The general layout of neuronal pathways can be discerned by a number of physiological and anatomical techniques. Mentioned earlier are the physiological studies in which evoked potentials were used and which served to describe the way in which the retina is projected onto the visual cortex (Chapter 2). A reliable anatomical method for sorting out the terminals that originate in a particular distant group of neurons is to damage the cell bodies; over the next few days their axon terminals undergo characteristic degenerative changes. The staining technique devised by Nauta and colleagues[4] allows ready identification of the degenerating nerve endings. The procedure, then, is to make a discrete lesion in a localized site—for example, in one layer of the geniculate—and to search in different regions of the cortex for degenerating terminals. For many years, this procedure and stimulation were the only methods of tracing the destinations of neurons.

Other methods (mentioned in Chapter 1) have proved very useful. For example, if horseradish peroxidase is injected into a region of the brain, the enzyme is taken up by nerve terminals and transported backward along the axons to the cell bodies of origin (retrograde transport). The cells are then distinctively labeled by allowing the enzyme to react with a substrate, which causes a dense end-product to be deposited intracellularly. Cells stained in this way appear black in light micrographs; the end-products of the reaction can also be seen by electron microscopy.[5] Instead of horseradish peroxidase, a fluorescent dye can be injected and, after retrograde transport, can be located in the neuronal cell bodies. A combination of horseradish peroxidase injection into one area and dye injection into another enables the identification of cells that project to both regions. Similar double labeling can be achieved by using two different fluorescent dyes.[6]

A further technique, used for exploring visual pathways, is similar in principle but makes use of radioactive amino acids. The amino acid

Methods for tracing interconnections of cells

[3]Van Essen, D. C. 1979. *Annu. Rev. Neurosci.* 2: 227–263.
[4]Nauta, W. J. H. and Gygax, P. A. 1954. *Stain Technol. 29:* 91–93.
[5]La Vail, J. H. and La Vail, M. M. 1974. *J. Comp. Neurol. 157:* 303–358.
[6]Kuypers, H. G. J. M., Catsman-Berrevoets, C. E. and Padt, R. E. 1977. *Neurosci. Lett.* 6: 127–135.

is injected into a structure, such as the vitreous of the eye, from which it is taken up by nerve cell bodies of the retina and incorporated into protein. The labeled protein is then transported from ganglion cells through the optic nerve fibers to their terminals within the lateral geniculate nucleus. There the label can be detected by autoradiography, a method that consists of cutting a thin section of the tissue and covering it with a photoemulsion. The radioactive disintegrations in the terminals of geniculate neurons cause silver grains to be reduced in the photoemulsion. The technique detects only label attached to insoluble molecules like proteins precipitated by the fixative; water-soluble amino acids tend to be washed out of the tissue during dehydration and embedding. An unforeseen feature discovered by Grafstein[7] was that the label is also transferred from neuron to neuron across synapses. Thus, the endings of geniculate fibers are labeled in layer IV. This finding has been put to good use in the mammalian visual system, where it permits entire pathways to be traced.

The information provided by the methods described above is a prerequisite for studying functional architecture and the overall representation, or destination in the brain, of sensory pathways. These techniques, however, do not yet provide the resolution to determine the fine grain of neuronal interconnections. To gain such a resolution, the branching patterns and interconnections must be analyzed at the level of individual cells. For many years the Golgi stain provided the only method for picking out individual cells and staining all their processes.[8] In recent years, an additional indispensable technique has been to label identified cells or axons by injecting them with fluorescent dyes[9] or horseradish peroxidase.[10] This method has the advantage of allowing the investigator to establish first the physiological performance of a neuron and then its geometry—for example, whether a simple or a complex cell is a stellate or a pyramidal neuron and where its processes run. Although the technique is difficult, particularly for axons and for small cells, it has produced a wealth of new information about cortical circuitry.[11]

Nevertheless, electrical recording from individual neurons with external electrodes remains a simple and effective first step in classifying individual cells according to their functions. Their location in the cortex can be established by marking the position of the electrode tip at the end of a penetration by an electrolytic lesion. In this way the entire path of the electrode can later be reconstructed after fixation, sectioning, and staining.

[7]Specht, S. and Grafstein, B. 1973. *Exp. Neurol. 41*: 705–722.
[8]Lund, J. S. 1980. In F. O. Schmitt, F. G. Worden and F. Dennis (eds.), *The Organization of the Cerebral Cortex*. MIT Press, Cambridge, pp. 105–124.
[9]Kelly, J. P. and Van Essen, D. C. 1974. *J. Physiol. 238*: 515–547.
[10]Muller, K. J. and McMahan, U. J. 1976. *Proc. R. Soc. Lond. B 194*: 481–499.
[11]Gilbert, C. D. 1983. *Annu. Rev. Neurosci. 6*: 217–247.

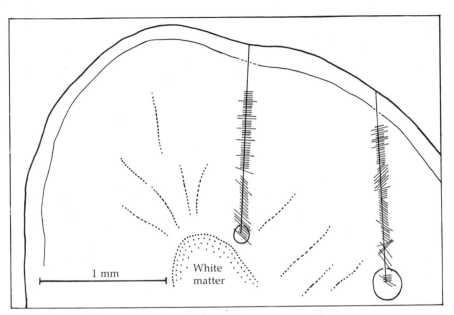

1 AXIS ORIENTATION of receptive fields of neurons encountered as an electrode traverses the cortex normal to its surface. Cell after cell tends to have the same axis orientation, indicated by the angle of the bar to the electrode track. The penetration to the right is more oblique; consequently the track crosses several columns and the axis orientations change frequently. The electrode track is reconstructed by making a lesion at the end of the penetration (circle) and cutting serial sections through the brain. Such experiments have established that cat and monkey cells with similar axis orientation are stacked in columns running at right angles to the cortical surface. (After Hubel and Wiesel, 1962.)

COLUMNS IN THE VISUAL CORTEX

To study the functional architecture of the visual cortex, Hubel and Wiesel made a systematic survey of cells by penetrating adjacent areas of cortex with microelectrodes and determining the properties of the various cells.[12-15] When the results found on repeated cortical penetrations were combined, a clear, unambiguous pattern emerged. With the electrode moving at right angles to the surface through the thickness of the gray matter, the cells that were encountered in sequence had the same field axis orientation. A sample experiment is shown in Figure 1. A microelectrode is inserted normal to the surface of the cortex in area 17 of the cat. Each bar indicates one cell and its preferred orientation in the progression through the cortex. The circle at the end marks the

Receptive field axis
orientation columns

[12]Hubel, D. H. and Wiesel, T. N. 1962. *J. Physiol. 160*: 106–154.
[13]Hubel, D. H. and Wiesel, T. N. 1963a. *J. Physiol. 165*: 559–568.
[14]Hubel, D. H. and Wiesel, T. N. 1968. *J. Physiol. 195*: 215–243.
[15]Hubel, D. H. and Wiesel, T. N. 1974. *J. Comp. Neurol. 158*: 267–294.

site of a lesion and shows the final position of the electrode tip. From this end point and the electrode track observed in the fixed brain after the end of the experiment, the following sequence is seen: At first, all the cells are optimally driven by bars or edges, at about 90° to the vertical, at one position in the visual field. After a penetration of about 0.6 mm, the axis of the receptive field orientation changes to about 45°. A second track to the right shows other cells, with slightly different receptive field positions and field axis orientations. With this more oblique penetration, the field axis changes repeatedly with only small movements of the electrode tip, as though a series of columns with different axis orientations is being traversed.

From a large series of such experiments in cats and monkeys, it became apparent that in area 17 simple and complex NEURONS WITH SIMILAR RECEPTIVE FIELD AXIS ORIENTATION ARE NEATLY STACKED ON TOP OF EACH OTHER IN DISCRETE COLUMNS that run perpendicular to the cortical surface. The columns receive their input from largely overlapping receptive fields on the retinal surface. Separate columns exist for each axis orientation.

Other functional variables are also grouped in columnar aggregates of cells; in cortical areas of the monkey beyond area 17, there exist columns of cells with well-defined color sensitivity and other columns in which the direction of movement of the visual stimulus is important.[16,17]

Ocular dominance columns

The COLUMNS FOR EYE PREFERENCE constitute a second system of cortical organization. Cortical neurons that are dominated, or more strongly influenced, by one eye or the other are grouped closely together. Figure 2 illustrates these characteristics of neurons in the striate cortex of the monkey. The cells (total 1116) are subdivided into seven groups. Groups 1 and 7 are driven exclusively by one of the two eyes; in groups 2,3 and 5,6 the effect of one eye is stronger than that of the other; and the cells in the middle, group 4, are equally influenced. The majority respond to both eyes (Chapter 2).

In the striate cortex of the monkey and the cat, the inputs from the two eyes are segregated at first (as in the geniculate) before being combined in an orderly manner. The majority of incoming geniculate fibers end in layer IV of area 17. In this clearly defined region, only simple cells and cortical neurons with concentric geniculatelike fields are encountered, and they are all stimulated by one eye only. Outside layer IV, the simple and complex cells are driven for the most part by both eyes. For example, in some of the penetrations of Figure 2B all the cells are better driven by the left eye, whereas in others the cells are better driven by the right eye. These groupings are called OCULAR DOMINANCE COLUMNS.

As the electrode moves away from layer IV, either deeper or more superficially, complex cells are found, all with the same field axis ori-

[16]Van Essen, D. C. and Zeki, S. M. 1978. *J. Physiol. 277*: 549–573.
[17]Michael, C. R. 1981. *J. Neurophysiol. 46*: 587–604.

entation, ocular dominance, and position in the visual field. This arrangement of cells detected physiologically supports the concept of organization through the thickness of the cortex. Thus, in general, the processes of cortical neurons pass up and down at right angles to the surface of the cortex and spread laterally only over short distances (see Figure 8).

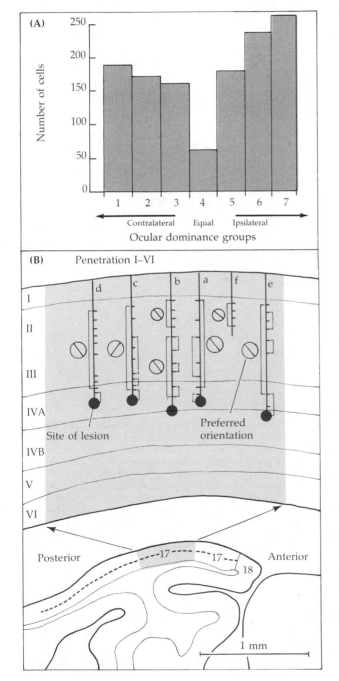

2

OCULAR DOMINANCE COLUMNS. (A) Eye preference of 1116 cells in 28 Rhesus monkeys. Most cells (groups 2 through 6) are driven by both eyes. (B) When penetrations (I through VI) are made through the cortex at right angles to the surface, the cells encountered in one track have similar eye preference, indicated by numbers, as well as similar orientation specificity, indicated by lines within open circles. (After Hubel and Wiesel, 1968; Wiesel and Hubel, 1974.)

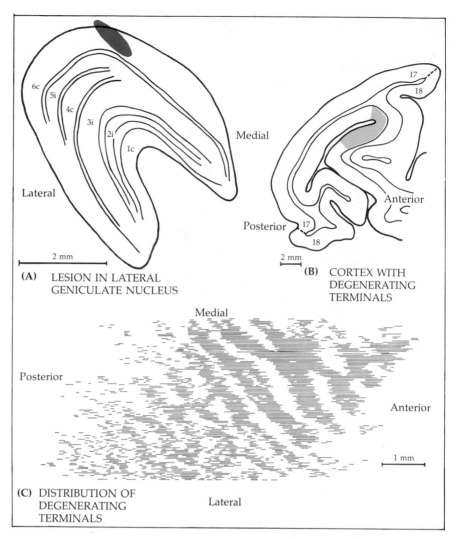

3 OCULAR DOMINANCE COLUMNS are demonstrated in the monkey by detection of degenerating axonal endings in layer IV of the cortex after lesions in the lateral geniculate nucleus. (A) The lesion (blue area) in layer 6, which is supplied by the contralateral eye. (B) Degenerating terminals are restricted to a small area (in blue) of the cortex. (C) The columns have been reconstructed from serial sections, 30 μm thick, cut through the cortex. Each short line represents the region over which degenerating terminals were found in layer IVC of a section. The overview in (C) was produced by aligning all the drawings appropriately according to their relative positions in adjacent sections. The columns actually have the shapes of slabs. (From Hubel and Wiesel, 1972.)

Shape and distribution of columns in the cortex

The discovery of the columnar arrangement through electrical recording methods was followed, some years later, by anatomical demonstration of ocular dominance columns. This development reinforced confidence in the reliability of the physiological approach and provided new opportunities for a glimpse into the layout of functional synaptic organization. An opening was provided by the observation just mentioned, that the simple cells in layer IV of area 17 are driven exclusively by one eye. As a result it has become possible to use a variety of procedures for demonstrating alternating groups of cells in layer IV supplied by one eye or the other.

One procedure is to make a small lesion in one layer of the lateral geniculate nucleus; degenerating terminals subsequently appear in layer IV. Degenerating endings of axons from the geniculate are distributed in a characteristic pattern of alternating strips or slabs (Figure 3C). These correspond to the areas driven by the one eye in whose line of connection the lesion is made. If a lesion is made so that axons from two layers of the geniculate are affected, carrying inputs from both eyes, a more evenly distributed pattern of degeneration is produced.[18] A further striking demonstration of the ocular dominance columns was provided by the transport of radioactive fucose or proline from one eye, as described earlier.[19,20] Figure 4 clearly shows the radioactivity around the terminals of geniculate neurons supplied by the injected eye. Zones that get their input from the other eye remain clear. The columnar arrangement demonstrated in this way is the same as that produced by lesions to the lateral geniculate nucleus, but now the columns are seen in all parts of the striate cortex. Finally, it is possible to observe the boundaries of the columns in layer IV of the monkey cortex by staining the cortex with a modified silver stain.[21]

At the cellular level, a similar pattern of bands has been revealed by backfilling or by injection of horseradish peroxidase into individual axons of the lateral geniculate nucleus as they approach the cortex.[22-24] The axon shown in Figure 5 is an off-center Y afferent (one that gives transient responses to dark moving spots). It ends in two distinct clumps of processes in layer IVab. The processes are separated by a blank area that presumably represents an ocular dominance column of the other eye. A more detailed description of the layers in which X and Y axons end is given later, in the discussion of microcircuitry. All the techniques described earlier indicate that the slabs for eye dominance are about 0.25 to 0.5 mm wide and run horizontally for long distances (Figures 4, 5, and 7). Strictly speaking, they are not shaped like columns, but the term has become established and is generally retained.

The form of the orientation columns has not been established, but it appears that, like the ocular dominance columns, they are shaped like slabs. This information about their arrangement in the visual cortex of monkeys and cats was obtained by making oblique or tangential electrode penetrations through the cortex. An example is shown in

[18]Hubel, D. H. and Wiesel, T. N. 1972. *J. Comp. Neurol. 146*: 421–450.
[19]Hubel, D. H. and Wiesel, T. N. 1977. *Proc. R. Soc. Lond. B 198*: 1–59.
[20]Shatz, C. J., Lindström, S. and Wiesel, T. N. 1977. *Brain Res. 131*: 103–116.
[21]LeVay, S., Hubel, D. H., and Wiesel, T. N. 1975. *J. Comp. Neurol. 159*: 559–576.
[22]Ferster, D. and LeVay, S. 1978. *J. Comp. Neurol. 182*: 923–944.
[23]Gilbert, C. D. and Wiesel T. N. 1979. *Nature 280*: 120–125.
[24]Blasdel, G. G. and Lund, J. S. 1983. *J. Neurosci. 3*: 1389–1413.

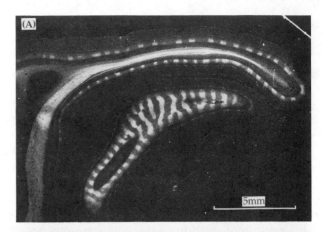

(A)

5mm

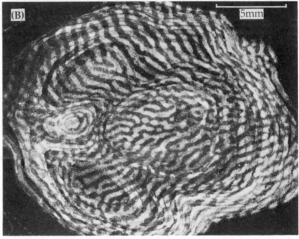

(B)

5mm

4

OCULAR DOMINANCE COLUMNS IN MONKEY CORTEX demonstrated by injection of radioactive proline into one eye. (A) and (B) are autoradiographs photographed with dark field illumination in which the silver grains appear white. (A) This horizontal section first passes through the visual cortex at right angles to the surface displaying columns cut perpendicularly, then in the center horizontally through layer IV cutting columns tangentially. (B) Reconstruction made from numerous horizontal sections of layer IVC in another monkey in which the ipsilateral eye had been injected (no single horizontal section can encompass more than a part of layer IV of the cortex because of its curvature). Dorsal is above, medial to the right. In both (A) and (B), the ocular dominance columns appear as stripes of equal width supplied by one eye or the other. (C) Reconstruction of the pattern of ocular dominance columns over the entire exposed part of layer IVc. Scale 5 mm. (A and B from LeVay, unpublished, photos by courtesy of S. LeVay; C from LeVay, Hubel, and Wiesel, 1975.)

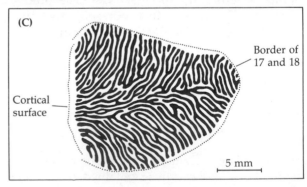

(C)

Border of
17 and 18

Cortical
surface

5 mm

Figure 6, which reveals once again the orderliness of the arrangement of cells (in a spider monkey named George[15]). Each 50-μm shift of the electrode along the cortex is accompanied by a change in field axis orientation of about 10°, in a regular sequence through 180°. The field axis orientation columns are narrower than those for ocular dominance—20 to 50 μm compared with 0.25 to 0.5 mm. The anatomical

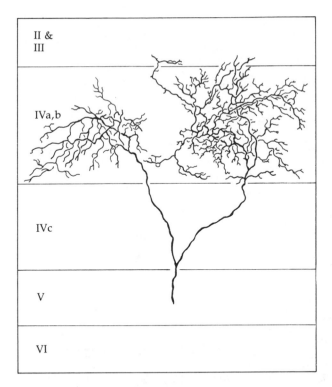

II &
III

IVa,b

IVc

V

VI

5

OFF-CENTER Y AXON FROM LATERAL GE-
NICULATE NUCLEUS terminating in layer IVab
of cat. The axon was injected with horseradish
peroxidase through a microelectrode. The terminals are grouped in two clusters separated by a
vacant zone supplied by the other eye. (After
Gilbert and Wiesel, 1979.)

demonstration of columns for orientation is more elusive than for ocular
dominance. Hints have, however, been provided by the use of 2-deoxy-
glucose, a technique pioneered by Sokoloff.[25] The principle (Chapter 1)
is that active cells take up radioactive deoxyglucose, as though it were
glucose. This molecule, however, cannot be metabolized or transported
out; as a result, metabolically active cells become labeled with radioactivity and their distribution can be seen by autoradiography. In monkeys
and cats, after the eyes were exposed to vertical stripes, bands of
radioactivity appeared in the cortex, possibly corresponding to orientation columns.[26] (In addition, discrete "blobs" of neurons were labeled;
their properties are discussed later.)[27] A particularly clear anatomical
demonstration of orientation columns has been made in the tree
shrew.[28] Deoxyglucose studies reveal regularly spaced, parallel zones
of uptake 150–350 μm wide following exposure of the animal to stripes
of one orientation. Labeling extends from the surface of the cortex to
white matter.

A scheme for the way in which the two sets of columns might be
arranged in the cortex is shown in Figure 7. For simplicity in elaborating
fields, one would expect the orientation and ocular dominance columns

[25]Sokoloff, L. 1977. *J. Neurochem. 29*: 13–26.
[26]Hubel, D. H., Wiesel, T. N. and Stryker, M. P. 1978. *J. Comp. Neurol. 177*: 361–380.
[27]Livingstone, M. S. and Hubel, D. H. 1982. *Proc. Natl. Acad. Sci. USA 79*: 6098–6101.
[28]Humphrey, A. L., Skeen, L. C. and Norton, T. T. 1980. *J. Comp. Neurol. 192*: 549–
566.

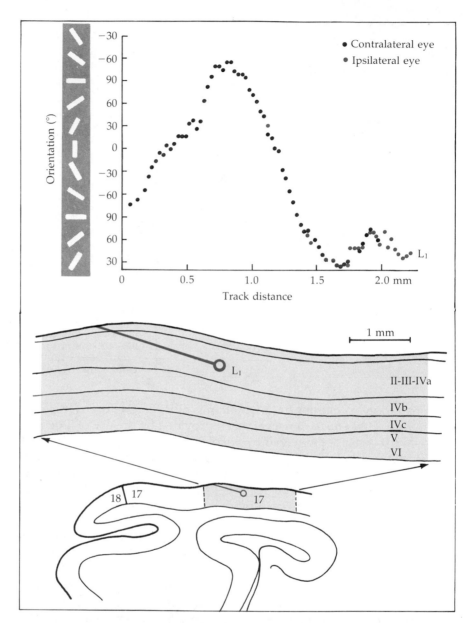

6 ARRANGEMENT OF ORIENTATION COLUMNS. As the electrode moves obliquely through the cortex, the orientation specificity of the cells encountered shifts systematically. The shift is about 10° for each 50 μm, as though a series of columns or slabs were being traversed in a regular sequence. (After Hubel and Wiesel, 1974.)

to run at right angles to each other. If they did, inputs from the two eyes could be combined to produce coherent binocular fields with similar orientation for the synthesis of simple or complex fields positioned in the same part of the visual field. The illustration provides a further clue to cortical organization and the relation of the orientation to the

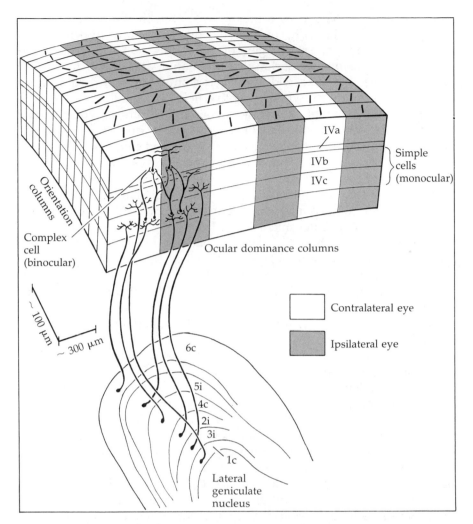

IVa

IVb

IVc

Simple
cells
(monocular)

Orientation
columns

Complex
cell
(binocular)

Ocular dominance columns

~ 100 μm

~ 300 μm

Contralateral eye

Ipsilateral eye

6c

5i

4c

2i

3i

1c

Lateral
geniculate
nucleus

7 **RELATION BETWEEN OCULAR DOMINANCE** and orientation columns. Scheme in which the ocular dominance and orientation columns
run at right angles to each other. An example of a complex cell is shown
in an upper layer, receiving its inputs from two simple cells that lie in
adjacent ocular dominance columns but share the same orientation columns. (From Hubel and Wiesel, 1972.)

ocular dominance columns. The functional unit of the visual cortex
appears to be a roughly cuboidal aggregate of cells in which all the
possible orientations are represented for a receptive field area in the
same place in each of the two eyes. Hubel and Wiesel have named this
unit the HYPERCOLUMN. An adjacent square of cortex would analyze
information in the same way for an adjacent but overlapping part of
the visual field, and so on. In parts of the cortex that deal with peripheral fields of vision, the receptive fields of individual cells become larger.
In such areas a hypercolumn of the same size deals with a relatively
large area of peripheral retina. And moving from one hypercolumn to

the next is accompanied by a much larger shift in the position of the receptive field—still with considerable overlap, many hypercolumns being influenced by a small region of retina. But the basic organization of the cortex remains similar, in that orientation and ocular dominance columns have the same width as their counterparts in cortical areas dealing with the fovea, where receptive fields are small.

Layering of cortex and microcircuitry

What processing of information goes on within a hypercolumn? Upon which cells do geniculate fibers end? What connections are made within the cortex to synthesize fields of simple and complex cells; which cells send signals out again and to which destinations? In the cat, exploration of such questions has now become possible through a combination of techniques already mentioned: Golgi staining, horseradish peroxidase injection of individual neurons, and recordings of synaptic activity with intracellular microelectrodes.

Figures 8 and 9 show the principal patterns of layering, the inputs and outputs for the various layers, as well as the predominant distributions of simple and complex cells.[11,29,30] As mentioned in Chapter 2, simple cells are found only in layers IV and VI and complex cells are not present in layer IV.

Inputs, outputs, and intracortical connections

As mentioned earlier, incoming geniculate fibers end for the most part, but not exclusively, in layer IV.[11,22,23] In both the cat and the monkey, layer VI is also supplied by geniculate afferents. In addition, more superficial layers receive inputs from C laminae of the lateral geniculate nucleus (containing Y and W cells) in the cat and from the pulvinar (a region of the thalamus) in the monkey.[31] Within layer IV itself, geniculate fibers of different types (parvocellular or magnocellular) end in discrete sublaminae; for example, in monkey cortex, cells in layers IVA, IVCβ and the upper part of layer VI receive inputs from parvocellular layers of the geniculate. (Layer IVC is itself subdivided into two sublayers: IVCα above and IVCβ below.) Cells in IVCα and the deeper part of VI are supplied by the magnocellular afferents. The degree to which separation of X and Y is maintained at subsequent stages of processing is not known. Numerous complex and simple cells are supplied by both types of input.[32,33] Some cortical cells, especially those in layers II, upper III, and V, receive inputs derived mainly from neurons within the cortex; cortical cells in layers IV and VI receive inputs both from geniculate afferents and from cortical cells. A specialized punctate projection of geniculate fibers to small clusters of cells, the "blobs" labeled with deoxyglucose, in layer III constitutes a quite separate system (page 92).

[29]Gilbert, C. D. and Wiesel, T. N. 1981. In F. O. Schmitt, F. G. Worden and F. Dennis (eds.), *The Organization of the Cerebral Cortex*. MIT Press, Cambridge, pp. 163–191.

[30]Gilbert, C. D. 1977. *J. Physiol. 268*: 391–421.

[31]Hendrickson, A. E., Wilson, J. R. and Ogren, M. P. 1978. *J. Comp. Neurol. 182*: 123-136.

[32]Malpeli, J. G., Schiller, P. H. and Colby, C. L. 1981. *J. Neurophysiol. 46*: 1102–1119.

[33]Movshon, J. A., Thompson I. D. and Tolhurst, D. J. 1978b. *J. Physiol. 283*: 101–120.

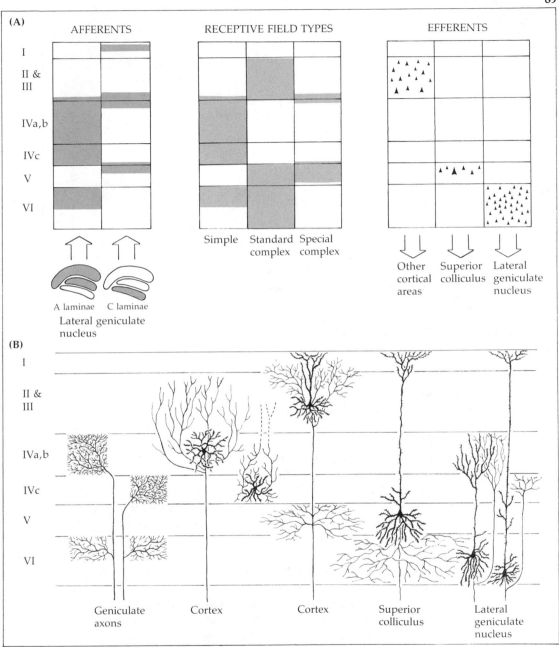

8   SCHEMATIC DIAGRAM OF CORTICAL CONNECTIONS IN CAT. (A)
Distribution of inputs to the various layers, properties of cells, and
outputs of cortex. (B) Principal arborizations of processes in layers I–
VI. (After Gilbert and Wiesel, 1981.)

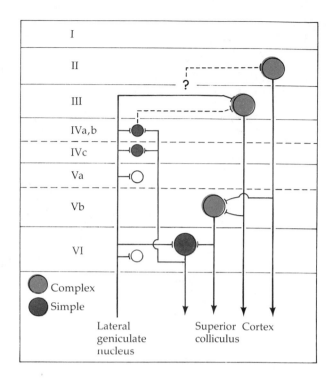

9

EXCITATORY SYNAPTIC CONNECTIONS IN AREA 17 OF CAT based on intracellular recordings from nerve cells in different layers while stimulating incoming geniculate fibers. Electrophysiological criteria are used to determine whether connections are direct (solid lines) or indirect (dashed lines). (After Ferster and Lindström, 1983.)

The OUTPUTS, which are derived from layers VI, V, IV, III, and II (Figures 8 and 9), are as follows:[11,29] Axons of cells in layer VI project back down to the lateral geniculate nucleus and to another deep structure, the claustrum. Cells in layer V (complex cells) also project down, principally to the superior colliculus, a structure in the midbrain concerned with eye movements (Appendix B). Cells in layer IV (simple cells) project to layers III and II; they also send axons out to end in other areas of cortex (not shown). Complex cells in layers II and III provide the major output to other cortical areas; they also project down to layer V. Within a given layer, discrete groups of cells give rise to these different projections. Thus, separate populations of cells in layers II and III supply the various areas of cortex. A single cell that carries efferent signals out of the cortex can also mediate intracortical connections from one layer to another.

A simplified scheme of intracortical connections between cells from layer to layer is shown for the cat in Figures 8 and 9. Simple cells in layer IV project to layers II and III; cells in II and III in turn project to layer V. Layers II, III, and V are the layers where the complex and specialized complex cells are abundant. Complex cells are also present in layer VI, which is the only layer apart from IV containing simple cells. Axons from layer VI, in addition to supplying the geniculate, end in layer IV. Unlike those in layer IV, simple cells in layer VI have greatly elongated receptive fields.

Certain correspondences with the original schemes of Hubel and Wiesel for constructing receptive fields within the cortex emerge from inspection of these pathways. Thus, simple cells in layer IV lie within the predominant geniculate afferent zone. Geniculate inputs and simple cells are also in proximity in layer VI. Moreover, the dynamic properties of geniculate axons and simple cells are nicely matched. In the monkey, on-center geniculate fibers supply cortical cells that respond to light bars or edges; off-center fibers supply those that respond best to dark bars or edges.[32] Similarly, as one might expect from the properties of X and Y geniculate axons (Box 2, Chapter 2), simple cells in layers where X-type fibers end have X-like characteristics and smaller receptive fields than Y-type simple cells.[22,23,33] Again, for complex cells it seems reasonable that complex cells in layers II and III should receive their predominant input from layer IV simple cells and that those in layer V should be supplied by layers II and III. At the same time, as already mentioned, there is not a simple through-line: Geniculate fibers also project to layers other than IV and VI. Even in the absence of input from layer IV, it has been reported that complex cells in layers II and III can still respond to visual stimuli.[34] Moreover, it is evident that nothing has been said of layer I, about which little is known.

A start has been made in analyzing how the receptive fields of simple and complex cells are synthesized. For this one needs to know about the neuron's response properties, the position of its cell body, the distribution of its dendrites, the inhibitory and excitatory inputs it receives, and the sources of those inputs—no small task. Some guides are already available from the findings reported above. Although the layer in which a cell is found provides essential clues, the shape of its cell body appears not to be useful diagnostically. Pyramidal cells, which provide the predominant output from the cortex, can be simple or complex. So can smooth and spiny stellate cells, which arborize for the most part within the cortex. A correlation has, however, been found between the size of the receptive field and the size of the cell. In general, large neurons have extensive, long dendrites and large receptive fields. For example, simple cells with X characteristics and small receptive fields have shorter dendrites grouped within a more restricted radius than Y-type simple cells or the simple cells with very long receptive fields in layer VI.[29] At the same time, the finding that processes end in one layer does not necessarily mean that they synapse with the cells in that layer: They may connect instead to processes of neurons whose cell bodies lie elsewhere. This is presumably the case in layer I, which contains relatively few cell bodies but abundant fibers from various sources.

Certain individual neurons have been found to send processes over surprisingly long distances through several hypercolumns, with endings clustered in discrete groups in various layers.[35] These long axonal projections might enable columns to intercommunicate and might also serve for synthesizing long receptive fields.

Cell types and synaptic interactions

[34]Malpeli, J. G. 1983. *J. Neurophysiol. 49*: 595–610.
[35]Gilbert, C. D. and Wiesel, T. N. 1983. *J. Neurosci. 3*: 1116–1133.

To trace connections in terms of their effectiveness, Ferster and Lindström[36] have recorded intracellularly with microelectrodes from individual cortical cells while stimulating geniculate axons. In this way they have been able to determine whether connections are excitatory or inhibitory and whether they are mediated directly or indirectly. Figure 9 shows the principal types of connections. Direct excitatory potentials were evoked in simple cells of layers IV and VI in the cat, and electrophysiological criteria confirmed that these arose from monosynaptic connections (i.e., by geniculate axons ending directly on cortical neurons with no interposed cells). Geniculate fibers also produced monosynaptic excitatory potentials in complex cells in layers III, Va (upper), and VI but not on complex cells in layers II and Vb (lower); the only excitation seen in those cells was mediated indirectly. In addition to these excitatory effects, inhibitory potentials (always indirect) were evoked by geniculate stimulation in all cortical cells recorded from. There are suggestions that these inhibitory potentials arise from activation of smooth stellate cells that use γ-aminobutyric acid as their transmitter (see Chapter 12). The role of these inhibitory connections in synthesizing receptive fields is not yet known. Several lines of evidence suggest that inhibitory inputs driven disynaptically or polysynaptically from separate geniculate fibers may give rise to the flanking "off" areas for cells responding to bright edges or bars.

Many other problems of circuitry within the columns still require elucidation. For example, simple cells in layer VI differ from those in IV by having much longer receptive fields. The way in which these are synthesized, what role is played by the descending inputs from layer V and what effects these layer VI cells have on layer IV simple cells are still not known. Other important questions concern the mechanisms by which end inhibition is produced and what contribution geniculate inputs make to the properties of complex cells.

Cytochrome oxidase-stained "blobs"

An unexpected development has been the discovery of regional specializations in the visual cortex of the monkey. These specialized regions constitute a system quite dissimilar from the columnar architecture described earlier. Discrete, circumscribed clusters of neurons have been found mainly in layers II and III, but also in layers V and VI and have distinctive inputs and properties. These patches or "blobs"[27,37] form a highly regular array ("a polka-dot pattern . . . as if the animal's brain had the measles"[38]) (Figure 10). They were first seen in cortical tissue stained for cytochrome oxidase, an enzyme indicative of high metabolic activity. Subsequently, it became apparent that some of the deoxyglucose labeling seen after visual stimulation in the monkey cortex was distributed over blobs. Injections of peroxidase or proline indicate that clusters of geniculate axons end in close association with the blobs.

In macaque monkeys, the patches of cytochrome oxidase stain are

[36]Ferster, D. and Lindström, S. 1983. *J. Physiol. 342*: 181–215.
[37]Wong-Riley, M. 1979. *Brain Res. 171*: 11–28.
[38]Hubel D. H. 1982b. *Nature 299*: 515–524.

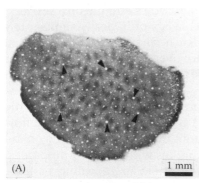
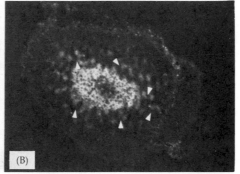

10 CYTOCHROME OXIDASE "BLOBS" IN VISUAL CORTEX of monkey. (A) Tangential section through area 17 in layer III stained for cytochrome oxidase, showing dark blobs. (B) Adjacent section stained for horseradish peroxidase. Horseradish peroxidase had been injected into the ipsilateral lateral geniculate nucleus. The section was photographed with dark-field illumination. The arrowheads represent corresponding points in the two sections. The blobs in the two sections are in precise register, showing that the blobs receive a special input from the lateral geniculate nucleus. (From Livingstone and Hubel, 1982.)

arranged in parallel rows about 0.5 mm apart. A remarkable feature is that they lie precisely in register with the ocular dominance columns. Removal of one eye causes every other row of blobs to shrink and stain less darkly. The properties of cells within the blobs are quite different from those surrounding them in the normal columnar arrangement. Thus, unlike the complex cells found in layers II and III, blob cells fire in response to small spots with no orientation specificity and have receptive fields that are concentric. Many of them are color-coded like geniculate cells, responding best to particular wavelengths of light in the center and surround.

The idea that the blobs represent a system separate from the columns but related to them is reinforced by the highly specific pattern of their intracortical connections. In area 18 are found aggregates of cells with properties resembling those of the blobs: they stain for cytochrome oxidase, become labeled with deoxyglucose, and are arranged in stripes. It has been shown that blobs of area 17 are connected to the homologous stripes in area 18 in a highly specific manner.[39] One intriguing speculation is that each blob may be concerned with processing information about color contrast within the field represented by the small area of the hypercolumn it is associated with.

The existence of separate columns for all the diverse orientations and their presumed use in providing components for processing information suggest why the cortex needs such a staggering number of cells. Look, for example, at just a very small segment of the primary visual cortex. Such a segment consists of columns with their rows of cells concerned exclusively with a small part of the visual field that is ana-

Significance of cell groupings

[39]Livingstone, M. S. and Hubel, D. H. 1983. *Nature (Lond.) 304*: 531–534.

lyzed for movement, orientation, color, and other stimulus parameters. Each part of the primary visual cortex consists of basically similar stereotyped repeating arrays of cells. And each small region of the retina is represented not once, but over and over again, in column after column, first for one receptive field axis orientation and then for another.

The columnar system enables the cortex to analyze each part of the visual world in terms of many different variables. As expected, columns are not entirely independent units; through their connections they hand on the result of their processing by linking up with other cortical areas, where further analysis takes place along a similar repeating pattern. There is direct anatomical evidence to show that fibers leave the primary visual cortex (area 17) to supply several adjacent areas, which in turn project to still other areas.

At this stage it is natural to wonder what other features of the perceptual world are organized in columns and how the rest of the cortex is arranged. The first demonstration of columns was made, as mentioned earlier, by Mountcastle[1] in the somatosensory cortex, where mechanosensory cells are grouped according to modality and receptive field position.[40] Within a column, all the cells respond either to touch or to deep pressure and joint position. Similarly, cells in the motor cortex are arranged in columns according to the muscles they innervate.[41] In the auditory cortex, the neurons in a column display similar binaural interactions.[42] Other types of groupings may occur in different areas. For example, in the face area of mouse and rat somatosensory cortex, there are barrel-shaped arrangements of neurons aligned in well-defined rows (see Chapter 17).[43] Each group of cells corresponds to one of the hairs, the vibrissae, on the animal's face. Here, however, the functional organization of the cortical cells is not understood; yet the principle remains the same—cells with similar functions are aggregated together. Within this small area, the cells receive inputs from various sources, process the information layer by layer, and send signals out to other parts of the brain.

An encouraging aspect of the recent studies of the functional organization of the cortex is that a few methodologies have already yielded so many advances. New approaches that enable areas of cortex to be identified as they become active seem particularly promising. These include positron emission tomography (PET scanning): the technique uses computers to reconstruct the distribution of radioisotopes in transverse sections of the living, functioning brain (fortunately without decapitation or brain slicing). Positron-emitting isotopes injected into the bloodstream give rise to signals that can be detected externally. Changes

[40]Kaas, J. H. 1983. *Physiol. Rev. 63*: 206–231.

[41]Asanuma, H. 1975. *Physiol. Rev. 55*: 143–156.

[42]Middlebrooks, J. C., Dykes, R. W. and Merzenich, M. M. 1980. *Brain Res. 181*: 31–48.

[43]Simons, D. J. and Woolsey, T. A. 1979. *Brain Res. 165*: 327–332.

in activity can be ascribed to different areas and measured as a function of time. Such measurements have provided information not obtainable in any other way about local glucose consumption, oxygen utilization, blood flow or tissue composition.[44] The resolution is at present too crude to resolve small groups of cells, but eventually procedures of this type that can be performed in conscious subjects may help to define structures involved in processing information as it moves through the cortex, area by area.

SUGGESTED READING

General reviews

Gilbert, C. D. 1983. Microcircuitry of the visual cortex. *Annu. Rev. Neurosci. 6*: 217–247.

Gilbert, C. D. and Wiesel, T. N. 1981. Laminar specialization and intracortical connections in cat primary visual cortex. In F. O. Schmitt, F. G. Worden and F. Dennis (eds.), *The Organization of the Cerebral Cortex*. MIT Press, Cambridge, pp. 163–191.

Hubel, D. H. and Wiesel, T. N. 1977. Ferrier Lecture. Functional architecture of macaque monkey visual cortex. *Proc. R. Soc. Lond. B 198*: 1–59.

Hubel, D. H. 1982b. Exploration of the primary visual cortex 1955–78. *Nature 299*: 515–524.

Lund, J. S. 1980. Intrinsic organization of the primate visual cortex, area 17, as seen in Golgi preparations. In F. O. Schmitt, F. G. Worden, G. Adelman and S. G. Dennis (eds.), *The Organization of the Cerebral Cortex*. MIT Press, Cambridge, pp. 105–124.

Raichle, M. E. 1983. Positron emission tomography. *Annu. Rev. Neurosci. 6*: 249–267.

Van Essen, D. 1979. Visual areas of the mammalian cerebral cortex. *Annu. Rev. Neurosci. 2*: 227–263.

Original papers

Ferster, D. and LeVay, S. 1978. The axonal arborizations of lateral geniculate neurons in the striate cortex of the cat. *J. Comp. Neurol. 182*: 923–944.

Ferster, D. and Lindström, S. 1983. An intracellular analysis of geniculocortical connectivity in area 17 of the cat. *J. Physiol. 342*: 181–215.

Gilbert, C. D. and Wiesel, T. N. 1979. Morphology and intracortical projections of functionally characterized neurons in the cat visual cortex. *Nature 280*: 120–125.

Hubel, D. H. and Wiesel, T. N. 1962. Receptive fields, binocular interaction and functional architecture in the cat's visual cortex. *J. Physiol. 160*: 106–154. (The original description of orientation and ocular dominance columns in cat cortex.)

Hubel, D. H. and Wiesel, T. N. 1968. Receptive fields and functional architecture of monkey striate cortex. *J. Physiol. 195*: 215–243.

Hubel, D. H. and Wiesel, T. N. 1972. Laminar and columnar distribution of geniculo-cortical fibers in the macaque monkey. *J. Comp. Neurol. 146*: 421–450.

[44]Phelps, M. E., Mazziotta, J. C. and Hueng, S.-C. 1982. *J. Cerebr. Blood Flow Metab. 2*: 113–162.

Hubel, D. H. and Wiesel, T. N. 1974. Sequence regularity and geometry of orientation columns in the monkey striate cortex. *J. Comp. Neurol, 158*: 267–294. (A number of basic concepts about columnar organization, receptive fields, and cortical function are incorporated in this paper.)

Humphrey, A. L. and Hendrickson, A. E. 1983. Background and stimulus-induced patterns of high metabolic activity in the visual cortex (area 17) of the squirrel and macaque monkey. *J. Neurosci. 3*: 345–358.

Humphrey, A. L., Skeen, L. C. and Norton T. T. 1980. Topographic organization of the orientation column system in the striate cortex of the tree shrew (*Tupaia glis*). II. Deoxyglucose mapping. *J. Comp. Neurol. 192*: 549–566.

Livingstone, M. S. and Hubel, D. H. 1982. Thalamic inputs to cytochrome oxidase-rich regions in monkey visual cortex. *Proc. Natl. Acad. Sci. USA 79*: 6098–6101.

Malpeli, J. G. 1983. Activity of cells in area 17 of the cat in absence of input from layer A of lateral geniculate nucleus. *J. Neurophysiol. 49*: 595–610.

Malpeli, J. G., Schiller, P. H. and Colby, C. L. 1981. Response properties of single cells in monkey striate cortex during reversible inactivation of individual lateral geniculate laminae. *J. Neurophysiol. 46*: 1102–1119.

Michael, C. R. 1981. Columnar organization of color cells in monkey's striate cortex. *J. Neurophysiol. 46*: 587–604.

Zeki, S. M. 1980. The response properties of cells in the middle temporal area (area MT) of owl monkey visual cortex. *Proc. R. Soc. Lond. B 207*: 239–248.

PART TWO
MECHANISMS FOR NEURONAL SIGNALING

ELECTRICAL SIGNALING

CHAPTER FOUR

The signals used by nerve cells to transmit information consist of potential changes produced by electrical currents flowing across their surface membranes. The currents are carried by ions such as sodium, potassium, and chloride. Compared with insulated wires, nerves are poor conductors of electricity. The internal ions are relatively low in concentration and much less mobile than electrons in a wire. Furthermore, the cell membrane separating the internal conducting medium from the external fluid is an imperfect insulator and permits leakage of ions into and out of the cell. It also has a relatively large electrical capacitance. These properties impose restrictions on the efficacy with which electrical signals can be conducted along a nerve.

Neurons carry only two types of signals: localized potentials and action potentials. The localized, graded potentials can spread only short distances, which are usually limited to 1 or 2 mm. They play an essential role at special regions, such as sensory nerve endings (where they are called receptor potentials) or at junctions between cells (where they are called synaptic potentials). The localized potentials enable individual nerve cells to perform their integrative functions and to initiate action potentials. The action potentials are regenerative impulses which are conducted rapidly over long distances without attenuation.

These two types of signals are the universal language of nerve cells in all animals that have been studied.

The signals used by the nervous system to process and transmit information are electrical in nature. Thus, nerve cells have RESTING POTENTIALS across their membranes; their membranes can be HYPERPOLARIZED or DEPOLARIZED from resting level by local potential changes; and impulses or ACTION POTENTIALS travel along their axons. The processes underlying these electrical events differ, however, from those underlying electrical signals encountered in everyday life in radios, TV sets, or computers. Signals in the nervous system, instead of being carried by electrons in metal wires, are caused by movements of ions such as

Current flow in nerve cells

sodium and potassium in the intracellular and extracellular fluids and across the cell membranes separating these fluid compartments.

In order to understand the processes underlying electrical signaling, it is useful to have in mind a picture of the relevant structural components of the nerve fiber which carries the signals. The fiber can be considered as a tube filled with a watery solution of salts and proteins separated from a similar extracellular solution by a membrane. The solutions are of the same ionic strength but of different ionic composition, and the membrane is relatively, but not totally, impermeable to the ions present on either side. Although the axoplasm is analogous to a copper wire and the membrane to a layer of insulation around a wire, the two systems are quantitatively quite different. First, the axoplasm is about 10^7 times worse than a metal wire as a conductor of electricity. This is because the density of charge carriers (ions) in the axoplasm is very much less than that of free electrons in a wire; and, in addition, their mobility is less. Second, movement of currents along the axon for any great distance is hampered by the fact that the membrane, although relatively impermeable to ions, is not a perfect insulator. Consequently, any current flowing along the axoplasm is gradually lost by leakage across the membrane to the outside solution. Finally, the fact that nerve fibers are extremely small (usually not exceeding 20 μm in diameter in vertebrates) further limits the amount of current they can carry. Hodgkin has provided a striking illustration of the consequences these factors have on the spread of electrical signals:[1]

> If an electrical engineer were to look at the nervous system he would see at once that signalling electrical information along the nerve fibres is a formidable problem. In our nerves the diameter of the axis cylinder varies between about 0.1μ and 10μ. The inside of the fibre contains ions and is a reasonably good conductor of electricity. However, the fibre is so small that its longitudinal resistance is exceedingly high. A simple calculation shows that in a 1μ fibre containing axoplasm with a resistivity of 100 ohm cm, the resistance per unit length is about 10^{10} ohm per cm. This means that the electrical resistance of a metre's length of small nerve fibre is about the same as that of 10^{10} miles of 22 gauge copper wire, the distance being roughly ten times that between the earth and the planet Saturn. An electrical engineer would find himself in great difficulties if he were asked to wire up the solar system using ordinary cables.

Electrical signals, then, are severely reduced, or "attenuated" over a relatively short length of nerve fiber. In addition, when such signals are brief in duration, their time course may be severely distorted and their amplitude further attenuated by the electrical capacitance of the cell membrane. These properties of membrane resistance and capacitance and axoplasmic resistance will be discussed in detail in Chapter 7. For now, it is sufficient to point out that the nervous system makes use of two kinds of electrical signals. The first kind are graded, passive

[1]Hodgkin, A. L. 1964. *The Conduction of the Nervous Impulse.* Liverpool University Press, Liverpool, p. 15.

or LOCALIZED POTENTIALS, which are subject to the attenuation and distortion discussed earlier; the second kind are impulses, or ACTION POTENTIALS, which involve active processes and can travel rapidly and without distortion from one end of a nerve to another.

The main characteristic of local potentials is that they can be graded continuously in size. In sensory endings, such potentials are known as generator potentials or RECEPTOR POTENTIALS. In a sensory nerve ending sensitive to pressure on the skin, for example, the size of the receptor potential increases in relation to the magnitude of the applied pressure. There are many types of such endings or receptors, each responsive to one type of physical stimulus such as bending of a hair, changes in temperature, changes in angle of a joint, or (in the retina of the eye) light. The job of such receptors is to change or "transduce" the physical stimulus into a receptor potential which then can be processed further by the nerve cell so that information about the stimulus eventually reaches the central nervous system. Other types of local potentials occur at synapses where they are known as POSTSYNAPTIC POTENTIALS. Postsynaptic potentials can be EXCITATORY (if they depolarize the cell and thus tend to give rise to nerve impulses) or INHIBITORY (if they tend to counteract or suppress such depolarization). (For a summary of nomenclature, see Box 1.) The size of a postsynaptic potential also can be graded and is a reflection of the number and rate of activity of excitatory or inhibitory presynaptic nerve terminals giving rise to it.

In contrast to local potentials, the action potential is a brief event that travels unattenuated along an axon. It is "all-or-none," which means that its amplitude remains relatively constant along the axon. Furthermore, action potential amplitudes do not vary very much from one kind of nerve fiber to the next, being just about the same in an optic nerve as in an auditory nerve or in a sensory fiber carrying information from the big toe to the spinal cord. What does vary, however, is the speed of propagation of the action potential: speed of propagation is greater in larger axons than in smaller ones. Initiation of the nerve impulse and its mechanisms of propagation are essentially the same in neurons from a wide range of invertebrates and vertebrates, as are the mechanisms underlying receptor potentials and postsynaptic potentials. Thus, the combination of local potentials and propagated action potentials constitutes the universal language of all known nervous systems.

To describe the electrical properties of nerve fibers more specifically, it is convenient to examine the potentials produced in a fiber by passing current across its membrane. In practice, muscle fibers are often used for such experiments for reasons of convenience, yielding results similar to those discussed here for nerve.

Two electrodes are used to measure the potential difference across the membrane, one in contact with the axoplasm and the other with the outside fluid. The difference in potential between them is amplified and fed into an oscilloscope. The connection to the intracellular fluid is usually made by a microelectrode, which is a fine glass capillary with an internal tip diameter of less than 1 μm. Such electrodes are placed

Types of signals

Measurement of membrane potentials

in the cell with the aid of a micromanipulator and are small enough to penetrate a cell membrane without causing damage. Their high electrical resistance (which may reach values greater than 200 MΩ) requires that special amplifiers be used to avoid distortion of the electrical potentials arising in the nerve.

When both recording electrodes are in the extracellular fluid, no potential difference is recorded; as the microelectrode is pushed through the cell membrane, a sudden jump of about 80 mV in the negative direction is registered on the oscilloscope (Figure 1). This potential difference is called the RESTING POTENTIAL. A reduction in the magnitude of the resting potential (toward zero) is called DEPOLARIZATION; an increase in magnitude is called HYPERPOLARIZATION. In most of the figures in this section, potential is plotted against time, as on an oscilloscope trace, and the potentials represent those at the microelectrode tip, recorded with respect to the indifferent electrode in the extracellular solution. Negativity is shown in the downward direction.

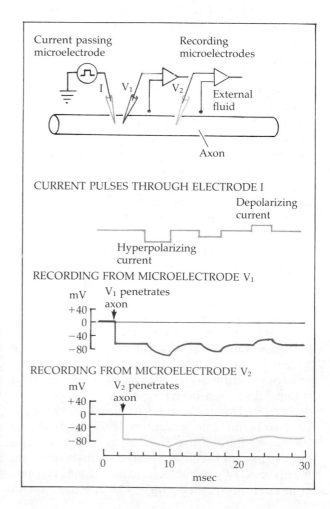

1

INTRACELLULAR RECORDING from a large axon with microelectrodes. One electrode (V_1) is inserted into the axon and records a resting potential of −70 mV (inside negative with respect to the outside). A second electrode (I), next to V_1, is used to pass pulses of current that produce localized graded potentials. The first two are hyperpolarizations and the third is a depolarization. Electrode V_2, about 1 mm away from V_1, also measures a resting potential of −70 mV when it penetrates the axon, but the localized graded potentials are smaller and slower than at V_1, owing to the passive electrical properties of the axon.

BOX 1   A NOTE ON NOMENCLATURE

The localized changes in membrane potential that depend on the passive electrical properties of the membrane are all basically similar, but they have been given a variety of names de- pending on the mechanism that generates the potential change, its effect, and the site at which it occurs. Defined here are some commonly used terms.

Designation	Potential Change
ELECTROTONIC POTENTIAL	Brought about by passing current through electrodes
SYNAPTIC POTENTIAL often abbreviated as epsp and ipsp (excitatory and inhibitory postsynaptic potentials); in muscle the excitatory postsynaptic potential is called an end plate potential (epp)	Brought about by chemical transmitters at synapses
RECEPTOR POTENTIAL (or generator potential)	In sensory receptors; brought about by the appropriate stimulus

In the experiment illustrated in Figure 1, a second microelectrode is inserted near the first and is used to pass current through the mem- brane. In addition, yet another electrode is placed in the axon about 2 mm away to record events at some distance from the point of current injection. The illustration shows the effect of passing hyperpolarizing or depolarizing current pulses into the axon. The local potentials pro- duced by the current pulses have the following properties:

1. They are graded. The amplitudes of the voltage changes at both the near and distant electrodes increase with the amount of injected current.
2. The duration of the potential changes also varies with the duration of the current pulse, but the potentials rise and fall more slowly because of the capacitance of the membrane (Chapter 7).
3. At the distant electrode, the change in potential is much smaller than at the point of current injection, and it rises and falls more slowly. At greater distances (several millimeters), little or no potential change is observed as a result of the current injection.

Localized, graded potentials of this type, which depend on the passive properties of the membrane, are therefore useless for long-range signaling. However, as we will see later, they are essential for the workings of the nervous system and underlie many basic neural pro- cesses, such as initiation of sensory signals and integration of signals arriving at synapses.

The localized, graded potentials that occur naturally in the nervous system—namely, receptor potentials and synaptic potentials—are able to influence more distant parts of the nervous system by giving rise to action potentials. We can initiate action potentials in an experiment by increasing the amount of depolarizing current we inject into the axon. The results of such depolarization are shown in Figure 2. In this part of the experiment, the second voltage-recording electrode (V_2) has been moved a greater distance from the point of current injection than previously—say, 2 cm rather than 2 mm. As before, a modest amount of depolarizing current injected into the axon produces an electrotonic potential which is recorded by the nearby voltage electrode. Now, however, the distant voltage-recording electrode sees nothing because it is well outside the range of spread of the electrotonic potential. When we inject more depolarizing current into the axon, a totally new type of event occurs: The membrane depolarizes rapidly so that the inside of the fiber becomes transiently positive and then returns rapidly to near its resting level. This event is the action potential, and it occurs whenever the nerve cell membrane is depolarized to a certain critical level, called the THRESHOLD. Once threshold is reached, the response is automatic and bears no relation to the form of the original stimulus. The

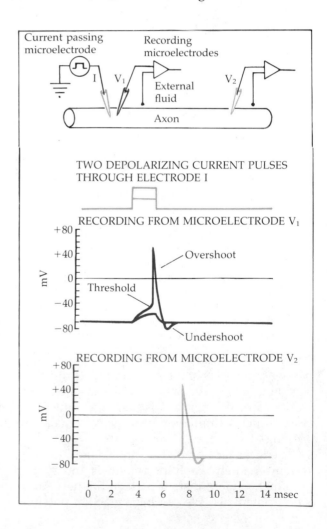

2

ACTION POTENTIALS recorded from a large axon by means of intracellular electrodes V_1 and V_2. The resting potential is −70 mV (inside negative); each trace shows two successive sweeps of the oscilloscope beam. During one sweep, a relatively small depolarizing current pulse is passed through microelectrode I, causing a small localized potential (recorded in V_1) that does not reach threshold. The second current pulse is larger, giving rise to a depolarization that reaches threshold and initiates an action potential that propagates rapidly along the axon and reaches V_2 at a distance of 2 cm. Unlike the graded localized potential, which cannot spread more than 1 to 2 mm and is therefore not recorded at V_2, the action potential is the same size all along the axon.

reversal of the membrane potential beyond 0 mV is called the OVER-SHOOT. In most nerve cells, the action potential is terminated by a brief hyperpolarization called the UNDERSHOOT (Figure 2). A most remarkable property of the action potential is that it is propagated along the axon and arrives at the distant voltage-recording electrode unaltered in size and form. The properties of the action potential may be summarized as follows:

1. The action potential is a triggered, explosive, all-or-nothing event. It has a distinct threshold, and, once initiated, its amplitude and duration are not determined by the amplitude and duration of the initiating event: Larger currents do not give rise to larger action potentials, and currents of longer duration do not prolong the action potential.

2. The entire action potential sequence must ensue before another action potential can be initiated. After each action potential, there is a period of enforced silence (the REFRACTORY PERIOD) during which a second impulse cannot be initiated. If depolarization of the nerve beyond threshold outlasts the refractory period, a second action potential may be initiated. In many neurons, prolonged depolarization may produce a train of action potentials that lasts as long as the depolarization. The frequency of the repeated action potentials is limited by the refractory period.

3. The action potential is propagated along the axon and does not decline with distance. In mammals, the fastest action potentials travel in the largest fibers at a rate of about 120 meters/sec (432 km/hr or 270 miles/hr) and are therefore capable of conveying information rapidly over a long distance. The mechanism whereby the action potential travels along the axon will be discussed in detail in Chapter 7.

SIGNALS USED IN A SIMPLE REFLEX

The use made by the nervous system of local potentials and action potentials can be illustrated in a simple form by the STRETCH REFLEX. A familiar reflex of this nature is the knee jerk, which is initiated by tapping the patellar tendon below the knee. The tap on the tendon stretches a group of muscles that extend the leg, and, as a result of this stretch, they undergo a reflex contraction. As only two types of neuron are involved in the reflex, it provides a simple example of the ways in which the nervous system performs its tasks.

The structural elements that subserve the stretch reflex are shown schematically in Figure 3. The first to be involved is a sensory neuron that has its cell body in a dorsal root ganglion near the spinal cord. In the periphery its sensory ending is in intimate contact with a specialized structure called the MUSCLE SPINDLE. Muscle spindles lie within the mass of the muscle and respond to muscle stretch by producing a receptor potential in the sensory nerve ending. The sensory neuron sends processes centrally to form synapses on many other cells in the spinal cord. Among such connections are excitatory synapses on motoneurons supplying the same muscle. These motoneurons, then, are the second type of neuron involved in the reflex; they complete the

Neurons involved in stretch reflex

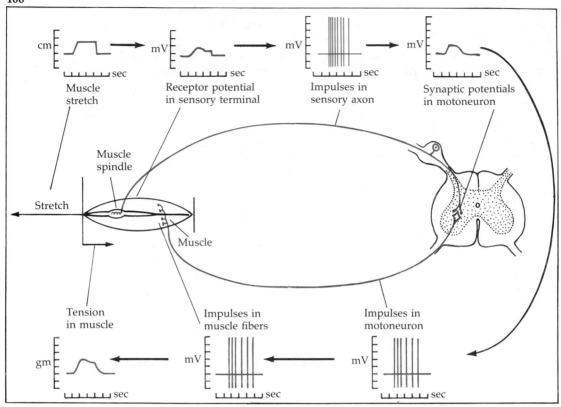

cm — Muscle stretch

mV — Receptor potential in sensory terminal

mV — Impulses in sensory axon

mV — Synaptic potentials in motoneuron

Muscle spindle

Stretch

Muscle

Tension in muscle

Impulses in muscle fibers

Impulses in motoneuron

gm — sec

mV — sec

mV — sec

3 STRUCTURES AND SIGNALS INVOLVED IN THE STRETCH RE-FLEX. Some important structural and functional features are omitted for clarity and will be described in detail in later chapters.

reflex arc by forming excitatory synapses on the muscle fibers. This reflex connection is another example of the remarkable specificity of neural connections discussed previously. The sensory fiber selects not just any motoneurons for its excitatory reflex connections but rather only those supplying the same muscles. This orderly anatomical arrangement is, of course, a prerequisite for the functioning of the reflex.

Figure 3 also depicts sequentially the excitatory events occurring during the reflex. The first event is, of course, the production of a receptor potential in the sensory ending as a result of the muscle stretch. The size and the duration of the receptor potential reflect the intensity and duration of the applied stretch (in this case, the magnitude and duration of the tap to the tendon). The receptor potential itself is confined to the first few millimeters of the sensory terminal but is of sufficient amplitude to depolarize the adjacent nerve fiber beyond threshold and initiate a series of action potentials. The frequency of action potentials in the train is related to the amplitude of the receptor potential (and hence to the intensity of the stretch) and limited, as noted previously, by the refractory period. The duration of the train reflects the duration of the stretch. This train of action potentials, then, is the means whereby information about the stretch reaches the central nervous system; intensity is coded by action potential frequency within the train and duration by the length of the train. The action potentials travel from the muscle toward the spinal cord at a velocity of about 100 meters/sec and arrive at the synapses formed on the motoneurons by the central terminals of the sensory nerve. Here each impulse releases from each terminal a small quantity of neurotransmitter, which causes a depolarizing synaptic potential in the motoneuron. The duration of each synaptic potential is relatively long, so that the train of excitatory postsynaptic potentials produced by the train of presynaptic action potentials results in a prolonged, relatively smooth depolarization of the motoneuron resembling the receptor potential that initiated the sequence (Figure 3). This depolarization of the motoneuron in turn initiates a train of action potentials that travel rapidly out again to the nerve–muscle synapses to produce depolarizing synaptic potentials (end plate potentials) in the muscle fibers. Here again the synaptic depolarization produces a train of action potentials in the muscle fiber, which causes muscle contraction. This whole sequence of events (receptor potential → sensory nerve impulses → synaptic potentials in motoneuron → motor nerve impulses → end plate potential in muscle → muscle impulses → muscle contraction) is rapid, taking less than 50 msec in humans, with a significant fraction of the time occupied by the last step, namely, initiation of the contractile process in the muscle. The reflex is therefore adequate for speedy adjustments in muscle tension, for example, to maintain a desired posture in the face of external perturbations.

The sensory fibers subserving the stretch reflex constitute only a very small fraction of the synaptic inputs to a motoneuron. Thousands of other neurons converge on a motoneuron to form many thousands of synaptic connections—some excitatory, some inhibitory—and the stretch reflex can be overruled in many ways. For example, a pin stuck in the toe will cause the knee to bend—a reaction opposite to that of the stretch reflex. This is because a painful stimulus to the foot causes contraction of other muscle groups, which produce bending of the knee, and at the same time gives rise to inhibition of the motoneurons responsible for the knee jerk. The process by which the myriad of excitatory and inhibitory influences on a neuron are sorted out was called by Sherrington[2] the INTEGRATING ACTION OF NEURONS. Integration at the cellular level is simply the way in which action potentials in fibers converging on a neuron become converted into postsynaptic potentials, the sum total of which determine its firing pattern. This firing pattern, then, is the synthesis of all the various inputs. Integration will be discussed in much more detail later on. The point to be made here is that in all the complex activities of the nervous system only two basic

How does a neuron take account of different converging influences?

[2]Sherrington, C. S. 1947. *The Integrative Action of the Nervous System.* Yale University Press, New Haven.

C. S. Sherrington with
one of his pupils
(J. C. Eccles) in the
mid 1930s.

types of signal are used to convey the abstractions of the surrounding
world and to implement actions. Integration by the motoneuron, which
adds up excitation and inhibition and then fires one or more impulses
or remains silent, is strikingly similar to integration by the nervous
system as a whole. The cell and the brain both decide whether or not
to act on the basis of information received from a wide variety of
sources.

Many of these general principles we owe to Sherrington, who discovered them through recording tension in skeletal muscle by the
stretch reflex before electrical recording from individual nerve cells was
possible. The following quotation is still a useful, concise description of
different neural signals:[3]

> The nerve nets are patterned networks of threads. The human brain is
> a vast example, offering immense numbers of determinate paths, and im-

[3]Sherrington, C. S. 1933. *The Brain and Its Mechanism*. Cambridge University Press,
London.

mense numbers of junctional points. At these latter the travelling signal so to speak hesitates and sets up a local gradable state which may have to accumulate before transmitting further, or indeed may there subside and fail. These junctional points are often convergent points for lines from several directions. Arrived there signals convergent from several lines may coalesce and thus reinforce each other's exciting power.

At such points too appears a process which instead of exciting, quells and precludes excitation. This inhibition, like its opposite process, excitation, does not travel. It is evoked, however, by travelling signals not distinguishable from those which call forth excitement. The travelling signals calling up excitement and those calling up inhibition never, however, reach the nodal point by the same path, never have paths in common.

The two are relatively antagonistic. Each can be neutralized gradually by a dosage of the other. The inhibition may be a temporary stabilization of the membrane at the nodal point, which is potentially a relay station. The inhibitory stabilization produced by a travelling signal is evanescent; a train of signals is required to maintain it. While it lasts, the nodal point is blocked to signals, or only transmits them slowly.

These two opposed processes, excitation and inhibition, cooperate at nodal point after nodal point in the nerve-circuits. Their joint operation at any moment settles what will be the conduction pattern, and so the motor outcome, of the signalling going forward to the brain.

IONIC BASIS OF RESTING AND ACTION POTENTIALS

CHAPTER FIVE

The electrical potential difference between the inside and the outside of a nerve cell membrane depends on the ionic concentration gradients across the membrane and on its relative permeability to the ions present. Simple principles of physical chemistry can be used to explain how resting potentials arise in a variety of excitable cells and, in particular detail, in the giant axon of the squid.

The distribution of ions on either side of a cell membrane is subject to two major constraints: (1) the bulk solutions inside and outside the cell must be electrically neutral; (2) the total osmotic concentration of intracellular ions and molecules in solution must be equal to that in the extracellular fluid. Each individual permeant species that is distributed across the membrane is subject to two additional gradients tending to drive it into or out of the cell—a concentration gradient and an electrical gradient. For potassium, to which the membrane has a relatively large permeability and which is much more concentrated inside the cell than out, these two gradients are nearly in balance. Thus, the inside surface of the membrane is negative with respect to the outside and the tendency of potassium ions to move out of the cell along their concentration gradient is opposed almost exactly by the potential gradient tending to prevent such movement. The membrane potential at which there is no net potassium flux is called the potassium equilibrium potential. The relation between concentration ratio and equilibrium potential for any ion is given by the Nernst equation.

Chloride ions are subject to the same rules. Chloride concentration is higher outside the cell than inside and this concentration gradient is again balanced by the membrane potential, which tends to oppose inward movement of the negatively charged ion. The resting membrane potential, however, is determined mainly by the potassium concentration because the internal chloride concentration tends to accommodate itself to the resting potential.

Sodium is much more concentrated in the extracellular fluid than in the cell cytoplasm. Thus, both its concentration gradient and the membrane potential favor movement of the ion into the cell. At rest, the membrane is only sparingly permeable to sodium. During the

111

action potential, the membrane becomes highly permeable to sodium; sodium ions cross the membrane and recharge the inner surface, causing the membrane potential to swing rapidly toward the sodium equilibrium potential (about 55 mV, inside positive).

In general, then, the actual value of membrane potential for a cell is determined by the relative permeabilities of the membrane to the major ions: Na, K, and Cl. The way in which ionic concentrations and permeabilities determine membrane potential is described by the Goldman equation. Electrical signals, which consist of changes in membrane potential, are generated by alterations in the selective permeability of the membrane. These changes in selective permeability result in ion movements without the direct intervention of metabolism. However, at rest and with each action potential, the cell gains minute amounts of sodium and loses corresponding amounts of potassium, which are ultimately restored by metabolic processes. Many action potentials can be generated before the metabolic debt must be paid.

In Chapter 4 a simple reflex arc was used to demonstrate that both localized and conducted electrical signals are generated in nerve cells and muscle fibers. The mechanisms by which such signals arise are now understood to a large extent; the membrane potential is controlled by the relative permeability of the membrane to certain ions and the electrochemical gradients acting upon them. Metabolic energy is expended only to maintain the gradients and has no immediate role in the generation of the localized and conducted signals we have discussed so far. These are generated instead by changes in membrane permeability, which allow ions to run downhill along established electrochemical gradients and, in so doing, to change the charge on the membrane. As we will see, in the case of the action potential the permeabilities themselves are in turn dependent on membrane potential. An early review by Hodgkin[1] first brought together many of the concepts described in this chapter.

Before describing the physiological experiments that have been done on neurons to explain the resting and action potentials, it is useful to start with a brief discussion of the principles of physical chemistry that underlie the distribution of ions across membranes and the relation between such ionic distributions and the membrane potential. The discussion will be related to the idealized cell shown in Figure 1. Some of the principles are intuitively understandable, whereas others in which several different ions are considered at once require algebraic manipulations (see Boxes 1–3).

The idealized cell in Figure 1 contains potassium, sodium, chloride, and a large anion and is bathed in a solution of sodium chloride and

[1]Hodgkin, A. L. 1951. *Biol. Rev. 26*: 339–409.

potassium chloride. The concentrations inside and outside the cell are similar to those found in nerve cells of a frog. In the model the volume of the extracellular fluid is very large compared with the volume of the cell, so that movements of ions and water into or out of the cell have no effect on the extracellular concentrations. One requirement for any such system is that both the intracellular and extracellular solutions be electrically neutral. For example, solutions of chloride ions alone do not exist; their charges must be balanced by an equal number of positive charges on cations such as sodium or potassium. A second requirement is that the cell be in osmotic balance. When the total concentration of particles on the inside is not equal to that on the outside, water will enter or leave the cell, causing it to swell or shrink, until osmotic balance is achieved. Both these requirements are fulfilled by the cell shown in Figure 1. Finally, it is convenient to start with a simplified system in which the cell membrane is impermeable to sodium and to the large internal anions, although in reality nerve cells have a small resting permeability to sodium ions.

When these three conditions of electrical neutrality, osmotic balance, and selective permeability are met, how are the ions distributed and what electrical potential is developed across the cell membrane? First, Figure 1 shows that potassium and chloride are distributed in reverse ratio; potassium is more concentrated on the inside of the cell, chloride on the outside. A question that arises is why the ions do not simply diffuse down their concentration gradients to equalize their concentrations on either side of the cell membrane. The answer is that they cannot because the instant they start to do so there is a charge separation across the membrane that prevents further diffusion. Thus, potassium ions leaving the cell and chloride ions entering both result in an excess accumulation of negative ions on the inner surface and positive ions on the outer surface of the membrane. The resulting membrane potential (negative on the inside) prevents further net diffusion of the ions, and the system is said to be in EQUILIBRIUM. Then the concentration gradient and potential gradients for each ion balance one another exactly (arrows). When such a condition occurs for a specific ion, the membrane

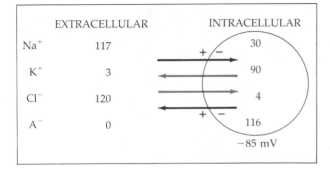

1

IONIC DISTRIBUTIONS IN A MODEL CELL. The cell membrane is impermeable to Na^+ and to the internal anion (A^-) and permeable to K^+ and Cl^-. The concentration gradient for K^+ tends to drive it out of the cell (color arrows); this is balanced by the potential gradient (black arrows) which tends to keep K^+ from leaving. Concentration and electrical gradients for Cl^- are in the reverse directions. Ionic concentrations are expressed as mM.

potential is equal to the EQUILIBRIUM POTENTIAL for that ion. The relation between the equilibrium potential and the concentration gradient is given by the Nernst equation, which is discussed in more detail in Box 1. The Nernst equation for potassium is

$$E_K = (RT/zF) \ln (K_o/K_i)$$

where R is the gas constant, T the absolute temperature, z the valence of the ion ($+1$ for potassium), and F the faraday. RT/zF is the potential needed to balance an e-fold concentration gradient across the membrane, and at room temperature it has a value of 25 mV. Hence the equilibrium potential for potassium (in millivolts) is $E_K = 25 \ln (K_o/K_i)$ or, if we convert from natural logarithms to base 10 logarithms:

$$E_K = 58 \log (K_o/K_i)$$

BOX 1   THE NERNST EQUATION

The Nernst equation describes the potential necessary to balance an ionic concentration gradient across the membrane so that the net flux of the ion in question is zero. In other words, it gives the equilibrium potential for that ionic species. The equation can be applied to passive ionic distributions irrespective of the mechanism whereby the ions cross the membrane. For example, a particular ion species might move across the membrane simply by diffusion along the potential gradient or, on the other hand, might hop along a series of binding sites within an aqueous channel; in either case the Nernst equation is applicable. The equation, being very general, can be derived from purely thermodynamic considerations without reference to any particular membrane model. It is useful, however, to illustrate how it arises in relation to ionic fluxes in a simple diffusional model. The first point to be made is that such fluxes are never zero; being at equilibrium simply means that the number of ions of a given species leaving the cell at any given time is equal to the number entering, so that the *net* flux is zero. We can write that the inward flux of potassium (*flux*$_{in}$; moles/cm^2 sec) is proportional to the outside concentration (K_o). That is,

$$flux_{in} = r_{in} K_o$$

where r_{in} is the rate constant for inward movement and has the dimensions centimeters per second. Similarly the outward flux is proportional to the inside concentration (K_i) and the rate constant for outward movement (r_{out}):

$$flux_{out} = r_{out} K_i$$

The net flux is the difference between the inward and outward fluxes:

$$net\ flux = r_{in} K_o - r_{out} K_i$$

Now, if the membrane were not charged, we might expect the rate constants for inward and outward movement to be equal, and this would indeed be true: both r_{in} and r_{out} would be equal to the permeability constant for potassium, p_K. In that case, because of the 30-fold difference between inside and outside concentrations, the outward flux for potassium in our idealized cell (Figure 1) would be 30 times faster than the inward flux. However, the fact that the membrane is negative on the inside with respect to the outside means that potassium ions enter the cell more readily than they leave. Because of this effect, the rate constants depend on membrane potential as well as on permeability. The exact effect of the potential depends on the mechanism of permeation. For simple diffusion down a potential gradient, it can be shown that the quantitative relation between potential and the rate constant for inward movement is given by

$$r_{in} = p_K z V'/(e^{zV'} - 1)$$

where z is the valence of potassium ($+1$) and V'

For the cell shown in Figure 1, the concentration ratio for potassium is 1/30 and E_K is therefore -85 mV.

What about chloride? Because for an anion $z = -1$, the equilibrium potential for chloride is given by

$$E_{Cl} = -58 \log (Cl_o/Cl_i)$$

The concentration ratio is 30 and E_{Cl} is therefore also -85 mV. As with potassium, the internal negativity balances exactly the tendency for chloride to move into the cell along its concentration gradient. Because potassium and chloride are the only two ions that can move across the membrane and because both are in equilibrium at -85 mV, a cell of this type can exist indefinitely without any net gain or loss of ions.

The charge separation that occurs at the membrane appears to violate the principle of electrical neutrality that we started out with; and, in-

---

is the membrane potential (V_m) divided by the thermodynamic potential RT/F, that is, $V' = V_mF/RT$, where R is the gas constant, T the absolute temperature, and F the faraday (the number of coulombs of charge carried by one mole of univalent ion). RT/F is the potential needed to balance an e-fold concentration ratio across the membrane and has a value of 25 mV at room temperature. For the idealized cell, which has a potential of -85 mV, V' is $-85/25 = -3.4$ and $e^{V'}$ is 0.033, so that $r_{in} = -3.4p_K/(0.033 - 1) = 3.5p_K$, or 3.5 times greater than if there were no voltage across the membrane.

For outward movement, the rate constant is given by an algebraically similar expression:

$$r_{out} = p_K zV'/(1 - e^{-zV'})$$

Again for the idealized cell, V' is -3.4 as before, $e^{-V'}$ is 30, and $r_{out} = -3.4p_K/(1 - 30)$, or about 8.5 times smaller than the rate constant with no membrane potential. In summary, the effect of the membrane potential is to increase the inward potassium flux by a factor of about 3.5 and reduce the outward flux by a factor of about 8.5. In other words, the effect of the potential is to produce a 30-fold difference in the rate constants for inward and outward movement, which balances the 30-fold difference in concentration.

We can now use the expressions for r_{in} and r_{out} to rewrite the equation for net flux of potassium in terms of its permeability constant and the membrane potential:

$$net\ flux = \frac{zV'}{e^{zV'} - 1} p_K K_o - \frac{zV'}{1 - e^{-zV'}} p_K K_i$$

If we multiply the numerator and denominator of the second term by $e^{zV'}$, the expression becomes

$$net\ flux = \frac{zV'}{e^{zV'} - 1} p_K K_o - \frac{zV'e^{zV'}}{e^{zV'} - 1} p_K K_i$$

or, upon rearranging,

$$net\ flux = p_K zV' \frac{K_o - K_i e^{zV'}}{e^{zV'} - 1}$$

When $K_o = K_i e^{zV'}$, the net flux is zero, which means that the potassium ion is in equilibrium. It follows that at equilibrium

$$zV' = \ln (K_o/K_i)$$

When we remember that $V' = V_mF/RT$, then the membrane potential required for potassium to be at equilibrium is given by

$$V_m = E_K = (RT/zF) \ln (K_o/K_i)$$

which is the Nernst equation. As RT/zF is close to 25 mV at room temperature, the equation can be written as $E_K = 25 \ln (K_o/K_i)$ or, when we convert from natural (base e) logarithms to base 10 logarithms,

$$E_K = 58 \log (K_o/K_i)$$

The dependence of
the resting potential
on extracellular
potassium

deed, it does so. Quantitatively, however, the ionic *concentration* changes in the solutions produced by the charge separation are negligible and could not possibly be measured. For example, if we consider a cell with a radius of 25 μm and a resting membrane potential of −85 mV, we can calculate that there are about 7×10^{-17} moles of excess negative charges on the inner surface of the membrane. If these were distributed throughout the bulk solution as univalent ions, they would represent a concentration of about 10^{-9}, or about one part in 10^8 of the total anionic concentration.

In neurons and in many other cells, the membrane potential is sensitive to changes in extracellular potassium concentration, but it is relatively unaffected by changes in extracellular chloride. To understand how this comes about, it is useful to consider the consequences of such changes in the model cell. Figure 2A shows the changes in intracellular composition and membrane potential that result from increasing extracellular potassium from 3 mM to 6 mM. This is done by replacing 3 mM NaCl with 3 mM KCl, thereby keeping the osmolarity constant at 240 mM. As a result of this change, the cell is depolarized from −85 mV to −68 mV, intracellular potassium concentration is increased slightly, and intracellular chloride concentration is almost doubled. How did this change come about? As we will see, it was achieved by potassium, chloride, and water entering the cell in response to the increased extracellular potassium concentration. First, when external potassium concentration is increased, potassium ions enter the cell and the resulting accumulation of positive charge inside the membrane causes depolarization. The depolarization in turn results in chloride being out of equilibrium, so chloride enters as well. This process of potassium and chloride entry continues until both ions are at the same concentration ratio and therefore have the same equilibrium potential. Except for the minute excess of cations required to charge the membrane to the new potential, the entry process must be electrically neutral, that is, potassium and chloride entry must be in equal amounts. In addition, water must accompany the ions to maintain osmotic balance. Now imagine for the moment that we can separate the processes of ionic diffusion and water entry, with ionic diffusion occurring first. Sufficient potassium chloride would enter the cell to increase the intracellular concentration by about 4.2 mM. As a result, intracellular potassium concentration would be 94.2 mM and chloride 8.2 mM. Intracellular osmolarity would be 248.4 mM. Now, if we let the water enter, all the intracellular constituents would be diluted by a factor of 240/248.4 (or by about 3.5%), and the final concentrations shown in Figure 2A would result. In particular, the internal potassium concentration would be diluted to 91 mM and chloride to 7.9 mM. Their concentration ratios would again be equal and would correspond to the new membrane potential of −68 mV.

Similar considerations apply to changes in extracellular chloride, but with a marked difference: No change occurs in membrane potential. The consequences of a 50 percent reduction are shown in Figure 2B, in

(A)

	NORMAL			HIGH POTASSIUM		
	Extracellular	Intracellular		Extracellular	Intracellular	
Na^+	117	30		114	29.0	
K^+	3	90		6	91.0	
Cl^-	120	4		120	7.9	
A^-	0	116		0	112.1	
Relative volume:		1.0			1.035	
Membrane potential:		-85 mV			-68 mV	

(B)

	NORMAL			LOW CHLORIDE		
	Extracellular	Intracellular		Extracellular	Intracellular	
Na^+	117	30		117	30.5	
K^+	3	90		3	89.5	
Cl^-	120	4		60	2.0	
A^-	0	116		60	118.0	
Relative volume:		1.0			0.98	
Membrane potential:		-85 mV			-85 mV	

2 **EFFECTS OF CHANGING EXTRACELLULAR IONIC COMPOSITION** on intracellular ionic concentrations and on membrane potential. In (A) extracellular K^+ is doubled, with a corresponding reduction in Na^+ to keep osmolarity constant. In (B) half the extracellular Cl^- is replaced by an impermeant anion, A^-. Ionic concentrations are in mM and extracellular volumes are assumed to be very large with respect to cell volume so that ionic movements do not change extracellular concentrations.

which 60 mM of chloride in the solution bathing the cell is replaced by an impermeant anion. Chloride leaves the cell because it is no longer in equilibrium and, as before, potassium and chloride must move together, accompanied by water. Equal quantities of the ions leave and those remaining are concentrated slightly as the cell shrinks.

From Figure 2 we can arrive at some general conclusions: In the model cell, and in most real cells, changes in external potassium concentration result in changes in membrane potential, with internal chloride concentration accommodating itself to the change; changes in external chloride result in a similar accommodating change in internal chloride without a major effect on membrane potential. This difference in the effects of potassium and chloride concentration changes arises from the relative concentrations of the two ions inside the cell and from the fact that it is the *concentration ratio* of each ion (not, for example, the concentration difference) that determines its equilibrium potential. As we have seen, the two ions move across the membrane in concert in response to an external change of either. Because the internal chloride concentration is relatively small, such movement has a much greater effect on the chloride concentration ratio than on the potassium concentration ratio. In other words, a change of a few mM in internal potassium chloride concentration can easily double the initial concentration of chloride but only change that of potassium by a few percent. As a consequence, when a change in the composition of the extracellular solution occurs, the two ions achieve equilibrium in different ways: The membrane potential changes to the new *potassium equilibrium potential* and the internal chloride concentration changes so that the chloride equilibrium potential is equal to the new membrane potential.

Experimental effect of
potassium on
membrane potential

The idea that the resting membrane potential is the result of an unequal distribution of potassium ions between the extracellular fluid and the intracellular fluid was first proposed by Julius Bernstein[2] in 1902. From chemical analysis he knew that the interior of nerve and muscle was rich in potassium but contained little sodium or chloride. He knew further that there were anions within the cell to which the membrane was not permeable, and the evidence at the time suggested that the membrane was impermeable to sodium ions. Thus, his conception of the cell was much the same as our idealized model, and he arrived at the same conclusion, namely, that the magnitude of the resting membrane potential should be given quantitatively by the Nernst equation. He could not test his hypothesis directly because there was no satisfactory method of measuring membrane potential accurately. One approximation was to place an electrode onto the cut end of a muscle—thereby making contact with the intracellular fluid—and to measure the resulting "injury potential." He found that the variation of the injury potential with temperature conformed to that expected from the Nernst relation, within the error of his experimental measurements.

In more recent years, more detailed tests have been made by varying external and internal potassium concentrations to see whether the membrane behaves as expected from the Nernst relation. Many of the experimental ideas were formulated by Boyle and Conway,[3] who also worked with muscle fibers; but the experiments we shall discuss were done for the most part on large nerve fibers that innervate the mantle of the squid. These "giant axons" are up to 1 mm in diameter (Figure 3), and their size makes them particularly useful for a number of tests that could not be done on smaller axons.[4] Many of the pioneering

[2]Bernstein, J. 1902. *Pflügers Arch. 92*: 521–562.
[3]Boyle, P. J. and Conway, E. J. 1941. *J. Physiol. 100*: 1–63.
[4]Young, J. Z. 1936. *J. Microsc. Sci. 78*: 367–386.

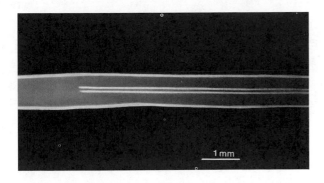

3

RECORDING ELECTRODE inside an isolated giant axon of the squid. (From Hodgkin and Keynes, 1956.)

TABLE 1
CONCENTRATIONS OF IONS INSIDE AND OUTSIDE FRESHLY ISOLATED AXONS OF SQUID

	Concentration (mM)		
Ion	Axoplasm	Blood	Seawater
Potassium	400	20	10
Sodium	50	440	460
Chloride	40—150	560	540
Calcium	0.1 μM	10	10

Modified from Hodgkin (1964); estimate of ionized intracellular calcium from Baker, Hodgkin, and Ridgway (1971).

investigations of mechanisms underlying the resting and action potentials were done on these fibers, and the results have been shown to be relevant to the function of many other excitable tissues such as vertebrate nerve fibers, vertebrate skeletal muscle, and even heart muscle. A. L. Hodgkin, who together with A. F. Huxley initiated many experiments on squid axon (for which they later received the Nobel prize), has said:[5]

> It is arguable that the introduction of the squid giant nerve fiber by J. Z. Young in 1936 did more for axonology than any other single advance during the last forty years. Indeed a distinguished neurophysiologist remarked recently at a congress dinner (not, I thought, with the utmost tact), "It's the squid that really ought to be given the Nobel Prize."

The concentrations of some of the major ions in squid blood and in the axoplasm of the squid nerves are given in Table 1 (several ions, such as magnesium and internal anions, are omitted). The ratio of potassium concentrations in the blood and axoplasm is 1:20, so that according to the Nernst equation the resting membrane potential should be −75 mV, provided that (1) the given intracellular concentration rep-

[5]Hodgkin, A. L. 1973. *Proc. R. Soc. Lond. B 183*: 1–19.

4

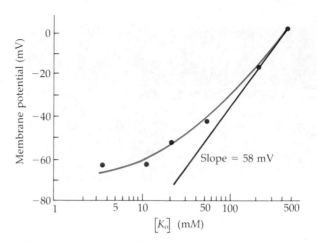

MEMBRANE POTENTIAL AND EXTERNAL POTASSIUM concentration in a squid axon, plotted on a semilogarithmic scale. The solid straight line is drawn with a slope of 58 mV per 10-fold change in concentration, according to the Nernst equation. Because the membrane is also permeable to sodium (and to chloride), the points deviate from the straight line, especially at low external potassium concentrations (see text). (After Hodgkin and Keynes, 1955.)

resents the concentration of free (as opposed to bound) potassium, (2) potassium is the major permeant species, and (3) potassium is at equilibrium. In fact, the measured membrane potential is usually slightly less negative (−70 mV) for a nerve in situ. On the other hand, the membrane potential is more negative than the chloride equilibrium potential, the latter being about −65 mV. The reason for these discrepancies will be discussed later in this chapter. After the axon is removed from the squid and placed in a chamber for recording, the resting potential is usually lower still (−65 to −60 mV). The lower resting potential is probably due to slight damage during dissection and electrode penetration.

Given the ability to measure the membrane potential of the squid axon with some accuracy, it is possible to test in more detail the original hypothesis of Bernstein by examining how changes in potassium concentration affect the potential. From the Nernst equation, changing the concentration ratio by a factor of 10 should change the membrane potential by 58 mV at room temperature. The results of such an experiment in which the external potassium concentration was changed are shown in Figure 4. As with our model cell, such changes would be expected to produce no significant change in internal potassium concentration. The external concentration is plotted on a logarithmic scale on the abscissa and the membrane potential on the ordinate. The expected slope of 58 mV per tenfold change in extracellular potassium concentration is realized only at relatively high concentrations (solid straight line), with the slope becoming less and less as external potassium is reduced. This result suggests that the potassium ion distribution is a major, but not the only, factor contributing to the membrane potential.

A further test of the role of potassium in determining membrane potential is to determine the effects of changing internal concentration. This experiment was carried out in a unique way on squid axon by Baker, Hodgkin, and Shaw.[6] They squeezed out the axoplasm by means

[6]Baker, P. F., Hodgkin, A. L. and Shaw, T. I. 1962. *J. Physiol. 164*: 330–354, 355–374.

of a rubber roller and perfused the interior through a cannula with solutions of their own choosing. This remarkable procedure is illustrated in Figure 5, together with the experimental observation that the ability of the surface membrane to generate action potentials was unimpaired by the procedure. In fact, when the axon was perfused with a solution containing normal sodium and potassium (but not the usual proteins and amino acids), it was capable of producing many thousands of action potentials when stimulated. This result emphasizes dramatically the self-sufficiency of the membrane in producing action potentials; in the short run, the normal axoplasm is not required for the process.

To return to the role of potassium in determining the membrane potential, it was found that when the internal concentration equaled that in the bathing solution the membrane potential was zero—as pre-

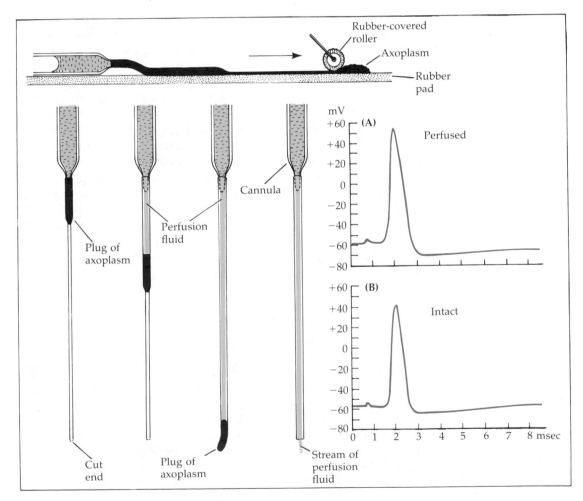

5 AXOPLASM BEING EXTRUDED from a squid axon, which is then cannulated and perfused internally. (A) Record of an impulse from a perfused axon. (B) Record of an impulse from a normal axon. (After Baker, Hodgkin, and Shaw, 1962.)

dicted by the Nernst equation—over a wide range of concentrations. When the internal concentration was less than that in the bathing solution, the membrane potential reversed—the inside becoming positive, as predicted by the Nernst equation. Furthermore, alterations in the ratio of sodium and chloride concentrations had very much smaller effects on membrane potentials.

From the experiments on squid axon, then, we can conclude that the hypothesis made by Bernstein in 1902 was almost correct; the membrane potential is strongly, but not exclusively, dependent on the potassium concentration ratio. How do we account for the deviation from the Nernst relation shown in Figure 4? First we have to abandon the assumption made in our model cell that the membrane is impermeable to sodium. In the model and in the squid axon, the concentration gradient and the membrane potential both tend to drive sodium into the cell. As the permeability of the membrane to sodium is low, the rate of entry is relatively slow; nevertheless, the consequences are important. As the sodium ions enter, the accumulation of positive charge depolarizes the membrane. This depolarization means that potassium is no longer in equilibrium and that potassium ions will leave the cell. Under these conditions a steady state will be reached when the potential is such that the inward leak of sodium is equaled by the outward leak of potassium, that is, when the net current across the membrane is zero. However, it turns out that in squid axon and in most other cells, chloride is not in equilibrium either, so that there is a chloride current across the membrane as well. Thus, for a steady state to exist, all three currents must add up to zero. This condition was first treated theoretically by Goldman,[7] who derived a relation describing the dependence of the membrane potential on the concentration ratios of potassium, sodium, and chloride and on their relative permeability constants. When K_o, K_i, Na_o, Na_i, Cl_o, and Cl_i are the external and internal ionic concentrations and p_K, p_{Na}, and p_{Cl} are their membrane permeability constants, then the membrane potential, V_m, is given by

$$V_m = 58 \log \frac{K_o + (p_{Na}/p_K)Na_o + (p_{Cl}/p_K)Cl_i}{K_i + (p_{Na}/p_K)Na_i + (p_{Cl}/p_K)Cl_o}$$

This is called the Goldman equation, or sometimes the GHK equation, because it was later used independently by A. L. Hodgkin and B. Katz.[8] It is also known as the constant field equation because one of the assumptions made in arriving at the expression was that the voltage gradient (or "field") across the membrane was uniform. The way in which the equation is derived is discussed in Box 2.

The Goldman equation describes with reasonable accuracy the relation between potassium concentration and membrane potential shown in Figure 4. By examining the equation we can see how this occurs. If we take the relative permeability constants for potassium, sodium, and

[7]Goldman, D. E. 1943. *J. Gen. Physiol. 27*: 37–60.
[8]Hodgkin, A. L. and Katz, B. 1949. *J. Physiol. 108*: 37–77.

BOX 2 THE GOLDMAN EQUATION

This relation can be derived from previous considerations. Recall that for simple diffusion along a potential gradient the net flux of potassium across a cell membrane was given by

$$net\ flux = p_K z V' \frac{K_o - K_i e^{zV'}}{e^{zV'} - 1}$$

where $V' = V_m F/RT$, z is the valence and p_K is the potassium permeability. This flux, which has the units of moles/cm^{2}sec, can be converted into current (i_K) in amperes/cm^2 (coulombs/cm^{2}sec) by multiplying by the faraday (F), which is the number of coulombs of charge in a mole of univalent ion, and by the valence of the ion in question (for potassium, +1). We can write similar expressions for sodium (i_{Na}) and chloride (i_{Cl}) currents (remembering that the valence for chloride is -1). The results are

$$i_K = p_K F V' \frac{K_o - K_i e^{V'}}{e^{V'} - 1}$$

$$i_{Na} = p_{Na} F V' \frac{Na_o - Na_i e^{V'}}{e^{V'} - 1}$$

$$i_{Cl} = p_{Cl} F V' \frac{Cl_o - Cl_i e^{-V'}}{e^{-V'} - 1}$$

$$= p_{Cl} F V' \frac{Cl_i - Cl_o e^{V'}}{e^{V'} - 1}$$

For a steady state to exist, the sum of the three currents must equal zero. When we take out the factor $FV'/(e^{V'} - 1)$ common to all three expressions, we can write

$$p_K(K_o - K_i e^{V'}) + p_{Na}(Na_o - Na_i e^{V'}) + p_{Cl}(Cl_i - Cl_o e^{V'}) = 0$$

or

$$p_K K_o + p_{Na} Na_o + p_{Cl} Cl_i = e^{V'}[p_K K_i + p_{Na} Na_i + p_{Cl} Cl_o]$$

Rearrangement yields

$$e^{V'} = \frac{p_K K_o + p_{Na} Na_o + p_{Cl} Cl_i}{p_K K_i + p_{Na} Na_i + p_{Cl} Cl_o}$$

Remembering that $V' = V_m F/RT$, with simple algebraic manipulation we arrive at the final expression:

$$V_m = \frac{RT}{F} \ln \frac{p_K K_o + p_{Na} Na_o + p_{Cl} Cl_i}{p_K K_i + p_{Na} Na_i + p_{Cl} Cl_o}$$

This equation can be written in a slightly different way by dividing both the numerator and the denominator by p_K and converting to base 10 logarithms:

$$V_m = 58 \log \frac{K_o + (p_{Na}/p_K)Na_o + (p_{Cl}/p_K)Cl_i}{K_i + (p_{Na}/p_K)Na_i + (p_{Cl}/p_K)Cl_o}$$

In deriving the Goldman equation, we have made two basic assumptions. The first is that the individual ionic species move across the membrane by simple diffusion along a potential gradient. This is sometimes known as the constant field assumption, although a constant (i.e., linear) potential gradient across the membrane is not a necessary condition for the expressions to be valid. The second assumption is made when the three ionic currents are combined to give the total current: It is assumed that they are independent of each other, that is, that sodium ions crossing the membrane, for example, do not interfere with potassium ions crossing at the same time.

chloride in squid axon to be roughly in the ratio 1.0:0.03:0.1, respectively,[8] we can use these ratios together with the ionic concentrations given in Table 1 for axoplasm and seawater to calculate the resting membrane potential:

$$V_m = 58 \log \frac{10 + (0.03)460 + (0.1)40}{400 + (0.03)50 + (0.1)540} = -70\ mV$$

This is in reasonable agreement with the values observed experimentally. Furthermore, if we look at the magnitude of the individual terms in the numerator (10, 13.8, and 4), we can see that the potassium term constitutes only about one-third of the total. Because of this, doubling the external potassium concentration will not double the numerator (as would happen in the Nernst equation), and the change in potential with extracellular potassium concentration will be less than would be expected if potassium were the only permeant ion. When the external potassium concentration is raised to a high level, however, the potassium term will dominate the numerator and the effect of potassium concentration on potential will approach the theoretical limit of 58 mV per tenfold change in concentration, as observed experimentally (Figure 4). In fact, the 58-mV slope is approached more rapidly than predicted by the Goldman equation because of a secondary effect: Depolarization of the membrane by increased extracellular potassium causes an increase in potassium permeability. This will be discussed in detail in Chapter 6. Because the permeability constant for sodium is only 3 percent of that for potassium, changing the extracellular sodium concentration will have a much smaller effect. For example, reducing extracellular sodium from 460 mM to 230 mM will reduce the numerator by about 25 percent and increase the membrane potential by only 7 mV. Finally, because of the relatively low chloride permeability, extracellular chloride makes only a minor contribution to the denominator of the expression, the individual terms of which have the magnitudes 400, 1.5, and 54. Consequently, we might expect changes in extracellular chloride to have little effect on membrane potential. For example, if we reduced extracellular chloride in the bathing solution from 540 mM to 270 mM, we would reduce the denominator by only about 6 percent. In addition, even if the permeability were significant, changes in external chloride would lead to changes in internal concentration as well, with the chloride concentration ratio tending to accommodate to the membrane potential rather than contributing to its regulation.

The role of active transport

In going from the idealized cell (with a membrane totally impermeable to sodium) to the squid axon, we have gone from a cell in equilibrium to one in steady state. The distinction is important. The idealized cell requires no metabolic energy to maintain the ionic distributions because the only permeant ions (potassium and chloride) are in equilibrium. On the other hand, in the squid axon and in neurons in general, the membrane is permeable to sodium, and neither sodium nor potassium is in equilibrium. This means that there is a constant leak of sodium into the cell and a constant loss of potassium. In the face of these ionic leaks, the internal concentrations are maintained by active transport processes requiring metabolic energy to pump sodium back out of the cell and to accumulate potassium. In addition, although chloride could, in theory, be in equilibrium in most cells, it is not. Thus, in addition to active transport of sodium and potassium, there is usually transport of chloride across the membrane as well. In the squid axon, chloride is transported into the cell; in other nerve cells, chloride trans-

125
IONIC BASIS
OF RESTING
AND ACTION
POTENTIALS

Ionic basis of the
action potential

port is outwardly directed. These processes will be considered in detail in Chapter 8.

At the same time as Bernstein proposed his hypothesis in 1902, Overton[9] made the important discovery that sodium ions are necessary for nerve and muscle cells to maintain their ability to produce action potentials and that lithium could substitute for sodium in this process. Overton suggested that the action potential might come about through sodium entering the cell, but pointed out that an excitable cell conducted literally millions of action potentials during the life of an animal without becoming enriched in intracellular sodium. The role of sodium ions could not be clarified further until two additional observations were made: (1) In 1939, Curtis and Cole[10] and Hodgkin and Huxley[11] showed that the action potential is actually larger than the resting potential, that is, that there is an overshoot during which the membrane potential becomes transiently positive on the inside. (2) In 1949, Hodgkin and Katz[8] showed that changes in external sodium concentration affect the amplitude of the action potential, as shown in Figure 6. Furthermore, they showed that the changes could be predicted accurately by the constant field equation by assuming that at the peak of the action potential the sodium permeability increased several hundredfold over

[9]Overton, E. 1902. *Pflügers Arch. 92*: 346–386.
[10]Curtis, H. J. and Cole, K. S. 1940. *J. Cell. Comp. Physiol. 15*: 147–157.
[11]Hodgkin, A. L. and Huxley, A. F. 1939. *Nature 144*: 710–711.

(A)

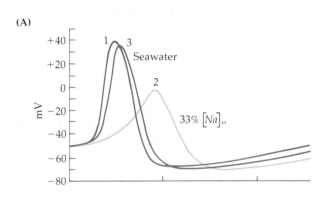

(B)

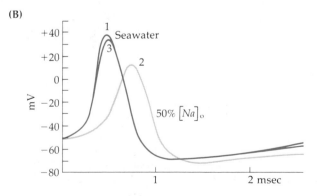

6

ROLE OF SODIUM in conduction of an action potential in a squid axon. In record 2 in (A), external sodium was reduced to one-third of normal and in (B) to one-half of normal. Records 1 and 3 in each case are control records in normal seawater before and after exposure to low sodium. (From Hodgkin and Katz, 1949.)

its resting value. Similar results were obtained later by Baker, Hodgkin, and Shaw,[6] who varied the internal sodium concentration in perfused squid axons. Together, these results show that during the action potential the selectivity of the membrane permeability changes so that the sodium permeability dominates the membrane potential. Suppose, for example, that the permeability ratios $p_K:p_{Na}:p_{Cl}$ change from 1:0.03:0.1 to 1:15:0.1 or, in other words, that the sodium permeability increases by a factor of 500. Then, if we use numerical values as before, the membrane potential predicted by the constant field equation becomes:[8]

$$V_m = 58 \log \frac{10 + (15)460 + (0.1)40}{400 + (15)50 + (0.1)540} = +44 \text{ mV}$$

We see, then, that the membrane potential is determined by the relative permeability of the membrane to the ions present, particularly sodium and potassium, and it is changes in permeability to these two ions that underlie the generation of the action potential. Its rise is due to a sudden increase in the permeability of the membrane to sodium, its fall to a subsequent marked increase in permeability to potassium. The ionic movements and changes in membrane potential that result from changes in selective permeability to these two ions are summarized schematically in Figure 7. If a membrane that is permeable primarily to potassium (Figure 7A) suddenly becomes permeable primarily to sodium (by an as yet undefined mechanism), then the membrane potential will become dominated by sodium concentration (Figure 7B). Starting at, say, -70 mV (near E_K), both the concentration gradient and the membrane potential favor sodium entry into the cell. Once the sodium permeability is increased, sodium ions enter across the cell membrane along their electrochemical gradient without any requirement for metabolic energy. As they enter, however, they recharge the membrane, the inside becoming more and more positive until a new steady state is reached. Because the membrane potential is now positive, sodium ions will enter less readily and there will be an outward movement of

 7

INFLUENCE OF POTASSIUM AND SODIUM PERMEABILITY on the membrane potential. (A) At rest, the potassium permeability is dominant and the membrane potential is near E_K (-75 mV). **(B)** An unspecified mechanism increases sodium permeability relative to that of potassium, and sodium ions enter the cell, recharging the membrane toward E_{Na} ($+55$ mV). The actual potential reached depends on the ratio of the permeabilities, and a steady state will be reached when the inward sodium current equals the outward potassium current. **(C)** When the original membrane properties are restored, the outward potassium current will restore the membrane potential to its original value. Chloride fluxes are very much smaller and have been ignored.

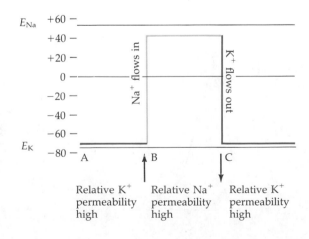

potassium ions and an inward movement of chloride ions. The new steady state will be reached when the sum of the potassium and chloride currents exactly equals the inward sodium current. Because the membrane is much more permeable to sodium than to the other two ions, this will occur at a potential near the sodium equilibrium potential.

The increase in sodium permeability that causes the membrane depolarization is followed by a return of the sodium permeability to normal and a marked increase in potassium permeability (Figure 7C). As both the concentration gradient and the potential gradient for potassium are now outward, potassium ions will leave the cell, and as they do so the inner surface of the membrane will become more and more negative until the membrane potential approaches the potassium equilibrium potential and a new steady state is reached. In real life, the changes in permeability are not stepwise as shown in Figure 7 and are not completely separated in time. The exact time course of the permeability changes and the mechanisms underlying them will be discussed in detail in the next chapter.

If the interior of the nerve gains sodium during the rising phase of the action potential and loses potassium on the falling phase, one might expect the sodium and potassium concentrations in the axoplasm to change. In fact, as we have already shown for charge separation associated with the resting potential, the charge movements required to discharge and recharge the membrane are negligible compared with the internal concentrations of the anions. This can be shown in two ways: by calculation and by direct measurement.

The calculation depends on the fact that the cell membrane behaves as a capacitor—which is simply another way of saying that there is a charge separation across it. The capacitance of the membrane is defined in the same way as that of any other capacitor, by the amount of charge it accumulates for each volt of potential across it. Thus, the capacitance (C) is given by $C = Q/V$, where Q is the charge on the membrane and V is the membrane potential. The membrane capacitance depends on membrane composition and thickness and for most nerve cells, including squid axons, is about 1 $\mu F/cm^2$. If we consider a length of 1 cm of axon with a diameter of 1 mm, then its surface area will be about 0.3 cm^2 and its capacitance 0.3 μF. With a membrane potential of -70 mV, the charge on the membrane will be given by $Q = CV = 0.3\ \mu F \times 0.07\ V = 0.02$ microcoulombs, or 2×10^{-8} coulombs. This can be converted into moles of univalent ion by dividing by the faraday, which is the number of coulombs per mole of univalent ion (96,501 or approximately 10^5). The result, then, is that the charge on the membrane is the equivalent of 2×10^{-13} moles of univalent ion. To discharge the membrane to zero potential, this number of moles of sodium ion would have to cross the membrane; to discharge it to $+40$ mV would require an additional 1.1×10^{-13} moles or a total of 3.1×10^{-13} moles of sodium. To return the membrane potential to its resting state would require the loss of an equal number of potassium ions. When we divide this number by the volume of 1 cm of axon (7.5×10^{-6} liters), the resulting change

How many ions enter and leave during an action potential?

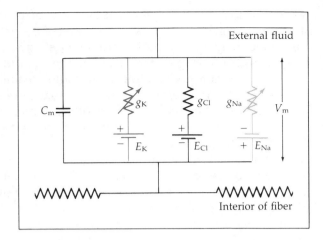

8

EQUIVALENT CIRCUIT for a length of excitable membrane. E_K, E_{Cl}, and E_{Na} are the Nernst potentials for the individual ions. The individual ionic conductances are represented by resistors (which have the value $1/g$ for each ion) and the arrows indicate that they are variable for potassium and sodium. C_m is the membrane capacitance and V_m the membrane potential. External and internal fluid connect this length of membrane with neighboring elements.

in concentration is 4×10^{-8} molar. This is about one ten-millionth of the internal potassium concentration.

These calculated values are supported satisfactorily by experimental measurements of the amount of radioactive sodium and potassium entering and leaving a squid axon for each nerve impulse. The measured value of 3×10^{-12} to 4×10^{-12} moles/cm^2 (or about 1×10^{-12} moles per centimeter of axon) is somewhat larger than the value predicted theoretically, mainly because the movements of the two ions overlap in time to some extent so that there are ions moving simultaneously in opposite directions. Such simultaneous movement will contribute to the measured fluxes, but their contributions to the change in membrane charge will tend to cancel.

Membrane conductance

Two principles relating to the control of membrane potential have been outlined so far: (1) The membrane potential is determined by the relative permeabilities of the membrane to the major ions present. (2) With respect to the resting and action potentials, the major ions we have to consider are sodium and potassium (we will see later that in some cells calcium ions are involved in action potential generation). These principles can be represented in a rather different way by an electrical model (Figure 8) in which membrane potential and ionic currents are related through membrane CONDUCTANCES to the ions present, rather than through membrane permeabilities. Conductance is a measure of the ease with which charge will flow in a system; it is the reciprocal of resistance and is measured in reciprocal ohms, or siemens. Thus, the current i flowing through a resistor of resistance R connected to a battery with voltage V is given by $i = V/R$ or, in terms of conductance, $i = gV$, where $g = 1/R$ is the conductance of the resistor. The conductance of the cell membrane to a given ion can be used to relate ionic current to the electrical driving force. For sodium, potassium, and chloride we can write

$$i_{Na} = g_{Na}(V_m - E_{Na})$$
$$i_K = g_K(V_m - E_K)$$
$$i_{Cl} = g_{Cl}(V_m - E_{Cl})$$

What the equations say is that the ionic current is proportional to the difference between the membrane potential (V) and the equilibrium potential for the particular ion (E_{Na}, E_K, or E_{Cl}). When the membrane potential is equal to the equilibrium potential for an ion, then that ionic current will be zero. When the membrane potential is more negative than the equilibrium potential, then the current will be negative, or inward, when more positive, outward. The constant of proportionality relating current and voltage (the ionic conductance, g) is a measure of the permeability of the membrane to the ion in question. This particular definition of conductance is called CHORD CONDUCTANCE and is not quite the same as SLOPE CONDUCTANCE. The latter is defined, not with reference to the equilibrium potential, but rather by the slope of the relation between ionic current and membrane potential at any point. The two are different because the relation between current and voltage is not usually linear. The relation between permeability and conductance and the difference between chord conductance and slope conductance are discussed in more detail in Box 3.

As in the derivation of the constant field equation (Box 2), we can say that the membrane potential is in a steady state when the total ionic current is zero. In other words, the membrane potential will remain constant as long as there is no net inward or outward movement of ions. In the steady state, then,

$$i_{Na} + i_K + i_{Cl} = 0$$

By algebraic manipulation, we then arrive at

$$V_m = \frac{E_K + (g_{Na}/g_K)E_{Na} + (g_{Cl}/g_K)E_{Cl}}{1 + (g_{Na}/g_K) + (g_{Cl}/g_K)}$$

This equation can be represented by the electrical model shown in Figure 8.[12] E_{Na} is represented as a battery with its positive pole pointing inward; for a squid axon the voltage of the battery would be +55 mV. Similarly, E_K and E_{Cl} are represented by batteries of the appropriate voltage with their negative poles pointing inward. Current will flow through each arm of the circuit as specified by each of the individual equations for ionic current presented above. The membrane potential (V_m) will, as in the final equation, depend on the relative values of the conductances. The arrows on the resistors representing the sodium and potassium conductances indicate that they are variable, but the chloride conductance is fixed. Also shown is the membrane capacitance C_m and the longitudinal resistance of the interior of the fiber, which is represented by resistors. Because the extracellular bathing solution is considered to be of infinitely large volume by comparison, its resistance is represented as being zero.

The principles outlined in the derivation of the constant field equation and represented in the electrical model presented in Figure 8 apply

[12]Hodgkin, A. L. 1964. *The Conduction of the Nervous Impulse.* Liverpool University Press, Liverpool.

BOX 3 PERMEABILITY AND CONDUCTANCE

One question that arises frequently with respect to ionic movements across cell membranes is, What is the relation between permeability and conductance? This question can be answered, at least in a qualitative way, by considering a graph of current as a function of membrane potential for a given ion. We will use as an example diffusion of potassium across the membrane along a constant field. As already derived (Box 2), the potassium current is given by

$$i_K = p_K F V' \frac{K_o - K_i e^{V'}}{e^{V'} - 1}$$

where p_K is the permeability constant for potassium, K_o and K_i are the outside and inside potassium concentrations, respectively. $V' = V_m F/RT$, where V_m is the membrane potential. Suppose now that we can do an experiment on the idealized cell (see figure) in which the membrane potential can be held at any desired value and the potassium current measured. The relation between current and voltage, given by the equation for i_K, will have the form:

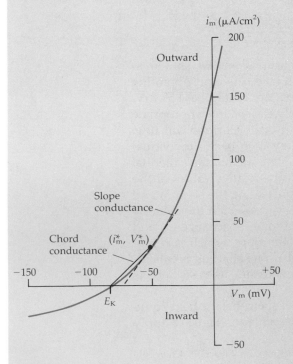

The current is, of course, zero at $V_m = E_K$, inward at more negative membrane potentials, and outward at more positive potentials. The slope of the relation becomes steeper and steeper as the potential becomes more and more positive. This is because at positive potentials the current is directed outward from a relatively concentrated solution of potassium (i.e., there are lots of ions available to carry the current). At negative potentials, on the other hand, the current is directed inward from a relatively low external concentration. At extreme positive and negative potentials, the current is directly proportional to the concentration of ions in the solution from which the current is flowing.

We can use the relation shown in the graph to illustrate the meaning of potassium conductance (g_K), as defined in the text:

$$g_K = i_K/(V_m - E_K)$$

At a particular membrane potential V_m^*, where the current is i_K^*, the potassium conductance is simply the slope of the solid straight line drawn from the point (V_m^*, i_K^*) to E_K, where the current is zero. Because the voltage–current relation is curved, the conductance (i.e., the slope of the line) depends on the membrane potential (V_m^*) at which it is measured. This definition of conductance is called CHORD CONDUCTANCE. It is important because it provides a measure of the ionic current through the membrane at any given potential. There is another measure of conductance called SLOPE CONDUCTANCE, which is defined in a different way—namely, by the slope of the voltage–current relation at a particular point: $g'_K = (di_K/dV_m)$. The slope conductance at V_m^* is indicated on the graph by the dashed line. Slope conductance does not provide an absolute measure of membrane current, but it is sometimes useful in more complicated analyses of membrane behavior. Like the chord conductance, slope conductance is also dependent on membrane potential; as V_m approaches E_K, the two measures of conductance approach the same value.

In summary, then, conductance is directly related to permeability, as can be seen by the equation. Increasing p_K increases the slope of the voltage–current relation and therefore increases the potassium conductance. But conductance depends on other factors as well, specifically ionic concentrations and membrane potential. For example, suppose we reduced the potassium concentration outside the cell to some very low value. Then, even though the potassium *permeability* would be unaffected, at large negative membrane potentials the inward current, and hence the conductance, would be very small because there would be few potassium ions available for inward movement.

to cells other than idealized cells and the squid axon, but there are many variations in detail between one cell and the next. For example, chloride permeability is relatively much higher in skeletal muscle fibers, and it therefore plays a relatively greater role in determining the membrane potential in these cells. Many neurons have quite low resting potentials, of the order of -45 mV or so, because of a high resting sodium conductance. Glial cells (the satellite cells of the brain; Chapter 13) have high resting potentials very close to E_K and behave as if their membranes were permeable almost solely to potassium. In spite of these variations, the cell membrane potentials either at rest or during activity can in each case be predicted from the concentration ratios of the major ions and their relative permeabilities or conductances.

SUGGESTED READING

General

Hodgkin, A. L. 1951. The ionic basis of electrical activity in nerve and muscle. *Biol. Rev. 26*: 339–409.

Katz, B. 1966. *Nerve, Muscle and Synapse.* McGraw-Hill, New York, Chapter 4. (Both provide useful descriptions of some of the concepts developed here and contain many key references to older literature.)

Selected original papers

Baker, P. F., Hodgkin, A. L. and Shaw, T. I. 1962a. Replacement of the axoplasm of giant nerve fibres with artificial solutions. *J. Physiol. 164*: 330–354.

Baker, P. F., Hodgkin, A. L. and Shaw, T. I. 1962b. The effects of changes in internal ionic concentrations on the electrical properties of perfused giant axons. *J. Physiol. 164*: 355–374.

Hodgkin, A. L. and Horowicz, P. 1959. The influence of potassium and chloride ions on the membrane potential of single muscle fibres. *J. Physiol. 148*: 127–160.

Hodgkin, A. L. and Katz, B. 1949. The effect of sodium ions on the electrical activity of the giant axon of the squid. *J. Physiol. 108*: 37–77. (This paper deals in detail with the general relation between permeability and membrane potential. The constant field equation is derived in Appendix A.)

CONTROL OF MEMBRANE PERMEABILITY

CHAPTER SIX

The ionic mechanisms responsible for generating the nerve impulse in squid axons have been described quantitatively, largely through the use of the voltage clamp method, which maintains the membrane potential at a set value. Using this method, it is possible to measure the membrane current flow resulting from a step change in voltage to a new steady level and to determine which components of the current are carried by individual ions. From these measurements one can estimate the magnitude and time course of changes in permeability of the membrane to individual ions as a function of membrane potential.

Such experiments have shown that the permeabilities of the membrane to sodium and potassium are influenced by the membrane potential. Depolarization increases sodium permeability and also, more slowly, permeability to potassium. The increase in sodium permeability produced by depolarization is transient, being turned off by a process of inactivation. The increase in potassium permeability persists as long as the depolarization is maintained. These sodium and potassium permeability mechanisms behave independently. For example, the poison tetrodotoxin blocks the increase in sodium permeability without any effect on the potassium permeability mechanism. Conversely, tetraethylammonium ion blocks the voltage-dependent potassium permeability selectively.

It is the dependence of sodium and potassium permeabilities on membrane potential that is responsible for the action potential. The magnitudes of the permeability changes to the two ions and their sequential timing account quantitatively for its rising and falling phases. The physical dimensions of the sodium permeability channels and their distribution in the membrane have been estimated from other kinds of experiments. In addition, evidence has been obtained that the opening of the channels by depolarization is associated with the movement of charged molecules within the membrane. Modern techniques have enabled investigators to measure directly the conductance and kinetic behavior of single sodium and potassium channels.

These findings have helped to explain many of the basic properties of excitable membranes, not only in squid axons but in other cells, such as vertebrate nerves, skeletal muscle fibers, and heart muscle.

In Chapter 5 it was shown that the potential of a cell membrane is determined by its relative permeabilities, particularly to sodium and potassium. In this chapter evidence will be presented that the membrane permeabilities are in turn determined primarily by the membrane potential. The discussion will be based on the initial experiments of Hodgkin and Huxley and on more recent work that has provided detailed information about the structure and properties of the membrane channels through which the ions move.

Two clues to the permeability change underlying the rising phase of the action potential were given in the previous chapter: Depolarization is required to trigger the potential, and its amplitude depends on extracellular sodium concentration. Suppose, then, that the permeability of the membrane to sodium is increased by depolarization: The immediate result will be an inward movement of sodium along its electrochemical gradient. The inward movement of sodium will depolarize further, producing a still greater increase in sodium permeability and more rapid sodium entry, and so on (Figure 1). If the increased sodium permeability is not turned off, the membrane will approach and remain near the sodium equilibrium potential. This kind of explosive process involves the principle of "positive feedback." Another example is a gunpowder explosion in which the heat generated by the initial chemical reaction accelerates the reaction itself, which in turn generates still more heat, and the explosion ensues.

What happens if the permeability to potassium is increased by depolarization? In this case the electrochemical gradient for potassium is

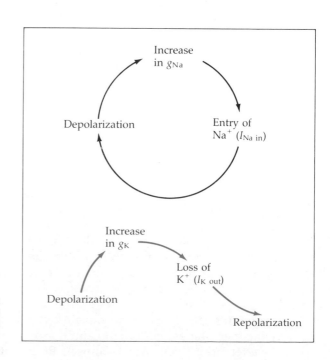

1

EFFECTS OF INCREASING g_{Na} AND g_K on membrane potential. Sodium entry reinforces depolarization; potassium efflux leads to repolarization.

outward, not inward, and potassium leaves the cell. The outward movement results in repolarization toward the initial membrane potential and a return of potassium permeability to normal; this restoration process is an example of "negative feedback."

In theory, then, the rising and falling phases of the action potential can be explained by assuming that depolarization causes an initial increase in permeability to sodium, which is then turned off and followed by an increase in permeability to potassium. Such a sequence will lead to rapid depolarization toward the sodium equilibrium potential followed by repolarization toward the potassium equilibrium potential. Depending on the concentration gradients for the two ions, this means that the action potential will approach about +50 mV at its peak and return to a resting level of about −75 mV. Hodgkin and Huxley developed these ideas and showed experimentally that in squid axon changes in sodium and potassium permeabilities occurred; the changes were timed correctly and were of the correct magnitude to account exactly for the action potential (see Suggested Reading at the end of this chapter). In their experiments, which will be described here in some detail, they determined how membrane permeabilities to potassium and sodium (or, more precisely, the membrane conductances, g_{Na} and g_K) varied with membrane potential and time. They were then able to use this information to reconstruct the entire action potential and, in addition, to account for many other properties of excitable membranes such as threshold, refractory period, and propagation.

At first thought, it might seem simple to make the appropriate measurements of conductance (g). All that is needed is to measure the amount of current (I) flowing inward or outward across the membrane at various levels of potential (V_m), since for each ion (i)

$$I_i = g_i(V_m - E_i)$$

However, three major problems arise with this approach and must be solved experimentally.

1. Current flowing across the membrane will change the membrane potential, which in turn will change the membrane conductance. How, then, can the system be held steady to determine systematically the way in which conductance depends on potential? The difficulty is severe as the action potential rises at a rate of several hundred volts per second; accordingly, a system with a response time of a few microseconds is needed. The solution, which will be described later, was to hold steady or "clamp" the membrane potential and to measure the magnitude and time course of the membrane current.
2. A fraction of the membrane current during an action potential is not ionic at all; it flows into or out of the membrane capacitance as the potential changes. How can this current be separated from the total current so that only the ionic currents are measured?
3. Given that the capacitative current can, in fact, be separated from the ionic currents, how can the currents carried by sodium and potassium be determined individually?

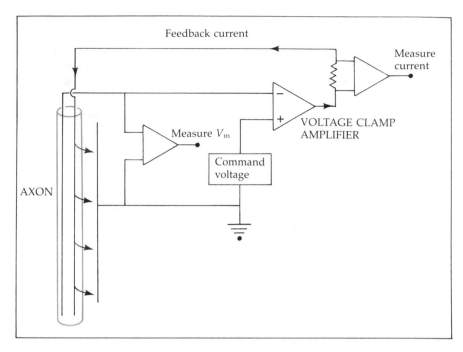

2 **VOLTAGE CLAMP TECHNIQUE for squid axon.** One longitudinal electrode measures membrane potential (V_m) and is connected to the voltage
clamp amplifier. When V_m is different from the command potential (set
by experimenter), the clamp amplifier passes current into the axon
through a second electrode (arrows). With this circuit, the membrane
potential can be displaced abruptly and held constant at a new value
while the current flowing into the axon and across the membrane can
be measured (see text and Figure 3).

Principle of the
voltage clamp

The technique of holding the membrane potential at a constant
value, the VOLTAGE CLAMP, was devised by Cole and his colleagues[1,2]
and was developed further by Hodgkin, Huxley, and Katz.[3] The experimental arrangement is shown in Figure 2. A squid axon is bathed in
seawater, and into one end two fine silver wires are inserted longitu-
dinally. One of the wires is connected to the "inverting" (−) input of
the voltage clamp amplifier and provides a measure of the potential of
the inside of the fiber with respect to the potential of the seawater
(which is grounded), or, in other words, a measure of the membrane
potential (V_m). The "noninverting" (+) input of the voltage clamp amplifier is connected to a variable voltage source, which can be set by
the person doing the experiment; the value to which it is set is known
as the COMMAND POTENTIAL (V_c). The output of the amplifier is con-

[1]Marmont, G. 1949. *J. Cell. Comp. Physiol. 34:* 351–382.
[2]Cole, K. S. 1968. *Membranes, Ions and Impulses.* University of California Press, Berkeley.
[3]Hodgkin, A. L., Huxley, A. F. and Katz, B. 1952. *J. Physiol. 116:* 424–448.

nected to the second fine silver wire, which is inside the axon and is used to pass current across the membrane. The current through the wire is measured by measuring the voltage drop across a small series resistor. As with any other amplifier, the voltage clamp amplifier will deliver current from its output whenever there is a voltage difference between the $(+)$ and $(-)$ inputs. The essential feature that makes it a voltage clamp is the way it is connected to the axon.

Now let us see how the system works. Suppose, to start with, that the resting membrane potential of the fiber is -70 mV and that the command potential has been set at -70 mV as well. Now imagine the occurrence of an event that would normally depolarize the membrane to -65 mV. This event will start to drive V_m positive with respect to V_c. Because V_m is connected to the inverting input of the amplifier, a change in V_m in the positive direction will result in negative current flow at the output of the amplifier; that is, current will flow from the axoplasm toward the amplifier output (that is another example of negative feedback). Thus, the membrane potential will be driven in the negative direction and current will continue to flow until V_m is again equal to V_c. If the circuitry is properly designed, this all happens within a few microseconds so that the membrane never really deviates from -70 mV; the instant it tries to do so, the clamping action of the circuit takes effect. In summary, the voltage clamp arrangement requires that the membrane potential be equal to the command potential at all times. This condition is realized by the clamp amplifier automatically delivering current of the appropriate magnitude and sign to the current electrode in the axoplasm any time a difference starts to develop between V_m and V_c.

In principle, the system is like a thermostatically controlled bath. The thermostat control is the command setting, and a thermometer records the actual bath temperature. When the two are equal, no heat flows into the bath from the heater; when the bath temperature falls below the command temperature or when the command temperature is suddenly increased, then additional heat is supplied until the two temperatures are again equal.

One point that requires further explanation is the meaning of "positive" and "negative" current. Positive current means that positive charge is flowing from the amplifier output into the axoplasm; negative current means that positive charge is flowing in the opposite direction, from the axoplasm to the amplifier. With respect to the ionic currents across the membrane, when there is inward current (carried, for example, by sodium ions), then the clamp amplifier current must be negative to hold the membrane potential constant; when the membrane current is outward, then the amplifier current will be positive. Consequently, it is conventional to represent outward ionic currents as positive and inward ionic currents as negative. The convention adopted by Hodgkin and Huxley in the original voltage clamp papers was the opposite, with inward currents positive. In addition, depolarization from the resting potential was shown as a negative voltage step. Results from those papers presented in this chapter have been redrawn to conform to modern convention.

Now consider what will happen if the membrane is first clamped at -70 mV and the command potential then stepped to -15 mV. V_m will then be negative with respect to V_c, and we would expect the amplifier to deliver positive current into the axon to drive V_m to -15 mV. This is indeed what happens, by way of a rapid surge of capacitative current (see later), but something more interesting occurs as well. The depolarization to -15 mV produces an increase in sodium conductance, and there is a consequent flow of sodium ions inward across the membrane. In the absence of the clamp, this would tend to depolarize the membrane still further (toward the sodium equilibrium potential); with the clamp in place, however, the amplifier provides just the correct amount of negative current to hold the membrane potential constant. In other words, the current provided by the amplifier is exactly equal to the current flowing across the membrane. Here, then, is the great power of the voltage clamp: In addition to holding the membrane potential constant, it provides an exact measure of the membrane current at all times.

Membrane currents produced by depolarization

As noted previously, the current measured by the voltage clamp is the *total* current across the membrane; it does not solve the problem of separating capacitative currents from ionic currents and ionic currents one from another. To discuss these problems it is useful to describe in more detail the magnitudes and time courses of the currents that occur when the command potential is suddenly stepped from the resting membrane potential (in this example, -65 mV) to a depolarized level (-9 mV). These are shown in Figure 3A. The current following the voltage step consists of three phases: (1) a brief positive surge lasting only a few microseconds while the membrane is moving to its new value, (2) a transient inward current, and (3) a delayed outward current.

Capacitative and leak currents

The first component is the capacitative current, which occurs because a step from one potential to another requires that the membrane capacitance be recharged from the old potential to the new. The magnitude of the current (I_c) depends on the magnitude of the capacitance (C) and

3 CURRENT FLOW ACROSS MEMBRANE during depolarization. (A) Membrane currents measured by voltage clamp during a 56-mV depolarization of a squid axon membrane. The resulting currents (lower tracing) consist of a brief outward capacitative current, a transient phase of inward current, and a delayed, maintained outward current. These are shown separately in (B), (C), and (D). The capacitative current (B) lasts for only a few microseconds. The small outward leakage current is due partly to the movement of chloride. The transient inward current (C) is due to sodium entry (see Figure 6) and the prolonged outward current (D) to potassium movement out of the fiber. In this and in other voltage clamp records from squid axons, the membrane potential is assumed to be -60 to -65 mV.

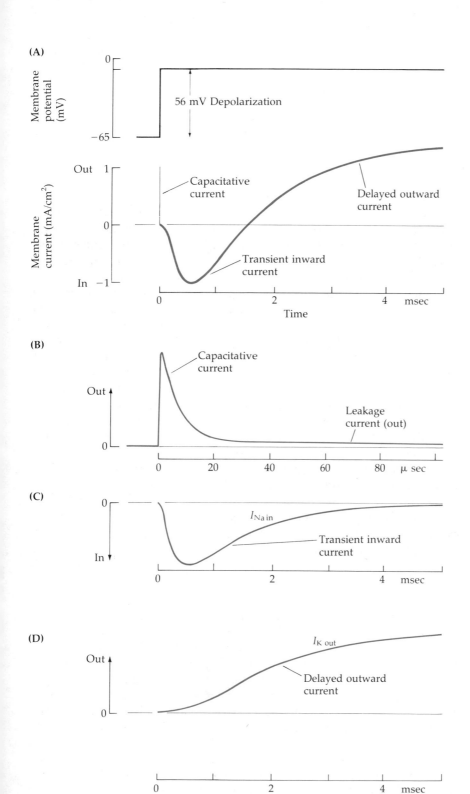

the rate of change of the potential during the step (dv_m/dt) (see Appendix A). The relation is simply

$$I_C = C(dv_m/dt)$$

If the clamp amplifier is capable of delivering a large amount of current, then the membrane can be recharged to its new potential rapidly, and the current will last only an infinitesimal time. Once the new potential is reached ($dV/dt = 0$), there is no more capacitive current. Note that the capacitive current is neither "inward" nor "outward" in the sense of ionic movements *through* the membrane. Instead it is associated with a change in charge separation at the membrane surfaces. The initial current is shown in more detail on an expanded time scale in Figure 3B. It can be seen that in practice the surge of capacitive current lasts about 20 μsec and is followed by a small steady outward current. This steady current is what one would expect if the resting membrane conductances were unaltered by the step depolarization. Because the cell is depolarized, the sum of all the ionic currents across the membrane will no longer be zero, as it was at rest, and there will be a net outward current. Consequently, the clamp amplifier will have to deliver an equal amount of positive current to hold the membrane potential constant. This outward ionic current is known as "leak" current and is carried largely by potassium and chloride ions. It varies linearly with voltage displacement from rest and lasts throughout the duration of the voltage step. However, it is obscured by the much larger ionic currents in the later phases of the response. Reducing extracellular chloride concentration results in a slight reduction in leak current but has no effect on the later currents. It is concluded that chloride ions contribute to the leak current but are not otherwise involved in the response.

Currents carried by sodium and potassium

Turning now to the second and third phases, Hodgkin and Huxley showed that they were due first to the entry of sodium and then to the exit of potassium across the cell membrane. By various experimental manipulations they were able to show that the ions moved independently and to deduce the time course of each of the ionic currents, as illustrated in Figure 3C and D.

Dependence of ionic currents on membrane potential

One way to obtain information about the nature of the early and late ionic currents is to determine how the magnitudes of the currents depend on the size of the depolarizing voltage step. Currents produced by various levels of depolarization from a holding potential of −65 mV are shown in Figure 4. First of all, a step *hyperpolarization* to −85 mV (lower record) produces only a small inward current, as would be expected from the resting properties of the membrane. As already shown in Figure 3, moderate depolarizing steps each produce an early inward current followed by a sustained outward current. With greater depolarizations, the early current becomes smaller, is absent at about +52 mV, and then reverses to become outward as the depolarizing step is increased still further. The current–voltage relations for the early and late currents are shown in Figure 5, in which the peak amplitude of the early current and the steady-state amplitude of the late current are plotted against the potential to which the membrane is stepped. With

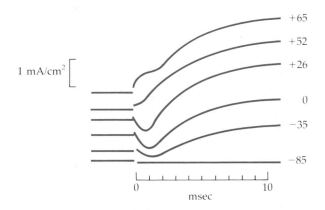

1 mA/cm^2

+65
+52
+26
0
−35
−85

0   10
msec

4

CURRENTS PRODUCED BY VOLTAGE STEPS from a holding potential of −65 mV to a hyperpolarized level (−85 mV) and to depolarized levels of increasing magnitude, as indicated. The early sodium current first increases, then decreases in magnitude as the depolarizing step increases, and is reversed in sign at +65 mV. The late potassium current increases monotonically with increasing depolarization. (After Hodgkin, Huxley, and Katz, 1952.)

hyperpolarizing steps, there is no separation of early and late currents; the membrane simply responds as a passive resistor, with a hyperpolarization producing the expected inward current. The late current also behaves as one would expect of a resistor in the sense that depolarization produces outward current, but the magnitude of the current is much larger than expected from the resting membrane properties. This larger current is the result of activation of the voltage-dependent potassium conductance. The early current behaves in a much more complex way. As already noted, it first increases and then decreases with increasing depolarization, becoming zero at about +52 mV and then reversing in sign. The reversal potential is, in fact, the equilibrium potential for sodium, a finding that provides one important piece of evidence that the early current is carried by sodium ions.

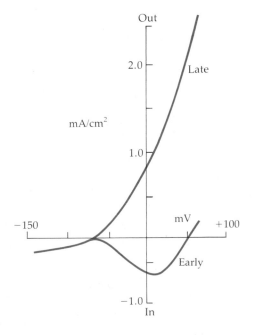

Out

2.0

Late

mA/cm^2

1.0

−150   mV   +100

Early

−1.0
In

5

AMPLITUDE OF EARLY AND LATE CURRENTS plotted against the potential to which the membrane is stepped. Late outward current increases rapidly with depolarization. Early inward current first increases in magnitude, then decreases, reversing to outward current at about +55 mV (the sodium equilibrium potential). (After Hodgkin, Huxley, and Katz, 1952.)

One point of interest in the current–voltage relation for the early current is that between about −50 mV and +10 mV the slope of the relation is negative. In this region of *negative slope conductance,* the magnitude of the current increases even though the driving force for sodium entry ($E_{Na} − V_m$) is decreasing, because sodium conductance increases with depolarization and the conductance increase overrides the effect of the decrease in driving force. It is this effect of potential on sodium conductance that, in the absence of the voltage clamp, gives the regenerative response shown schematically in Figure 1. Under voltage clamp conditions, the membrane is prevented from producing a regenerative response, but its ability to do so appears in the voltage–current relation as a region of negative slope conductance.

One further procedure used by Hodgkin and Huxley to test the idea that the early current was carried by sodium ions was to vary the external sodium concentration and determine how this influenced the

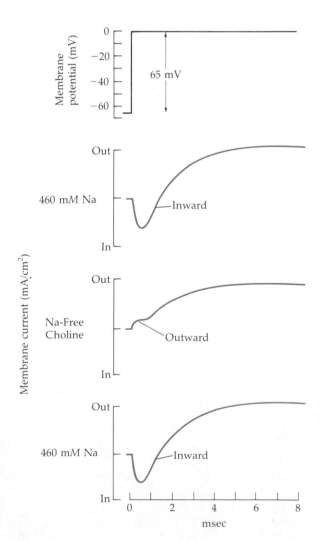

6

MEMBRANE CURRENT contributed by an inward flow of sodium when the membrane of a squid axon is depolarized by 65 mV. In sodium-free solution, the inward current disappears. It is replaced by a small transient outward current as sodium moves out of the fiber. The inward current is restored (last record) when sodium is reintroduced. (After Hodgkin and Huxley, 1952a.)

currents. When the external sodium concentration was reduced to zero (by substituting choline chloride for sodium chloride in the solution bathing the axon), the initial phase of inward current disappeared and was replaced by a small outward current (Figure 6). The delayed phase of outward current was unchanged. This result was consistent with the idea that the initial current was due to an increase in sodium conductance; when all the extracellular sodium was removed, the electrochemical gradient for sodium was outward and an increase in conductance would be expected to result in outward current rather than no current or the inward current seen in the normal bathing solution. The fact that the late potassium current was unaffected by the magnitude and direction of the sodium current was evidence for the idea that the ions moved independently.

Other experiments were done to determine the time course of the individual ionic currents. One approach was to displace the membrane potential to near E_{Na}. At this potential, which in Figure 4 was +52 mV, an increase in sodium conductance should result in no net movement of sodium, and, as expected, there was no initial phase of inward current. Instead there was a single S-shaped phase of outward current. At E_{Na}, therefore, it was possible to observe the magnitude and time course of the late potassium current. It would be more useful, however, to measure the magnitude and time course of the potassium current at all membrane potential levels, not just at +52 mV. Then, in addition, the magnitude and time course of the sodium current could be obtained by subtracting the potassium current from the total. Hodgkin and Huxley accomplished this in an ingenious way by experiments like that illustrated in Figure 7. First they stepped from the resting potential to a new holding potential (for example, from −65 to −9 mV) and measured the resulting current (Figure 7, curve A). Then, after returning to the resting potential, they replaced about 90 percent of the extracellular sodium with choline so that E_{Na} was −9 mV. On repeating the step from −65 to −9 mV, they now observed the potassium current in isolation (curve B). Subtracting the potassium current from the previous

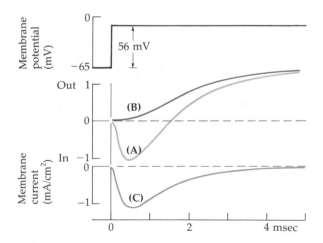

7

SODIUM AND POTASSIUM COMPONENTS of membrane current. With a displacement of 56 mV of the membrane potential in a squid axon in normal seawater, one sees the normal sodium and potassium membrane current components in (A). After replacing most of the extracellular sodium so that E_{Na} is at −9 mV, the pure potassium current (B) remains. The difference between (A) and (B) is the normal sodium current (C). (After Hodgkin and Huxley, 1952a.)

total current gave a measure of the magnitude and time course of the sodium current (curve C). Using this kind of procedure, then, it was possible to observe the potassium current at a selected step potential by appropriate adjustment of the extracellular sodium concentration.

The idea that the late outward current is carried by potassium was tested directly by Hodgkin and Huxley. They loaded an axon with radioactive potassium and clamped the membrane at a depolarized level. The total charge movement across the membrane during the maintained outward current was then calculated and compared with the efflux of radioactive potassium into the bathing solution over the same period. The agreement was excellent, providing direct evidence that the outward current was indeed carried almost entirely by potassium ions.

Selective poisons for sodium and potassium channels

Since the original experiments of Hodgkin and Huxley, new and convenient pharmacological methods have been found for blocking sodium and potassium currents selectively. Tetrodotoxin (TTX) in particular has turned out to be useful for a wide range of experiments. Tetrodotoxin is a virulent poison that is concentrated in the ovaries of certain fish. Its potent effects have given rise to the Chinese proverb, "To throw away life eat blowfish" (puffer fish). Kao[4] has reviewed the fascinating history of TTX, beginning with the discovery of its effects by the Chinese emperor Shun Nung (2838–2698 B.C.), who personally tasted 365 drugs while compiling a pharmacopoeia and lived (for an amazingly long time) to tell the tale.

One great advantage of TTX for neurophysiological studies is that its action is highly specific. Working with squid axons, Moore, Narahashi, and colleagues[5] have shown that it blocks the voltage-sensitive sodium conductance selectively. Thus, when a TTX-poisoned axon is subjected to a depolarizing voltage step, no inward sodium current is seen, only the delayed outward potassium current. The latter is unchanged in amplitude and time course by the poison. Application of TTX to the inside of the membrane by adding it to an internal perfusing solution has no effect. The toxin appears to bind to a site in the outer mouth of the channel through which sodium ions move (see later, Figure 15), thereby physically blocking ionic current through the channel. Many other excitable cells are affected by TTX is a similar manner, including vertebrate myelinated and unmyelinated axons and skeletal muscle fibers. The effect of TTX on sodium currents in a myelinated nerve fiber is shown in Figure 8A and B.[6] There are other substances that block the sodium current; of clinical importance are local anesthetics such as procaine. Another toxin whose action, like that of TTX, is highly specific for voltage-sensitive sodium channels is saxitoxin (STX). It is produced by microscopic marine dinoflagellates and accumulated by

[4]Kao, C. T. 1966. *Pharmacol. Rev. 18:* 977–1049.

[5]Moore, J. W., Blaustein, M. P., Anderson, N. C. and Narahashi, T. 1967. *J. Gen. Physiol. 50:* 1401–1411.

[6]Hille, B. 1970. *Prog. Biophys. Mol. Biol. 21:* 1–32.

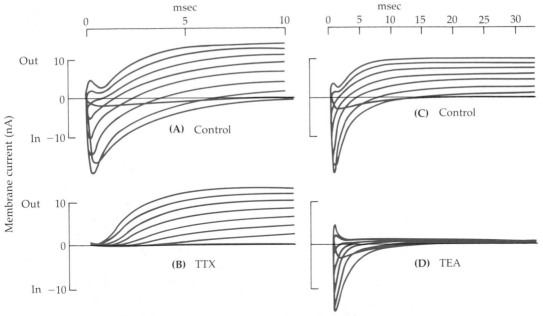

8 PHARMACOLOGICAL SEPARATION of membrane currents into sodium and potassium components. The membrane potential of a frog myelinated nerve fiber is displaced to various levels between −60 and +75 mV. (A) and (C) are controls in normal fluid. In (B), addition of 300 nM tetrodotoxin (TTX) causes the sodium currents to disappear while the potassium currents remain. In (D), addition of tetraethylammonium (TEA) blocks the potassium currents, leaving the sodium currents intact. (After Hille, 1970.)

filter feeding shellfish, including the clam *Saxidomus*, after which the poison is named.

Just as TTX has a selective action on the sodium channels, a number of substances that have been found have a similar effect on the potassium channels. For example, in squid axons and in frog myelinated axons, Armstrong, Hille, and others have shown that the voltage-sensitive potassium conductance is blocked selectively by tetraethylammonium ion (TEA),[7] as shown in Figure 8C and D. In contrast to the effect of TTX, TEA is most effective when added to the axoplasm or to an internal perfusate, and it exerts its action on the inner surface of the membrane. Another quaternary ammonium compound, 4-aminopyridine (4-AP), is effective in blocking the potassium conductance from either the inside or the outside of the membrane.

The three problems posed earlier in this chapter—namely, how to hold the membrane potential constant, how to separate ionic and capacitative currents, and how to separate ionic currents from one another—were all solved in a satisfactory manner by the voltage clamp experiments. Using this technique, Hodgkin and Huxley were able to

[7]Armstrong, C. M. and Hille, B. 1972. *J. Gen. Physiol. 59:* 388–400.

deduce the magnitude and time course of sodium and potassium currents as a function of the membrane potential V_m and to determine the equilibrium potentials E_{Na} and E_K. It was then a straightforward matter to deduce the magnitude and time courses of the sodium and potassium conductance changes, using the relations

$$g_{Na} = I_{Na}/(V_m - E_{Na})$$
$$g_K = I_K/(V_m - E_K)$$

The results for five different voltage steps are shown in Figure 9. The first point to be made is that g_{Na} and g_K are both increased by depolarizing the membrane. The relations between the peak conductance and membrane potential are shown for sodium and potassium in Figure 10. The curves are remarkably similar. In contrast, the *time courses* of the conductance changes for the two ions are quite different. The increase in potassium conductance is very much delayed, compared to the time course of the sodium conductance, and remains high through the duration of the step. The sodium conductance increase, on the other hand, rises much more rapidly and then decays to normal, even though the membrane is still depolarized. This decline of the sodium current is called INACTIVATION.

Inactivation of the sodium current

Hodgkin and Huxley did additional experiments to determine the nature of the inactivation process. In particular, they investigated the voltage dependence of inactivation by examining the effect of hyperpolarizing and depolarizing prepulses on the peak amplitude of the sodium current produced by a subsequent standard depolarization. Records from such an experiment are presented in Figure 11. In Figure 11A the membrane is stepped from a holding potential of −65 mV to −21 mV, producing a peak sodium current of about 1 mA/cm². When the step is preceded by a hyperpolarizing prepulse of 31 mV, the peak sodium current is increased (B). Depolarizing prepulses, on the other hand, cause a decrease in the sodium current (C, D). These effects of hyperpolarizing and depolarizing prepulses are time dependent; brief

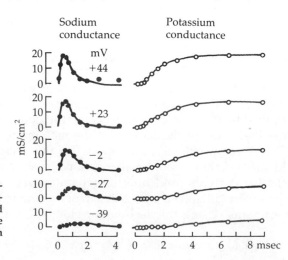

9

MEMBRANE CONDUCTANCE CHANGES produced by voltage steps from −65 mV to the indicated potentials. Peak sodium conductance and steady-state potassium conductance both increase with increasing depolarization. (After Hodgkin and Huxley, 1952a.)

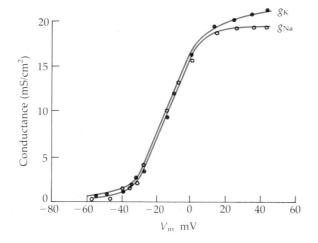

10

**SODIUM AND POTASSIUM CONDUC-
TANCES as a function of membrane potential.
Peak sodium conductance and steady-state potassium conductance are plotted against the potential to which the membrane is stepped. Both conductances increase steeply with depolarization
between −20 and +10 mV. (After Hodgkin and
Huxley, 1952a.)**

pulses with a duration of only a few milliseconds have little effect. In the experiment shown here, the prepulses were of sufficient duration for the effects to reach their maximum. The results were expressed quantitatively by plotting the peak sodium current after a conditioning step as a fraction of the control current (I_{Na},step/I_{Na},no step) against the amplitude of the conditioning step, as shown in Figure 11E. With a depolarizing prepulse of about 30 mV, the subsequent sodium current was reduced to zero; i.e., inactivation was complete. The increase in sodium current produced by hyperpolarizing prepulses reached a maximum of about 70 percent, also with about a 30-mV prepulse. Hodgkin and Huxley represented this range of sodium currents from zero to their maximum value with a single parameter h, varying between zero (complete inactivation) to 1 (no inactivation), as indicated on the right-hand ordinate of Figure 11E.

The experiments with prepulses suggested that inactivation was a distinct phenomenon, separable from the activation process. A more recent experimental observation that gives credence to this idea is that Pronase, a mixture of proteolytic enzymes, when perfused through the inside of a squid axon, leads to reduction and eventual abolition of inactivation[8] without affecting activation of either the sodium or potassium conductance. In other words, the axon behaves as if the enzyme had removed from the inner surface of the membrane some molecular entity associated specifically with the inactivation process. In contrast, the enzyme is quite ineffective when applied in the same concentration to the outer surface of the axon.

A number of neurotoxins act on both activation and inactivation of the sodium conductance.[9] Lipid-soluble steroidal alkaloids, such as veratridine and batrachotoxin (from the skin of Colombian frogs) cause

[8]Armstrong, C. M., Bezanilla, R. and Rojas, E. 1973. *J. Gen. Physiol.* 62: 375–391.
[9]Catterall, W. A. 1980. *Annu. Rev. Pharmacol. Toxicol.* 20: 15–43.

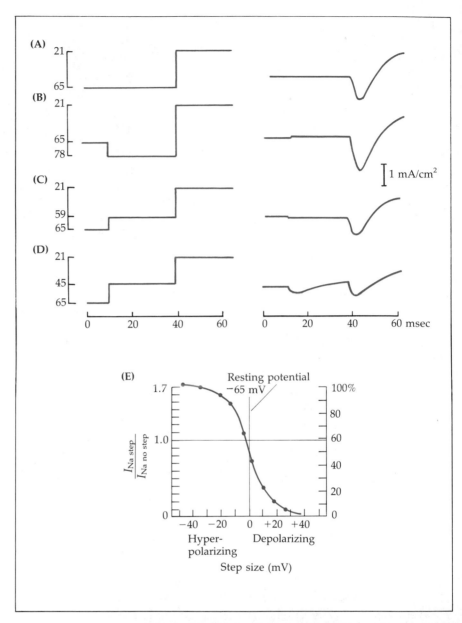

11 EFFECT ON SODIUM CURRENTS of depolarization and hyperpolarization. If a depolarizing step from −65 to −21 mV (A) is preceded by a hyperpolarizing conditioning step (B), the sodium current produced by the later step is increased. Depolarizing conditioning steps of increasing amplitude (C and D) cause a progressive reduction in sodium current. Graph in (E) shows fractional increase or reduction of sodium current as a function of the amplitude of the conditioning steps. Maximum current is about 1.7 times control with a hyperpolarizing step of −40 mV. Depolarizing step of +40 mV reduces subsequent response to zero. Full range of sodium current is scaled from zero to unity by the "h" ordinate (see text). (After Hodgkin and Huxley, 1952c.)

slow activation at rest by shifting the activation to more negative membrane potentials. At the same time, inactivation is blocked so that the poisoned membrane becomes highly permeable to sodium. Repetitive stimulation during exposure to the toxins enhances their actions, indicating that they bind selectively to channels in the open state. A number of scorpion toxins act similarly.

In summary, the results obtained by Hodgkin and Huxley indicated that depolarization of the nerve membrane led to three distinct processes: (1) activation of a sodium conductance mechanism, (2) subsequent sodium inactivation, and (3) activation of a potassium conductance mechanism. These processes can be represented by a physical model in which sodium and potassium channels in the membrane are controlled by voltage-sensitive "gates" (Figure 12). The sodium channel has two such gates, one of which (the "m-gate") is closed at rest and opens in response to depolarization. The other ("h-gate") closes on depolarization and is responsible for sodium inactivation. The potassium channel, as it does not inactivate, has only one "n-gate" which, like the m-gate, opens upon depolarization. In response to a depolarizing pulse, the m-gates open rapidly, allowing the entry of sodium ions. The h-gates respond much more slowly to the depolarization, closing well after the m-gates have opened and subsequently blocking the sodium channels. Similarly, the response of the n-gates to depolarization is relatively slow so that the outward potassium current is delayed.

After obtaining the experimental results, Hodgkin and Huxley proceeded to develop a mathematical description of the precise time courses

Physical and mathematical models of the voltage-dependent changes in sodium and potassium conductance

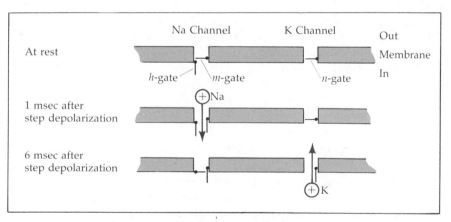

12 **RESPONSE OF MEMBRANE CHANNELS to depolarization. Membrane is represented schematically as containing separate channels for sodium and potassium. The sodium channel is controlled by voltage-sensitive m- and h-gates, the potassium channel by the n-gate. Upon depolarization the m-gates open quickly, allowing sodium entry; the h- and n-gates respond more slowly, closing the sodium channel and opening the potassium channel after a delay.**

of the sodium and potassium conductance changes produced in response to the depolarizing voltage steps. To deal first with the potassium conductance, one might imagine that the effect of a sudden change in membrane potential would be to provide a driving force for the translocation of charged particles in the membrane or for some other change in molecular conformation that would then lead to the opening of n-gates. If the opening of each gate were associated with the translocation of a single particle, then the change in the overall potassium conductance might be expected to be governed by ordinary first-order kinetics; that is, its rise after the onset of the voltage step would be exponential. Instead it is S-shaped. Hodgkin and Huxley were able to account for the S-shaped onset of the potassium conductance increase by assuming that opening of an n-gate required the simultaneous activation of four first-order processes, for example, the translocation of four particles in the membrane. In other words, the S-shaped time course of activation could be fitted by an exponential raised to the fourth power. The expression for the increase in potassium conductance for a given voltage step, then, is

$$g_K = g_{K_{max}} \, n^4$$

where $g_{K_{max}}$ is the maximum conductance reached for the particular voltage step and n is a rising exponential function varying between zero and unity, given by $n = 1 - \exp(- t/\tau_n)$. $g_{K_{max}}$ varies with voltage as shown in Figure 10, and the exponential time constant τ_n is also voltage dependent, ranging between about 4 msec for small depolarizations and 1 msec for depolarization to zero at 10°C. The time course of the rise in sodium conductance is also S-shaped and was fitted by an exponential raised to the third power. Again, this would be the case if the simultaneous translocation of three charged particles were required to open each m-gate. In contrast, the fall in sodium conductance due to inactivation was consistent with a simple exponential decay process.

The overall time course of the sodium conductance change, then, was given by the product of the activation and inactivation processes:

$$g_{Na} = g_{Na_{max}} \, m^3 h$$

where $g_{Na_{max}}$ is the maximum level to which g_{Na} would rise if there were no inactivation and $m = 1 - \exp(-t/\tau_m)$. The inactivation process is a falling, not a rising, exponential and is given by $h = \exp(-t/\tau_h)$. As with the expression for the potassium conductance, $g_{Na_{max}}$ is voltage dependent, as are the activation and inactivation time constants. The activation time constant τ_m is much shorter than that for potassium, having a value at 10°C of approximately 0.6 msec near the resting potential and decreasing to about 0.2 msec at zero potential. τ_h, on the other hand, is much longer, being similar in magnitude to τ_n.

Once the theoretical expressions were obtained for sodium and potassium conductances as a function of voltage and time, Hodgkin and Huxley were then able to predict the entire time course of the action

Reconstruction of the
action potential

potential. Starting with a depolarizing step to just above threshold, they calculated what the subsequent potential changes would be at successive intervals of 0.01 msec. Thus, during the first 0.01 msec after the membrane had been depolarized to, say, −45 mV, they calculated how g_{Na} and g_K would change, what increments of I_{Na} and I_K would result, and then the amount of additional depolarization produced by the net current. Knowing the change in V_m at the end of the first 0.01 msec, they then repeated the calculations for the next time increment, and so on all through the rising and falling phases of the action potential (a laborious exercise to undertake in the days before electronic computers were readily available). The result was that the procedure duplicated with remarkable accuracy the naturally occurring action potential in squid axon. An example of such a calculated action potential and the time courses of the underlying sodium and potassium conductance changes are shown in Figure 13A. Theoretical and observed action potentials produced by brief depolarizing pulses at three different stim-

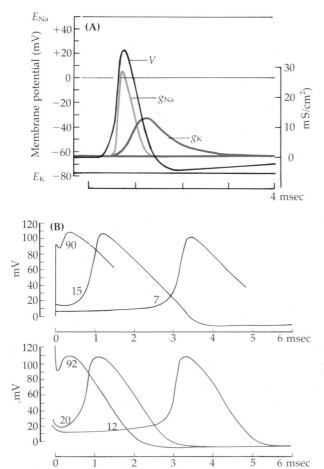

13

(A) RECONSTRUCTION OF ACTION POTENTIAL (curve *V*) and the underlying changes in sodium and potassium conductance, using results from voltage clamp experiments. **(B)** Observed action potentials following brief shocks to the axon at three different intensities (lower tracings) and theoretical solutions for three similar shocks (upper tracings). (After Hodgkin and Huxley, 1952d.)

ulus strengths are compared in Figure 13B. In order to appreciate fully the magnitude of this accomplishment it is necessary to keep in mind that the calculations used to duplicate the action potential were based on current measurements made under completely artificial conditions, with the membrane potential clamped first at one value, then at another, and with various sodium and potassium concentrations in the external fluid. In addition to describing the action potential accurately, Hodgkin and Huxley were able to explain in terms of ionic conductance changes the propagation of the action potential and many other properties of excitable axons, such as refractory period, threshold, and anode break excitation. Further, their findings have been found to be applicable to a wide variety of other excitable tissues.

Threshold and refractory period

How do the findings of Hodgkin and Huxley explain the threshold membrane potential at which the impulse takes off, especially when it might seem that a discontinuity like threshold would require a discontinuity in g_{Na} or g_K? The explanation is that when the membrane is depolarized to threshold the resulting outward current carried by potassium (plus the small leak current) is exactly equal to the inward current carried by sodium. This balance is the same as that at rest, but with an important difference: The sodium conductance is now unstable. When an extra sodium ion enters the cell, the depolarization is increased, g_{Na} increases, more sodium enters, and the regenerative process explodes. When, on the other hand, an extra potassium ion leaves the cell, the depolarization is decreased, g_{Na} decreases, sodium current decreases, and the excess potassium current causes repolarization. As the membrane potential approaches its resting level, the potassium current decreases until it again equals the resting inward sodium current. Depolarization above threshold results in an increase in g_{Na} sufficient for inward sodium movement to swamp outward potassium movement immediately. Subthreshold depolarization fails to increase g_{Na} sufficiently to override the resting potassium conductance.

And how is the refractory period explained? Two changes that develop make it difficult for the nerve fiber to produce a second action potential immediately following a first: (1) Inactivation, which prevents any increase in g_{Na}, is maximal during the falling phase of the action potential and requires several more milliseconds to decay to zero; (2) g_K is now very large, thus requiring a very large increase in g_{Na} to initiate any regenerative depolarization. These two factors result in an ABSOLUTE REFRACTORY PERIOD that lasts throughout the falling phase of the action potential and during which no amount of externally applied depolarization can initiate a second regenerative response. Following the action potential, there is a RELATIVE REFRACTORY PERIOD during which the residual inactivation of the sodium conductance and the relatively high potassium conductance combine to produce an increase in threshold for action potential initiation.

Calcium ions and excitability

An ion of considerable importance, not yet discussed, is calcium, which plays a key role in a variety of processes. For example, a transient increase in intracellular calcium is responsible for secretion of chemical transmitters by neurons and for contraction of muscle fibers. With respect to excitation, changes in extracellular calcium result in changes in threshold for initiation of the action potential; and in some excitable cells it is calcium, rather than sodium, that carries the inward current on the rising phase of the action potential. In the squid axon, and in other nerve and muscle cells having sodium action potentials, a reduction in extracellular calcium lowers the threshold for initiation of impulses; conversely, increasing extracellular calcium raises the threshold. Frankenhaeuser and Hodgkin[10] examined these effects and found not only a change in threshold but also a change in the amount of sodium and potassium current produced by a given voltage step. When, for example, extracellular calcium was lowered, the ionic currents produced by a depolarizing voltage step were increased. The increases in the sodium and potassium conductances produced by a fivefold decrease in extracellular calcium were similar to those that would have been produced by an increase of 10–15 mV in the depolarizing voltage step. In summary, the overall effect of extracellular calcium is to influence both the threshold and the current–voltage relations. At normal levels, calcium serves to stabilize the membrane by maintaining a margin of safety between the resting potential and the threshold potential for activation of the voltage-sensitive conductance channels.

In addition to its action on excitability, calcium itself enters squid axons and other nerve fibers during the impulse. This ionic movement was first demonstrated by measuring accumulation of radioactive calcium and subsequently by using a more rapid optical technique.[11] The optical technique involves the use of aequorin, a protein that is obtained from the jellyfish *Aequoria*, is luminescent in the presence of ionized calcium, and is a reliable and sensitive indicator of calcium activity. Accordingly, when the protein is injected, for example, into the axoplasm of a squid axon, it provides a method for measuring the background concentration of intracellular calcium and changes in concentration during activity. In the squid axon, as in other neurons, the internal calcium concentration is very low, about 0.1 μM or less. This means that the equilibrium potential for calcium is greater than +200 mV, so that there is a large driving force for calcium entry. The internal concentration is kept at a low level by transport mechanisms that extrude calcium across the cell membrane. In the presence of aequorin, trains of action potentials are accompanied by an increased emission of light, a reaction indicating an increase in intracellular calcium concentration.

Voltage clamp experiments established that calcium enters the axon in two phases. The early phase is blocked by tetrodotoxin and has a time course similar to that of the sodium current. This phase represents leak through the sodium channels; the permeability of the sodium channel to calcium is about 1 percent of the permeability to sodium.

[10]Frankenhaeuser, B., and Hodgkin, A. L. 1957. *J. Physiol. 137:* 218–244.
[11]Baker, P. F., Hodgkin, A. L. and Ridgway, E. B. 1971. *J. Physiol. 218:* 709–755.

Calcium action
potentials

Effect of intracellular
calcium on potassium
conductance

The later phase of calcium entry is not blocked by TTX or by TEA and, unlike the sodium current, does not inactivate. The presence of the second phase indicates a separate voltage-sensitive calcium conductance with properties similar to those of the voltage-sensitive potassium channels. We will see later (Chapter 10) that the late calcium entry is responsible for secretion of neurotransmitters at nerve terminals.

In some muscle fibers and some neurons, the voltage-dependent calcium conductance becomes sufficiently large for calcium current to contribute significantly to or even be solely responsible for the rising phase of the action potential. Because g_{Ca} increases with depolarization, the process is a regenerative one, entirely analogous to that discussed for sodium. The participation of calcium in the action potential process was first studied in vertebrate cardiac muscle and in invertebrate muscle fibers.[12] Calcium action potentials have now been shown to occur in a wide variety of invertebrate neurons, vertebrate autonomic neurons, and neurons in the vertebrate central nervous system. Such action potentials occur in nonneural cells as well, including a number of endocrine cells and some invertebrate egg cells. The voltage-dependent calcium channels can be blocked by adding millimolar concentrations of cobalt, manganese, or cadmium to the extracellular bathing solution. Barium can enter the channels and substitute for calcium in the action potential process; magnesium, on the other hand, cannot substitute for calcium. One particularly striking example of the coexistence of sodium and calcium action potential mechanisms in the same cell is the mammalian cerebellar Purkinje cell,[13] which generates calcium action potentials in branches of its dendritic tree and sodium action potentials in its soma. These two processes interact to produce complex discharge patterns in response to steady depolarization (Figure 14).

It is now known that, in addition to the delayed increase in g_K ("delayed rectification") responsible for the late current in the Hodgkin and Huxley voltage clamp experiments, there are a number of other conductance pathways for potassium, some of which contribute to the resting conductance of the membrane.[14] Among these are the "anomalous" or "inward rectification" channels that are seen, for example, in skeletal and cardiac muscle and are turned *off* by depolarization. In addition, "A-channels," seen in many neurons, are like delayed rectification channels in that they are activated by depolarization, but activation is seen only after a preceding hyperpolarization; that is, they are inactivated at rest.[15] "M-channels" are also seen in a variety of neurons; these also are similar to delayed rectification channels but are inactivated by acetylcholine.[16] They will be discussed in more detail later in relation to synaptic transmission (Chapter 9). Yet another potassium channel is

[12]Hagiwara, S. and Byerly, L. 1981. *Annu. Rev. Neurosci. 4:* 69–125.
[13]Llinás, R. and Sugimori, M. 1980. *J. Physiol. 305:* 171–195, 197–213.
[14]Latorre, R. and Miller, C. 1983. *J. Memb. Biol. 71:* 11–30.
[15]Connor, J. A. and Stevens, C. F. 1971. *J. Physiol. 213:* 21–30.
[16]Adams, P. R., Brown, D. A. and Constanti, A. 1982b. *J. Physiol. 330:* 537–572.

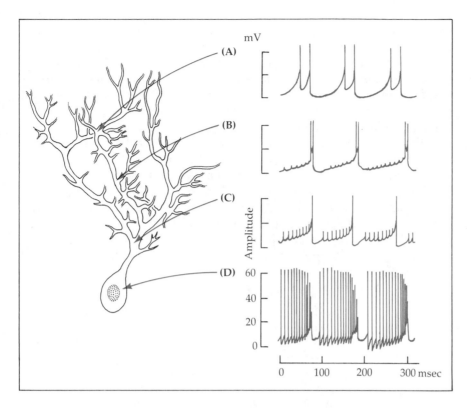

14 SODIUM AND CALCIUM ACTION POTENTIALS in cerebellar Purkinje cell. Records were obtained by impaling the cell at various locations and passing a depolarizing current through the recording electrode. When the electrode was in a remote portion of the dendritic tree (A), depolarization produced long-duration calcium action potentials. In the cell soma (D), steady depolarizing current produced high-frequency sodium action potentials, interrupted periodically by calcium action potentials, followed by a transient hyperpolarization. Sodium action potentials spread passively from the soma into intermediate portions of the dendritic tree (B,C). (From Llinás and Sugimori, 1980b.)

controlled by the level of intracellular calcium.[17] This can be shown experimentally by raising intracellular calcium above normal, for example, by injection from an intracellular micropipette. Following such an injection, the membrane resistance of the cell decreases rapidly and the resting membrane potential approaches the equilibrium potential for potassium. The resistance and potential then return to their control levels as the excess calcium is removed from the cytoplasm by internal buffering mechanisms and active transport. The calcium-dependent potassium channels occur in a wide variety of cells (including nonexcitable cells such as erythrocytes); in those with calcium action potentials, the influx of calcium stimulates an increase in potassium conductance,

[17]Schwarz, W. and Passow, H. 1983. *Annu. Rev. Physiol. 45:* 359–374.

which contributes to the recovery phase of the action potential and produces a subsequent hyperpolarization.

Since the experiments of Hodgkin and Huxley in 1952, a number of lines of evidence have come together to provide more detailed information about the sodium channel. Electrophysiological and biochemical evidence can be combined to provide a physical picture of the channel similar to that shown in Figure 15. In biochemical experiments, a protein that binds to TTX with high affinity and that probably represents part or all of the functional sodium channel has been isolated from membranes of eel electroplax.[18] The molecular size of the protein is approximately 250,000 daltons, and electron microscopy indicates that it is a rodlike structure about 17 nm in length and 4.1 nm in diameter.[19] The channel itself, which presumably extends through the core of the rod, is considered to be an aqueous pore. By measuring its conductance to ions of various sizes and charges, Hille[20] has proposed that an essential

[18]Agnew, W. S., Levinson, S. R., Brabson, J. S. and Raftery, M. A. 1978. *Proc. Natl. Acad. Sci. USA 75:* 2606–2610.

[19]Ellisman, M. H., Agnew, W. S., Miller, J. A. and Levinson, S. R. 1982. *Proc. Natl. Acad. Sci. USA 79:* 4461–4465.

[20]Hille, B. 1971. *J. Gen. Physiol. 58:* 599–619.

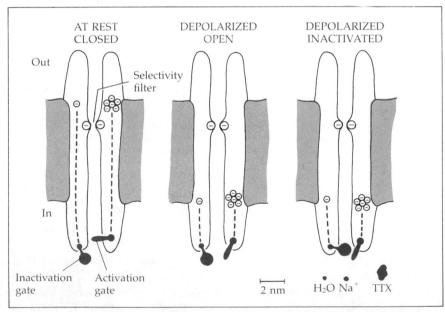

15 **VOLTAGE-SENSITIVE SODIUM CHANNEL,** drawn schematically to scale according to biochemical, electron microscopic, and electrophysiological information. Ionic selectivity is provided by a constriction lined with negative charges near the outer surface of the membrane. The activation gate near the inner surface opens in association with translocation of negative charges across the membrane from out to in. The inactivation gate blocks the inner mouth of the channel and prevents closing of the activation gate. Water molecule, hydrated sodium ion, and tetrodotoxin molecule are drawn to scale for comparison.

part of the pore is a "selectivity filter" consisting of a 0.3 × 0.5 nm constriction ringed by carbonyl oxygens that exclude anions and attract cations. An ionized carboxyl group may also be part of the selectivity mechanism. Measurements of the effects of pH on the channel permeability and the voltage-dependence of the pH effects are consistent with the model and suggest that the negatively charged region is about 25 percent of the way into the membrane from the outside surface. The outer dilatation of the channel accommodates the binding of tetrodotoxin (and saxitoxin), which is thought to occur within the mouth of the channel. The inner dilatation represents the fact that the open channel can be blocked from the inside by a variety of compounds with diameters up to at least 0.8 nm.

Much less is known about the gating structure. Hodgkin and Huxley proposed that the activation process might be associated with translocation of charged structures, or particles, within the membrane. As has already been discussed, the time course of activation suggested that movement of three such *m* particles was required to open the gating structure. The steep voltage dependence of activation (Figure 10) indicated that activation was accompanied by a charge movement equivalent to the translocation of six electronic charges across the membrane, or two per particle. This is represented in the model as a cloud of six monovalent particles that move inward across the membrane upon depolarization. The nature of the charge movement, if any, associated with inactivation is less clear; the voltage-dependence of the inactivation gate may arise in ways other than from associated charge movement.[21] The *h*-gate itself is represented as being at the inner mouth of the pore because of the fact that it can be removed by intracellular Pronase. The *m*-gate is placed on the inner surface as well because some molecules appear to enter and block the channel from the cytoplasm only during activation (when the *m*-gate is open). In addition, inactivation appears to interfere with closing of the activation gate, as discussed in the following section.

One consequence of the proposed gating mechanism is that the postulated charge movements should appear as capacitative currents directed outward in response to a depolarizing voltage step. After a number of technical difficulties were resolved, such "gating currents" were finally seen.[22] An example of such a gating current, produced by a step depolarization, of an internally perfused squid axon is shown in Figure 16A, followed by inward sodium current. The sodium current is much smaller than usual because extracellular sodium concentration was reduced to 5 percent of normal. How is the gating current separated from other capacitative currents produced by a depolarizing step? The total capacitative current has a number of possible components, as indicated in Figure 16B. One is simply charge displacement from the membrane surfaces in response to the change in potential (i) another is

Gating currents

[21]Armstrong, C. M. 1981. *Physiol. Rev. 61:* 644–683.
[22]Armstrong, C. M. and Bezanilla, F. 1974. *J. Gen. Physiol. 63:* 533–552.

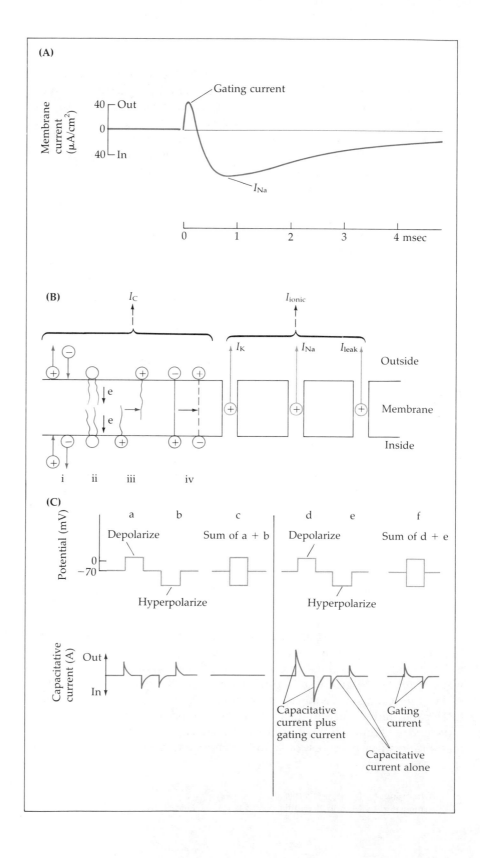

electron displacement within the membrane (ii), and finally movements of charged molecules (iii) or dipole reorientation (iv) may occur. Some or all of these currents may vary linearly and symmetrically with voltage steps of either polarity. Those associated with sodium channel activation, however, should appear upon depolarization from a holding potential of, say, -70 mV, but not upon hyperpolarization. In other words, if the channels are already closed, there should be no gating current upon further hyperpolarization. Similarly, gating currents associated with channel *closing* might be expected at the "off" of a brief depolarizing pulse but not after a hyperpolarizing pulse. One experimental way of recording gating currents, then, is to sum the currents produced by two identical voltage steps of opposite polarity, as shown in Figure 16C. If there were no asymmetrical currents, the result of such a summation would be as shown in a, b, and c of the figure: The current at the start of the depolarizing pulse is exactly equal and opposite to that at the start of the hyperpolarizing pulse, and their sum is zero. Similarly, the sum of the currents is zero when the step is terminated. Asymmetries due to gating currents show up as in b, c, and d of the figure: Both the "on" and "off" of the depolarizing pulse are larger than the corresponding portions of the hyperpolarization because of the additional charge movement associated with gating of the sodium channel. In such experiments, ionic currents are often suppressed by replacing sodium and potassium in the external and internal solutions with impermeant cations, or by adding TTX and TEA to the solutions. The effect of TTX is of interest: It appears to eliminate voltage-sensitive sodium conductance by physically blocking the mouth of the channel, not by interfering with the gating process.

The evidence that the asymmetry currents produced in the manner just described are, in fact, associated with sodium channel activation has been summarized by Armstrong.[21] First of all, the current observed at the onset of the pulse has kinetics similar to those of the opening of the sodium channels, and the current increases with the amplitude of the depolarizing step as expected. In addition, the maximum charge movement observed in squid axon represents about 2000 electronic charges/μm^2. When six monovalent charges are assigned to each chan-

16 GATING CURRENT IN SQUID AXON. (A) Brief outward gating current precedes inward sodium current. Gating current is a small component of the total capacitative current produced by depolarizing the axon. (B) In response to a depolarizing step, charge is displaced from the surface of the membrane (i); within the membrane, electrons are redistributed (ii); and charged (iii) or bipolar (iv) molecules become reoriented. To separate charge movements associated with gating from the rest of the capacitative currents, the procedure shown in (C) is used. Symmetrical depolarizing (a) and hyperpolarizing (b) pulses are applied. For a perfect capacitance, the sum of the resulting currents should be zero (c). If, however, a gating current is produced by depolarization (d) but not by hyperpolarization (e), then such a current should appear when the currents produced by the two pulses are summed (f). (After Armstrong and Bezanilla, 1974.)

nel, a channel density of $333/\mu m^2$ is calculated, a value that is in fair agreement with the channel density estimated by other means, such as binding of TTX (discussed later). This kind of evidence, then, indicates that the magnitude and time course of the asymmetry currents are consistent with those expected of gating currents. In addition, when the sodium current and the gating current are eliminated or attenuated by a number of other experimental procedures, subsequent recovery of the two processes is concurrent. For example, perfusion with internal zinc removes both sodium currents and gating currents reversibly and both recover with the same time course.

One point of particular interest is that during sodium inactivation the gating currents associated with repolarization are attenuated, as shown in Figure 17. Thus, after a brief depolarization of 0.3 msec (A), the areas under the "on" and "off" gating currents are equal, a result indicating opening and closing of an equal number of gates. On the other hand, when repolarization is delayed for 5 msec (B), which is sufficient time for inactivation to develop, the gating current associated with repolarization is reduced considerably. After Pronase treatment (which removes inactivation), this effect is abolished (C). Experiments such as this suggest that inactivation consists of a "foot in the door" process that interferes with closing of the m-gates, as indicated schematically in Figure 15.

Density of sodium channels and single channel conductance

Given this schematic model of the sodium channel, a number of additional questions can be asked experimentally; for example, What is the conductance of an individual open channel? How many channels

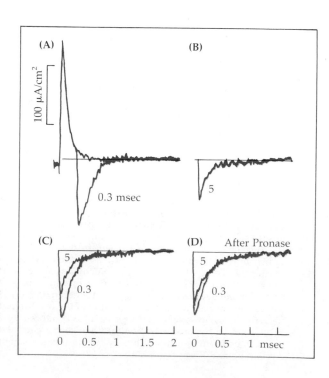

17

EFFECT OF INACTIVATION ON "OFF" GAT-ING CURRENT. (A) Depolarizing pulse pro-duces "on" gating current (upward deflection); repolarization results in "off" current with same area. If repolarization is delayed for 5 msec, "off" current is reduced (B). "Off" currents are superimposed in (C). After removal of inactiva-tion with Pronase (D), "off" currents are similar. (From Armstrong, 1981.)

are there in a unit area of membrane? The last question has been answered in a number of tissues simply by measuring the density of TTX binding sites. Using tritiated tetrodotoxin, Levinson and Meves[23] estimated that in squid axon an average of 553 molecules were bound to each square micrometer of membrane. Values in other tissues have been found to range from a low of $35/\mu m^2$ in garfish olfactory nerve to $12,000/\mu m^2$ at the node of Ranvier in rabbit sciatic nerve.[24] Knowing the sodium channel density, one can easily calculate the conductance of a single channel. For example, Hodgkin and Huxley's results indicated that the voltage-dependent sodium conductance had a maximum value of the order of 120 mS/cm^2, or 1200 pS/μm^2 at 20°C. Dividing this by 553 channels/μm^2 gives a single channel conductance of 2.2 pS. Similar calculations for frog muscle indicate a single channel conductance of 8.6 pS at 13°C.[25,26] One further point of interest with respect to frog muscle is that sodium channels appear to be distributed over the sarcolemma in clusters rather than uniformly. This distribution has been demonstrated by Almers and his colleagues,[27] who found that when small patches of the membrane were depolarized with a focal extracellular pipette, inward sodium currents varied markedly from one patch to the next, a finding indicating an underlying variation in density of sodium channels.

Other techniques have been used to estimate the conductance and density of single sodium channels. One of these is noise analysis, an experimental technique that will be described in detail later in relation to ion channels at synapses (Chapter 10). Briefly, when sodium channels are activated by depolarization, the number of open channels fluctuates in a statistical manner. This causes a fluctuation in the membrane current or, in other words, "current noise." Analysis of the magnitude and frequency composition of this noise then provides information about the individual channel characteristics. In squid axon,[28] the sodium channel conductance estimated in this way was 4 pS and the channel density $330/\mu m^2$. At frog nodes of Ranvier,[29] corresponding measurements resulted in an estimated single channel conductance of about 6 pS and a channel density of 10^5/node or $2000/\mu m^2$.

More direct measures of channel conductance are possible by other experimental procedures. In particular, the technique of patch clamping has been used to measure currents through a variety of individual membrane channels, including sodium channels. The technique was developed by Neher and Sakmann and their colleagues[30] to record from

Single channel conductance measured by patch clamping

[23]Levinson, S. R. and Meves, H. 1975. *Phil. Trans. R. Soc. Lond. B 270:* 349–352.
[24]Ritchie, J. M. and Rogart, R. B. 1977. *Rev. Physiol. Biochem. Pharmacol. 79:* 1–50.
[25]Almers, W. and Levinson, S. R. 1975. *J. Physiol. 247:* 483–509.
[26]Hille, B. and Campbell, D. T. 1976. *J. Gen. Physiol. 67:* 265–293.
[27]Almers, W., Stanfield, P. and Stuhmer, W. 1983. *J. Physiol. 336:* 261–284.
[28]Conti, F., De Felice, L. J. and Wanke, E. 1975. *J. Physiol. 248:* 45–82.
[29]Conti, F., Hille, B., Neumcke, B., Nonner, W. and Stämpfli, R. 1976. *J. Physiol. 262:* 729–742.
[30]Hamill, O. P., Marty, A., Neher, F., Sakmann, B. and Sigworth, F. J. 1981. *Pflügers Arch. 391:* 85–100.

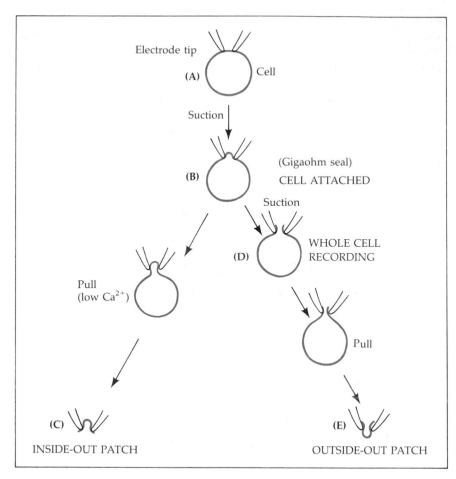

Electrode tip

(A) Cell

Suction

(B) (Gigaohm seal)
CELL ATTACHED

Suction

(D) WHOLE CELL
RECORDING

Pull
(low Ca^{2+})

Pull

(C) INSIDE-OUT PATCH

(E) OUTSIDE-OUT PATCH

18 **PATCH CLAMP RECORDING** configurations, represented schematically. **The electrode forms a seal on contact with the cell membrane (A),**
which is converted to a gigaohm seal by gentle suction (B). Records
may then be made from the patch of membrane within the electrode tip
(cell-attached patch) or a cell-free, inside-out patch may be made by
pulling (C). Alternatively, membrane within the electrode tip may be
ruptured by further suction to obtain a whole cell recording (D) or, by
pulling, to obtain an outside-out patch (E). (After Hamill, Marty, Neher,
Sakmann, and Sigworth, 1981.)

small cells and from cell-free membrane patches. The four basic types
of recording possible with the technique are illustrated in Figure 18. A
small (0.5–3.0 μm inside diameter) micropipette with a heat-polished
tip is brought up to the surface of a cell and forms a low-resistance seal
to the membrane (A). Under appropriate conditions, slight suction applied to the pipette results in the formation of a "gigaohm seal" (B),
that is, a seal to the external surface of the membrane so tight that
electrically the resistance between the inside of the pipette and the
bathing solution is tens of gigaohms or greater (1 GΩ = 10^9 Ω). Once
such a seal is obtained, currents flowing through the cell-attached patch
of membrane can be recorded, or a cell-free, "inside-out" patch may be

formed by pulling the electrode away from the cell in low-calcium solution (C). Alternatively, the membrane inside the pipette can be broken by further suction and recordings made from the whole cell (D). Such recordings are equivalent to those made with an intracellular micropipette. Finally, an "outside-out," cell-free patch can be made by first rupturing the initial patch and then pulling the electrode away from the cell (E).

The patch clamp technique has been used to observe directly current through individual sodium channels,[31] using cell-attached patches of cultured rat muscle. Such patches typically contain only two or three active channels. The recording arrangement is shown in Figure 19A. The amplifier used to record from the patch (a current-to-voltage converter) has two essential properties: (1) The output is a direct measure of *current*, I_P, flowing through the patch; and (2) the potential of the outside of the patch membrane is the same as the command potential V_c. Thus, when V_c is zero, the membrane potential of the patch is the same as that of the rest of the cell. When V_c is stepped to, say, -40 mV, then the outer surface of the patch is made negative by this amount; that is, the patch is depolarized.

[31]Sigworth, F. J. and Neher, E. 1980. *Nature 287:* 447–449.

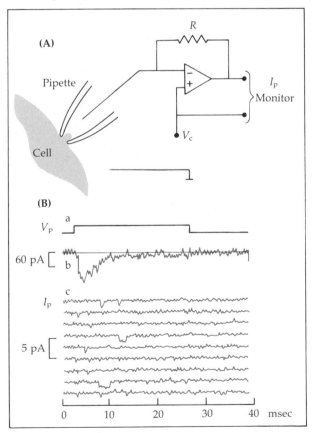

19

SODIUM CHANNEL CURRENTS recorded with a patch clamp. **(A) Recording arrangement. A patch clamp amplifier enables recording of small currents from the cell-attached patch and provides a method of supplying depolarizing pulses. (B) Successive voltage pulses applied to the patch with the waveform shown in (a) result in current pulses from individual channels (downward deflections) in the nine successive records shown in (c). The sum of 300 such records (b) shows that most channels open in the initial 1–2 msec, after which the probability of channel openings decays with the time constant of inactivation. (From Sigworth and Neher, 1980.)**

The results of one such experiment are shown in Figure 19B. The electrode was held at a command potential of +30 mV, so that the membrane patch was hyperpolarized to −100 mV from its normal resting potential of about −70 mV. A 40 mV depolarizing step was then applied to the electrode for about 23 msec (a). During this period, as shown in the nine individual sweeps in part (c), small pulses of inward current 1–2 pA in amplitude and of various durations could be seen, tending to occur most frequently near the onset of the pulse. These pulses of current are the result of opening and closing of single sodium channels. The mean channel current in the experiment shown was 1.6 pA and the mean channel open time was 0.7 msec. When one assumes the sodium equilibrium potential to be +30 mV, the driving potential is 90 mV; thus, the single channel conductance is about 18 pS. This value is somewhat larger than the value of 8.6 pS calculated for frog muscle, but the higher temperature in these experiments (about 22°C) accounts for much of the difference. When 300 individual sweeps were summed (part b), the total current reached a maximum in 1–2 msec and then declined as inactivation occurred. This result illustrates an important point about the channel behavior, namely, that the overall time course of the sodium current does *not* reflect the time course of the current through individual channels. Instead, the individual channel currents are all-or-nothing, and the magnitude of the summed current at any time after the step is determined by the number of channels open at that particular time. This number, in turn, is a reflection of the probability that an individual channel will be open. Depolarization, then, produces an increase in the probability of individual channel openings; as inactivation occurs, this probability is reduced.

Potassium and calcium channel characteristics

Rather less is known about the detailed structure and characteristics of the voltage-sensitive potassium channels. However, single channel conductance has been measured in squid axon both by noise analysis[28] and with the patch clamp.[32] Patch clamp experiments were particularly ingenious, as the squid axon membrane was patch clamped from the *inside*! Both methods gave a single channel conductance of about 10 pS. The estimated channel density was about $60/\mu m^2$. Thus, the potassium channels appear to have a somewhat higher conductance and lower density than the sodium channels. In frog muscle, potassium channels, like sodium channels, are distributed nonuniformly,[27] and the distributions of the two types of channel are not correlated. At nodes of Ranvier in rabbit, myelinated nerve depolarization produces no late outward current, and the potassium channels are therefore presumably absent.[33] During the action potential, repolarization is achieved by a large leak current after relatively rapid inactivation of the sodium channels.

Currents from single voltage-sensitive calcium channels have been

[32]Conti, F. and Neher, E. 1980. *Nature 285:* 140–143.

[33]Chiu, S. Y., Ritchie, J. M., Rogart, R. B. and Stagg, D. A. 1979. *J. Physiol. 292:* 149–166.

recorded in a number of cells, including bovine chromaffin cells,[34] snail ganglion cells,[35] and pituitary tumor cells.[36] In most such experiments, single channel currents were maximized by bathing the external surface of the membrane in isotonic barium, which passes through the channels. Under such conditions, channel conductances are approximately 7 pS. In normal solution (e.g., 5 mM calcium outside), conductances are very much lower, possibly near 0.05 pS.

SUGGESTED READING

Books and reviews

Armstrong, C. M. 1981. Sodium channels and gating currents. *Physiol. Rev. 61:* 644–683.

Hille, B. 1976. Ionic basis of resting and action potentials. In E. Kandel (ed.), *Handbook of the Nervous System,* Vol. I. American Physiological Society, Bethesda, MD, Chap. 3.

Hille, B. In press. *Ionic Channels of Excitable Membranes.* Sinauer, Sunderland, MA.

Hodgkin, A. L. 1964. *The Conduction of the Nervous Impulse.* Liverpool University Press, Liverpool.

Katz, B. 1966. *Nerve, Muscle and Synapse.* McGraw-Hill, New York, Chap. 5.

Latorre, R. and Miller, C. 1983. Conduction and selectivity in potassium channels. *J. Memb. Biol. 71:* 11–30.

Tsien, R. W. 1983. Calcium channels in excitable cells. *Annu. Rev. Physiol. 45:* 341–358.

Original papers

Adams, P. R., Brown, D. A. and Constanti, A. 1982. M-currents and other potassium currents in bullfrog sympathetic neurons. *J. Physiol. 330:* 537–572.

Frankenhaeuser, B. and Hodgkin, A. L. 1957. The action of calcium on the electrical properties of squid axons. *J. Physiol. 137:* 218–244.

Hamill, O. P., Marty, A., Neher, F., Sakmann, B. and Sigworth, F. J. 1981. Improved patch-clamp techniques for high-resolution current recording from cells and cell-free membrane patches. *Pflügers Arch. 391:* 85–100.

Hodgkin, A. L. and Huxley, A. F. 1952a. Currents carried by sodium and potassium ions through the membrane of the giant axon of *Loligo. J. Physiol. 116:* 449–472.

Hodgkin, A. L. and Huxley, A. F. 1952b. The components of the membrane conductance in the giant axon of *Loligo. J. Physiol. 116:* 473–496.

Hodgkin, A. L. and Huxley, A. F. 1952c. The dual effect of membrane potential on sodium conductance in the giant axon of *Loligo. J. Physiol. 116:* 497–506.

Hodgkin, A. L. and Huxley, A. F. 1952d. A quantitative description of membrane current and its application to conduction and excitation in nerve. *J. Physiol. 117:* 500–544.

[34]Fenwick, E. M., Marty, A. and Neher, E. 1982. *J. Physiol. 331:* 599–635.
[35]Lux, H. D. and Nagy, K. 1981. *Pflügers Arch. 391:* 252–254.
[36]Hagiwara, S. and Harunori, O. 1983. *J. Physiol. 336:* 649–661.

Hodgkin, A. L., Huxley, A. F. and Katz, B. 1952. Measurement of current–voltage relations in the membrane of the giant axon of *Loligo*. *J. Physiol. 116:* 424–448.

Meech, R. W. 1974. The sensitivity of *Helix aspersa* neurones to injected calcium ions. *J. Physiol. 237:* 259–277.

A. L. Hodgkin, 1949

A. F. Huxley, 1974

CHAPTER SEVEN

NEURONS
AS CONDUCTORS
OF ELECTRICITY

Impulses propagate along axons by the longitudinal spread of current. As each region of the membrane generates an all-or-nothing action potential, it depolarizes and excites the adjacent not-yet-active region and gives rise to a new regenerative impulse. For an understanding of impulse propagation, as well as of synaptic transmission and integration, one must know how an electric current spreads passively along a nerve.

As current spreads along a nerve fiber, it becomes attenuated with distance. This attenuation depends on a number of factors, principally the diameter and membrane properties of the fiber. A longitudinal current spreads further along a fiber with large diameter and high membrane resistance. The electrical capacitance of the membrane influences the time course of the electrical signals and usually their longitudinal distribution as well. Thus, to estimate how far a subthreshold potential change will spread, one needs to know the geometry and membrane characteristics of the nerve and, in addition, the waveform of the potential change.

In many vertebrate nerve cells, segments of the axon are covered by a high-resistance, low-capacitance myelin sheath. This sheath acts as an effective insulator and forces currents associated with the nerve impulse to flow through the membrane at intervals where the sheath is interrupted (nodes of Ranvier). The impulse jumps from one such node to the next, and thereby its conduction velocity is increased. Such myelinated nerves occur in pathways in the nervous system where speed of conduction is important.

The permeability properties of nerve cell membranes and the way in which these properties behave to produce regenerative electrical responses have been discussed in the preceding chapters. In this chapter we will describe in more detail how currents spread along nerve fibers to produce local graded potentials. The "passive" electrical properties of nerves that underlie such current spread are essential for signaling in the nervous system. At sensory end organs they are the link between the stimulus and the production of impulses; along axons

Passive electrical properties of nerve and muscle membranes

167

they allow the impulse to spread and propagate; at synapses they enable the postsynaptic neuron to add and subtract synaptic potentials that arise in it from numerous converging inputs. The discussion that follows will deal primarily with spread of current along nerve fibers of uniform diameter, that is, along cylindrical conductors. The same principles apply to more complex systems such as tapered dendrites and to the convergence of currents from a number of such dendrites into a cell body.

A cylindrical nerve fiber has the same formal components as an undersea cable, namely, a central or core conductor and an insulating sheath surrounded by a conducting medium. However, as already noted in Chapter 4, the two systems are quantitatively quite dissimilar. In a cable, the core conductor is usually copper, which has a very high conductance, and the surrounding sheath is neoprene, plastic, or some other material of very high resistance. In addition, the sheath is usually relatively thick, so it has a very low capacitance. Voltage applied to one end of such a cable will spread a very long distance because the resistance to longitudinal current flow along the copper conductor is relatively low and virtually no current is lost through the insulating sheath. In a nerve fiber, on the other hand, the core conductor is a salt solution similar in concentration to that bathing the nerve and (compared with copper) is a poor conductor. Furthermore, the plasma membrane of the fiber is a relatively poor insulator and, being thin, has a relatively high capacitance. A voltage signal applied to one end of a nerve fiber, then, will fail to spread very far for two reasons: (1) The core material has a low conductance so that resistance to current flow down the fiber is high; (2) current that starts off flowing down the axoplasm is progressively lost along the fiber by lateral current flow through the poorly insulating plasma membrane. The analysis of current flow in cables was developed by Lord Kelvin for application to transatlantic telephone transmission and refined by Oliver Heaviside in the late nineteenth century. Heaviside was the first to consider the effect of resistive leak through the insulation, equivalent to membrane resistance in nerve, and he made many other contributions to cable theory, including the concept of what he called impedance (it was, incidentally, Heaviside and not Maxwell, who formulated Maxwell's equations in their modern form). Cable theory was first applied in detail to nerve fibers by Hodgkin and Rushton,[1] who used extracellular electrodes to measure the spread of applied current along lobster axons. Later on, intracellular electrodes were used in a number of nerve and muscle fibers for similar studies.

The factors that determine the spread of current and potential along a fiber are demonstrated by the experiment illustrated in Figure 1A. A microelectrode is placed in the middle of a large fiber (for example, a lobster axon), and a small amount of positive current is passed into the axoplasm. As shown by the arrows, some of the current will immediately flow outward across the membrane adjacent to the electrode; the

[1]Hodgkin, A. L. and Rushton, W. A. H. 1946. *Proc. R. Soc. Lond. B 133*: 444–479

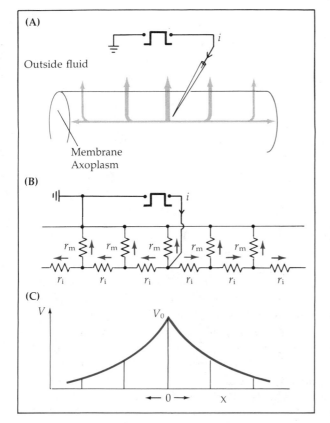

1

PATHWAYS FOR CURRENT FLOW in an axon.
(A) Current flow across the surface membrane
produced by injection of current from a micro-
electrode. Thickness of arrows indicates the cur-
rent density at various distances from the point
of injection. (B) An equivalent electrical circuit,
assuming zero resistance in the external fluid and
ignoring membrane capacitance; r_i, longitudinal
resistance of axoplasm per unit length; r_m, mem-
brane resistance of a unit length. (C) Decrement
of potential (V) along the axon from its amplitude
at the point of current injection (V_0).

remainder will spread laterally for various distances along the axon
before leaving the cell. The relative amounts of current flowing across
the membrane are indicated roughly by the thickness of the arrows.
The *potential change* produced across the membrane at a given distance
from the electrode will be proportional to the current flow across the
membrane at that point, in accordance with Ohm's law, provided the
potential changes themselves are well below threshold, that is, provided
there are no changes in sodium and potassium permeability. Figure 1B
shows an equivalent circuit for an axon in which only the resistive
components are considered; the membrane capacitance is, for now,
ignored. The circuit is obtained by imagining that the axon is cut along
its length into a series of rings. The transverse resistance r_m then represents the resistance outward across such a ring of membrane; the
longitudinal resistance r_i represents the internal resistance along the
axoplasm from the midpoint of one ring to the midpoint of the next.
Because nerves are normally bathed in a large volume of fluid, the
extracellular longitudinal resistance from one ring to the next is represented as being zero. Any length can be selected for the rings themselves; however, the resistances r_m and r_i usually refer to a 1-cm length
of axon. Thus, the dimensions of r_m are Ωcm and those of r_i are Ω/cm.
The dimensions for r_m seem strange until one realizes that this resistance

decreases as the length of axon under consideration increases (more channels are available for current to leak through the membrane). Thus, the resistance in ohms of a given length of axon membrane is the resistance of a 1-cm length (r_m, in Ωcm) *divided* by the length (in centimeters).

Suppose now that a current is passed into such a resistive network at a point $x = 0$, as indicated in Figure 1B. This current will produce a potential, V_0, at the point of injection. On either side the potential will be smaller. When the potential in successive segments is plotted as in Figure 1C, it is seen to fall exponentially with distance. In other words, the relation between the potential (V_x) and the distance (x) from the point of injection is given by

$$V_x = V_0 e^{-x/\lambda}$$

The shape of the curve, then, is determined by two parameters, V_0 and λ. Both depend on r_m and r_i. λ is known as the SPACE CONSTANT of the fiber and is given by $\lambda = (r_m/r_i)^{1/2}$. It is the distance at which the potential has fallen to $1/e$ of its maximum value. Note that λ has the dimensions of centimeters as required. In addition, the expression fulfills the intuitive expectation that the distance over which the potential change will spread should increase with increasing membrane resistance and decrease with increasing internal resistance. The peak potential change V_0 is, of course, proportional to the magnitude of the injected current. The constant of proportionality is known as the INPUT RESISTANCE of the fiber, r_{input}. In other words, if the amount of current injected is given by i, then $V_0 = ir_{input}$. The input resistance is related to the transverse membrane resistance and to the internal longitudinal resistance by $r_{input} = 0.5(r_m r_i)^{1/2}$. Again the expression has the required dimensions (Ω) and tells us that the input resistance increases with both membrane resistance and internal resistance. The factor 0.5 arises because the cable extends both ways from the point of current injection; each half has an input resistance $(r_m r_i)^{1/2}$. Note that the input resistance has the same value as the resistance of a cylinder of membrane extending a distance λ in both directions from the point of current injection (i.e., $r_{input} = r_m/2\lambda$).

The spread of current in a cable is analogous to the spread of heat along a metal rod surrounded by insulation and immersed in a conducting material (such as water). When one end of the rod is heated, heat is conducted along the rod and, at the same time, is lost to the surrounding medium. At progressively greater distances from the heated end, the temperature becomes less and less; similarly, the heat loss decreases progressively with distance as the rod itself becomes less hot. The fall of temperature with distance is exponential, like the fall of voltage with distance in a nerve. Assuming the surrounding medium is a good heat conductor (just as the medium surrounding a nerve is a good electrical conductor), then the distance the heat spreads depends primarily on (1) the conductivity of the rod, (2) its diameter, and (3) the effectiveness of the insulation in preventing heat loss.

For practical purposes the cable parameters r_{input} and λ characterize the resistive properties of a given nerve fiber: Once calculated, they predict how much potential change will be produced by injection of a given amount of current and how far the potential will spread along the fiber. Conversely, r_{input} and λ can be measured experimentally and used to calculate r_m and r_i. This can be done, as in Figure 1, by injecting a known amount of current into the fiber and measuring the resulting potential change at various positions along it. Using the relations already given, it is easy to show that $r_m = r_{input}\lambda$ and $r_i = r_{input}/\lambda$. It is clear, however, that these parameters depend on the fiber diameter: Both the longitudinal resistance and the membrane resistance per unit length will be larger in a small fiber than in a large one. Consequently, r_m and r_i do not provide a direct measure of the SPECIFIC RESISTANCES of the membrane and the axoplasm. The specific resistances, R_m and R_i, are measures of the conducting properties of the cell membrane and the axoplasm and are independent of fiber geometry. Knowledge of specific membrane resistance is important, therefore, if one wishes to compare one nerve cell membrane with another.

The resistance (r) of any segment of conducting material is given by the relation $r = \rho L/A$, where L is the length of the segment and A is its cross-sectional area. ρ is the SPECIFIC RESISTANCE of the material, which is the resistance of a volume 1 cm in length and 1 cm^2 in cross-sectional area; it has the dimensions Ωcm. For a unit length of axon, the longitudinal resistance is $r_i = R_i/\pi a^2$, where R_i is the specific resistance of the axoplasm and a is the fiber radius. If r_i and a have been measured experimentally, the specific resistance of the axoplasm can be calculated from the inverse relation: $R_i = \pi r_i a^2$. For squid nerve R_i is about 30 Ωcm, or about 10^7 greater than copper. For mammals, in which the ionic strength of the cytoplasm is lower, estimated values of the specific internal resistance are higher (about 125 Ωcm); for frogs, with still lower ionic strength and lower temperature than mammals, the specific internal resistance is further increased (about 250 Ωcm). Neither the specific resistance of the membrane material nor the membrane thickness is usually measurable with any accuracy. Consequently, the resistive property of the membrane is represented by the single parameter R_m, which is the resistance of a unit area of membrane and has dimensions Ωcm^2. The membrane enclosing a unit length of axon has a transverse resistance that can be calculated by dividing the specific membrane resistance by the membrane area, that is, $r_m = R_m/2\pi a$. Again, R_m can be calculated from the inverse expression $R_m = 2\pi a r_m$. R_m is inversely related to the resting conductance of the membrane in question, which in turn is related primarily to the resting permeability to potassium and chloride; these vary considerably from one cell to the next. The average value for R_m reported by Hodgkin and Rushton for lobster axon was about 2000 Ωcm^2; in other preparations measurements range from less than 1000 Ωcm^2 for membranes with a large number of channels through which ions can leak to more than 50,000 Ωcm^2 for membranes with relatively few such channels.

Given a specific resistance R_i for the axoplasm and a specific membrane resistance R_m, how are the cable parameters r_{in} and λ influenced by fiber diameter? The answer can be obtained quantitatively from the relations presented in the preceding paragraphs. First, one calculates input resistance as

$$r_{input} = 0.5(r_m r_i)^{1/2} = 0.5(R_m R_i / 2\pi^2 a^3)^{1/2}$$

Thus, the input resistance increases as fiber radius decreases; specifically the resistance varies inversely with the 3/2 power of the radius. The space constant is given by

$$\lambda = (r_m / r_i)^{1/2} = (aR_m / 2R_i)^{1/2}$$

Other properties being equal, then, λ increases with the square root of the fiber radius. A squid axon of 1-mm diameter with a specific internal resistance of 30 Ωcm and a specific membrane resistance of 2000 Ωcm^2 would have a space constant of almost 13 mm. The space constant of a frog muscle fiber with the same specific membrane resistance and a diameter of 100 μm would be 1.5 mm and that of a 1-μm diameter mammalian nerve fiber only 0.2 mm. In summary, once the cable parameters r_{in} and λ are known, the steady-state distribution of potential produced by injection of current into the cable as shown in Figure 1 can be described quantitatively.

Qualitatively, the current flow in a cable can be described in ionic terms: Positive charges flowing into the axoplasm from the tip of the microelectrode repel other cations and attract anions. By far the most abundant intracellular ion is potassium, which, therefore, carries most of the current away from the electrode. Positive charge accumulates at the membrane and spreads laterally along the axon; as it spreads some is lost by ion movement through the membrane as well, carried by potassium or chloride. In reality, no one ion migrates very far through the axoplasm; the displacement of ions resembles more closely collisions along a series of billiard balls. The extent to which the potential spreads depends on the membrane resistance relative to the internal longitudinal resistance. In a cell with low membrane resistance, the potassium and/or chloride conductance is large, and charge can leak across the membrane without spreading very far. In contrast, in a cell with high membrane resistance, a greater portion of the charge is displaced laterally before leaking out. Along the nerve, less and less charge accumulates so that the quantity leaving each adjacent ring of membrane decreases in an exponential manner with distance from the electrode. In a small fiber the resistance to lateral current flow relative to the resistance across the membrane is higher than in a large fiber, so that the decrease in potential with distance is steeper.

Physical considerations indicate that the cell membrane should have a relatively large electrical capacitance. A capacitor consists of two conducting sheets or plates separated by a layer of insulating material; usually the conducting sheets are metallic foil and the insulator is mica or a plastic such as Mylar. The closer together the plates are, the greater their ability to separate and store charge. In the case of a nerve cell, the "plates" are the conducting fluids on either side of the membrane and the insulating material is the lipoprotein of the membrane itself. Because the membrane is only about 7 nm thick, it is capable of storing a relatively large amount of charge. As noted in Chapter 5, the capacitance (C) of a capacitor is defined by how much charge (Q) it will accumulate for each volt of potential (V) applied to it, that is, $C = Q/V$. C has the units coulombs/volt or farads (F). Typically nerve cell membranes have a capacitance of the order of 1 $\mu F/cm^2$. When we turn the equation around, the charge stored in a capacitor is given by $Q = CV$. Thus, when a cell has a resting potential of 80 mV, the amount of charge separated by the membrane will be $(1 \times 10^{-6}) \times (80 \times 10^{-3}) = 0.08$ μcoulombs/cm^2. The *current* flowing into or out of a capacitor can be deduced from the relation between charge and voltage by remembering that current is the rate of change of charge with time, that is, that *amperes = coulombs per second*. It follows that

$$i = dQ/dt = C(dV/dt)$$

The relations between current and voltage in circuits with resistors and capacitors "in parallel" are illustrated in Figure 2. When a rectangular current pulse of amplitude i is applied to a resistor (R), it simply produces a voltage pulse across the resistor of amplitude $V = iR$ (Figure 2A). When the same pulse is applied to a capacitor (C), the voltage on the capacitor builds up as it accumulates charge. Because the current is constant (or, in other words, the rate of delivery of charge is constant), the voltage will build up at a constant rate given by $dV/dt = i/C$ (Figure 2B). When the two elements R and C are combined in parallel (Figure 2C), the current initially flows entirely into the capacitor, charging it at a rate i/C. However, the instant any voltage develops across the capacitor, current starts to flow through the resistor as well. Because some of the current is now being diverted from the capacitor, the rate at which it is charged will decrease. Eventually all of the applied current flows through the resistor, producing a potential $V = iR$, and the capacitor is fully charged to this potential. In other words, the capacitor initially shorts out the resistor; the short then becomes less and less as the capacitor is charged closer and closer to the final voltage. When the current pulse is terminated, the charge stored in the capacitor leaks off through the resistor and the voltage falls to zero.

The rise and the fall of the potential in Figure 2C are described by exponential functions. If we take $t = 0$ at the beginning of the pulse, the rising phase during the pulse is described by

$$V = iR(1 - e^{-t/\tau})$$

The exponential time constant τ is given by the product of the resistance and capacitance in the circuit; that is, $\tau = RC$. τ is the time for the potential to rise to a fraction $(1 - 1/e)$, or 63 percent, of its final value. At the termination of the current pulse, the applied current is, of course, zero. The capacitor then discharges through the resistor until the voltage

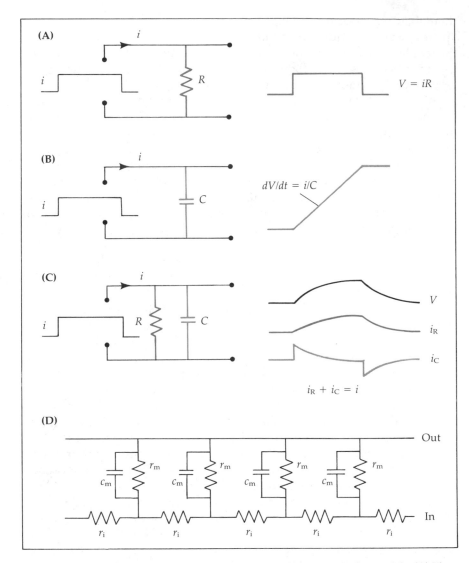

2 **EFFECT OF CAPACITANCE ON TIME COURSE** of potentials. **(A)** Time course of potential (V) resulting from current (i) in a purely resistive network. Voltage is proportional to, and has the same time course as, the applied current. **(B)** In a purely capacitative network the *rate of change* of voltage is proportional to the applied current. **(C)** In a combined RC network, the initial surge of current is through the capacitor (i_C); finally all the current flows through the resistor (i_R). Voltage rises to final value iR exponentially with a time constant $\tau = RC$. At termination of the current pulse, the capacitance discharges through the resistance with the same time constant, and i_C and i_R are equal and opposite. **(D)** Electrical model of a cable, as in Figure 1, but with membrane capacitance per unit length (c_m) added.

returns to zero as well. The fall in voltage is again exponential with the same time constant τ. Just as the voltage rises and falls exponentially, so must the resistive current, i_R. On the rising phase, then, the resistive current starts at zero and rises exponentially toward its final value i. Conversely, the capacitative current starts at i and falls with the same time constant. After the termination of the pulse, because external current is no longer being applied, the only current flowing across the resistor is that flowing out of the capacitor as it is being discharged. Consequently, the two currents must be equal and opposite as shown.

The circuit just described, with a resistor and capacitor in parallel, can be used to represent a spherical nerve cell without dendrites and with an axon so small that it makes only a negligible contribution to the electrical properties of the cell. In the equivalent circuit for an axon or muscle fiber, however, the membrane capacitance must be distributed along the length of the fiber in company with the elements representing the membrane resistance, as shown in Figure 2D. The membrane capacitance per unit length, c_m ($\mu F/cm$), is related to the specific capacitance per unit area, C_m ($\mu F/cm^2$) by $c_m = 2\pi a C_m$, where a is the fiber radius. The membrane time constant is independent of fiber radius as

$$\tau = r_m c_m = (R_m/2\pi a)(2\pi a C_m) = R_m C_m$$

The time constant, then, is the third parameter specifying the behavior of the cable, the other two being the input resistance and the space constant. Time constants in nerve and muscle cells range from 1 to 20 msec.

How does the time constant affect current flow in a cable? As with the simple RC circuit, the rise and fall of the potential change produced by a rectangular current pulse are delayed by the presence of the distributed capacitance. The effects are more complicated, however, because current no longer flows into a single capacitor; instead, each segment of the circuit with its capacitative and resistive elements interferes with the others. Because of this interaction, the rising and falling phases of the potential changes are not exponential and, in addition, the growth and decline of the potentials become increasingly prolonged as records are made farther and farther from the point of current injection (Figure 3). As the waveform of the electrotonic potential is not exponential, the membrane time constant, $\tau = r_m c_m$, cannot be obtained by measuring the time to rise to 63 percent of its final value (see Box 1).

A further consequence of the membrane capacitance is that brief signals do not spread as far as signals of long duration. For sufficiently long pulses, the spread of the steady-state potential along the axon is unaffected by capacitance, that is, $V_x = V_0 e^{-x/\lambda}$ as before. However, for brief events, such as synaptic potentials, the current flow giving rise to the signal may be over before the membrane capacitances become fully charged. This has the effect of reducing the spread of the potential along the fiber. In other words, for brief signals the effective space constant is less than for longer duration ones. In addition, such signals are distorted as they spread along the fiber, the peaks becoming more rounded and occurring progressively later with increasing distance.

Again, the effect of membrane capacitance can be explained in terms of ionic movements. When positive charges are injected into an axon,

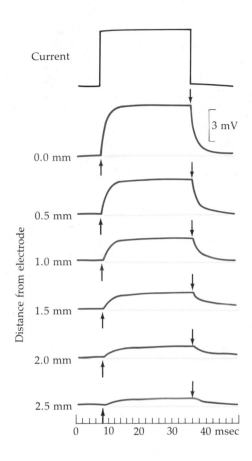

Current

3 mV

Distance from electrode

0.0 mm

0.5 mm

1.0 mm

1.5 mm

2.0 mm

2.5 mm

0   10   20   30   40 msec

3

SPREAD OF POTENTIAL along a lobster axon, recorded with a surface electrode. A rectangular current pulse is applied at 0 mm, producing a large electrotonic potential. With increasing distance from the site of current injection, the rise time of the potential change is slowed and the height of the plateau attenuated. (After Hodgkin and Rushton, 1946.)

potassium ions on the inside move radially toward the membrane and spread laterally. As the ions accumulate, they recharge the membrane capacitance. Such ionic movements take time and constitute a capacitative current that does not pass through the membrane but alters the membrane charge distribution. As the capacitance is charging, the resulting change in potential also causes current flow through the membrane, carried primarily by potassium and chloride ions. Eventually the membrane potential reaches its steady-state distribution with the distributed capacitances fully charged and a constant ionic current through the membrane. The time required to reach the steady state is determined by the membrane time constant.

It was mentioned previously that the change in temperature along a metal rod heated at one point is a good analogy for the longitudinal spread of potential along a fiber; it also provides a good analogy for the time course of the potential change. The temperature of a metal rod rises more slowly at a distance from the heated end because of the capacity of the metal to store heat. Thus, each segment must be heated up before passing heat to the next. In the nerve, the capacitance is in the membrane whereas in the rod it is within the metal; nevertheless the effects and the equations used to describe them are the same.

BOX 1 ELECTROTONIC POTENTIALS AND MEMBRANE TIME CONSTANT

The electrotonic potentials shown in Figure 3, recorded at various distances along an axon from the point of current injection, do not rise and fall exponentially. Instead their waveforms are described by complicated functions of both time and distance. These functions rise more rapidly than exponentials near the current electrode and more slowly farther away. Consequently, the membrane time constant, $\tau = RC$, cannot be obtained by measuring the time for an electrotonic potential to rise to 63 percent of its final value. At the point of current injection, the potential, in fact, rises to 84 percent of its maximum amplitude in one time constant. At a distance of two space constants from the current electrode, on the other hand, it reaches only 37 percent of its final value in the same time.

How, then, does one measure τ in a cable? To do this, it is necessary to know the separation of the current-passing and voltage-recording electrodes as a fraction of the space constant λ. One then consults a table of values to find out how far the electrotonic potential will rise in an interval equivalent to one time constant at that particular electrode separation. An abbreviated version of such a table (modified from Hodgkin and Rushton, 1946) is presented here. The numbers give the amplitude reached by an electrotonic potential at a time τ after the onset of the current pulse (V_τ), as a fraction of the final steady-state amplitude (V_∞), for various electrode separations. The distance between the electrodes (d) is expressed as fractions or multiples of one space constant (λ).

Separation (d/λ)	0	0.2	0.4	0.6	0.8	1.0	1.5	2.0
Amplitude (V_τ/V_∞)	0.84	0.81	0.77	0.73	0.68	0.63	0.50	0.37

Conduction of impulses in a nerve is influenced considerably by the nerve's cable properties (r_{input}, λ, and τ). The effects of these passive properties on conduction are important because conduction velocity plays a significant role in the scheme of organization of the nervous system. It varies by a factor of more than 100 in nerve fibers that transmit messages of different information content. In general, nerves that conduct most rapidly (more than 100 m/sec) are involved in mediating rapid reflexes, such as those used for regulating posture. Slower conduction velocities are associated with less urgent tasks such as regulating the distribution of blood flow to various parts of the body, controlling the secretion of glands, or regulating the tone of visceral organs.

Propagation of action potentials

To illustrate the factors involved in impulse propagation, the action potential can be "frozen" at an instant in time and its spatial distribution along the axon plotted as shown in Figure 4. The distance occupied instantaneously by an action potential depends on its duration and conduction velocity. For example, if the action potential duration is 2 msec and if it is conducted at 10 m/sec, then the potential will be spread over a 2 cm length of axon. At the peak of the action potential, there is a transient reversal of the membrane potential, the inside being charged positively with respect to the outside. Along the leading edge of the action potential, where the membrane is depolarized above threshold, there is a rapid influx of sodium ions, carrying positive charge into the cell. This charge spreads ahead of the active region,

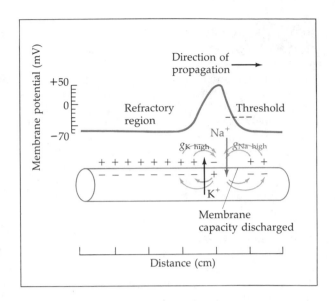

4

CURRENT FLOW DURING A NERVE IMPULSE
at an instant in time. Influx of positive current
in the active region spreads ahead of the impulse
to depolarize the membrane toward threshold
and produce an outward current. An outward
potassium current from the depolarized segment
behind the active region causes rapid repolarization.

depolarizing the adjacent membrane toward threshold. Behind the peak
of the action potential, on its falling phase, the conductance to potassium is high and there is outward potassium movement in the depolarized region, restoring the membrane to its resting level. Thus, the
two outward currents have very different effects: The current spread
ahead of the action potential discharges the membrane capacitance,
causing depolarization of the membrane and outward current; behind
the active region, outward potassium current from the depolarized
membrane is responsible for repolarization.

When an action potential is set up in the middle of an axon, it
propagates in both directions from the point of excitation. Normally
this does not happen in the body, and impulses move in one direction
only. The action potential cannot double back on itself, reversing the
direction of propagation, because of the refractory period. In the refractory region, indicated in Figure 4, the sodium conductance is still
inactivated and the potassium conductance is high so that a backward-
conducting regenerative response cannot be initiated.

The conduction velocity of the action potential depends largely on
the rate at which the membrane capacitance ahead of the active region
is discharged to threshold by the spread of positive charge. This, in
turn, depends on the amount of current generated in the active region
and on the cable properties of the fiber, in particular on the membrane
capacitance (c_m) and the internal longitudinal resistance through which
the current must flow to discharge the capacitance (r_i). Because of the
dependence on the longitudinal resistance of the axon core, conduction
velocity is higher in large fibers than in small fibers. In theory, other
factors being equal, the velocity of propagation should vary directly
with the square root of the fiber diameter.[2]

[2]Hodgkin, A. L. 1954. *J. Physiol. 125*: 221–224.

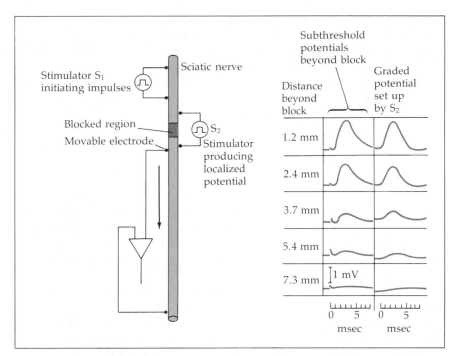

5 ROLE OF CURRENT FLOW in the propagation of impulses. Impulses in a sciatic nerve generate currents that spread across a blocked region, where they cause a depolarization that declines with distance. When a small current is passed through electrode S_2, the resulting subthreshold potential declines with distance in an identical manner. (After Hodgkin, 1937.)

The idea that action potential propagation depended on flow of current through the passive membrane elements ahead of the active region (i.e., in "local circuits") was confirmed by Hodgkin[3] in 1937. In his first series of experiments on the sciatic nerve of the frog, he blocked conduction by applying pressure or by cooling a short segment of the nerve and showed that impulses did not spread through the blocked region. However, when impulses arrived at the block, current that spread through the blocked area produced subthreshold depolarization extending several millimeters beyond the block. This result is illustrated in Figure 5, which shows records taken at five different positions beyond the blocked region. The decay of the depolarization with distance was identical with that of a subthreshold potential produced by passing current from a second stimulating electrode (S_2) into the region distal to the block. In other words, current spread from the action potential through the blocked region could be mimicked by extrinsic current injected from an electrode. Hodgkin also showed that excitability at various points beyond the blocked region changed in parallel with the spread of potential produced either by a blocked action potential or by

Evidence of involvement of local circuits in the conduction of action potentials

[3]Hodgkin, A. L. 1937. *J. Physiol. 90*: 183–210, 211–232.

extrinsic current pulses. Thus, the amount of current needed for excitation during the spread of depolarization in the axon distal to the block was less than normal near the blocked region and increased with distance as the subthreshold potentials decayed.

In another series of experiments,[4] local circuit theory was tested further by showing that alterations in the external longitudinal resistance affect conduction velocity in the expected way. A large axon from crab was stimulated at one end and the arrival of the action potential recorded at the other. The experiment was done first in seawater, in which the external resistance (r_o) was low; then a high external resistance was produced by replacing the bathing solution with mineral oil, leaving only a thin film of seawater around the axon. As expected, the increase in external resistance resulted in a reduction in conduction velocity. Finally, in the converse experiment, the external resistance was reduced below that in seawater by laying the nerve along a series of silver plates. When the plates were shorted together, thereby decreasing the external resistance, the conduction velocity increased.

Myelinated nerves and saltatory conduction

In the vertebrate nervous system, the larger nerve fibers are myelinated. In peripheral nerve, myelin is formed by Schwann cells, which during development wrap themselves tightly around axons. With each wrap the Schwann cell cytoplasm between the membrane pair is squeezed out so that the final result is a spiral of tightly packed membranes. The number of wrappings (lamellae) ranges from a minimum of between 10 and 20 to a maximum of about 160.[5] A wrapping of 160 lamellae means that there are more than 300 membranes in series between the plasma membrane of the axon and the extracellular fluid. Thus, the effective membrane resistance is increased by a factor of more than 300 and the membrane capacitance reduced by a similar factor. In terms of dimensions, the myelin occupies 20–40 percent of the diameter of the fiber; that is, the diameter of the axon is 60–80 percent of the overall fiber diameter. The myelin sheath is interrupted periodically by nodes of Ranvier, exposing patches of axonal membrane. The internodal distance is usually approximately 100 times the external diameter of the fiber, ranging between 200 μm and 2 mm. The effect of the myelin sheath is to restrict current flow largely to the node, as ions cannot flow into or out of the high-resistance internodal region and the internodal capacitative currents are very small as well. It follows that only the restricted portion of the axon membrane at the nodes becomes involved in impulse propagation and that relatively small charge movements are required to displace the membrane potential. As a result, the impulse jumps from node to node, thereby greatly increasing the conduction velocity. Such impulse propagation is called SALTATORY CONDUCTION (from Latin *saltare*, to jump, leap, or dance).

An additional consequence of myelination is that during impulse propagation fewer sodium and potassium ions enter and leave the axon,

[4]Hodgkin, A. L. 1939. *J. Physiol. 94*: 560–570.
[5]Arbuthnott, E. R., Boyd, I. A. and Kalu, K. U. 1980. *J. Physiol. 308*: 125–157.

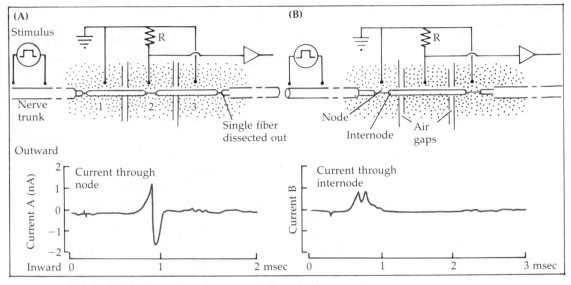

6 **CURRENT FLOW THROUGH A MYELINATED AXON.** A single myelinated axon passes through two air gaps, which create three compartments that are not linked by extracellular fluid. During the propagated
action potential, radial currents flow through the resistor R; the voltage
drop across the resistor is then a measure of the current. In (A) a node
of Ranvier is in compartment 2. The initial upward deflection is due to
depolarization of the node and consequent outward current; this reverses to a large inward current when threshold is reached. In (B) an
internode is in the center compartment, and only outward capacitative
currents are seen as the impulse approaches and then leaves the internodal segment. (After Tasaki, 1959.)

as regenerative activity is restricted to the nodes. Consequently, less
metabolic energy is required by the sodium–potassium exchange pump
(Chapter 8) to restore the intracellular concentrations to their resting
levels. Myelinated axons not only conduct more rapidly than unmyelinated ones but also are capable of firing at higher frequencies for
more prolonged periods of time.

Experiments to demonstrate saltatory conduction were first made in
1941 by Tasaki[6] and later by Huxley and Stämpfli,[7] who recorded current
flow at nodes and internodes. Such an experiment on a single myelinated axon is illustrated in Figure 6. The nerve is placed in three saline
pools, the central pool being narrow and separated from the others by
air gaps of very high resistance. Electrically, the pools are connected by
the external recording circuitry as shown, so that during impulse propagation current flow into or out of the central pool is through the resistor
R. When the central pool contains a node of Ranvier, stimulation of the
nerve trunk results first in outward current through the node as it is

[6]Tasaki, I. 1959. In H. W. Magoun (ed.), *Handbook of Physiology*, Vol. 1, American
Physiological Society, Bethesda, pp. 75–121.

[7]Huxley, A. F. and Stämpfli, R. 1949. *J. Physiol.* 108: 315–339.

depolarized by the oncoming excitation, followed by inward current after threshold is reached and a regenerative response initiated. In contrast, when the central pool contains an internode, there is only a small outward capacitative current as the impulse in the first pool depolarizes the internodal region; as this begins to decay, there is a second small surge of capacitative current as the impulse depolarizes the node in the third pool. These kinds of experiments, then, confirmed that there is no inward current, and hence no regenerative activity, in the internodal region. More recently, sophisticated recording techniques have been developed by Bostock and Sears[8] for recording saltatory conduction in undissected mammalian axons in situ. With such techniques it is possible to measure inward currents at the nodes and longitudinal internodal currents and thereby to estimate accurately the positions of the nodes and the distances between them.

Theoretical calculations suggest and experimental recordings confirm that in myelinated fibers the conduction velocity is proportional to the diameter of the fiber. The relation between fiber diameter and conduction velocity was first considered theoretically by Rushton[9] and has been examined in peripheral nerve in some detail by Boyd and his colleagues.[5] One theoretical point of interest is the best thickness for the myelin sheath if conduction velocity is to be optimized. Obviously the increase in membrane impedance in the myelinated region will be greater with a thick sheath than with a thin one. On the other hand, for a fiber of given diameter, the cross-sectional area of the axon decreases with increasing myelin thickness, thereby increasing internal longitudinal resistance. The first effect would be expected to increase conduction velocity, the second to decrease it. It turns out that the optimal compromise between these opposing effects is achieved when the axon diameter is about 0.7 times the fiber diameter. As already noted, this ratio in mammalian peripheral nerve ranges between 0.6 and 0.8. For large fibers the conduction velocity in meters per second is approximately 6 times the diameter in micrometers. For smaller fibers below 11 μm in diameter, the constant of proportionality is about 4.5.

The optimum internodal length for conduction is approximately that found in practice, namely, about 100 times the external fiber diameter. Longer internodal distances would allow the excitation to jump farther, tending to increase conduction velocity. On the other hand, with increased longitudinal resistance between the nodes, the depolarization produced by activity in a preceding node would be smaller and rise more slowly so that excitation at a node would be slowed, tending to decrease conduction velocity. Because of these opposing factors, variations in internodal length around the optimum values have little effect on conduction velocity. Eventually, however, as the internodal length is increased, conduction is blocked as the depolarization from activity in the preceding node no longer reaches threshold.

[8]Bostock, H. and Sears, T. A. 1978. *J. Physiol. 280*: 273–301.
[9]Rushton, W. A. H. 1951. *J. Physiol. 115*: 101–122.

In myelinated peripheral nerve, the safety factor for conduction is about five; that is, the depolarization produced at a node by excitation of a preceding node is approximately five times larger than necessary to reach threshold. This safety factor can be reduced considerably in several morphological circumstances. For example, when a myelinated nerve branches, the current supplied by the single node at the branch point is divided between *two* nodes beyond the branch and the safety factor for conduction along one or other or both may be reduced, depending on the geometry. Similarly, when the myelin sheath terminates—for example, near the end of a motor nerve—the current from the last node is distributed over a large area of unmyelinated nerve terminal membrane and, as a consequence, provides less overall depolarization than would occur at a node. It is perhaps for this reason that the last few internodes before an unmyelinated terminal are shorter than normal.[10]

In recent experiments Ritchie and his colleagues[11] have examined the properties of the axon membrane in the paranodal region normally covered by the myelin sheath. To do this, the myelin is loosened by enzyme treatment or osmotic shock. Voltage clamp studies are then made of currents in the region of the node and compared with those obtained before the treatment. Rabbit nodes of Ranvier normally display only inward sodium current upon excitation. Repolarization is caused, not by an increase in potassium conductance (as in the squid axon), but rather by a relatively large leak current and rapid sodium inactivation. After exposure of axon membrane in the paranodal region, there is no increase in inward current, but excitation produces a delayed outward potassium current as well. It is concluded that the axon under the myelin, at least in the region adjacent to the nodes, has no voltage-sensitive sodium channels but does have voltage-sensitive potassium channels. In other words, its properties are the reverse of those of the nodal membrane. As would be expected, one of the consequences of extensive demyelination is conduction block. However, mammalian axons that have been demyelinated chronically with diphtheria toxin can develop continuous conduction through a demyelinated region,[8] implying that after demyelination voltage-sensitive sodium channels appear in the exposed axon membrane. As in the acutely demyelinated paranodal membrane, voltage-sensitive potassium channels are present as well.[12]

For much physiological work in the central and peripheral nervous systems, extracellular electrodes are used to stimulate or to record from axons of various diameters. When two extracellular electrodes are used to stimulate a nerve trunk, much of the current flow is through the extracellular fluid; the remainder enters individual axons under the positive electrode, flows along the axoplasm, and exits under the neg-

Stimulating and recording from nerves with external electrodes

[10]Quick, D. C., Kennedy, W. R. and Donaldson, L. 1979. *Neuroscience* 4: 1089–1096.
[11]Chiu, S. Y. and Ritchie, J. M. 1981. *J. Physiol.* 313: 415–437.
[12]Bostock, H., Sears, T. A. and Sherratt, R. M. 1981. *J. Physiol.* 313: 301–315.

ative electrode. The membrane under the positive electrode is hyper-polarized (the outside is made more positive with respect to the inside), that under the negative electrode depolarized. The amount of current required to reach threshold depolarization depends on fiber diameter: Large fibers require less current than small fibers. The reason for this becomes apparent upon consideration of the voltage drops produced by the current flow in the axon. First, a larger fraction of the total current will enter a large axon than will enter a small one. Second, in a large axon a larger fraction of the voltage drop produced by the current occurs across the membrane and a correspondingly smaller fraction along the internal longitudinal resistance: A current with the magnitude i entering the axon hyperpolarizes the membrane by an amount $\Delta V_m = i r_{input}$ and produces an equal depolarization when it leaves; in addition, it produces a voltage drop $\Delta V_l = i r_l \, l$ along the core of the fiber, where l is the longitudinal distance between the electrodes. In an axon with a large cross-sectional area, ΔV_l is relatively small compared with ΔV_m, so that most of the voltage drop is across the membrane where it is needed for excitation. This relation between size and threshold for stimulation by external electrodes is fortunate for physiological and clinical purposes; threshold and conduction velocity can be tested, for example, in nerves supplying muscles, which are relatively large, without exciting pain fibers, which are very much smaller. Just as the largest fibers are the easiest to stimulate, they are also the most difficult to block, for example, by cooling, localized pressure, or local anesthetics. Again, this means that pain fibers can be blocked with anesthetic without interfering with conduction in larger sensory and motor fibers. When records are made with extracellular electrodes, the recorded potentials are generated by current flow in the extracellular fluid. Because of their larger membrane area and lower core resistance, larger fibers generate more current than smaller ones. As a result, they produce a larger signal at the recording electrodes.

Optical recording of membrane potential changes

A promising new technique for recording from cells or processes too small to study with conventional electrodes has been developed by Cohen, Salzberg, Grinvald, and colleagues.[13] The method involves dyes that bind to the cell membrane and the optical characteristics (absorbance or fluorescence) of which then change with membrane potential. Such optical signals can give reliable measures of both action potentials and subthreshold potential changes, such as synaptic potentials. Moreover, by using extremely fine laser microbeams for excitation, it is possible to record changes in fluorescence from processes much too small to be impaled with microelectrodes.[14] An example is shown in Figure 7. The records are from a neuroblastoma cell grown in culture, the soma of which has been impaled with two microelectrodes, one for passing current and the other for measuring the resulting potential

[13]Cohen, L. B., Salzberg, B. B. and Grinvald, A. 1978. *Annu. Rev. Neurosci.* 1: 171–182.

[14]Grinvald, A. and Farber, I. C. 1981. *Science* 212: 1164–1167.

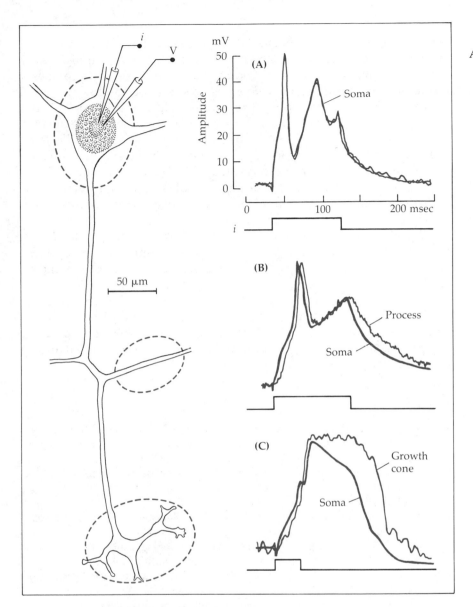

7 OPTICAL RECORDING from a neuroblastoma cell in culture. Two microelectrodes are placed in the soma of the cell, one to pass depolarizing current and the other to record the resulting electrical activity. In addition, the cell has been treated with a dye which fluoresces when excited by a laser microbeam, and whose fluorescence changes with membrane potential. In (A) the microbeam encompasses the cell soma, as indicated by the dashed outline. Depolarization produces a fast, sodium-dependent and a slower, calcium-dependent action potential (thin trace). Optical record of potential change (noisy trace) is identical. (B) The microbeam is focused on an axonal process, as indicated. Excitation produces optically recorded potential change similar to that recorded electrically from soma. (C) After treatment with TTX and TEA to block voltage-sensitive sodium and potassium conductances, depolarization produces a calcium-dependent action potential recorded electrically from the cell body and optically from the growth cone at the termination of the axonal process. (Courtesy of A. Grinvald.)

change. The current and voltage records are shown in A. Superimposed on the voltage record is the optical record from the field indicated by the dashed oval around the cell soma. Such superposition provides a means of calibrating the optical record in terms of absolute potential change. In Figure 7B the optical recording field has been moved to a small process branching from the axon; the action potential is delayed slightly but is similar to that recorded by the microelectrode in the soma. In Figure 7C the optical field now encompasses a growth cone. Tetrodotoxin and TEA have been added to the bathing medium to block voltage-sensitive changes in sodium and potassium conductance, and excitation produces a calcium-dependent action potential recorded electrically from the soma and optically from the growth cone.

A further extension of the optical recording technique is to use an array of 100 or more photodiodes, each about 1.5 mm square, to record activity in organized groups of neurons, such as invertebrate ganglia[15] or thin slices of rat brain.[16] This method of recording provides information about interactions between large numbers of cells; this information could otherwise be obtained only with great difficulty—for example, the sequential order of firing and whether, during a particular pattern of activity, ongoing synaptic interactions between particular cells are excitatory or inhibitory.

SUGGESTED READING

Arbuthnott, E. R., Boyd, I. A. and Kalu, K. U. 1980. Ultrastructural dimensions of myelinated peripheral nerve fibres in the cat and their relation to conduction velocity. *J. Physiol. 308*: 125–157.

Grinvald, A., Hildesheim, R., Farber, I. C. and Anglister, L. 1982. Improved fluorescent probes for the measurement of rapid changes in membrane potential. *Biophys. J. 39*: 301–308.

Hodgkin, A. L. and Rushton, W. A. H. 1946. The electrical constants of a crustacean nerve fibre. *Proc. R. Soc. Lond. B 133*: 444–479.

Huxley, A. F. and Stämpfli, R. 1949. Evidence for saltatory conduction in peripheral myelinated nerve fibers. *J. Physiol. 108*: 315–339.

Ritchie, J. M. 1982. On the relation between fibre diameter and conduction velocity in myelinated nerve fibres. *Proc. R. Soc. Lond. B 217*: 29–35.

Rushton, W. A. H. 1951. A theory of the effects of fibre size in medullated nerve. *J. Physiol. 115*: 101–122.

[15]Grinvald, A., Cohen, L. B., Lesher, S. and Boyle, M. B. 1981. *J. Neurophysiol 45*: 829–840.

[16]Grinvald, A., Manker, A. and Segal, M. 1982. *J. Physiol. 333*: 269–291.

ACTIVE TRANSPORT OF IONS

In resting neurons there is a constant inward leak of sodium and an outward leak of potassium. This process is accelerated greatly when nerves conduct impulses for long periods. To maintain the appropriate ionic distributions, energy must be expended by the cells to pump sodium out and potassium in. A continuously active restorative pumping mechanism is particularly important for small cells because they have a small volume in which to store ions, relative to their surface area.

Experiments on a variety of invertebrate nerve cells and unmyelinated mammalian axons have shown that the transport mechanism gives rise to greater outward movement of sodium than inward movement of potassium. Activity of the pump, therefore, produces a net outward movement of positive charge, the effect of which is to add to the resting membrane potential. The rate of sodium extrusion increases with intracellular sodium accumulation so that following trains of impulses there is increased pumping activity, which in turn results in hyperpolarization of the membrane. Such hyperpolarization can change the signaling properties of neurons.

The energy source for the sodium–potassium pump is derived from the hydrolysis of adenosine triphosphate (ATP), and it has been shown that the ATPase is itself part of the pump. Both the pumping activity and the enzymatic activity depend on the presence of sodium and potassium; both are blocked by the cardiac glycoside ouabain.

In neurons and in other cells such as red blood cells and epithelial cells, there are transport mechanisms not only for sodium and potassium but also for other ions such as calcium and chloride and, in order to regulate intracellular pH, for hydrogen ions and bicarbonate. In squid axon, for example, there is a pump that extrudes calcium from the axoplasm and another that transports chloride into the axoplasm, both coupled to inward sodium movement. In other cells, such as snail neurons, chloride is pumped out of the cytoplasm in exchange for bicarbonate, with the exchange mechanism again being coupled to inward movement of sodium.

All the electrical signals of the nervous system described so far are due to changes in membrane permeability that result in ions moving into or out of the cell along their electrochemical gradients, without requiring any metabolic energy. During action potentials sodium and potassium ions run rapidly into and out of the cell. Receptor potentials are also associated with sodium ion entry, and synaptic potentials, depending on whether they are excitatory or inhibitory (Chapter 9), are associated with movements of sodium, potassium, or chloride. At the resting potential, as noted in Chapter 5, none of the major ions are at equilibrium, so that similar ionic leaks proceed continuously, although much more slowly than during activity. Inevitably, such leaks would dissipate the concentration gradients across the membrane (and destroy the osmotic equilibrium of the cell) if no restitution were made.

Until information became available about ion transport in other types of cells, it was not obvious how nerve cells could adjust their internal concentrations of ions. As early as 1902, Overton[1] recognized that though the problem applied to all types of cells in the body, it was especially serious in nerve and muscle fibers that conduct millions of action potentials throughout the life of an animal but still maintain high potassium and low sodium concentrations in their cytoplasm. The mechanism for achieving this must be quite different from the permeability mechanisms governing passive ionic movements because an energy source is required. For example, for sodium to move out of a cell, it must be transported UPHILL against an electrochemical gradient.

Early recordings from nerve fibers provided electrical signs of active transport processes in the form of slow changes in membrane potential following trains of impulses. It was known that these AFTERPOTENTIALS were markedly reduced by metabolic poisons; but their possible function and underlying mechanism were not appreciated, and for many years they were ignored. Their significance was understood only after more became known about processes of transport of ions against electrochemical gradients. Such transport mechanisms are referred to as PUMPS, and their mode of operation is of interest for a number of reasons: (1) As already noted, pumps are essential for maintaining the long-term viability of nerve cells; (2) their effects can lead to relatively large changes in membrane potential, which can persist for seconds or even minutes, far longer than the millisecond time scale associated with usual electrical signaling; (3) pump-generated potentials inevitably influence signaling—for example, the threshold depolarization required for excitation increases when the membrane is hyperpolarized. Such hyperpolarization, then, can influence in a significant manner frequency of action potential activity, conduction at axonal branch points, and synaptic integration.

It has been demonstrated that sodium–potassium exchange is driven by hydrolysis of adenosine triphosphate (ATP) and that the phosphatase

---

[1]Overton, E. 1902. *Pflügers Arch.* 92: 346–386.

itself is an integral part of the pump.[2] Before describing the experimental evidence for such metabolically driven ion transport, it is convenient to discuss the changes in ionic concentrations inside and outside cells produced by impulse activity.

It was noted in Chapter 5 that during an action potential in a squid axon about 3 picomoles of sodium and potassium move into and out of a cell through each square centimeter of membrane. In a large cell like the squid axon, changes in concentration resulting from such fluxes are insignificant over short periods, especially since the natural frequency of firing is low, occurring in short bursts of only 8 to 10 impulses per second. In contrast, action potential activity in small nerve fibers might be expected to lead to larger changes in ion concentration because of their smaller reservoir of axoplasm in relation to membrane surface area. For example, if 3 picomoles/cm^2 of sodium entered a fiber 1 μm in diameter, the change in internal concentration would be 120 μM, or 1000 times greater than in a 1-mm diameter squid axon. If there were no transport mechanism, the sodium gradient in the small fiber would be dissipated after only 1000 impulses.

Ionic concentration changes resulting from nerve impulses

As sodium accumulates in the cytoplasm, potassium is lost to the extracellular space. Almost all neurons within the central nervous system are surrounded by narrow clefts that are only a few tens of nanometers wide and in which potassium can accumulate. During action potential activity, potassium concentration in this extracellular space can increase sufficiently to produce a significant change in E_K and hence in membrane potential. Thus, both sodium and potassium transport mechanisms are required, not only to maintain intracellular concentrations, but also to limit ion accumulation and loss in extracellular spaces. Such transport mechanisms necessarily require expenditure of energy (Figure 1).

Evidence for sodium–potassium transport in squid axon was provided by a series of experiments by Hodgkin and Keynes and their colleagues.[3,4] The principle of the experiments was to load an axon with ^{24}Na, either by stimulating it in seawater containing the radioactive isotope or by intracellular injection, and then to measure the rate of efflux. The results (Figure 2) indicated that at rest there was a continuous movement of radioactive sodium out of the axon, which was later shown to have the following properties:

Active transport of sodium and potassium in axons

1. It is rapidly and reversibly reduced by a number of poisons that interfere with the oxidative metabolism of the cell, notably dinitrophenol (DNP), cyanide, and azide.
2. It is prevented by the digitalis glycosides, particularly ouabain and strophanthidin, which specifically block the sodium–potassium pump but do not interfere with the oxidative metabolism of the cell. Interestingly,

[2]Skou, J. C. 1957. *Biochim. Biophys. Acta* 23: 394–401.
[3]Hodgkin, A. L. and Keynes, R. D. 1955. *J. Physiol. 128*: 28–60.
[4]Baker, P. F., Blaustein, M. P., Keynes, R. D., Manil, J., Shaw, T. I. and Steinhardt, R. A. 1969. *J. Physiol. 200*: 459–496.

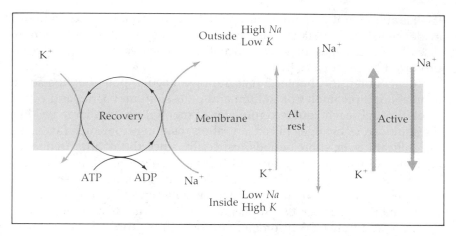

1 MOVEMENT OF IONS through a nerve membrane. The downhill movements that occur during the impulse are shown on the right by heavy arrows, and the net downhill movements at rest by the light arrows. Uphill movements driven by ATP are shown on the left. Pump ratio and resting leak ratio are 3:2 (Na:K).

ouabain acts only on the outside of the axon and is ineffective when introduced into the axoplasm.

3. It is blocked by removal of potassium from the external bathing fluid.

4. Agents that block active transport cause little immediate change in the resting or action potentials,[4] which continue for long periods; the squid axon is so large that ion accumulation over short periods produces only small changes in intracellular concentration.

5. Lithium ions (which can substitute for sodium in producing action potentials) are not extruded from the cell by the sodium pump.

In addition to extruding sodium actively, the axon also accumulates potassium by a process blocked by metabolic inhibitors and by ouabain. The two active transport processes are coupled and depend on hydrolysis of adenosine triphosphate (ATP). The interdependence of the two transport mechanisms is illustrated by the finding that the extrusion of sodium requires the presence of potassium in the extracellular fluid. When potassium is removed from the bathing solution, outward sodium transport is greatly reduced. Measurements made on squid axons have shown that the coupled sodium–potassium transport is not a simple one-to-one exchange; instead, under normal experimental conditions, the coupling ratio of sodium extruded to potassium accumulated is 3:2.

Evidence that the sodium–potassium pump is an ATPase How are metabolic processes of the nerve cell used to drive sodium and potassium against their electrochemical gradients? Numerous experiments have shown that the immediate energy source is ATP. In experiments on squid axons, for example, first metabolism was blocked by cyanide or by DNP—to block the transport mechanism. Next, a variety of compounds were injected into the cytoplasm to determine whether or not they could serve as a substrate for the pump in order

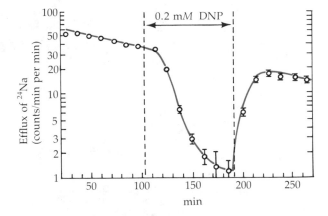

2

SODIUM EFFLUX from a cuttlefish (*Sepia*) axon before, during, and after treatment with a metabolic inhibitor, dinitrophenol (DNP). The axon was loaded with ^{24}Na before the experiment and was maintained in artificial seawater. The rate at which ^{24}Na leaves the axon (ordinate) is markedly reduced by DNP. Vertical lines are ±2 SE of the mean. (From Hodgkin and Keynes, 1955.)

to restore its activity. In early experiments[5] it appeared that a number of energy-rich compounds were able to reactivate the pump. Later, however, Mullins and Brinley[6] showed that in axons depleted of small organic ions by internal dialysis only replacement of ATP could provide a direct source of energy for the exchange mechanism (Figure 3). This observation was later confirmed in internally perfused axons.[7]

The finding that ATP consumption was a necessary condition for sodium–potassium transport suggested that the pump itself might be an enzyme, hydrolyzing ATP. This idea was reinforced by Skou's demonstration that in crab nerve an ATPase fulfilled many of the requisite criteria for the pump.[8] As with the pump, the enzyme activity depended on the presence of sodium and potassium; lithium would not substitute for sodium, and activity was inhibited by ouabain.

The identity between the sodium–potassium exchange pump and sodium–potassium ATPase was demonstrated more directly by three types of evidence obtained in red blood cells: (1) The optimal sodium and potassium concentrations required for enzyme activity are the same as those for the exchange pump; (2) the ionic specificity for activation of the enzyme and the pump are the same and similar to that observed in nerve (e.g., lithium will not substitute for sodium); (3) ouabain is a highly specific blocker of both the enzyme and the pump.

In addition to these similarities, in red blood cells the enzyme has the same polarity as the pump with respect to effects of changes in intracellular and extracellular composition. Both the ATPase and the transport mechanism are activated by high concentrations of sodium inside and potassium outside, but not by the reverse ratios. These

[5]Caldwell, P. C., Hodgkin, A. L., Keynes, R. D. and Shaw, T. I. 1960. *J. Physiol. 152*: 561–590.

[6]Mullins, L. J. and Brinley, F. J. 1967. *J. Gen. Physiol. 50*: 2333–2355.

[7]Baker, P. F., Foster, R. F., Gilbert, D. S. and Shaw, T. I. 1971. *J. Physiol. 219*: 487–506.

[8]Skou, J. C. 1964. *Prog. Biophys. Mol. Biol. 14*: 133–166.

3

EFFECT OF ATP ON SODIUM EFFLUX from an internally dialyzed squid axon. The composition of the axoplasm is regulated by perfusion through an intracellular dialysis capillary that removes substrates used for metabolism. While "fuel-free" fluid flows through the tubing, the rate of sodium extrusion is low. When ATP in a concentration of 5 mM is added internally, sodium efflux increases. (After Mullins and Brinley, 1967.)

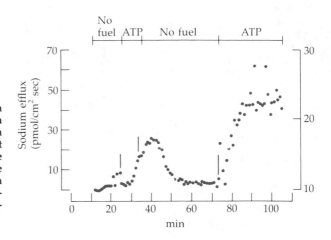

properties were demonstrated by Glynn[9] and his colleagues, using red cell "ghosts." Ghosts are obtained by immersing normal red cells in hypotonic solution, a process causing them to lose their hemoglobin through a disrupted membrane. They are then reconstituted with the selected intracellular ionic contents (including radioactive tracers, if desired) by immersion in an isotonic solution of appropriate composition. Fluxes then can be measured with the cells bathed in a variety of extracellular solutions. In this way, then, the effects on ATPase activity of a wide range of extracellular and intracellular sodium and potassium concentrations could be studied in detail. In the presence of high internal sodium and external potassium, ATPase activity of reconstituted ghosts was stimulated, a result indicating that the enzyme, like the pump, was sensitive to concentration differences on either side of the membrane. Furthermore, with unusually high concentrations of sodium outside and potassium inside the cells, the reversed gradients could be increased sufficiently to drive the pump backward, reversing its enzymatic activity and thus leading to a net synthesis of ATP.[10]

Highly purified sodium–potassium ATPase has been isolated from cell membranes from a wide variety of tissues, including mammalian kidney and brain, human red cells, dogfish rectal gland, and eel electroplaque, and has been characterized biochemically.[11] The biochemical properties of the enzyme and their relation to sodium–potassium exchange have been summarized briefly by Post.[12] The stoichiometry of the enzyme is as expected from the pump characteristics: Three sodium and two potassium ions are bound for each molecule of ATP hydrolyzed. The requirement for sodium is remarkably specific. The only

[9]Glynn, I. M. 1968. *Br. Med. Bull. 24*: 165–169.

[10]Garrahan, P. J. and Glynn, I. M. 1967. *J. Physiol. 192*: 237–256.

[11]Schuurmans Stekhoven, F. and Bonting, S. L. 1981. *Physiol. Rev. 61*: 1–76.

[12]Post, R. L. 1981. In T. P. Singer and R. N. Ondarza (eds.), *Molecular Basis of Drug Action.* Elsevier, New York, pp. 299–331.

substrate accepted for net outward transport is sodium; conversely, the only monovalent cation *not* accepted for inward transport is again sodium. Thus, for example, lithium, ammonium, rubidium, cesium, and thallium are all able to substitute for potassium in the external solution. The requirement for external potassium is not absolute. Without it, the pump will run at a maximum of about 10 percent of capacity in an "uncoupled" mode. There is a corresponding sodium-dependent, ouabain-sensitive ATPase activity of the purified enzyme in the absence of potassium.

In theory, the sodium–potassium exchange pump need not generate any current flow as a result of its activity. If the transport ratio were 1:1, then the net charge transfer would be zero. Turning off the pump would, in that case, have no immediate effect on membrane potential, although in the long run the potential would decline because of the gradual loss of internal potassium and accumulation of sodium. In short, the pump would be electrically neutral. As has already been noted, however, the transport ratio is in fact 3:2; that is, two potassium ions are transported inward for each three sodium ions transported outward. Thus, the pump produces a net outward current equivalent to one ionic charge per cycle. This outward charge movement contributes to the internal negativity of the cell; and when the pump is stopped there is an immediate fall in resting potential because of removal of the hyperpolarizing current. Such a pump is said to be ELECTROGENIC.

Electrogenic pumps and resting membrane potential

The effect of the electrogenic pump on membrane potential can be derived, using constant field theory as in Chapter 5. To understand how the resting potential is influenced by the pump, it is worth noting that in the *steady state* the net transmembrane currents must be zero for both sodium and potassium. Otherwise the intracellular ionic concentrations would not remain constant. In other words, the leak current across the membrane for each ion must be exactly equal and opposite to the current generated by the pump (Box 1). The resting potential of the cell is then given by[13]

$$V_m = 58 \log \frac{rK_o + bNa_o}{rK_i + bNa_i}$$

where r is the pump ratio (in this case, 1.5) and b is the ratio of sodium to potassium permeability (p_{Na}/p_K). The result may seem unexpected, as the equation takes account neither of the pump *rate* nor of the overall permeability of the membrane across which the electrogenic current flows. One might expect, for example, that if the pump rate were increased (without changing r) the additional current flow would result in a further increase in membrane potential. Conversely, if the pump rate were kept constant and the permeability of the membrane increased (without changing b), the membrane potential might be expected to fall because the electrogenic component would now be shorted out by the increase in passive ion movements. In fact, neither of these considera-

[13]Mullins, L. J. and Noda, K. 1963. *J. Gen. Physiol.* 47: 117–132.

BOX 1 CONTRIBUTIONS OF THE ELECTROGENIC PUMP TO THE RESTING MEMBRANE POTENTIAL

Expressions were derived in Chapter 5 for passive sodium and potassium currents across the cell membrane, using the constant field assumption. The expressions were that

$$i_K = p_K FV' \frac{K_o - K_i e^{V'}}{e^{V'} - 1}$$

$$i_{Na} = p_{Na} FV' \frac{Na_o - Na_i e^{V'}}{e^{V'} - 1}$$

where i_K and i_{Na} are the sodium and potassium currents, p_K and p_{Na} the membrane permeabilities to the ions, and K_o, K_i, Na_o, and Na_i the extracellular and intracellular concentrations of the ions. F is the faraday and V' is the membrane potential as a multiple of the thermodynamic potential RT/F, that is, $V' = V_m F/RT$. In the steady state, the passive sodium and potassium currents ("leak currents") must be exactly equal and opposite to the corresponding currents carried by the pump. In other words, if t_K and t_{Na} are the potassium and sodium currents generated by the transport mechanism, then $t_K = -i_K$ and $t_{Na} = -i_{Na}$. It follows that if the transport ratio (t_{Na}/t_K) is $-r$ (negative because the ions are being transported in opposite directions), then

the ratio of the passive currents must also be $-r$, or $i_{Na} = -r i_K$.

Remove the factor $FV'/(e^{V'} - 1)$ common to both expressons; then,

$$p_{Na}(Na_o - Na_i e^{V'}) = -r p_K(K_o - K_i e^{V'})$$

It follows that

$$e^{V'} = \frac{r p_K K_o + p_{Na} Na_o}{r p_K K_i + p_{Na} Na_i}$$

Recall that $V' = V_m F/RT$ and write $p_{Na}/p_K = b$; then,

$$V_m = \frac{RT}{F} \ln \frac{r K_o + b Na_o}{r K_i + b Na_i}$$

The resting membrane potential, then, depends on both the pump ratio and the permeability ratio for sodium and potassium. Note that no assumptions were made about chloride current in deriving the relation. In the steady state, net chloride current must, of course, be zero as well, but whether this is achieved by passive distribution or by an active transport mechanism has no effect on the result obtained for sodium and potassium.

tions appears in the equation because in the *steady state* the rates of transport of sodium and potassium each must be matched exactly to their individual passive leaks (Figure 4A). Thus, to maintain constant intracellular sodium and potassium concentrations, the leak ratio, like the pump ratio, must be 3:2. The membrane potential at which this occurs depends, in turn, on the relative permeability of the membrane to the two ions (i.e., on b). If the membrane is relatively leaky to cations, then this requires that the pump rate be correspondingly high to maintain the steady state; conversely, a low conductance membrane requires a low pump rate. In other words, once a steady state is specified the pump rate and membrane conductance are irrevocably linked.

How much does the electrogenic pump contribute to the resting membrane potential? If, for a squid axon, we take K_o and K_i to be 10 mM and 400 mM, respectively, and Na_o and Na_i to be 440 mM and 50 mM, respectively, and assign a value of 0.03 to b (Chapter 5), then with a pump ratio of 1.5 the theoretical membrane potential is -88 mV. With no pump, or an electrically neutral pump ($r = 1$), the correspond-

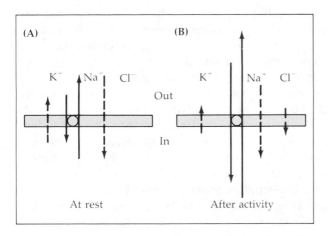

4 CURRENT FLOW in the membrane associated with the electrogenic sodium–potassium pump. The magnitudes of individual ionic currents are represented by the lengths of the arrows. **(A)** At rest, net passive sodium and potassium currents are equal and opposite to currents generated by the pump, and net chloride flux is zero. **(B)** After activity, increased pump rate causes hyperpolarization, little change in inward sodium current, a reduction in outward potassium current, and an inward chloride current. In both situations the net membrane current is zero, but in **(B)** the net fluxes of individual ions are not zero, so the cell is not in a steady state.

ing value is -82 mV, so the pump adds 6 mV or about 7 percent to the resting potential. In cells with a relatively higher sodium permeability (and hence b), the pump makes a greater contribution. For example, in some mammalian cells b might be as large as 0.1. If K_o and K_i are 5 mM and 120 mM, respectively, and Na_o and Na_i 150 mM and 20 mM, respectively, then V_m would be -53 mV; on stopping the pump, the potential would fall to -46 mV. In this case, then, the pump adds 7 mV, or 15 percent to the resting potential.

The constraints on the contribution of the electrogenic pump to membrane potential in the resting cell do not apply to transient conditions when the pump and leak rates are not matched. The ATPase itself is exquisitely sensitive to increases in internal sodium concentration so that sodium accumulation after a train of impulses in most nerve cells results in an acceleration of the pump rate and consequent hyperpolarization. This sequence of events is accompanied by changes in both active and passive current flow (Figure 4B). Suppose, to begin with, that there is no change in pump rate. Because the intracellular sodium concentration is relatively low, it might easily be doubled by moderate accumulation during activity, thus reducing E_{Na} and therefore the passive inward sodium leak. The loss of the same amount of potassium, on the other hand, would produce only a small *fractional* change in intracellular potassium concentration and hence little change in E_K. Because of the small change in E_K and the relatively small effect of

Electrogenic pumps and hyperpolarization

intracellular sodium on membrane potential, the resting potential will change very little (unless there is potassium accumulation in *extracellular* spaces around the nerve). In summary, the changes in ionic concentration, by themselves, have little effect on the resting potential or on the outward potassium flux but reduce the inward sodium flux.

If the pump rate is now increased, the primary effect is to increase the membrane potential due to the net outward movement of positive charge. At the same time, this hyperpolarization has the secondary effect of increasing the inward leak of sodium and decreasing the outward leak of potassium. The potassium current may even be reversed if the membrane is hyperpolarized beyond E_K. Finally, if there is no net chloride current at rest, hyperpolarization will result in outward chloride movement (inward current) as well. A steady membrane potential will be reached when the net inward current through the passive permeability channels is equal in magnitude to the net outward pump current (i_{pump}). Thus,

$$-i_{pump} = g_{Na}(V_m - E_{Na}) + g_K(V_m - E_K) + g_{Cl}(V_m - E_{Cl})$$

It follows that

$$V_m = \frac{-i_{pump} + g_{Na}E_{Na} + g_KE_K + g_{Cl}E_{Cl}}{g_{Na} + g_K + g_{Cl}}$$

If there were no pump current, the first term in the numerator would disappear and the equation would be the same as that given in Chapter 5 for membrane potential in terms of equilibrium potentials and conductances. The increment that is added by the pump, then, is simply $i_{pump}/(g_{Na} + g_K + g_{Cl})$. Thus, a cell with a high membrane resistance (low conductance for Na, K, and Cl) will be hyperpolarized more than one with a more leaky membrane. Although the membrane potential may remain constant for some time at the high pump rate, the cell is *not* in a steady state, as the individual ionic leak currents are not equal and opposite to those carried by the pump; for example, both the pump and the leak currents for potassium may be inward. The steady state is restored when the ionic concentrations again reach their resting values and the pump returns to its resting rate.

Experimental evidence
for an electrogenic
pump
An example of these effects is provided by neurons in snail ganglia, which become markedly hyperpolarized when their internal sodium concentration is increased either by repetitive activity or by direct sodium injection from a micropipette. The relation between internal sodium concentration, pump current and membrane potential was studied in detail by Thomas.[14,15] Two intracellular pipettes were used to deposit ions into the cell, one filled with sodium acetate and the other with potassium acetate; a third intracellular pipette was used as an electrode to record membrane potential. To inject sodium by ionophoresis, the sodium pipette was made positive with respect to the potassium pipette (Figure 5A). Current flow in the injection system, then, was between

[14]Thomas, R. C. 1969. *J. Physiol.* 201: 495–514.
[15]Thomas, R. C. 1972b. *J. Physiol.* 220: 55–71.

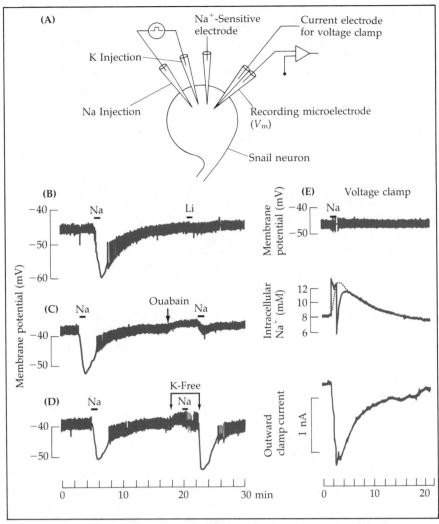

5 EFFECT OF SODIUM INJECTION. Changes in intracellular sodium concentration, membrane potential, and membrane current following injection of sodium into snail neurons. **(A)** Sodium injected by passing current between two electrodes filled with sodium acetate and potassium acetate (see text). A Na^+-sensitive electrode measures Na_i; two other electrodes measure membrane potential and pass current through the cell to obtain voltage clamp records in (E). **(B)** Hyperpolarization of the membrane following intracellular injection of sodium. (The small rapid deflections are action potentials, reduced in size because of the poor high-frequency response of the pen recorder.) Injection of lithium does not produce hyperpolarization. **(C)** After application of ouabain (20 µg/ ml), which blocks the sodium pump, hyperpolarization by sodium injection is greatly reduced. **(D)** Removal of potassium from the extracellular solution blocks the pump so that sodium injection produces no hyperpolarization until extracellular potassium is restored. **(E)** Voltage clamp records. Sodium injection results in increased intracellular sodium concentration and outward current across the cell membrane. The sharp peaks on the sodium concentration record are artifacts due to coupling with the injection system. The time course of the concentration change is indicated by the dashed line. Note that in the lower record downward deflection corresponds to *outward* current. (After Thomas, 1969.)

the two pipettes, the current being carried out of the sodium pipette by sodium ions and into the potassium pipette by negatively charged acetate ions *leaving* the tip. As a result, no current from the injection circuit flowed through the cell membrane. The result of such a sodium injection is shown in Figure 5B. After a brief injection, the cell became hyperpolarized by about 20 mV and gradually recovered over several minutes. The injection of lithium (Figure 5B), acetate, or potassium did not produce hyperpolarization.

Several lines of evidence showed that the potential change after sodium injection was due to the action of a sodium pump and not to changes in membrane permeability. For example, the input resistance of the cell did not decrease as might be expected if hyperpolarization were the result of an increased permeability to potassium or chloride. The hyperpolarization could, however, be greatly reduced or abolished by addition of ouabain to the bathing solution (Figure 5C), as would be expected of a sodium–potassium exchange pump. Similarly, removal of potassium from the external solution markedly reduced the effect of sodium injection (Figure 5D).

Quantitative estimates of the pump rate and the exchange ratio were obtained by voltage clamp measurements (Figure 5A). A fourth micropipette, connected to a voltage clamp circuit (Chapter 6), was used for passing current through the cell membrane. At the same time, the sodium concentration inside the cell was monitored by a fifth electrode, made of sodium-sensitive glass. The potential developed at the tip of the electrode provides a measure of the sodium concentration of the fluid in which it is immersed. The results of the experiment (Figure 5E) showed that sodium injection gave rise to an outward surge of current, the amplitude and duration of which followed the intracellular sodium concentration. The total charge carried out of the cell, measured by integrating the total clamp current, was only about one-third of the charge injected in the form of sodium ions. The evidence, then, was consistent with the idea that two out of three sodium ions pumped out of the cell were exchanged for potassium ions carried inward.

Effects of the electrogenic pump on signaling

Long-lasting hyperpolarizing potentials that can be attributed to sodium–potassium pump activity have now been found in a number of vertebrate and invertebrate nerve cells. These include unmyelinated axons in the vagus nerve of the rabbit, neurons in the cat's spinal cord, sympathetic ganglion cells, crustacean stretch receptors, and neurons in the nervous system of the leech and a variety of mollusks.[16] In such cells the hyperpolarization following trains of impulses can be as large as 35 mV and persist for many minutes. Such hyperpolarizations can have important consequences for synaptic integration and action potential propagation. Most significant is that hyperpolarization moves the membrane away from threshold and therefore causes a decrease in sensitivity of neurons to excitatory stimulation or, at sensory nerve endings, to depolarizing receptor potentials produced by external stimuli.[17] In leech neurons it has been demonstrated that pump activity can

[16]Thomas, R. C. 1972a. *Physiol. Rev. 52*: 563–594.
[17]Nakajima, S. and Onodera K. 1969a. *J. Physiol. 200*: 161–185.

cause conduction block at axonal branch points so that after repetitive activity conduction is favored along certain pathways over others.[18] The point to be emphasized is that because of activation of the electrogenic pump, previous activity of neurons can exert a major influence over their subsequent performance over relatively long periods.

A way of estimating the density of voltage-sensitive sodium channels by binding radioactive tetrodotoxin to the cell membrane was discussed in Chapter 6. A similar approach has been used to measure the number of sodium–potassium pump sites in neurons. The results of such analyses reinforce the concept of an inhomogeneous membrane with a mosaic arrangement of different fundamental properties. Landowne and Ritchie[19] measured the binding of tritiated ouabain to desheathed rabbit vagus nerve, which contains largely unmyelinated fibers. When just sufficient ouabain was added for complete inhibition of the sodium pump, the density of ouabain binding sites was estimated at about $750/\mu m^2$. If one assumes that each ATPase molecule binds one ouabain molecule, $750/\mu m^2$ would then be the density of pump sites. The results of a more recent study of the kinetics of ouabain binding to frog skeletal muscle[20] were consistent with a one-to-one binding ratio and provided an estimate of 2500 pump sites/μm^2.

Changes in intracellular concentration of calcium play a fundamental role in a number of aspects of neural signaling: (1) In many nerve and muscle cells, calcium entry either dominates or contributes to the inward current during the action potential; (2) calcium entry into nerve terminals during depolarization is an essential link in the processes leading to secretion of neurotransmitters; and (3) calcium entry into the myoplasm is an intermediate step in initiation of muscle contraction. In addition, changes in cytoplasmic calcium may be crucial in generating the electrical responses of photoreceptors and in processes involving the activation of cyclic AMP. Because of its involvement in such processes, it is important that intracellular calcium concentration be regulated very closely in nerve and muscle.

During the action potential in a squid axon, the permeability of the cell membrane to calcium is increased; and, like sodium, calcium enters the cell down its electrochemical gradient. Estimates of free calcium in the axon, using aequorin, range from about 0.3 μM[21] to as low as 0.02 μM,[22] and for this low concentration to be maintained it is clear that the cell must transport calcium outward against a very steep gradient.

Sarcoplasmic reticulum of vertebrate skeletal muscle has an active transport mechanism that sequesters calcium during the relaxation process. As with the sodium–potassium transport system, the pump is an ATPase, but it is activated by calcium instead of by sodium and potassium; and magnesium is a necessary cofactor for ATP binding. The

Density of pump sites

Regulation of intracellular calcium

[18]Yau, K.-W. 1976. *J. Physiol. 263:* 513–538.

[19]Landowne, D. and Ritchie, J. M. 1970. *J. Physiol. 207:* 529–537.

[20]Venosa, R. A. and Horowicz, P. 1981. *J. Memb. Biol. 59:* 225–232.

[21]Baker, P. F., Hodgkin, A. L. and Ridgway, E. B. 1971. *J. Physiol. 218:* 709–755.

[22]Dipolo, R., Requena, J., Brinley, F. J., Mullins, L. J., Scarpa, A. and Tiffert, T. 1976. *J. Gen. Physiol. 67:* 433–467.

enzyme, then, is known as calcium–magnesium ATPase. Although it has been studied extensively only in muscle, the enzyme is found in other tissues as well, including red blood cells, salivary gland, renal tubular cells, brain microsomes, and mammalian nerves.[12] It is responsible for maintaining the low basal level of intracellular calcium in squid axon and possibly in other nerves as well.

A second mechanism for removing calcium that leaks into cells is different in principle from the ATP-driven transport systems just discussed. Experimental evidence suggests that in squid axons, when intracellular calcium is sufficiently high, a fraction of the outward movement of calcium, instead of being driven directly by ATP, is coupled to inward flux of sodium. This mechanism is shown schematically in Figure 6A. The energy derived from the inward flux of three sodium ions down their electrochemical gradient is used to finance the uphill movement of a calcium ion. Such a scheme is known as coupled transport. The sodium gained by the process must, of course, be removed from the axoplasm in order for the electrochemical gradient to be maintained; this is taken care of by the sodium–potassium pump, the activity of which, in the long run, is necessary for the system to operate. This transport mechanism for calcium, then, is dependent only indirectly on metabolism insofar as it requires the maintenance of an adequate inward gradient for sodium movement.

In the squid axon at rest, there is a resting efflux of calcium equal to the rate at which it leaks in along its electrochemical gradient. The uphill, outward movement is reduced when extracellular sodium concentration is reduced. The effect of extracellular calcium and sodium on the concentration of ionized calcium in the axoplasm, measured by aequorin luminescence (Chapter 6), is illustrated in Figure 6B. Initially the intracellular calcium concentration is relatively high, as the axon is bathed in a high (112 mM) calcium solution. When the extracellular calcium is reduced to 11 mM, the intracellular calcium concentration, indicated by the level of luminescence, is reduced and a new steady-state distribution is reached. Reducing extracellular sodium, first to one-half and then to one-quarter normal, results in a progressive increase in intracellular calcium concentration, thereby reflecting a reduction in the rate of calcium extrusion. Replacement of extracellular sodium by lithium increases intracellular calcium still further; and, finally, return to normal extracellular sodium is followed by a return of intracellular calcium to its control level as the excess accumulated during the previous period is extruded by the exchange mechanism.

Coupling between transmembrane movement of sodium and calcium has been demonstrated in other experiments in which *outward* movement of sodium from the axoplasm into sodium-free bathing solution was found to be dependent on extracellular calcium.[23] Thus, the system behaves as though there were a carrier in the membrane with

[23]Baker, P. F., Blaustein, M. P., Hodgkin, A. L. and Steinhardt, R. A. 1969. *J. Physiol.* *200*: 431–458.

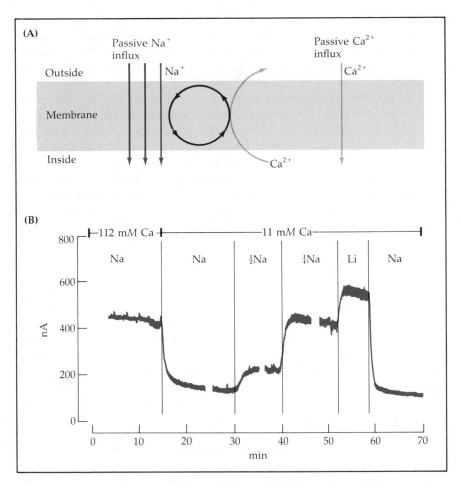

6 **TRANSPORT OF CALCIUM IONS** from nerve fibers. (A) Scheme for coupling sodium entry to calcium efflux. Influx of three sodium ions, which move passively downhill along their electrochemical gradient, is coupled to uphill extrusion of calcium. (B) Effect of changes in external calcium and sodium on intracellular calcium activity in a squid axon. Changes in calcium activity are measured as changes in intensity of luminescence of injected aequorin, indicated on the ordinate as nA of current from the photodetector. Increased readings mean increased intracellular, free calcium. Reducing external calcium from 112 to 11 mM reduces the intracellular concentration. Reducing extracellular sodium reduces outward calcium transport and hence increases the intracellular concentration. Lithium ions do not substitute for sodium in producing calcium extrusion (From Baker, Hodgkin, and Ridgway, 1971.)

an affinity for sodium and calcium that transports the two ions in opposite directions. The stoichiometry (one calcium transported out for three sodium in) is supported by a number of experimental observations;[24] for example, the exchange rate is reduced by depolarization, suggesting that there is a net inward movement of positive charge. In

[24]Blaustein, M. P. 1977. *Biophys. J. 20*: 79–111.

addition, direct measurements of sodium-dependent calcium efflux and calcium-dependent sodium influx provide direct evidence for a ratio of about 3:1. Finally, it is clear from energy considerations that the entry of at least three sodium ions is required for the extrusion of one calcium ion along its electrochemical gradient.

The energy calculations necessary to arrive at the stoichiometric relation between sodium entry and calcium extrusion are (1) how much energy is dissipated by the entry of one sodium ion down its concentration gradient and (2) how much energy is required to transport one calcium ion uphill out of the cell. The energy dissipated per mole of sodium entering the axon is given by the product of the charge carried by one mole of the ion and the voltage gradient down which it moves, or $zF(E_{Na} - V_m)$. Similarly, the work that is required to transport one mole of calcium out of the cell is given by the analogous expression: $zF(E_{Ca} - V_m)$. When, for squid axon, E_{Na} is taken as 58 mV and V_m as -77 mV, then the driving force for sodium entry is 135 mV and the energy dissipated on sodium entry is 1.3×10^4 joules/mole. When extracellular and intracellular calcium concentrations are taken as 10 mM and 300 nM, respectively, then E_{Ca} is 130 mV and the ion must be transported against a total gradient of 207 mV; the work that is required is 4×10^4 joules/mole. Consequently, the entry of at least three sodium ions is necessary to provide enough energy for the extrusion of one calcium ion. With lower levels of intracellular calcium, the mechanism cannot work with a 3:1 exchange ratio and calcium extrusion is dependent upon calcium–magnesium ATPase.

The free cytoplasmic calcium concentration in resting squid axon may be as low as 100 nM, but may rise to several times this value immediately inside the membrane during transient periods of calcium entry. The total intracellular calcium concentration is very much greater—about 50 μM. The bulk of the calcium, then, is buffered by intracellular constituents, about 3 μM in mitochondria and the remainder in other organelles, principally endoplasmic reticulum.[25] This buffered calcium is presumably available for release under appropriate physiological conditions and the metabolically labile fraction, held in mitochondria, is released by metabolic inhibitors. This mechanism explains the apparently anomalous effect of cyanide on calcium extrusion.[21,26] In the presence of cyanide, the outward transport of calcium against its electrochemical gradient is dramatically *increased*. Paradoxically, a process that requires energy seems to function better when the ATP level of the cell has fallen too low to maintain other transport activities. The effect is due to inhibition of mitochondrial enzymes, allowing calcium to escape into the cytoplasm; the resulting increase in free calcium concentration then stimulates the sodium–calcium exchange mechanism (Figure 6A).

[25]Brinley, F. J. 1980. *Fed. Proc. 39*: 2778–2782.
[26]Blaustein, M. P. and Hodgkin, A. L. 1969. *J. Physiol. 200*: 497–528.

The intracellular concentration of hydrogen ions, like that of most other ionic constituents of the cytoplasm, is not at equilibrium with the concentration in the extracellular fluid. Instead, it is less than the equilibrium concentration, implying that there exists an active transport mechanism for extrusion of hydrogen ions or, alternatively, for inward transport of HCO_3^-. Thomas[27] used pH-sensitive microelectrodes to measure intracellular hydrogen ion concentration in snail neurons and found an average internal pH of 7.4 when the pH of the bathing solution was 8.0. If hydrogen ions were in equilibrium at a resting potential of -55 mV, the pH would have been 7.05; thus, the intracellular hydrogen ion concentration was about one-half that which would be expected if there were no active transport mechanism for the regulation of pH. To determine what mechanisms might be important, Thomas[28] acidified the cytoplasm of the neurons by intracellular injection of HCl or exposure to CO_2 and then measured the rate of recovery to normal pH under various experimental conditions. Recovery times were prolonged when extracellular bicarbonate concentration was reduced or when intracellular chloride was depleted, a finding suggesting that one mechanism for pH regulation was a chloride–bicarbonate exchange pump, carrying bicarbonate into the cell in exchange for chloride. Such a chloride–bicarbonate exchange mechanism exists in a variety of cells; it is inhibited by 4-acetamido-4'-isothiocyanostilbene-2,2'-disulfonic acid (SITS) and the related compound DIDS and is not an ATPase.[12] Both chloride and bicarbonate are carried against their electrochemical gradient by the exchange mechanism; the energy required appears to be obtained by coupling to inward sodium movement. This idea is supported by the observation that in snail neurons recovery from acidification was greatly prolonged by removal of external sodium. Finally, with respect to internal ionic concentration, there was a fall in intracellular chloride and an increase in intracellular sodium during recovery from intracellular acidification. The results, then, were consistent with pH regulation by a transport mechanism that carries sodium and bicarbonate into the cell in exchange for chloride. As there was no change in resting potential during the recovery period, the system did not appear to be electrogenic; therefore, the stoichiometry would have to be consistent with electrical neutrality, for example, a $Na:Cl:HCO_3^-$ ratio of 1:1:2. Alternatively, as proposed by Thomas largely on grounds of symmetry, a hydrogen ion could be transported outward instead of the extra bicarbonate inward. The latter view is supported by similar experiments on crayfish neurons.[29] More than 90 percent of the recovery from acidification was abolished when sodium was removed from the bathing solution. SITS was shown to slow the recovery by about 45 percent and abolish its dependence on bicarbonate, but the remaining recovery was still dependent on extracellular sodium. It was concluded that two

[27]Thomas, R. C. 1974. *J. Physiol. 238*: 159–180.
[28]Thomas, R. C. 1977. *J. Physiol. 273*: 317–338.
[29]Moody, W. J. 1981. *J. Physiol. 316*: 293–308.

Chloride transport

separate mechanisms were involved in pH regulation, a bicarbonate-independent sodium–hydrogen ion exchange and a chloride–bicarbonate exchange, both driven by the electrochemical gradient for sodium.

As noted in Chapter 5, chloride is usually not at its equilibrium concentration in nerve cells or in muscle cells. One mechanism for outward transport of chloride has already been discussed in the previous paragraph, namely, the SITS-sensitive, sodium-driven, chloride–bicarbonate exchange mechanism. Other cells, however, *accumulate* chloride, among them vertebrate skeletal muscle and squid axon. It has been shown by Russell[30] that in squid axon inward chloride transport requires intracellular ATP and both sodium and potassium in the extracellular bathing medium. It appears then that chloride is carried into the cell together with the two cations. The co-transport system is insensitive to ouabain and DIDS and is blocked by furosemide and bumetamide, substances shown to block chloride transport in other tissues. The transport system is electrically neutral, the Na:K:Cl ratio being 1:1:2.

In summary, none of the major ions in the cytoplasm of nerve cells is in equilibrium. The bulk of the energy required to maintain the intracellular ionic concentrations is expended in the operation of the sodium–potassium pump. The inward gradient for sodium thus maintained serves to drive the sodium–calcium and bicarbonate–chloride exchange systems, maintaining low intracellular calcium and regulating intracellular pH. Finally, although the chloride accumulation mechanism requires ATP, the requirement appears to be indirect, as the sodium concentration gradient is more than sufficient to drive the transport of the other two ions.

Lithium

An ion of interest mentioned earlier is lithium. Although this ion can enter the voltage-sensitive sodium channels, it cannot be transported in place of sodium by the sodium–potassium exchange pump. Nevertheless, neurons and other cells are somehow able to extrude it. This is important for higher functions of the central nervous system. It is known that patients with a mental disorder—manic-depressive illness—can be helped with small doses of lithium given orally. The improvement in both mania and depression can be dramatic. Recent studies have shown that lithium–sodium exchange occurs in excitable cells, transporting lithium out of the cytoplasm.[31]

SUGGESTED READING

Reviews

Hobbs, A. S. and Albers, R. W. 1980. The structure of proteins involved in active membrane transport. *Annu. Rev. Biophys. Bioeng. 9*: 259–291.

Post, R. L. 1981. The sodium and potassium ion pump. In T. P. Singer and R. N. Ondarza (eds.), *Molecular Basis of Drug Action*. Elsevier, New York, pp. 299–331.

[30]Russell, J. M. 1983. *J. Gen. Physiol. 81*: 909–925.
[31]Ehrlich, B. E. and Diamond, J. M. 1980. *J. Memb. Biol. 52*: 187–200.

Schuurmans Stekhoven, F. and Bonting, S. L. 1981. Transport adenosine tri-phosphatases: Properties and functions. *Physiol. Rev. 61*: 1–76.

Thomas, R. C. 1972a. Electrogenic sodium pump in nerve and muscle cells. *Physiol. Rev. 52*: 563–594.

Original papers

Baker, P. F., Blaustein, M. P., Keynes, R. D., Manil, J., Shaw, T. I. and Stein-hardt, R. A. 1969. The ouabain-sensitive fluxes of sodium and potassium in squid giant axons. *J. Physiol. 200*: 459–496.

Blaustein, M. P. 1977. Effects of internal and external cations and of ATP on sodium–calcium and calcium–calcium exchange in squid axons. *Biophys. J. 20*: 79–111.

Brinley, F. J. 1980. Regulation of intracellular calcium in squid axons. *Fed. Proc. 39*: 2778–2782.

Hodgkin, A. L. and Keynes, R. D. 1955. Active transport of cations in giant axons from *Sepia* and *Loligo*. *J. Physiol. 128*: 28–60.

Moody, W. J. 1981. The ionic mechanism of intracellular pH regulation in crayfish neurons. *J. Physiol. 316*: 293–308.

Mullins, L. J. and Brinley, F. J. 1967. Some factors influencing sodium extrusion by internally dialyzed squid axons. *J. Gen. Physiol. 50*: 2333–2355.

Russell, J. M. 1983. Cation-coupled chloride influx in squid axon. Role of potassium and stoichiometry of the transport process. *J. Gen. Physiol. 81*: 909–925.

Thomas, R. C. 1969. Membrane current and intracellular sodium changes in a snail neurone during extrusion of injected sodium. *J. Physiol. 201*: 495–514.

Thomas, R. C. 1972b. Intracellular sodium activity and the sodium pump in snail neurons. *J. Physiol. 220*: 55–71.

Thomas, R. C. 1977. The role of bicarbonate, chloride and sodium ions in the regulation of intracellular pH in snail neurons. *J. Physiol. 273*: 317–338.

CHAPTER NINE

SYNAPTIC TRANSMISSION

Synaptic transmission—the transfer of signals from one cell to another—can be treated as an extension of the principles that have been discussed in relation to the resting membrane potential and the conduction of nerve impulses. Two distinct modes of transmission are known, one electrical and the other chemical.

At electrical synapses, currents generated by an impulse in the presynaptic nerve terminal spread directly to the next neuron through a low-resistance pathway. The sites for electrical communication between cells have been identified in electron micrographs as gap junctions, in which the usual intercellular space of several tens of nanometers is reduced to about 2 nm.

At chemical synapses, the fluid-filled gap between presynaptic and postsynaptic membranes prevents a direct spread of current. Instead, the nerve terminals secrete a specific substance, the neurotransmitter. This compound diffuses across the synaptic gap to the postsynaptic cell, where it changes the permeability of the membrane to specific ions, producing either an excitatory or an inhibitory synaptic potential. Whether the response is excitatory or inhibitory depends on the ion species carrying the synaptic current and their equilibrium potentials relative to the threshold for action potential initiation. In the best-studied case of synaptic excitation—the vertebrate neuromuscular junction—the chemical transmitter acetylcholine (ACh) produces a simultaneous increase in the permeability of the postjunctional membrane to sodium and potassium, leading to a depolarizing synaptic potential.

At chemical inhibitory synapses, the permeability to chloride or potassium is increased, driving the membrane potential away from threshold. Nerve cells in the central nervous system receive both excitatory and inhibitory synapses, the effects of which are graded. The generation of impulses depends on the balance between excitatory and inhibitory influences.

Slow synaptic potentials, both excitatory and inhibitory, have a delayed onset and may last for several minutes. They are produced by either increases or decreases in membrane permeability. The very slow time course suggests that several intermediate steps may intervene between the arrival of transmitter at the cell membrane and activation of the conductance change, steps possibly involving protein phosphorylation.

A second mode of inhibition is presynaptic; the inhibitory nerve terminal secretes its transmitter onto a neighboring excitatory nerve terminal. The inhibitory transmitter, in turn, causes a reduction in the output of excitatory transmitter. Presynaptic inhibition serves to eliminate selectively specific pathways converging on a cell without influencing the efficacy of others.

Initial approaches

In the second half of the nineteenth century, a vigorous discussion took place between proponents of the CELL THEORY, who considered that neurons were independent units, and those who believed that nerve cells were interconnected by protoplasmic bridges. It was not until the early part of the present century that the cell theory won general acceptance and most biologists started to think of nerve cells as being similar to other cells in the body. Although in retrospect the opponents of the cell theory may seem to have been unnecessarily stubborn, one should remember that at the time it was difficult to obtain convincing evidence to show whether or not there was continuity between neurons. It remained for electron microscopy to show that each neuron was surrounded completely by its own plasma membrane. Even so, it turns out that continuity still occurs, but in the form of tenuous connections of molecular dimensions between certain cells, connections that permit intercellular flow of ions and small molecules.

Over the past 20 years, a wide variety of modes of synaptic transmission have been discovered, in addition to simple, chemically mediated increases in permeability leading to excitation and inhibition. Such modes of transmission include electrical excitation and inhibition, combined chemical–electrical synapses, chemical synaptic changes produced by *reductions* in membrane permeability, and prolonged synaptic potentials mediated by chemical reactions in the postsynaptic cell. Because of this complexity, it is useful to review briefly the development of some of the relevant ideas.

Concept of chemical transmission between cells

In 1843 Du Bois-Reymond showed that flow of electric current was involved not only in muscle contraction but also in nerve conduction, and it required only a small extension of this idea to arrive at the conclusion that transmission of excitation between nerve and muscle was also brought about by current flow.[1] Du Bois-Reymond himself proposed, and in fact favored, an alternative explanation—the secretion by nerve of an excitatory substance that then caused contraction. However, the idea of animal electricity had a potent hold on people's thinking, and for almost 100 years no contrary evidence could budge the assumption of electrical transmission between nerve and muscle and, by extension, between nerve cells in general.

One reason that the idea of chemical transmission seemed unattractive was the fact that transmission between nerve and muscle and

---

[1]Du Bois-Reymond, E. 1848. *Untersuchungen über thierische Electricität*. Reimer, Berlin.

between nerve cells in the central nervous system was extremely rapid. No such difficulties existed in theory for autonomic nerves that innervate glands and blood vessels because they act relatively slowly. Accordingly, results of experiments on these tissues carried little weight in the argument. For example, Langley proposed that synaptic transmission in autonomic ganglia was chemical; his proposal was based on the selective action of nicotine in blocking transmission through the mammalian ciliary ganglion.[2] Eliot[3] in 1904 pointed out that an extract from the adrenal gland, adrenaline (epinephrine), mimicked the action of sympathetic nerves and suggested cautiously that it might be secreted by nerve terminals as a transmitter. He was discouraged from even publishing his results fully. Interesting sidelights on the men and their ideas on this subject are contained in the writings of Dale,[4] for several decades one of the leading figures in British physiology and pharmacology. Among his many contributions were the clarification of the action of acetylcholine (ACh) at synapses in autonomic ganglia and the

Henry Dale (left) and Otto Loewi, mid 1930s. (Courtesy of Lady Todd and W. Feldberg.)

establishment of its role in neuromuscular transmission. As one of our colleagues has delicately put it, "He walked in a cow pasture and trod only on daisies."

In 1921 Otto Loewi did a direct and simple experiment that established the chemical nature of transmission between the vagus nerve

[2]Langley, J. N. and Anderson, H. K. 1892. *J. Physiol.* *13*: 460–468.
[3]Elliot, T. R. 1904. *J. Physiol.* *31*: 20–26.
[4]Dale, H. H. 1953. *Adventures in Physiology.* Pergamon Press, London.

and the heart.[5] He perfused the heart of a frog and stimulated the vagus nerve, thereby causing stopping or slowing of the heartbeat. When the fluid from the inhibited heart was transferred to a second unstimulated heart, it too began to beat more slowly. Apparently the vagus nerve had released, upon stimulation, an inhibitory substance into the perfusion fluid. Loewi and his colleagues then showed by various bioassays that the substance was mimicked in every way by ACh. It is an amusing sidelight that Loewi had the idea for the experiment in a dream, wrote it down in the middle of the night, but could not decipher the writing the next morning. Fortunately, the dream returned and the second time Loewi took no chances; he rushed to the laboratory and performed the experiment. In a personal account he reflected:

> On mature consideration, in the cold light of morning, I would not have done it. After all, it was an unlikely enough assumption that the vagus should secrete an inhibitory substance; it was still more unlikely that a chemical substance that was supposed to be effective at very close range between nerve terminal and muscle be secreted in such large amounts that it would spill over and, after being diluted by the perfusion fluid, still be able to inhibit another heart.

Subsequently, in the early 1930s, support for the role of ACh as a chemical synaptic transmitter was extended to other autonomic synapses, particularly in sympathetic ganglia by Feldberg and his colleagues.[6] Then, in 1936, Dale and his colleagues demonstrated that stimulation of the motor nerve to mammalian skeletal muscle caused release of ACh.[7] In addition, when ACh was injected into arteries supplying the muscle, it caused a large synchronous twitch.

Pharmacological approaches were indispensable for such experiments. For example, the hydrolysis of ACh was prevented by a drug (eserine) that inhibits the enzyme cholinesterase. Another indispensable tool was curare (Figure 2A), an Indian arrow poison prepared at that time as a crude extract of cinchona bark; curare was used in the nineteenth century by Claude Bernard, who showed that it blocked neuromuscular transmission, and by Langley, who demonstrated a similar action in autonomic ganglia.

Synaptic potentials at the neuromuscular junction

A new phase in the study of synaptic transmission started with the use of improved electrical recording techniques on mammalian and amphibian nerve–muscle preparations. Figure 1 illustrates diagrammatically a vertebrate neuromuscular junction and a synapse on a nerve cell. The area of nerve–muscle contact is called an END PLATE, and the potential change recorded in that region following nerve stimulation is called an END PLATE POTENTIAL (epp).[8,9] It is now clear that there is no essential difference between the epp and the excitatory SYNAPTIC PO-

[5]Loewi, O. 1921. *Pflügers Arch. 189*: 239–242.
[6]Feldberg, W. 1945. *Physiol. Rev. 25*: 596–642.
[7]Dale, H. H., Feldberg, W. and Vogt, M. 1936. *J. Physiol. 86*: 353–380.
[8]Göpfert, H. and Schaefer, H. 1938. *Pflügers Arch. 239*: 597–619.
[9]Eccles, J. C. and O'Connor W. J. 1939. *J. Physiol. 97*: 44–102.

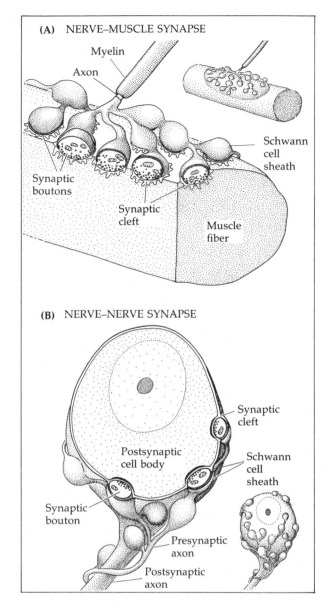

(A) NERVE–MUSCLE SYNAPSE

Myelin

Axon

Schwann
cell
sheath

Synaptic
boutons

Synaptic
cleft

Muscle
fiber

(B) NERVE–NERVE SYNAPSE

Synaptic
cleft

Postsynaptic
cell body

Schwann
cell
sheath

Synaptic
bouton

Presynaptic
axon

Postsynaptic
axon

1

STRUCTURAL FEATURES OF CHEMICAL SYNAPSES. (A) The end plate formed by a motor nerve on a skeletal muscle fiber of the snake consists of a collection of synaptic boutons. Boutons contain synaptic vesicles and mitochondria and are always separated from the postsynaptic cell by a cleft about 50 nm wide. They are covered by fine lamellae of Schwann cells. (B) Similar synaptic boutons are distributed on a ganglion cell in the heart of the frog. (After McMahan and Kuffler, 1971.)

TENTIAL seen in nerve cells. A synaptic potential that excites a postsynaptic cell is usually referred to as an epsp (excitatory postsynaptic potential) and one that inhibits as an ipsp.

Electrical recording techniques were used in experiments such as that illustrated in Figure 2 to test in some detail the idea that neuromuscular transmission was mediated by the release of ACh from the nerve terminals and its subsequent action on the postsynaptic membrane.[10] Stimulating electrodes were placed on the nerve and a pair of

[10]Eccles, J. C., Katz, B. and Kuffler, S. W. 1942. *J. Neurophysiol. 5*: 211–230.

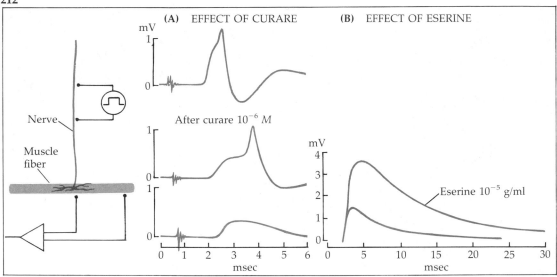

2 **EFFECT OF CURARE AND ESERINE on end plate potential. At the neuromuscular junction, nerve stimulation leads to a synaptic potential that initiates a muscle impulse (upper record in A). In the presence of curare, the separation of these two events is more pronounced (middle record), and eventually only the end plate potential remains (lower record). In (B) the curarized end plate potential (lower record) is increased in size and prolonged by eserine, which prevents hydrolysis of ACh. The potentials were recorded with extracellular electrodes. (A after Kuffler, 1942; B after Eccles, Katz, and Kuffler, 1942.)**

recording electrodes on the muscle, one of these very close to the end plate region. Nerve stimulation then produced a diphasic action potential, recorded extracellularly from the muscle fibers (A, upper record). When curare was added to the solution bathing the muscle, the action potential was delayed and was preceded by a slower depolarization at the end plate region (middle record). Eventually, in the presence of curare, the action potential failed completely, leaving only the slow depolarization (the end plate potential). The addition of eserine to the bathing solution (B) produced a marked increase in the size and duration of the potential, consistent with its anticholinesterase action; that is, because of the action of eserine in preventing hydrolysis, ACh is allowed to remain longer in the synaptic cleft and therefore produces a larger and more prolonged action on the postsynaptic membrane. Experiments such as this, then, provided important evidence in favor of chemical transmission at the neuromuscular junction. In addition, it was shown by Kuffler[11] that currents generated by impulses in the presynaptic nerve terminals did not spread effectively into the postsynaptic region of the muscle and could not be responsible for the end plate potential. More detailed information became available after the

[11]Kuffler, S. W. 1948. *Fed. Proc.* 7: 437–446.

Bernard Katz (right),
John Eccles and friend
in Australia, about
1941.

introduction of fine-tipped glass micropipettes for intracellular record-ing.[12] The technique was used by Fatt and Katz[13] to study in detail the time course and spatial distribution of the epp in frog muscle fibers. They stimulated the nerve in a curarized muscle and recorded the epp intracellularly at various distances from the end plate (Figure 3). At the end plate region of the muscle, the potential rose rapidly to a peak and then declined slowly over the next 10–20 msec. As they moved the recording microelectrode farther and farther away from the end plate, the epp amplitude became progressively smaller and its time to peak progressively longer. By analyzing the decay of the potential with dis-tance, Fatt and Katz were able to show that it is similar to the decay of an electrotonic potential with increasing distance from the point of current injection (Chapter 7) and is determined solely by the passive electrical characteristics of the muscle fiber. In other words, the synaptic potential is generated only at the end plate region and decays passively with distance, rather than being propagated actively along the muscle fiber. In addition, Fatt and Katz showed that after the epp reaches its peak, its falling phase is also passive. In summary, the synaptic potential at the end plate is generated by a brief surge of current which flows into the membrane and causes a rapid depolarization. The potential then decays passively, spreading outward in both directions as it dies away. Fatt and Katz proposed that the inward current underlying the rising phase of the potential was produced by a general increase in membrane permeability, as if ACh opened nonselective holes or pores in the membrane.

[12]Ling, G. and Gerard, R. W. 1949. *J. Cell. Comp. Physiol. 34*: 383–396.
[13]Fatt, P. and Katz, B. 1951. *J. Physiol. 115*: 320–370.

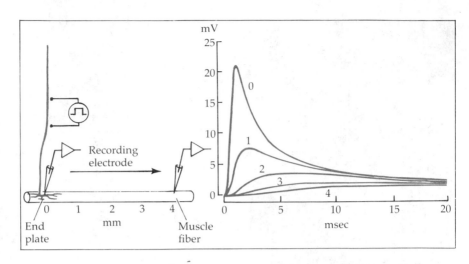

3 SYNAPTIC POTENTIALS recorded at various distances from the end plate show a decline in size and slowing of rise time with distance from the site of release of ACh. An intracellular electrode was inserted into the muscle fiber successively at distances of 1, 2, 3, and 4 mm from the end plate. (After Fatt and Katz, 1951.)

Shortly after the introduction of the microelectrode, glass micropipettes were used also for discrete application of ACh (and later of other drugs) to the end plate region of muscle.[14] The technique is illustrated in Figure 4A. A microelectrode is inserted into the end plate of a single muscle fiber for recording while an ACh-filled micropipette is held just outside the cell. To apply ACh to a localized region of the membrane, a brief current is passed through the pipette, causing positively charged ACh ions to be ejected from the tip by IONOPHORESIS. With this type of experiment, it was shown that ACh depolarized the muscle fiber membrane only when applied to the end plate region and only from the outside; intracellular injection was without effect.[15] The ionophoretic technique made it possible to map in detail the distribution of postsynaptic receptors for ACh in nerve cells[16] (Figure 4B) and muscle fibers.[17] The effect of ionophoretic application of ACh to a parasympathetic ganglion cell is compared with the effect of preganglionic stimulation in Figure 4C. Acetylcholine application from a micropipette produces a depolarization (upper left) that is indistinguishable from that produced by stimulation of the preganglionic nerve trunk (upper right). Applications of larger doses of drug and of stronger stimuli to the preganglionic nerve trunk both result in action potential initiation in the cell (lower records). The close correlation between ionophoresis and stim-

[14]Nastuk, W. L. 1953. *Fed. Proc. 12*: 102.

[15]del Castillo, J. and Katz, B. 1955. *J. Physiol. 128*: 157–181.

[16]Dennis, M. J., Harris, A. J. and Kuffler, S. W. 1971. *Proc. R. Soc. Lond. B 177*: 509–539.

[17]Miledi, R. 1960. *J. Physiol. 151*: 24–30.

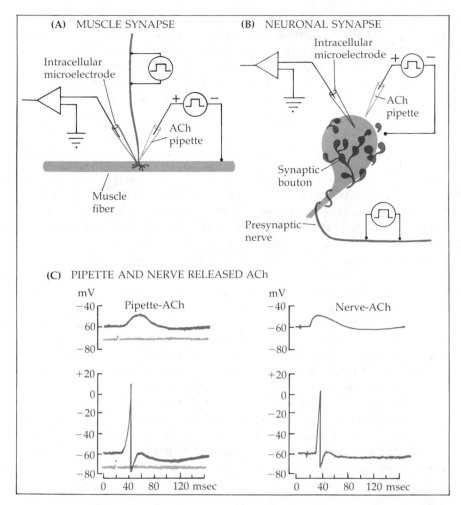

4 IONOPHORESIS OF ACh. ACh applied by a pipette to a neuromuscular junction (A) and to a synapse on a ganglion cell (B) mimics nerve stimulation. (C) shows a nerve-evoked synaptic potential (upper right) that in the lower record initiates an impulse. The records on the left were obtained from the same cell and show responses to brief pulses through an ACh-filled micropipette. Potentials were recorded with an intracellular electrode. (After Dennis, Harris, and Kuffler, 1971.)

ulation is obtained only when the ACh-filled pipette is in the immediate vicinity of a synaptic bouton; application more than a few micrometers away from a bouton results in only a very small, prolonged depolarization, presumably as a result of diffusion of ACh from the pipette into the synaptic region.

The experiments with intracellular recording and ionophoresis served to establish firmly the idea of chemical synaptic transmission, and the alternate concept of electrical transmission at synapses almost disappeared. However, the definite establishment of chemical transmission came just as electrical interactions were being discovered between nerve cells and in time for the first direct demonstration of electrical synaptic transmission by Furshpan and Potter in 1959.[18] This was an instructive and, in a sense, amusing event: The long-championed hypothesis of electrical synaptic transmission, which had been accepted almost universally for about 100 years on inadequate evidence, was finally forsaken by its supporters for good experimental reasons, only to be shown valid after all at a different kind of synapse.

ELECTRICAL SYNAPTIC TRANSMISSION

The principle of electrical synaptic transmission is simply that the current flow generated by an action potential and responsible for conduction along an axon is also capable of transferring depolarization directly across a synapse. Passage of current from one cell to the next requires minimal interruption of electrical continuity at the contact area. This principle is illustrated in Figure 5. In the first model (Figure 5A), the adjoining cell membranes of two neurons, A and B, are closely apposed. For electrical transmission to work, the resistance to current flow through the contact area (r_c) must be relatively low compared with the transverse membrane resistance. If current is passed into cell A from a microelectrode, some of that current will flow outward across the membrane of cell A and some through the apposed membranes into cell B. If electrical transmission is to be effective, a significant fraction of the current must flow through the area of apposition and outward across the surface membrane of cell B. How will the current be divided between the cells? This can be determined by considering the resistances involved. If cells A and B are considered to be cylindrical conductors as drawn, they each represent a cable extending in one direction only. Their input resistances (Chapter 7) will be given by

$$r_{input} = (R_m R_i/2\pi^2 a^3)^{1/2}$$

where R_m and R_i are the specific resistances of the cell membrane and the cytoplasm, respectively, and a is the fiber radius. If the area of contact of the cells consists of two intimately apposed surface membranes, then the resistance across the contact will be

$$r_c = 2R_m/\pi a^2$$

Suppose now that R_m is 2000 Ωcm^2 for both cells and R_i is 100 Ωcm (both reasonable values for nerve cells) and that a is 10 μm. The r_{input} will be 3.2 MΩ and r_c 1300 MΩ. The pathway for current outward through the membrane of cell A, then, has a resistance of 3.2 MΩ; the resistance for current flow through the junction and outward through cell B is 1303.2 MΩ. Consequently, virtually all the injected current will flow out through cell A and none through the junction. Furthermore, even if a significant amount of current did flow into cell B, most of the

---

[18]Furshpan, E. J. and Potter, D. D. 1959. *J. Physiol. 145*: 289–325.

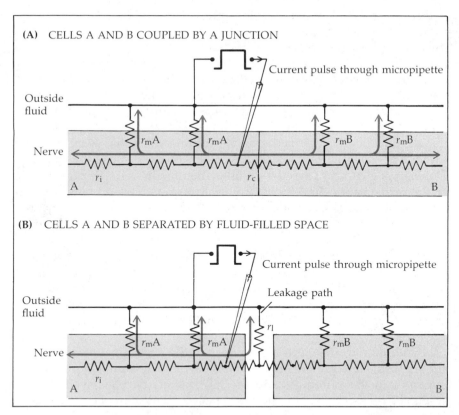

(A) CELLS A AND B COUPLED BY A JUNCTION

Current pulse through micropipette

Outside fluid

Nerve

r_mA r_mA r_mB r_mB

r_i

r_c

A

B

(B) CELLS A AND B SEPARATED BY FLUID-FILLED SPACE

Current pulse through micropipette

Leakage path

Outside fluid

r_l

Nerve

r_mA r_mA r_mB r_mB

r_i

A

B

5 **PATHWAYS FOR CURRENT FLOW during electrical synaptic transmission. (A) A model of two cells in close contact; in this model, current can flow from cell A to cell B. The coupling resistance (r_c) must be low compared with the other membrane resistances (r_m). (B) The cells are separated by a fluid-filled gap. Current flow between cells A and B is excluded.**

voltage drop would be produced across the high-resistance junction rather than across the relatively low-resistance surface membrane where depolarization is needed to initiate an action potential. The situation between most nerve cells is even more unfavorable. As shown in Figure 5B, the membranes in the synaptic region are separated by a fluid-filled gap of the order of 30 nm, providing a low-resistance pathway to the extracellular fluid. Thus, any current leaving the end of cell A is shunted away from cell B. Clearly, electrical transmission is excluded unless two requirements are met: The intercellular resistance must be made very much smaller and current must be prevented from escaping from the junctional region. As will be discussed later, this is accomplished by a special structural arrangement allowing electrical continuity to be established between one neuron and the next. An additional requirement is that there must be a reasonable match between the sizes of the presynaptic and postsynaptic elements for effective transmission to occur. A small presynaptic bouton—say, 1 μm in diameter—could not be expected to produce enough current to depolarize a 50-μm diameter cell to any significant degree. To return to the earlier analogy of the conduction of heat (Chapter 7), a hot knitting needle cannot heat up a cannon ball.

In their detailed study of electrical synaptic transmission, Furshpan and Potter used the synapse between a large nerve fiber (the lateral giant fiber, Figure 6) in the abdominal nerve cord of the crayfish and a motor nerve that innervates fast-acting flexor muscles of the tail (the giant motor fiber). Stimulating and recording electrodes were inserted into the presynaptic and postsynaptic fibers on each side of the synapse. Depolarizing current passed with little attenuation from the presynaptic to the postsynaptic fiber (orthodromically), and an action potential initiated in the lateral giant fiber was conducted into the giant motor fiber (Figure 6A) without the characteristic delay associated with chemical synapses. Another interesting property of the synapse is that flow of depolarizing current was permitted in the orthodromic direction but not antidromically from postsynaptic element to presynaptic element (Figure 6B). In other words, the synapse shows RECTIFYING properties that allow current to flow only in one direction. This allows transmission to go forward but not antidromically; like a chemical synapse, it is polarized.

In addition to the electrical synapse in the abdominal nerve cord of the crayfish, electrical transmission has been demonstrated at numerous synapses in the vertebrate and the invertebrate nervous systems. For example, cells coupled to each other electrically have been found in nervous systems of annelids, mollusks, and arthropods.[19] In the medulla of the puffer fish, groups of large cells are coupled electrically.[20] Electrical coupling has been demonstrated between motoneurons in the spinal cord of the frog[21] and a corresponding ultrastructural specialization for the site of coupling has been described.[22] Electrical synaptic transmission has also been described in the central nervous system of mammals.[23] Unlike the crayfish giant synapse, most electrical synapses do not exhibit rectification, but conduct equally well in both directions.

Shortly after the demonstration of electrical synaptic transmission, yet another type of synapse was discovered in which electrical synaptic transmission and chemical synaptic transmission were combined. This dual type of synapse was first described in the ciliary ganglion of the chick,[24] the neurons of which regulate the accommodation of the lens of the eye and the diameter of the iris, and has since been found in a number of cells, including spinal interneurons of the lamprey[25] and frog spinal motoneurons.[26] Perhaps more frequently, postsynaptic cells re-

[19]Bennett, M. V. L. 1974. *Synaptic Transmission and Neuronal Interaction.* Raven Press, New York, pp. 153–178.

[20]Bennett, M. V. L. 1973. *Fed. Proc. 32*: 65–75.

[21]Grinnell, A. D. 1970. *J. Physiol. 210*: 17–43.

[22]Sotelo, C. and Taxi, J. 1970. *Brain Res. 17*: 137–141.

[23]Llinás, R., Baker, R. and Sotelo, C. 1974. *J. Neurophysiol. 37*: 560–571.

[24]Martin, A. R. and Pilar, G. 1963. *J. Physiol. 168*: 443–463.

[25]Rovainen, C. M. 1967. *J. Neurophysiol. 30*: 1024–1042.

[26]Shapovalov, A. I. and Shiriaev, B. I. 1980. *J. Physiol. 306*: 1–15.

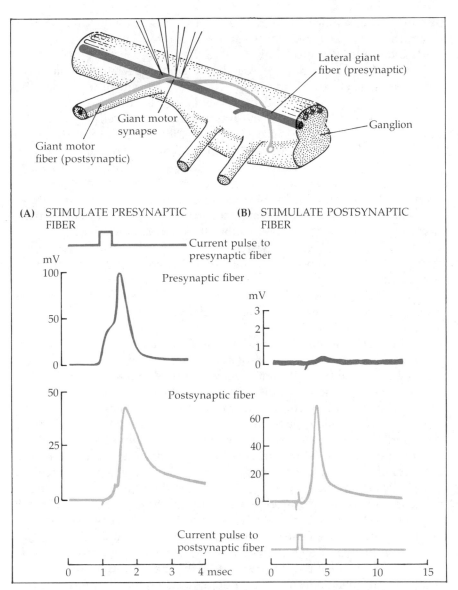

6 **ELECTRICAL TRANSMISSION** between two nerve fibers in the nerve cord of the crayfish. Stimulating and recording electrodes are inserted into the presynaptic and postsynaptic axons. The current flow generated by an impulse in the presynaptic axon (A) spreads to produce an impulse in the postsynaptic axon. Because of the electrical continuity between the two, no signficant synaptic delay occurs between the impulses. In contrast, postsynaptic depolarization by a nerve impulse (B) does not spread effectively into the presynaptic axon. (After Furshpan and Potter, 1959.)

ceive chemical and electrical synaptic inputs separately from different inputs converging upon them. This dual arrangement was seen originally in crayfish neurons and later in goldfish Mauthner cells and in the central nervous system of the leech. Some of these connections will

Structural basis of
electrical transmission:
The gap junction

be considered in more detail in Chapter 18 to illustrate the diverse ways in which converging influences can be integrated.

The discovery of electrical coupling between a variety of excitable cells stirred up an intensive search for some specialized relation of membranes at electrical junctions. There is now strong evidence that such coupling is represented morphologically by the GAP JUNCTION.[27] The gap junction is a region of close apposition of two cells characterized by aggregates of particles in each of the adjoining membranes (Figure 7). The particles range from 6 to 15 nm in diameter, depending on the tissue, and may be distributed in a hexagonal array or irregularly, depending on fixation procedures. The particles are exactly paired to span the 2- to 3-nm gap in the region of contact. The unit constituted by a particle pair has been termed a CONNEXON.[28] The individual particles are composed of six protein subunits about 2.5 nm in diameter and 7.5 nm long in a slightly twisted circular array with a 2-nm central opening.[29] The opening, then, presumably represents the pathway for current flow and movement of small molecules between cells. Electrophysiological evidence has been obtained that each particle of the pair making up the connexon behave as independently regulated channels in series.[30]

Significance of gap
junctions

Cell-to-cell channels are widely distributed phylogenetically from sponges to humans in tissues of mesenchymal and epithelial origin. Generally a given cell within a tissue is coupled to its neighbors so that whole organs or subdivisions of organs are coupled from within. In addition to permitting the passage of current between cells, the intercellular channels allow passage of small molecules up to about 2 nm in diameter. Coupling may serve a variety of functions, such as tissue homeostasis among large numbers of cells and transmission of regulatory signals from one cell to the next.[27] Coupling is widespread in embryos,[31] even between cells with quite different future functional roles, and such cells uncouple as development progresses. One example of embryonic coupling in the nervous system is between sensory cells (Rohon-Beard cells) in the spinal cord of tadpole larvae.[32]

Uncoupling can occur under various circumstances. Embryonic cells in the tadpole uncouple naturally at about the time they develop the ability to produce sodium action potentials. Other cells can be uncoupled by depolarization of one or both cells on either side of the junction or by changes in the potential across the junction. In addition, changes

[27]Loewenstein, W. 1981. *Physiol. Rev. 61*: 829–913.

[28]Caspar, D. L. D., Goodenough, D. A., Makowski, L. and Phillips, W. C. 1977. *J. Cell Biol. 74*: 605–628.

[29]Unwin, P. N. T. and Zampighi, G. 1980. *Nature 283*: 545–549.

[30]Obaid, A. L., Socolar, S. J. and Rose, B. 1983. *J. Memb. Biol. 73*: 68–89.

[31]Sheridan, J. D. 1978. In *Intercellular Junctions and Synapses*. Chapman and Hall, London, pp. 39–59.

[32]Spitzer, N. C. 1982. *J. Physiol. 330*: 145–162.

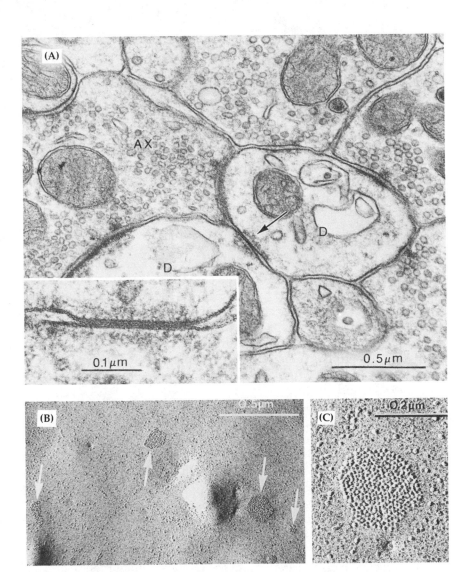

7   GAP JUNCTIONS between neurons. (A) Two dendrites (labeled D) in
the inferior olivary nucleus of the cat are joined by a gap junction
(arrow). To the left is an axon terminal (Ax) that contains numerous
vesicles and makes chemical synapses on both dendrites. The inset
shows a high magnification of a gap junction. The usual space between
the cells is almost obliterated in the contact area, which is traversed by
cross bridges. (B) Freeze-fracture through the presynaptic membrane of
a nerve terminal that forms gap junctions with a neuron in the ciliary
ganglion of a chicken. A broad area of the cytoplasmic fracture face is
exposed, showing numerous specialized areas that correspond to sites
of gap junctions (arrows). (C) Higher magnification of fracture face; a
cluster of closely packed particles, about 9 nm in diameter, is seen
within the membrane. The particles are assumed to form part of the
channel system between the presynaptic terminal and the postsynaptic
cell. For freeze-fracture details, see Chapter 11, Figure 3. (A from Sotelo,
Llinás, and Baker, 1974; B, C from Cantino and Mugnaini, 1975.)

in the chemical composition of the cytoplasm, such as increased intracellular calcium or decreased intracellular pH, cause uncoupling.[30]

In terms of synaptic transmission between nerve cells, there are several possible functions for electrical coupling. One advantage of an electrical synapse is the absence of the synaptic delay of 0.5 to 1.0 msec associated with chemical synaptic transmission. Electrical synapses give a greater speed of transmission and possibly a greater certainty that an impulse will be transmitted from the presynaptic to the postsynaptic cell. For example, an electrical synapse located on the Mauthner cell of the goldfish brain is involved in a rapid reflex reaction whereby the fish gives a strong tail flip, suitable for escape when the water is disturbed by a predator. In such a case, a saving of a fraction of a millisecond may be important for survival. Electrical coupling can also produce subthreshold or integrative actions between nerve cells. The effects depend on the degree of electrical coupling between cells. This is usually expressed as a COUPLING RATIO, a ratio of 1:2 meaning that half the total presynaptic voltage change appears in the postsynaptic cell. Thus, stronger or weaker coupling between nerve cells represents one way to grade and control the amount of current flowing between neurons and hence their interactions with each other. In addition, multiple electrical synapses converging on a neuron have simple additive effects with little fluctuation, compared with chemical synaptic processes to be discussed in the next section.

Electrical inhibition   In addition to producing excitation, current flow between cells can, surprisingly, produce inhibition at at least one specialized site. The one known example of electrical inhibition was demonstrated in a series of rigorous experiments by Furukawa and Furshpan.[33] The observations were made on the Mauthner cell in the medulla of the goldfish. During electrical inhibition, the fluid space just OUTSIDE the Mauthner cell is made more POSITIVE by current flow from the presynaptic terminals. This occurs in the initial segment of the axon where impulses are initiated. The effect is analogous to that of an external electrode placed in the region of the initial segment and made positive with respect to a remote electrode on another part of the cell. The phenomenon appears to require specialized anatomical features, particularly an axon cap surrounding the initial segment of the axon and enclosing numerous presynaptic fibers.

CHEMICAL SYNAPTIC TRANSMISSION

Some features of the mechanism of chemical synaptic transmission have already been described. Certain obvious questions arise when one considers the elaborate scheme for transmission that entails the secretion of a specific chemical by a nerve terminal. How does the terminal liberate the chemical? Does the action potential with its characteristic

[33]Furukawa, T. Y. and Furshpan, E. J. 1963. *J. Neurophysiol.* 26: 140–176.

sequence of permeability changes have some specific feature that is specialized to cause secretion of transmitter? Or is secretion activated by any depolarization of the terminal? The release process will be considered in detail in Chapter 10; the present discussion is concerned with the question of how neurotransmitters act on the postsynaptic cell to produce excitation or inhibition. Many of the pioneering studies of chemical synaptic transmission were done on relatively simple preparations, such as the neuromuscular junction of the frog. This particular preparation has the great advantage that the neurotransmitter has been definitely identified as acetylcholine. It has allowed a detailed examination of the mechanism of action of ACh on the postsynaptic membrane and of the way in which ACh is released from the motor nerve terminals. Most chemical synapses between neurons, particularly in the brain, are far less easy to study. In fact, in only a few specific examples can the transmitters be identified and rarely can the synaptic mechanisms be analyzed in detail.

Once electrical transmission at the neuromuscular junction has been ruled out, the question that then arises is how does ACh produce a depolarization in the postjunctional membrane? One possibility considered by Fatt and Katz[13] is that ACh itself enters the membrane. If a sufficient number of positively charged ACh ions moved along their electrochemical gradients into the cell, the membrane would be depolarized. However, calculations show that ACh itself cannot be the source of the synaptic current. The amount liberated by each impulse is many thousands of times too small to account for the size of the synaptic potential. In other words, for each ACh molecule liberated, several thousand ions cross the postsynaptic membrane. This illustrates one property of chemical synapses: They provide a very large amplification mechanism.

General mechanism of the action of ACh

Experiments by Fatt and Katz led them to the important conclusion that ACh produces a marked, nonspecific increase in permeability of the postsynaptic membrane to small ions. It was found later that the permeability increase was only to cations,[34,35] the major ones present being, of course, sodium and potassium. What are the consequences of assuming that the postsynaptic membrane becomes equally permeable to sodium and potassium under the influence of ACh? Such an increase in permeability would tend to drive the membrane potential to a level between E_K and E_{Na}, close to zero. Thus, when the membrane potential is at its resting level (for example, -90 mV), ACh would cause inward current and consequent depolarization. If, on the other hand, the membrane were already depolarized (for example, to $+30$ mV), ACh would then cause outward current and repolarization. At some intermediate potential there would be no net current produced by ACh; that is, the outward potassium current would be exactly equal to the inward sodium current. This potential is called the REVERSAL POTENTIAL (V_r). Fatt

Ionic permeabilities and reversal potential

[34]Takeuchi, A. and Takeuchi, N. 1960. *J. Physiol. 154*: 52–67.
[35]Jenkinson, D. H. and Nicholls, J. G. 1961. *J. Physiol. 159*: 111–127.

and Katz were able to estimate the reversal potential in a very ingenious way. They initiated successive action potentials in a muscle fiber by direct stimulation and then stimulated the nerve so that the release of ACh occurred at various times in relation to the arrival of the action potential at the end plate region. When the release of ACh was timed to occur at the peak of the action potential, the peak amplitude was reduced. Release of ACh on the rising or falling phase of the action potential produced depolarization when the membrane potential at the time was more negative than about −15 mV or repolarization when the membrane was more positive than −15 mV (Figure 8). They concluded, then, that the reversal potential was about −15 mV.

Ionic conductance changes produced by ACh

Two techniques were used to access the permeability changes produced by ACh. One involved the use of radioactive isotopes, which showed that the permeability of the postsynaptic membrane was increased to both sodium and potassium (and also calcium) but not to chloride.[35] This experiment provided convincing evidence concerning the ion species involved but revealed nothing about details of the conductance changes, such as their timing or voltage dependence. This information was provided by voltage clamp experiments, first performed by the Takeuchis.[34] The experimental arrangement is shown in Figure 9A. Two microelectrodes were inserted into the end plate region of a frog muscle fiber, one for recording membrane potential (V_m) and the other for current injection to clamp the membrane potential at the desired level. The nerve was then stimulated to release ACh, or, in later experiments, ACh was applied directly by ionophoresis. The Takeuchis found that ACh produced a transient inward current at the normal resting potential, which, when the membrane was depolarized, decreased toward zero at an estimated reversal potential of −15 mV. Changing the concentrations of sodium, potassium, or calcium in the bathing solution resulted in changes in reversal potential, but changes in extracellular chloride did not. They concluded that the effect of ACh was to produce an increase in permeability of the postsynaptic membrane to cations.

Subsequently, more detailed studies of the effect of membrane potential on synaptic current were carried out in a number of laboratories.

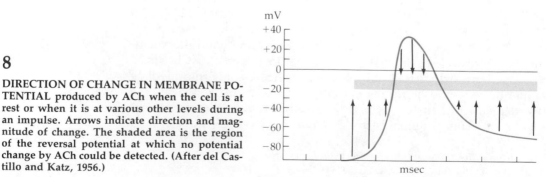

8

DIRECTION OF CHANGE IN MEMBRANE POTENTIAL produced by ACh when the cell is at rest or when it is at various other levels during an impulse. Arrows indicate direction and magnitude of change. The shaded area is the region of the reversal potential at which no potential change by ACh could be detected. (After del Castillo and Katz, 1956.)

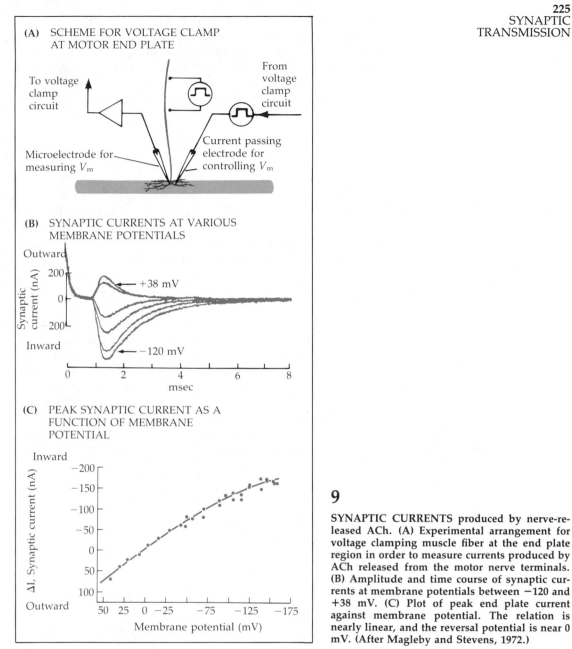

(A) SCHEME FOR VOLTAGE CLAMP AT MOTOR END PLATE

To voltage clamp circuit

From voltage clamp circuit

Microelectrode for measuring V_m

Current passing electrode for controlling V_m

(B) SYNAPTIC CURRENTS AT VARIOUS MEMBRANE POTENTIALS

Outward

Synaptic current (nA)

200

0

200

Inward

+38 mV

−120 mV

0   2   4   6   8

msec

(C) PEAK SYNAPTIC CURRENT AS A FUNCTION OF MEMBRANE POTENTIAL

Inward

ΔI, Synaptic current (nA)

−200

−150

−100

−50

0

50

100

Outward

50 25 0 −25   −75   −125   −175

Membrane potential (mV)

9

SYNAPTIC CURRENTS produced by nerve-released ACh. (A) Experimental arrangement for voltage clamping muscle fiber at the end plate region in order to measure currents produced by ACh released from the motor nerve terminals. (B) Amplitude and time course of synaptic currents at membrane potentials between −120 and +38 mV. (C) Plot of peak end plate current against membrane potential. The relation is nearly linear, and the reversal potential is near 0 mV. (After Magleby and Stevens, 1972.)

An example from experiments by Magleby and Stevens[36] is shown in Figure 9B. The muscle was treated with hypertonic glycerol, which disables the contractile mechanism and leaves the fibers in a somewhat depolarized state. At the resting potential (−40 mV), nerve stimulation produces an inward current of about 150 nA. At more negative holding

[36]Magleby, K. L. and Stevens, C. F. 1972a. *J. Physiol.* 223: 151–171.

potentials, up to −120 mV, the current increases. At depolarized levels of +21 and +38 mV, the synaptic current is outward. When current is plotted against holding potential, the reversal potential is found to be near zero (Figure 9C). The relation between voltage and current is approximately linear, so that the peak current (ΔI) is roughly proportional to the difference between the membrane potential (V_m) and the reversal potential. Thus,

$$\Delta I = \Delta g(V_m - V_r)$$

where Δg is the increase in membrane conductance produced by the ACh.

Whereas the peak change in conductance produced by ACh is little influenced by membrane potential, the same does not apply to the time course of the conductance change. At the neuromuscular junction, the decay of the synaptic current is much more rapid at depolarized than at hyperpolarized levels of membrane potential (Figure 9B); at other synapses the time course of decay of the synaptic current may either decrease or increase with membrane potential, or it may be unaffected by membrane potential changes.

The conclusion so far, then, is that ACh produces an increase in conductance to cations, particularly sodium and potassium. Because of the relatively low concentration of calcium normally present in extracellular fluids and its almost negligible intracellular concentration, the contribution of calcium to the overall synaptic current can be ignored, as can that of other cations, such as magnesium. In the electrical model shown in Figure 10, the resting membrane, with the usual sodium, potassium, and chloride channels, is shunted by ACh-activated channels for sodium and potassium. Note that although the ACh-activated

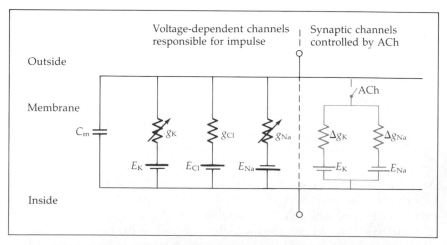

10 ELECTRICAL MODEL of the synaptic membrane activated by ACh in parallel with the remaining cell membrane, which has resting and voltage-sensitive conductances.

conductances for sodium and potassium are represented separately, current flows through them as a single unit. In other words, ACh does not activate *separate* sodium and potassium channels. However, it is assumed that sodium and potassium move independently through such channels, so that separate expressions can be written for the sodium and potassium currents (ΔI_{Na} and ΔI_K):

$$\Delta I_{Na} = \Delta g_{Na} (V_m - E_{Na})$$

$$\Delta I_K = \Delta g_K (V_m - E_K)$$

These equations are useful in that they provide a means of determining the relative conductance change to sodium and potassium produced by ACh. This can be determined by considering current flow at the reversal potential, V_r. This is the potential at which the current is zero or, in other words, the potential at which the inward sodium current is exactly equal in magnitude to the outward potassium current. Thus,

$$\Delta g_{Na}(V_r - E_{Na}) = -\Delta g_K(V_r - E_K)$$

It follows that

$$\Delta g_{Na}/\Delta g_K = -(V_r - E_K)/(V_r - E_{Na})$$

The Takeuchis calculated that with V_r at -15 mV, $\Delta g_{Na}/\Delta g_K$ was approximately 1.3. The equation can also be turned around so that, when the relative conductances are known, the expected reversal potential can be calculated:

$$V_r = \frac{(\Delta g_{Na}/\Delta g_K)E_{Na} + E_K}{\Delta g_{Na}/\Delta g_K + 1}$$

Many investigators have found that this equation predicts reasonably well the changes in reversal potential produced by changes in extracellular concentrations of sodium and potassium and, hence, in the equilibrium potentials E_{Na} and E_K. In other words, when the conductance ratio $\Delta g_{Na}/\Delta g_K$ is determined in a bathing solution of normal ionic composition, that value can then be used to predict the reversal potential when the ionic composition is changed, for example, in reduced extracellular sodium. This result implies that the conductance ratio is constant over quite a wide range of ionic concentrations, a conclusion that is useful for predicting reversal potentials but is otherwise mysterious.

In theory, the conductance ratio $\Delta g_{Na}{:}\Delta g_K$ should *not* remain constant when the ionic composition of the extracellular bathing solution is changed, because conductances should depend on concentrations (Chapter 5). Therefore, if, for example, the extracellular sodium concentration is reduced, Δg_{Na} (and, hence, the conductance ratio) would be expected to decrease, unless for some reason there were a compensatory increase in the sodium permeability (Δp_{Na}) at the same time. The fact that in many experiments sodium conductance did not appear to decrease with concentration over quite a wide range may be a fortuitous consequence of other experimental factors. For example, in many of the experiments sodium chloride was replaced by sucrose, thus reducing the ionic strength of the bathing solution. This, in turn, might in some way have increased the apparent sodium permeability of the end plate channels, possibly by unmasking negative charges around the channel openings and thereby increasing the local sodium concentration.

More detailed studies of the effects of acetylcholine on the eel electroplaque[37] and the mouse end plate[38] have shown that when various cations are substituted for sodium and potassium, the conductance channels behave as if the ionic permeabilities are constant. For example, when a large, virtually impermeant cation (glucosamine) was used to replace sodium, the reversal potential was predicted by the constant field relation (Chapter 5),

$$V_r = \frac{RT}{F} \ln \frac{(\Delta p_{Na}/\Delta p_K)Na_o + K_o}{(\Delta p_{Na}/\Delta p_K)Na_i + K_i}$$

Variations in reversal potential with concentration were predicted accurately with a permeability ratio of about 0.9.

In summary, chemical synaptic transmission involves a new type of channel, one activated by a chemical rather than by changes in potential. In the case of the epp, ACh activates a cationic channel that is roughly equally permeable to sodium and potassium. At other excitatory synapses, the permeability ratio, and hence the reversal potential, may be quite different, ranging between −30 and +30 mV in various preparations. In Table 1 some of the properties of the ACh-activated channels are compared with those of the voltage-sensitive channels discussed previously in relation to the action potential. The permeability characteristics and kinetics of individual synaptic channels have been studied in detail at the molecular level, using methods of noise analysis and patch clamping; these experiments will be discussed in Chapter 11.

SYNAPTIC INHIBITION

The process of synaptic inhibition involves the same principles as those underlying synaptic excitation; only the details are different. Both excitatory and inhibitory neurotransmitters act by changing the permeability of the postsynaptic membrane, but excitation is achieved by driving the membrane potential toward threshold, whereas inhibition is achieved by holding the membrane below threshold. The effect of the inhibitory transmitter is to increase the membrane permeability, not to sodium, but instead to chloride or to potassium, both of which have equilibrium potentials near the resting level. Synaptic inhibition has been studied in detail in a number of preparations, most notably the spinal motoneuron of the cat,[39] the crustacean neuromuscular junc-

[37]Lassignal, N. and Martin, A. R. 1977. *J. Gen. Physiol. 70*: 23–36.
[38]Linder, T. M. and Quastel, D. M. J. 1978. *J. Physiol. 281*: 535–556.
[39]Coombs, J. S., Eccles, J. C. and Fatt, P. 1955. *J. Physiol. 130*: 326–373.

TABLE 1
PROPERTIES OF IMPULSE AND SYNAPTIC
POTENTIAL AT NERVE MUSCLE SYNAPSE

Property	Impulse	Synaptic potential
Initiated by	Depolarization	ACh
Changes of membrane conductance		
During rising phase	Specific increase in g_{Na}	Simultaneous increase in g_{Na} and g_K
During falling phase	Specific increase in g_K	Passive decay
Reversal potential of active response	E_{Na} approx. $+50$ mV	Close to 0 mV
Other features	Regenerative ascent, followed by refractory period	No evidence for regenerative action or refractoriness
Pharmacology	Blocked by TTX, not influenced by curare	Blocked by curare, not influenced by TTX

Modified from del Castillo and Katz (1956).

tion,[40,41] and the crayfish stretch receptor.[42] The motoneurons are inhibited by sensory inputs from antagonist muscles by way of inhibitory interneurons in the spinal cord. The effects of the inhibitory fibers can be studied by impaling a cell with two micropipettes, one to record potential changes and the other to pass current through the cell membrane, as in the experiment illustrated in Figure 11A. At the normal resting potential (about -75 mV), stimulation of the inhibitory fibers causes a slight hyperpolarization of the cell—the ipsp. When the cell is depolarized to -64 mV by passing positive current into the cell from the current passing electrode, the amplitude of the inhibitory potential is increased. When the cell is hyperpolarized to -82 mV, the synaptic potential is very small and reversed in sign, and at -100 mV, the reversed ipsp is very much larger. It appears, then, that in this experiment the reversal potential for the inhibitory response is about -80 mV. A similar experiment on the crayfish stretch receptor is shown in Figure 11B. In this preparation it is not necessary to use a current electrode to change the resting potential of the cell; because the cell is a stretch receptor, it can be depolarized or hyperpolarized by increasing or decreasing the tension on the muscle fiber. The inhibitory potential reverses between -67 and -70 mV. Crustacean muscle, unlike frog muscle, receives both excitatory and inhibitory inputs. It is one of the few preparations in which the inhibitory neurotransmitter has been identified. The presynaptic inhibitory terminals release γ-aminobutyric acid

[40]Dudel, J. and Kuffler, S. W. 1961. *J. Physiol. 155*: 543–562.
[41]Takeuchi, A. and Takeuchi, N. 1967. *J. Physiol. 191*: 575–590.
[42]Kuffler, S. W. and Eyzaguirre, C. 1955. *J. Gen. Physiol. 39*: 155–184.

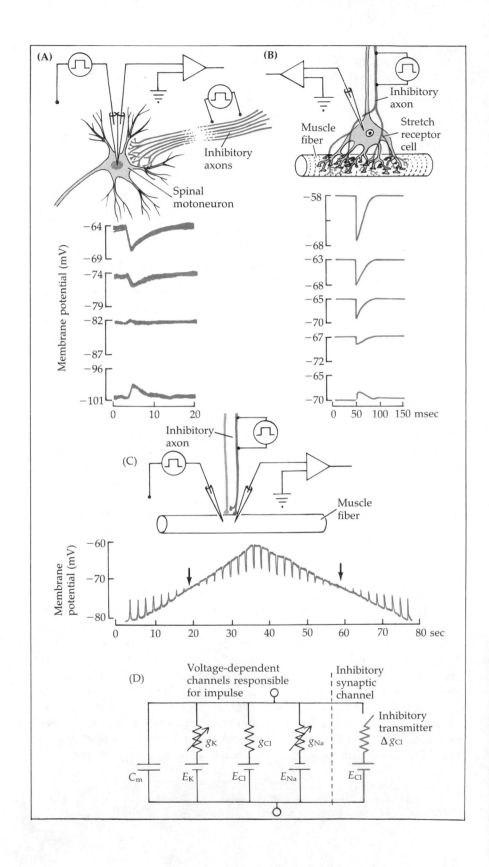

(GABA).[43] In the experiment shown in Figure 11C, a current-passing electrode was used to change the membrane potential continuously while the inhibitory nerve was being stimulated. The inhibitory synaptic potential, recorded with the voltage recording electrode, reverses in sign at about -70 mV.

At the crustacean inhibitory synapses, in spinal motoneurons, and indeed in most nerve cells, there is strong evidence that the type of inhibition illustrated in Figure 11 is associated with an increase in membrane permeability to chloride. For example, in spinal motoneurons, injection of chloride into the cell from a micropipette results in a shift of the ipsp reversal potential in the positive direction, and bathing the other preparations in solutions with altered chloride concentrations produces shifts in reversal potential expected of a chloride conductance pathway. The reversal potential for the ipsp, then, is the chloride equilibrium potential

$$V_r = E_{Cl} = - \frac{RT}{F} \ln \frac{Cl_o}{Cl_i}$$

The equivalent electrical circuit is illustrated in Figure 11D. The cell membrane with its resting and voltage-dependent conductances is shunted by a chloride pathway activated by the inhibitory transmitter. Just as the excitatory synaptic channel is permeable to cations, the chloride inhibitory channel is permeable to a wide variety of anions, the relative permeabilities depending upon the hydrated radius. The permeability characteristics of both excitatory and inhibitory synaptic channels have been reviewed by Edwards.[44]

Whether a particular neurotransmitter is excitatory or inhibitory depends only upon the ionic channels it opens in the postsynaptic membrane, and the same transmitter can, in fact, be inhibitory at one synapse and excitatory at another. For example, ACh, although excitatory

[43]Otsuka, M., Iverson, L. L., Hall, Z. W. and Kravitz, E. A. 1966. *Proc. Natl. Acad. Sci. USA 56*: 1110–1115.

[44]Edwards, C. 1982. *Neuroscience 7*: 1335–1366.

11 EFFECT OF INHIBITORY TRANSMITTER on a cat spinal motoneuron (A), a crayfish stretch receptor (B), and a muscle fiber in a crayfish (C). Potential changes recorded with an intracellular microelectrode. In the motoneuron and the muscle fiber, the membrane potential is set to different levels by passing current through a second intracellular electrode; in the stretch receptor, it is altered by adjusting the amount of stretch on the dendrites. Each cell has a reversal potential at which inhibitory stimulation causes no potential change. In (A), it is between -74 and -82 mV; in B, between -67 and -70 mV. In C, membrane potential is varied continuously from -80 to -60 mV and back, with inhibitory stimulation every 2 sec. Inhibitory potentials reverse at -72 mV (arrows). (D) Equivalent circuit of inhibitory conductance channel in parallel with the rest of the membrane. (A after Coombs, Eccles, and Fatt, 1955; B after Kuffler and Eyzaguirre, 1955; C after Dudel and Kuffler, 1961.)

at the vertebrate skeletal neuromuscular junction, is inhibitory on heart muscle. Here the nerve terminals of parasympathetic ganglion cells release ACh, which acts on the muscle by increasing the postjunctional permeability to potassium.[45] It thereby hyperpolarizes the cardiac muscle fibers and slows or stops the heartbeat. In the sea hare *Aplysia*, Kandel and his colleagues[46] showed that ACh produced two different postsynaptic actions on the same cell, one excitatory and the other inhibitory. Later Kehoe[47] showed that the cells had three receptors for ACh, one excitatory and two inhibitory. In addition, the cells have separate excitatory and inhibitory responses to dopamine.[48] The variety of substances known or suspected to be neurotransmitters and their postsynaptic actions will be discussed in Chapter 12. For the moment the point to be made is that since a given transmitter may have different actions it is misleading to refer to a substance as "excitatory" or "inhibitory" without reference to the particular receptor upon which it acts.

SLOW SYNAPTIC RESPONSES

Molluscan neurons

It has already been mentioned that cells in *Aplysia* have multiple responses to ACh. The channels involved in these responses subserve a conventional fast epsp of the type already discussed, a fast ipsp produced by an increase in chloride permeability, and a much slower inhibitory potential produced by an increase in potassium permeability. The fast synaptic potentials have time courses of tens of milliseconds, the slow inhibitory potential lasts several seconds. Other studies on *Aplysia* neurons have shown no fewer than six responses to applied 5-hydroxytryptamine (5-HT).[49] Two of these are depolarizations with a fast and a slow time course, both involving increases in sodium permeability; one is a slow hyperpolarizing response produced by an increase in potassium permeability and another a faster hyperpolarizing response due to a chloride permeability increase. The remaining two are produced by CONDUCTANCE DECREASES. In other words the potential changes are accompanied by an increase in the input resistance of the cell. In one of these, the conductance to potassium is decreased; because the membrane potential is determined by the ratio of sodium to potassium permeabilities, a reduction in the latter means that sodium becomes more dominant and the membrane is depolarized as its potential moves closer to the sodium equilibrium potential. Conversely, a hyperpolarizing response is observed that is due to a reduction in sodium permeability, causing the membrane to move toward the potassium equilibrium potential. One curious feature of such potentials is that they seem to behave anomalously with imposed changes in membrane po-

[45]Harris, E. J. and Hutter, O. F. 1956. *J. Physiol. 133:* 58P–59P.
[46]Wachtel, H. and Kandel, E. R. 1971. *J. Neurophysiol. 34:* 56–68.
[47]Kehoe, J. 1972. *J. Physiol. 225:* 85–114, 115–146, 147–172.
[48]Ascher, P. 1972. *J. Physiol. 225:* 173–209.
[49]Gerschenfeld, H. M. and Paupardin-Tritsch, D. 1974. *J. Physiol. 243:* 427–456.

tential. The depolarizing response, for example, becomes *smaller* if the membrane is hyperpolarized and, because it involves a change in potassium conductance, reverses at E_K. In summary, slow synaptic responses in molluscs have a number of unusual features: As their name implies, they are of much longer duration than the synaptic potentials we have discussed so far; one of the slow inhibitory responses involves an increase in potassium rather than in chloride permeability; and other excitatory and inhibitory responses are produced by decreases rather than by increases in membrane permeability.

Both mammalian and frog sympathetic ganglion cells have a variety of synaptic responses; this variety indicates that they are much more complex than simple relay stations transmitting messages unaltered from the central nervous system to peripheral organs. At least four separate synaptic responses can be seen under appropriate experimental conditions,[50] as shown in Figure 12: (1) a fast epsp, lasting up to 50 msec; (2) a slow inhibitory potential of about 2 seconds duration; (3) a slow excitatory potential lasting 30 to 60 seconds. All three are mediated by ACh. The fourth is a late slow depolarization that can persist 5–10 minutes and is caused by the release of a peptidergic transmitter (LHRH).[51] The characteristics of the four synaptic potentials are summarized in Table 2.

Autonomic ganglion cells

The fast epsp was first recorded intracellularly by Nishi and Koketsu[52] and was studied in detail by Ginsborg and his colleagues.[53] It is analogous to the end plate potential seen in skeletal muscle and to other fast excitatory synaptic potentials. The potential is produced by an increase in sodium and potassium permeability and is blocked by curare. The slow excitatory potential appears to be related to a reduction in potassium conductance.[54] In particular, there are voltage-sensitive potassium channels (M-channels) in the cells, some of which (unlike the more familiar delayed rectification channels) are open at normal resting potentials.[55] As additional channels are opened when the cell is depolarized, they tend to counteract any such depolarization and to stabilize the membrane at a relatively large resting potential. Consequently, their inactivation by ACh causes depolarization. This action of ACh is blocked by atropine and not by curare.

The slow inhibitory potential is also activated by ACh, both in the frog sympathetic ganglion[56] and in the parasympathetic cardiac ganglion of the mud puppy (*Necturus maculosis*).[57] It is the result of an increase

[50]Kuffler, S. W. 1980. *J. Exp. Biol. 89*: 257–286.

[51]Jan, Y. N., Jan, L. Y. and Kuffler, S. W. 1979. *Proc. Natl. Acad. Sci. USA 76*: 1501–1505.

[52]Nishi, S. and Koketsu, K. 1960. *J. Cell. Comp. Physiol. 55*: 15–30.

[53]Blackman, J. G., Ginsborg, B. L. and Ray, C. 1963. *J. Physiol. 167*: 355–373.

[54]Weight, F. and Votava, J. 1970. *Science 170*: 755–758.

[55]Adams, P. R., Brown, D. A. and Constanti, A. 1982b. *J. Physiol. 330*: 537–572.

[56]Dodd, J. and Horn, J. P. 1983. *J. Physiol. 334*: 271–291.

[57]Hartzell, H. C., Kuffler, S. W., Stickgold, R. and Yoshikami, D. 1977. *J. Physiol. 271*: 817–846.

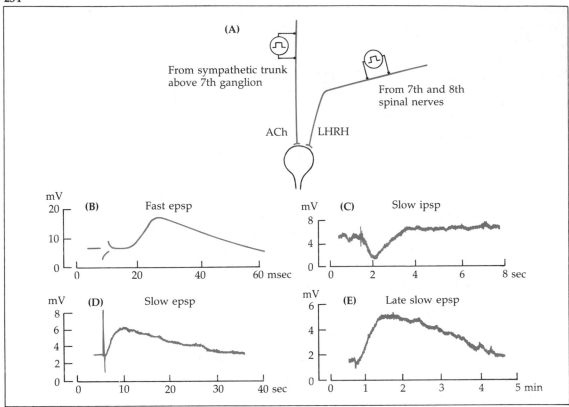

12 FOUR TYPES OF SYNAPTIC RESPONSES from sympathetic neurons of the frog. (A) Innervation of a sympathetic B neuron in the ninth ganglion of the paravertebral chain of the bullfrog; the diagram shows the separation of cholinergic and noncholinergic innervation. (B) A single preganglionic stimulus produces a fast epsp. (C) Repetitive stimulation produces a slow ipsp lasting about 2 sec; the fast epsp is blocked with a nicotinic blocking agent. (D) Repetitive stimulation also produces a slow epsp which occurs after the first two responses and lasts about 30 sec. (E) The late slow epsp, produced by stimulating preganglionic peptidergic fibers, lasts more than 5 min after repetitive stimulation. (Modified from Kuffler, 1980.)

in potassium permeability that is extremely slow in onset (more than 100 msec at room temperature) and that reaches its peak long after the ACh concentration in the synaptic region has dropped to zero. The possibility that such long latency and duration is due to some kind of delayed diffusion of ACh between the nerve terminal and the postsynaptic receptors is highly unlikely, and such an explanation is ruled out by the fact that the delay is extremely temperature sensitive; a 10°C decrease in temperature slows the rate of rise by a factor of ten or more. The time course and temperature sensitivity led Kuffler and his colleagues[57] to propose that there were at least three intermediate chemical steps between interaction of ACh with the postsynaptic receptor

TABLE 2
SYNAPTIC POTENTIALS IN SYMPATHETIC GANGLION CELLS

Property	Fast excitatory	Slow inhibitory	Slow excitatory	Late slow excitatory
Time course	Tens of milliseconds	Seconds	Tens of seconds	Minutes
Change in conductance	Increased Na, K	Increased K	Reduced K	?
Transmitter	ACh	ACh	ACh	LHRH
Blocked by	Curare	Atropine	Atropine	Some LHRH analogues

and activation of the conductance change. The potassium channels associated with the slow inhibitory response are voltage dependent, their conductance increasing with hyperpolarization. Thus, when the cell is hyperpolarized by passing current through the membrane, the amplitude of the slow ipsp first increases (because of the increase in conductance) and then decreases toward zero as the membrane potential approaches E_K. Because of the initial increase in amplitude with hyperpolarization, the response can be mistakenly attributed to a decrease in sodium conductance. This behavior illustrates the dangers of attempting to deduce the conductance changes underlying a synaptic response without examining its dependence upon membrane potential over a wide range that includes the reversal potential.

The late slow epsp was first reported by Nishi and Koketsu,[58] who correctly deduced that it was noncholinergic; and Kuffler and his colleagues[51] found that the response could be elicited by analogues of luteinizing hormone releasing hormone (LHRH). A number of additional tests confirmed that the peptide was the likely transmitter responsible for the response (see Chapter 12).

Fast synaptic potentials are produced by the direct interaction of the neurotransmitter substance with postsynaptic receptors responsible for opening membrane channels. The mechanism by which slow excitatory and inhibitory synaptic potentials are produced is clearly much more complex. As already noted, the potentials persist long after the neurotransmitter itself has disappeared from the synaptic region, such persistence suggesting that activation of the conductance changes involves a number of intermediate steps. The chemical reactions involved in these intermediate steps have not been determined in detail.

There is evidence that in *Aplysia* cells one of the steps leading to slow changes in potassium conductance involves protein phosphorylation by a cyclic AMP-dependent protein kinase. Such phosphorylation

Mechanisms underlying slow responses

[58]Nishi, S. and Koketsu, K. 1968. *J. Neurophysiol. 31*: 109–118.

might be expected to alter the conformation of channel proteins in the membrane, thus altering membrane conductance. The proposed sequence of events[59] begins with activation by the transmitter of an adenylate cyclase, which produces adenosine-3',5'-phosphate (cyclic AMP); the cyclic AMP, in turn, activates the protein kinase. Although detailed evidence relating to such a sequence of events is still to be obtained, it has been shown that injections of cyclic AMP, of a catalytic subunit of cyclic AMP-dependent protein kinase, or of protein kinase inhibitor have the expected effects on some of the potassium conductance changes. For example, injection of a protein kinase inhibitor reversibly blocks the increase in potassium conductance produced by 5-HT.[60] The available evidence implicates cyclic AMP in activation of a number of other potassium channels in *Aplysia* cells as well; whether protein phosphorylation is the final step leading to an increase or decrease in potassium conductance remains to be determined.

The same cyclic AMP mechanism has been proposed by Greengard to account for the slow hyperpolarization produced by catecholamines in mammalian sympathetic ganglia.[61] In accordance with this proposal, theophylline, which inhibits the hydrolysis of cyclic AMP by phosphodiesterase, increased the hyperpolarization by dopamine of rabbit superior cervical ganglion cells. In addition, in the same experiments, exogenous cyclic AMP was shown to hyperpolarize the cells.[62] However, subsequent experiments on rat superior cervical ganglia[63] showed that the hyperpolarization produced by cyclic AMP was replicated by other adenosine compounds and that they all acted through membrane receptors different from those activated by the catecholamines. Thus, the exact role of cyclic AMP in mediating these slow inhibitory potentials has not yet been determined.

PRESYNAPTIC INHIBITION

In the examples of synaptic transmission described in this chapter, the postsynaptic effects of neurotransmitters have been stressed. Over the years, however, a number of experiments have indicated that it is difficult to account for inhibition solely in terms of postsynaptic permeability changes.[64,65] Eventually, in 1961, an additional inhibitory mechanism was described in the mammalian spinal cord by Eccles and his colleagues[66] and at the crustacean neuromuscular junction by Dudel and Kuffler.[40] As shown in Figure 13, the action of the inhibitory nerve at the crustacean neuromuscular junction is exerted not only on the

[59]Kuo, J. F. and Greengard, P. 1969. *Proc. Natl. Acad. Sci. USA 64*: 1349–1355.
[60]Adams, W. B. and Levitan, I. B. 1982. *Proc. Natl. Acad. Sci. USA 79*: 3877–3880.
[61]Greengard, P. 1976. *Nature 260*: 101–108.
[62]McAfee, D. A. and Greengard, P. 1972. *Science 178*: 310–312.
[63]Brown, D. A., Caulfield, M. P. and Kirby, P. J. 1979. *J. Physiol. 290*: 441–451.
[64]Fatt, P. and Katz, B. 1953. *J. Physiol. 121*: 374–389.
[65]Frank, K. and Fuortes, M. G. F. 1957. *Fed. Proc. 16*: 39–40.
[66]Eccles, J. C., Eccles, R. M. and Magni, F. 1961. *J. Physiol. 159*: 147–166.

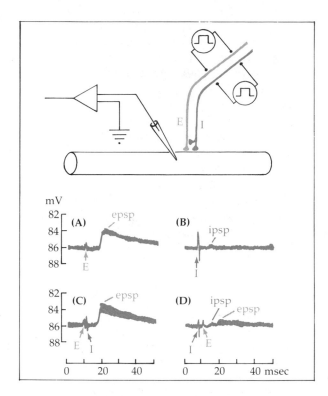

mV

(A) epsp

(B) ipsp

(C) epsp

(D) ipsp epsp

0   20   40   0   20   40 msec

13

PRESYNAPTIC INHIBITION in a crustacean muscle fiber innervated by one excitatory and one inhibitory axon. At the resting potential (-86 mV), the excitatory impulse E causes a 2-mV epsp (A) while the inhibitory impulse I depolarizes by about 0.2 mV (B). If, as in (C), the inhibitory impulse arrives at the synapse just after the excitatory one, it is too late to exert an inhibitory effect. If, however, I precedes E by several milliseconds, as in (D), it greatly reduces the epsp. (After Dudel and Kuffler, 1961.)

muscle fibers but also on the excitatory terminals, reducing the output of transmitter. In Chapter 10 we shall see that the transmitter is released in packets or quanta, each containing several thousand molecules. The presynaptic effect of the inhibitory transmitter is to reduce the number of quanta released from the excitatory terminal. The presynaptic effect of an inhibitory impulse is relatively brief, reaching a peak in a few milliseconds and declining to zero after a total of 6 to 7 msec. To exert their maximum effect, inhibitory impulses must arrive at the nerve terminal several milliseconds before the excitatory impulse. The importance of accurate timing is illustrated in Figure 13, where A and B show the excitatory and inhibitory potentials following separate stimulation of each of the two nerves. In C the inhibitory impulse follows the excitatory one by 1.5 msec and arrives too late to exert its effect; in D, however, it precedes the excitatory impulse and arrives in time to reduce the size of the epsp markedly. There is evidence that GABA, the transmitter responsible for postsynaptic inhibition at the crustacean neuromuscular junction, is also responsible for the presynaptic effect.[40,67]

In the mammalian spinal cord, as at the crayfish neuromuscular junction, presynaptic inhibition results in the reduction in the number of quanta of transmitter released from the excitatory nerve terminals.[68]

[67]Takeuchi, A. and Takeuchi, N. 1966. *J. Physiol. 183*: 433–449.
[68]Kuno, M. 1964. *J. Physiol. 175*: 100–112.

The relative importance of presynaptic and postsynaptic inhibition in the nervous system has not yet been established. However, one can see readily how presynaptic inhibition could be useful for information processing in a neuron innervated by many converging pathways. For such a cell it may be important that under certain circumstances some inputs act by suppressing others selectively. This cannot be done by the use of a postsynaptic conductance change that affects the whole cell.

Structural basis of presynaptic inhibition

The experiments demonstrating presynaptic inhibition naturally imply the existence of axoaxonic synapses between inhibitory and excitatory terminals. These have been demonstrated directly by electron microscopy at the crustacean neuromuscular junction[69] and at numerous locations in the mammalian central nervous system.[70] Inhibitory neurons themselves can be inhibited presynaptically, as synapses have been demonstrated electron microscopically on inhibitory terminals that end on stretch receptors.[71]

SUGGESTED READING

Reviews

Eccles, J. C. 1964. *The Physiology of Synapses.* Springer-Verlag, Berlin.

Hall, Z. W., Hildebrand, J. G. and Kravitz, E. A. 1974. *Chemistry of Synaptic Transmission.* Chiron Press, Newton.

Katz, B. 1966. *Nerve, Muscle and Synapse.* McGraw-Hill, New York.

Kehoe, J. and Marty, A. 1980. Certain slow synaptic responses: Their properties and possible underlying mechanisms. *Annu. Rev. Biophys. Bioeng. 9*: 437–465.

Kuffler, S. W. 1980. Slow synaptic responses in autonomic ganglia and the pursuit of a peptidergic transmitter. *J. Exp. Biol. 89*: 257–286.

Loewenstein, W. 1981. Junctional intercellular communication: The cell-to-cell membrane channel. *Physiol. Rev. 61*: 829–913.

Takeuchi, A. 1977. Junctional transmission I. Postsynaptic mechanisms. In E. Kandel (ed.), *Handbook of the Nervous System,* Vol. 1. American Physiological Society, Baltimore, pp. 295–327.

The Synapse. 1976. *Cold Spring Harbor Symp. Quant. Biol., Vol. 40.* (A collection of about 60 articles.)

Original papers

Coombs, J. S., Eccles, J. C. and Fatt, P. 1955. The specific ion conductances and the ionic movements across the motoneuronal membrane that produce the inhibitory post-synaptic potential. *J. Physiol. 130*: 326–373.

Dennis, M. J., Harris, A. J. and Kuffler, S. W. 1971. Synaptic transmission and its duplication by focally applied acetylcholine in parasympathetic neurones in the heart of the frog. *Proc. R. Soc. Lond. B 177*: 509–539.

---

[69]Atwood, H. L. and Morin, W. A. 1970. *J. Ultrastruct. Res. 32*: 351–369.

[70]Schmidt, R. F. 1971. *Ergeb. Physiol. 63*: 20–101.

[71]Nakajima, Y., Tisdale, A. D. and Henkart, M. P. 1973. *Proc. Natl. Acad. Sci. USA 70*: 2462–2466.

Dudel, J. and Kuffler, S. W. 1961. Presynaptic inhibition at the crayfish neuro-muscular junction. *J. Physiol. 155*: 543–562.

Eccles, J. C., Eccles, R. M. and Magni, F. 1961. Central inhibitory action attrib-utable to presynaptic depolarization produced by muscle afferent volleys. *J. Physiol. 159*: 147–166.

Fatt, P. and Katz, B. 1951. An analysis of the end-plate potential recorded with an intracellular electrode. *J. Physiol. 115*: 320–370.

Furshpan, E. J. and Potter, D. D. 1959. Transmission at the giant motor synapse of the crayfish. *J. Physiol. 145*: 289–325.

Furukawa, T. Y. and Furshpan, E. J. 1963. Two inhibitory mechanisms in the Mauthner cell neurons of the goldfish. *J. Neurophysiol. 26*: 140–176.

Hartzell, H. C., Kuffler, S. W., Stickgold, R. and Yoshikami, D. 1977. Synaptic excitation and inhibition resulting from direct action of acetylcholine on two types of chemoreceptors on individual amphibian parasympathetic neu-rones. *J. Physiol. 271*: 817–846.

Takeuchi, A. and Takeuchi, N. 1960. On the permeability of the end-plate membrane during the action of transmitter. *J. Physiol. 154*: 52–67.

Takeuchi, A. and Takeuchi, N. 1966. On the permeability of the presynaptic terminal of the crayfish neuromuscular junction during synaptic inhibition and the action of γ-aminobutyric acid. *J. Physiol. 183*: 433–449.

CHAPTER TEN

RELEASE OF CHEMICAL TRANSMITTERS

The mechanism whereby the impulse arriving at a presynaptic nerve terminal causes transmitter release has been studied in great detail in a number of preparations, particularly the frog neuromuscular junction and the giant synapse of the squid. The stimulus for secretion of the transmitter is depolarization of the presynaptic ending, produced either by an impulse or by artificially applied current. Invariably, a delay (about 0.5 msec at 22°C) intervenes between the voltage change and secretion. For release to occur, calcium ions must be present in the bathing fluid at the time of the depolarization, and there is good evidence that calcium enters the terminal to trigger transmitter release. In low-calcium medium, the release can be reduced or completely eliminated.

Changes in the postsynaptic membrane potential provide a convenient assay for estimating the timing and the quantity of transmitter release. A key finding is that transmitter is secreted in multimolecular packets, or quanta, each containing several thousand molecules. These constitute the fundamental physiological units of release.

The synaptic potential at the sketal neuromuscular junction is normally made up of about 200 quanta of acetylcholine (ACh), which are released almost synchronously. Variations in release usually occur through more or fewer quanta being secreted by the terminal. At rest, the nerve also releases quanta of transmitter spontaneously, an action giving rise in the postsynaptic cell to miniature synaptic potentials with amplitudes in the millivolt range. Quantal release appears to be a general mechanism for secretion of neurotransmitters at synapses in the central and peripheral nervous systems of both vertebrates and invertebrates.

A number of questions arise concerning the way in which the presynaptic neuron releases transmitter. Experimental analysis of this process requires a highly sensitive, quantitative, and reliable measure of the amount of transmitter released, with a time resolution in the millisecond scale. In the experiments described below, the membrane potential of

241

the postsynaptic cell is used as a sensitive detector for the transmitter liberated by the presynaptic terminal. Once again, the vertebrate neuromuscular junction, where acetylcholine (ACh) is known to be the transmitter, offers many advantages. However, to obtain more complete information about the release process, it is useful to be able to record from the presynaptic endings as well; for example, such recordings are needed to establish how membrane potential affects transmitter release. The presynaptic terminals at the neuromuscular junction are too small to be impaled by a microelectrode, but this can be done at a number of other synapses, most notably the giant fiber synapse in the stellate ganglion of the squid.[1] This is a chemical synapse,[2] but the identity of the transmitter is not yet firmly established. As at the neuromuscular junction, an action potential in the presynaptic fiber normally gives rise, after a delay, to a large depolarizing potential in the postsynaptic membrane, which usually reaches threshold and produces an action potential.

The stellate ganglion of the squid was used by Katz and Miledi to determine the precise relation between the membrane potential of the presynaptic terminal and the amount of transmitter release.[3] The preparation and the arrangement for recording from the presynaptic terminal and the postsynaptic fiber simultaneously are shown in Figure 1A. When tetrodotoxin (TTX) was applied to the preparation, the presynaptic action potential failed gradually over the next 15 minutes (Figure 1B). As it did so, the postsynaptic action potential failed as well, leaving an excitatory postsynaptic potential (epsp). Further reduction in the amplitude of the presynaptic action potential led to a continued loss of epsp amplitude until the later failed completely. When the epsp amplitude is plotted against the amplitude of the failing presynaptic impulse, as in Figure 1C, the synaptic potential decreases rapidly as the presynaptic action potential amplitude falls below about 75 mV, and at amplitudes less than about 45 mV there is no postsynaptic response. Tetrodotoxin has no effect on the sensitivity of the postsynaptic membrane, so the reduction in epsp amplitude indicates a reduction in the amount

[1]Bullock, T. H. and Hagiwara, S. 1957. *J. Gen. Physiol. 40*: 565–577.
[2]Hagiwara, S. and Tasaki, I. 1958. *J. Physiol. 143*: 114–137.
[3]Katz, B. and Miledi, R. 1967c. *J. Physiol. 192*: 407–436.

1 **PRESYNAPTIC IMPULSE AND POSTSYNAPTIC RESPONSE at squid giant synapse. (A) Sketch of the stellate ganglion in the squid emphasizing the two large axons that form the chemical synapse. The fibers can be impaled with microelectrodes as shown. (B) Simultaneous recordings from the presynaptic (lower records) and postsynaptic axons during the development of conduction block following TTX application. (C) The relation between the amplitude of the presynaptic impulse and the postsynaptic response plotted on linear and semilogarithmic (inset) scales. Closed circles from results shown in (B); open and half-filled circles obtained after complete TTX block by applying depolarizing current pulses to presynaptic terminals. (A after Bullock and Hagiwara, 1957; B, C after Katz and Miledi, 1967c.)**

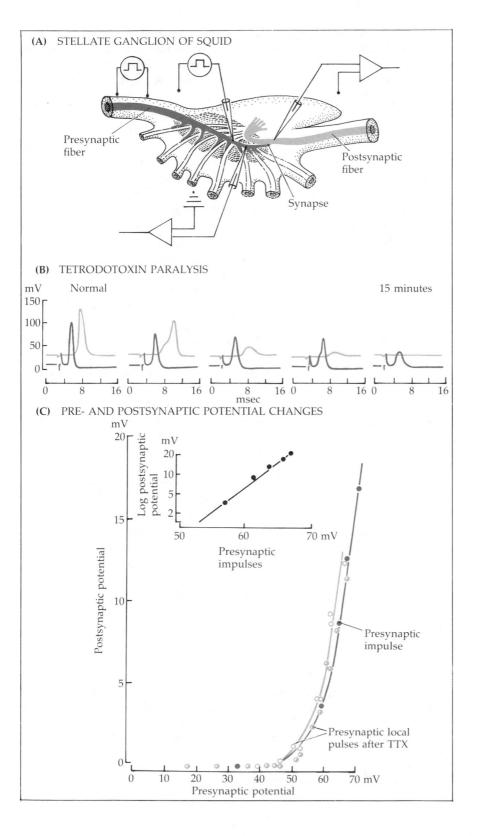

(A) STELLATE GANGLION OF SQUID

Presynaptic fiber

Postsynaptic fiber

Synapse

(B) TETRODOTOXIN PARALYSIS

mV   Normal   15 minutes

150

100

50

0

0   8   16 0   8   16 0   8   16 0   8   16 0   8   16
msec

(C) PRE- AND POSTSYNAPTIC POTENTIAL CHANGES

mV

20

Log postsynaptic potential (mV)

20

10

5

2

50   60   70 mV

Presynaptic impulses

15

Postsynaptic potential

10

Presynaptic impulse

5

Presynaptic local pulses after TTX

0

0   10   20   30   40   50   60   70 mV
Presynaptic potential

of transmitter released from the presynaptic terminal. This result suggests, then, that there is a threshold for transmitter release at about 45 mV depolarization, after which the amount released, and hence the epsp amplitude, increases rapidly with presynaptic action potential amplitude.

Katz and Miledi used an additional procedure to explore the relation further: They placed a second electrode in the presynaptic terminal, through which they applied brief (1–2 msec) depolarizing current pulses, thus mimicking a presynaptic action potential. The relation between the amplitude of the artificial action potential and the epsps was the same as that obtained with the failing action potential immediately after TTX poisoning. This result indicates that depolarization alone is sufficient to release transmitter from the terminal and that the normal sequence of permeability changes to sodium and potassium responsible for the action potential are not involved. The results obtained with the depolarizing current pulses are plotted on a semilogarithmic scale in the inset of Figure 1C; the linear relation indicates that the initial increase in epsp amplitude with presynaptic depolarization is exponential.

Synaptic delay   One characteristic of the transmitter release process evident in Figure 1B is the delay between onset of the presynaptic action potential and that of the synaptic potential. In Katz and Miledi's experiments, which were done at about 10°C, the delay was 3–4 msec. Detailed measurements at the frog neuromuscular junction showed a delay of at least 0.5 msec at room temperature between depolarization of the presynaptic terminal and the onset of the end plate potential (epp). The time is too long to be accounted for by diffusion of ACh across the synaptic cleft (a distance of about 50 nm), which should take no longer than about 50 μsec. With ACh applied ionophoretically from a micropipette, delays of as little as 150 μsec can be achieved, though the pipette, even with the most careful placement, can never contact the postsynaptic membrane as intimately as does the nerve terminal. Furthermore, synaptic delay is much more sensitive to temperature than would be expected if it were due to diffusion. Cooling the frog nerve–muscle preparation to 2°C increases the delay to as long as 7 msec, whereas the delay in the response to ionophoretically applied ACh is not perceptibly altered.[4] Several explanations for the synaptic delay are suggested by its marked temperature sensitivity. Two of these, which might act exclusively or in combination, are (1) a metabolic process intervening between depolarization and release and (2) the entry of some substance across the terminal membrane as a necessary requirement for release.

Evidence that calcium entry is required for release   Calcium has long been known as an essential link in the process of synaptic transmission. When its concentration in the extracellular fluid is decreased, release of ACh at the neuromuscular junction is reduced and eventually abolished.[5] The importance of calcium for release has

[4]Katz, B. and Miledi, R. 1965. *Proc. R. Soc. Lond. B 161*: 483–495.
[5]del Castillo, J. and Stark, L. 1952. *J. Physiol. 116*: 507–515.

been established at all synapses where it has been tested, irrespective of the nature of the transmitter. Its role has been generalized further to other secretory processes such as liberation of hormones by cells of the pituitary gland, release of epinephrine from the adrenal medulla, and secretion by salivary glands.[6] As discussed below, transmitter release is preceded by calcium entry into the terminal. Its effect in promoting release is antagonized by magnesium, which blocks such entry. Transmitter release can be reduced, then, either by removing calcium from the bathing solution or by adding magnesium.

Experiments by Katz and Miledi on the neuromuscular junction showed not only that calcium is essential for transmitter release to occur but also that it must be present at the time of depolarization of the presynaptic terminal. The experiment in which this was demonstrated is illustrated in Figure 2. Calcium was removed from the bathing solution so that release of ACh in response to nerve stimulation was virtually abolished (A). Calcium was then applied to the nerve terminal by ionophoresis, from a micropipette close to the terminal, just before the nerve was stimulated (B); application of calcium in this way restored transmitter release, producing an epp with each trial (B). Calcium pulses alone were not effective in producing release (C), nor were calcium pulses applied *after* stimulation, during the synaptic delay period (D). Similar results were obtained in the presence of TTX, when the nerve terminal was depolarized by current pulses. Again, calcium application was effective just before the depolarizing pulse. These experiments were taken as evidence that for release to occur calcium must be available during the period of depolarization.

Other experiments on nerve fibers have indicated that the calcium conductance (g_{Ca}) of the membrane is increased by depolarization (Chapter 6) and that some calcium enters with each action potential. This idea was supported by an extension of the experiment shown in Figure 1. Katz and Miledi reasoned that if the presynaptic terminal were depolarized to E_{Ca} or beyond, then there would be no calcium entry during the pulse and hence no transmitter release. To produce large and relatively long-lasting depolarizations of the squid presynaptic terminal, it was necessary to block delayed rectification; this was accomplished by intracellular injection of tetraethylammonium ion (TEA; see Chapter 6). The results of one such experiment are shown in Figure 3. A current pulse with a duration of about 17 msec produces a steady depolarization of the nerve terminal and a brief epsp when the depolarization is small (A). At a moderate level of depolarization (B), the initial epsp is suppressed, and there is an additional response at the end ("off") of the pulse. With a very large depolarization (C) the "on" response is abolished and a large epsp occurs at the end of the pulse. This result is consistent with the idea that the release of transmitter depends on calcium entry during depolarization. The number of calcium channels that are opened increases with depolarization, but the calcium

[6]Douglas, W. W. 1978. *Ciba Fdn. Symp. 54*: 61–90.

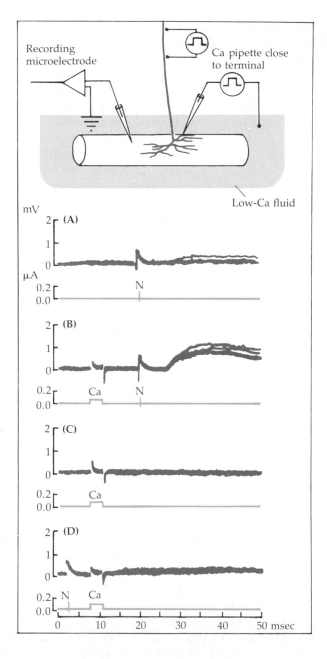

2

**TIMING OF CALCIUM ACTION ON TRANS-
MITTER RELEASE. (A)** With repeated stimuli in
low-calcium solution, a motor nerve impulse (N)
causes little or no transmitter release at the skeletal neuromuscular junction. **(B)** If the nerve impulse is preceded by application of calcium (Ca)
from a pipette to the terminal, transmitter is released. **(C)** Calcium application alone has no effect. **(D)** If calcium is applied after the arrival of
the presynaptic impulse in the terminal, but before the expected epp, it has no influence on
transmitter release. (After Katz and Miledi,
1967b.)

current is suppressed when the membrane potential during the pulse
approaches E_{Ca}. When the pulse is terminated, the calcium channels
are still open and calcium can then enter the terminal down its electrochemical gradient to trigger the transmitter release process.

The experiment illustrated in Figure 3, in addition to providing
evidence about the role of calcium in the release process, also gives us
some clue about the nature of the synaptic delay. At low and moderate
depolarization, the epsp at the beginning of the pulse appears with a

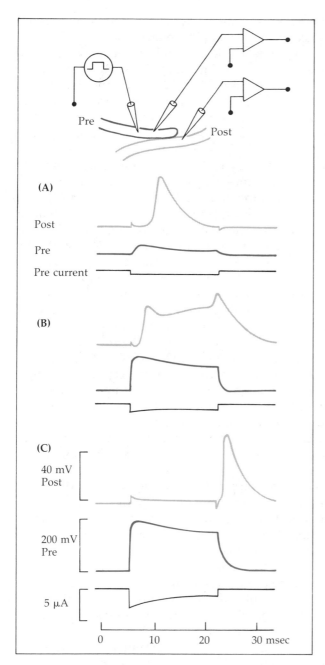

(A)

Post

Pre

Pre current

(B)

(C)

40 mV
Post

200 mV
Pre

5 µA

0   10   20   30 msec

3

EFFECT OF PROLONGED DEPOLARIZATION
of the presynaptic terminal in the squid giant
synapse (preparation treated with TTX and TEA).
(A) Moderate depolarization of the presynaptic
terminal (Pre) produces a postsynaptic epsp
(Post) with a delay of about 3.5 msec. (B) Increased depolarization reduces the amplitude of
the initial epsp slightly and depolarization due
to continuing transmitter release is maintained
throughout the pulse. Upon termination of the
pulse, an "off" response occurs as a result of
additional transmitter release. (C) Very large presynaptic depolarization suppresses the initial
epsp completely, leaving only the "off" response. Suppression occurs because the presynaptic terminal is depolarized to near the calcium
equilibrium potential; consequently, there is no
calcium entry until after the pulse. Note that the
delay of the "off" response (C) is considerably
less than that of the "on" response (A). (After
Katz and Miledi, 1967c.)

delay of about 3.5 msec. At the end of the pulse, however, the "off"
response occurs with a delay of less than 1.0 msec. This difference
presumably arises because the "on" response cannot occur until a sufficient number of calcium channels have opened, whereas at the termination of the pulse the "off" response can occur with little delay
because the calcium channels are already open and calcium can enter
immediately upon repolarization. These observations suggest that a

significant fraction of the normal synaptic delay is due to the time taken for activation of the calcium channels by depolarization.

Calcium entry into the presynaptic terminal was demonstrated directly by Llinás and his colleagues, using the luminescent dye aequorin (Chapter 6). In addition, they measured the magnitude and time course of the calcium current produced by presynaptic depolarization, using voltage clamp techniques.[7] An example is shown in Figure 4A. The sodium and potassium conductances associated with the action potential had been blocked by TTX and TEA so that only the voltage-sensitive calcium channels remained. A presynaptic depolarizing pulse from −70 to −18 mV (upper record) produced an inward calcium current that increased slowly in magnitude to about −200 nA (middle record). The onset of calcium entry was followed, after a delay of more than a millisecond, by a large synaptic potential in the postsynaptic cell (lower record). When the depolarizing pulse was increased to +50 mV (B), the calcium current was suppressed during the pulse. On repolarization, however, there was an immediate calcium current, accompanied by a postsynaptic potential with a delay of less than 0.2 msec. This experiment shows directly that much of the synaptic delay is associated with the time taken for activation of the calcium conductance. The effect of an artificial action potential is shown in Figure 4C. A presynaptic action potential, recorded before the addition of TTX and TEA to the preparation, is "played back" through the voltage clamp circuit to produce exactly the same voltage change in the terminal, but now without the normal changes in sodium and potassium conductance. The postsynaptic potential is indistinguishable from that produced by a normal presynaptic action potential, showing that only the voltage change, and not the changes in sodium and potassium conductances that normally accompany the action potential, is necessary for normal transmitter release. The voltage clamp technique also enabled Llinás and his colleagues to deduce the magnitude and time course of the calcium current produced by the artificial action potential (black curve). The current began about 0.5 msec after the onset of the depolarization, rose rapidly to a peak and then declined over the next 1.0 msec.

Calcium currents related to synaptic transmission have also been observed at the neuromuscular junction of the mouse. Using extracellular recording with precisely placed microelectrodes, combined with discrete application of TTX and TEA to confined regions of the nerve terminal, Brigant and Mallart[8] were able to deduce the sequence of permeability changes associated with nerve terminal depolarization. Their experiments show that as the impulse arrives there is inward sodium current restricted to the preterminal membrane (where the myelin ends), followed by inward calcium current throughout the terminal and repolarization by inward potassium current. Thus, the actual terminal appears to have no voltage-sensitive sodium channels and is

[7]Llinás, R. 1982. *Sci. Am.* 247: 56–65.
[8]Brigant, J. L. and Mallart, A. 1982. *J. Physiol.* 333: 619–636.

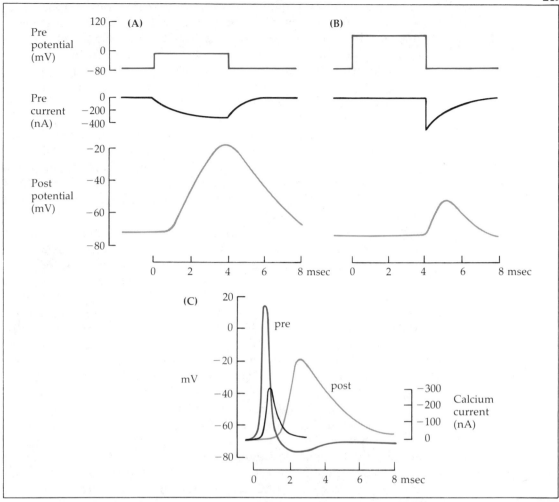

4 PRESYNAPTIC CALCIUM CURRENTS and transmitter release. The presynaptic terminal is voltage clamped and treated with TTX and TEA to abolish voltage-sensitive sodium and potassium currents. (A) A voltage pulse from −70 to −18 mV (upper trace) results in a slow, inward, calcium current (middle trace). Calcium entry is followed by a postsynaptic potential (lower trace) with a delay of about 1 msec. (B) During a larger voltage step (to +50 mV), inward calcium current is suppressed and no synaptic potential occurs until the pulse is terminated. A surge of calcium current at the end of the pulse is followed by a synaptic potential within 0.2 msec. (C) A voltage waveform identical in shape to a normal action potential (pre) produces a synaptic potential (post) indistinguishable from that produced by a normal presynaptic impulse. The black curve gives the magnitude and time course of the inward calcium current (negative upward). (After Llinás, 1982.)

specialized to promote calcium entry. On the other hand, the frog motor nerve terminal, which is much longer, has a propagating sodium action potential along most of its length.[9]

[9]Katz, B. and Miledi, R. 1968b. *J. Physiol. 199*: 729–741.

So far the general scheme can be summarized as presynaptic depolarization → calcium entry → transmitter release. Once this general framework has been established, it remains to be shown how the transmitter is secreted from the terminals. In experiments on the frog neuromuscular junction Fatt and Katz[10] showed that ACh is released from the terminals in MULTIMOLECULAR PACKETS, which they called QUANTA. A packet, or quantum, simply describes the smallest unit in which the transmitter is normally secreted. To take an arbitrary number, quantal release means that only 0, 3000, 6000, 9000, and so on, molecules can be released, but not intermediate amounts such as 2056 or 3762. In general, the number of quanta released by depolarization (the QUANTUM CONTENT of a synaptic response) may vary, but the number of molecules in a quantum (QUANTAL SIZE) is relatively fixed. Some exceptions to this rule will be discussed later.

The first evidence for packaging of ACh in multimolecular quanta was the observation by Fatt and Katz that at the motor end plate, but not elsewhere in the muscle fiber, spontaneous depolarizations of about 1 mV occurred irregularly (Figure 5). They had the same time course as the epps evoked by nerve stimulation. The spontaneous potentials were decreased in amplitude and eventually abolished by increasing concentrations of curare in the bathing solution, and they were increased in amplitude and time course by acetylcholinesterase inhibitors such as prostigmine. These two pharmacological tests suggested that the potentials were produced by the spontaneous release of discrete amounts of ACh from the nerve terminal. The possibility that they might have been due to the release of *single* molecules of ACh was ruled out by several lines of reasoning: (1) Curare and prostigmine had graded effects on miniature epp amplitude; if the events were due to the action of single molecules their amplitude should have been fixed and the drugs would have been expected to affect their frequency of occurrence. Prostigmine, by inhibiting hydrolysis of ACh and thereby increasing its concentration in the synaptic cleft, should have increased miniature potential frequency. Curare, by occupying postsynaptic receptors and thereby decreasing the number available for interaction with ACh, should have decreased the frequency. (2) Addition of ACh to the solution bathing the muscle caused a graded depolarization of the end plate region without any increase in miniature epp frequency. Again, if the spontaneous miniature epps were caused by single molecules, then addition of ACh to the bathing solution should have increased the number of molecular collisions with the postsynaptic membrane and hence the miniature potential frequency. Moreover, with small applications of ACh, a depolarization smaller in amplitude than a miniature potential could be produced.

Fatt and Katz made an additional important observation on the behavior of the stimulus-evoked epp after synaptic transmission had been reduced by lowering the extracellular calcium concentration. Un-

[10]Fatt, P. and Katz, B. 1952. *J. Physiol. 117*: 109–128.

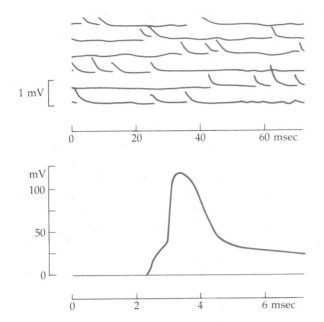

5

MINIATURE SYNAPTIC POTENTIALS occurring spontaneously at the frog neuromuscular junction. The potentials, which are due to spontaneous release of quanta of ACh from the presynaptic terminal, are less than 1 mV in amplitude, in contrast to the large nerve-evoked epp giving rise to the action potential, shown in the lower record (note difference in time scale). Spontaneous miniature epps are confined to the end plate (synaptic) region of the muscle fiber. (From Fatt and Katz, 1952.)

expectedly, they found that at very low levels of calcium the responses to stimulation fluctuated in a stepwise manner. Some stimuli produced no response at all, some a response of about 1 mV in amplitude, identical in size and shape to a spontaneous miniature potential; some responses were twice the unit amplitude, and some three times as large. This remarkable observation led Fatt and Katz to propose that the single quantal events observed to occur spontaneously also represented the building blocks for the stimulus-evoked epps, and that the effect of reduced extracellular calcium was to reduce the quantum content of the epp in a graded fashion, without reducing the quantal size. When extracellular calcium was sufficiently low, only a few quanta were released and stepwise fluctuations in the release process could then be observed.

Subsequent evidence confirmed in a variety of different ways that the spontaneous miniature epps are indeed due to ACh liberated by the nerve terminal. For example, depolarization of the nerve terminals by passing a steady current through them causes an increase in frequency of the spontaneous activity,[11] whereas muscle depolarization has no effect on frequency. Botulinum toxin, which blocks release of ACh in response to nerve stimuli, also abolishes the spontaneous activity.[12] Shortly after denervation of the muscle and degeneration of its motor nerve, the miniature potentials disappear.[13] After an interim period, spontaneous potentials reappear in denervated frog muscle, but

[11]del Castillo, J. and Katz, B. 1954c. *J. Physiol. 124*: 586–604.
[12]Brooks, V. B. 1956. *J. Physiol. 134*: 264–277.
[13]Birks, R., Katz, B. and Miledi, R. 1960. *J. Physiol. 150*: 145–168.

convincing evidence has been obtained that these arise because of ACh release from Schwann cells that have engulfed segments of the degenerating nerve terminals by phagocytosis (see Chapter 13).

In summary, release of ACh from motor nerve terminals at the neuromuscular junction occurs in packets, or quanta, each quantum containing several thousand molecules. Quantal release occurs spontaneously at a low rate, and a large number of quanta are released in response to nerve stimulation, depending on the extracellular calcium concentration. Normally the epp is made up of about 200 quantal units; in low calcium, when the number is very small, the quantum content fluctuates from trial to trial. This fluctuation in release results in stepwise fluctuations in the amplitude of the epp, as shown in Figure 6.

Statistical fluctuations of the epp Two questions arise with respect to quantal release: (1) Do the fluctuations really occur in quantal steps? (2) Why do the fluctuations occur? These questions were addressed by del Castillo and Katz, who studied the nature of the fluctuations in some detail.[14] Their experiments consisted of reducing transmitter release by reducing extracellular calcium or by adding magnesium to the bathing solution and then recording a large number of epps evoked by nerve stimulation, as well as a large number of spontaneous miniature potentials. They plotted histograms of the amplitude distributions, as shown in Figure 7. First, it is clear that the spontaneous miniature potentials (inset) are not all the same size; they are distributed around their mean amplitude with a variance of about 10 percent of the mean. Because of this variance, evoked epps

[14]del Castillo, J. and Katz, B. 1954a. *J. Physiol. 124*: 560–573.

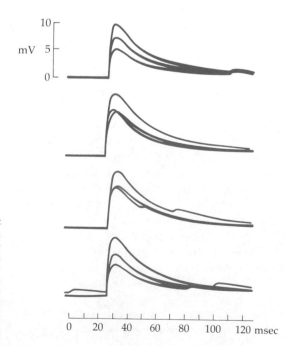

6

FLUCTUATIONS IN SYNAPTIC RESPONSE at a neuromuscular junction. The output of ACh was reduced by adding 10 m*M* MgCl₂ to the bathing solution (reducing extracellular calcium would have had the same effect). Stepwise fluctuations in amplitude are due to variations in the number of quanta of ACh released on successive shocks to the motor nerve. Occasional spontaneous miniature epps can be seen. (From del Castillo and Katz, 1954a.)

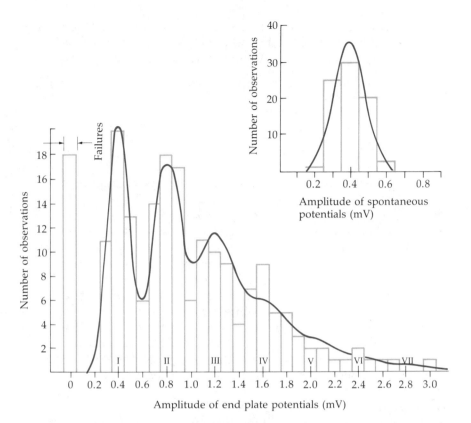

7 AMPLITUDE DISTRIBUTION of stimulus-evoked and spontaneous
(inset) epps at the mammalian neuromuscular junction in which release
of ACh has been reduced. The histogram shows the numbers of epps
observed at each amplitude. The peaks in the histogram occur at one,
two, three, and four times the mean amplitude of the spontaneous
miniature epps. The solid line represents the theoretical distribution of
synaptic potential amplitudes calculated according to the Poisson equation and allowing for the spread in amplitude of the miniature potentials
(see text). Arrows indicate the predicted number of failures of response.
(From Boyd and Martin, 1956.)

of all amplitudes can be seen (main histogram), but there are pronounced peaks in the distribution occurring at precisely one, two, three,
and four times the mean amplitude of the spontaneous potentials.
Further, the variance of the first peak is the same as that of the spontaneous potentials, and the multiple responses have correspondingly
increasing variances. This kind of experiment has been done at a large
number of synapses[15] and the results are always the same: Evoked
release occurs in multiples of the quantal unit. The answer to the first
question, then, is that release is indeed quantized.

[15]Martin, A. R. 1977. In E. Kandel (ed.), *Handbook of the Nervous System*, Vol. 1.
American Physiological Society, Baltimore, pp. 329–355.

To answer the second question, del Castillo and Katz[14] proposed that the fluctuations were statistical in nature and formulated what has come to be known as the quantum hypothesis:

> Suppose we have, at each nerve–muscle junction, a population of n units capable of responding to a nerve impulse. Suppose, further, that the average probability of responding is $\bar{p}$ . . . , then the mean number of units responding to one impulse is $m = n\bar{p}$.

What they proposed, then, was that release of individual quanta from the nerve terminal was similar to shaking marbles out of a box. If such a box has a small hole in the top and contains a large number (n) of marbles and if, when the box is turned over and shaken, there is a small probability p that any given marble will fall out of the hole, then sometimes, on successive tries, one marble will fall out, other times two, other times none, and so on. In a large number of trials, the mean number falling out per trial will be given by np (from now on p rather than $\bar{p}$ will be used for mean release probability). This kind of hypothesis not only explains *why* the fluctuations occur but also provides a way of predicting the distribution of events (i.e., how many failures, singles, doubles, and so forth to expect in a series of trials). Statisticians have shown that under such circumstances the relative occurrence of multiple events is predicted by the BINOMIAL DISTRIBUTION (see Box 1). If n and p are known, the binomial equation can then be used to predict how many failures, single responses, and multiple responses to expect in a series of trials. (Note that the *sequence* in which the various responses occur is not predictable, only their total number.) In summary, the fact that release is quantized tells us *where* the peaks should occur in the distribution shown in Figure 7; the quantum hypothesis tells us *how high* the peaks should be relative to one another.

If, in a total of N trials, the number of responses containing x quanta is called n_x, where x takes on the values, 0, 1, 2, . . . , n, then according to the binomial distribution n_x is given by

$$n_x = N \frac{n!}{(n - x)!x!} p^x q^{(n-x)}$$

where $q = (1 - p)$. To test whether the binomial hypothesis is applicable, the number of failures, single, and multiple unit responses produced by a series of stimuli are compared with those expected theoretically. The theoretical prediction, however, requires that one know n (the number of marbles in the box) and p, although the only experimental parameter that can be observed directly is their product, m. In order to deal with this difficulty, del Castillo and Katz went on:

> Under normal conditions, $\bar{p}$ may be assumed to be relatively large, that is a fairly large part of the synaptic population responds to an impulse. However, as we reduce the Ca and increase the Mg concentration, the chances of responding are diminished and we observe mostly complete failures with an occasional response of one or two units. Under these conditions, when $\bar{p}$ is very small, the number of units which make up the epp in a large series

BOX 1 THE BINOMIAL DISTRIBUTION

It may not be obvious why the quantal hypothesis, as proposed by del Castillo and Katz, implies that fluctuations in release should obey binomial statistics. How this arises can be illustrated by a simple numerical example. Suppose there are three units of the type proposed by del Castillo and Katz and that the probability of any one releasing a quantum of transmitter is 0.1. One might think of the units as release sites in the presynaptic terminal membrane and of the release probability as the probability that such a site will interact with a synaptic vesicle (Chapter 11). In any case, it is necessary to assume that n stays constant at 3 and p at 0.1. The probability that a site will *not* release a quantum is $q = (1 - p) = 0.9$. Now suppose the nerve is stimulated 1000 times. How many failures, single unit responses, doubles, and triples would be expected to occur? To begin with failures, the probability that on any given trial all three sites fail to release a quantum is simply q^3 or 0.729. So in the 1000 trials we would expect 729 failures. The probability of a single release is the product of the probabilities that one site releases a quantum and the other two do not (q^2p). As there are three possible ways for this to happen (site a releases, b and c do not; c releases, a and b do not; and so on), the total probability of a single release is $3q^2p = 0.243$. By similar reasoning, the probability of a double release is $3qp^2 = 0.027$. In the 1000 trials then, we would expect 243 single releases and 27 doubles. Finally, the probability of all three sites releasing a quantum is p^3, or 0.001, so only one triple release would be expected.

In this example there are 243 single quanta released in the 1000 trials; an additional 54 are released as double responses (27 × 2); and 3 are released in the single triple event, for a total of 300 quanta, or an average of 0.3 per trial. The mean number released, then, is $m = np$, as would be expected. In addition, the sum of the four probabilities, $q^3 + 3q^2p + 3qp^2 + p^3$, is unity, as there are no additional combinations. The individual probabilities are the successive terms of the expansion of the expression $(q + p)^3$. Because the expression has two variables, this is called a binomial expansion. For any number of units, n, the probabilities of failures, single responses, and so forth are given by the successive terms of the expansion of $(p + q)^n$. In other words, if such probabilities are designated p_x, where $x = 0, 1, 2, \ldots$, then,

$$p_x = \frac{n!}{(n - x)!x!} p^x q^{(n - x)}$$

As noted by del Castillo and Katz, this expression is simplified if the assumption is made that p becomes infinitely small. As the product np remains finite, this means that n becomes infinitely large. In that case, it can be shown mathematically that the above expression approaches the limit

$$p_x = (m^x/x!)e^{-m}$$

This is the same as the Poisson distribution, which normally arises from quite different considerations. When the release probability is small, then, the distribution of synaptic potential amplitudes should follow the Poisson distribution quite closely. But this is only an approximation; the hypothesis is a binomial one and the amplitude distribution should be consistent with binomial statistics for all values of p. The point is that for small values (p less than about 0.1) there is no practical difference between the two distributions.

of observations should be distributed in the characteristic manner described by Poisson's law.

In their experiments, then, del Castillo and Katz tested the applicability of the POISSON DISTRIBUTION to the observed fluctuations in epp amplitude. In the Poisson equation there is no n or p, only m. The expected numbers of epps containing x quanta are given by

$$n_x = N(m^x/x!)e^{-m}$$

For the Poisson distribution to be applicable requires only that the quanta be released independently so that the release of one has no influence on the probability of release of the next. In general, the distribution is used to predict the frequency of occurrence of discrete events in a continuum, for example, the distribution in time of airplane crashes or the distribution of breaks along a telephone line. One of the best-known applications is an analysis of the number of Prussian cavalry officers hurt each year by horse kicks. Some years were failures—none was kicked; in other years, one or two were kicked. Over a long period the distribution was described closely by the Poisson equation, using only the mean number of kicks per year (m) to describe the theoretically expected distribution. Another convenient analogy, in which the Poisson distribution is used as an approximation to the binomial distribution, is the dime slot machine that contains a very large number of dimes but gives them back only with great reluctance. Most of the time the player, on putting a dime in the machine, will suffer a failure, but sometimes he will receive one, two, or (in rare cases) a large number of dimes in return. Again, if the mean number of dimes paid out by the machine per play (m) is known, then during a long period of play it is possible to predict the number of times there is no payoff, the number of times the player receives only one dime in return, and so on. Two points to be noted about the Poisson distribution are (1) n and p have no relevance; the characteristics of the distribution depend on the single parameter m. (2) There is no restriction on m; in the unlikely event that the machine paid out an average of 100 dimes per play, the distribution would still be applicable (provided the content of the machine was sufficiently large). In that case, however, the frequency of failures would be vanishingly small, as would the profits of the establishment owning the machine.

To test the applicability of the Poisson distribution, then, it is necessary to have a measure of m, the mean number of units released per trial. For the slot machine, this could be obtained by calculating the average amount of money paid out per trial and dividing by the size of the unit (10¢). An analogous approach at the neuromuscular junction is to take advantage of the fact that the mean amplitude of the spontaneous miniature potential ($\bar{v}_1$) is the same as that of the individual units making up the epp evoked by nerve stimulation. The mean quantum content (m) is then given by the mean amplitude of the evoked response ($\bar{v}$) divided by the mean unit amplitude:

$$m = \bar{v}/\bar{v}_1$$

A second method is to determine m from the number of failures. When $x = 0$ in the Poisson equation, $n_0 = Ne^{-m}$. It follows, then, that

$$m = \ln (N/n_0)$$

The agreement between m estimated in these two ways is excellent. Put the other way around, if m is calculated by the first method, the value so obtained predicts accurately the observed number of failures.

A more stringent test of the applicability of the Poisson equation is to predict the *entire* distribution of epp amplitudes using only m and the mean amplitude of the unit potential, that is, to predict the height of the individual peaks in Figure 7 as well as their position. To do this, m is calculated from the ratio of the mean response amplitude to the mean spontaneous potential amplitude, and the number of expected failures is calculated (arrows at zero amplitude). The expected number of single responses is then calculated and these are distributed about the mean unit size with the same variance as the spontaneous events (inset). Similarly, the predicted multiple responses are distributed around their means with proportionately increasing variances. The individual distributions are then summed to give the theoretical distribution shown by the continuous curve. The agreement with the experimentally observed distribution (bars) is remarkably good.

The agreement between the overall amplitude distribution and that predicted by Poisson statistics provided strong evidence that the quanta were released independently. Such experiments, however, provided no information about the validity of the quantum hypothesis. Although adherence of the release process to Poisson statistics was demonstrated at a large variety of vertebrate and invertebrate preparations,[15] it was several years before the applicability of the binomial distribution was confirmed, first at the crayfish neuromuscular junction[16] and then in a number of other preparations. In summary, there is now ample evidence that transmitter is released in packets, or quanta, and that the fluctuations in release from trial to trial can be accounted for by binomial statistics, as predicted by del Castillo and Katz. When the release probability p is small, as in a low-calcium medium, the Poisson distribution provides an equally good description of the fluctuations.

The quantal nature of the release process bears directly on the mechanisms that operate during integration in the nervous system because the quanta provide the units that make up synaptic signals. One striking feature of the vertebrate nervous system is the variation in mean quantum content of the response as one moves from the neuromuscular junction where the safety factor for transmission is high (m in the range of 100 to 300) to synapses on cells in the central nervous system, such as the spinal motoneuron. The motoneuron is concerned mainly with integrating a myriad of incoming information, and no one synapse has a dominating influence on the cell. A primary afferent fiber from a muscle spindle, for example, operates with a mean quantum content of only about one quantum per incoming presynaptic action potential.[17] Inhibitory inputs to the motoneuron have correspondingly low quantum contents.[18] Autonomic ganglion cells occupy an intermediate position with quantum contents of 1–3 in mammalian sympathetic ganglia[19] and about 20 in the ciliary ganglion of the chicken.[20]

General significance of quantal release

[16]Johnson, E. W. and Wernig, A. 1971. *J. Physiol. 218:* 757–767.
[17]Kuno, M. 1964a. *J. Physiol. 175:* 81–99.
[18]Kuno, M. and Weakly, J. N. 1972. *J. Physiol. 224:* 287–303.
[19]Blackman, J. G. and Purves, R. D. 1969. *J. Physiol. 203:* 173–198.
[20]Martin, A. R. and Pilar, G. 1964. *J. Physiol. 175:* 1–16.

The quantal nature of the release process is of interest for another reason: Changes in synaptic efficacy are almost always due to changes in quantum content. For example, at most synapses, including the neuromuscular junction of the frog and crayfish, a second impulse to a presynaptic fiber delivered shortly after the first gives rise to a larger postsynaptic potential. Similarly, a train of impulses can lead to a continuing growth of the response, as shown in Figure 8. This effect, called FACILITATION, has been shown to be due to an increase in the number of quanta of transmitter released by the presynaptic terminal.[21-23] Further statistical analyses at the crayfish neuromuscular junction have led to the suggestion that this increase in m is due, in turn, to an increase in the probability of release, p.[24,25] Similarly, presynaptic inhibition (Chapter 9) has been shown to be due to a reduction in the number of quanta released from the affected terminal, both at the crayfish neuromuscular junction[22] and at primary afferent synapses on spinal motoneurons.[23]

Role of calcium in facilitation

Facilitation is the earliest of several components of increase in quantal release following repetitive stimulation. These have been classified according to their time course,[26] with facilitation lasting a few hundred milliseconds. Two later components can be seen following longer periods of stimulation: These are augmentation, lasting seconds, and potentiation, which may last minutes to hours, depending on the preparation and experimental conditions. Experimental evidence was obtained by Katz and Miledi[27] that facilitation was related to a residue of calcium that enters the terminal during the nerve impulse. With repetitive stimulation, as in Figure 8, continuing calcium accumulation then leads to a progressive increase in transmitter release. Similarly, augmentation and potentiation are believed to be related to increases in cytoplasmic calcium in the presynaptic nerve terminal, increases due to both accumulation during activity and release from internal stores.[28] Mathematical models relating facilitation to the kinetics of calcium entry and removal from the terminal have been developed by H. Parnas and shown to be applicable to facilitation at the crayfish neuromuscular junction.[29]

Throughout this discussion it has been implied that the quantal size is constant. This is not always true. For example, during nerve terminal regeneration, the amplitudes of spontaneous miniature epps, rather than being distributed normally (Figure 6), are skewed into the baseline

[21]del Castillo, J. and Katz, B., 1954b. *J. Physiol. 124*: 574–585.
[22]Dudel, J. and Kuffler, S. W. 1961. *J. Physiol. 155*: 543–562.
[23]Kuno, M. 1964b. *J. Physiol. 175*: 100–112.
[24]Wernig, A. 1972. *J. Physiol. 226*: 751–759.
[25]Zucker, R. S. 1973. *J. Physiol. 229*: 787–810.
[26]Magleby, K. L. and Zengel, J. E. 1982. *J. Gen. Physiol. 80*: 613–638.
[27]Katz, B. and Miledi, R. 1968a. *J. Physiol. 195*: 481–492.
[28]Lev-Tov, A. and Rahamimoff, R. 1980. *J. Physiol. 309*: 247–273.
[29]Parnas, H., Dudel, J. and Parnas, I. 1982. *Pflügers Arch. 393*: 1–14.

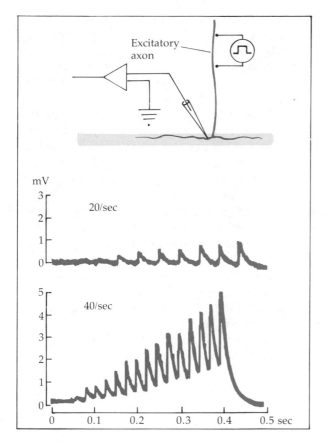

Excitatory axon

mV

20/sec

40/sec

8

FACILITATION at the crayfish neuromuscular junction brought about by the liberation of increased numbers of quanta with successive nerve stimuli at 20/sec and 40/sec. The extent of facilitation depends on the frequency of stimulation. (From Dudel and Kuffler, 1961.)

noise; that is, there are large numbers of very small spontaneous potentials. The small quantal units, however, do not appear to be released in response to stimulation.[30] Small miniature epps are also seen in normal frog and mouse muscle. It has been suggested that these might represent subunits of the usual miniature epp,[31] on the grounds that under some experimental conditions the amplitude distribution of the large miniature potentials appears to be fragmented into multiples of the small miniature amplitude. However, the existence of such fragmentation has been disputed,[32] and the significance of the small miniature potentials is not clear. Conversely, spontaneous synaptic potentials larger than the usual miniature potentials are seen occasionally. In some preparations these appear to be due to the spontaneous release of two or more quanta simultaneously; in others their amplitude shows no clear relation to the normal quantal amplitude.[15]

[30]Dennis, M. J. and Miledi, R. 1974b. *J. Physiol.* 239: 571–594.
[31]Kriebel, M. E. and Gross, C. E. 1974. *J. Gen. Physiol.* 64: 85–103.
[32]Magleby, K. L. and Miller, D. C. 1981. *J. Physiol.* 311: 267–287.

In addition to leaving the motor nerve terminal in the form of individual quanta, ACh also leaks continuously from the cytoplasm into the extracellular fluid. In other words, there is a steady nonquantal "ooze" of ACh from the presynaptic terminal. This was first detected electrophysiologically by Katz and Miledi at frog muscle end plates treated with anticholinesterase.[33] Under these conditions, ACh that would normally have been hydrolyzed by the cholinesterase at the synapse instead built up a steady concentration in the synaptic cleft, producing a slight depolarization. This depolarization was detected by applying large pulses of curare to the end plate region by ionophoresis from a focally placed pipette. Such application resulted in a hyperpolarization of about 50 μV. The concentration of ACh in the cleft required to account for a potential of this magnitude was calculated to be about 10 pM. The rate of leak of ACh from the terminals required to maintain such a concentration was estimated to be larger by two orders of magnitude than the spontaneous quantal release of ACh. The leak itself is of no significance with respect to the immediate processes involved in synaptic transmission, as its postsynaptic effect is small and is seen only when cholinesterase is inhibited. It does, however, explain the biochemical finding that the amount of ACh collected from resting neuromuscular preparations is about 100 times greater than that which can be accounted for by spontaneous miniature epps.[34]

SUGGESTED READING

General

Douglas, W. W. 1978. Stimulus-secretion coupling: Variations on the theme of calcium-activated exocytosis involving cellular and extracellular sources of calcium. *Ciba Fdn. Symp. 54*: 61–90.

Katz, B. 1969. *The Release of Neural Transmitter Substances*. Liverpool University Press, Liverpool.

Llinás, R. 1982. Calcium in synaptic transmission. *Sci. Am. 247*: 56–65.

McLachlan, E. M. 1978. The statistics of transmitter release at chemical synapses. *International Review of Physiology, Neurophysiology III, 17*: 49–117.

Martin, A. R. 1977. Junctional transmission II. Presynaptic mechanisms. In E. Kandel (ed.), *Handbook of the Nervous System*, Vol. 1. American Physiological Society, Baltimore, pp. 329–355.

Original papers

Boyd, I. A. and Martin, A. R. 1956. The end-plate potential in mammalian muscle. *J. Physiol. 132*: 74–91.

del Castillo, J. and Katz, B. 1954a. Quantal components of the end-plate potential. *J. Physiol. 124*: 560–573.

del Castillo, J. and Katz, B. 1954b. Statistical factors involved in neuromuscular facilitation and depression. *J. Physiol. 124*: 574–585.

[33]Katz, B. and Miledi, R. 1977. *Proc. R. Soc. Lond. B 196*: 59–72.
[34]Vizi, S. E. and Vyskočil, F. 1979. *J. Physiol. 286*: 1–14.

Dodge, F. A. and Rahamimoff, R. 1967. Cooperative action of calcium ions in transmitter release at the neuromuscular junction. *J. Physiol. 193*: 419–432.

Dudel, J. and Kuffler, S. W. 1961. Presynaptic inhibition at the crayfish neuromuscular junction. *J. Physiol. 155*: 543–562.

Johnson, E. W. and Wernig, A. 1971. The binomial nature of transmitter release at the crayfish neuromuscular junction. *J. Physiol. 218*: 757–767.

Katz, B. and Miledi, R. 1967b. The timing of calcium action during neuromuscular transmission. *J. Physiol. 189*: 535–544.

Katz, B. and Miledi, R. 1967a. The release of acetylcholine from nerve endings by graded electrical pulses. *Proc. R. Soc. Lond. B 167*: 23–38.

Katz, B. and Miledi, R. 1967c. A study of synaptic transmission in the absence of nerve impulses. *J. Physiol. 192*: 407–436.

Zucker, R. 1973. Changes in the statistics of transmitter release during facilitation. *J. Physiol. 229*: 787–810.

CHAPTER ELEVEN

MICROPHYSIOLOGY OF CHEMICAL TRANSMISSION

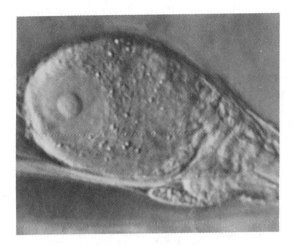

1

LIVING NEURON from the heart of the frog. (From McMahan and Kuffler, 1971.)

Several steps in the release of chemical transmitter and its postsynaptic action have been correlated with subcellular structures and with molecular constituents of synapses, particularly at neuromuscular junctions. Transmitters are concentrated within nerve terminals in membrane-bounded vesicles, which can be depleted by sustained neural activity. Electron microscopic evidence supports the hypothesis that vesicles fuse with the cell membrane to release their contents into the synaptic cleft.

Once acetylcholine (ACh) molecules have been released by motor terminals, they diffuse across the synaptic cleft and combine with receptor molecules in the subsynaptic membrane. This is the area where the ACh receptors are concentrated. In snake and frog muscles, the sensitivity to ACh decreases by a factor of about 100 within a distance of a few micrometers from the edge of the nerve terminal. Acetylcholine receptors have been extracted from various membranes, purified, and characterized biochemically.

Electrical recording techniques have provided insight into the molecular mechanisms involved in the interactions of ACh and other neurotransmitters with postsynaptic chemoreceptors and in the subsequent opening of membrane channels. The events associated with the opening and closing of single postsynaptic channels can be deduced from fluctuations in membrane current (membrane noise) produced by transmitter application. Such analyses have revealed that individual ACh-activated channels open for 1–2 msec and allow a net entry of about 2×10^4 ions. Other synaptic channels stay open for tens of milliseconds and pass 10^5 ions or more. Much of the information obtained by noise analysis has been confirmed by experiments in which the current through individual molecular channels has been recorded, using "patch clamp" techniques.

Quanta, the physiological units of presynaptic transmitter release, each consist of only a few thousand molecules. The postsynaptic current produced by a nerve impulse is due to the linear summation of responses to synchronously released quanta. The enzyme acetylcholinesterase (AChE) hydrolyzes ACh molecules and thereby restricts the spread of the transmitter within the synaptic clefts. When AChE is inhibited, however, molecules in the quanta are free to diffuse into adjacent regions so that they cover overlapping areas and potentiate each other's effects.

Analysis of quantal release, membrane noise, and currents in single postsynaptic channels have advanced our understanding of the mechanism of action of transmitters and drugs that affect synaptic transmission. The techniques have been applied to a wide variety of preparations, including the vertebrate central nervous system, and the experimental results indicate that the underlying principles are the same at all synapses.

Chapters 9 and 10 have treated the synapse as a simple functional unit made up of two components: the presynaptic terminal and the post-synaptic membrane. This concept provides an adequate framework for discussing the basic principles of chemical synaptic transmission, a discussion focusing chiefly on the secretion of transmitters and the permeability changes they cause in the postsynaptic membrane. However, at the subcellular, and even at the molecular level, there is now much more detailed information about the sites where transmitter is released and about the localization and function of postsynaptic receptors. Correlation of fine structure and molecular behavior with function has revealed much about the detailed process of synaptic transmission.

This chapter emphasizes studies on the vertebrate neuromuscular junction. As with the more general aspects of synaptic transmission discussed so far, the details worked out at this synapse are, to a large extent, applicable to other synapses as well, including, for example, inhibitory synapses in the vertebrate brain. Progress in the last ten years or so has been remarkable in a variety of areas. Transmission electron microscopy and freeze-fracture have provided much information about synaptic structure. Improved optical methods have made it possible to map with considerable accuracy the distribution of chemo-receptors on the surface membranes of living nerve and muscle cells. Improved biochemical techniques have made possible the isolation and characterization of acetylcholine receptors. New analytical techniques applied to electrophysiological methods have resulted in the ability to deduce the characteristics of individual membrane channels from the current fluctuations they produce in response to transmitter application. These deductions have, in turn, been confirmed by direct measurements of currents through individual molecular channels. With information gleaned from such techniques, it is now possible to discuss with some degree of certainty such questions as, What is the ACh concentration in the synaptic cleft when a quantum is released? How many molecules make up a quantum? What are the dimensions of a postsynaptic channel?

A consistent feature of chemical synapses is the cleft—a few tens of nanometers wide—that separates the presynaptic nerve terminal from the postsynaptic membrane. The internal architecture of the nerve terminals at many diverse synapses shows a number of common elements such as mitochondria, neurofilaments, microtubules, an agranular endoplasmic reticulum, and synaptic vesicles. This discussion will concentrate chiefly on SYNAPTIC VESICLES, which were tentatively linked with chemical synaptic transmission shortly after their discovery. This correlation of structure and function has been well supported, and the presence of vesicles is one of the single most useful criteria for morphological recognition of a chemical synapse.[1,2] The vesicles range from 40 to 200 nm in diameter, and the characteristics of their morphology are often associated with a particular transmitter. For example, nerves that release catecholamines invariably contain large vesicles, each with an electron-dense granule at its core,[3] whereas nerve terminals known to be cholinergic contain smaller, clear (agranular) vesicles. The presence of agranular vesicles does not, however, indicate that ACh is the transmitter at a synapse where the transmitter is unknown. In fact, agranular vesicles are ubiquitous and probably represent the means of packaging a number of neurotransmitters.

The neuromuscular junction in the frog, illustrated in Figure 2, contains all of the major features typical of chemical synapses. As shown in Figure 2A, the incoming motor nerve loses its myelin sheath and then gives off branches that run in shallow grooves on the surface of the muscle. On the muscle cell side of the cleft are POSTSYNAPTIC FOLDS, which radiate from the cleft at regular intervals. Within the cleft is basement membrane, which follows the contours of the muscle surface. Schwann cell lamellae cover the terminal, sending fingerlike processes around it, thereby creating regularly spaced subdivisions. Within the cytoplasm of the terminal are mitochondria and synaptic vesicles. Many of the latter, instead of being randomly distributed, are lined up in double rows along narrow transverse bars of electron-dense material attached to the presynaptic membrane. This region is known as the ACTIVE ZONE; much evidence indicates that it represents the region

Structural elements of synapses

[1]Palay, S. L. and Palade, G. E. 1955. *J. Biophys. Biochem. Cytol.* 1: 69–88.
[2]DeRobertis, E. 1967. *Science 156*: 907–914.
[3]Bloom, F. E. 1972. *Handbook Exp. Pharmacol. 33*: 46–78.

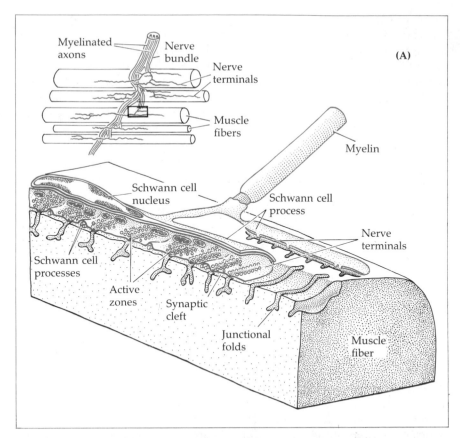

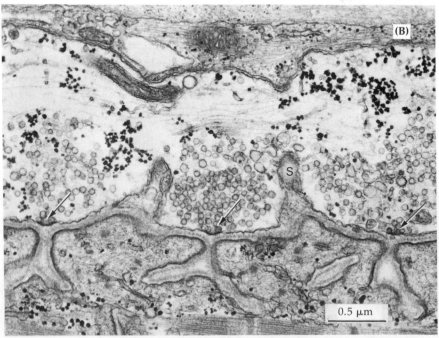

2 NEUROMUSCULAR JUNCTION of the frog. (A) Low-power view of several muscle fibers and their innervation. Below is a three-dimensional sketch of part of a synaptic contact area. Synaptic vesicles are clustered in the nerve terminal in special regions opposite the opening of the postsynaptic folds. These active zones are the sites where transmitter is released into the synaptic cleft. Processes of Schwann cells usually extend between the nerve terminal and the postsynaptic membrane, separating active zones. (B) Electron micrograph of a portion of the motor nerve terminal, showing many of the features seen in the sketch, including active zones (arrows) and Schwann cell processes (S). Within the synaptic cleft and the junctional folds, basement membrane is visible. (Courtesy of U. J. McMahan.)

where transmitter is released.[4-6] These details are illustrated in Figure 2B, which is an electron micrograph of a longitudinal section through a presynaptic terminal and the adjacent muscle. The principal subcellular elements shown also occur in nerve–nerve synapses (Figure 5C).

The technique of freeze-fracturing has revealed many additional features of cell membranes. This procedure consists of freezing the tissue and breaking it apart before preparing replicas for scanning electron microscopy. When the tissue is broken, the lines of cleavage generally do not occur along the spaces between adjacent cells; instead, freeze-fracture planes occur between the bilayers of plasma membranes. This creates two artificial surfaces, or faces, one belonging to the cytoplasmic half of the membrane leaflet, the other to the external half. Thus, with scanning electron microscopy, broad areas of the interior of the membranes can be viewed on either face.

The nature of the fracture planes is shown in detail in Figure 3. Figure 3A shows the main features of two active zones in a presynaptic terminal and the adjacent postsynaptic structures. If the presynaptic and postsynaptic membrane bilayers are fractured along the planes indicated by the arrows, the separations occur as shown in the three-dimensional drawing of Figure 3B. (In practice a fracture would occur in one membrane or the other, but not in both at the same time.) The upper portion of the figure shows the exposed surface of the cytoplasmic half of the presynaptic membrane with vesicles lined up on the cytoplasmic side. Some of the vesicles are represented as fusing with the membrane and spilling their contents into the synaptic cleft. This process, called EXOCYTOSIS, is believed to be the mechanism whereby the individual quanta of acetylcholine are released from the terminal. There are, in addition, intramembranous particles protruding from the fracture face of the cytoplasmic leaflet; pits corresponding to the particles appear on the fracture face of the outer leaflet. Similar particles and pits also occur on the corresponding postsynaptic fracture faces.

[4]Couteaux, R. and Pécot-Dechavassine, M. 1970. *C. R. Acad. Sci. (Paris)* 271: 2346–2349.
[5]Heuser, J. E., Reese, T. S. and Landis, D. M. D. 1974. *J. Neurocytol.* 3: 109–131.
[6]Peper, K., Dreyer, F., Sandri, C., Akert, K. and Moor, H. 1974. *Cell Tissue Res.* 149: 437–455.

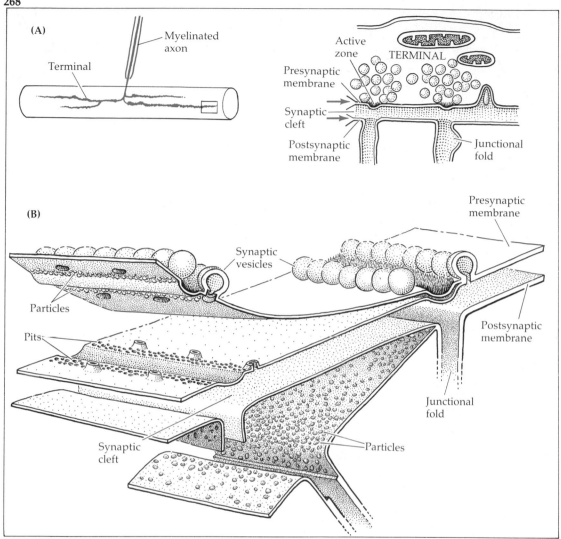

(A)

Myelinated axon

Terminal

Active zone

TERMINAL

Presynaptic membrane

Synaptic cleft

Postsynaptic membrane

Junctional fold

(B)

Presynaptic membrane

Synaptic vesicles

Particles

Pits

Postsynaptic membrane

Junctional fold

Synaptic cleft

Particles

3 **SYNAPTIC MEMBRANE STRUCTURE. (A) Entire frog neuromuscular junction (left) and longitudinal section through a portion of the nerve terminal (right). Arrows indicate planes of cleavage during freeze-fracture. (B) Three-dimensional view of presynaptic and postsynaptic membranes with active zones and immediately adjacent rows of synaptic vesicles. Plasma membranes are split along planes indicated by the arrows in (A) to illustrate structures observed upon freeze-fracturing. The cytoplasmic half of the presynaptic membrane at the active zone shows on its fracture face protruding particles whose counterparts are seen as pits on the fracture face of the outer membrane leaflet. Vesicles which fuse with the presynaptic membrane give rise to characteristic protrusions and pores in the fracture faces. The fractured postsynaptic membrane in the region of the folds shows a high concentration of particles on the fracture face of the cytoplasmic leaflet; these are probably ACh receptors. (Courtesy of U. J. McMahan.)**

A conventional transmission electron micrograph of a horizontal section through an active zone is shown in Figure 4A. It is equivalent to looking down through the active zone from above. An orderly row of synaptic vesicles is lined up along either side of the band of dense material. Figure 4B shows a corresponding image of the fracture face of the cytoplasmic leaflet. It is as if one were now viewing the same region from below the fracture plane. A row of particles, each about 10 nm in diameter, flanks the active zone on either side; and indentations, which are believed to be openings resulting from fusion of synaptic vesicles with the membrane, appear more laterally. A lower power freeze-fracture image is shown in Figure 4C. At the upper left of the photograph, the first fracture face seen is that of the outer leaflet of the presynaptic terminal (T). The fracture crosses the synaptic cleft (C) and exposes the face of the cytoplasmic leaflet of the postsynaptic membrane. Populations of particles are seen around the borders of the postsynaptic folds (F); these are believed to correspond to ACh receptors that are concentrated in this region.[5-7]

When brain tissue is homogenized and fractionated, some of the fractions contain large numbers of membrane-enclosed structures called SYNAPTOSOMES. These are pinched-off nerve terminals, with attached fragments of postsynaptic membrane. Synaptosomes are rich in vesicles and in neurotransmitters. When synaptosomes from preparations like the electric organ of *Torpedo*, where transmission is cholinergic, are further fractionated, ACh is found in the vesicular fraction of the cytoplasm.[8] Furthermore, short periods of stimulation before fractionation of the tissue have been shown to lead to depletion of vesicles in the fraction and to a parallel reduction of acetylcholine content.[9] During prolonged low-frequency stimulation in the presence of radioactive precursors of ACh (acetate or choline), the vesicular fraction accumulates radioactive ACh. Similarly, electron-dense markers such as dextran, added to the bathing solution, are taken up by the vesicles during stimulation. These kinds of observations suggest that vesicles are recycled during activity, first discharging their contents into the synaptic cleft by exocytosis and then reforming and reaccumulating ACh.[10]

In intact nerve terminals, vesicle depletion by stimulation has been demonstrated with electron microscopy,[11-14] and after application of black widow spider venom, which causes a massive outpouring of transmitter at the neuromuscular junction, motor nerve terminals are

Vesicles as the sites of transmitter storage and release

[7]Porter, C. W. and Barnard, E. A. 1975. *J. Membr. Biol. 20*: 31–49.
[8]Whittaker, V. B., Essman, W. B. and Dowe, G. H. C. 1972. *Biochem. J. 128*: 833–846.
[9]Dowdall, M. J., Boyne, A. F. and Whittaker, V. P. 1974. *Biochem. J. 140*: 1–12.
[10]Zimmerman, H. 1979. *Neuroscience 4*: 1773–1804.
[11]Atwood, H. L., Lang, F. and Morin, W. A. 1972. *Science 176*: 1353–1355.
[12]Heuser, J. E. and Reese, T. S. 1973. *J. Cell Biol. 57*: 315–344.
[13]Birks, R. I. 1974. *J. Neurocytol. 3*: 133–160.
[14]Dickenson-Nelson, A. and Reese, T. S. 1983. *J. Neurosci. 3*: 42–52.

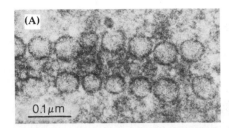

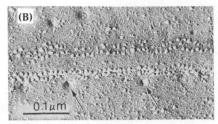

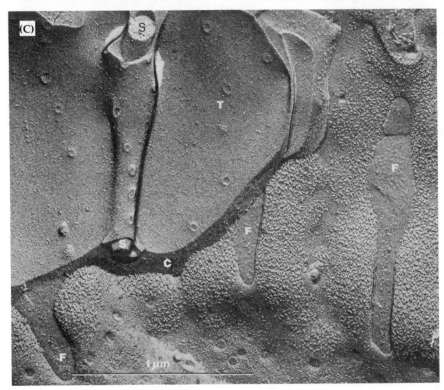

4 REGION OF TRANSMITTER RELEASE and postsynaptic action at the neuromuscular junction of the frog. (A) Electron micrograph of a section through the nerve terminal parallel to an active zone in the presynaptic membrane. (B) Fracture face of the cytoplasmic half of the presynaptic membrane in the active zone, obtained by freeze-fracture. The region of the active zone is delineated by membrane particles (about 10 nm in diameter); further to the side are deformations (arrows) caused by fusion of synaptic vesicles with the membrane. (C) Lower power view of fractured synaptic region. The fracture passes first through the presynaptic membrane of the terminal (T), the outer fracture face of which is visible; it then crosses the synaptic cleft (C) and enters the postsynaptic membrane. On the fracture face between the synaptic folds (F), one sees on the postsynaptic cytoplasmic leaflet particles which are characteristic of regions of concentration of ACh receptors. A Schwann cell process (S) passes between the terminal and the muscle. (A from Couteaux and Pécot-Dechavassine, 1970; B, C from Heuser, Reese, and Landis, 1974.)

virtually devoid of vesicles.[15] Recycling of vesicles in frog motor nerve terminals was studied in detail by Heuser and Reese,[12] who found that when nerve–muscle preparations were stimulated in the presence of horseradish peroxidase (HRP; an enzyme that gives rise to precipitation of electron-dense material—Chapters 1 and 3), the HRP appeared sequentially in subcellular particles in the presynaptic terminals. When electron micrographs of terminals fixed after short periods of electrical stimulation were examined, HRP was found primarily in coated vesicles around the outer margins of the synaptic region, suggesting that these vesicles were being recovered by endocytosis from the terminal membrane and, in the process, were capturing the HRP from the extracellular space. Horseradish peroxidase also appeared in larger cisternae within the terminal and, after longer periods of stimulation, in the synaptic vesicles. Synaptic vesicles loaded in this way with HRP could then be depleted of the substance by stimulation in HRP-free medium, an experimental result supporting the idea that upon such stimulation they discharged their contents into the extracellular space. The sequence that has been proposed from this and similar studies, then, is that quanta are released by exocytosis and that the vesicle membrane is recovered in the form of coated vesicles at the lateral margins of the synapses. These then lose their coats and are refilled with ACh to become synaptic vesicles. The cisternae appear to be formed in the synaptic region at the same time as the coated vesicles by a separate parallel process.[16]

An important experimental innovation that was developed by Heuser and Reese and their colleagues[17] enabled frog muscle to be quick-frozen within milliseconds after a single shock to the motor nerve and then to be prepared for freeze-fracture. With such an experiment it was possible to obtain scanning electron micrographs of vesicles in the act of fusion with the presynaptic membrane and to determine with some accuracy the time course of such fusion. To do this, the muscle is mounted on the under surface of a falling plunger, with the motor nerve attached to stimulating electrodes. As the plunger falls, a stimulator is triggered, shocking the nerve at a selected interval before the muscle smashes into a copper block cooled to 4°K with liquid helium. An essential additional part of the experiment is that the duration of the presynaptic action potential is increased by addition of 4-aminopyridine (4-AP; Chapter 6) to the bathing solution. This treatment greatly increases the number of quanta released by a single shock and hence the number of vesicle openings seen in the electron micrographs. Two important observations were made: First, the maximum number of vesicle openings occurred when stimulation preceded freezing by 3–5 msec. This corresponded to the peak of the postsynaptic current re-

[15]Ceccarelli, B., Hurlburt, W. P. and Mauro, A. 1973. *J. Cell Biol. 57*: 499–524.

[16]Heuser, J. E. and Reese, T. S. 1981. *J. Cell Biol. 88*: 564–580.

[17]Heuser, J. E., Reese, T. S., Dennis, M. J., Jan, Y., Jan, L. and Evans, L. 1979. *J. Cell Biol. 81*: 275–300.

corded from curarized, 4-AP-treated muscles in separate experiments. In other words, the maximum number of openings coincided in time with the peak postsynaptic conductance change determined physiologically. Second, the number of vesicle openings increased with 4-AP concentration, and the increase was related linearly to the estimated increase in quantum content of the end plate potentials by 4-AP, again obtained from separate physiological experiments. Thus, vesicle openings were correlated both in number and in time course with quantal release. In later experiments, Heuser and Reese[16] characterized the time course of the vesicle openings in greater detail, showing that openings first increase during a 3- to 6-msec period after stimulation and then decrease over the next 40 msec.

In summary, there is now much evidence that synaptic vesicles are the morphological correlate of the quantum of transmitter, each vesicle containing a few thousand molecules of the transmitter in question. Vesicles can release their contents by exocytosis both spontaneously at a low rate (producing miniature synaptic potentials) and in response to presynaptic depolarization. This view is not universally held,[18] but other mechanisms proposed for quantal release, such as calcium-activated quantal gates in the presynaptic membrane, have rather less extensive experimental support.

CHEMORECEPTORS AND THEIR DISTRIBUTION

A neurotransmitter opens specific permeability channels in the postsynaptic membrane, an effect produced by interaction with specific chemoreceptor molecules. This type of interaction raises some general questions about the manner in which the cell surface is constructed. For example, are the chemoreceptors and the ionic channels they activate sharply confined to structurally distinct synaptic areas? If so, receptors should be concentrated on the cell surface in restricted areas according to the distribution of synapses. On the cell bodies and dendrites of neurons, such specialization is presumably even more complex; patches of receptors underlying various boutons should have properties distinct from one another, depending on the nature of the synapse (inhibitory or excitatory, fast or slow). In other words, each transmitter substance released onto the cell should control specific permeability channels. Questions of receptor characteristics and distribution have been studied most extensively in synapses at which ACh is the transmitter.

The existence of special properties of skeletal muscle fibers in the region of innervation has been known since the beginning of the century. For example, Langley assumed the presence of a receptive substance around motor nerve terminals, where a localized sensitivity exists for various chemical agents, such as nicotine.[19] Gradually, it has become

[18]Tauc, L. 1982. *Physiol. Rev. 62*: 857–893.
[19]Langley, J. N. 1907. *J. Physiol. 36*: 347–384.

accepted that the term RECEPTOR means a site where a transmitter, hormone, or drug exerts its effects. Isolation of ACh receptors has been helped greatly by the discovery of snake toxins that bind specifically and with high affinity to ACh receptors.[20] Once occupied by the toxin, the receptors cannot react with ACh, which accounts for the paralyzing action of the bite of certain snakes. One of the most useful toxins has been α-bungarotoxin, obtained from *Bungarus* snakes. When labeled with radioactive iodine, the toxin can be used as a means of identifying and assaying the isolated ACh receptor. The richest sources of ACh receptor are the electric organs of the fishes *Electrophorus* and *Torpedo*, and the receptor has been isolated from these and other preparations in a number of laboratories.[21-23] The receptor is a glycoprotein with a molecular weight of 255,000 (255 K), and is composed of five subunits, the complete amino acid sequences of which are now known.[24-26] There are two α-subunits of about 40 K, a β-subunit of 50 K, a γ-subunit of 60 K, and a δ-subunit of 65 K. The α-subunits each carry a binding site for α-bungarotoxin and cholinergic ligands, so there are two sites per receptor monomer. Electron microscopic studies indicate that the monomer is about 11 nm long by 8.5 nm in diameter.[27] The diameter of the ion channel spanning the core (when open) can be estimated by its permeability to organic cations of different sizes[28] and seems to be about 0.65 nm.

The chemosensitivity of the surface membrane of the postsynaptic cell can be determined by ionophoretic application of transmitter substances, such as ACh, from a micropipette onto restricted regions of membrane (Chapter 9). The method is particularly useful with thin preparations in which the presynaptic and postsynaptic structures can be resolved with interference contrast optics[29] and the position of the ionophoretic pipette in relation to the synapse can be determined with some precision. One such preparation is the INTERATRIAL SEPTUM of the heart. The septum is a thin, transparent sheet in which are embedded nerve cells that receive their synaptic innervation from the vagus nerve. Usually one or more vagal axons distribute 12 to 15 synaptic boutons on a cell body (Figure 5A). The boutons are sites where ACh is released

Chemical specialization of the postsynaptic membrane

[20]Lee, C. Y. 1972. *Annu. Rev. Pharmacol. 12*: 265–286.

[21]Heidmann, T. and Changeux, J.-P. 1978. *Annu. Rev. Biochem. 47*: 317–357.

[22]Karlin, A. 1980. In *Cell Surface and Neuronal Function*. Elsevier–North Holland, New York, pp. 191–260.

[23]Conti-Tronconi, B. M. and Raftery, M. A. 1982. *Annu. Rev. Biochem. 51*: 491–530.

[24]Noda, M. *et. al.* 1982. *Nature 299*: 793–797.

[25]Noda, M. *et. al.* 1983. *Nature 301*: 251–255.

[26]Ballivet, M., Patrick, J., Lee, J. and Heinemann, S. 1982. *Proc. Natl. Acad. Sci. USA 79*: 4466–4470.

[27]Kistler, J., Stroud, R. M., Klymkowsky, M. W., Lalancett, R. A. and Fairclough, R. H. 1980. *Biophys. J. 37*: 371–383.

[28]Dwyer, T. M., Adams, D. J. and Hille, B. 1980. *J. Gen. Physiol. 75*: 469–472.

[29]McMahan, U. J., Spitzer, N. C. and Peper, K. 1972. *Proc. R. Soc. Lond. B 181*: 421–430.

274

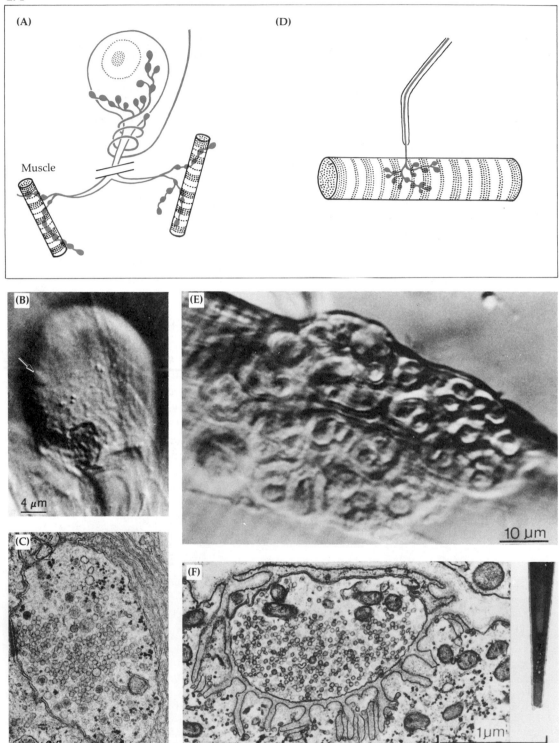

NEURONAL AND NEUROMUSCULAR SYNAPSES. (A) A ganglion
cell in the interatrial septum sends its axon to innervate the heart muscle
and receives synapses from the vagus nerve. (B) A live ganglion cell
viewed with interference contrast optics. A conspicuous synaptic bouton
is marked by the arrow; others can be observed elsewhere on the cell
by changing the focus. (C) Cross section through a synaptic bouton that
contains all of the elements typical of a chemical synapse. (D) Sketch
of an end plate on a skeletal muscle of a snake. (E) End plate on a live
muscle fiber. Individual synaptic boutons rest in craters in the muscle
surface. (F) Cross section through a bouton. To the right is an electron
micrograph of a micropipette used for ionophoresis of ACh; it has an
outer diameter of 100 nm and an opening of about 50 nm. (B, C from
McMahan and Kuffler, 1971; E, F from Kuffler and Yoshikami, 1975a.)

from the presynaptic fibers, thereby causing excitatory postsynaptic
potentials that lead to the initiation of impulses in the nerve cells. A
synaptic bouton can be seen in Figure 5B (arrow). It is possible to locate
other boutons on the cell surface by focusing the microscope to different
depths. An electron micrograph of a cross section through such a bouton (Figure 5C) shows the typical constituents of a chemical synapse,
particularly the synaptic vesicles and the presynaptic and postsynaptic
membrane thickenings.

Figure 5D, E, and F shows similar pictures of synapses on a snake
muscle fiber. The end plates in snake muscle resemble in their com-
pactness those in mammals; they are about 50 μm in diameter with 50
to 70 terminal swellings that are analogous to synaptic boutons. The
swellings rest in craters sunk into the surface of the muscle fiber (Figure
5E); an electron micrograph of such a synapse is shown in Figure 5F,
again illustrating the characteristic features observed at all chemical
synapses. The inset to the right shows an electron micrograph (at the
same magnification as the rest of Figure 5F) of a typical ionophoretic
micropipette. The outer diameter of the tip is about 100 nm and the
opening is about 50 nm, similar in size to a synaptic vesicle.

As expected, the neuronal membrane is highly sensitive to application of ACh in the synaptic regions, where the transmitter is normally
released from the presynaptic boutons. In contrast, the extrasynaptic
membrane is relatively insensitive, even as close as a few micrometers
from the edge of a bouton.[30] For technical reasons, muscle fibers have
been more suitable for demonstrating this sharp delineation of sensitivity. The motor nerve terminals can be removed by applying to the
synaptic region the enzyme collagenase, which frees the nerve terminals
without damaging the muscle fibers.[31] The process of lifting off the
terminals from a snake muscle is shown in Figure 6A. Each of the
boutons leaves behind a circumscribed crater lined with the exposed
subsynaptic membrane. This is shown in more detail in Figure 6B

[30]Harris, A. J., Kuffler, S. W. and Dennis, M. J. 1971. *Proc. R. Soc. Lond. B 177*: 541–
553.

[31]Betz, W. J. and Sakmann, B. 1973. *J. Physiol. 230*: 673–688.

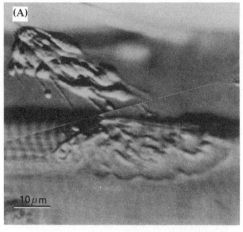

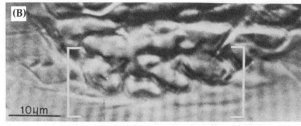

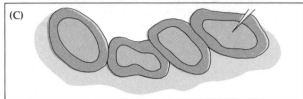

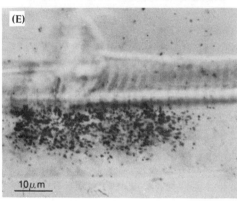

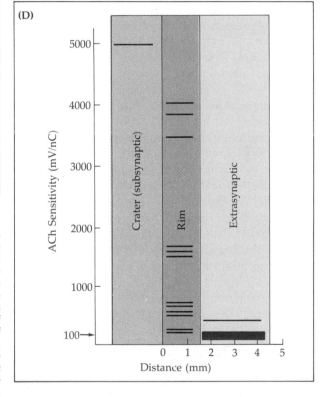

6

ACETYLCHOLINE RECEPTOR DISTRIBU-
TION at skeletal neuromuscular junctions of the
snake. (A) Process of removal of nerve terminal
with boutons from a muscle treated with collagenase. (B) The remaining empty synaptic craters
show exposed subsynaptic membrane. An ACh-
filled pipette (entering from upper right) points
to the floor of a crater. (C) Enlarged drawing of
the bracketed area in (B). (D) Determination of
sensitivity to ACh. In the subsynaptic region,
sensitivity (indicated by horizontal lines) is uniformly high (5000 mV/nC); in the extrasynaptic
region beyond the rims of the craters, it is uniformly low (about 100 mV/nC). In the rim areas,
the sensitivity is variable. (E) An end plate in a
boa constrictor muscle. Receptors are labeled with radioactive α-bungarotoxin, which is confined almost exclusively to the synapses. (A, B,
C, D from Kuffler and Yoshikami, 1975a; E from Burden, Hartzell, and
Yoshikami, 1975.)

(redrawn in Figure 6C), in which an ACh-filled micropipette points at an empty crater. There is a distinct rimlike border area about 1.5 μm wide between the subsynaptic and extrasynaptic membrane, which is probably an optical effect of the synaptic folds lining the walls of the crater (Figure 5F). If the tip of an ACh-filled pipette is placed on the subsynaptic membrane, 1 pC of charge passed through the pipette releases enough ACh to cause, on the average, a 5-mV depolarization. The sensitivity of the membrane is then said to be 5 mV/pC, or 5000 mV/nC (Figure 6D). In contrast, at a distance of about 2 μm, just beyond the rim, the same amount of ACh applied to the extrasynaptic membrane produces a response that is 50 to 100 times smaller.[32] In the rim itself, the sensitivity fluctuates over a wide range. We shall see later that the number of molecules of ACh carried by 1 pC of charge ejected from the pipette can be determined.

These experiments, and similar studies on other preparations, lead to the conclusion that there exists a gradient of chemosensitivity that falls off within micrometers from the edge of the subsynaptic membrane. The chemosensitivity measured by ionophoresis provides a good index of the relative densities of chemoreceptors, and such findings are fully supported by an independent chemical method. When radioactive α-bungarotoxin is used to label cholinergic receptors, the receptor sites can then be located by autoradiography (Figure 6E). By counting the sites of silver deposition in the emulsion, density of receptors can be estimated; in snake muscle the density is about $10^4/\mu m^2$. The density is likely to be greatest around the lip or the crest of the postjunctional folds (Figure 4C). Similar experiments on other neuromuscular preparations have yielded similar estimates of receptor density and, in contrast, far lower densities in the extrasynaptic region (about $5/\mu m^2$).[33,34]

The sharp boundaries between the subsynaptic and extrasynaptic membrane with respect to ACh sensitivity are not immutable. If the input from the nerve is removed by motor nerve section or by blocking the presynaptic nerve impulses, a relatively high density of new receptors appears in the extrasynaptic regions of the postjunctional membrane (Chapter 19).

COMPONENTS OF THE SYNAPTIC RESPONSE

Having discussed the chemical specialization of the postsynaptic membrane, we can now reexamine more closely the signals that are generated there. The ionophoretic technique has been particularly useful in answering a number of questions about both presynaptic and postsynaptic processes involved in transmission. For example, it has enabled

[32]Kuffler, S. W. and Yoshikami, D. 1975a. *J. Physiol.* 244: 703–730.
[33]Fambrough, D. M. 1974. *J. Gen. Physiol.* 64: 468–572.
[34]Fertuck, H. C. and Salpeter, M. M. 1974. *Proc. Natl. Acad. Sci. USA* 71: 1376–1378.

Artificial synaptic
response and the
number of molecules
in a quantum

investigators to estimate the number of molecules of ACh involved in the quantal response and, in combination with an analytical technique known as noise analysis, has provided a means of estimating the conductance and kinetics of single, ACh-activated channels.

There are a number of methods for estimating the number of molecules in one quantum of acetylcholine. The most accurate of these was devised by Kuffler and Yoshikami,[35] who used very fine pipettes (inset, Figure 5F) for ionophoresis of ACh onto the postsynaptic membrane of snake muscle. By careful placement of the pipette, they were able to produce a response to a brief pulse of ACh that mimicked almost exactly the spontaneous miniature end plate potential. An example of such a response is shown in Figure 7. The amplitude of the response to ACh application is the same as that of the miniature synaptic potential, and the time course was only slightly slower.

Having mimicked the synaptic response by ionophoresis, Kuffler and Yoshikami then set out to measure the number of molecules released from the pipette. One method they tried was to fill the pipette with radioactive ACh, place the pipette tip in a small volume of fluid, and apply a large number of pulses. In theory, the amount of ACh released per pulse could then be determined by measuring the total amount of radioactivity released into the fluid. This method was not satisfactory for a number of technical reasons, and a second method was adopted. The ACh was released by repetitive pulses into a small (about 0.5 ml) droplet of saline under oil. The droplet was then applied to the end plate of a snake muscle fiber (Figure 8) and the resulting depolarization measured. The response was compared with responses to droplets of the same size containing known concentrations of ACh. In this way the concentration of ACh in the test droplet was determined and the number of ACh molecules released per pulse was calculated. A pulse of ACh identical to that required to mimic the miniature synaptic potential contained fewer than 10,000 molecules. This represents an upper limit to the number of molecules in a quantum, as the iono-

Stephen W. Kuffler
in 1975

7

SYNAPTIC RESPONSES to artificial and nerve-released ACh in frog muscle. The synaptic potential caused by the spontaneous release of a quantum of ACh from the nerve terminal (miniature excitatory postsynaptic potential) and a postsynaptic response resulting from a 0.5-msec pulse of ACh are virtually identical. Rise time of the response to artificial ACh application is slightly slower because the pipette is not as intimately apposed as the nerve terminal to the postsynaptic membrane. (From Kuffler and Yoshikami, 1975a.)

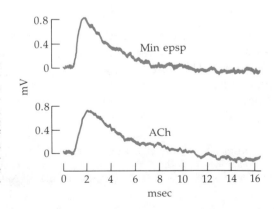

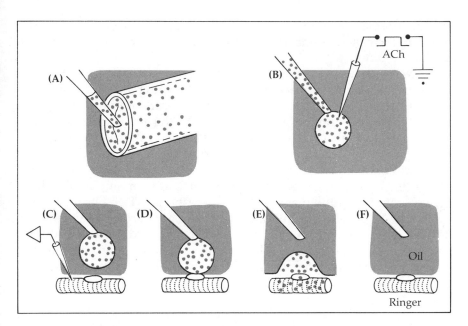

8 ASSAY OF ACh released from a micropipette by ionophoresis. (A) A droplet of fluid (stippled) is removed by a transfer pipette from the dispensing capillary under oil (shaded). (B) ACh is then injected into the droplet from an ionophoresis pipette by repetitive pulses. (C)–(F) An ACh-loaded droplet is touched against the oil–Ringer interface at the end plate and discharges its contents into the aqueous phase. The resulting depolarization is compared with that produced by droplets of known ACh concentration to determine the amount of ACh released by the ionophoresis pipette; the amount of ACh ejected per pulse is then calculated. (After Kuffler and Yoshikami, 1975b.)

phoretic pipette is in a considerably less favorable position with respect to the subsynaptic membrane than is the nerve terminal itself and therefore must release more ACh to achieve the same effect.

The effectiveness of transmission at any synapse is regulated by the number of quanta secreted by the presynaptic nerve terminals. It is of obvious interest, therefore, to know whether two or more quanta interact at the postsynaptic membrane or whether the full synaptic response represents the simple sum of the effects of the individual quantal components. This question was examined by Kuffler and his colleagues[36] at end plates of snake muscle, using the voltage clamp technique (Chapter 9); with this technique, end plate current rather than end plate potential is measured, and the responses provide a direct indication of the magnitude and time course of the postsynaptic conductance change. To obtain synaptic responses with a variable number of quanta, quantal release was first blocked completely by reducing the calcium and increasing the magnesium concentration in the bathing

Quantal interaction

[36]Hartzell, H. C., Kuffler, S. W. and Yoshikami, D. 1975. *J. Physiol.* 251: 427–463.

solution; normal solution was then reintroduced slowly so that progressively increasing numbers of quanta were released with successive stimuli.

Figure 9A shows successive synaptic currents corresponding to the release of 5 to 12 quanta, each quantum contributing a peak current of about 4 nA. Eventually, when the calcium concentration in the bathing solution reaches its normal level (Figure 9B), the response reaches a peak amplitude of about 1250 nA, representing the release of more than 300 quanta distributed over the entire end plate. A comparison of the records in Figure 9A and B shows that the half-decay time (1.2 msec at arrows) does not change as the response increases in amplitude. The only difference is in the magnitude of the response. Two observations in this type of experiment indicate that the quanta act independently: (1) The increment of peak current added by an additional quantum remains constant (in this example, at 4 nA) as the response increases; (2) the time course is independent of the size of the response.

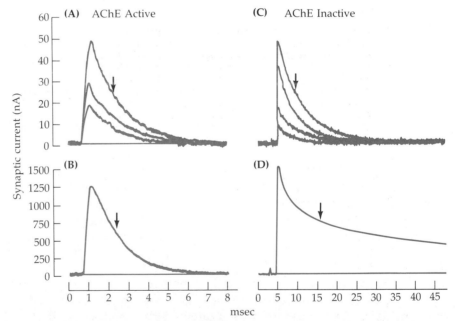

9 INDEPENDENT ACTION AND INTERACTION among quanta. The number of quanta released is regulated by adjusting the calcium and magnesium content of the bathing solution. (A) Nerve-evoked synaptic currents in an end plate of the snake due to the release of 5 to 12 quanta of ACh. (Successive sweeps are superimposed.) (B) Synchronous secretion of about 300 quanta. The half-decay times in (A) and (B) are all close to 1.2 msec (arrows). (C) Release of 2 to 12 quanta when the hydrolyzing action of AChE is inhibited. The half-decay times (arrows) of the synaptic currents range from 2 to 5 msec (note difference in time scale) and increase progressively with quantum content, a finding indicating quantal interaction. (D) Release of 300 quanta is accompanied by marked prolongation of the half-decay times to 9.5 msec. (After Hartzell, Kuffler, and Yoshikami, 1975.)

In contrast to the functional independence of the quanta normally seen, clear interaction is observed when acetylcholinesterase (AChE) is inhibited. Under these circumstances ACh survives in the synaptic cleft and acts repeatedly on receptors as it diffuses out of the region. As a result, the synaptic currents are prolonged significantly, and the prolongation increases with increasing response amplitude. This effect is shown in Figure 9C and D. In Figure 9C the responses increase from 2 to 12 quanta. The half-time of decay for the smallest response is about 2 msec (note change in time scale from Figure 9A and B), increasing to about 5 msec for the largest. When the full synaptic current (corresponding to about 300 quanta) is reached (Figure 9D), the half-decay time is further increased to almost 10 msec.

The difference between the responses in the presence and absence of AChE is illustrated pictorially in Figure 10, which shows the relative density of activated ionic channels in the postsynaptic membrane at different times after release of ACh. Three quanta, out of the normal total of about 300, are shown as examples. The x,y-plane represents the subsynaptic membrane under the presynaptic release sites, and the z-axis represents the density of activated channels. When AChE is fully active (A), each quantum acts independently in a punctate manner on the subsynaptic membrane. At 0.2 to 0.3 msec after release (T_1), when the synaptic current is at its peak, most of the ACh molecules are likely to be bound to receptors. At later times (T_2 and T_3), all of the ACh has been hydrolyzed and the current decays as the ionic channels close (see later). When AChE is inhibited (Figure 10B), the initial situation at T_1 is not much different from normal, except that in addition to receptor-bound ACh, free ACh is also present in the cleft. Subsequently, at T_2 and T_3, the free ACh spreads laterally, during which time the diffusing molecules bind repeatedly to postsynaptic receptors before escaping into open space.

In summary, the AChE normally ensures that quanta act rapidly and independently of each other by confining the area over which each quantal cloud can spread. In other words, each quantum of transmitter has an independent territory on the postsynaptic membrane. In the absence of the esterase, ACh diffusing in the cleft interacts repeatedly with postjunctional receptors, greatly prolonging the response.

The general scheme usually assumed to underlie the action of ACh or of other neurotransmitters on postjunctional membranes is that one or more transmitter molecules combine with a postjunctional receptor molecule, which then undergoes a conformational change resulting in opening of an ionic channel. This idea can be represented schematically:

Single ACh-activated channels

$$n\mathrm{A} + \mathrm{R} \rightleftharpoons \mathrm{A}_n\mathrm{R} \underset{\alpha}{\overset{\beta}{\rightleftharpoons}} \mathrm{A}_n\mathrm{R}^*$$

where A represents the transmitter ("agonist"), R the postsynaptic receptor, n the number of molecules interacting with the receptor, and $\mathrm{A}_n\mathrm{R}^*$ the open state of the agonist–receptor molecule corresponding to an open postsynaptic channel. β and α are the rate constants for the

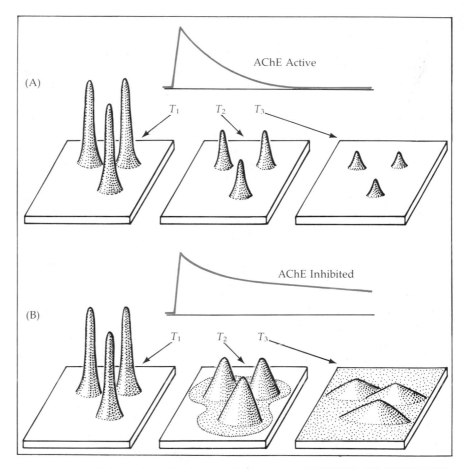

10 **HYPOTHETICAL DISTRIBUTION OF ACTIVATED ACh RECEPTORS** after release of three quanta onto neighboring sites on the subsynaptic membrane. Density of activated receptors (open ionic channels) is shown on the z-axis; the x,y-plane represents the postsynaptic membrane. The progressive change in the number of open channels is drawn at three different times: T_1 at the peak of the synaptic current, T_2 and T_3 during its falling phase. (A) When AChE is fully active, there is little opportunity for lateral diffusion from the point of release, and within 0.3 msec the ACh is bound to the receptors. The declining phase of the current is due to the closing time of the ionic channels. (B) When AChE is inhibited, ACh is able to diffuse along the synaptic membrane so that at T_2 and T_3 it covers overlapping areas, leading to interaction and prolongation of the synaptic current. (From Hartzell, Kuffler, and Yoshikami, 1975.)

transitions between the open and closed states. It is of obvious physiological interest to determine the characteristics of the individual molecular channels; for example, How much current does each carry? How long do the channels stay open? How are their open times distributed? Two relatively recent experimental techniques have been used to answer such questions. One—single channel recording with the patch clamp

283
MICRO-
PHYSIOLOGY
OF CHEMICAL
TRANSMISSION

Noise analysis

technique—has already been described (Chapter 6). The other is a technique known as NOISE ANALYSIS.

The noise analysis technique was first used by Katz and Miledi to study the behavior of postsynaptic channels.[37] This was an original and unusual approach, as electrical noise is not as a rule regarded as a welcome phenomenon in neurophysiological experiments. The method is based on the observation that when ACh is applied to a muscle end plate by ionophoresis, it causes not only depolarization but also an increase in the baseline noise. This extra noise arrives because the number of channels activated by the ACh is not constant. Instead, the number of open channels fluctuates in a statistical manner as the ACh molecules collide with the receptors. The amplitude of the noise relative to the mean depolarization gives a direct measure of the size of the elementary voltage produced by each individual channel. Information about the time course of the elementary voltage contributions can also be obtained from the frequency components present in the noise. However, the time course of voltage fluctuations is determined largely by the membrane time constant and not by the behavior of the elementary channels. For this reason it is more informative to use the voltage clamp technique and examine CURRENT NOISE. This approach also provides a direct measure of individual channel currents and, hence, of channel conductance.

Current noise produced by application of ACh to the neuromuscular junction was first studied in detail by Anderson and Stevens.[38] The end plate region of a frog muscle fiber was voltage clamped and ACh applied ionophoretically. Two characteristics of the response were recorded: (1) the amount of current flowing into the end plate region, recorded on a relatively low-amplification DC trace, and (2) the increase in baseline noise, recorded at much higher amplification and using AC coupling to eliminate the steady DC component. Records of this nature are shown in Figure 11. The lower records show the DC oscilloscope traces at rest (no current) and during a steady application of ACh to the end plate, resulting in a maintained inward current of about 120 nA. Above are shown the corresponding high-amplification AC traces. At rest, the baseline fluctuations are less than 0.1 nA, except for a miniature end plate current caused by a spontaneous quantal release. During ACh application, baseline fluctuations approach 1 nA in peak-to-peak amplitude.

Given experimental observations such as those shown in Figure 11, what information can we obtain from them? First, it might be expected intuitively that the amplitude of the noise should be related directly to the size of the molecular channels. If, for example, the average current of 120 nA were due to ACh opening an average of 60 channels, each carrying 2 nA of current, we would expect rather large fluctuations in

[37]Katz, B. and Miledi, R. 1972. *J. Physiol.* 224: 665–699.
[38]Anderson, C. R. and Stevens, C. F. 1973. *J. Physiol.* 235: 655–691.

11

CURRENT NOISE PRODUCED BY ACh at the frog neuromuscular junction. The lower two records show membrane current at the end plate; the upper two were taken simultaneously at higher amplification to show baseline noise. At rest there is no current across the membrane, and the high-gain record shows little baseline noise (the large deflection is a spontaneous miniature synaptic current). Application of ACh produces about 120 nA of inward current and a marked increase in noise level. (After Anderson and Stevens, 1973.)

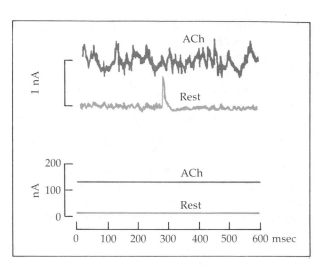

the baseline as the channels opened and closed at random. On the other hand, if the 120 nA mean current were due to an average of 1200 open channels, each carrying 0.1 nA, much smaller fluctuations would be expected. This intuitive expectation is borne out theoretically. The increase in baseline variance (Var I = SD^2) due to fluctuations in the ACh-induced current is directly proportional to the mean current (I) and the size of the single-channel current (c); that is, var I = cI. In deriving this relation, we assume (1) that the channels open and close independently of each other; (2) that each channel undergoes instantaneous transitions between the closed state and an open state of fixed conductance (i.e., that each opening produces a rectangular pulse of current); and (3) that only a small fraction of the postsynaptic receptors are activated. When the equation is rearranged, the single-channel current is given by

$$c = \text{var } I/I$$

To determine the single-channel current, then, it is necessary only to measure the increase in baseline variance and to divide it by the mean current. Once a value for the single-channel current is obtained, the SINGLE-CHANNEL CONDUCTANCE (γ) can be calculated from the current and the driving potential. The driving potential (ΔV) is the difference between the reversal potential for the ACh response (V_r, about -15 mV; Chapter 9) and the potential at which the membrane is clamped (V_m). The relation is

$$\gamma = c/\Delta V$$

Anderson and Stevens obtained single-channel conductances in the range of 20 to 30 pS for channels activated by ACh. This means that each channel passes about 1.6×10^4 ions/msec (see below).

If the amplitude of the single-channel current can be deduced from the amplitude of the baseline fluctuations, it is reasonable to expect that information about the duration of channel opening might be contained in the frequency composition of the extra noise. If most of the channels stayed open for a relatively long time (for example, for 30 msec), the noise might be expected to consist mainly of low-frequency components. If, on the other hand, most of the channels remained open for only a brief period (say, 1 msec), the noise might be expected to contain higher frequency components. Again, this intuitive expectation is borne out theoretically. In practice, the frequencies contained in the extra noise are determined by computer analysis and presented in the form of a SPECTRAL DENSITY DISTRIBUTION, such as that shown in Figure 12. This is a plot of the power density, $S(f)$, as a function of frequency, usually presented on double logarithmic coordinates. What is power density? It is the amount of power contained in each increment of frequency; that is, the amount of power between 0 and 1 Hz, 1 and 2 Hz, 2 and 3 Hz, and so on. It has the units A^2/Hz, or (since $Hz = sec^{-1}$) A^2sec. If all the incremental bits of power are added up, their sum is a measure of the total power contained in the noise, in units of A^2. This is identical to the total variance (Var I).

The exact form expected for the spectral density distribution depends on how one imagines the channels to behave. If the transition from the open to the closed state is a first-order process, determined by the single rate constant α, the channel open times will be distributed exponentially. This means that if a large number (N_0) of channels were opened simultaneously, the number (N) remaining open at any later

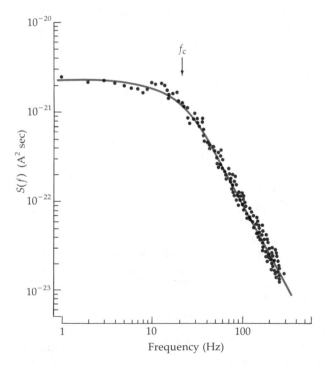

12

SPECTRAL DENSITY DISTRIBUTION of extra noise produced by ACh application (Figure 11). Ordinate, $S(f)$, is the amount of power, or variance, contained in each frequency increment in A^2sec; abscissa is frequency in hertz (Hz). Measurements expressed in this form provide information about the frequency components of the ACh noise and thereby enable estimates to be made of channel open times. Experimental points agree with the relation $S(f) = S(0)/[1 + (2\pi f\tau)^2]$ (solid curve), a result suggesting that ACh-activated channels close with a single time constant, τ. The value for τ is obtained from the corner frequency f_c (see text). (After Anderson and Stevens, 1973.)

time would decrease according to the relation

$$N = N_0 e^{-t/\tau}$$

where the closing time constant $\tau = 1/\alpha$. It is a mathematical property of an exponential distribution of this nature that the time constant is numerically the same as the mean open time. Consequently, the terms *closing time constant* and *mean channel open time* tend to be used interchangeably. If the channel open times are indeed distributed exponentially, it can be shown that the spectral density distribution will have the form

$$S(f) = S(0)/[1 + (2\pi f\tau)^2]$$

This relation is shown in Figure 12. Its shape is determined by the two constants in the equation, $S(0)$ and τ. $S(0)$ determines the maximum value of the the curve where it begins at the low-frequency end of the spectrum. Its value depends on channel conductance. As already noted, large channels produce large current fluctuations; hence, the power density increases with channel size. The mean channel open time, τ, determines how quickly the curve bends over as the frequency increases. τ is usually calculated from the so-called corner frequency (f_c), shown on the graph. This is the frequency at which the power density is one-half that at the start of the curve or, in other words, where $S(f) = 0.5S(0)$. It can be seen from the equation that for this to be true $2\pi f_c\tau$ must be equal to unity, or $\tau = 1/2\pi f_c$.

The spectral density distribution in Figure 12 contains two important pieces of information. One is that the experimental points fit the theoretical curve reasonably well. Thus, the experimental results are consistent with the idea that the channel open times are distributed exponentially. Given that the fit is reasonable, we are then able to determine the mean channel open time. The corner frequency is 21 Hz, so that the calculated mean open time is 7.2 msec. This particular experiment was at 8°C; at room temperature (20°C), the mean channel open time is 1–2 msec.

In summary, analysis of current noise provides information about the conductance of single molecular channels in the membrane and about the kinetic behavior of the channels. At the frog neuromuscular junction, the channels activated by ACh have a conductance of about 25 pS and a mean open time of about 1.5 msec. At the normal resting potential ($V_m = 90$ mV), the current flowing through the channel is about 2 pA, so the total charge transfer through the channel is 2 pA × 1.5 msec, or 3×10^{-15} C. This is the amount of current carried by approximately 20,000 univalent ions. Another calculation of interest is the number of channels activated by a quantum of ACh released from the nerve terminal. At normal resting potential, a miniature end plate current has a peak amplitude of about 3 nA; this is equivalent to 1500 elementary channels or, if two ACh molecules are required to open a channel, 3000 molecules of ACh. This is in good agreement with the experiments discussed earlier (p. 278) in which it was shown that release by ionophoresis of well under 10,000 molecules was required to mimic the effect of a single quantum.

Noise analysis has revealed a number of important properties of the postsynaptic conductance change at the neuromuscular junction. For example, the time constant for channel closing (τ) is the same as the time constant of decay of the miniature end plate current. Furthermore, the time constant of decay of the end plate current is voltage-dependent, the decay being accelerated by depolarization and slowed by hyperpolarization;[39] the time constant for channel closing displays an identical voltage dependence.[38] These observations indicate that, when the molecules of ACh in a quantum produce a relatively synchronous opening of about 1500 postsynaptic channels, the subsequent decay of the response is determined by channel kinetics and not, for example, by the rate of removal of ACh from the synaptic cleft. When ACh removal is slowed—for example, by inhibition of AChE (Figure 9)—the latter becomes important as individual molecules have a chance to interact repeatedly with the channels, thereby prolonging the response. The closing time constant itself depends on the agonist molecule; mean open times are shorter with carbamylcholine than with acetylcholine and longer with suberyldicholine.[37] Another point of interest is that the major action of drugs like curare on individual channels is, as expected, all-or-none. Thus, by competing with ACh for receptor sites, curare produces a reduction in the *number* of channels available for activation by ACh rather than a reduction in individual channel conductances.[37] The blocking action of atropine, on the other hand, is associated with a marked reduction in channel open time and, hence, in the amount of charge entry through the channels.[40]

Deductions about the behavior of single, ACh-activated channels made from noise analysis have been largely confirmed by direct observation of channel currents, using the patch clamp technique.[41] As described in Chapter 6, a high-resistance seal is made between the tip of a glass micropipette and the cell membrane; single channel currents can then be measured in the sealed-off membrane within the tip. This technique has revealed a number of additional features of channel activation and blocking not apparent from noise analysis. An example is the effect on the channels of local anesthetics. These substances have a curare-like action in the sense that they block the postsynaptic effect of ACh. However, the blocking action is accompanied by a striking alteration in the time course of decay of the end plate current. At the single channel level, patch clamp records show that the effect of the drugs is to chop the single channel openings into a series of multiple bursts.[42] The channel behavior is consistent with the idea that the blocking

Single channel currents revealed by patch clamping

[39]Magleby, K. L. and Stevens, C. F. 1972a. *J. Physiol.* 223: 151–171.

[40]Katz, B. and Miledi, R. 1973. *Proc. R. Soc. Lond. B 184*: 221–226.

[41]Hamill, O. P., Marty, A., Neher, E., Sakmann, B. and Sigworth, F. 1981. *Pflügers Arch. 391*: 85–100.

[42]Neher, E. and Steinbach, J. H. 1978. *J. Physiol 277*: 153–176.

molecules enter and block open channels and that the flickering of the channel within a burst represents repeated occurrences of this reaction.

Patch clamp records have also shown that the assumption of a single conductance state is not always valid. In cultured embryonic rat muscle, channels activated by ACh display a fully open state and, in addition, a substate of relatively low conductance.[43] An example is shown in Figure 13, in which two channels are present in an "outside-out" patch

[43]Hamill, O. P. and Sakmann, B. 1981. *Nature* 294: 462–464.

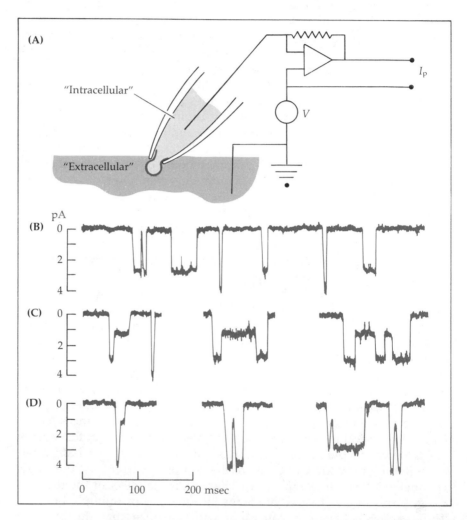

13 PATCH CLAMP RECORDS of single channel currents (I_p) activated by ACh in cultured embryonic rat muscle. (A) Recording arrangement. Outside-out patches were formed with artificial intracellular solution in the pipette and normal bathing solution on the outside containing ACh (2 μM). (B) Two channels in the patch are activated by ACh, one with a peak current of about 2.7 pA, the other 4 pA. (C, D) Examples of each of the channels relaxing to a substate of about 1.3 pA. (After Hamill and Sakmann, 1981.)

(the outside of the membrane faces the external solution; see Figure 18, Chapter 6). Both channels show relaxations to the same substate from which they either close or reopen. Another observation of interest is that ACh channels in frog muscle activated by suberyldicholine (an ACh analogue with which channel open times are relatively long) may have their open states interrupted by one or more very brief closures.[44] Unlike the observations with local anesthetics discussed earlier, the transient closures are not dependent on agonist concentration and therefore do not seem to represent channel blockade by the agonist molecules themselves; the observations with agonists suggest that a channel can close and reopen repetitively while occupied by agonist molecules. These and other observations are continuing to provide important information about the detailed behavior at the molecular level of transmitter-activated channels.

Noise analysis and patch clamp techniques have been applied to a variety of other excitatory and inhibitory synapses. In locust muscle, excitatory synaptic channels activated by glutamate have conductances of approximately 150 pS[45,46] (approximately six times larger than the ACh-activated channel conductance), and mean open times of the order of 2 msec at room temperature. Chloride channels activated by GABA and by glycine at inhibitory synapses have been studied in a number of preparations[47-51] and their characteristics are summarized in Table 1. In the central nervous system, noise analysis techniques have been used to examine the characteristics of glycine-activated inhibitory syn-

Other synaptic channels

[44]Colquhoun, D. and Sakmann, B. 1981. *Nature 294*: 464–466.
[45]Patlak, J. B., Gration, K. A. F. and Usherwood, P. N. R. 1979. *Nature 278*: 643–645.
[46]Cull-Candy, S. G., Miledi, R. and Parker, I. 1980. *J. Physiol. 321*: 195–210.
[47]Dudel, J., Finger, W. and Stettmeier, H. 1980. *Pflügers Arch. 387*: 167–174.
[48]Cull-Candy, S. G. and Miledi, R. 1981. *Proc. Roy. Soc. Lond. B 211*: 527–535.
[49]Jackson, M. B., Lecar, H., Mathers, D. A. and Barker, J. L. 1982. *J. Neurosci. 2*: 889–894.
[50]Barker, J. L., McBurney, R. N. and MacDonald, J. F. 1982. *J. Physiol. 322*: 365–387.
[51]Gold, M. R. and Martin, A. R. 1983. *J. Physiol. 342*: 85–98, 99–117.

TABLE 1
CHARACTERISTICS OF INHIBITORY CHANNELS

Preparation	Agonist	Conductance (pS)	Open time (msec)
Crayfish muscle[47]	GABA	14	7
Locust muscle[48]	GABA	23	4
Cultured mouse spinal neurons[49,50]	GABA	21	35
	Glycine	25	8
Lamprey brain stem[51]	Glycine	75	30

aptic channels in reticulospinal neurons, using the intact brain of the lamprey.[51] The channels, which are permeable to chloride, have large conductances (about 75 pS) and open times of about 30 msec at 5°C. The conductance of the channels is dependent in an anomalous way on intracellular chloride concentration, decreasing rapidly as the concentration increases. Thus, it appears that intracellular chloride interacts with the channel in some way to reduce its conductance. As at the neuromuscular junction, the time constant of decay of miniature inhibitory synaptic currents is consistent with the time constant for channel closing. In addition, the miniature synaptic currents decrease in amplitude with increasing intracellular chloride concentration, as would be expected if the synaptic channels and the glycine-activated channels were identical. With normal intracellular chloride, the peak conductance increase associated with the miniature synaptic current is slightly greater than 100 nS; thus, the single quantum event corresponds to the activation of about 1500 postsynaptic channels.

SUGGESTED READING

General reviews

Conti-Tronconi, B. M. and Raftery, M. A. 1982. The nicotinic cholinergic receptor: Correlation of molecular structure with functional properties. *Annu. Rev. Biochem. 51*: 491–530.

Kistler, J., Stroud, R. M., Klymkowsky, M. W., Lalancett, R. A. and Fairclough, R. H. 1980. Structure and function of an acetylcholine receptor. *Biophys. J. 37*: 371–383.

Peper, K., Bradley, R. J. and Dreyer, F. 1962. The acetylcholine receptor at the neuromuscular junction. *Physiol. Rev. 62*: 1271–1340.

Verveen, A. A. and DeFelice, L. J. 1974. Membrane noise. *Prog. Biophys. Mol. Biol. 28*: 189–265.

Zimmerman, H. 1979. Vesicle recycling and transmitter release. *Neuroscience 4*: 1773–1804.

Original papers

Anderson, C. R. and Stevens, C. F. 1973. Voltage clamp analysis of acetylcholine produced end-plate current fluctuations at frog neuromuscular junction. *J. Physiol. 235*: 655–691.

Dennis, M. J., Harris, A. J. and Kuffler, S. W. 1971. Synaptic transmission and its duplication by focally applied acetylcholine in parasympathetic neurons in the heart of the frog. *Proc. R. Soc. Lond. B 177*: 509–539.

Gold, M. R. and Martin, A. R. 1983b. Analysis of glycine-activated inhibitory post-synaptic channels in brain-stem neurones of the lamprey. *J. Physiol. 342*: 99–117.

Hamill, O. P., Marty, A., Neher, E., Sakmann, B. and Sigworth, F. J. 1981. Improved patch-clamp techniques for high-resolution current recording from cells and cell-free membrane patches. *Pflügers Arch. 391*: 85–100.

Heuser, J. E., Reese, T. S., Dennis, M. J., Jan, Y., Jan, L. and Evans, L. 1979. Synaptic vesicle exocytosis captured by quick freezing and correlated with quantal transmitter release. *J. Cell Biol. 81*: 275–300.

Heuser, J. E. and Reese, T. S. 1981. Structural changes after transmitter release at the frog neuromuscular junction. *J. Cell Biol. 88*: 564–580.

Katz, B. and Miledi, R. 1972. The statistical nature of the acetylcholine potential and its molecular components. *J. Physiol. 224*: 665–699.

Kuffler, S. W. and Yoshikami, D. 1975a. The distribution of acetylcholine sensitivity at the post-synaptic membrane of vertebrate skeletal twitch muscles: Iontophoretic mapping in the micron range. *J. Physiol. 244*: 703–730.

Kuffler, S. W. and Yoshikami, D. 1975b. The number of transmitter molecules in a quantum: An estimate from iontophoretic application of acetylcholine at the neuromuscular junction. *J. Physiol. 251*: 465–482.

Magleby, K. L. and Stevens, C. F. 1972b. A quantitative description of endplate currents. *J. Physiol. 223*: 173–197.

THE SEARCH FOR CHEMICAL TRANSMITTERS

CHAPTER TWELVE

Several substances have been shown to act as transmitters at chemical synapses. They include acetylcholine, norepinephrine, epinephrine, and γ-aminobutyric acid (GABA) as well as a variety of other amino acids, amines, and peptides. Two examples will be described to illustrate the steps required for unequivocal identification of the transmitter molecule acting at a particular synapse: first, GABA, an inhibitory transmitter in the periphery of invertebrates and also within the mammalian central nervous system; and second, a peptide that produces slow potentials in autonomic ganglia of the frog.

Action potentials in inhibitory nerves supplying crustacean muscle release GABA, which opens specific channels to increase the chloride permeability of the postsynaptic membrane. Application of GABÀ mimics the action of the neural transmitter, including its presynaptic inhibitory effects. Analysis of individual neurons has shown that GABA is highly concentrated in inhibitory axons, whereas little or none is found in excitatory or sensory neurons. The enzymes that synthesize and degrade GABA have been extracted from inhibitory and excitatory axons. Both types of cells contain similar amounts of GABA transaminase, the enzyme that degrades GABA to succinate. Glutamic decarboxylase, however, which synthesizes GABA from glutamate, is far more concentrated in inhibitory nerves. This apparently accounts for the higher level of GABA in inhibitory cells. γ-Aminobutyric acid also acts as a transmitter within the mammalian brain. It is found in relatively high concentrations in various regions including the retina, the olfactory bulb, and the cerebral cortex. In the cerebellum, high concentrations of GABA are found in isolated Purkinje cells, which are known to be inhibitory. Their synaptic actions on identified postsynaptic cells are mimicked by ionophoretically applied GABA.

Peptides have been implicated as transmitters at various sites within the central nervous system, where they can produce long-lasting effects that persist for seconds, minutes, or hours. At most synapses it has not been possible to establish their role with certainty. In frog autonomic ganglia, however, a decapeptide (resembling luteinizing hormone releasing hormone, or LHRH) has been shown to be the transmitter. This peptide is present in the presynaptic terminals, is released by stimulation, and mimics the action of the presynaptic nerve.

Numerous other peptides that are candidates for transmitters have been identified. Some have actions resembling those of morphine and bind to the same receptors within the central nervous system. Peptides occur in neurons situated not only within the brain but also in the gut. Since the amino acid composition and sequence of peptides can be analyzed, it may become possible to study regulatory mechanisms at the level of the gene. In addition to peptides, amines such as norepinephrine, 5-hydroxytryptamine, and dopamine have been shown to exert profound effects at synapses within the central nervous system. Often these are "modulatory" in the sense that the amine or peptide modifies the effect of another transmitter, for example, by making it less effective or preventing its release. Moreover, two different transmitters, such as an amine and a peptide, can coexist in the same terminal. The mode of action of such transmitters can be quite different from the punctate actions of GABA or acetylcholine: Release sites may not be closely applied to the postsynaptic membrane. Transmitter may diffuse widely to affect distant targets and thereby influence large numbers of neurons rather uniformly (like a garden sprinkler). Small groups of nerve cells in discrete circumscribed locations within the central nervous system are the principal, or even the only, sources of axons that contain 5-hydroxytryptamine, dopamine, or norepinephrine. Those axons, however, branch extensively to supply widespread areas of the brain, with important functional consequences.

The preceding account of synaptic transmission has emphasized the role of acetylcholine as a transmitter, its release from nerve terminals, its interaction with receptors, and the conductance changes that follow. Transmitter physiology and chemistry continue to be developing and expanding areas of research; conspicuous gaps still exist in the knowledge about the identity of transmitters and their actions. This chapter discusses the general problem of identifying chemical transmitters and their distribution in excitatory and inhibitory neurons. Instead of reviewing comprehensively the large number of transmitters now being studied, we have chosen to restrict the discussion to two examples: (1) γ-aminobutyric acid (GABA) and (2) a decapeptide resembling luteinizing hormone releasing hormone, or LHRH—a hormone that is liberated by neurons in the hypothalamus and that acts on cells in the anterior pituitary gland.

Chemical studies on GABA and the LHRH peptide are well suited to the cellular approach we stress in this book. Thus, for GABA the enzymatic steps in its synthesis and degradation can be studied by microchemical techniques in large nerve cells that can be identified and isolated. Although GABA was first shown to be a transmitter at crustacean neuromuscular junctions, similar approaches have proved fruitful in demonstrating its role in the mammalian brain. Similarly, the LHRH-like peptide is of interest in that it represents a class of transmitters that produce dramatic but little understood effects on the functions of the central nervous system as well as of the peripheral nervous system. As for GABA, an account of the search for the identity of LHRH peptide at synapses in frog autonomic ganglia illustrates well the general problems encountered in establishing with certainty that a particular molecule is a transmitter at a particular synapse.

Little direct information exists about the identity of the transmitters at the majority of synapses within the vertebrate brain. There are obvious technical reasons for this gap in knowledge. Nevertheless, it is possible to determine whether transmission across a particular synapse operates by a chemical mechanism even in higher centers. The criteria are morphological and physiological. For example, electron microscopy reveals clusters of vesicles and electron-dense material in the presynaptic terminal, and across the synaptic cleft postsynaptic thickenings are generally visible (Chapter 11). Physiological testing can identify a reversal potential and specific changes in ionic conductance that, together with pharmacological tests, provide reliable evidence for a chemical synapse. This is fortunate, because even before a transmitter is identified, the knowledge that certain cells have chemical rather than electrical synapses fashions the experimental approach (Chapter 9).

Establishing the identity of a chemical transmitter

To obtain firm evidence for the identity of the substance liberated by the presynaptic terminals, a number of different experimental approaches are required. It is not enough to know that a particular chemical mimics the action of the presynaptic nerve. To conclude that it is the transmitter, one must demonstrate in addition that the terminals (1) normally contain it, (2) release it at the right time in response to stimulation, and (3) release sufficient quantities to produce the appropriate effects in the postsynaptic cell. These procedures require the microanalysis of very small amounts (often as small as 10^{-14} moles) of a number of candidate compounds, together with physiological recordings from individual cells. The methodological difficulties are so great that it has not yet been determined with certainty which chemical is responsible for transmission at synapses for which much detailed physiological information is available, such as the squid giant synapse and synapses on the mammalian spinal motoneuron. Even in the peripheral nervous system, particularly in autonomic ganglia and the gut, the identification of transmitters and their mode of action are still not fully understood, as will become apparent in the later discussion of peptides as transmitters.

In contrast to many peripheral structures, from which most information about transmitters is derived, the central nervous system is composed of an inhomogeneous population of nerve cell bodies and terminals that are closely apposed and intermingled. This makes the analysis much more difficult than that described for ACh at neuromuscular junctions (Chapters 9, 10, and 11). The transmitter that is liberated

Identification of transmitters at synapses within the brain

from neurons in the brain cannot easily be collected from the release sites but often has to diffuse over large distances before it appears at the surface of the brain or in the cerebrospinal fluid. It is often difficult to stimulate selectively a particular homogeneous group of fibers known to be excitatory or inhibitory; and further difficulties are associated with attempts to record intracellularly from the postsynaptic cells and to mimic the action of nerves by applying chemicals through a micropipette. Nevertheless, as a working hypothesis, demonstrating the presence of a substance in presynaptic terminals as well as appropriate postsynaptic effects can provide strong evidence for its presumed role as a transmitter. To this group belong such substances as 5-hydroxytryptamine (5-HT, or serotonin), dopamine, octopamine, adenosine triphosphate (ATP), the amino acids glutamate and glycine, and various peptides including LHRH, substance P, enkephalins, and β-endorphin.

A convenient feature of the mammalian brain is that discrete pools of neurons sharing the same transmitter are often grouped together. It is perhaps somewhat surprising that the populations of neurons containing 5-hydroxytryptamine, norepinephrine, and dopamine are aggregated in a few separate clusters (see later). These three substances and epinephrine are known as biogenic amines; the term CATECHOLA-MINE is used to designate collectively the substances dopa, dopamine, epinephrine, and norepinephrine, all of which contain a catechol nucleus—a benzene ring with two adjacent hydroxyl groups (as shown in Figure 1). It is remarkable that the same molecules are found so widely distributed in the animal kingdom, from leeches and insects to amphibians and mammals. At the same time it seems highly likely that the list is by no means complete and that more transmitter substances remain to be discovered.

A variety of different techniques have now been devised for making maps of the distribution of neurons with special chemical properties. A particularly significant advance has been the development by Falck and Hillarp of the fluorescence method that enables the recognition under the light microscope of axons or terminals containing epinephrine, norepinephrine, dopamine, and 5-HT.[1] In the terminals of such nerves, electron micrographs show characteristic vesicles containing granules about 40 to 140 nm in diameter that can be readily identified; these vesicles have been shown to be rich in biogenic amines (Chapter 11). When condensed with formaldehyde, each of these substances emits light of a characteristic wavelength under ultraviolet light. These fluorescence methods have paved the way for striking advances because they have opened up for easier exploration the entire central nervous system. The techniques enable the tracing in the brain of many pathways and groups of neurons whose terminals are likely to liberate amines. High specificity for staining selectively neurons that contain peptides as well as amines is possible by the use of fluorescence-labeled

---

[1]Falck, B., Hillarp, N.-A., Thieme, G. and Thorp, A. 1962. *J. Histochem. Cytochem. 10:* 348–354.

1   CHEMICAL STRUCTURES OF TRANSMITTER SUBSTANCES.

antibodies to the various transmitter molecules. Had such fluorescence-labeling techniques been available for ACh and norepinephrine at a time when their functions as transmitters were first being tested, the analysis would have been greatly simplified.

A different approach is to take advantage of the specialized metabolic machinery of different neurons to make and degrade neurotransmitters. During these activities neurons use special enzymes. For example, cells having GABA as a transmitter are known to possess high levels of the enzyme glutamic decarboxylase. This enzyme has been purified and an antibody made. When appropriately tagged, the antibody can be used to localize the enzyme in tissue sections; thus, GABA-containing neurons can be visually identified.[2,3] Similarly, for identifying cholinergic neurons, antibodies have been made against the synthesizing enzyme choline acetyltransferase.[4] In addition, cells that use a transmitter may selectively take up and recapture the molecule or its precursor from the external fluid. For example, there is evidence that the

[2]Matsuda, T., Wu, J-Y. and Roberts, E., 1973. *J. Neurochem.* 21: 159–166, 167–172.

[3]Hendrickson, A. E., Ogren, M. P., Vaughn, J. E., Barber, R. P. and Wu, J.-Y. 1983. *J. Neurosci.* 3: 1245–1262.

[4]Levey, A. I., Armstrong, D. M., Atweh, S. F., Terry, R. D. and Wainer, B. H. 1983. *J. Neurosci.* 3: 1–9.

neurons in the cerebral cortex that take up radioactive GABA and become labeled with it use it as a transmitter. Identifying the receptors on postsynaptic cells can also serve to provide clues about the transmitter liberated by presynaptic terminals. Thus, a hallucinogenic plant alkaloid known as muscimol binds to GABA receptors with high specificity; similarly the distribution of peptide binding sites has provided essential clues about the sites where peptides may be liberated by presynaptic fibers.

From a combination of the various approaches outlined above it becomes possible to identify with confidence transmitters at certain synapses within the central nervous system, even though all the ideal criteria may not be satisfied. For example, a variety of physiological tests have supported the idea that motoneurons, which release acetylcholine from their terminals on skeletal muscles, also release the same transmitter onto interneurons within the spinal cord (Renshaw cells; Chapter 16).[5] At other synapses, hints may be provided by staining selectively for particular transmitters in presynaptic terminals.

The following section deals with GABA, which on physiological and chemical grounds is one of the relatively few substances known to be a transmitter at specific synapses in the peripheral as well as the central nervous system and the first to be identified by a systematic chemical survey of a nervous system.

GABA: AN INHIBITORY TRANSMITTER

GABA was first found in extracts of the mammalian brain in 1950 almost simultaneously in the laboratories of Awapara, Udenfriend, and Roberts. At that time it was of interest mainly as a possible metabolite in the glutamic acid cycle. The role of GABA as a transmitter was first suggested by Florey, Elliott, and their colleagues, who used a sensory nerve cell in the peripheral nervous system of a crustacean to assay inhibitory substances.[6] This cell responds to stretch with a steady discharge of impulses (Chapter 15). It was shown that the firing is inhibited by GABA, by extracts of the mammalian brain that contained GABA, and by extracts of crustacean peripheral nerves.

Florey's physiological finding of the inhibitory effect of GABA on the crustacean stretch receptor was confirmed in 1958 by Kuffler and Edwards,[7] who showed in addition that GABA mimics closely the action of the neurally released transmitter on the membrane. On its own this was not adequate to establish GABA as a transmitter. Extracts of the entire nervous system of lobsters were therefore systematically fractionated and the extracts tested for inhibitory activity, using crustacean nerve–muscle junctions as well as stretch receptor neurons for the bioassay.

[5]Kuno, M. and Rudomin, P. 1966. *J. Physiol. 187*: 177–193.
[6]Florey, E. 1961. *Annu. Rev. Physiol. 23*: 501–528.
[7]Kuffler, S. W. and Edwards, C. 1958. *J. Neurophysiol. 21*: 589–610.

The initial chemical work was carried out by Kravitz and his colleagues[8,9] on a large sample of the pooled central and peripheral nervous systems of about 500 lobsters. After a long series of fractionations, ten amino acids that produce inhibition were found. Of these, the following eight were identified (in order of potency): GABA, taurine, betaine, alanine, β-alanine, aspartate, glutamine, and homarine. Taurine was present in the highest concentration, accounting in some extracts for one-half the inhibitory activity. Yet, in its specific inhibitory action, GABA (mole for mole) was about ten times more potent. Since then, evidence has accumulated for the role of taurine as a transmitter.

At this stage it became essential to track the various inhibitory substances down to their cells of origin, because it was expected that the neural transmitter would be preferentially concentrated in inhibitory neurons. Fortunately, the locations of excitatory and inhibitory axons innervating crustacean leg muscles were known, chiefly through the studies of Wiersma.[10] As a result, one could isolate either an individual excitatory axon that on electrical stimulation makes the muscles contract or an inhibitory axon that inhibits contraction (Chapter 9). A pair of such fibers, each about 50 μm in diameter, is shown in Figure 2.

Large numbers of individual inhibitory and excitatory axons were identified, dissected, extracted, and fractionated. The results showed that the concentrations of inhibitory amino acids, with the notable exception of GABA, were roughly equal in both excitatory and inhibitory axons. In contrast, GABA was present in the surprisingly high concentration of approximately 100 mM in the inhibitory axon; this corresponds to about 0.5 percent of the wet weight. But in excitatory axons less than 1 mM of GABA could be detected.

The great asymmetry in distribution of transmitter in the axons of excitatory and inhibitory cells suggests that the inhibitory nerve cell bodies should also contain much larger amounts of GABA than do the excitatory cells. This distribution was found for single cells dissected from ganglia of lobsters, in which cells can be reliably identified as excitatory or inhibitory by physiological criteria.[11] With some practice the cells can also be recognized by their shapes and positions in different ganglia (Figure 3A). The inhibitory cell bodies contain 13 to 15 mM of GABA; excitatory motoneurons contain less than 1 mM. The procedure for isolating cells is shown in Figure 4; a single inhibitory cell body is shown first outside the ganglion and then in the microtube used for chemical analysis. From such a direct microchemical analysis, a partial chemical map of excitatory and inhibitory neurons of a lobster ganglion can be constructed (Figure 3B).

It seems surprising that GABA in the inhibitory axon should be six

[8]Kravitz, E. A., Kuffler, S. W., Potter, D. D. and van Gelder, N. M. 1963. *J. Neurophysiol. 26*: 729–738.

[9]Kravitz, E. A., Kuffler, S. W. and Potter, D. D. 1963. *J. Neurophysiol. 26*: 739–751.

[10]Wiersma, C. A. G. and Ripley, S. H. 1952. *Physiol. Comp. Oecol. 2*: 391–405.

[11]Otsuka, M., Kravitz, E. A. and Potter, D. D. 1967. *J. Neurophysiol. 30*: 725–752.

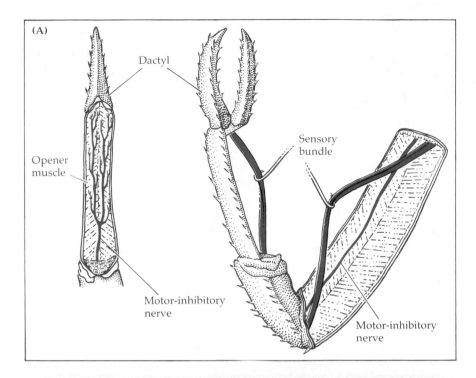

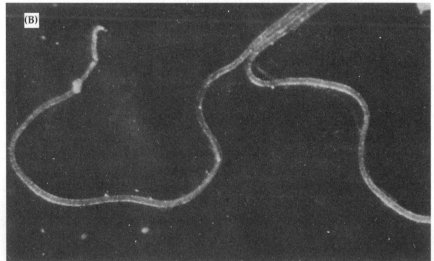

2   **WALKING LEG OF LOBSTER and its various nerve bundles. (A)** Some
muscles are cut away. The opener muscle of the dactyl is supplied by
one excitatory and one inhibitory fiber. Long segments of these can be
dissected out during their course before they reach their target and hence
are available for stimulation or chemical analysis. **(B)** The excitatory
axon is marked by a knot. The axons are about 50 μm in diameter. (A
from Hall, Bownds, and Kravitz, 1970; B from Kravitz, Kuffler, Potter,
and van Gelder, 1963.)

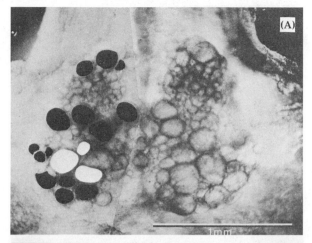

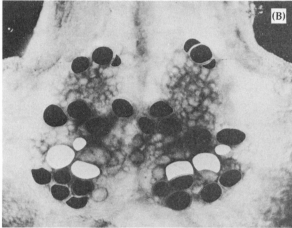

3

CHEMICAL AND PHYSIOLOGICAL ARCHI-
TECTURE OF NEURONS in a ganglion of the
lobster. (A) Inhibitory neurons are marked in
white; black markings indicate nerve cells that
excite muscles. (B) Chemical map of homologous
neurons in both halves of the same ganglion.
Inhibitory neurons (white) contain more than
2×10^{-11} moles of GABA. Excitatory nerves containing no detectable GABA are indicated in
black. (After Otsuka, Kravitz, and Potter, 1967.)

times more concentrated than in the cell body. One possible reason for
the relatively low GABA estimate in cell bodies may be trivial. The
calculations were made from the total volume of the cell. However,
organelles occupy a large fraction of the cell body but not of the axon;
the difference in concentration between the cytoplasm of axon and cell
bodies may therefore be smaller than the figures indicate.

In addition to demonstrating the presence of GABA in inhibitory
neurons, Kravitz and his colleagues made tests to see what other transmitters, if any, these cells could synthesize.[12] The principle was first to
bathe a lobster ganglion in a solution containing radioactive precursors
of various transmitters and then to dissect out individual cells and see
what they had synthesized. Since the precursors are common metabolites, they are taken up by all cells and metabolized. Only in cells using

[12]Hildebrand, J. G., Barker, D. L., Herbert, E. and Kravitz, E. A. 1971. *J. Neurobiol.*
2: 231–246.

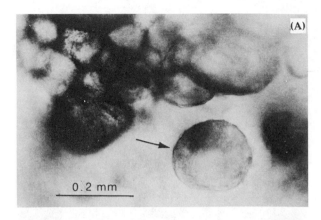

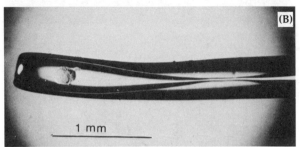

4

SINGLE NEURON ANALYSIS. (A) Single cell body (arrow) isolated from a lobster ganglion. (B) Cell is transferred for chemical analysis or histological examination. (C) Stained cell. (From Otsuka, Kravitz, and Potter, 1967.)

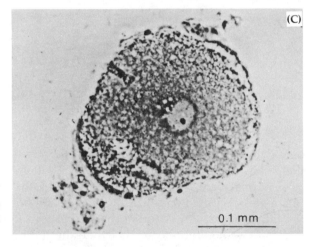

a particular transmitter is the precursor converted to the transmitter product. For example, cholinergic cells incorporate choline and make acetylcholine, whereas noradrenergic cells take up tyrosine for eventual production of norepinephrine, and other cells convert glutamate to GABA. In the ganglia of the lobster, as expected, inhibitory cells make radioactive GABA but not acetylcholine or catecholamines. Similarly, cells that liberate transmitters such as GABA, 5-HT, and norepinephrine have specific reuptake mechanisms for recapturing the molecule from extracellular fluid.[13]

The virtual restriction of GABA to inhibitory neurons greatly increased the likelihood of its being a transmitter. There still remained the problem of showing that GABA is in fact released from nerve terminals. An experiment similar to that originally performed by Loewi on the heart was therefore performed on lobster muscles.[14]

The inhibitory nerve to the opener muscle of the claw was stimulated while the muscle was perfused. The perfusate was collected and analyzed for GABA using an enzymatic assay. (Testing the activity on a stretch receptor or on a nerve–muscle preparation would have been less specific and too insensitive.) Only small amounts of GABA were released (about $1-4 \times 10^{-14}$ moles per nerve impulse) from all inhibitory terminals in the opener muscle of the claw. In solutions containing low calcium concentrations, nerve conduction was unimpaired, but transmission was blocked and extra GABA no longer appeared in the perfusate following stimulation. This result supports the assumption that in normal fluid nerve stimulation causes the release of GABA from the inhibitory terminals.

At present the actual amount of GABA released cannot be estimated because in lobster preparations much of the GABA that is liberated from the terminals is taken up into tissues before it can be collected. It is intriguing that most of the uptake is by Schwann cells and connective tissue around the inhibitory nerve fibers rather than by the nerve terminals themselves, as might be expected by analogy with the norepinephrine system or with GABA in the vertebrate central nervous system (Chapter 13).[15]

Several experiments show that GABA acts on the postsynaptic membrane in a manner similar to that of the neurally released transmitter.

Experiments by several groups have demonstrated that applied GABA, like nerve stimulation, produces its effects by increasing permeability to chloride, so that both procedures shift the postsynaptic membrane potential toward E_{Cl}, the equilibrium potential for chloride. At E_{Cl}, the synaptic current and potential reverse (Chapter 9). Potassium ions, on the other hand, do not contribute significantly to the inhibitory actions of applied or nerve-released GABA since changing the potassium concentration in the bathing fluid does not alter the reversal potential.[16–18] As expected, however, removal of chloride ions abolishes inhibitory synaptic potentials without disturbing excitatory transmission.

Similarity between the actions of GABA and the nerve-released transmitter

[13]Cooper, J. R., Bloom, F. E. and Roth, R. H. 1982. *The Biochemical Basis of Pharmacology,* 4th Ed. Oxford University Press, New York.

[14]Otsuka, M., Iversen, L. L., Hall, Z. W. and Kravitz, E. A. 1966. *Proc. Natl. Acad. Sci. USA 56*: 1110–1115.

[15]Orkand, P. M. and Kravitz, E. A. 1971. *J. Cell Biol. 49*: 75–89.

[16]Boistel, J. and Fatt, P. 1958. *J. Physiol. 144*: 176–191.

[17]Dudel, J. and Kuffler, S. W. 1961. *J. Physiol. 155*: 543–562.

[18]Takeuchi, A. and Takeuchi, N. 1967. *J. Physiol. 191*: 575–590.

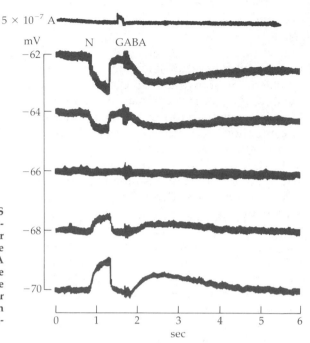

5×10^{-7} A

mV   N   GABA

−62

−64

−66

−68

−70

0   1   2   3   4   5   6

sec

5

SIMILARITY OF CONDUCTANCE CHANGES resulting from the action of nerve-released transmitter and from applied GABA in a muscle fiber of the crayfish. Brief trains of inhibitory nerve stimuli (N) and ionophoretically applied GABA appear on each record. As the resting membrane is depolarized from −70 mV the synaptic and the GABA potentials reverse at the same level, near −66 mV. Trace at top indicates current through a GABA-filled pipette (From Takeuchi and Takeuchi, 1965.)

Figure 5 shows an example in which the membrane of a muscle fiber in the crayfish claw was depolarized from its resting level of −70 mV by passing current through a microelectrode. At various potentials, the inhibitory axon was stimulated by a short train of pulses at 30/sec and its effect compared with that of GABA applied ionophoretically. There was a unique value of membrane potential—in this case, −66 mV—at which no potential change was caused by the neural transmitter or by GABA. This potential is called the REVERSAL POTENTIAL.

It has also been shown that when GABA is released from the micropipette, its action is confined to the region of the postsynaptic membrane where the inhibitory terminals release their transmitter and where the inhibitory potentials are set up.[18,19] Apparently, nerve-released transmitter and artificially applied GABA act on the same synaptic receptors and cause identical conductance changes. The similarity has been further demonstrated in tests using a pharmacological agent—picrotoxin—that blocks inhibitory synapses in crustacean muscle and also abolishes the response to GABA.[20] In conclusion, at crustacean neuromuscular synapses, GABA fulfills many of the same criteria as acetylcholine at the vertebrate skeletal neuromuscular synapse.

Inhibitory neuromuscular junctions in the earthworm and in several insects also seem to be operated by GABA, although the studies have not been so extensive as those on lobsters and crayfish. There are cells,

[19]Cull-Candy, S. G. and Miledi, R. 1981. *Proc. R. Soc. Lond. B 211:* 527–535.
[20]Takeuchi, A. and Takeuchi, N. 1969. *J. Physiol. 205:* 377–391.

however, in which GABA has an excitatory depolarizing effect. These include mammalian dorsal root ganglion cells[21] and cells from the superior cervical ganglion in culture.[22]

The inhibitory action of crustacean peripheral nerves is not confined exclusively to the postsynaptic muscle membrane. Inhibitory nerve impulses can also reduce the amount of transmitter released by excitatory nerve terminals. Identical presynaptic inhibitory effects are observed when GABA is applied to the bathing fluid in low concentrations that do not significantly affect the postsynaptic membrane (Chapter 9).[17,23]

GABA action on excitatory presynaptic terminals

In parallel series of experiments, the groups of Roberts[24] and of Kravitz[25] have shown that the pathways of GABA metabolism are similar in crustaceans and in the mammalian brain. In crustaceans, both excitatory and inhibitory axons possess large stores of glutamate (the precursor of GABA), but only the inhibitory axons are able to accumulate GABA. Figure 6 illustrates the components of the metabolic pathways in individual excitatory and inhibitory neurons. All the metabolites and enzymes have been isolated and purified. A conspicuous difference between excitatory and inhibitory neurons is the high glutamic decarboxylase activity in inhibitory axons. An interesting property of the decarboxylase in lobster is that the product of the reaction—GABA—inhibits the activity of the eynzyme. These features have enabled Kravitz and his colleagues to propose an explanation for the regulation of GABA accumulation: According to their scheme, the inhibitory axon synthesizes GABA until the concentration reaches about 100 mM, a stage at which GABA inhibition of the decarboxylase prevents further accumulation. If the GABA level falls, synthesis increases until the concentration once again reaches 100 mM. The turnover of transmitter, therefore, depends upon the amount of activity in the axon, a process common to other transmitters such as norepinephrine.

Metabolic basis of GABA accumulation

The scheme, based on the difference in the amount of decarboxylase, does not explain the presence of a small, but apparently real, amount of GABA in some excitatory axons or cell bodies in which little or no decarboxylase is detected. The explanation may lie in a failure to assay small amounts of the enzyme. Alternatively, the small amounts of GABA may be contained in cells that surround neurons in the lobster.[15] Schwann cells and connective tissue in peripheral nerves have been shown to possess an active GABA transport system, and some neurons may take up GABA in relatively small amounts. Similarly, glial cells in the mammalian brain also have an active GABA-uptake mechanism in areas where many GABA-containing neurons are found (Chapter 13).

GABA was originally discovered in the mammalian brain, where its distribution is inhomogeneous, a finding suggesting that some cells

GABA in vertebrate central nervous system

[21]De Groat, W. C. 1972. *Brain Res. 38*: 71–88.
[22]Obata, K. 1974. *Brain Res. 73*: 71–88.
[23]Takeuchi, A. and Takeuchi, N. 1966. *J. Physiol. 183*: 433–449.
[24]Susz, J. P., Haber, B. and Roberts, E. 1966. *Biochemistry 5*: 2870–2877.
[25]Hall, Z. W., Bownds, M. D. and Kravitz, E. A. 1970. *J. Cell Biol. 46*: 290–299.

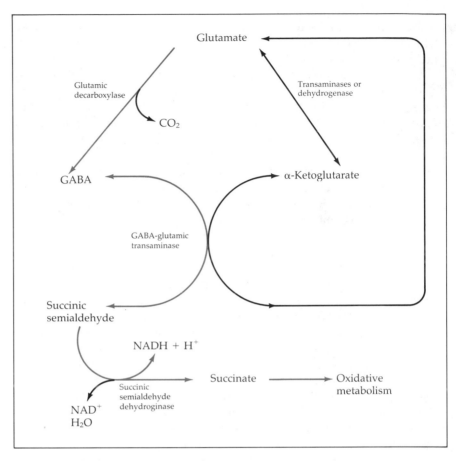

6 METABOLISM OF GABA. Main pathways are indicated in color. (After Kravitz, Kuffler, and Potter, 1963.)

contain relatively high concentrations. In addition, Curtis,[26] Krnjević,[27] and their colleagues have shown that many nerve cells in various regions of the brain are inhibited from discharging when GABA is applied diffusely or by ionophoresis. These observations, together with the finding that GABA can be collected from fluid on the surface of the cortex or from ventricles after electrical stimulation,[28] encouraged studies at a more detailed cellular level, mainly in cats.

Tests made by Otsuka, Ito, Obata, and their colleagues are of particular interest because they focused on specific, identified cells. These investigators examined the inhibitory action of the Purkinje cells of the cat cerebellum. These are large, conspicuous neurons with a wide arborization onto which converges most of the circuitry of the cerebellum

[26]Curtis, D. R. and Johnston, G. A. R. 1974. *Ergeb. Physiol. 69*: 97–188.
[27]Krnjević, K. 1974. *Physiol. Rev. 54*: 418–540.
[28]Jasper, H. and Koyama, I. 1969. *Can. J. Physiol. Pharmacol. 47*: 889–905.

(Figure 2, Chapter 1). The entire output of the cerebellum is by way of the Purkinje axons, which form inhibitory synapses in several distinct nuclei. One of these includes a collection of large cells called the lateral vestibular nucleus of Deiters. This direct pathway provides an exceptional opportunity to study an inhibitory neuron that can be identified, together with its target cells (to be discussed later).

To estimate the GABA content of Purkinje cells, single cells were dissected out and analyzed by an enzymatic assay that measured quantities down to 2×10^{-14} moles. The concentration found in Purkinje cells was 5 to 8 mM. In control experiments, cholinergic motoneurons were found to contain less GABA—about 1 mM.[29]

Two considerations arise in relation to these analyses on isolated cell bodies: (1) As in crustaceans, the cell body may have a lower concentration of transmitter than the axon or the terminals where it is released; (2) some GABA is present in cells that are not inhibitory (for example, in spinal motoneurons), and in these the GABA may be contributed by presynaptic nerve terminals. Pursuing such considerations, Otsuka and co-workers analyzed neurons in the ventral and dorsal portions of the lateral vestibular nucleus of Deiters. Most Purkinje cells make synapses within the dorsal portion of the nucleus. Neurons dissected from this region had associated with them a GABA concentration as high as that in Purkinje cells (about 6 mM). Neurons from the ventral nucleus, which receive no innervation from the Purkinje cells, had about one-half the GABA concentration of Purkinje cells. When Purkinje axons were cut and degenerated, the GABA levels in cells dissected from the dorsal nucleus were greatly reduced, whereas those in the ventral nucleus were not affected. This suggests that the GABA in the nucleus of Deiters is not derived from the cell bodies but from the synaptic endings upon them. The results reinforce the view that GABA (and other transmitters) may be highly concentrated in terminals.[30]

If the cerebellar output is stimulated electrically and recordings are made with an intracellular electrode from a Deiters neuron, inhibitory hyperpolarizing synaptic potentials appear. Ionophoretic application of GABA to the postsynaptic Deiters cells mimics these synaptic potentials, a finding similar to results with the crustacean neuromuscular junction.[31] Thus, hyperpolarizing potentials, set up neurally and electrophoretically, decrease and reverse at the same membrane potential when the postsynaptic membrane is hyperpolarized by current. Furthermore, with cerebellar stimulation, GABA appears in the fluid that diffuses out of the medulla into the nearby fourth ventricle (Chapter 14). As expected, localization of glutamic decarboxylase to the Purkinje cells by the antibody technique (see earlier) provides further confirmation for a role of GABA as the inhibitory transmitter at this synapse.

[29]Obata, K. 1969. *Experientia 25*: 1283.
[30]Otsuka, M., Obata, K., Miyata, Y. and Tanaka, Y. 1971. *J. Neurochem. 18*: 287–295.
[31]Obata, K., Takeda, K. and Shinozaki, H. 1970. *Exp. Brain Res. 11*: 327–342.

Several lines of evidence have suggested a role for GABA as a transmitter in other, widespread areas of the brain. Specific GABA receptors have been extracted from mammalian brain and purified. One promising approach for studying the properties of the receptors in detail has been to inject messenger RNA that codes for GABA and that has been extracted from chick brain into frog oocytes. This procedure causes the GABA receptor to appear in vitro on the cell membrane.[32] The synthetic enzymes for GABA have been found in the cortex and in deeper structures such as the substantia nigra (see later). The visual system provides one example in which both physiological and morphological evidence exists. Thus, in the retina, GABA is associated with certain types of horizontal and amacrine cells.[33,34] Agents that block the action of GABA have effects on signaling:[35] Bicuculline and picrotoxin change the receptive field organization of ganglion cells of the Y type (transient responses to moving spots; Chapter 2) by decreasing surround inhibition. In particular, Y cells driven by the rods used for dim vision were affected whereas those driven by cones were not.[36,37] In the geniculate nucleus, cells are found that contain glutamic decarboxylase.[3] In the visual cortex, GABA uptake and glutamic decarboxylase are associated with smooth stellate cells (see Chapter 3).[38] Local application of bicuculline can modify the responses of complex cells to oriented light stimuli.[39]

However, these experiments are neither quantitative nor comprehensive enough to be certain that GABA is the transmitter released by particular cells and acting at particular synapses. Many actions resembling those of GABA can be produced by the amino acid glycine, which is also a candidate as an inhibitory transmitter, particularly in the spinal cord. Like GABA, it is unevenly distributed, and if applied ionophoretically it mimics potentials set up by inhibitory stimulation. A useful distinction is in the agents that block inhibition pharmacologically, such as strychnine for glycine or picrotoxin and bicuculline for GABA.[13] At present, the main questions concern the specificity of each of these drugs and whether their actions provide diagnostic criteria for recognizing the transmitter by selectively and reliably blocking either GABA or glycine. At some synapses such as those on neurons in the spinal cord of the cat and those in the lamprey central nervous sytem, sufficiently detailed tests can be made to indicate that glycine rather than GABA is the more likely candidate.[40] A technique of growing importance used for these experiments is to measure the characteristics of

[32]Miledi, R., Parker, I. and Sumikawa, K. 1982. *Proc. R. Soc. Lond. B 216*: 509–515.
[33]Sterling, P. 1983. *Annu. Rev. Neurosci. 6*: 149–185.
[34]Lam, D. M.-K. and Ayoub, G. S. 1983. *Vision Res. 23*: 433–444.
[35]Caldwell, J. H. and Daw, N. W. 1978. *J. Physiol. 276*: 299–310.
[36]Kirby, A. W. and Enroth-Cugell, C. 1976. *J. Gen. Physiol. 68*: 465–484.
[37]Kirby, A. W. and Schweitzer-Tong, D. E. 1981. *J. Physiol. 320*: 303–308.
[38]Gilbert, C. D. 1983. *Annu. Rev. Neurosci. 6*: 217–247.
[39]Sillito, A. M. 1979. *J. Physiol. 289*: 33–53.
[40]Gold, M. R. and Martin, A. R. 1983. *J. Physiol. 342*: 85–98.

conductance channels activated by transmitter candidates, either by noise analysis or by patch clamp techniques (Chapter 11). For the substance to be a viable candidate, the channels must have characteristics identical to those of channels activated synaptically.

In conclusion, GABA plays an important role as a transmitter mediating inhibition at many crustacean and insect synapses. The evidence for its role as a transmitter in mammals is also good. In addition, the finding that GABA can depolarize and excite certain neurons, such as dorsal root ganglion cells, raises the possibility that it may also act as an excitatory transmitter.

PEPTIDES AS TRANSMITTERS

It was in the gut that the first hormone—secretin—was discovered by Bayliss and Starling in 1902.[41] Since then numerous additional intestinal hormones have been isolated and characterized, including gastrin, bradykinin, somatostatin, cholecystokinin (CCK), and others that were later shown to be peptides. It had also been known since the 1950s that certain neurons within the brain could secrete peptide hormones into the local circulation: For example, nerve cells in the hypothalamus were shown to secrete releasing factors that reached the gland cells of the anterior lobe of the pituitary and caused them in turn to secrete other hormones into the general circulation.[42] What was quite unexpected was the finding in the 1970s that the peptide hormones identified in the gut were also widely distributed in the brain. Advances in immunological, cytochemical, and physiological detection techniques made it possible to demonstrate the presence of cholecystokinin, gastrin, vasoactive intestinal polypeptide (VIP), and other gut hormones in widespread regions of the central nervous system. What is more, the peptides already known to occur in the hypothalamus were later located in structures such as the pancreas and the gut, where they exerted profound effects.

An early hint of this unity of peptides occurring in brain and gut is a transmitter known as substance P.[43] This was isolated in 1931 by Von Euler and Gaddum from gut and brain and was shown to cause contractions of smooth muscles. Since then the structure of the peptide has been worked out and its distribution in various neurons established. In particular, it is present in small-diameter sensory axons concerned with nociception or pain endings in the dorsal layers of the spinal cord, where it may act as a transmitter (see Box 1, Chapter 17; "P" does not refer to "pain" or "peptide" but was the name used by Von Euler and Gaddum to designate the crude "preparation" that contained the active extract). Like many other peptides, substance P has clear actions on the gut, on smooth muscles, and on neurons in the central nervous system, but its role is still not fully understood.

[41]Bayliss, W. M. and Starling, E. H. 1902. *J. Physiol.* 28: 325–353.
[42]Harris, G. W., Reed, M. and Fawcett, C. P. 1966. *Br. Med. Bull.* 22: 266–272.
[43]Nicoll, R. A., Schenker, C. and Leeman, S. E. 1980. *Annu. Rev. Neurosci.* 3: 227–268.

Interest in brain peptides increased further in the mid-1970s as a result of two sets of experiments made by Kosterlitz, Hughes, Goldstein, Snyder, and their colleagues:[44-46] First, receptors were found in the brain and in the gut that bind with high specificity to morphine and other derivatives of opium (opiates). Second, peptides that were identified within the brain had actions similar to those of opiates. It seemed clear that the characterization of the brain's own peptides could be important not only for determining their possible function as transmitters but also for understanding mechanisms involved in the control of pain and the horrors of drug addiction. Other key findings were that stimulation of specific regions of the brain known to contain neurons with opioid peptides (peptides with opiate activity) could produce analgesia;[47] these effects were reversed by naloxone, a drug that blocks opiate receptors. Interest was still further spurred by the finding that the terminals of opioid neurons were related to those of the selfsame substance P containing terminals supposed to mediate pain sensation in the spinal cord[48] (see Box 1, Chapter 17).

Extensive searches have been made for peptides with biological activity in the brain and the gut.[13] These additional peptides are (1) the enkephalins, which are pentapeptides; one enkephalin is known as met-enkephalin and the other as leu-enkephalin, depending on whether the molecule contains a methionine or a leucine group. They occur in the gut and the adrenal gland as well as in the brain. (2) A second peptide with opiate activity is β-endorphin; it is found in the pituitary gland, the brain, the pancreas, and the placenta. This 31-residue peptide is generated from a large molecule that also acts as a precursor for other hormones such as ACTH. β-Endorphin and the enkephalins produce widespread effects such as rigidity, analgesia, and profound changes of behavior. Naloxone, the morphine antagonist, antagonizes the actions of these opioid peptides.

Numerous other peptides have been found in brain and gut, including dynorphin, dermorphins, bradykinin, somatostatin, and bombesin (first isolated from the skin of a frog, *Bombina bombina*, mentioned here for its romantic name). In many instances, peptides can be shown to be released by stimulating appropriate regions of the brain or of slice preparations.[49] At the same time, the functional role played by each of these various peptides in the working of the brain and the gut remains an open question. It is often not clear whether they are acting as transmitters in the normal sense of the word, or as hormones.

[44]Hughes, J., Smith, T. W., Kosterlitz, H. W., Fothergill, L. A., Morgan, B. A. and Morris, H. R. 1975. *Nature (Lond.) 258*: 577–579.

[45]Teschemacher, H., Ophein, K. E., Cox, B. M. and Goldstein, A. 1975. *Life Sci. 16* 1771–1776.

[46]Pert, C. B. and Snyder, S. H. 1973. *Science 179*: 1011–1014.

[47]Fields, H. L. and Basbaum, A. I. 1978. *Annu. Rev. Physiol. 40*: 217–248.

[48]Jessel, T. M. and Iversen, L. L. 1977. *Nature (Lond.) 268*: 549–551.

[49]Iversen, L. L., Lee, C. M., Gilbert, R. F., Hunt, S. and Emson, P. C. 1980. *Proc. R. Soc. Lond. B 210*: 91–111.

SYNAPTIC TRANSMISSION MEDIATED BY A PEPTIDE IN FROG SYMPATHETIC GANGLIA

In a series of experiments, Kuffler and his colleagues have unequivocally identified one peptide that acts as a transmitter in sympathetic ganglia of the frog.[50,51] As for GABA, a peripheral synapse presented the opportunity for the stepwise analysis of the chemical involved in the transmission process. It was known from earlier studies by Libet, Nishi, Koketsu, Weight, and others that slow synaptic potentials with unusual properties occurred in autonomic ganglia of frogs and mammals.[50] Some of these, like the conventional fast excitatory potentials, were mediated by acetylcholine. Others were not. In particular, a late excitatory potential of long duration (several minutes) was not blocked by agents such as curare or atropine that eliminated the earlier synaptic potentials.

A convenient feature of the innervation of frog sympathetic ganglia is that separate nerves can be stimulated to evoke selectively the fast cholinergic synaptic potentials and the late slow excitatory potentials (Figure 7; see also Figure 12 in Chapter 9). No known transmitters other than ACh and norepinephrine had been observed in these ganglia. Trials of GABA, octopamine, ADP, and 5-hydroxytryptamine could not mimic the slow potentials produced by stimulating the appropriate nerves. Agents that blocked the action of 5-hydroxytryptamine or norepinephrine were without effect. Similarly, a variety of peptides including substance P, neurotensin, bombesin (again, from *Bombina bombina*), somatostatin, and others were ineffective. Of hundreds of components studied, only one decapeptide (LHRH) produced a small, slow depolarization in a concentration of 1 μM or more. (As mentioned earlier, LHRH is a hormone released into the local circulation by hypothalamic neurons. It causes cells in the anterior pituitary gland to secrete luteinizing hormone, an essential hormone involved in the ovarian cycle and in the secretion of testosterone.) This dose seemed too high for LHRH to be a serious transmitter candidate. With this as a clue, analogues of LHRH were tested and some were indeed found to be 100 times more potent.

Figure 8 shows an experiment made by Kuffler and Sejnowski[50] in which the synaptic potentials evoked by nerve stimulation and by an LHRH analogue are highly similar. Further evidence for transmitter action was provided by the following experiments: (1) Specific blocking agents simultaneously reduced the effects of nerve stimulation and of the LHRH analogue. (2) Removal of calcium or increases of magnesium in the bathing fluid blocked the effects of nerve stimulation but not those produced by application of the peptide directly. This result suggested that the peptide was released from presynaptic terminals by a calcium-dependent mechanism and that the site of action of the peptide

Evidence for LHRH-like peptide as a transmitter

[50]Kuffler, S. W. 1980. *J. Exp. Biol. 89*: 257–286.

[51]Jan, Y. N., Jan, L. Y. and Kuffler, S. W. 1980. *Proc. Natl. Acad. Sci. USA 77*: 5008–5012.

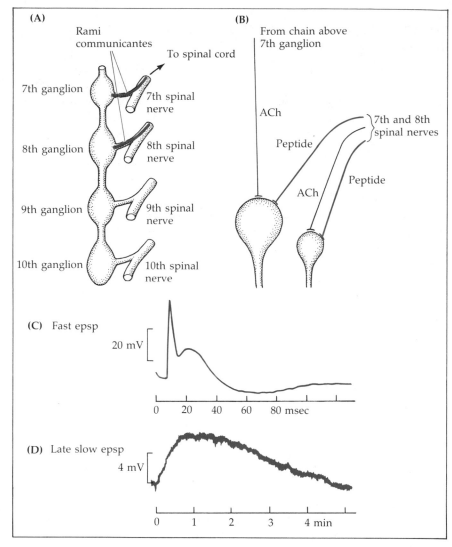

7 **INNERVATION OF SYMPATHETIC GANGLION CELLS in bullfrog.**
(A) Diagram of sympathetic ganglia 7, 8, 9, and 10. Nerve cells receive
inputs from the spinal cord via *Rami communicantes* and also from the
chain above. (B) The larger B cells in caudal ganglia (9 and 10) receive
cholinergic input from roots rostral to ganglion 7 and receive peptide
input from roots 7 and 8. Accordingly, the different inputs can be
selectively stimulated. (C, D) Fast and slow excitatory potentials evoked
by stimulating above the 7th ganglion (single shock) or spinal nerves 7
and 8 (10/sec for 5 sec). The large, fast cholinergic epsp lasts a few
milliseconds and gives rise to an impulse. The slow depolarization lasts
for several minutes. (After Kuffler, 1980.)

was postsynaptic. (3) The presence of an LHRH-like substance in the
appropriate presynaptic nerves and in the ganglion was shown by a
highly specific and sensitive immunological technique—radioimmune
assay, or RIA. By this method one can measure minute amounts of a

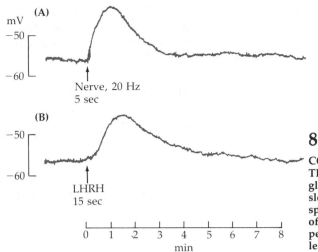

8

COMPARISON OF EFFECTS OF LHRH AND TRANSMITTER in bullfrog sympathetic ganglion cells. (A) Intracellular recording of late slow synaptic potential evoked by stimulating spinal nerves 7 and 8. (B) Same cell, application of LHRH by pressure from a micropipette. The peptide mimics the action of the naturally released transmitter. (After Kuffler, 1980.)

substance. [The principle is to bind a known quantity of the radioactive antigen (say, LHRH) to a fixed amount of the antibody. This labeled antigen–antibody combination is added to an unknown amount of the same nonradioactive antigen. The unlabeled antigen competes with the labeled antigen for the binding sites on the antibody. As a result, the amount of radioactive label bound to the antibody decreases quantitatively and provides a measure of the amount of antigen in the extract.] As expected, after cutting the presynaptic nerves the LHRH activity disappeared from the ganglia. (4) With fluorescence-labeled antibody, the LHRH terminals could be seen by light microscopy. Additional tests strengthened the idea of the LHRH analogue as the transmitter. (5) The peptide was released by nerve stimulation or by raising the potassium concentration to depolarize the neurons. The release was calcium dependent. (6) Extracts of the ganglion and nerves could mimic the effects of nerve stimulation and showed a variety of properties similar to those of LHRH, including their molecular weights and their sensitivity to degradation by proteolytic enzymes. (7) Kuffler and Sejnowski[52] showed further that the amplitude and duration of conductance changes produced by the neurally released transmitter and by the peptide on the postsynaptic cell were similar. The conductance changes themselves have been studied in detail by voltage clamp and display interesting characteristics including the closing as well as the opening of channels exhibiting voltage sensitivity.[52,53] A further feature of the peptide is that it can diffuse over considerable distances to produce its effect on the postsynaptic targets. Thus, peptide liberated by a neuron may act on more than one postsynaptic cell, a feature discussed later.

[52]Kuffler, S. W. and Sejnowski, T. J. 1983. *J. Physiol. 341*: 257–278.
[53]Adams, P. R., Brown, D. A. and Constanti, A. 1982a. *J. Physiol. 332*: 223–262.

These findings, while clearly establishing the transmitter role of the peptide, do not reveal what function the slow synaptic potentials perform in autonomic ganglia. One may speculate that prolonged depolarization with a concomitant increase in excitability could facilitate transmission via the normal acetylcholine pathway. Thus, the peptide would "modulate" the excitability. It is tempting also to wonder whether such prolonged changes might not under certain circumstances be responsible for disorders of functions that are controlled by the autonomic nervous system including blood pressure, gastrointestinal motility, and glandular secretion.

NOREPINEPHRINE, DOPAMINE, AND 5-HYDROXYTRYPTAMINE AS TRANSMITTERS IN THE NERVOUS SYSTEM

Norepinephrine is a transmitter that has been unequivocally identified in the autonomic nervous system by the criteria already mentioned. Of all transmitters, its chemistry has been most thoroughly studied. It is released by nerves in internal organs, including gut, spleen, and heart. The processes involved in the synthesis, storage, and release of this substance have been worked out in great detail by von Euler, Axelrod, Udenfriend, and their colleagues.[54] Much detailed information is also available about the mechanism of action of norepinephrine at synapses. At some junctions—for example, those on the smooth muscle of the gut—norepinephrine produces miniature synaptic potentials and excitatory synaptic potentials that are large enough to activate contractions. On other targets, slower potentials, usually inhibitory, are produced indirectly by norepinephrine with changes in the intracellular concentration of cyclic AMP. Similarly, for 5-hydroxytryptamine and dopamine, the metabolic pathways are well understood and their roles as transmitters have been established at certain synapses in the periphery in vertebrates and in the central nervous system of invertebrates. Like norepinephrine, these transmitters can give rise to slow potentials, which often arise after a prolonged delay and are accompanied by changes in intracellular cyclic AMP.[55] As mentioned earlier (page 296), dense-core vesicles and characteristic fluorescence with the Falck-Hillarp technique have served as diagnostic criteria for determining pathways and synapses at which the amines are involved.

Within the mammalian central nervous system, there is evidence that norepinephrine, epinephrine, dopamine, and 5-hydroxytryptamine act as transmitters. They are found in pathways essential for sensory and motor performance as well as for higher functions. However, out of the total cells in the human brain, relatively few appear to contain these transmitters—*thousands* rather than millions or billions. What is more, most of the cells containing one of these transmitters are clustered together in a discrete region of the brain. Thus, in the rat central nervous

[54]Axelrod, J. 1971. *Science 173*: 598–606.
[55]Kehoe, J. and Marty, A. 1980. *Annu. Rev. Biophys. Bioeng. 9*: 437–465.

system the majority of cells containing norepinephrine—only about one thousand five hundred or so in number—are located in a small nucleus known as the *locus coeruleus*.[56,57] Those for dopamine lie for the most part in the *substantia nigra*, and those for 5-hydroxytryptamine in groups of cells called *raphe nuclei*. From these regions, which are shown in Figure 9 and in Appendix B, fibers spread out to supply virtually all areas of the brain. For example, norepinephrine fibers supply the cerebellar Purkinje cells, the cerebral cortex, and the thalamus. A single cell can branch to supply widely different areas. Dopamine-containing fibers supply structures such as the basal ganglia in addition to the cortex. 5-Hydroxytryptamine-containing fibers are almost ubiquitous. Often no synaptic specializations with defined postsynaptic targets are seen, as though the transmitter were liberated at sites from which it diffused to populations of cells in the vicinity bearing the appropriate receptors.

Numerous comprehensive monographs and reviews give full accounts of the actions of the amines, which are beyond the scope of this book.[13,57] Here we may cite a few, well-known examples. Destruction of the norepinephrine system of the locus coeruleus in young kittens leads to marked changes in the development of the visual cortex, particularly in its susceptibility to damage by sensory deprivation (see Chapter 20). Local replacement of norepinephrine in the visual cortex can reverse the effects of the lesion.[58] For the dopamine system, damage to the substantia nigra or a reduction in its dopamine content leads to disorders of movement.[59] These disorders resemble the condition known as Parkinson's disease, which is characterized by rigidity, weakness, and a tremor at rest that progressively deteriorate. Diffuse replacement of transmitter activity can alleviate the symptoms of parkinsonism to some extent. In patients, the precursor of the transmitter (L-dopa) rather than dopamine is given by mouth. Dopamine will not work because it cannot cross the blood-brain barrier to enter the brain (Chapter 14). Presumably the dopamine neurons that do remain are able to synthesize and release sufficient transmitter if provided with extra precursor. Dopamine neurons are also implicated in schizophrenia: Phenothiazine drugs that antagonize dopamine actions are used to treat the disease. At the same time these drugs have as a side effect the production of symptoms resembling those of Parkinson's disease, which can resolve when the drugs are withdrawn. A third example of the importance of amines is provided by 5-hydroxytryptamine, which is liberated from fibers of the raphe nuclei and has profound effects on the level of wakefulness and on pain sensation.

How can so few neurons give rise to all-encompassing changes in the central nervous system? At present, one attractive hypothesis is

[56]Foote, S. L., Bloom, F. E. and Aston-Jones, G. 1983. *Physiol. Rev. 63*: 844–914.
[57]Moore, R. Y. and Bloom, F. E. 1979. *Annu. Rev. Neurosci. 2*: 113–168.
[58]Kasamatsu, T., Pettigrew, J. D. and Ary, M. 1981. *J. Neurophysiol. 45*: 254–266.
[59]Baldessarini, R. J. and Tarsy, D. 1980. *Annu. Rev. Neurosci. 3*: 23–41.

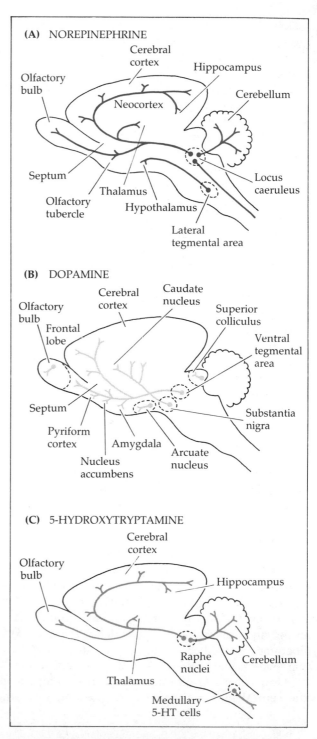

9 NEURONS CONTAINING NOREPINEPHRINE, DOPAMINE, AND 5-HT. Simplified scheme of principal locations and projections. (After Shepherd, 1983.)

that the axons containing amines do not always have a defined structural relationship to the postsynaptic targets. Rather, it may be diffuse release of transmitter that modulates the functioning of pathways to increase or decrease their effectiveness. Examples of such diffuse modulatory effects are particularly obvious in the central nervous systems of invertebrates. In some pathways, well-defined synaptic structures and one-to-one transmission are found. As general actions, however, 5-hydroxytryptamine can increase the swimming activity of leeches and modify synaptic effectiveness in sensory motor pathways of the sea slug *Aplysia* (see Chapter 18). A diffuse "garden sprinkler" distribution could explain how substitution of transmitter can restore function: In this way 5-hydroxytryptamine, norepinephrine, or dopamine (and perhaps peptide transmitters) need only reach an appropriate average concentration in the vicinity of neurons with the appropriate receptors. In contrast, one could not imagine diffuse application of acetylcholine restoring appropriate contractions of skeletal muscles in coordinated movements. In Chapter 19 we discuss further how implantation of embryonic neurons containing dopamine or 5-hydroxytryptamine can be used to correct deficits of these transmitters in adult mammalian brain.

Nerve terminals continually secrete transmitter, and the questions arise whether it is shipped to them ready-made, whether they assemble it from parts provided by the cell body, or whether they themselves synthesize some or many of the essential components. It is known that transmitter can be assembled in terminals, provided the precursors are available. The terminals cannot, however, manufacture the enzymes required either for transmitter synthesis or for the making of new parts of the cell membrane; this seems to be so, because no ribosomes have been found in either axons or terminals, and it is unlikely that mitochondria could provide the needed proteins. These are apparently transported from the cell body to the rest of the neuron. In many instances, the distances are so great that if simple diffusion accounted for transport, it might take years for a substance to move, for example, from the spinal cord to the toe of a large animal. A relatively rapid transport mechanism must therefore be postulated.

Axoplasmic flow and transport of transmitters

There is now abundant evidence for a continuous movement of substances along axons in both directions.[60,61] For forward transport of materials there are slow and fast components, from 1 to about 400 mm/day. Examples of this are mentioned earlier in connection with the visual system, where movement of labeled amino acids along neurons from the eye to the lateral geniculate nucleus and the cortex is demonstrated. This occurs far more rapidly than simple diffusion. Similarly, the reverse movement of a protein—horseradish peroxidase—from nerve terminal to cell body is now used to mark specific pathways.[62]

[60]Schwartz, J. H. 1979. *Annu. Rev. Neurosci.* 2: 467–504.
[61]Grafstein, B. and Forman, D. S. 1980. *Physiol. Rev. 60*: 1167–1283.
[62]La Vail, J. H. and La Vail, M. M. 1974. *J. Comp. Neurol. 157*: 303–358.

A variety of approaches have been used to demonstrate movement of transmitters along the axon. For example, Schwartz and his colleagues[63,64] injected radioactive choline or 5-HT into large neurons in *Aplysia*. The movement of acetylcholine or 5-HT could be followed down the nerve to the terminals. Similarly, Dahlström,[65] Geffen,[66] and their colleagues have observed the movement of transmitters and enzymes along sympathetic neurons.

What at present is not clear is the mechanism for movements faster than diffusion. Simple bulk flow is difficult to understand in view of the two-way traffic in axons. The idea that microtubules may be involved is attractive but incompletely defined. Further, what happens to products once they reach the nerve endings or the cell bodies? Materials do not continue to accumulate. Presumably a mechanism exists that regulates the turnover of the amount of various substances and that is sensitive to the metabolic needs of neurons in various physiological states.[67]

Long-term regulation of transmitters

An important variable in considering the neurochemistry of transmitters is that the amount of transmitter synthesized and secreted by a neuron is not constant. In Chapter 19 it will be shown that specific proteins and the amount of physiological activity, as well as pharmacological agents, can significantly influence the level of synthetic enzymes in presynaptic neurons. This, in turn, can influence the amounts of transmitter released at the nerve terminals. Even the identity of the transmitter manufactured by an individual neuron can be switched. Thus, neurons isolated from the sympathetic ganglia of neonatal rats and destined to synthesize norepinephrine can be induced in culture to synthesize acetylcholine instead.[68] Contrary to previous ideas, it has also been shown that a neuron can synthesize and contain in its terminals more than one transmitter; for example, VIP (vasoactive intestinal peptide) and acetylcholine can coexist in individual nerve cells.[69] It is interesting that VIP produces an enhancement of ACh release at neuromuscular junctions of the frog.[70] Figure 10 shows the localization by antibody techniques and fluorescence of substance P and 5-HT within individual neurons in the central nervous system of the guinea pig. It is not yet known whether under normal conditions a modulator as well as a conventional transmitter can be released by a single terminal so as to influence synaptic efficacy.

At present there is little information about how the genes for a transmitter or a transmitter-synthesizing enzyme can be turned on or off so that appropriate regulation can be made during development,

[63]Koike, H., Kandel, E. R. and Schwartz, J. H. 1974. *J. Neurophysiol.* 37: 815–827.
[64]Shkolnik, L. J. and Schwartz, J. H. 1980. *J. Neurophysiol.* 43: 945–967.
[65]Dahlström, A. 1971. *Phil. Trans. R. Soc. Lond. (Biol. Sci.)* 261: 325–358.
[66]Geffen, L. B. and Livett, B. G. 1971. *Physiol. Rev.* 51: 98–157.
[67]Collier, B. and MacIntosh, F. C. 1969. *Can. J. Physiol. Pharmacol.* 47: 127–135.
[68]Patterson, P. H. and Chun, L. L. Y. 1977. *Devel. Biol.* 56: 263–280.
[69]Fahrenkrug, J. and Emson, P. C. 1982. *Br. Med. Bull.* 38: 265–270.
[70]Gold, M. R. 1982. *J. Physiol.* 327: 325–335.

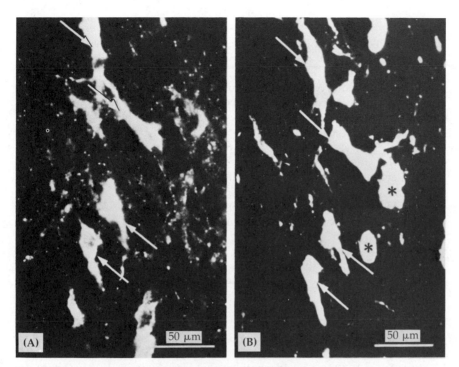

10 5-HYDROXYTRYPTAMINE AND SUBSTANCE P in individual nerve cells of rat medulla oblongata. (A) Immunofluorescence micrograph of section with antiserum to substance P. (B) Adjacent section with antiserum to 5-hydroxytryptamine. Comparison of A with B reveals many cells that contain both transmitters (marked by arrows) and others that do not (asterisks). Bars indicate 50 μm. (From Schultzberg, Hökfelt, and Lundberg, 1982.)

following damage or under the influence of physiological activity or hormones. With modern techniques involving recombinant DNA, it has become possible to isolate the genes responsible for producing various peptides with known amino acid sequences. Accordingly, the search for the specific genes controlling the synthesis of peptide transmitters is a burgeoning research area. Once the gene has been isolated and cloned, one can hope to study regulatory mechanisms at the molecular level.

SUGGESTED READING

General reviews

Cooper, J. R., Bloom, F. E. and Roth, R. E. 1982. *The Biochemical Basis of Neuropharmacology*, 4th Ed. Oxford University Press, New York.

Schwartz, J. H. 1982. (10) Chemical basis of synaptic transmission. (11) Biochemical control mechanisms in synaptic transmission. In E. R. Kandel and J. H. Schwartz (eds.). *Principles of Neuroscience*. Elsevier, New York, Chapters 10 and 11.

GABA

Hall, Z. W., Bownds, M. D. and Kravitz, E. A. 1970. The metabolism of gamma aminobutyric acid in the lobster nervous system. *J. Cell Biol. 46*: 290–299.

Kravitz, E. A., Kuffler, S. W. and Potter, D. D. 1963. Gamma-aminobutyric acid and other blocking compounds in crustacea. III. Their relative concentrations in separated motor and inhibitory axons. *J. Neurophysiol. 26*: 739–751.

Obata, K. 1980. Biochemistry and physiology of amino acid transmitters. In *Handbook of Physiology. The Nervous System*, Vol. 1, Part 1. American Physiological Society, Bethesda, pp. 625–650.

Otsuka, M., Kravitz, E. A. and Potter, D. D. 1967. Physiological and chemical architecture of a lobster ganglion with particular reference to γ-aminobutyrate and glutamate. *J. Neurophysiol. 30*: 725–752.

Otsuka, M., Obata, K., Miyata, Y. and Tanaka, Y. 1971. Measurement of γ-aminobutyric acid in isolated nerve cells of cat central nervous system. *J. Neurochem. 18*: 287–295.

Peptides and biogenic amines

Dahlström, A. and Fuxe, K. 1964. Evidence for the existence of monoamine containing neurons in the central nervous system. I. Demonstration of monoamines in the cell bodies of brain stem neurons. *Acta Physiol. Scand. 62* (Suppl. 232): 1–55.

Gregory, R. A. (ed.) 1982. Regulatory peptides of gut and brain. *Br. Med. Bull. 38*: 219–318.

Hughes, J. (ed.). 1983. Opioid peptides. *Br. Med. Bull. 39*: 1–106. (These are excellent review articles by various authors and cover the history, distribution, and mode of action of peptides.)

Hughes, J., Beaumont, A., Fuentes, J. A., Malfroy, B. and Unsworth, C. 1981. Opioid peptides: Aspects of their origin, release and metabolism. *J. Exp. Biol. 89*: 239–255.

Kuffler, S. W. 1980. Slow synaptic responses in autonomic ganglia and the pursuit of a peptidergic transmitter. *J. Exp. Biol. 89*: 257–286.

Kuffler, S. W. and Sejnowski, T. J. 1983. Peptidergic and muscarinic excitation at amphibian sympathetic synapses. *J. Physiol. 341*: 257–278.

Moore, R. Y. and Bloom, F. E. 1978. Central catecholamine neuron systems: Anatomy and physiology of the dopamine systems. *Annu. Rev. Neurosci. 1*: 129–169.

Axoplasmic flow

Grafstein, B. and Forman, D. S. 1980. Intracellular transport in neurons. *Physiol. Rev. 60*: 1167–1283.

PART THREE

THE SPECIAL ENVIRONMENT OF NERVE CELLS IN THE BRAIN FOR SIGNALING

CHAPTER THIRTEEN

PHYSIOLOGY OF NEUROGLIAL CELLS

Most nerve cells in the central and the peripheral nervous systems are surrounded by satellite cells. These satellites are divided according to anatomical criteria into (1) neuroglial cells in the brain, which are further subdivided into oligodendrocytes and astrocytes, and (2) Schwann cells in the periphery. Taken together, the neuroglial cells make up almost one-half the volume of the brain, and they greatly outnumber neurons. To some glial cells can be assigned a definite functional role; for example, the oligodendrocytes and Schwann cells form myelin around the larger axons and speed up conduction of nerve impulses. Although much experimental evidence is now available, definite roles for the other glial cells remain to be clarified.

The basic membrane properties of glial cells differ in several essential aspects from those of neurons. Glial cells behave passively in response to electric current and, unlike neurons, their membranes do not generate conducted impulses. The glial membrane potential is higher than that of neurons and depends primarily on the distribution of potassium, the principal intracellular cation. Further, glial cells are linked by low-resistance connections that permit the direct passage of ions and small molecules, such as fluorescent dyes, between the cells. Neurons and glial cells, on the other hand, are separated from each other by narrow, fluid-filled, extracellular spaces that are about 20 nm wide and prevent currents generated by nerve impulses from spreading into neighboring glial cells.

Neurons transmit signals to glial cells by releasing potassium into the intercellular spaces during the conduction of impulses, thereby depolarizing the glial membrane. There is no difference in the potassium-mediated signaling that occurs between excitatory or inhibitory neurons and glial cells. Such signaling is nonsynaptic.

Potassium-mediated glial depolarization creates potential changes that can be recorded from the surface of tissue; as a result glial cells contribute to the electroencephalogram and the electroretinogram.

Some of the intriguing questions that await further elucidation include how glial cells respond metabolically to potassium-mediated depolarization, whether they are important for supplying materials to neurons, and what role they play during development.

Nerve cells in the brain are intimately surrounded by satellite cells called NEUROGLIAL CELLS. From counts of cell nuclei, it has been estimated that they outnumber neurons by at least 10:1 and make up about one-half of the bulk of the nervous system. Studies of glial cells are in a peculiar state. The importance of these cells is stressed frequently; yet, except for their role in speeding conduction, no clear-cut essential function has been firmly established for them. It is remarkable that one should have to discuss the performance of the brain in terms of neurons only, as if glial cells did not exist. Even if in the end this neglect should turn out to be correct, which we doubt, one ought to be able to justify the reasons for ignoring the most numerous cell types in the nervous system.

Glial cells were first described in 1846 by Rudolf Virchow, who later gave them their name. He clearly recognized that they differed fundamentally from neurons and from interstitial tissue elsewhere in the body. Several excerpts from a paper by Virchow give the flavor of his approach and thinking.[1] He pointed out many aspects of glial tissue that later became important for formulating various hypotheses.

> Hitherto, considering the nervous system, I have only spoken of the really nervous parts of it. But . . . it is important to have a knowledge of that substance also which lies *between the proper nervous parts,* holds them together and gives the whole its form. (our italics)

Speaking of the ependyma (see later), he continued,

> This peculiarity of the membrane, namely, that it becomes continuous with interstitial matter, the real cement, which binds the nervous elements together, and that in all its properties it constitutes a tissue different from the other forms of connective tissue, has induced me to give it a new name, that of *neuro-glia.* (nerve glue; our italics)

Later on he stated,

> Now it is certainly of considerable importance to know that in all nervous parts, in addition to the real nervous elements, a second tissue exists, which is allied to the large group of formations, which pervade the whole body, and with which we have become acquainted under the name of connective tissues. In considering the pathological or physiological conditions of the brain or spinal marrow, the first point is always to determine how far the tissue which is affected, attacked or irritated, is nervous in its nature, or merely an interstitial substance. . . . Experience shows us that this very interstitial tissue of the brain and spinal marrow is one of the most frequent seats of morbid change, as for example, of fatty degeneration. . . . Within the neuroglia run the vessels, which are therefore nearly everywhere *separated from the nervous substance* by a slender intervening layer, and are not in immediate contact with it. (our italics)

In the subsequent 100 years, neuroglial cells were intensively studied, chiefly by neuroanatomists and also by pathologists, who knew

[1]Virchow, R. 1859. *Cellularpathologie.* (F. Chance, trans.). Hirschwald, Berlin. Excerpts are from pp. 310, 315, and 317.

them to be the most common source of tumors in the brain. This is perhaps not surprising, because normally—unlike neurons—glial cells can still divide in the mature animal.

This chapter considers glial cells from a cellular standpoint. The morphology and the various functions attributed to glial cells are reviewed, but the main emphasis is on their physiological properties, about which a good deal is known. Without background knowledge about their membrane potentials and how they are influenced by neuronal signaling, it appears difficult to deal with the wider issues that inevitably arise.

The experiments on glial cells illustrate once again the basic unity of the principles by which the nervous system works in higher and lower animals. Thus, for technical reasons the membrane properties of glial cells were most easily determined in especially large cells of the leech brain. Once this had provided a clue to what to look for, it became simpler to investigate amphibian and then mammalian glial cells, which share many key properties with those in the more modest nervous system of the leech. Far from being a roundabout approach, this turned out to be a shortcut.

One of the most distinct structural features of neuroglial cells as compared with neurons is the absence of axons, but many other differences have been demonstrated by light and electron microscopy. A representative picture of mammalian neuroglial cells is shown in Figure 1. The cytoplasmic contents suggest that glial cells are metabolically active structures containing the usual organelles, including mitochondria, endoplasmic reticulum, ribosomes, lysosomes, and often deposits of glycogen and fat.

Appearance and classification of glial cells

Glial cells are classified on histological grounds, and in the vertebrate central nervous system they are usually subdivided into two main groups (astrocytes and oligodendrocytes) and several subgroups.

ASTROCYTES can be classified into two subgroups: (1) fibrous astrocytes, which contain filaments and are more prevalent among bundles of myelinated nerve fibers, the white matter of the brain; and (2) protoplasmic astrocytes, which contain less fibrous material and are more abundant in the gray matter around nerve cell bodies, dendrites, and synapses. Both types of astrocytes make contacts with capillaries and neurons (see later).

OLIGODENDROCYTES are predominant in the white matter, where they form myelin around the larger axons. This is a wrapping of glial cell processes with practically all the cytoplasm squeezed out in between so that the membranes are tightly apposed as they spiral around the axon (Figure 2A). The large numbers of smaller diameter axons (1 μm or less) that are unmyelinated are also surrounded by glial cells, singly or in bundles.[2,3]

EPENDYMAL CELLS that line the inner surfaces of the brain, in the

[2]Bunge, R. P. 1968. *Physiol. Rev. 48*: 197–251.
[3]Morell, P. and Norton, W. T. 1980. *Sci. Am. 242*: 88–118.

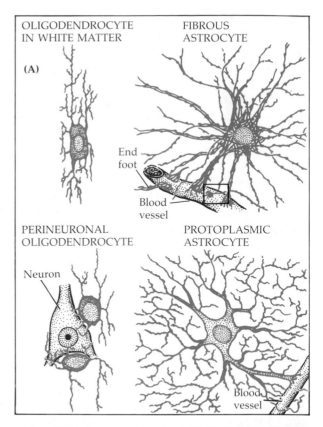

1

NEUROGLIAL CELLS in the mammalian brain.
(A) Neuroglial cells stained with silver impregnation. Oligodendrocytes and astrocytes represent the principal neuroglial cell groups in the vertebrate brain. They are closely associated with neurons and form end feet on blood vessels. (B) Electron micrograph of glial cells in the optic nerve of the rat. In the lower portion is the lumen of a capillary (CAP) lined with endothelial cells (E). The capillary is surrounded by end feet formed by processes of fibrous astrocytes (AS). Between the end feet and the endothelial cells is a space filled with collagen fibers (COL). In the upper portion is part of a nucleus of an oligodendrocyte (OL) and to the right are axons surrounded by myelin wrapping (see Figure 2). (A after Penfield, 1932; B from Peters, Palay, and Webster, 1976.)

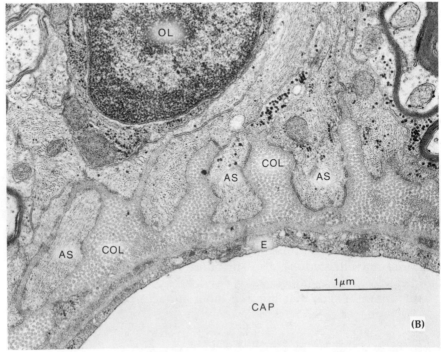

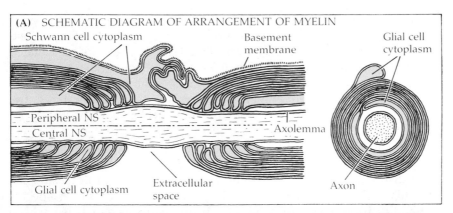

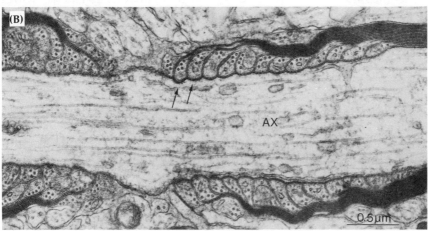

2 **MYELIN AND NODES OF RANVIER. Oligodendroglial cells and Schwann cells form wrappings of myelin around segments of axons. (A) At periodic intervals, at the nodes of Ranvier, the myelin covering is interrupted and the axon is exposed. The upper half of the nodal region, with a loose covering of processes, is typical of the arrangement in peripheral nerves. To the right is a transverse section through a myelin-covered axon portion. (B) An electron micrograph of a nodal region in the optic nerve of the rat. Compare with the lower portion of the drawing in (A). At the edge of the node, where the Schwann cell lamellae terminate, there occurs a specialized close contact area between the membrane of the axon (AX) and the membrane of the myelin wrapping (arrows). (A after Bunge, 1968; B from Peters, Palay, and Webster, 1976.)**

ventricles, are also usually classed as glial cells. No physiological role has been assigned to them.

In invertebrates, the classification of glial cells into distinct groups is not well established. However, there is no question about the functional analogy of the various glial structures.[4]

[4]Lane, N. J. 1981. *J. Exp. Biol. 95*: 7–33.

In vertebrate peripheral nerves, the SCHWANN CELLS are analogous to oligodendrocytes in that they form MYELIN around the larger, fast-conducting axons (up to 20 μm in diameter). Smaller, nonmedullated axons (usually below 1 μm in diameter), as in the brain, have a Schwann cell envelope without myelin. The glial cells of the brain and of peripheral nerves have different embryological origins: Glial cells in the central nervous system are derived from precursor cells that line the inner surface of the brain; Schwann cells arise from the neural crest.

The various types of satellite cells can now be further characterized by immunological techniques (see Box 1, Chapter 19). For example, a clear distinction has been made between fibrous and protoplasmic astrocytes.[5] Fibrous astrocytes contain a protein against which specific antibodies have been prepared. When this antibody is labeled with a fluorescent marker, the fibrous astrocytes, to which it binds selectively, can be clearly distinguished in micrographs. The protein is present in fibrous astrocytes in all vertebrates that have been examined and also in glial cells of the enteric neural plexus.[5,6] An example of stained fibrous astrocytes in rat brain is shown in Figure 3. The role of this fibrillary protein in the function of the astrocytes is not yet known.

Other antibodies have been found that bind specifically to oligodendrocytes and to Schwann cells.[6–8] Such labels can make it possible to study, not only different molecular components of glial cells in adult animals, but also the precursor cells in the embryonic brain that will

[5]Bignami, A. and Dahl. D. 1974. *J. Comp. Neurol. 153:* 27–38.

[6]Mirsky, R. 1982. In J. Brockes (ed.). *Neuroimmunology.* Plenum Press, New York, pp. 141–181.

[7]Schachner, M., Hedley-Whyte, E. T., Hsu, D. W., Schoonmaker, G. and Bignami, A. 1977. *J. Cell Biol. 75:* 67–73.

[8]Schachner, M. 1982. *J. Neurochem. 39:* 1–8.

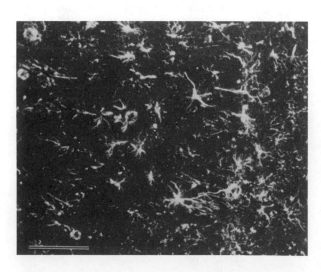

3

FIBROUS ASTROCYTES specifically labeled in the brain of a rat. These cells contain a protein against which an antibody is formed. The antibody is then made to fluoresce. Scale, 0.1 mm. (After Bignami and Dahl, 1974.)

eventually give rise to astrocytes, oligodendrocytes, and Schwann cells.[9]

A glance at almost any electron micrograph of brain tissue brings home the difficulty of making physiological and chemical studies of neuroglial cells. Figure 4 shows an example from the cerebellum of a

Structural relation between neurons and glia

[9]Schachner, M., Sommer, I., Lagenaur, C. and Schnitzer, J. 1982. In T. A. Sears (ed.). *Neuronal-Glial Cell Interrelationships*. Springer-Verlag, New York, pp. 321–336.

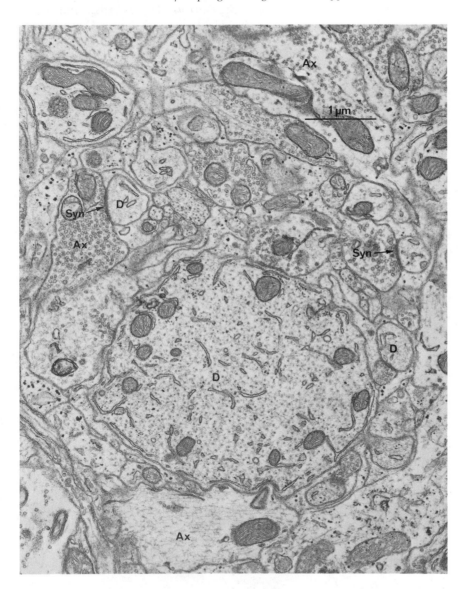

4 NEURONS AND GLIAL PROCESSES in the cerebellum of the rat. The glial contribution is lightly colored. The neurons and glial cells are always separated by clefts of about 20 nm in width. The neural elements are dendrites (D) and axons (Ax). Two synapses (Syn) are marked by arrows. (After Peters, Palay, and Webster, 1976.)

rat. The section is filled with neurons and glial cells, which can be distinguished from each other only after considerable experience. To make the task simpler, the glial contribution is highlighted. The extracellular space is restricted to narrow clefts, about 20 nm wide, that separate all cell boundaries. Astrocytic processes generally surround neurons except where synaptic contacts are made. Many of the axons are typically grouped together and instead of each having an individual covering, entire bundles of axons are surrounded by a common glial envelope. This arrangement is usual in the central nervous system.

From a comparison of neurons and glial cells, one sees that in some regions the cross-sectional area is about equally divided between neurons and astrocytes, whereas in others, as in Figure 4, the glial contribution is smaller. Glial processes tend to be thin, at times less than 1 μm thick. Only around the glial nuclei are there larger volumes of nonlamellated glial cytoplasm.

Electron microscopy has clarified the neuron–glia relation by demonstrating the intimate apposition of their membranes. Special connections, such as gap junctions, are not seen between the two cell types.[10] The clefts (several tens of nanometers wide) seen in Figure 4 always intervene between the surface membranes of neurons and glial cells. In spite of occasional reports of direct communication,[11] the visible separation agrees with results of physiological tests. These fail to reveal direct low-resistance pathways between the neurons and glial cells. Such pathways do, however, link glial cells. The special connections are made by gap junctions (Chapter 9). The relation of glial cells, neurons, and extracellular space is shown diagrammatically in Figure 5. One interesting exception, mentioned earlier (Figure 2), is the close apposition of myelin to axons at the edge of the node; specialized contacts that limit longitudinal spread of current are seen between the two structures in freeze-fracture preparations.[12,3] Other contacts between neurons and glia occur during development and after injury, when transient specializations reminiscent of synapses have been found.[10,13]

Basic to the difficulty of studying glial function is the lack of a method for separating neurons from glia. The tissues are so interwoven (Figure 4) that it is difficult to separate brain tissue into pure glial and neuronal fractions. Some promising immunological methods, however, can now be used to distinguish various glial and neuronal components, such as nuclei or membrane fractions and their chemical constituents. Examples are the proteins associated with fibrous astrocytes and the surface antigens of oligodendrocytes. In tissue culture, pure lines of glial and Schwann cells have been grown.

[10]Mugnaini, E. 1982. In T. A. Sears (ed.). *Neuronal–Glial Cell Interrelationships*. Springer-Verlag, New York, pp. 39–56.

[11]Peracchia, C. 1981. *Nature 290*: 597–598.

[12]Livingston, R. B., Pfenninger, K., Moor, H. and Akert, K. 1973. *Brain Res. 58*: 1–24.

[13]Hendrickson, C. K. and Vaughn, J. E. 1974. *J. Neurocytol. 3*: 659–679.

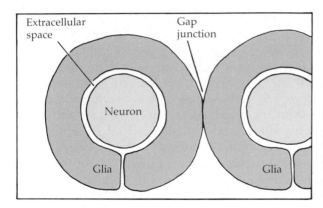

5

NEURONS AND GLIA. Diagrammatic presentation of the relationship between neurons and glia and between glial cells. While neurons are always separated from glia by a continuous cleft, the interiors of glial cells are linked by gap junctions.

One well-established role for oligodendrocytes and Schwann cells is the production of a myelin sheath around axons—a high-resistance covering akin to insulating material around wires. The myelin is interrupted at the nodes of Ranvier (Figure 2), which occur at regular intervals of about 1 mm in most nerve fibers. Since the ionic currents associated with the conducted nerve impulse cannot flow across the myelin, the ions move in and out at the nodes between the insulation (Chapter 7). This leads to an increased conduction velocity and seems an ingenious solution for acquiring speed. The alternative is to make nerve fibers larger, but this is less effective. For example, in the squid, the largest axons are between 0.5 and 1.0 mm in diameter and conduct at rates no faster than about 20 m/sec; a myelinated axon 20 μm in diameter conducts at 120 m/sec.

Myelin and the role of neuroglial cells in axonal conduction

The fact that myelin is not a static substance was shown in studies that followed the uptake and movement of radioactive cholesterol into Schwann cells and then along the lamellae of myelin.[3,14] The idea of a dynamic role of the satellites is reinforced by observations in diseases such as allergic encephalomyelitis. Following demyelination, which need not be accompanied by axonal disintegration, conduction is greatly slowed or blocked.[15] With recovery, remyelination can occur and, thereby, restore normal conduction. Alternatively, as described in Chapter 7, continuous conduction may develop along the demyelinated axon.[16]

The association of satellite cells with axons to form myelin raises a number of interesting problems. For example, what are the genetic or environmental factors that enable glial cells to select the appropriate axons, to surround at the right time, and to maintain myelin sheaths around them? What are the characteristics of some of the neurological disorders caused by genetic abnormalities?

[14]Rawlins, F. 1973. *J. Cell Biol. 58*: 42–53.
[15]Rasminsky, M. and Sears, T. A. 1972. *J. Physiol. 227*: 323–350.
[16]Bostock, H. and Sears, T. A. 1978. *J. Physiol. 280*: 273–301.

These problems have been studied in mice by Aguayo and his colleagues.[17,18] The principle of the experiments was to remove a segment of the sciatic or the sural nerve and replace it with a graft. The graft consists of a segment of peripheral nerve from either the same or a different mouse. Within the grafted segment, axons, which have all been separated from their cell bodies, degenerate, leaving a chain of Schwann cells. Axons from the proximal stump regenerate and grow through the grafted Schwann cells to reach the distal stump, which contains the animal's own Schwann cells. This is shown schematically in Figure 6. The experiment enables a test to be made of how effective the Schwann cells of the graft are in forming myelin, compared to those normally present in the sciatic nerve. For example, a segment of unmyelinated sympathetic nerve from the neck was grafted into a leg nerve that is normally myelinated. As axons grew into the graft they became myelinated. Thus, a population of Schwann cells that do not make myelin in situ can do so if they come into contact with different axons of the appropriate type.[19]

In other studies, made on mutant mice, the genetic defects that are responsible for deficiency of myelin formation were analyzed. One such mutant mouse, called "Trembler," exhibits a gross deficiency of myelin. In this inherited disorder, myelin in the peripheral nervous system is abnormally thin or totally absent. Does the defect result from Schwann cells that are unable to manufacture myelin or from axons that do not provide adequate signals to the Schwann cells? Grafts were made in normal and Trembler peripheral nerves, as shown in Figure 6. When a graft of Trembler nerve was inserted into a normal mouse sciatic nerve, regenerating axons were myelinated above and below the graft but not within it. Conversely axons of the mutant mouse growing through a graft containing normal Schwann cells became myelinated in the graft segment but not above it or below it. In Trembler mice, therefore, the defect resides in the Schwann cells and not in the axons; these are capable of becoming myelinated if they come into contact with healthy satellite cells. By such techniques it has become possible to assess the sites of defects in various other disorders of myelination, including those affecting patients suffering from demyelinating diseases.[20]

HYPOTHESES FOR FUNCTIONAL ROLES
OF NEUROGLIAL CELLS

In the past dozen years, glial cells have been burdened by almost every nervous system task for which no other obvious explanation has been

[17]Aguayo, A. J. and Bray, G. M. 1982. In T. A. Sears (ed.). *Neuronal-Glial Cell Inter-relationships*. Springer-Verlag, New York, pp. 57–75.

[18]Bray, G. M., Rasminsky, M. and Aguayo, A. J. 1981. *Annu. Rev. Neurosci.* 4: 127–162.

[19]Aguayo, A. J., Charron, L. and Bray, G. M. 1976. *J. Neurocytol.* 5: 565–573.

[20]Aguayo, A. J., Kasarjian, J., Skamene, E., Kongshavn, P. and Bray, G. M. 1977. *Nature* 268: 753–755.

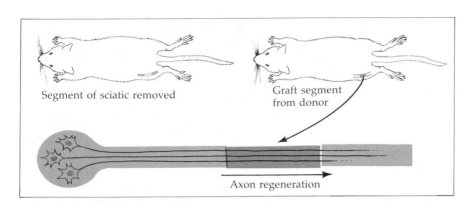

Segment of sciatic removed

Graft segment
from donor

Axon regeneration

4 months

Proximal	Graft	Distal

N-N-N

N-T-N

T-N-T

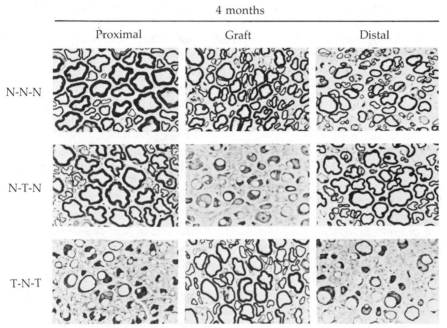

6 MYELINATION OF REGENERATING AXONS BY SCHWANN CELLS.
A segment of sciatic nerve is removed from the leg of a mouse. In its
place is grafted a segment of sciatic nerve from a donor mouse. Axons
in the graft and in the host's sciatic nerve distal to the graft degenerate,
leaving only Schwann cells and connective tissue. Axons grow back
into the graft and the distal stump. N-N-N represent axons in cross
section from a normal mouse in which the grafted segment was from a
normal mouse: the axons are myelinated above, within, and beyond the
graft. In N-T-N the grafted segment was from a Trembler mouse and
was inserted into a normal animal; the grafted Schwann cells did not
form myelin, although distal parts of the axon became myelinated. In
T-N-T a normal segment was grafted into a Trembler mouse sciatic;
only the part of the axon within the graft, surrounded by normal
Schwann cells, became myelinated. These results show that the defect
in Trembler resides in the Schwann cells, not the axons. (After Aguayo,
Bray, and Perkins, 1979.)

found. For example, they have been implicated in the processes of learning and memory and in specifically regulating the chemical environment of nerve cells. The principal difficulty until relatively recently has been the lack of physiological and chemical methods adequate for testing the various proposals. Understandably, the majority of the long-established hypotheses relied primarily on histological observations, but these can hardly on their own reveal the physiological interaction between glial cells and neurons that involve biochemical changes or signaling. In the following paragraphs, we summarize and comment briefly on possible functions of glial cells.

At the beginning of this century, several hypotheses assigning various roles for glial cells were formulated, and some of them have gained wide acceptance. The proposals are still current in most treatises on the subject.

Structural support

Virchow, around 1850, recognized that neuroglial cells are interspersed among neurons and therefore form part of the "structure" of the brain. In one sense, physical support is provided, but neuroglial cells may not necessarily "bind" the brain structure together.

Isolation and insulation of neurons

Ramón y Cajal very early proposed that glial cells prevent "cross talk" by current spread during conduction of nerve impulses. This plausible suggestion has now been tested in several instances, but it lacks force. The fluid-filled clefts between cells seem adequate to prevent effective current flow between cells. However, a role as a spatial barrier for the spread of various substances, such as potassium and transmitters, has found some experimental support (see later).

Role in repair and regeneration

Unlike neurons, glial cells retain the ability to divide throughout the life of an animal; they take part in the formation of scar tissue and have phagocytotic properties. When neurons disappear during aging or after injury, glial cells divide and occupy the vacant spaces. Studies made in tissue culture have shown that growth factors that cause proliferation of Schwann cells can be extracted from neural tissue.[21] Moreover, damage to peripheral nerves causes the synthesis of novel proteins by Schwann cells.[22] At synapses in autonomic ganglia and in the spinal cord, characteristic changes occur after section of the postsynaptic axons.[23,24] In electron micrographs glial cells can be seen to invade the synaptic cleft; at the same time electrical recordings indicate that transmission is impaired (see Chapter 19). These results demonstrate that glial cells can be influenced by changes in the properties of the neurons they surround.

In the course of regeneration, peripheral axons can grow back to their destinations along a route marked by residual Schwann cells. This can be interpreted as a sign of chemical affinity that guides the regenerating axon to its proper site. Regeneration of synaptic connections

[21]Brockes, J. P., Fryxell, K. J. and Lemke, G. E. 1981. *J. Exp. Biol. 95*: 215–230.
[22]Skene, J. H. P. and Shooter, E. M. 1983. *Proc. Natl. Acad. Sci. USA 80*: 4169–4173.
[23]Hunt, C. C. and Nelson, P. 1965. *J. Physiol. 177*: 1–20.
[24]Purves, D. 1975. *J. Physiol. 252*: 429–463.

can, however, still occur in the absence of glia. For example, in the central nervous system of the leech (Chapter 18), axons grow back to reform their original connections after their processes have been severed by an injury, and such regeneration can still be achieved after the glial cells that surround them have been killed.[25]

A role for glial cells in the growth of neurons and the formation of connections has frequently been suggested. In a comprehensive series of experiments, Rakič and Sidman[26–28] have studied the development of the cerebral cortex, the hippocampus, and the cerebellum in monkeys and in man. The formation of the various cell types and their migration to their final destinations have been followed by light and electron microscopy and by labeling of the glial cells with specific antibodies. The neurons appear to move along the glial processes during development. The close association of the two cell types suggests that glial cells provide an initial framework around which subsequent neuronal organization takes place. A generalization along such lines for nonlaminated brain structures, however, is not at present warranted.

Neuroglial cells and
the development of
the nervous system

The role of glial cells in synapse formation is less certain. In the embryonic spinal cord of the monkey, formation of chemical synapses is in progress before glial cells appear in the area of synapses at motoneurons.[29] It is also known that functioning synapses can be formed in tissue culture without satellite cells.[30]

There is now evidence from autoradiographic studies that glial cells can take up γ-aminobutyric acid (GABA) in mammalian dorsal root ganglia, in the spinal cord, in autonomic ganglia, and at crustacean neuromuscular junctions.[31,32] An example is shown in Figure 7. The preparation was bathed in radioactive GABA, which then is visualized in the form of silver grains over the cytoplasm of the satellite cells. Intriguing also is the suggestion that GABA may be accumulated preferentially in the brain by glial cells in areas in which the neurons are rich in GABA and presumably release it.[33] The possibility arises that transmitters may be taken up and stored in glial cells, to be handed back again as the need arises.

Uptake of chemical
transmitters by glia

Such a role was clearly suggested by Nageotte[34] in 1910. He wrote that "la nevroglie est une glande interstitielle annexée au système nerveux." One example has now been well documented for the secretion of neurotransmitter by Schwann cells. In chronically denervated skeletal

Secretory function

[25]Elliott, E. J. and Muller, K. J. 1983. *J. Neurosci. 3:* 1994–2006.

[26]Rakič, P. 1971. *J. Comp. Neurol. 141:* 283–312.

[27]Sidman, R. L. and Rakič, P. 1973. *Brain Res. 62:* 1–35.

[28]Rakič, P. 1982. In T. A. Sears (ed.). *Neuronal-Glial Cell Interrelationships.* Springer-Verlag, New York, pp. 25–38.

[29]Bodian, D. 1966. *Bull. Johns Hopkins Hosp. 119:* 129–149.

[30]Fischbach, G. D. and Dichter, M. A. 1974. *Dev. Biol. 37:* 100–116.

[31]Orkand, P. M. and Kravitz, E. A. 1971. *J. Cell Biol. 49:* 75–89.

[32]Schon, F. and Kelly, J. S. 1974. *Brain Res. 66:* 275–288.

[33]Ljungdahl, A. and Hökfelt, T. 1973. *Brain Res. 62:* 587–595.

[34]Nageotte, J. 1910. *C. R. Soc. Biol. (Paris) 68:* 1068–1069.

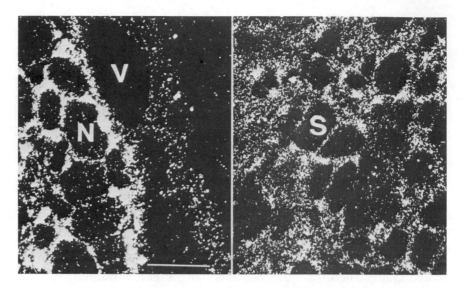

7   GLIAL CELLS take up labeled γ-aminobutyric acid (GABA). Silver
grains appear white in this autoradiograph observed with dark-field
illumination. Left: Schwann cells surrounding neurons (N) in a dorsal
root ganglion of a rat have preferentially accumulated GABA labeled
with [3]H. Only sparse labeling occurs around the vessels (V). Right:
Similar situation in a sympathetic ganglion in which the neurons (S,
dark roundish structures) are practically devoid of radioactivity but the
satellite cells are heavily labeled. White bar is 25 μm long. (Courtesy
of F. Schon and J. S. Kelly.)

muscles, Schwann cells come to occupy the site of the degenerated
nerve terminals. At rest they secrete quanta of acetylcholine (ACh) that
produce miniature synaptic potentials in the muscle. Release of acetylcholine can also be provoked by passing large currents through
Schwann cells.[35] Unfortunately, it is not known whether Schwann cells
can normally secrete transmitter or acquire the ability only after the
nerve has degenerated (see later and Chapter 19).

A second example of transmitter release from glial cells is provided
by the experiments of Brown, Iversen, Kelly, and their colleagues.[36-39]
They have shown that radioactive GABA is released from satellite cells
in sympathetic ganglia and dorsal root ganglia when the potassium
concentration of the bathing fluid is increased. Glial cells are depolarized
by potassium concentration changes of the same order as those following nerve impulses (see later). The functional role of GABA release by

[35]Dennis, M. J. and Miledi, R. 1974a. J. Physiol. 237: 431–452.
[36]Bowery, N. G. and Brown, D. A. 1972. Nature New Biol. 238: 89–91.
[37]Minchin, M. C. W. and Iversen, L. L. 1974. J. Neurochem. 23: 533–540.
[38]Currie, D. N. and Kelly, J. S. 1981. J. Exp. Biol. 95: 181–193.
[39]Bowery, N. G., Brown, D. A., White, R. D. and Yamini, G. 1979. J. Physiol. 293:
51–74.

glial cells is not yet clear, but it is likely to be significant in view of the physiological occurrence of potassium buildup in the central nervous system (see later).[40]

Glial cells in culture have also been shown to contain and secrete molecules that influence the survival, growth, and differentiation of neurons. For example, they contain nerve growth factor and other proteins that act to promote growth of sympathetic and sensory nerve cells (see Chapter 19).[41,42]

The hypothesis that neurons need glial cells in their vicinity to synthesize transmitters is supported by experiments in cultured dissociated nerve cells from sympathetic and dorsal root ganglia. In the absence of satellite cells, neurons reversibly lose their ability to synthesize acetylcholine.[43] Although these results suggest one more secretory role for glia, the effect is not specific. Other cells, such as fibroblasts, produce similar results on transmitter levels.

This nutritive role of neuroglia has been the most popular hypothesis up to the present. It was proposed by Golgi in about 1883. He wrote,[44]

Nutritive role and transfer of substances between glial cells and neurons

> I find it convenient to mention that I have used the term connective tissue with regard to neuroglia. I would say that "neuroglia" is a better term, serving to indicate a tissue which, although connective because it connects different elements and for its own part *serves to distribute nutrient substances,* is nevertheless different from ordinary connective tissue by virtue of its morphological and chemical characteristics and its different embryological origin. (our italics)

Coupled with Golgi's histological staining methods, these ideas seemed so reasonable and had such force that they were hardly questioned through the years. Recent studies with freeze-fracture have further demonstrated aggregates of uniform small particles packed in orthogonal array on astrocytic membranes apposed to blood vessels.[45] The function of these assemblies is unknown, but their specific positioning reinforces the idea that they may play some part in transporting materials into or out of the brain.

Although the proposal for glial transfer of nutrients is attractive, it still requires experimental support. Figure 8 shows graphically the considerations that relate to the Golgi hypothesis, which implies the following events. A molecule that leaves a capillary is either preferentially taken up by a glial cell or diffuses into the cell's interior. Once inside the glial cell, the molecule must be conveyed across the glial cytoplasm by some directional internal transport mechanisms to the region where it meets the neuron; the molecule must then leave the glial cell for

[40]Bowery, N. G., Brown, D. A. and Marsh, S. 1979. *J. Physiol. 293*: 75–101.

[41]Longo, A. M. and Penhoet, E. E. 1974. *Proc. Natl. Acad. Sci. USA 71*: 2347–2349.

[42]Thoenen, H., Barde, Y.-A. and Edgar. D. 1982. In J. G. Nicholls (ed.). *Repair and Regeneration of the Nervous System.* Springer-Verlag, New York, pp. 173–185.

[43]Patterson, P. 1978. *Annu. Rev. Neurosci. 1*: 1–17.

[44]Golgi, C. 1903. *Opera Omnia*, Vols. I and II. U. Hoepli, Milan.

[45]Landis, D. M. and Reese, T. S. 1981. *J. Exp. Biol. 95*: 35–48.

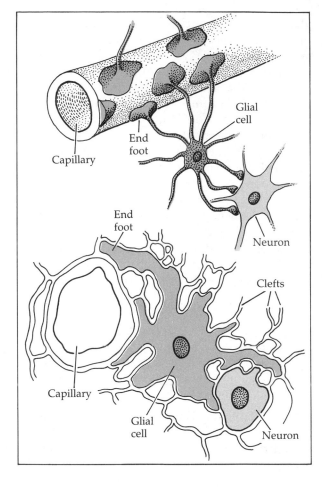

8

RELATIONS OF CAPILLARY, GLIA, AND
NEURONS as seen in the light and electron mi-
croscopes. The most direct pathway from the capillary to the neuron is through the aqueous intercellular clefts that are open for diffusion. Cell
dimensions are not in proportion. (After Kuffler
and Nicholls, 1966.)

uptake by the neuron. The nutritive hypothesis also assumes implicitly
that no adequate direct pathways for free diffusion of substances exist
that would provide easier and preferential access to the neuron. Rather
than crossing glial cell boundaries that intervene between capillary and
neuron, perhaps a molecule could diffuse straight to the target. In fact,
the glial membrane has a relatively high electrical resistance and the
cell has its own particular internal environment (see later). Materials
cannot, therefore, diffuse readily into the interior of glial cells. Further,
free channels that are open for diffusion exist all around the neurons
and glial cells, so that the most direct access to nerve cells is through
the aqueous cleft system (Chapter 14).

Experiments have been done to demonstrate the transfer of materials
between glial cells and neurons. The principle is to label a protein or
lipid component of one type of cell and measure its appearance in the
other. If a label appears "simultaneously" in nerve and glia, one assumes that it was made or taken up independently by the two tissues.
If, however, a distinct time sequence occurs, such as label turning up

first in glia and then in neurons, it may suggest that material was handed on from the first to the second cell type. For example, Droz and his colleagues[46] injected labeled glycerol into the brain of a chick. The label was incorporated into lipids that were transported along axons and then several hours later appeared in the surrounding myelin, first in the innermost layers and then farther away. Transfer in the opposite direction, from Schwann cells to axons, has been measured in the squid axon. Axons are devoid of ribosomes and cannot synthesize proteins. When radioactive amino acids were applied to squid nerve, labeled proteins appeared in the sheath cells and then in the axon.[47,48] Injection or perfusion of the amino acids into the axon itself did not lead to the appearance of labeled protein. Moreover, certain specific proteins seem to be preferentially transferred from the surrounding glia to the axon.[49] Similar experiments using radioactive amino acids, dyes, and horseradish peroxidase have been done in crustacea.[50] That transfer of material from cell to cell can occur is clear (see also Chapter 3 for a description of transsynaptic transfer of protein in the visual system). It remains to be determined whether such transfer is quantitatively significant, what the mechanism of transfer is, and what role it might play in the normal functioning of the nerve cell.

PHYSIOLOGICAL PROPERTIES OF NEUROGLIAL CELL MEMBRANES

The study of the physiological properties of neuroglia has been greatly aided by two preparations in which the glial cells are large and accessible and in which they can be conveniently investigated in their normal relation to nerve cells. One is the CENTRAL NERVOUS SYSTEM OF THE LEECH, the other is the OPTIC NERVE OF THE MUD PUPPY, *Necturus*. In both preparations glial cells can be impaled with microelectrodes and their membrane properties studied. Such studies have allowed physiological criteria to be defined by which glial cells can be recognized within the mammalian brain.

The leech nervous system—described more fully in Chapter 18—consists of a chain of ganglia joined by connectives and lies within a vascular sinus; it is less than 1 mm in diameter, contains no blood vessels, and is quite transparent. Ganglia can readily be removed from the animal and survive well in Ringer's solution. A photograph of a leech ganglion is shown in Figure 9. The relative sizes of neurons and glia in this nervous system are the reverse of what is seen in vertebrates, for in the leech the glial cells are larger than the nerve cell bodies and

Simple preparations used for intracellular recording from glia

[46]Droz, B., Di Giamberardino, L., Koenig, N. J., Boyenval, J. and Hassig, R. 1978. *Brain Res.* 155: 347–353.

[47]Lasek, R. J., Gainer, H. and Barker, J. L. 1977. *J. Cell Biol.* 74: 501–523.

[48]Gainer, H., Tasaki, I. and Lasek, R. J. 1977. *J. Cell Biol.* 74: 524–530.

[49]Lasek, R. J. and Tytell, M. A. 1981. *J. Exp. Biol.* 95: 153–165.

[50]Bittner, G. D. 1981. *Comp. Biochem. Physiol.* 68A: 299–306.

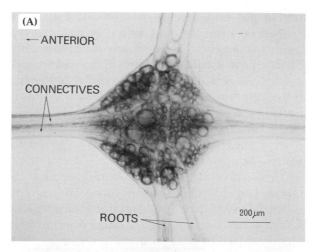

9

LEECH GANGLION. (A) Living isolated ganglion in the central nervous system of the leech
under transmitted illumination. The shapes and
sizes of nerve cells can be seen; the glial cytoplasm between the neurons is not visible because
it is transparent. Connectives and roots run to
neighboring ganglia and body wall, respectively.
(B) The cytoplasm of one glial cell is traced in
two histological sections and appears white. It
fills all the spaces not occupied by neurons (the
enclosed black areas). Scale, 50 μm. (A from
Nicholls and Baylor, 1968; B from Kuffler and
Potter, 1964.)

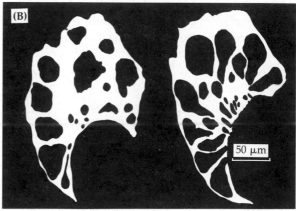

are few in number.[51] Ten large glial cells are present in each ganglion
and its connectives. Six of these are associated with the nerve cell bodies
in each ganglion, two with the synaptic regions, and two with several
thousand axons running in the connectives. The glial cells in the ganglion are transparent and therefore appear under the dissecting microscope as spaces between nerve cells. They can be impaled consistently
by a microelectrode inserted next to a neuron.

Another simple central nervous system preparation that has been
studied extensively is the optic nerve of *Necturus*.[52] It is about 0.15 mm
in diameter and is covered by a layer of connective tissue containing
blood vessels that run parallel to the surface; no blood vessels occur
within the tissue. The glial cells are large and intimately surround the
nerve fibers. As in the leech, the neurons and glia receive their nutrients
principally from the surface. Figure 10A presents a cross section of the

[51]Coggeshall, R. E. and Fawcett, D. W. 1964. *J. Neurophysiol.* 27: 229–289.
[52]Kuffler, S. W., Nicholls, J. G. and Orkand, R. K. 1966. *J. Neurophysiol.* 29: 768–787.

Necturus optic nerve stained with toluidine blue. The glial nuclei are dark and prominent, whereas the glial cytoplasm remains unstained. The outlines of the bundles of nonmedullated axons (fiber diameters 0.1 to 1.0 μm) are too small to be seen in the light micrograph. The blood vessels run close to the surface of the nerve within the loose connective tissue. The neuron–glia relation is illustrated in Figure 10B and C; spaces about 20 nm wide exist between the cells, as in the mammalian (Figure 4) and the leech nervous systems. Also visible is part of a cluster of tightly packed axons where glia surround the bundle but not the individual axons.

Glial cells in both the central nervous system of the leech and the optic nerve of *Necturus* occupy 35 to 55 percent of the cross-sectional area. For emphasis the glial cells with their processes between the nervous elements are shaded in Figure 10C, and the tortuous intercellular clefts are traced in black. Note that in two places (arrows) the clefts open to the outside. The fine structure of the leech nervous system appears strikingly similar in many respects, with narrow spaces intervening between neuronal and glial membranes (Chapter 18).

Recordings of the membrane potentials in the central nervous system of the leech show directly that glial cells have higher resting potentials than the neurons they surround; about −75 mV compared with −50 mV.[53] In vertebrates, including the frog, mud puppy, cat, and rat, the highest membrane potentials recorded for neurons are about −70 to −75 mV, whereas the values for glial cells consistently approach −90 mV. Their membrane resistance (R_m, Chapter 7) is about 1000 Ωcm^2 or more, which is comparable to the value measured in neurons.

Glial membrane potentials

To study the origin of the high resting potentials, the membrane potentials of glial cells have been measured in different potassium concentrations (Figure 11). In isolated optic nerves of *Necturus*, the potassium concentrations in the bathing fluid can be varied over a wide range. The results are unexpected, because the glial membrane behaves like a perfect potassium electrode and accurately follows the Nernst equation (Chapter 5):

Dependence of membrane potential on potassium

$$E = \frac{RT}{F} \ln \frac{K_o}{K_i} = 58 \log \frac{K_o}{K_i}$$

Changes in sodium and chloride concentrations do not produce significant changes in potential. One can conclude that ions other than potassium make a negligible contribution to the membrane potential, which is determined by the ratio K_o/K_i. Figure 11C shows a series of membrane potential measurements plotted against K_o on a logarithmic scale. The solid line is the theoretical slope of 59 mV per tenfold change in concentration predicted by the Nernst equation (at 24°C), and it agrees excellently with the experimental points. A particular feature of the relation is the good fit at low concentrations of K_o, down to 1.5 m*M*. In this respect glial cells differ significantly from most neurons,

[53]Kuffler, S. W. and Potter, D. D. 1964. *J. Neurophysiol.* 27: 290–320.

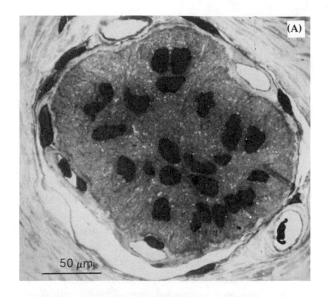

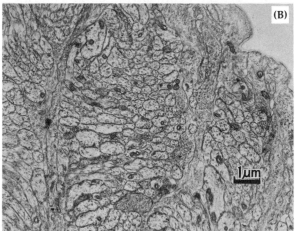

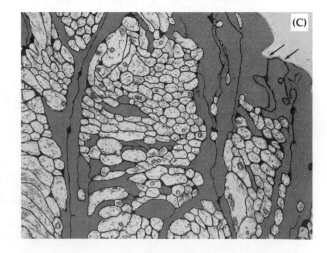

10

OPTIC NERVE of the mud puppy (*Necturus*) is shown in cross section (A). Nuclei of glial cells are stained black. The outlines of the glial cytoplasm and bundles of nonmedullated axons cannot be resolved. The nerve is surrounded by connective tissue containing capillaries. (B, C) Two identical electron micrographs of part of the optic nerve. In one the glial processes are lightly shaded and the clefts that separate the cells are traced in black. Two cleft openings reaching the surface are marked by arrows. Axons run in closely packed bundles, as in the mammalian system (Figure 4). (From Kuffler, Nicholls, and Orkand, 1966.)

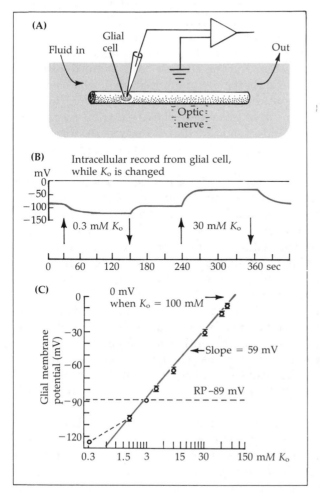

11

GLIAL MEMBRANE POTENTIAL depends on potassium concentration. (A) System of perfusing the optic nerve while recording from a glial cell. (B) Reducing the potassium concentration from the normal 3.0 mM to 0.3 mM increases the normal membrane potential of -89 mV to -125 mV, while increasing the potassium concentration to 30 mM decreases it by 59 mV. (C) Various values of potassium plotted against membrane potential show that the relation predicted by the Nernst equation (solid line) accurately fits the experimental results over a wide concentration range. The membrane potential is zero when the internal and external potassium concentrations are equal to 100 mM. (After Kuffler, Nicholls, and Orkand, 1966.)

which deviate from the Nernst prediction in the physiological range of 2 to 4 mM K_o (Chapter 5).[52]

The experiments illustrated in Figure 11 provide a good estimate of the internal concentration of potassium (K_i). The Nernst equation indicates that when K_o is the same as K_i, the membrane potential is zero. This occurred when the outside potassium concentration was increased to 100 mM. Similar determinations of K_i (110 mM) were made in glial cells of the leech, but there the conclusions were further strengthened by direct flame photometric measurements.[54]

Glial cells, therefore, contain a high potassium concentration and have a negligible ionic permeability for ions other than potassium. Although sodium channels are present in glial membranes, they are

[54]Nicholls, J. G. and Kuffler, S. W. 1965. *J. Neurophysiol. 28*: 519–525.

scarce and contribute little to the total conductance.[55] A sodium–potassium coupled pump has been demonstrated in glia and has properties similar to those of pumps described for neurons in Chapter 8.[56,57]

A salient feature of glial cells is their widespread distribution throughout the nervous system and the absence of an axonlike process. This indicated to many of the earlier workers that glial cells were unlikely to conduct impulses in the manner of axons. However, the possibility remained that they might give some active electrical responses, possibly slow ones. It should be added that no synapses, formed by axons, have as yet been seen on glial cells in electron microscopic studies of adult nervous system.[10]

In the leech, frog, mud puppy, and mammals, it has been shown that identified glial cells do not produce impulses. The results in the leech and in *Necturus* are particularly clear-cut, because the glial membrane behaves passively even when its potential is displaced over a range of about 200 mV. The cell membrane resistance remains constant throughout, like an ohmic resistance.[52,53] "Active" regenerative responses are, therefore, excluded. As expected, such properties make glial cells fundamentally different from the excitable neurons.

Schwann cells that surround squid axons have been impaled with electrodes, and their membranes also behave passively; that is, they are inexcitable. Their measured resting potentials (about 40 mV) are, however, lower than those found in glial cells. The low membrane potentials may be only apparent, the result of leakage current during electrode penetration of the cells that form a very thin layer—mostly several micrometers thick.[58]

Physiological properties of glial cells in the mammalian brain

When intracellular recordings are made from the mammalian central nervous system, certain cells appear to have properties quite different from those of conventional neurons. Their resting potentials are high, often near 90 mV, without the fluctuations typical of continued background synaptic bombardment, and they appear to be inexcitable by electrical currents.[59,60] These characteristics are basically the same as those of glial cells in the leech and *Necturus*. A further unambiguous physiological criterion for identifying glial cells is that they become depolarized through potassium accumulation when neurons in their vicinity discharge impulses (see later). Similar properties have been found in astrocytes recorded from in biopsy material from human brain.[61] Such "silent" cells have been injected with various dyes, in particular with the fluorescent dye Lucifer Yellow or with horseradish peroxidase, and have been shown to be glia by subsequent histological

[55]Tang, C.-M. Strichartz, G. R. and Orkand, R. K. 1979. *J. Gen. Physiol. 74*: 629–642.
[56]Tang, C.-M. Cohen, M. W. and Orkand, R. K. 1980. *Brain Res. 194*: 283–286.
[57]Orkand, R. K., Orkand, P. M. and Tang, C.-M. 1981. *J. Exp. Biol. 95*: 49–59.
[58]Villegas, J. 1981. *J. Exp. Biol. 95*: 135–151.
[59]Karahashi, Y. and Goldring, S. 1966. *Electroencephalogr. Clin. Neurophysiol. 20*: 600–607.
[60]Dennis, M. J. and Gerschenfeld, H. M. 1969. *J. Physiol. 203*: 211–222.
[61]Picker, S., Pieper, C. F. and Goldring, S. 1981. *J. Neurosurg. 55*: 347–363.

examination.[62,63] The slope of the relationship between potassium concentration and membrane potential follows that predicted by the Nernst equation. One exception is human reactive astrocytic glia in which the sensitivitiy to potassium is somewhat less.[61]

In the vertebrate brain only relatively few neurons are linked by the low-resistance gap junctions that enable current to pass readily from one nerve cell to its neighbor (Chapter 9). Adjacent glial cells, however, including those of mammals, are linked to each other by gap junctions through which dyes as well as current can pass.[64,65] In this respect they resemble epithelial and gland cells and heart muscle fibers. In the leech, frog, and *Necturus*, glial cells are always found to be coupled electrically.

The functional significance of the electrical coupling of glial cells is not known. Clearly, ions can exchange directly between cells without passing through the extracellular space, and such interconnections may be useful for equalizing concentration gradients that may arise. The possibility is open that there exists some metabolic interaction between coupled cells, linked with demand that is induced by activity. It is shown later that the low-resistance coupling between the cells is essential for generation by glial cells of current flow that can be recorded from the surface of nervous tissues by extracellular electrodes.

As mentioned earlier, no gap junctions between nerve and glia have been detected. This is of physiological interest because it is natural to wonder about the manner in which neurons and glial cells may interact. Direct tests have been made in the leech nervous system, where the potentials of neurons can be changed in a controlled manner by passing currents through them while recording from adjacent glial cells.[53] The reverse procedure has also been done—recording from nerve cells while changing the membrane potential of glial cells. Analogous tests in the optic nerve of *Necturus* show similar results: Current flow around glial cells created by synchronized conducting nerve impulses has no significant effect on the neighboring glial membrane, and therefore an electrical interaction between nerve and glia seems quite unlikely.

The simplest explanation of the failure of current to flow from one cell to another is their separation by intercellular channels (several tens of nanometers wide) filled with low-resistance fluid (about 100 Ωcm) (Figures 4 and 10). These channels provide pathways for currents to flow out to the bathing fluid without passing through the relatively high-resistance cell membranes of neurons or glia.

A SIGNALING SYSTEM FROM NEURONS TO GLIAL CELLS

Most speculation about the role of glial cells entails some kind of interaction with neurons. Such a mutual influence might be expected be-

[62]Kelly, J. P. and Van Essen, D. C. 1974. *J. Physiol. 238*: 515–547.
[63]Takato, M. and Goldring, S. 1979. *J. Comp. Neurol. 186*: 173–188.
[64]Brightman, M. W. and Reese, T. S. 1969. *J. Cell Biol. 40*: 648–677.
[65]Gutnick, M. J., Connors, B. W. and Ransom, B. R. 1981. *Brain Res. 213*: 486–492.

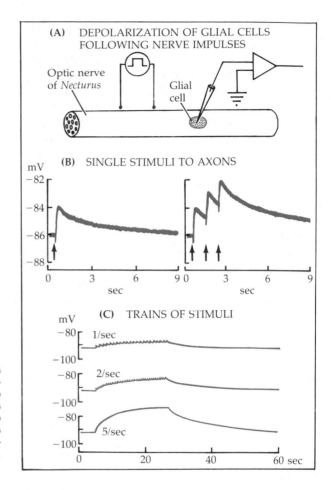

12

EFFECT OF NEURAL ACTIVITY on glial cells
in the optic nerve of the mud puppy. Synchronous impulses in nerve fibers cause glial cells to
become depolarized. Each volley of impulses
leads to a depolarization that takes seconds to
decline. The amplitude of the potentials depends
on the number of axons activated and the frequency of stimulation, as shown in (B) and (C).
(After Orkand, Nicholls, and Kuffler, 1966.)

tween two cell types that are so intimately apposed to each other. The
effect of nerve activity on glial cells can be most simply illustrated by
experiments made in the brain of *Necturus*; similar results have been
obtained in the leech and in mammals.

The basic observation is illustrated in Figure 12. Recordings are made
from a glial cell in the optic nerve of the mud puppy, and a volley is
set up in the nerve fibers so that impulses travel past the impaled glial
cell. Each volley of impulses is followed by a depolarization of the glial
cell, rising to a peak in about 150 msec and declining slowly over several
seconds. The size of the potential is graded, depending on the number
of nerve fibers activated. With repeated stimulation, the potentials in
the glial cells sum, depending on the frequency of the stimulation
(Figure 12B and C). If stimulation is maintained, surprisingly large glial
membrane depolarizations of up to 48 mV can be seen. At the end of
a train of stimuli, a residue of such large potentials may persist for 30
seconds or longer.[66]

[66]Kuffler, S. W. and Nicholls, J. G. 1966. *Ergeb. Physiol. 57:* 1–90.

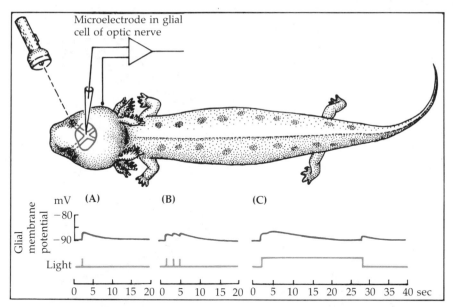

13 **EFFECT OF ILLUMINATION** of the eye on the membrane potential of glial cells in the optic nerve of an anesthetized mud puppy, with intact circulation. (A) Single flash of light 0.1 sec long. (B) Three flashes. (C) Light stimulus maintained for 27 seconds. During such prolonged illumination the initial glial depolarization declines as the nerve discharge adapts. At the end of illumination, a burst of "off" discharges initiates a renewed glial depolarization. Lower beams monitor light. (After Orkand, Nicholls, and Kuffler, 1966.)

To test whether potential changes recorded in glial cells are a physiological occurrence, experiments were done with natural stimulation of optic nerve fibers in anesthetized mud puppies with intact circulation. A single brief flash of light caused a glial depolarization of about 4 mV (Figure 13). Repeated flashes added distinct, but smaller, potentials. The glial potential declined progressively during maintained illumination but reappeared when the light was turned off. These results are in good agreement with the conclusion that discharges in the optic nerve are responsible for the glial potentials, since the discharge rate declines during maintained illumination but a renewed burst follows when the light is turned off.

In the cortex, glial cells become depolarized when neurons in their vicinity are activated by the stimulation of neural tracts, peripheral nerves, or the surface of the cortex. The magnitude of the potentials recorded from glia once more depends on the strength of stimulation as additional axons conducting into the region are activated. An example from experiments by Ransom and Goldring[67] is shown in Figure 14. The slow time course of the glial potentials and their magnitude

[67]Ransom, B. R. and Goldring, S. 1973. *J. Neurophysiol.* 36: 869–878.

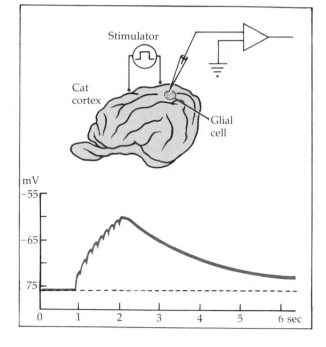

14

GLIAL DEPOLARIZATION resulting from stimulation of cortical neurons (at 8/sec). The amplitude of the glial depolarization is graded according to the strength and frequency of stimulation. (After Ransom and Goldring, 1973.)

resemble those seen in *Necturus.* Similar results have now been obtained from mammalian glial cells by a number of investigators.[63,65]

These results suggest that glial cells within the visual cortex should become depolarized only when certain specific patterns of light are shone into the eye to activate groups of neighboring neurons. Kelly and Van Essen[62] found such an effect in the visual cortex of the cat. Glial cells identified by physiological and morphological criteria became significantly depolarized only when a bar of light with an appropriate orientation was shone into the eye (Figure 15). Illumination of both eyes was effective if corresponding areas were illuminated, but diffuse lights or inappropriate orientations produced no appreciable change in potential. These results are in good agreement with the assumption that glial cells that respond only to selected stimuli are situated in a column of neurons whose receptive fields are all oriented at one particular angle to the vertical. Still unknown is whether glial cells are specifically arranged in relation to neuronal columns (Chapter 3).

Potassium release as mediator of the effect of nerve signals on glial cells

The first possibility to be considered as the mechanism for the action of nerve on glial cells is the flow of current during nerve impulses. This is excluded because the time courses of the events are greatly mismatched. The peak current flow caused by impulses, such as a synchronous volley in the optic nerve, has already declined when the glial depolarization starts to rise. Further, passing current directly into nerve cells in the leech does not produce detectable responses in neighboring glial cells.

A more likely hypothesis is suggested by experiments on squid

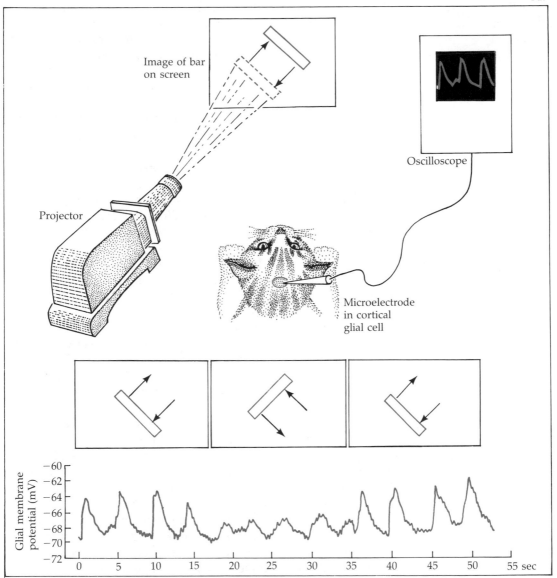

15 DEPOLARIZATION OF GLIAL CELL in the cortex of a cat as a result of visual stimulation. The greatest depolarization was produced by moving slits of light having a certain orientation and confined to a part of the visual field. Neurons in the region near the glial cell had the same orientation preference. (After Kelly and Van Essen, 1974.)

axons by Frankenhaeuser and Hodgkin,[68] who showed that following nerve impulses potassium accumulated in the spaces between axons and the surrounding Schwann cells. To check the hypothesis that glial cells are depolarized by potassium leakage from axons, use was made

[68]Frankenhaeuser, B. and Hodgkin, A. L. 1956. *J. Physiol. 131*: 341–376.

of the observation that the glial membrane potential provides a good quantitative assay for potassium in the environment of the cells (Figure 11). If potassium leaves axons and accumulates in the intercellular clefts, it changes the K_o/K_i ratio and alters the membrane potential of glial cells in a predictable way. The glial cells should depolarize as if potassium liberated by the axons into the clefts adds logarithmically to the K_o already present in the bathing fluid.

Figure 16 shows results that agree quantitatively with this prediction. The membrane potential was measured at three different external concentrations of potassium: normal Ringer's solution (K_o = 3 mM); 1.5 times normal (K_o = 4.5 mM); and one-half normal (K_o = 1.5 mM). These are the three solid circles of Figure 16C that lie on the curve (solid line) predicted by the Nernst equation. When, at the normal membrane potential (89 mV) and in 3 mM potassium solution, a brief train of nerve stimuli was given, the membrane potential was reduced by 12.1 mV (plotted on the ordinate, middle open circle). This depolarization is equivalent to an increase of 1.8 mM in potassium in the bathing fluid. How much depolarization would 1.8 mM potassium produce if it were added to a bathing solution that contained only 1.5 mM instead of 3 mM potassium? This value is indicated by the lower horizontal broken line, which is within 0.4 mV of the value observed with a train of stimuli (lowest open circle). The upper horizontal dotted line shows the depolarization predicted if 1.8 mM potassium is added to a solution that already contains 4.5 mM. Once more the depolarization (upper open circle) caused by the standard nerve stimulation is in excellent agreement with the prediction.[69]

These and other results can be explained by the simple assumption that in the physiological range of K_o nerve impulses release constant amounts of potassium into the intercellular clefts. As a result of the potassium that transiently accumulates, the membrane potential of the glial cell becomes depolarized, and its potential returns to normal as the potassium disappears by uptake and diffusion.

The introduction of potassium-sensitive glass electrodes has enabled several groups of investigators to measure directly the accumulation of potassium in the extracellular spaces of the brain during neuronal activity (Chapter 8). With repetitive stimulation of neurons, the potassium concentration increases, the value being comparable to that assumed from the glial cell depolarization.[70-72]

Effect on signaling of removal of glial cells

Do glial cells contribute to the signaling activity of those neurons they surround? A ready answer is obtained by cutting open ganglia in the leech and washing away the glial cytoplasm. This procedure leaves the neuronal surface exposed to the bathing fluid (Figure 17); yet these naked neurons continued to give signals for many hours.[53] The exper-

[69]Orkand, R. K., Nicholls, J. G. and Kuffler, S. W. 1966. *J. Neurophysiol. 29:* 788–806.
[70]Syková, E., Shirayev, B., Kriz, N. and Vycklický, L. 1976. *Brain Res. 106:* 413–417.
[71]Heineman, U. and Lux, H. D. 1977. *Brain Res. 120:* 231–249.
[72]Somjen, G. G. 1979. *Annu. Rev. Physiol. 41:* 159–177.

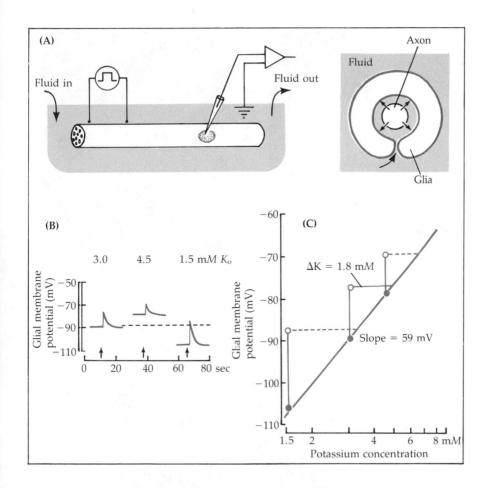

16 RELEASE OF POTASSIUM from axons depolarizes glia. (A) System of perfusing the optic nerve of the mud puppy and (right) a schematic representation of an optic nerve fiber surrounded by a glial cell. (B) Three different concentrations of potassium in the bath shift the membrane potential to three different levels (Figure 10). When a standard train of nerve impulses is set up, it causes a smaller depolarizing effect when K_0 is higher (4.5 mM) and a larger effect when K_0 is lower (1.5 mM). (C) The observations in (B) plotted on a semilogarithmic scale. The solid circles indicate the membrane potential at K_0 of 1.5, 3.0, and 4.5 mM, in quantitative agreement with prediction by the Nernst equation (solid line). The nerve-evoked peak depolarizations (open circles) are equivalent (dotted lines) to adding 1.8 mM of potassium to the bathing fluid. (After Orkand, Nicholls, and Kuffler, 1966.)

iment demonstrates that glia are not essential for the short-term survival of neurons or for impulse conduction. However, close scrutiny of the impulses in neurons deprived of glia shows that their manner of signaling is different when they are excited by a train of stimuli. The difference can be attributed to the absence of potassium accumulation. The analysis is based on the technique originally described by Frankenhaeuser and Hodgkin,[68] in which the undershoot of a nerve impulse is used to measure small changes of potassium concentrations in the

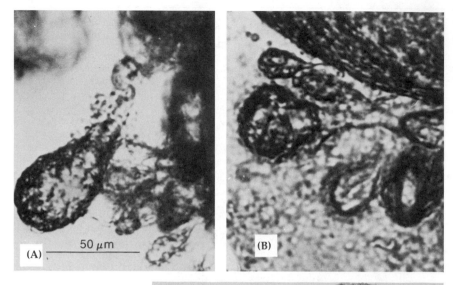

17

NEURONS DEPRIVED OF GLIA by making a cut in the connective tissue capsule surrounding a leech ganglion. (A) A single nerve cell body and its initial axon segment after the glial cytoplasm was rinsed away. (B) Same ganglion, just after the cut, before glial cytoplasm had been removed. (C) Electron micrograph made by D. E. Wolfe, of part of a neuron that is devoid of its glial environment. Such "naked" neurons continue to conduct impulses. (From Kuffler and Potter, 1964.)

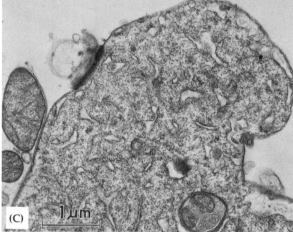

bathing medium. During this phase of the action potential, g_K is high and the cell is more sensitive to small changes in K_o than it is at rest (see below). As the potassium (K_o) in the environment increases, the undershoot becomes smaller. The undershoots of impulses can, therefore, be used as a bioassay for potassium concentration, just as the glial membrane potential was used in the tests described in Figure 16.

In the experiment illustrated in Figure 18, a train of nerve impulses was repeated with a neuron first in its normal environment with the glia and clefts intact, and then after the glial cytoplasm had been removed. In its normal environment, the undershoots of the nerve impulses decrease as more and more potassium accumulates during a train of impulses, whereas the impulse peaks are but little altered. After removal of the glia and in the absence of potassium accumulation, each

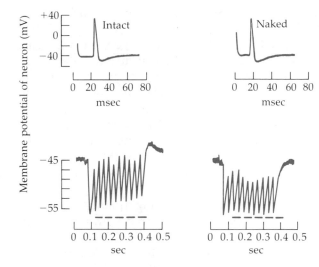

18

EFFECT OF GLIA ON NEURONAL SIGNALS in a leech ganglion. Records are from the same neuron before and after removal of the surrounding glia (Figure 17). In the intact ganglion the potassium accumulates in the clefts surrounding the neuron during a train of impulses, and the potassium-sensitive undershoot becomes progressively smaller (B, left record). In the naked neuron the undershoot does not change during a train of impulses (B, right record). (From Baylor and Nicholls, 1969.)

succeeding impulse has the same undershoot.[73]

These experiments bring out the significant fact that potassium accumulation during normal activity does not interfere with conduction but does affect the slower afterpotentials and, therefore, alters signaling activity. Accumulation of potassium must, therefore, be considered in any analysis of the pattern of signaling. Secondarily, the results also mean that the mere presence of glial cells is significant, by delineating the clefts and at the same time providing a spatial buffer around neurons.

GLIAL CELLS AS A SOURCE OF POTENTIALS RECORDED WITH SURFACE ELECTRODES

Recordings made with electrodes placed on the surface of a tissue measure currents flowing through extracellular fluids. Such currents are generated by a cell if various regions of its surface are at different potentials. Nerve cells, of course, use this principle as the mechanism for conduction. Thus, current flows from inactive regions of an axon into the part that is occupied by a nerve impulse. Positive charges are thereby drawn away from the uninvolved area in front of a nerve impulse so that the region ahead becomes depolarized and eventually "active." Glial cells do not extend over long distances like axons but are linked to each other by low-resistance connections. The conducting properties of such contiguous, coupled cells are therefore much the same as they would be if the cells were a syncytium. As a result, if several glial cells become depolarized by increased potassium concentrations in their environment, they draw current from the unaffected cells, thereby creating current flow. This, in turn, gives rise to a potential

General considerations

[73]Baylor, D. A. and Nicholls, J. G. 1969. *J. Physiol.* 203: 555–569.

difference in the external fluid, a potential difference which can be recorded with external electrodes.

One would expect the patterns of current flow generated by glial cells in the brain to be complex because of their diffuse distribution in the tissue. Each cell has an elaborate arborization and makes contact at various points with other glial cells. We emphasize that the situation in glial cells would be quite different if they were electrically independent of each other and uniformly depolarized. If that were the case, no current would flow in the external circuits and the cells would be isolated from each other in the same way as two neighboring skeletal muscle fibers or nerve fibers that change their potentials independently without reference to each other.

Correlation between glial potential changes and current flow

A model situation suitable for demonstrating quantitatively the amount of current generated by glial cells compared with neurons occurs in the optic nerve of *Necturus*. The nerve's geometry is simple: A population of nerve fibers runs in parallel through the cylindrical nerve. The glial cells that are in electrical continuity, each coupled to its neighbors, also run in parallel through the nerve. For present purposes, therefore, the nerves can be represented as a single continuous structure, and so can the glial cells (Figure 19).

The method of choice for measuring the relative contribution of neurons and glia to current flow is the sucrose gap technique. The chamber that holds the preparation is divided into three compartments (Figure 19) so that a short segment of nerve is surrounded by isotonic sucrose. Since sucrose, which is not a conductor of current, fills all the extracellular spaces, the only electrical connection for current between the two Ringer's-solution-filled compartments is through the interior of the axons and glial cells. Under these conditions, two surface electrodes across the sucrose gap register the voltage drop produced by current flow through the interior of all the cells. At the same time, an electrode within a glial cell near the boundary between the Ringer's solution and the sucrose measures the membrane potential across the cell membrane.

In a series of experiments by Cohen,[74] the glial cells were depolarized by an intracellular current electrode while the glial membrane potential at the gap was registered by another electrode. External electrodes across the gap recorded a potential about one-half as great as that recorded by the intracellular microelectrode (Figure 19). This result is expected because in the optic nerve of *Necturus* the glial cells and axons are seen by electron microscopy to occupy about the same volume. The potentials recorded from glial cells with external electrodes are therefore reduced by 50 percent compared with intracellularly recorded potentials. In other nerves, a variable contribution from glia is expected, depending on the volume ratio.

The conclusion that glial cells generate current flow is further strengthened by results of experiments in which the optic nerve was cut. After axons had degenerated, the glia:neuron ratio was changed

[74]Cohen, M. W. 1970. *J. Physiol.* 210: 565–580.

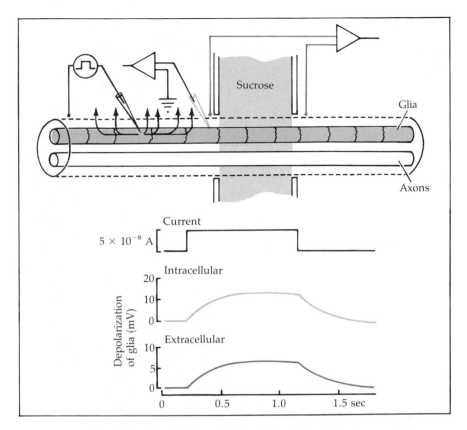

19 CURRENT FLOW generated by glial cells in the optic nerve of *Necturus*. In the sketch at top the axon bundles and glial cells are each represented as a continuous structure. Current was passed into the glial cells by an intracellular electrode, and the resulting potential change was recorded by another intracellular electrode as well as by electrodes across a sucrose gap. The extracellular potentials were about half the size of the intracellular ones and had a similar time course. This shows that glia and neurons make similar contributions to current flow. (After Cohen, 1970.)

so that a preparation of almost pure glial cells was left.[57,74] In these nerves, extracellular leads across the sucrose gap recorded a potential that was 80 to 90 percent of that recorded by an intracellular electrode in glia at the boundary between the sucrose and the Ringer's solution.

Such experiments show clearly that GLIAL CELLS DO CONTRIBUTE CURRENTS that can be measured by surface electrodes. By themselves, however, they do not enable prediction of the amount or time course of the resulting potential changes at other sites in the nervous system.

Electroencephalography is one of the avenues available for obtaining objective information about activity in the human brain. Electroencephalograms (EEGs) are made routinely in fully conscious subjects or in animals under a variety of experimental conditions. Therefore, any method that separates the contribution of various elements, such

Glial contribution
to the
electroencephalogram

as neurons and glia, is of potential interest. The same applies to recordings from the eye—electroetinograms (ERGs)—in which one gross electrode is placed on the cornea and another indifferent electrode elsewhere on the body. In each case the potential changes represent the summed electrical activity of the underlying mass of neurons and glial cells.

How much is contributed by glial depolarizations to surface recordings from the brain? A quantitative answer about the contribution of glial cells to the EEG is difficult to obtain for a number of reasons: (1) Both neurons and glial cells generate current flow, but depending on their distribution in the volume of the brain the potentials recorded on the surface of the tissue may be either positive or negative. (2) The contributions of currents by neurons and glia sum algebraically. If they occur simultaneously, their potentials may be additive; or if they are in the opposite direction, they may cancel each other. (3) Slow potentials are by no means confined to glia but are common in neurons as well; thus, neurons may be hyperpolarized through the activity of an electrogenic pump (Chapter 8) at the same time as glia are depolarized by the buildup of potassium.[73]

One approach to assessing the contribution of glial cells to the EEG is provided by experiments in which slow glial membrane potentials are recorded with intracellular electrodes in the cortex of cats following stimulation of neurons. At the same time, potentials are monitored with extracellular electrodes on the cortical surface and with other electrodes inserted at different depths within the cortex.[67] Potential changes, fast as well as slow, that can be correlated with neuronal activity change their polarity as the probing extracellular electrode passes through different depths of the tissue. On the other hand, potentials attributed to the glial cells are more evenly distributed throughout the cortex and do not change their polarity.

Glial contribution to
the electroretinogram

The ERG has been a useful tool in physiological studies of the eye and in clinical diagnoses. Literature going back over 100 years concerns slow potentials recorded from the eye. Intracellular and intraretinal recordings have helped greatly to determine the cellular origin of some of the surface potentials in the ERG. The main contributing elements are the photoreceptors, the neurons, and the glial cells. The glial cells, called Müller cells, extend through the whole depth of the retina. They have been impaled with microelectrodes and, as in the brain, have been shown to have higher membrane potentials (up to −85 mV) than the neurons. Illumination always depolarizes them, the response slowly declining if the light is kept on; when illumination is turned off, they give "off" responses rather like the records from a glial cell in the optic nerve of *Necturus* (Figure 13). Interestingly, the Müller cell potentials have a time course similar to that of one of the components (the b waves) in the ERG obtained with extracellular recordings from the eye of the mud puppy. They are likely to be responsible for most of that component of the standard ERG.[75]

[75]Miller, R. F. and Dowling, J. E. 1970. *J. Neurophysiol. 33*: 323–341.

The simplest interpretation of the Müller cell responses is that they are a secondary consequence of potassium leakage as neurons become active. Recall, however, that retinae of various fishes and turtles contain nerve cells that give maintained potentials that will contribute to slow waves in the ERG; for example, the receptor cells and horizontal cells hyperpolarize with illumination and the bipolar cells can give maintained responses in either direction (Chapter 2).

MEANING OF SIGNALING FROM NEURON TO GLIA

In general terms, glial cells indicate the level of impulse traffic in their environment. For example, the depolarization of a glial cell in the optic nerve is graded according to the number of axons and the frequency at which they carry nerve impulses. Impulse numbers are converted into potassium concentrations in the clefts, and these, in turn, into changes in glial cell membrane potentials.

A nonspecific system of communication

In many respects signaling between neurons and glia is radically different from specific synaptic activity. Synaptic action is confined to small specialized regions on the neuron cell body and the dendrites and may be excitatory or inhibitory. In contrast, signaling by the potassium mechanism is not confined to special structures, such as synapses, but occurs along the entire length of a neuron wherever an impulse liberates an excitatory or an inhibitory transmitter. In this respect the NEURON–GLIA SIGNALING IS NONSYNAPTIC AND NONSPECIFIC, except that one particular glial cell is influenced preferentially by a discrete population of neurons that are in close proximity to it. One would therefore expect the physiological role of glial cells to be a generalized rather than a discriminating one. One might speculate that potassium-induced depolarization leads somehow to stimulation of enzymes in glial cells, causing them to produce a product the nerves need for their activity or for recovery afterward. Other mechanisms of interaction that have been proposed involve transmitter-mediated actions of neurons on glia, an interaction resulting in release by glia of transmitter that acts back on its own membrane.[58]

The idea that external potassium affects the metabolic activity of glial cells has found support in experiments in which the overall metabolism was measured by studying changes in the fluorescence of a pyridine nucleotide (NADH).[76] The experiments were made on a pure glial cell preparation obtained from optic nerves of *Necturus* in which the axons had been previously cut and allowed to degenerate. Concentrations of cyanide, which decrease oxidative metabolism, produced marked reversible changes in NADH fluorescence. Interestingly, increasing the external potassium concentrations to 15 mM also increased the uptake and metabolism of [^{14}C]glucose by glial cells.[77] Similarly, in leech ganglia, repetitive firing by neurons or small increases in external potassium

Effects of increased potassium on glial metabolism and transmitter release

[76]Orkand, P. M., Bracho, H. and Orkand, R. K. 1973. *Brain Res. 55*: 467–471.
[77]Salem, R. D., Hammerschlag, R., Bracho, H. and Orkand, R. K. 1975. *Brain Res. 86*: 499–503.

(4 mM) have been shown to increase the synthesis of glycogen by glial cells.[78] Clearly it is of interest to determine what significance these effects may have for neuron–glia interactions.

One specific possibility is that increased concentrations of potassium may lead to secretion of materials from satellite cells. Already mentioned is the fact that GABA, which is stored in glial cells, can be released by raised concentrations of potassium (20 mM or more). Glial cells may, therefore, play a role in regulating the concentrations of transmitter in the intercellular spaces—both by taking it up and by secreting it again in the presence of increased potassium produced by signaling.[39,40] Equally possible is the secretion of other, as yet undetermined, materials under the influence of altered potassium concentrations in the clefts.

It has also been suggested that the current flow produced by glial cells may affect the signaling of neighboring neurons. This does not seem a likely possibility because clefts have been shown to be effective barriers to current spread between cells.

Another idea is a role for glia as a reservoir of electrolytes or as a device for rapid uptake of potassium from the clefts to regulate and preserve the constancy of the environment. Clearly, if glial cells are depolarized nonuniformly, current flow into and out of the cells will result in the movement of potassium ions. It is, however, not simple to estimate quantitatively how much potassium would move or how such movements would alter extracellular potassium concentrations. A series of experiments on the retina of the bee provide evidence for potassium uptake by glia following activation of photoreceptors and for subsequent breakdown of glycogen by the glial cells.[79]

Consequences of potassium accumulation on synaptic and integrative activity of neurons

A single brief flash of light into the eye can depolarize glial cells by about 4 mV (Figure 13). This potential corresponds to an average increase of about 0.5 mM of potassium in the clefts around these cells. It seems reasonable to assume that similar fluctuations also occur around terminals near synapses. Although such small changes have a negligible effect on nerve conduction, the consequences on synapses may be important. The accumulation of potassium becomes increasingly significant with synchronous activity in groups of axons. Thus, under rather artificial conditions when axons are driven electrically by stimuli at relatively high frequencies, the potassium concentration in the clefts may rise from the normal 2–3 mM to 20 mM. In fact, with prolonged synchronous activity, conduction can block itself through potassium accumulation.[69]

Fluctuations in potassium concentrations might produce effects at two sites. The first is at presynaptic nerve terminals where depolarization acts to modify the release of transmitter. In the squid stellate ganglion, a depolarization of 2 mV depresses the release of transmitter and reduces the effect of a nerve impulse by about 15 percent.[80] Repetitive stimulation can produce similar effects by increasing the potassium

[78]Pentreath, V. M. and Kai-Kai, M. A. 1982. *Nature* 295: 59–61.
[79]Coles, J. A. and Tsacopoulos, M. 1981. *J. Exp. Biol.* 95: 75–92.
[80]Katz, B. and Miledi, R. 1967a. *J. Physiol.* 192: 407–436.

concentration in the synaptic region.[81] At rat neuromuscular junctions, raising K_0 by 1 mM increases the spontaneous release of quanta of transmitter by about 25 percent.[82] The effects of changes in K_0 on transmitter release have also been studied extensively in the vertebrate central nervous system.[83] At the second site of its action—on the postsynaptic neuron—potassium accumulation brings the cell closer to the firing level by depolarizing it. Under certain experimental conditions at frog neuromuscular synapses, Katz and Miledi have shown that potassium liberated from presynaptic terminals can give rise to an end plate potential.[84]

In view of these considerations, one might think that increments of potassium concentration by even 2 to 4 mM, which is a physiological range, would alter signaling and integrative activity in neurons. It is, therefore, interesting that the membrane potential of nerve cells, in contrast to that of glial cells, is relatively insensitive to changes in the potassium concentration in its environment. For example, in neurons of the leech, a fivefold increase in K_0 (from 4 to 20 mM) results in a depolarization of only 5 mV, whereas the glial membrane potential changes by 25 mV.[85] The neuron's membrane potential, therefore, is at least in part protected from fluctuations, presumably because it has a significant resting permeability to other ions (chiefly sodium) besides potassium, in contrast to glial cells in which potassium practically determines the full membrane potential. Nevertheless, even in axons, appreciable changes occur in conduction velocity and in excitability as a result of potassium accumulation following trains of impulses.[86,87]

In spite of the relative insensitivity of neurons, the effects of potassium increments on synaptic transmission may be considerable, especially since sensitivity to potassium may increase after prolonged activity owing to calcium-activated increases in potassium permeability (Chapters 9 and 18). Moreover, one would expect that in the central nervous system, with its continued interactions between large numbers of neurons, even a very small change in synaptic efficiency might have profound consequences.

SUGGESTED READING

General reviews

Treherne, J. E. (ed.). 1981. *Glia–Neurone Interactions. J. Exp. Biol.*, Vol. 95.
Sears, T. A. (ed.). 1982. *Neuronal–Glial Cell Interrelationships.* Springer-Verlag, New York.
(These two volumes contains reviews and original articles)

[81]Erulkar, S. D. and Weight, F. F. 1977. *J. Physiol. 266*: 209–218.
[82]Cooke, J. D. and Quastel, D. M. J. 1973. *J. Physiol. 228*: 435–458.
[83]Syková, E. 1981. *J. Exp. Biol. 95*: 93–109.
[84]Katz, B. and Miledi, R. 1982. *Proc. R. Soc. Lond. B 216*: 497–507.
[85]Nicholls, J. G. and Kuffler, S. W. 1964. *J. Neurophysiol. 27*: 645–671.
[86]Kocsis, J. D., Malenka, R. C. and Waxman, S. 1983. *J. Physiol. 334*: 225–244.
[87]Grossman, Y., Parnas, I. and Spira, M. E. 1979b. *J. Physiol. 295*: 307–322.

Bray, G. M., Rasminsky, M. and Aguayo, A. J. 1981. Interactions between axons and their sheath cells. *Annu. Rev. Neurosci. 4*: 127–162.

Kuffler, S. W. 1967. Neuroglial cells: Physiological properties and a potassium mediated effect of neuronal activity on the glial membrane potential. *Proc. R. Soc. Lond. B 168*: 1–21.

Kuffler, S. W. and Nicholls, J. G. 1966. The physiology of neuroglial cells. *Ergeb. Physiol. 57*: 1–90.

Morell, P. and Norton, W. T. 1980. Myelin. *Sci. Am. 242*: 88–118.

Orkand, R. K. 1982. Signalling between neuronal and glial cells. In T. A. Sears (ed.). *Neuronal–Glial Cell Interrelationships*. Springer-Verlag, New York, pp. 147–157.

Rakič, P. 1982. The role of neuronal–glial cell interaction during brain development. In T. A. Sears (ed.). *Neuronal–Glial Cell Interrelationships*. Springer-Verlag, New York, pp. 25–38.

Schachner, M., Sommer, I., Lagenaur, C. and Schnitzer, J. 1982. Developmental expression of antigenic markers in glial subclasses. In T. A. Sears (ed.). *Neuronal–Glial Cell Interrelationships*. Springer-Verlag, New York, pp. 321–336.

Somjen, G. G. 1979. Extracellular potassium in the mammalian central nervous system. *Annu. Rev. Physiol. 41*: 159–177.

Original papers

Baylor, D. A. and Nicholls, J. G. 1969. Changes in extracellular potassium concentration produced by neuronal activity in the central nervous system of the leech. *J. Physiol. 203*: 555–569.

Bowery, N. G., Brown, D. A., White, R. D. and Yamini, G. 1979. (^{3}H)-γ-Aminobutyric acid uptake into neuroglial cells of rat superior cervical sympathetic ganglia. *J. Physiol. 293*: 51–74.

Cohen, M. W. 1970. The contribution by glial cells to surface recordings from the optic nerve of an amphibian. *J. Physiol. 210*: 565–580.

Kuffler, S. W., Nicholls, J. G. and Orkand, R. K. 1966. Physiological properties of glial cells in the central nervous system of amphibia. *J. Neurophysiol. 36*: 855–868.

Kuffler, S. W. and Potter, D. D. 1964. Glia in the leech central nervous system: Physiological properties and neuron–glia relationship. *J. Neurophysiol. 27*: 290–320.

Levitt, P. and Rakič, P. 1980. Immunoperoxidase localization of glial fibrillary acidic protein in radial glial cells and astrocytes of the developing rhesus monkey brain. *J. Comp. Neural. 193*: 815–840.

Mugnaini, E. 1982. Membrane specializations in neuroglial cells and at neuronglial contacts. In T. A. Sears (ed.). *Neuronal–Glial Cell Interrelationships*. Springer-Verlag, New York, pp. 39–56.

Orkand, R. K., Nicholls, J. G. and Kuffler, S. W. 1966. Effect of nerve impulses on the membrane potential of glial cells in the central nervous system of amphibia. *J. Neurophysiol. 29*: 788–806.

Picker, S., Pieper, C. F. and Goldring, S. 1981. Glial membrane potentials and their relationship to $[K^+]_o$ in man and guinea pig. A comparative study of intracellularly marked normal, reactive, and neoplastic glia. *J. Neurosurg. 55*: 347–363.

Takato, M. and Goldring, S. 1979. Intracellular marking with Lucifer Yellow CH and horseradish peroxidase of cells electrophysiologically characterized as glia in the cerebral cortex of the cat. *J. Comp. Neurol. 186*: 173–188.

REGULATION OF THE COMPOSITION OF THE FLUID SPACES IN THE BRAIN

CHAPTER FOURTEEN

A homeostatic system controls the fluid environment of nerve and glial cells in the brain and keeps its chemical composition relatively constant, compared with that of the blood plasma. This is necessary for the proper integrative activity of the neural signaling system.

Within the brain are three different types of fluid: (1) the blood supplied to the brain through a dense network of capillaries; (2) the cerebrospinal fluid (CSF) that surrounds the bulk of the nervous system and is also contained in the internal cavities (the ventricles); and (3) the fluid in the intercellular clefts. The constitution of the blood plasma and of the CSF is known from direct chemical analysis. This chapter deals principally with the determination and regulation of the composition of the fluid in the intercellular spaces that are generally no more than 20 nm wide. The contents of these spaces provide the immediate environment of nerve and glial cells in the brain.

The intercellular clefts are open for diffusion of ions and certain large molecules. They constitute the main channels through which materials are distributed to reach neurons and glial cells. The membrane potential of neuroglial cells within the brain can be used as a quantitative bioassay for determining concentrations of potassium ions within the intercellular clefts and also for estimating the rates of diffusion of various ions and small molecules. By the use of electron-dense markers and electron microscopy, the distribution of larger molecules has been traced through the cleft system of the brain.

The homeostatic control that keeps the environment of neural elements relatively constant, and different from blood plasma, is provided by two cell systems: (1) the endothelial cells of capillaries and (2) the epithelial cells surrounding capillaries of the choroid plexus. These epithelial cells secrete cerebrospinal fluid and act as a barrier for ions and various molecules. In addition, their regulating action is brought about by an active transport system.

Once substances have found their way into the CSF, they are free to diffuse into the tissues of the brain. Although the CSF dominates the composition of the intercellular spaces, there exist different regional chemical microclimates within the brain.

The problem

Nerve cells in the brain function in an environment that differs significantly from the fluid in which peripheral organs are bathed. It has long been known that the bulk of the brain and spinal cord is surrounded by a specially secreted, clear fluid called the CEREBROSPINAL FLUID (CSF) (Figure 1). This solution is almost devoid of protein containing only about 1/200 of the amount present in blood plasma. Chemical substances such as metabolites move relatively freely from the alimentary canal into the bloodstream, but not into the CSF. As a result, the levels in the blood plasma of sugar, amino acids, or fatty acids fluctuate over a wide range while their concentrations in the CSF remain relatively stable. The same is true for hormones, antibodies, certain electrolytes, and a variety of drugs. Injected directly into the bloodstream, they act rapidly on peripheral tissues such as the muscles, heart, or glands, but they have little or no effect on the central nervous system. When administered by way of the CSF, however, the same substances exert a prompt and strong action. The conclusion is that the substances injected into the bloodstream do not reach the CSF and the brain with sufficient rapidity and in effective concentrations.

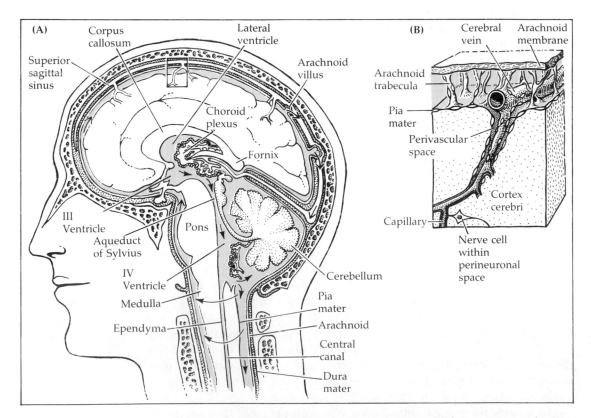

1 DISTRIBUTION OF CEREBROSPINAL FLUID (CSF) and its relation to larger blood vessels and to structures surrounding the brain (A). All spaces containing CSF communicate with each other. CSF is drained into the venous system through the arachnoid villi (B).

363
REGULATION
OF THE
COMPOSITION
OF THE
FLUID SPACES
IN THE BRAIN

It is hard to see how neurons could perform their essential signaling precisely unless the chemical environment were kept relatively constant. Constancy would seem particularly important in a system where the activity of many cells is integrated and where small variations may change the balance of delicately poised excitatory and inhibitory influences. In contrast, at myoneural junctions and in autonomic ganglia, where information is generally transmitted with a good safety margin, fluctuations in the environment are of relatively small consequence. Thus, homeostasis does not mean absolute constancy, but rather a setting of limits to fluctuations that are tolerable and cause no disruption of function. It therefore seems reasonable that the brain should have a special and more closely adjusted homeostatic control than the rest of the body.

The way in which the brain keeps its environment constant is frequently discussed in terms of a BLOOD-BRAIN BARRIER. The reasoning is generally based on the following considerations. The sustenance of the brain is derived from its rich blood supply. The vertebrate brain must discriminate against the entry of some materials from the blood, and therefore it has developed certain barriers. This term "barrier," however, which represents structural impediments to the access of substances, should not obscure the fact that other mechanisms also regulate the fluid environment.

If materials enter, however slowly, without participation and control by an active cell transport system, the concentration of fluids in the brain will eventually equilibrate with that in the plasma. The fallacy of an exclusively mechanical barrier concept is readily illustrated by the proposal of a blood-urine barrier—which cannot explain the mechanism by which glucose is normally kept out of the bladder, by reuptake from the kidney tubules. It therefore follows that linked with structural barriers there must exist active processes that regulate the environment of nerve cells in the brain.

Most of the general conclusions mentioned so far about access of materials to the brain have been derived from experiments determining differences between the CSF and blood plasma, two fluids that are readily available in relatively large quantities for chemical analysis. Frequently the assumption is made implicitly that the CSF actually provides the "true" environment of nerve cells, because of the convincing evidence that there are fewer obstacles to an exchange between the CSF and neurons than between the blood and the CSF. Yet, the immediate environment of neurons is neither the blood plasma nor the CSF, which is contained in the large spaces, such as the ventricles from which the samples are taken. The crucial sites providing the fluid environment of neurons are the intercellular spaces between neurons and glial cells. These are, as discussed in Chapter 13, narrow clefts several tens of nanometers in width. They are too small to be accessible for direct chemical analysis, and studies of their fluid composition have depended on a prior determination of the total extracellular spaces in the brain, a task beset by difficulties of interpretation.

The main problem discussed in this chapter involves the ways in which one can determine in the brain the composition of the extracellular spaces that surround nerve cells, for comparison with the composition of the blood plasma and the CSF. To set the stage, it is necessary to discuss first the gross distribution of CSF and blood in the brain and then the various pathways from capillaries or CSF to neurons.

Distribution of CSF
and blood

The brain and spinal cord are shock-mounted in a jacket of CSF. Between the mass of the brain and the skull are two layers of connective tissue, the pia-arachnoid and the dura mater, which is attached to the bone. The pia-arachnoid forms a trabecular meshwork providing fluid spaces on the surface of the brain (called subarachnoid spaces). A schematic illustration is shown in Figure 1. There is contact but no bulk fluid between the dura and the arachnoid. The CSF around the brain surface communicates with the CSF within the cavities of the brain, the ventricles, and the spinal canal. In man, the CSF is continuously replaced at an estimated rate of six to seven times a day. It drains into the sinus venosus through the arachnoid villi, which constitute a valve system. There is no lymphatic outflow from the brain.

The brain is supplied by many large blood vessels that run in the CSF-filled subarachnoid spaces before diving from the surface into the parenchyma, the bulk of the brain tissue, taking with them their surrounding connective tissue covering. Bulk CSF around the blood vessels thereby penetrates the brain at multiple places. As the blood vessels become progressively smaller, the space around them narrows until finally it is only a fraction of a micrometer around the capillaries (Figure 1B).

In considerations of the chemical environment of nerve cells, one should bear in mind that the vascular bed in the brain is very dense and permeates the entire structure, so that each nerve cell is probably no farther than 40 to 50 μm from a capillary.[1]

INTERCELLULAR CLEFTS AS CHANNELS
FOR DIFFUSION IN THE BRAIN

Nerve cells and neuroglia are so closely packed everywhere in the brain that the spaces between them are reduced to about 20 nm. For some time, around the early 1960s, it was commonly thought that such narrow spaces were not functional pathways and could not play a part in the distribution of materials. It was even postulated that neuroglial cells actually constitute the extracellular spaces through which materials must pass to reach neurons, much as originally suggested by Golgi (Chapter 13).

Several questions about intercellular clefts must be considered: (1) Do the intercellular spaces, if open, allow the diffusion of ions, such as sodium, or of molecules of various dimensions, such as sucrose or proteins? (2) What is the rate of movement of materials through the

[1]Scharrer, E. 1944. *Q. Rev. Biol. 19*: 308–318.

365
REGULATION
OF THE
COMPOSITION
OF THE
FLUID SPACES
IN THE BRAIN

nervous system? (3) What is the evidence that certain substances do not move through glial cells? If they do, what is the relative importance of the alternative pathways through the intercellular spaces and through the cytoplasm?

Theoretical considerations of diffusion through narrow channels

From the size of a molecule and the size of a fluid-filled channel, one can calculate rates of diffusion from simple diffusion equations. It is also possible to estimate the limiting size of the channels, if one assumes them to be tubes. The relative coefficient of diffusion within a tube, D'/D is given by

$$D'/D = \frac{(1 - a/r)^2}{1 + 2.4a/r}$$

where D is the coefficient of diffusion, r is the radius of the tube, and a is the size of the particle.[2] As an example, if one takes a value of 0.44 nm for the radius of sucrose and 15 nm as the diameter of a tube, the relative coefficient of diffusion turns out to be only 18 percent less than that in free solution. The size becomes significantly limiting only for molecules having a radius of more than about 15 percent of the tube width. Hence, the actual dimensions of the clefts would not necessarily prevent movement of ions or of small molecules. These considerations would not apply if the intercellular clefts were (1) filled with material that slowed or prevented diffusion or (2) closed off at certain points (see later).

The distribution of a number of substances of different molecular weights through the brain tissues can be detected by electron microscopy. One tracer molecule frequently used in the past is the protein ferritin, which has a diameter of about 10 nm and a molecular weight of 900,000. Another is the enzyme horseradish peroxidase, with a diameter of about 4 nm and a molecular weight of 43,000. More recently, microperoxidase (molecular weight near 2000) has become available. With such an enzyme as a marker, the product deposited as the result of the reaction of enzyme can be detected. The substances are generally injected into the CSF of animals, and their distribution is determined after fixation.

Diffusion of electron-dense molecules into the clefts

The results clearly demonstrate that horseradish peroxidase and even ferritin can enter intercellular clefts from the CSF if given adequate time for diffusion.[3,4] The electron-dense molecules, deposited by the peroxidase reaction, are then lined up in the clefts and fill the extracellular spaces. Tracer materials may also be taken up into the interior of cells, usually enclosed in pinocytotic vesicles. Further, one can determine the location of sites where clefts are occluded or reduced in diameter because tracers stop spreading when they reach these regions. An example is provided in Figure 2A, which shows that when horseradish peroxidase is injected into the ventricles, diffusion takes place

[2]Pappenheimer, J. R. 1953. *Physiol. Rev. 33*: 387–423.
[3]Brightman, M. W. 1965. *Am. J. Anat. 117*: 193–219.
[4]Brightman, M. W. and Reese, T. S. 1969. *J. Cell Biol. 40*: 648–677.

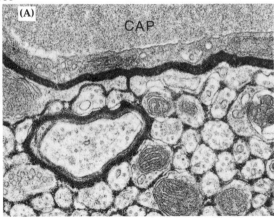
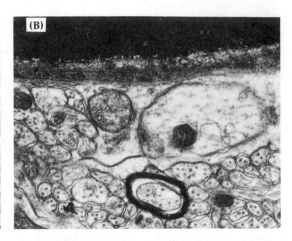

2 PATHWAYS FOR DIFFUSION IN THE BRAIN. (A) Demonstration in the mouse that the enzyme microperoxidase diffuses freely from the cerebrospinal fluid into the intercellular spaces of the brain which are filled with the dark reaction product. No enzyme is seen in the capillary (CAP). (B) The enzyme, injected into the circulation, fills the capillary but is prevented by the capillary endothelium from escaping into the intercellular spaces. (From Brightman, Reese, and Feder, 1970.)

throughout the clefts. The opposite result is shown in Figure 2B, in which the enzyme was injected into the circulation. The capillary is filled with the enzyme, but no significant deposits appear in the intercellular spaces. As explained later (Figure 7), the junctions between the capillary endothelial cells provide the barrier.

Although the electron microscopic method provides clear evidence that intercellular spaces are open for diffusion, it is difficult for technical reasons to establish accurately the rates of movement of substances by means of electron microscopy.

Rates of movement through intercellular clefts

Physiological measurements have established the rates at which sucrose, sodium, and potassium move through the clefts in the central nervous system of the leech[5] and of the mud puppy (*Necturus*).[6] These studies provide a better time resolution than do electron-dense tracers; and since the effects are reversible, controls can be made in live preparations.

The technique for measuring diffusion times depends on the use of neurons as indicators of the sodium and potassium in their environment. If the sodium around a neuron is reduced or replaced by sucrose or choline, the action potential becomes smaller (Chapter 5). Increased potassium concentration, on the other hand, reduces the membrane potential. The experimental arrangement sketched in Figure 3 has been used in the connectives of the leech and in the optic nerve of *Necturus*. The preparation lies in a chamber consisting of three compartments

[5]Nicholls, J. G. and Kuffler, S. W. 1964. *J. Neurophysiol.* 27: 645–671.
[6]Kuffler, S. W., Nicholls, J. G. and Orkand, R. K. 1966. *J. Neurophysiol.* 29: 768–787.

367
REGULATION
OF THE
COMPOSITION
OF THE
FLUID SPACES
IN THE BRAIN

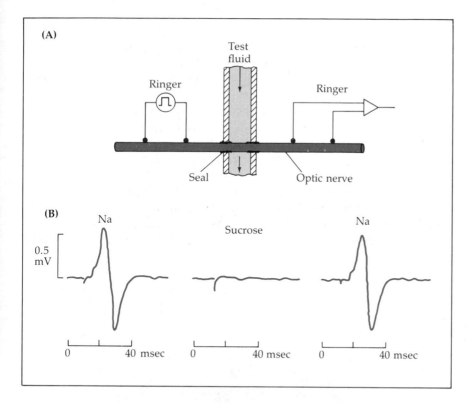

3 **RATE OF DIFFUSION THROUGH CLEFTS.** Movement of sodium and sucrose through the intercellular clefts of the optic nerve in *Necturus*. Sodium necessary for impulse conduction is replaced within 12 seconds by sucrose perfused through the middle compartment of a chamber shown in (A). As a result, conduction through the sucrose-filled region is blocked (middle record in B) and is restored to its original size after sodium is reintroduced (right record). (From Kuffler, Nicholls, and Orkand, 1966.)

sealed off from one another. The side compartments contain Ringer's fluid and are used for stimulating the nerve and for recording its action potential; the narrow central chamber can be perfused with various test solutions. Figure 3 shows that when the solution perfusing the central chamber is changed to sodium-free Ringer's solution that contains sucrose, conduction rapidly fails in the *Necturus* optic nerve. Within 12 seconds after sodium-free fluid reaches the central compartment, all the impulses are blocked. If sodium is introduced again into the perfusion fluid, the conducted potential is fully restored to its control size after 10 seconds (Figure 3). The results in the leech connectives are very similar when either sucrose or choline is used to replace sodium.

The above experiments can be interpreted as follows. When sodium is replaced by sucrose, the ion moves out of the extracellular spaces and sucrose takes its place. Block of conduction results because sodium, which is needed for the impulse mechanism, is lost from the immediate environment of the axons. Testing different concentrations of sodium shows that complete block occurs when 60 to 75 percent of the sodium in the central perfusion compartment has been replaced by sucrose. One can therefore conclude that within 12 seconds in sodium-free solution, at least 60 percent of the sodium in the optic nerve is exchanged for an equivalent amount of sucrose.

A more accurate measure of the rate of movement through the nervous system can be obtained by recording intracellularly from neurons in leech ganglia.[5] To measure the rate of penetration of potassium, for example, the relation between potassium concentration and the membrane potential of a neuron is first established (Figure 4B). From this curve, any membrane potential can be translated into an equivalent potassium concentration. Next, a high concentration of potassium is applied to the ganglion, and the cell becomes depolarized in a few seconds (Figure 4C). The equivalent potassium concentration around

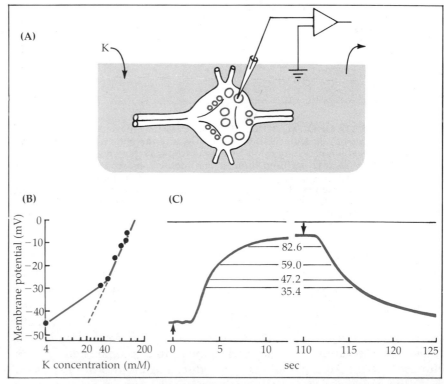

4   **RATE OF DIFFUSION THROUGH A GANGLION. (A)** Movement of potassium within a leech ganglion. An intracellular electrode measures the rate of depolarization of a neuron after introduction of a high potassium concentration into the bath, replacing sodium. **(B)** Membrane depolarization can be directly related to K_o. **(C)** The rate of movement of potassium through the ganglion can be obtained. The half-time for exchange of sodium and potassium is about 4 seconds. The numbers under the curve indicate the potassium concentration in mM during the depolarization. About 100 seconds were cut out of the records during the gap. (After Nicholls and Kuffler, 1964.)

369
REGULATION
OF THE
COMPOSITION
OF THE
FLUID SPACES
IN THE BRAIN

the neuron at any instant after the potassium has been applied to the bathing fluid can now be estimated. In this way, the influx and efflux of potassium have been plotted. By the same principle, the rates of movement of sodium, sucrose, and choline have been measured, using the size of the action potential as an indicator of concentration. These results have the advantage that a complete diffusion curve can be constructed. The points fall on a reasonably straight line when plotted on a logarithmic scale against time, as would be expected for a diffusion process. The half-time for the exchange of sodium with potassium through the leech ganglion is about 4 seconds; for the exchange of sodium and sucrose it is 10 seconds.

It is of interest to compare these times with the values predicted on the assumption that simple diffusion is occurring through narrow channels. Knowing the diffusion coefficients for sodium chloride and sucrose, the half-time for diffusion within an individual cleft in the nervous system can be calculated from the formula for linear diffusion.[7]

$$\frac{C}{C_0} = 1 - \frac{2}{\sqrt{\pi}} \int_0^y e^{-y^2} dy$$

In the equation, C_0 is the initial concentration and $y = x/2(Dt)^{1/2}$, x being the distance (in centimeters) and D the coefficient of diffusion (square centimeters per second). The distance (x), estimated by measuring the length of the clefts (mesaxons), is probably not more than 50 μm in a ganglion; in the connectives, it is probably not more than 30 μm. Calculations using these values for maximum distance yield values for the half-time for the exchange of sodium and sucrose of about 3.6 seconds in a leech ganglion packet and 1.3 seconds in a connective. In each case the half-times measured experimentally were considerably larger (by a factor of over two) than those predicted from the equation. This is not surprising, since the calculations at best yield only a rough approximation. They do indicate, however, that simple diffusion would probably be rapid enough to account for the rates of movement of sodium, potassium, and sucrose observed in the nervous system.

Do sodium and sucrose (or choline) exchange by taking a pathway through the glial cells, or around them through the clefts? The initial expectation is that movement through intercellular channels would leave the glial membrane potential unchanged, whereas passage through the cells would inevitably alter it. When glial membrane potentials are measured while isotonic sodium and sucrose move through the nervous system, the potentials remain practically unaffected by the exchange of sodium and sucrose. One can conclude that the electrolytes and sucrose diffuse through the clefts around the glial cells rather than through their cytoplasms.[5,6]

Exclusion of glia as a pathway for rapid diffusion

[7]Hitchcock, D. I. 1945. In R. Höber (ed.). *Physical Chemistry of Cells and Tissues.* Blakiston, Philadelphia.

Homeostatic
regulation of
hydrogen ions in
mammalian CSF

IONIC ENVIRONMENT OF NEURONS IN THE BRAIN

Preceding chapters show that the movement of ions and the charges they carry provide the mechanism for signals in nerve cells. It is therefore clear that changes in the ionic environment affect signaling and information processing within the brain. We argue, on teleological grounds, that a "constant" environment must be maintained for constancy of signaling, and that is the reason for a special homeostatic regulation in the brain.

A question that must be decided experimentally is the extent to which changes can be tolerated in the ionic environment of the nervous system without disrupting integrative activity. This problem has been extensively studied by Pappenheimer and his colleagues on the respiratory system of the goat.[8] The composition of the blood plasma and of the CSF was altered in unanesthetized goats whose ventricles were perfused with artificial CSF of known composition and whose blood plasma concentrations were also controlled by intravenous injection of a variety of substances. The respiration of the animal served as a sensitive index, or bioassay, of changes in pH in the vicinity of certain respiratory neurons. These cells are situated in the medulla and are very sensitive to small fluctuations in pH; their activity is immediately detected through changes in the rate of breathing of the animal.

In goats whose body fluid composition was artifically changed, small alterations in the content of H^+, HCO_3^-, or Cl^- within the ventricles had a large effect on respiration. In contrast, greater changes in the blood plasma had far less effect. These results show that activity in respiratory neurons is related to the fluctuations of concentrations within the CSF rather than in the blood. Even during prolonged modification of the plasma, such as maintained alkalosis or acidosis, a distinct environment in the brain was maintained. The simplest assumption is that a sharp discontinuity in the concentration profile of H^+ exists between the intercellular fluid around neurons and the blood in the nearby capillaries. It also follows that the differences are maintained by various active processes, such as pumps in the vicinity of the capillaries.

The conclusions derived from the experiments by Pappenheimer and his colleagues would not be affected if it turned out that respiratory neurons send processes (so far not noted) directly into the CSF within the ventricles. Neurons partially protruding into the spinal canal have already been described in reptiles.[9] In any event, there exist direct connections to neurons from the CSF via the intercellular clefts (see later). The mammalian experiments so far mentioned have been extended to amphibians and elasmobranchs (dogfish, shark).[10]

[8]Pappenheimer, J. R. 1967. *Harvey Lect. 61*: 71–94.

[9]Vigh, B., Vigh-Teichmann, I., Koritsánszky, S. and Aros, B. 1970. *Z. Zellforsch. 109*: 180–194.

[10]Cserr, H. and Rall, D. P. 1967. *Comp. Biochem. Physiol. 21*: 431–434.

371
REGULATION
OF THE
COMPOSITION
OF THE
FLUID SPACES
IN THE BRAIN

As explained in Chapter 13, the membrane potential of a glial cell provides a quantitative measure of the potassium concentration in its environment, since it behaves like a perfect potassium electrode. This finding has provided a technique for exploring the composition of fluids in spaces that are only several tens of nanometers wide and are therefore not accessible for conventional chemical analyses. Thus, the membrane potentials of glial cells can be used to register maintained changes in the fluid composition in the intercellular clefts. These have been produced by changing the concentration of potassium in the blood plasma of live mud puppies with intact circulations.

Homeostatic regulation of potassium in intercellular clefts

There exists in the brain of *Necturus,* as in mammals and fishes, a homeostatic control system.[11] Mud puppies were kept for long periods in tanks containing increased amounts of potassium; as a result the potassium concentrations in their body fluids increased. Figure 5 illustrates the changes occurring in the CSF when the potassium concentration in the blood had increased up to fivefold. The results demonstrate that the CSF maintained its potassium concentration within the relatively narrow range of 2 mM, in the face of plasma potassium increases of 7 to 8 mM.

Examination of the structural features of the optic nerve of the mud puppy (Figure 9, Chapter 13) shows clearly that the only separation between the blood within the capillaries and the CSF surrounding them is a single layer of endothelial cells. A sharp gradient must therefore exist in potassium concentration across capillary endothelial cells. With this background knowledge, we can consider how the potassium concentration changes in intercellular clefts around the neurons during long-term alterations of blood plasma and CSF levels. Do cleft concentrations in the optic nerve follow fluctuations in the CSF or in the plasma within the capillaries? Or do they perhaps maintain an in-between

[11]Cohen, M. W., Gerschenfeld, H. M. and Kuffler, S. W. 1968. *J. Physiol. 197*: 363–380.

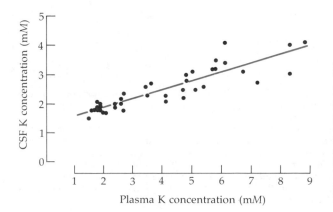

5

POTASSIUM CONCENTRATION in blood plasma and CSF. Determinations in mud puppies maintained in a high-potassium environment. Each point represents a measurement in a different animal. Potassium in the plasma can be chronically elevated from 1.5 mM to 9 mM, while in the CSF the increase is restricted to only 2 mM. This shows the existence of a homeostatic mechanism. (From Cohen, Gerschenfeld, and Kuffler, 1968.)

concentration of their own? Once more, the glial membrane potential, behaving like a potassium electrode, provides the answer.

If potassium in the intercellular clefts does faithfully follow fluctuations in the blood plasma, the membrane potential of glial cells should vary, as predicted by the Nernst relation (Chapter 13). The results plotted in Figure 6C show that although the glial membrane potential changes when potassium concentrations fluctuate in the blood, the slope is only 22 mV instead of the 59 mV predicted for a tenfold change in potassium concentration. On the other hand, the relation theoretically predicted by the Nernst equation is faithfully followed when potassium concentration in the CSF is plotted against the membrane potential, as in Figure 6B, where the slope is 59 mV.

These experiments show that in the mud puppy, the potassium concentration in the clefts closely reflects that in the CSF rather than that in the blood plasma. Such results support the longstanding view that the principal effect on neurons of fluctuations in ionic concentrations in the blood is indirect; the ionic composition of the CSF must be

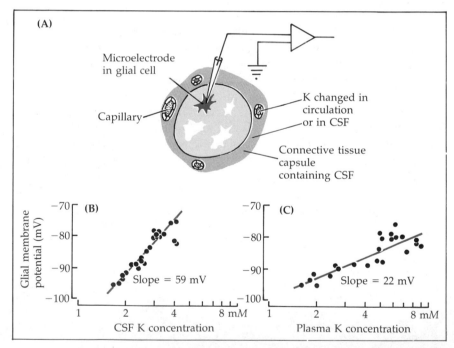

6   **GLIAL CELL AS A POTASSIUM ELECTRODE. Glial membrane potentials measured in mud puppies with intact circulations. Each point in (B) and (C) represents data from one animal maintained in a high-potassium environment. (B) The solid line gives the slope predicted by the Nernst equation for a tenfold change in potassium. The determinations show that potassium within the intercellular clefts follows concentration changes within the CSF. (C) In contrast, changes in the potassium in the blood plasma are not reflected by the changes in the potassium within the intercellular clefts. (After Cohen, Gerschenfeld, and Kuffler, 1968.)**

373
REGULATION
OF THE
COMPOSITION
OF THE
FLUID SPACES
IN THE BRAIN

7 EXCHANGE OF MATERIALS in the brain. Schematic presentation of cells involved in the exchange of materials between blood, CSF, and intercellular spaces. Molecules are free to diffuse through the endothelial cell layer lining capillaries in the choroid plexus. They are, however, restrained by circumferential junctions between the choroid epithelial cells which secrete CSF. There are no barriers between the bulk fluid of CSF and the various cell layers such as ependyma, glia, and neurons. The endothelial cells lining brain capillaries are occluded by a circumferential seal, as are the choroid epithelial cells. This prevents free diffusion of molecules out of the blood.

altered before the effect can be felt by the nerve cells. This shifts the emphasis onto the processes that control the production and composition of the CSF.

Within the brain certain blood vessels protrude into the ventricles to form a rich capillary network called the choroid plexus. Here the CSF is formed by an active-transport secretory mechanism.[12-14] Figure 7 shows the anatomical arrangement both in the brain parenchyma and in the choroid plexus. The choroid capillaries are lined with a fenestrated endothelium that is structurally different from endothelia in most other regions of the brain (exceptions are certain small areas such as the area postrema and the median eminence). These fenestrated endothelial cells apparently permit molecules, including peroxidase, to leak out. The next cell layer is made up of choroidal epithelial cells. They have the electron microscopic appearance of a secretory epithelium and resemble other active cells in which a regulated exchange occurs (such as those that line the renal tubules). Adjoining choroid epithelial

The choroid plexus and formation of CSF

[12]Ames, A., Higashi, K. and Nesbett, F. B. 1965. *J. Physiol. 181*: 506–515.
[13]Cserr, H. 1971. *Physiol. Rev. 51*: 273–311.
[14]Wright, E. M. 1978. *Rev. Physiol. Biochem. Exp. Pharmacol. 83*: 1–34.

cells are closed off by a circumferential girdle of tight junctions, so that the usual intercellular spaces are obliterated over a continuous beltlike area of contact (Figure 7). This layer of cells has the dual role of (1) preventing proteins from escaping and (2) regulating the newly formed CSF by active transport.

Sites other than the choroid plexus also contribute to the composition of CSF. The problem is quantitative because some leakage is bound to occur from various capillaries into their environment, either directly into the CSF or into the intercellular spaces. These in turn drain into the ventricular and subarachnoid spaces containing CSF (see later). A further contribution to the CSF comes from scattered leakage points, which permit blood plasma to enter the CSF directly, bypassing the barriers provided by the choroid epithelial cells. The nascent CSF secreted by choroid plexuses is therefore not identical in composition to the bulk of the CSF, although it dominates the makeup of the latter.

Sites of regulation of the ionic and chemical environment of neurons

In the mammalian brain the differences between CSF and blood plasma are brought about by the endothelium of the capillaries within the brain and by the choroid epithelium. In *Necturus*, physiological results suggest that the clefts communicate freely with the CSF. The conclusion that a barrier exists at the level of the capillaries is greatly strengthened by tracer studies, similar to those demonstrating that clefts are open for diffusion. In particular, the injection of horseradish peroxidase into the circulation of mice[4,15] and of *Necturus*[16] has shown that this enzyme, with a molecular weight of about 43,000, cannot leak out from brain capillaries because these cells are closed off by a continuous belt of tight junctions in the same way as the choroid epithelial cells (Figure 7). However, once materials have reached either the CSF or any of the intercellular spaces, they are free to diffuse throughout the bulk of the brain. Brain capillaries behave differently from those supplying skeletal or heart muscles, which do allow some proteins to escape.[17,18] Sharks and skates provide an interesting exception. In these animals the brain capillaries are leaky. The barrier occurs at the level of the glial end feet, which invest the vessels and which are linked by tight junctions.[19]

In conclusion, the main cellular sites of regulation of the fluid environment reside in the endothelial cells of the capillaries in the brain and in the choroid epithelial cells. Much of the physiological evidence presented above comes from lower vertebrates. There is good structural evidence that homeostasis around neurons and glia is brought about in a similar manner in the brains of the rat and the mouse.

[15]Reese, T. S. and Karnovsky, M. J. 1967. *J. Cell Biol. 34*: 207–217.
[16]Bodenheimer, T. S. and Brightman, M. W. 1968. *Am. J. Anat. 122*: 249–268.
[17]Karnovsky, M. J. 1967. *J. Cell Biol. 35*: 213–236.
[18]Clementi, F. and Palade, G. E. 1969. *J. Cell Biol. 41*: 33–58.
[19]Bundgaard, M. and Cserr, H. F. 1981. *Brain Res. 226*: 61–73.

375
REGULATION
OF THE
COMPOSITION
OF THE
FLUID SPACES
IN THE BRAIN

SUGGESTED READING

General reviews

Bradbury, M. 1979. *The Concept of a Blood-Brain Barrier.* John Wiley & Sons, Chichester.

Cserr, H. F. and Bundgaard, M. 1984. Blood-brain interfaces in vertebrates—a comparative approach. *Am. J. Physiol.* (in press).

Fleischhauer, K. 1972. Ependyma and subependymal layer. In G. H. Bourne (ed.). *The Structure and Function of Nervous Tissue.* Vol. IV. Academic, New York, pp. 1–46.

Kuffler, S. W. and Nicholls, J. G. 1966. The physiology of neuroglial cells. *Ergeb. Physiol. 57*: 1–90.

Leusen, I. 1972. Regulation of cerebrospinal fluid composition with reference to breathing. *Physiol. Rev. 52*: 1–56.

Wright, E. M. 1978. Transport processes in the formation of the cerebrospinal fluid. *Rev. Physiol. Biochem. Exp. Pharmacol. 83*: 1–34.

Original papers

Brightman, M. W., Klatzo, I., Olsson, Y. and Reese, T. S. 1970. The blood-brain barrier to proteins under normal and pathological conditions. *J. Neurol. Sci. 10*: 215–239.

Brightman, M. W. and Reese, T. S. 1969. Junctions between intimately apposed cell membranes in the vertebrate brain. *J. Cell Biol. 40*: 648–677.

Brightman, M. W., Reese, T. S. and Feder, N. 1970. Assessment with the electronmicroscope of the permeability to peroxidase of cerebral endothelium in mice and sharks. In E. H. Thaysen (ed.). *Capillary Permeability.* Alfred Benzon Symposium II. Munskgaard, Copenhagen.

Bundgaard, M. and Cserr, H. F. 1981. A glial blood-brain barrier in elasmobranchs. *Brain Res. 226*: 61–73.

Cohen, M. W., Gerschenfeld, H. M. and Kuffler, S. W. 1968. Ionic environment of neurons and glial cells in the brain of an amphibian. *J. Physiol. 97*: 363–380.

Nicholls, J. G. and Kuffler, S. W. 1964. Extracellular space as a pathway for exchange between blood and neurons in the central nervous system of the leech: Ionic composition of glial cells and neurons. *J. Neurophysiol. 27*: 645–671.

Reese, T. S. and Karnovsky, M. J. 1967. Fine structural localization of a blood-brain barrier to exogenous peroxidase. *J. Cell Biol. 34*: 207–217.

Van Deures, B. 1980. Structural aspects of brain barriers, with special reference to the permeability of the cerebral endothelium and choroidal epithelium. *Int. Rev. Cytol. 65*: 117–191.

Zeuthen, T. and Wright, E. M. 1981. Epithelial potassium transport: Tracer and electrophysiological studies in choroid plexus. *J. Membr. Biol. 60*: 105–128.

PART FOUR

HOW NERVE CELLS TRANSFORM INFORMATION

INTEGRATION BY INDIVIDUAL NEURONS

The central nervous system is faced incessantly with the task of making decisions on the basis of information about the outside world provided by sensory end organs situated in skin, muscles, internal organs, eyes, ears, and nose. In Chapter 15 sensory receptors in muscle are used to describe how electrical signals arise as the result of stretch or contraction of muscles and how these responses can be modified by commands descending from higher centers in the central nervous system.

At any one instant incoming signals from diverse sources in the periphery bombard the brain, some tending to reinforce and others to counteract each other. The mechanisms by which the various types of information are taken into account and assigned priorities is called integration. It is carried out by the nervous system as a whole, so that actions appropriate to the particular set of circumstances can be taken by the animal. It may not be too farfetched to picture individual cells within the central nervous system engaged in a similar task as they integrate the myriad of incoming signals. Each cell must perform an integrating function according to its particular position in the central nervous system. Examples already mentioned include ganglion cells in the retina and simple, complex, and special complex cells in the visual cortex. Each sifts and sorts out signals as visual processing progresses (Chapters 2 and 3).

To look at integration from a different angle, consider the consequences of an act such as an area of skin being tickled. This may give rise to a scratching movement by one hand while other groups of muscles perform different movements to make the limbs flex or extend to maintain steady posture. A smooth, coordinated movement has been initiated in the face of all the other external and internal stimuli to which the body is being subjected. This response may, however, be overridden at any instant by another stronger input arising at a different place in the skin or through some other sensory system. In our own brains an action of this type involves the confluence of literally millions of conflicting messages toward one decisive action. The question then is, How are such diverse influences handled by the brain?

Chapter 16 describes the manner in which individual cells integrate signals—combining excitation and inhibition into new instructions. We can go a long way toward understanding how the response of a single cell with known properties comes about. Out of the medley of incoming signals, every neuron makes its own synthesis and comes up with one piece of information that is typically its own. In other words, the cell does not simply hand the message on, but transforms it. To demonstrate these processes in action, three familiar cells are chosen: the crustacean muscle fiber, the Mauthner cell in the goldfish, and the mammalian spinal motoneuron. One can hardly imagine a more diverse set of excitable cells; yet, they all exemplify similar basic principles of summed synaptic action. Chapter 17 deals with problems concerning the production of coordinated movements involving the whole animal—in particular, locomotion and respiration. Breathing and panting, walking and running depend in mammals on the sequential activation of concerted groups of motoneurons. It has become apparent that groups of neurons within the central nervous system are responsible for initiating and programming such rhythmical excitatory and inhibitory drives to the motoneurons. The role of sensory inputs is to provide a constant flow of information, so that the movements can be modulated effectively. Chapter 18 carries the analysis on to a description of individual cells and their relation to behavior in simple invertebrates. There are many invertebrate nervous systems in which reflexes and integrative mechanisms can be studied at the cellular level. Each type of animal offers some advantages and disadvantages. Instead of providing a comprehensive review, we have again chosen to follow in some detail selected examples, the nervous systems of the medicinal leech and of the mollusk *Aplysia*. In these animals much is now known about the individual neurons and the circuitry, as well as about behavior and development.

HOW SENSORY SIGNALS ARISE AND THEIR CENTRIFUGAL CONTROL

CHAPTER FIFTEEN

In all sense organs external energy is translated into electrical signals. The receptor structures are specialized to respond to particular stimuli, such as sound, light, touch, or odors. Their properties determine the type of stimulus that can be handled and define the limits of what we can perceive. As examples of how sensory receptors function, two sense organs are compared—stretch receptors in the crayfish and muscle spindles in mammals. Both types of receptors measure the rate and extent of muscle stretch. They function in a basically similar manner and demonstrate the general principles that operate in sensory structures. The terminals of the receptor neurons that are deformed by stretch are inserted into specialized muscle fibers. The crustacean receptors provide the great technical advantage of allowing intracellular recording of events in their nerve endings.

The first electrical signal in nerve terminals in response to extension is a localized depolarization—the receptor potential—which is graded and reflects in size and time course the applied stimulus. When the receptor potential exceeds the threshold of the sensory axon, impulses propagate toward the central nervous system; their frequency is determined by the size of the receptor potential. The contractile and viscoelastic properties of muscle strands into which sensory terminals insert, combined with the properties of the sensory terminals, determine the character of the sensory discharges; some receptors register preferentially the rate of change and others the maintained phase of stretch.

In addition to responding to passive extension, muscle spindles and crustacean receptors are under the control of the central nervous system, which can increase or decrease their sensitivity to additional stretch. The muscle strands in which the receptor neurons are embedded receive motor innervation that makes them contract; thereby the sensory endings are stretched. Further, the crustacean stretch receptor neurons receive on their sensory terminals direct inhibitory innervation that can decrease or silence the sensory discharges in the face of continued stretch-excitation. The principle of centrifugal control, by which the central nervous system adjusts the information it receives, has been observed in a variety of sense organs, including the eye and the ear.

In addition to being subject to centrifugal control and sending information to the nervous system, stretch receptors have direct effects on motoneurons supplying the muscles in which they are embedded. This feedback system plays an essential role in regulation of muscle movement.

The sensory receptors are the gateways through which the outside world enters our minds. Right at the outset they set the stage for all the analysis of sensory events that is subsequently made by the central nervous system. They define the limits of sensitivity and determine the range of stimuli that can be perceived and acted upon. With rare exception, the nerve endings themselves are specialized in both their anatomical and physiological properties to respond preferentially to only one type of external energy. Nevertheless, the stimulus, whatever its form (light, sound waves, temperature changes, a chemical compound, or mechanical deformation), always gives rise to an electrical signal that acts as the symbol that can be used by the central nervous system. Furthermore, a great degree of amplification occurs at the receptor level, so that very small amounts of energy provide a trigger to release stored charges that appear in the form of electrical potentials. For example, a few quanta of light trapped by pigment molecules of a receptor in the retina give rise to a photochemical reaction that leads to an electrical potential change, to impulses, and even to a sensation (Chapter 2). Similarly, hair cells in the ear that are activated by sound waves respond to small displacement of a few tenths of a nanometer.

Each type of sensory receptor (like a photoelectric cell or a strain gauge) has a well-defined range of stimuli of a particular modality that it responds to. For example, sensory receptors in our ears respond to a restricted bandwidth of sound from about 20 to 20,000 Hz. Outside this range, sound waves produce no response. Our visual system cannot perceive infrared or ultraviolet light because light at those wavelengths does not activate receptors in the retina.

Animals with appropriately tuned transducers can take account of influences in the external world that we cannot perceive. Though the underlying sensory mechanisms are generally similar in different species, there are wide variations in the spectrum of sensitivity. Thus, dogs can hear higher frequencies and respond to sounds, such as a whistle, that are inaudible to us; bats emit and detect high-pitched sounds to locate with precision external objects in their vicinity and certain fishes use weak electrical pulses for a similar purpose. Some snakes can detect radiation in the infrared, whereas moths respond to ultraviolet light. Odors caused by a few molecules of a specific substance (pheromones) act as sex attractants for moths or ants. An involuntary extension of our own range of color perception is not uncommon after removal of a lens that has become cloudy (cataract formation). As a questionable bonus for such inconvenience, one becomes aware of the near-ultraviolet wavelengths that are normally filtered by the lens. After removal

381
HOW SENSORY
SIGNALS ARISE
AND THEIR
CENTRIFUGAL
CONTROL

of a cataract a person is able to read (although poorly) a chart that is illuminated in the 365-nm range and is not visible to the normal eye.

In some receptors, such as rods and cones that do not have long processes, the electrical signal generated by the transduction process spreads electrotonically from the sensory region to the synaptic region of the cell. Here the sensory information is passed on to the next cell in line. In other receptors such as cutaneous or muscle receptors, information must be sent much farther (for example, from the big toe to the spinal cord). In order to accomplish this the cell must perform a second transformation process: The receptor potentials give rise to trains of action potentials whose duration and frequency contain information about the duration and intensity of the original stimulus.

The way in which sensory receptors generate electrical signals in response to external stimuli can be studied conveniently and directly in stretch receptor neurons that register the length of muscles in crustaceans and in vertebrates. The stretch receptors in crayfish, first described by Alexandrowicz,[1] are particularly useful because their cell bodies lie in isolation—not within the central nervous system, but in the periphery, where they can be seen in live preparations (Figure 1). The cell bodies insert their dendrites into a fine muscle strand nearby and send an axon centrally to a segmental ganglion. Stretching the muscle strand deforms the dendrites where the transduction of energy takes place. A microelectrode inserted into the cell body of the neuron is close enough to the dendrites to obtain information about the transduction process and about the impulses initiated in the axon.

[1]Alexandrowicz, J. S. 1951. *Q. J. Microsc. Sci.* 92: 163–199.

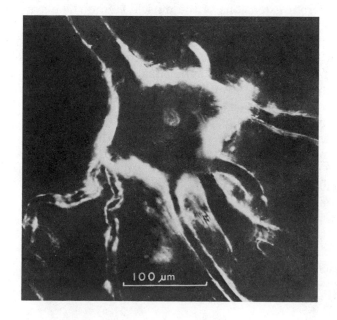

1

LIVING, UNSTAINED STRETCH RECEPTOR NEURON of the lobster viewed with dark-field illumination. Distal portions of six dendrites insert into the receptor muscle, which is not visible. (From Eyzaguirre and Kuffler, 1955.)

The crustacean stretch receptors are analogous to the vertebrate muscle spindles (described briefly in Chapter 4 in relation to the stretch reflex). The adequate stimulus for both types of receptors is mechanical deformation of the sensory nerve terminals produced by stretch of the muscle. These two types of mechanoreceptors serve to illustrate a number of general principles. It will become clear that the apparent simplicity of their function is deceptive. Impinging upon them is an elaborate control system, originating in the central nervous system, which precisely tunes their sensitivity to mechanical stimulation. These sensory cells provide one of the best examples of efferent or centrifugal control in which the central nervous system does not merely receive sensory information but reacts to it and modifies the performance of the receptors. Moreover, knowledge of the sensory end organs—their properties, control, and reflex connections—is a prerequisite for understanding integrative mechanisms.

Mechanoelectrical transduction in stretch receptors: The receptor potential

There are two types of crustacean stretch receptor organs with distinct structural and physiological characteristics. Each type of sensory nerve cell has a characteristic appearance, and its dendrites are embedded in a different type of receptor muscle. One neuron responds well to maintained stretch, whereas the other fires chiefly at the beginning of a stretch (see later). This decrease in response of a sensory nerve to a steady stimulus is called ADAPTATION.

The basic anatomical arrangement of stretch receptors in the crayfish or in the lobster is shown schematically in Figures 2 and 5. When the receptor muscle is stretched, the dendrites become deformed and their membrane potential is reduced.[2,3] This depolarization is the RECEPTOR POTENTIAL (also known as the "generator potential"), a graded, localized event, originating in the dendrites. Its amplitude and duration reflect, in electrical terms, the intensity and duration of the stretch, as shown in Figure 2B with different extensions of the muscle strand. Like synaptic and other localized potentials, the receptor potential depends on the passive electrical properties of the membrane and cannot spread far along the neuron without being attenuated. As the stretch is increased, more current flows, until eventually the depolarization reaches threshold and an all-or-nothing impulse is initiated, which propagates along the axon toward the central nervous system.

The conducted impulses start in a special region of the axon, called the initial segment or axon hillock, close to the cell body (Figure 2).[4] There the neuronal membrane has a lower threshold for initiation of a regenerative impulse than does the cell body, whose dendrites may not conduct impulses at all. Once initiated, the impulses propagate not only toward the central nervous system but also back into the cell body. It appears to be a common property of neurons that impulses are often initiated in a special region where the axon emerges from the soma.

[2]Eyzaguirre, C. and Kuffler, S. W. 1955. *J. Gen. Physiol. 39*: 87–119.
[3]Tao-Cheng, J.-H., Hirosawa, K. and Nakajima, Y. 1981. *J. Comp. Neurol. 200*: 1–21.
[4]Edwards, C. and Ottoson, D. 1958. *J. Physiol. 143*: 138–148.

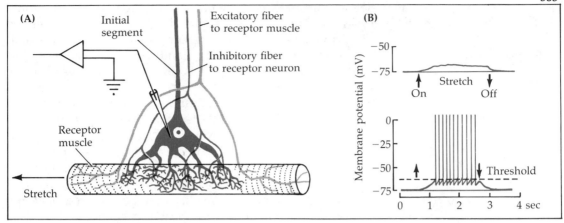

2   CRUSTACEAN STRETCH RECEPTOR. (A) Features of innervation; two additional inhibitory fibers are omitted. **(B)** A weak stretch for about 2 seconds causes a subthreshold receptor potential. With stronger stretch (lower record), a larger receptor potential sets up a series of impulses. (After Eyzaguirre and Kuffler, 1955.)

As in other conventional receptors, the intensity of the stimulus is expressed in the frequency of impulses. The relation between stimulus intensity and impulse frequency is established through the interaction of the maintained generator current in the dendrites and the conductance changes associated with the action potential (Figure 2B). At the end of each nerve impulse, the increased potassium conductance that occurs during the recovery phase drives the membrane potential in the hyperpolarizing direction, toward E_K. Although the increase in the potassium conductance is transient, the generator current is maintained by the stretch and depolarizes the membrane once more to the firing level. The stronger the current, the sooner the firing level is reached again, and the higher the impulse frequency.

As described above, not all sensory receptors generate action potentials. Instead the receptor potential spreads passively to the synaptic region, sometimes over a surprisingly long distance. For example, in a particular crustacean mechanoreceptor and in photoreceptors in the barnacle eyes, the receptor potential spreads passively over a distance of several millimeters. In such cells the membrane resistance is unusually high, thereby increasing the space constant for spread of passive depolarization.[5,6]

A second point of difference is that in some sensory cells the receptor potential is hyperpolarizing rather than depolarizing, for example, in the vertebrate retina (Chapter 2). Cochlear hair cells have both hyperpolarizing and depolarizing receptor potentials, depending upon which direction the hair is deflected.

[5]Roberts, A. and Bush, B. M. H. 1971. *J. Exp. Biol. 54*: 515–524.
[6]Hudspeth, A. J., Poo, M. M. and Stuart, A. E. 1977. *J. Physiol. 272*: 25–43.

The basic mechanism through which deformation leads to a permeability change is not known. Neither are the permeability changes in sensory endings that give rise to the receptor potential well understood. One would expect an increased sodium conductance to play a role, as in other excitatory depolarizing potentials. In accord with this idea, the receptor potential produced by stretch is reduced when sodium ions are replaced in the bathing fluid. It is interesting that, as for the end plate potential, tetrodotoxin does not affect the receptor potential.[7] The sodium channels are therefore different from those used for the production of the action potential; in addition, they are less specific for sodium. Thus, experiments made under voltage clamp have shown that large ions (such as Tris) and an amino acid (arginine) can pass through the channels. Ions other than sodium that may be involved have not yet been identified. As with other permeability changes, there is a characteristic reversal potential for the generator process that results from the combined equilibrium potentials of the different ions involved. In the stretch receptor, this appears to be about +15 mV.[8]

In their physiological behavior, mammalian muscle spindles (most frequently studied in cats) resemble, in many respects, the simple crustacean stretch receptors. As in crustaceans, the sensory nerve endings deformed by stretch are associated with specialized muscle fibers. Once again, there are two characteristic types of endings, one of which adapts rapidly and the other slowly. The general outline of the way in which muscle spindles respond was first worked out by B. H. C. Matthews in the early 1930s. For many years his experiments provided one of the best single attempts to describe comprehensively a sensory end organ and its control. Matthews was able to detect impulses in single nerve fibers that arose in individual spindles in frogs and cats. Recordings were made by an oscilloscope he designed for that purpose (no mean feat in 1930).[9–11] The existence of the receptor potential and its properties were first demonstrated in frog spindles by Katz.[12] As in the crustacean stretch receptor, the spindle receptor potential is not abolished by tetrodotoxin and depends on increases in permeability to Na, Ca, and K.[13] Since then, numerous other studies have revealed the detailed and elaborate specialization of the sensory apparatus and its control, particularly in snakes, rabbits, and cats.

An essential basis for the physiological experiments is analysis of the structure of the sensory and motor nerves and their terminations within the spindle. Much of this basic information is owed to the systematic tracing of fibers in the laboratories of Barker,[14] Boyd,[15] and

[7]Nakajima, S. and Onodera, K. 1969a. *J. Physiol. 200*: 161–185.

[8]Brown, H. M., Ottoson, D. and Rydqvist, B. 1978. *J. Physiol. 284*: 155–179.

[9]Matthews, B. H. C. 1931. *J. Physiol. 71*: 64–110.

[10]Matthews, B. H. C. 1931. *J. Physiol. 72*: 153–174.

[11]Matthews, B. H. C. 1933. *J. Physiol. 78*: 1–53.

[12]Katz, B. 1950. *J. Physiol. 111*: 261–282.

[13]Hunt, C. C., Wilkinson, R. S. and Fukami, Y. 1978. *J. Gen. Physiol. 71*: 683–698.

[14]Barker, D., Stacey, M. J. and Adal, M. N. 1970. *Philos. Trans. R. Soc. Lond. (Biol. Sci.) 258*: 315–346.

[15]Boyd, I. A. 1962. *Philos. Trans. R. Soc. Lond. (Biol. Sci.) 245*: 81–136.

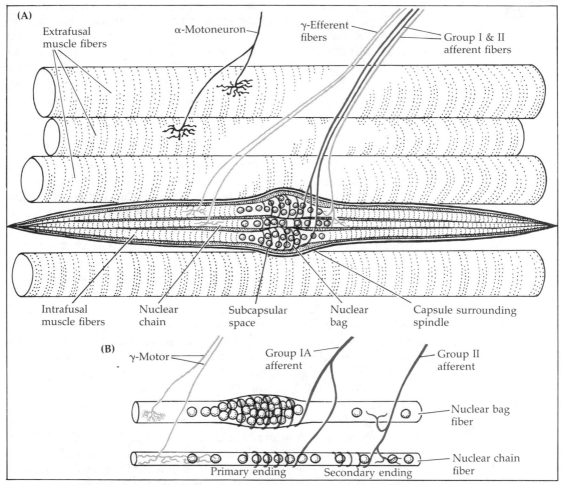

3 MAMMALIAN MUSCLE SPINDLE. (A) Scheme of mammalian muscle spindle innervation. The spindle is embedded in the bulk of the muscle, which is made up of large extrafusal muscle fibers and supplied by α-motoneurons. (B) Simplified diagram of intrafusal muscle fiber types and their innervation (see also Figure 10). (B from Matthews, 1964.)

Laporte.[16] Figure 3 illustrates schematically the sensory apparatus of spindles in leg muscles of the cat. The spindle consists essentially of eight to ten modified muscle fibers (called intrafusal fibers) running within a capsule. In the central, or equatorial region, there is in each fiber a large aggregation of nuclei. Their arrangement provides a basis for the classification of intrafusal bag or chain fibers, depending on whether the nuclei are grouped together in a swollen protuberance or are arranged linearly. Recently this classification has been refined further by subdividing the nuclear bag intrafusal muscle fibers into two

[16]Emonet-Dénand, F., Jami, L. and Laporte, Y. 1980. *Progr. Clin. Neurophysiol.* 8: 1– 11.

groups on the basis of structural and functional differences (see later).[17]

Two types of sensory neurons innervate each muscle spindle. The large nerve fibers (called group Ia afferents) have a diameter of 12 to 20 μm and conduct impulses at velocities up to 120 m/sec. These are the largest, most rapidly conducting nerve fibers in the mammal; their terminals are coiled around the central parts of both bag and chain fibers to form the primary endings. Smaller sensory nerves (called group II) are about 4 to 12 μm in diameter and conduct more slowly; their terminals are mainly in the less central region of the chain fibers, where they form the secondary endings. Such a classification according to fiber diameter and conduction velocity provides a convenient approximation for subdividing sensory fibers.[18]

When a rapid stretch is applied to a muscle and thereby to the spindles within it, receptor potentials and bursts of impulses arise in both types of sensory fibers. There is, however, a clear difference in the characteristics of the discharges in the two endings. The primary endings, connected to the larger group I axons, are sensitive mainly to the rate of change of stretch. The frequency of discharge is therefore maximal during the dynamic phase while stretch is increasing and subsides to a lower steady level while stretch is maintained. The secondary endings connected to the smaller group II fibers behave differently; they are relatively unaffected by the rate of stretch but are sensitive to the level of static tension.[17] This behavior is illustrated in Figure 4. Both types of nerve fibers contribute to the stretch reflex by exciting motoneurons in the spinal cord (see Chapter 16 and later).

Adaptation in sensory neurons

It is a common experience that the perception of a maintained stimulus tends to fade. One gradually becomes less aware of a constant pressure, an increased temperature on the skin, or contact with clothing or shoes. At least part of such sensory adaptation occurs through a decline in the frequency of firing in the primary receptors during a constant stimulus. This property is common to all receptors despite great differences in the degree and rate of adaptation. For example, certain mammalian mechanoreceptors, called Pacinian corpuscles, fire only a few impulses at the beginning of a maintained pressure but can follow oscillatory displacements at high frequencies.[19] Conversely, only little adaptation occurs in joint receptors that continually signal information to the central nervous system about the position of our limbs in space. Still other receptors start at a high rate but maintain a lower rate of impulses almost indefinitely.

A number of factors can contribute to the adaptation of sensory nerves. When bright light shines on photoreceptors, the amount of pigment diminishes as it becomes bleached. In addition, however, a neural component in visual adaptation is well established. In stretch

[17]Matthews, P. B. C. 1981. *J. Physiol. 320*: 11–30.

[18]Lloyd, D. P. C. and Chang, H. T. 1948. *J. Neurophysiol. 11*: 199–207.

[19]Gray, J. A. B. 1959. In J. Field (ed.). *Handbook of Physiology*, Vol. I. American Physiological Society, Washington, DC. 123–145.

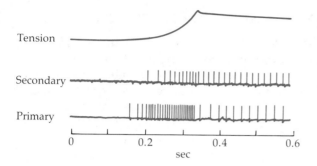

Tension

Secondary

Primary

0   0.2   0.4   0.6
sec

4

DIFFERENCES IN SPINDLE RESPONSES. Recordings from single primary (group I) and secondary (group II) sensory axons originating in a cat muscle spindle. The larger, fast-conducting primary ending greatly increases its response rate during development of tension; afterward, it slowly adapts. The smaller, more slowly conducting secondary ending is little affected by the change in tension but maintains its discharge well during continued tension. (From Jansen and Matthews, 1962.)

receptors, the viscoelastic properties of intrafusal muscle fibers or slippage in the attachment of the nerve terminals to the various receptor muscles may contribute to adaptation.[20] During maintained extension, such mechanical properties could allow the deformation of the terminals to decrease gradually.

At present it is often difficult to assess precisely the relative contribution of mechanical and electrical factors in an individual case. In the Pacinian corpuscles, however, adaptation has been shown to depend largely on mechanical factors. The nerve terminal is surrounded by a special structure made of a series of resilient lamellae. It has been observed that after removal of the outer lamellae, the receptor potential is better maintained and does not decline to this same extent during continued pressure.[21] For crustacean stretch receptors, a variety of processes occur.[22-24] The rapidly adapting receptor shows prompt adaptation of its firing rate to steady injected currents. In the slowly adapting receptor, in addition to mechanical factors, trains of impulses lead to an increase in internal sodium concentration and activation of the sodium pump. The net outward transport of positive charges by the pump produces a hyperpolarization, driving the membrane potential away from threshold (Chapter 8). With time, a steady stimulus to the slowly adapting neuron becomes less effective in initiating impulses because in a hyperpolarized cell a greater receptor potential is needed to initiate impulses. Decreases in firing rate with maintained stimulation can also occur through processes that involve changes in ionic permeability. For example, in the crustacean stretch receptor as in other neurons, calcium entry leads to a maintained increase in potassium conductance and this tends to cause repolarization. In addition, partial inactivation of the sodium conductance may play a role.

In conclusion, a number of electrical and mechanical factors seem to contribute to adaptation in different sense organs.

[20]Fukami, Y. and Hunt, C. C. 1977. *J. Neurophysiol. 40*: 1121–1131.
[21]Loewenstein, W. R. and Mendelson, M. 1965. *J. Physiol. 177*: 377–397.
[22]Nakajima, S. and Onodera, K. 1969b. *J. Physiol. 200*: 187–204.
[23]Nakajima, S. and Takahashi, K. 1966. *J. Physiol. 187*: 105–127.
[24]Sokolove, P. G. and Cooke, I. M. 1971. *J. Gen. Physiol. 57*: 125–163.

CENTRIFUGAL CONTROL OF SENSORY RECEPTORS

The central nervous system not only receives information from sensory receptors but also acts back upon them to modify their responses. The brain, therefore, has the built-in ability not only to edit or censor but also to adjust the flow of information that reaches it. This is usually called a feedback control and is executed through pathways leading centrifugally to a variety of peripheral sensory organs. Although centrifugal control has been most fully analyzed in mammalian muscle spindles, it is here discussed first in the crustacean stretch receptor, in which the elements of the system are most easily demonstrated.

Excitatory and inhibitory neural control of stretch receptors

The muscle strands (or receptor muscles) of the crustacean stretch receptors are innervated by excitatory and inhibitory motor axons that can be stimulated electrically. Excitation causes the ends of the muscles to contract more and thereby to stretch the centrally located dendrites of the sensory cells. The time course of the receptor potential initiated in this way thus mirrors the contraction of the muscle strand. The frequency of the sensory impulses therefore depends on the receptor potential produced in part by external stretch and in part by active contraction of the receptor muscles.[2,25] Figure 5A shows the principle in a simplified drawing; and, for comparison, Figure 5B presents a sketch of a mammalian spindle. Many essential features are similar in both systems. In the typical crustacean stretch receptor, the sensory neuron is embedded in a muscle strand that contracts when the motor nerves are stimulated. One sensory neuron inserts into a muscle strand that gives graded slow contractions with nerve stimulation, and when stretched the discharges adapt slowly; the dendrites of the other neuron insert into a muscle strand that gives twitchlike contractions, and when stretched the discharges adapt rapidly.

The effects of contractions of fast (twitch) bundles are shown in Figure 6B and C. The receptor potential faithfully reflects the transient rapid individual contractions. Figure 6D shows that sensory discharge is greatly accelerated by a contraction in the slow receptor muscle (note different time scales). There exists, therefore, a dual excitatory mechanism for initiating a sensory discharge: (1) passive stretch of the muscle, in which the receptor terminals are embedded, as discussed earlier; and (2) control originating in the central nervous system that makes the receptor muscle itself contract and that deforms the endings. Thus, with such a dual control, receptor neurons can be made to signal even when the muscles in which they are embedded are slack (Figure 6B, C), or their ongoing signals can be accelerated when they are already discharging (Figure 6D).

The control of receptor function is capable of even finer modulation than is indicated by the dual excitatory mechanism. The sensory neurons of both types of stretch receptors are innervated by efferent inhib-

[25]Kuffler, S. W. 1954. *J. Neurophysiol.* 17: 558–574.

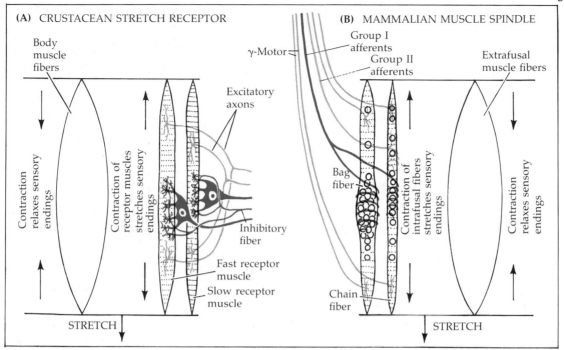

(A) CRUSTACEAN STRETCH RECEPTOR

Body muscle fibers

Excitatory axons

Contraction relaxes sensory endings

Contraction of receptor muscles stretches sensory endings

Inhibitory fiber

Fast receptor muscle

Slow receptor muscle

STRETCH

(B) MAMMALIAN MUSCLE SPINDLE

γ-Motor

Group I afferents

Group II afferents

Extrafusal muscle fibers

Bag fiber

Contraction of intrafusal fibers stretches sensory endings

Contraction relaxes sensory endings

Chain fiber

STRETCH

5 CENTRIFUGAL CONTROL of stretch receptors. (A) Excitatory and inhibitory innervation of the crustacean stretch receptor. The slow and fast receptor muscle strands can be made to contract by excitatory motor axons. A second excitatory mechanism is passive stretch of the receptor. The sensory receptor neurons are also innervated by inhibitory fibers that counteract excitation. (B) The scheme of excitation is basically similar in the mammalian spindle, which, however, lacks inhibitory innervation. For clarity, the main muscle mass is drawn separately in both preparations. Its contraction reduces stretch on sensory endings.

itory fibers.[26,27] Jansen and his colleagues[28] have shown that as many as three inhibitory axons form synapses on the dendrites and on cell bodies of these neurons. An example of inhibitory action is shown in Figure 7, in which only one inhibitory axon is indicated. In the face of a maintained stretch that makes the receptor fire at a rate of 11/sec, a train of inhibitory impulses (at 34/sec) keeps the membrane potential below the threshold level (Figure 7B). Once inhibitory stimulation is stopped, the sensory impulses are promptly resumed. In Figure 7C the individual contribution of each inhibitory impulse, this time at 21/sec and at higher amplification, can be resolved as a separate transient hyperpolarization. At a much greater frequency (150/sec), inhibition practically "clamps" the membrane potential at a steady level (Figure

[26]Kuffler, S. W. and Eyzaguirre, C. 1955. *J. Gen. Physiol. 39*: 155–184.

[27]Nakajima, Y. and Reese, T. S. 1983. *J. Comp. Neurol. 213*: 66–73.

[28]Jansen, J. K. S., Njå, A., Ormstad, K. and Walloe, L. 1971. *Acta Physiol. Scand. 81*: 273–285.

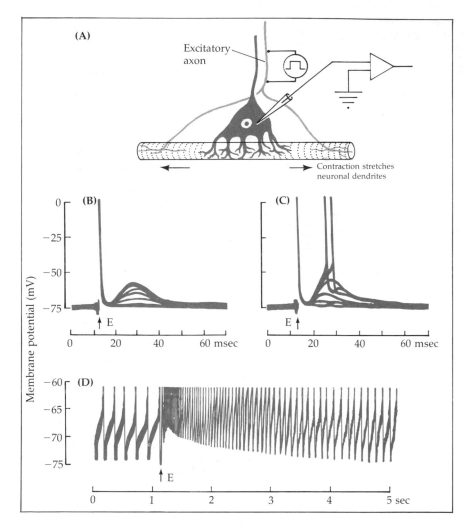

6 **CENTRIFUGAL EXCITATORY CONTROL OF SENSORY IMPULSES.**
(A) Excitatory stimulation of the motor nerve to the receptor muscle
while registering potential changes in terminals and cell body of the
receptor neuron. **(B)** Successive superimposed records during stimulation at 4/sec of a fast receptor muscle. The time course of the contraction
is reflected in the subthreshold receptor potentials. **(C)** With stimulation
(E) at 10/sec, the contractions build up until threshold is reached and
sensory impulses are set up. **(D)** A slow receptor muscle gives a maintained sensory discharge that is accelerated by two closely spaced motor
stimuli (arrow). The duration of the increased discharge rate reflects the
time course of contraction. Note different amplifications and time scales.
In (B) and (C) the large rapid deflections preceding the generator potentials are from the sensory axon which happened to be excited by the
excitatory stimulus. In (D) only the lower portions of impulses are seen.
(After Eyzaguirre and Kuffler, 1955.)

7D). Note that as far as the higher centers are concerned it makes no
difference whether the stimulus is 34/sec or 150/sec because both frequencies suppress all the sensory signals. However, a weaker inhibition

391
HOW SENSORY
SIGNALS ARISE
AND THEIR
CENTRIFUGAL
CONTROL

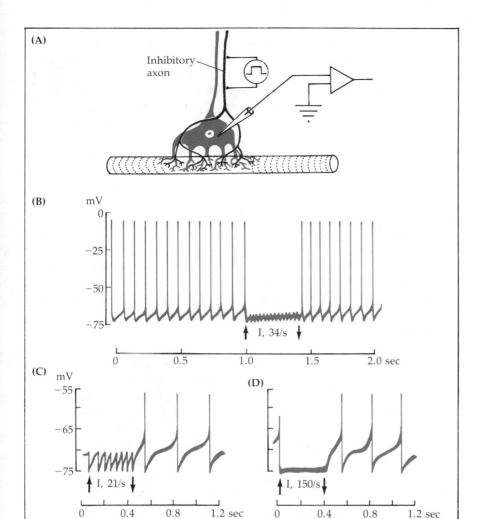

7 INHIBITORY CONTROL OF SENSORY DISCHARGES. (A) Stimulating electrodes on the inhibitory axon. (B) Maintained sensory signals are suppressed for the duration of inhibitory stimulation (I). (C) At higher amplification each inhibitory impulse at 21/sec is seen to hyperpolarize the membrane transiently. (D) High-frequency inhibitory impulses keep the membrane potential steady in the face of continued stretch. (From Kuffler and Eyzaguirre, 1955.)

can be more readily overruled or canceled by stronger excitation.

The stretch receptor provides an example of multiple excitatory and inhibitory actions characteristic of neurons within the central nervous system. The depolarizing generator action drives the membrane potential beyond its firing level while the inhibitory synaptic action tends to keep the membrane potential below threshold. The balance of these two competing influences determines whether the membrane potential of a receptor neuron is "set" above or below the critical threshold level, so that the cell keeps discharging or remains quiescent. The effectiveness of synaptic inhibitory modulation depends, as expected, on the frequency of the inhibitory impulses. But, in addition, the crustacean receptor in this example has a choice of three neurons with different pathways. The largest inhibitory axon exerts a more powerful inhibitory effect than the two smaller ones, which require a greater frequency of stimuli to produce an equivalent reduction in the rate of sensory discharges. They all act by the same ionic mechanism, producing conductance increases to chloride; they also use the same transmitter, γ-aminobutyric acid (GABA; Chapter 12).[29]

On closer scrutiny, then, the information that leaves the simple crustacean stretch receptors is highly controlled and adjustable over a wide range. The signals are determined by the excitatory action of graded amounts of external stretch, by contraction (shortening) of the receptor muscles, and by inhibitory synapses that apply a variable negative bias. These processes interact and exert their influence on the initial segment of the axon, where impulses start. The end result of impulses in the stretch receptors is to initiate a well-defined reflex by activating motoneurons within the central nervous system.[30]

Centrifugal control of muscle spindles

The motor control of mammalian muscle spindles is similar to that of crustacean stretch receptors. Stretch of the muscle in which they are embedded (Figures 3 and 4) gives rise to sensory discharges, as does activation of the intrafusal muscle fibers. In the original experiments by B. H. C. Matthews,[11] stimulation of the motor nerves innervating a muscle produced two effects. First, moderate stimulation caused contraction of the main mass of ordinary muscle fibers and a cessation of the afferent discharge arising in muscle spindles (Figure 11). This occurred because the spindles lie in parallel with the contracting fibers, so that their shortening reduces the tension on the sensory endings (Figure 5B). Stronger motor nerve stimulation added a burst of sensory impulses that he suspected to be caused by stimulation of small-diameter motor fibers, initiating contraction of the intrafusal muscle fibers.

In 1945, Leksell,[31] a neurosurgeon, studied the efferent nerve supply to skeletal muscles. This efferent system consists of a distinct group of small motor nerves (2 to 8 μm in diameter) and was originally described by Eccles and Sherrington.[32] Stimulation of these fibers, now called fusimotor or γ fibers, caused no increase in the tension recorded from the whole muscle but did increase the sensory activity originating in muscles. The role of fusimotor fibers was soon firmly established on limb muscles of the cat in a technically difficult and definitive series of experiments by Kuffler, Hunt, and Quilliam.[33] The procedure was to record in the dorsal root the activity in an individual afferent fiber coming from a spindle in an anesthetized cat while stimulating a fusimotor fiber in the ventral root going to the same spindle (Figure 8).

[29]Hagiwara, S., Kusano, K. and Saito, S. 1960. *J. Neurophysiol.* 23: 505–515.
[30]Fields, H. L., Evoy, W. H. and Kennedy, D. 1967. *J. Neurophysiol.* 30: 859–874.
[31]Leksell, L. 1945. *Acta Physiol. Scand.* 10 (Suppl. 31): 1–84.
[32]Eccles, J. C. and Sherrington, C. S. 1930. *Proc. R. Soc. Lond. B 106*: 326–357.
[33]Kuffler, S. W., Hunt, C. C. and Quilliam, J. P. 1951. *J. Neurophysiol.* 14: 29–54.

393
HOW SENSORY
SIGNALS ARISE
AND THEIR
CENTRIFUGAL
CONTROL

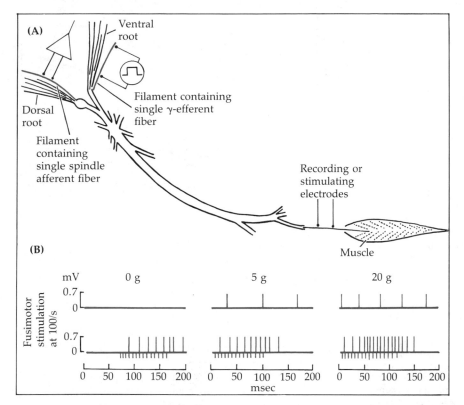

8 **CENTRIFUGAL CONTROL OF MUSCLE SPINDLE.** (A) Setup for recording from a single dorsal root axon arising in a muscle spindle of an
anesthetized cat; a single fusimotor (γ) axon in the ventral root innervating the same spindle is stimulated. (B) Upper records show sensory
discharges when the muscle is slack (0-g tension) or is lightly stretched
(5-g and 20-g tension). A brief train of 14 to 15 stimuli at 100/sec either
initiates sensory discharges (lower left) or accelerates them. Extrafusal
muscle fibers remain inactive. (After Kuffler, Hunt, and Quilliam, 1951.)

Such stimulation caused an increase in the frequency of the sensory
spindle discharge without visible muscle contraction or impulses in
skeletal muscle fibers. Trains of impulses in the fusimotor neurons
initiated a sensory discharge even in the absence of passive stretch in
a completely slack muscle (0 g tension in Figure 8B). If the initial load
(or stretch) caused a background discharge, this was accelerated during
a train of γ fiber impulses.

Further studies on the effects of intrafusal contraction in the cat have
established the existence of two main classes of efferent motor fibers to
mammalian spindles.[17,34] One produces an increase in the dynamic
discharge; the other affects mainly the static component (Figure 9). The
two kinds of fusimotor axons are called the DYNAMIC EFFERENT axon
(γ_d) and the STATIC EFFERENT axon (γ_s); they selectively innervate bag

[34]Jansen, J. K. S. and Matthews, P. B. C. 1962. *J. Physiol.* 161: 357–378.

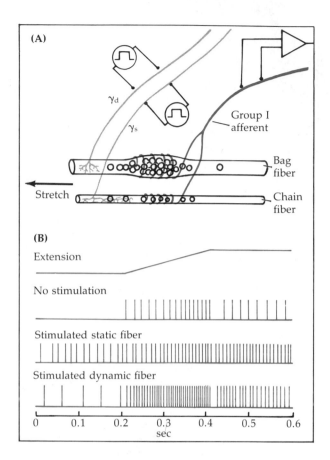

(A)

γ_d

γ_s

Group I
afferent

Bag
fiber

Stretch

Chain
fiber

(B)

Extension

No stimulation

Stimulated static fiber

Stimulated dynamic fiber

| 0 | 0.1 | 0.2 | 0.3 | 0.4 | 0.5 | 0.6 |

sec

9

CHARACTERISTIC ACTIONS OF STATIC AND DYNAMIC γ FIBERS on firing of primary ending. (A) Simplified scheme of pattern of innervation, modified from Matthews, 1964. (See also Figure 10.) (B) Different effects of stimulating two fusimotor fibers on the discharge arising in a primary (group I) sensory ending of a muscle spindle in the cat. During extension, the sensory fiber responds with an increased frequency. Stimulation of the static fusimotor fiber (γ_s), starting near the beginning of the second record, causes a steady increase in the rate of sensory discharges. Stimulation of the dynamic fusimotor fiber (γ_d) causes a larger increase in sensory discharge during stretch. (After Crowe and Matthews, 1964.)

and chain intrafusal muscle fibers according to the pattern shown in Figure 10. Moreover, contrary to earlier schemes, it is also clear that two distinct types of bag intrafusal muscle fibers exist: one called "bag₁" is responsible for the characteristic dynamic action. When bag₁ fibers contract, the sensitivity of Ia endings to the rate of stretch is increased. Contraction of the other (bag₂) fiber gives rise to the static component.

The types of bag fibers can be distinguished by their histological characteristics, by contractions they undergo and by the type of γ efferent nerve supply they receive. This has been shown by Boyd,[35] Barker, Laporte,[36] and their colleagues, who have made direct observation of the intrafusal fibers and of the effects of γ efferent stimulation in living spindles, dissected out of the animal and maintained in vitro. By using high resolution Nomarski microscopy, it has become possible to measure precisely the deformation of the primary and secondary endings as one γ efferent axon causes intrafusal contraction of a dy-

[35]Arbuthnott, E. R., Ballard, K. J., Boyd, I. A., Gladden, M. H. and Sutherland, F. I. 1982. *J. Physiol.* 331: 285–309.

[36]Barker, D., Emonet-Dénand, F., Laporte, Y. and Stacey, M. J. 1980. *Brain Res.* 185: 227–237.

395
HOW SENSORY
SIGNALS ARISE
AND THEIR
CENTRIFUGAL
CONTROL

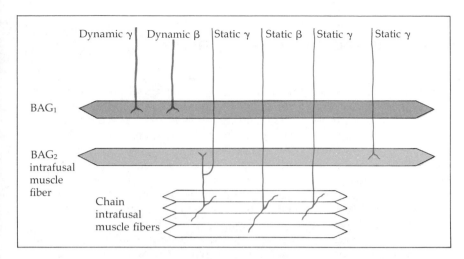

10 FUSIMOTOR INNERVATION OF BAG AND CHAIN fibers. The intrafusal bag$_1$ fiber is supplied by a dynamic γ axon, the bag$_2$ fiber by a separate static γ axon. Other static γ axons also supply bag$_2$ fibers; still others, both bag$_2$ fibers and chain fibers. In addition, β axons, dynamic and static, end on bag$_1$ and chain fibers, respectively. The evidence for this pattern was obtained by observing contractions of individual intrafusal muscle fibers following stimulation of single fusimotor axons in living spindles. (After Arbuthnott et al., 1982; Matthews, 1981.)

namic bag$_1$, a static bag$_2$, or a chain muscle fiber. In general, stimulation of a single γ fiber leads to a graded, local depolarization and contraction of the intrafusal muscle fiber, rather than to an action potential accompanied by a twitch. Simplest are the dynamic γ axons, which end exclusively on bag$_1$ muscle fibers at discretely localized motor end plates; their activity leads to localized shortening and stiffening of the intrafusal muscle fiber. Static γ axons end with distributed "trail" end plates on both bag$_2$ and chain fibers and produce more widespread contractions.

Figure 10 shows a further type of motor supply to the spindle. It has long been known that in frogs and toads extrafusal α-motoneurons that cause contraction of the main mass of the muscle also supply intrafusal fibers.[37] Laporte and his colleagues have now shown that such fibers also exist in the cat and the rabbit.[16] Some are dynamic, supplying bag fibers, others static. Such fibers have been called "Beta" because their diameters and conduction velocities correspond to the smaller and slower range of extrafusal motor axons. (In fact, this term Beta, although generally used, is a misnomer because they technically fall in the α range—see Box 1.) Beta fibers represent a system for automatically maintaining a spindle discharge when muscles contract.

What is the role of crustacean stretch receptors and muscle spindles? With the detailed, but still growing, knowledge about the peripheral

Stretch receptors and the control of movement

[37]Katz, B. 1949. *J. Exp. Biol.* 26: 201–217.

BOX 1 A NOTE ON THE CLASSIFICATION OF NERVE FIBERS IN MAMMALS

It is convenient to distinguish among different nerves according to their diameter and conduction velocity. If one stimulates a nerve bundle electrically at one end and records from it some distance away, one sees a series of potential peaks on the oscilloscope tracing. These are the result of dispersion of nerve impulses that travel at different speeds and therefore arrive at the recording electrodes at different times. On the basis of their electrical properties, the fibers in mammalian nerves were subdivided into groups named A, B (autonomic myelinated), and C. Next, the myelinated, rapidly conducting A fibers were grouped into four subdivisions designated α, β, γ, and δ, in order of decreasing conduction velocity. The motor efferent fibers that supply the intrafusal muscles correspond to the γ group, but the term "fusimotor," introduced at a later stage, is meant to include any

axon that innervates intrafusal fibers. When, in subsequent years, extensive recordings were made from group A sensory nerves supplying muscle, these nerves were classified according to a different convention as group I, group II, and group III. Group Ia fibers correspond to Aα and form the primary endings on muscle spindles; group Ib fibers also correspond to Aα but end instead in stretch receptors called "Golgi tendon organs"; group II fibers correspond to Aβ and form secondary endings. The still smaller myelinated fibers of group III are not discussed here. The term "β fiber" is used to designate α-motoneurons that supply intrafusal as well as extrafusal muscle fibers. Originally it was thought that these occurred only in frogs and in toads, but they have now been found in mammals.

mechanisms that govern the behavior of these sense organs, the search is shifting more and more to their central connections and the control that higher centers exercise. The magnitude of the task of working out the central organization becomes apparent: about one-third of all the motor nerves that leave through the ventral roots are concerned with the control of muscle spindles.

The fusimotor neurons together with their cell bodies in the spinal cord and attached connections constitute a cell system that is impressively large in its mass and numbers. It is therefore not surprising that knowledge of the actual workings of these neurons is incomplete and that discussion is still couched in terms of "general outlines" and "basic roles" rather than in the desired details of the wiring diagram for function that would give a comprehensive view. Besides, spindles are just one part of a larger interconnected sensory system and must be considered together with other peripheral sensory components from the muscles themselves, the joints, and the skin.

Muscle spindles are sensing elements that provide information about the state of muscles, their length, and the rate at which their length is changing. The skeletal muscles that do the work are directly regulated by groups of large motoneurons (usually called α-motoneurons) in the ventral part of the spinal cord. All the neural apparatus that influences movement must converge on these cells. Considering numbers once more, this is an apparatus in which one-third of all the motoneurons, the γ-motoneurons or fusimotor neurons, are concerned with regulation

397
HOW SENSORY
SIGNALS ARISE
AND THEIR
CENTRIFUGAL
CONTROL

rather than with execution of movement. For example, the soleus muscle in the leg of the cat contains about 25,000 muscle fibers and 50 muscle spindles.[38] The motor control consists of 100 α-motoneurons, whereas 50 γ-motoneurons supply the spindles that contain a total of 300 intrafusal muscle fibers. The sensory fibers comprise 50 Ia and 50 II spindle afferents, in addition to 40 Ib fibers supplying Golgi tendon organs (see later). The more finely honed the movement, the more nervous machinery is devoted to its control. Thus, the muscles of our hands are more densely supplied with spindles and γ fibers than are the larger limb muscles or the diaphragm.

One role played by the efferent innervation of intrafusal muscle fibers in the spindle becomes apparent when sensory discharges during muscle movement are considered. As the muscle contracts, the tension on the sensory element is reduced and the spindle is unloaded (Figure 11). During the shortening, the rate of sensory discharges is reduced or the signals stop. Therefore, in the absence of γ efferent or fusimotor impulses to the intrafusal muscle elements, the spindle would temporarily send fewer impulses, or if the muscle stayed in a shortened state, sensory signals would cease altogether. In such a situation, stimulation of the γ efferents shortens the intrafusal muscle elements, takes up the slack, and restores the tension on the sensory terminals (Figure 11C).[39] The part played by efferent nerves is to adjust the sensitivity of the measuring instrument—the spindles—so that they can perform over a wide range of muscle lengths. Spindles, therefore, can be made to maintain their discharge frequencies even when the external stretch on them is reduced during contraction of muscles (Figures 11 and 13).

The simplified diagram in Figure 12 outlines connections used for the stretch reflex and for innervation of extensor and flexor muscles in the cat. It is the simplest of many pathways and represents only a small part of what is known about spinal reflexes. Assume that the sketch in Figure 12 illustrates an extensor muscle—for example, a muscle that straightens the leg. If the muscle is stretched—for example, by tapping the tendon as in the knee-jerk reflex—the primary sensory endings are deformed and initiate impulses; a somewhat stronger stretch also brings in the secondary spindle endings. The first consequence of these impulses is a monosynaptic excitation of the α-motoneurons going back to the muscle that had been extended; discharges from the smaller sensory fibers (secondary group II, not shown) reinforce the effect shortly thereafter, both monosynaptically and by way of interneurons.[40] The reflex extension of the leg is further helped by a simultaneous inhibition of the α-motoneurons that innervate the antagonistic flexors. This principle of one group of muscles being excited while its antagonists are inhibited was first described by Sherrington, who called it reciprocal innervation.

Spinal connections of stretch receptors

[38]Matthews, P. B. C. 1972. *Mammalian Muscle Receptors and Their Central Action.* Edward Arnold, London.

[39]Hunt, C. C. and Kuffler, S. W. 1951. *J. Physiol. 113*: 283–297.

[40]Kirkwood, P. A. and Sears, T. A. 1974. *Nature 252*: 243–244.

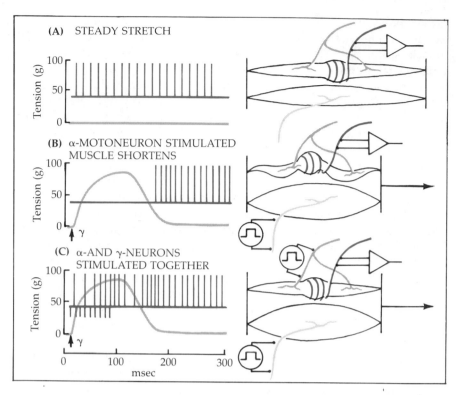

11 **EFFECT OF FUSIMOTOR STIMULATION** during contraction. Sensory discharges from a muscle spindle in the leg of a cat, recorded from an afferent fiber in the dorsal root. (A) During steady stretch the discharges are maintained. Tension is registered on the lower trace. (B) During stimulation of the large α-motoneurons, the muscle shortens and the spindle ceases to fire. (C) If the α-motoneurons and the γ-efferent (fusimotor) fiber are stimulated together, the spindle maintains its discharges during the muscle shortening. (After Hunt and Kuffler, 1951.)

If, instead of a brief tap on the tendon of an extensor muscle, steady tension is exerted, the muscle contracts in response to discharges from primary afferents (marked Ia), and the load is taken off the spindles. Thereby the drive on the α-motoneurons is reduced. This, in turn, leads to a renewed lengthening of the muscle and a renewed sensory acceleration of spindle afferent discharges. Such an alternate lengthening and shortening, an oscillation, would occur in the absence of some form of "damping" and would be in phase with the fluctuations of the discharges from spindles.

In practice, however, the spindles do not respond only to lengthening of the muscle. In addition, fusimotor discharges serve to regulate their firing. In the cat, γ-motoneurons lie in the same region of the spinal cord as the α-motoneurons but differ from them in size and morphology.[41] The γ-motoneurons have smaller cell bodies (as well as

[41]Westbury, D. R. 1982. *J. Physiol.* 325: 79–91.

399
HOW SENSORY
SIGNALS ARISE
AND THEIR
CENTRIFUGAL
CONTROL

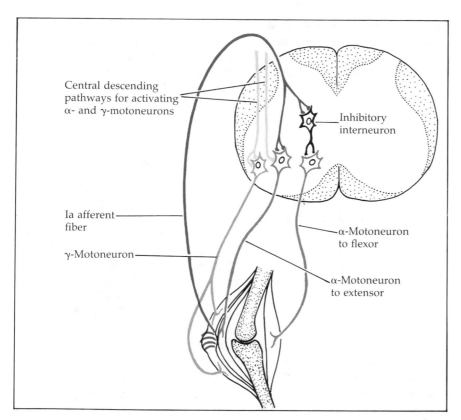

Central descending
pathways for activating
α- and γ-motoneurons

Inhibitory
interneuron

Ia afferent
fiber

γ-Motoneuron

α-Motoneuron
to flexor

α-Motoneuron
to extensor

12 **CONNECTIONS FOR STRETCH REFLEX.** Sketch of two antagonistic muscles, an extensor and a flexor, together with some of their connections in the spinal cord that are activated through sensory discharges from a spindle. When the extensor muscle is briefly stretched, the α-motoneuron is excited monosynaptically (the stretch reflex) while the flexor is inhibited via interneurons (reciprocal innervation). Thereby antagonistic muscles cooperate. Central descending pathways activate α-motoneurons to initiate voluntary movements and excite γ-motoneurons to maintain sensory discharges during contraction (see text).

smaller diameter axons), with fewer dendrites and axon collaterals. Unlike α-motoneurons, γ-motoneurons receive no spindle afferents.

For the sake of simplicity, a number of the known pathways and connections are omitted from Figure 12. These include the group II afferents, which are within the spindle and play a part in the excitatory drive for the stretch reflex, and the connections of the various interneurons that occur within the spinal cord; these connections have been worked out in detail by Lundberg, Jankowska, and their colleagues.[42,43] One type of interneuron, called the Renshaw cell, mediates recurrent inhibitory interactions between the population of α-motoneurons sup-

[42]Lundberg, A. 1979. *Progr. Brain. Res. 50*: 11–28.
[43]Czarkowska, J., Jankowska, E. and Sybirska, E. 1981. *J. Physiol. 310*: 367–380.

plying the same muscle. Firing by a motoneuron produces excitatory synaptic potentials (mediated by acetylcholine) and impulses in the Renshaw interneuron, which in turn causes inhibition of the motor neurons. The Renshaw cells, like other interneurons involved in coordination, also receive descending inputs from higher centers. Renshaw cells do not produce inhibition of γ-motoneurons. The precise role played by recurrent inhibition in regulating movements is not understood.[44]

Tendon organs

Other sensory receptors exist within muscles and play a role in stretch reflexes. The most important of these are usually found near the tendon–muscle junctions and are called the Golgi TENDON ORGANS; they lie in series with the contracting regular skeletal muscles and can be made to discharge impulses by passive stretch. Contraction of muscle fibers to which they are extremely sensitive is the principal stimulus that activates firing. Thus, a contraction of one or two muscle fibers that leads to a tension increase of less than 100 mg can cause a brisk discharge.[45,46] Tendon organs have no centrifugal control. Their axons, classified as Ib, activate interneurons that inhibit the α-motoneurons innervating the muscle from which the axons of the tendon organ arise.[38,43] Together, the tendon organs in a muscle provide a measure of the force exerted by it as it contracts.

Coactivation of γ
efferent axons

A number of schemes have assigned roles to spindles for movement and for the maintenance of posture. For example, it was proposed that fusimotor fibers might be used to initiate contractions by producing sensory discharges that in turn lead to reflex excitation of α-motoneurons. This view has been shown to be incorrect by recordings made in man and in cats. For example, when a voluntary movement of the thumb is made, the α-motoneurons fire before the γ-motoneurons.[47,48] Probably the greatest weight of evidence now suggests that the γ efferent system is used chiefly to maintain the synaptic drive during muscular contraction in the face of varying external loads. Imagine, for example, that during a planned movement an unexpected opposition that slows extrafusal muscle shortening is encountered. Coactivation of α- and γ-motoneurons will mean that intrafusal muscle fibers go on shortening, an action increasing the afferent discharge from spindles and thereby increasing the α-motoneuron discharge to overcome the opposition to the movement.

A favorable situation for demonstrating a role for γ efferent fibers in normal movement has been found in respiratory muscles. Sears,[49]

[44]Kirkwood, P. A., Sears, T. A. and Westgaard, R. H. 1981. *J. Physiol. 319*: 111–130.

[45]Crago, P. E., Houk, J. C. and Rymer, W. Z. 1982. *J. Neurophysiol. 47*: 1069–1083.

[46]Fukami, Y. 1982. *J. Neurophysiol. 47*: 810–826.

[47]Taylor, A. and Prochazka, A. 1981. *Muscle Receptors and Movement*. Macmillan, London.

[48]Vallbo, Å. B., Hagbarth, K.-E., Torebjörk, H. E. and Wallin, B. G. 1979. *Physiol. Rev. 59*: 919–957.

[49]Sears, T. A. 1964a. *J. Physiol. 174*: 295–315.

401
HOW SENSORY
SIGNALS ARISE
AND THEIR
CENTRIFUGAL
CONTROL

(A) NORMAL

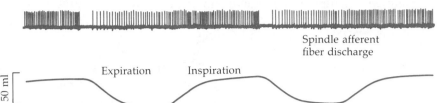

Spindle afferent
fiber discharge

Expiration   Inspiration

(B) AFTER FUSIMOTOR PARALYSIS

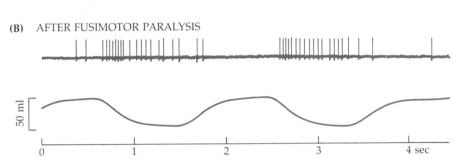

13 **REFLEX CONTROL OF SPINDLE ACTIVITY.** Evidence in the cat for the simultaneous efferent activation of the motor innervation of an intercostal respiratory muscle and of its muscle spindles during breathing. (A) Normally the sensory discharge frequency from an inspiratory muscle is highest during inspiration, even though the muscle is shortening. (B) After the fusimotor fibers have been selectively blocked by procaine, the spindle behaves passively and its discharges cease during inspiration when the muscle contracts. This is the response expected for a muscle spindle afferent fiber in the absence of γ-control. In each pair, upper trace registers spindle afferent discharge, lower trace registers respiration. (After Critchlow and von Euler, 1963.)

von Euler,[50] and their colleagues have recorded the discharges in afferent fibers from spindles in the intercostal muscles used for inspiration and expiration. Figure 13 shows that under normal conditions the afferent discharge from an inspiratory muscle is greatest during inspiration. This at first is surprising because during inspiration that muscle actually shortens (Figure 13A). A simple explanation is that the fusimotor fibers are being excited together with the α-motoneurons to the same muscle and thereby overcome the slack created by contraction. This is confirmed in Figure 13B, which shows the responses of the afferent fiber after the γ efferents to the inspiratory muscle were paralyzed by applying the local anesthetic procaine to the intercostal nerve. This procedure blocks the relatively small diameter γ fibers, but not the larger α fibers, allowing the extrafusal muscle fibers to contract as before. Under these conditions the afferent discharge occurs, as expected, but only during expiration while the inspiratory muscle is being stretched. Other examples of the parallel activation of γ- and α-moto-

[50]Critchlow, V. and von Euler, C. 1963. *J. Physiol. 168*: 820–847.

neurons are provided by experiments on recording spindle activity accompanying finger movements in man.[48] During voluntary movements, there is little change in the frequency of Ia or II fiber discharges even though the muscle is shortening.

A problem that remains concerns the mechanisms by which the central nervous system is able to distinguish between sensory discharges from spindles evoked on the one hand by stretch and on the other by intrafusal contraction. It seems evident that commands descending from the central nervous system to shape a movement must be registered in the centers analyzing the sensory responses to the load on the muscle and its movements. By analogy, no purpose would be served by continuously altering the gain of a measuring device without keeping track of its sensitivity at all times. The timing, location, and properties of such postulated corollary discharges have not yet been examined within the central nervous system. Presumably the analysis involves comparison of (1) the α signals causing contraction, (2) the γ discharges initiating sensory responses, and (3) the incoming signals actually received from the spindles, tendon organs, and joint receptors.

It is clear, however, that whatever the mechanism, the efferent fusimotor fibers emerging from the spinal cord play an essential role in controlling and modulating the reflexes used in maintaining posture and in executing smooth movements. Clinical studies, particularly those on injuries to dorsal roots, support this view.

Do spindle discharges contribute to our sense of position or movement? For many years it was thought that sensory information from muscle spindles was concerned solely with reflex regulation of muscles and did not reach the cerebral cortex. Position sense was attributed entirely to receptors situated in joints. Recently, however, it has been shown that information from spindles does reach the cerebral cortex and consciousness.[51,52] Patients and volunteers have reported sensations of movements, of changes in position of a limb or digit, or of stiffness when muscles were vibrated or stretched after all sensation had been abolished in joint receptors.

OTHER RECEPTORS AND CONTROL OF ASCENDING PATHWAYS

Although the principle of centrifugal control has been discussed for spindles and stretch receptors only, similar mechanisms also operate in other sense organs. For example, an inhibitory efferent innervation to the first-order sensory cells in the auditory system has long been known in mammals and in frogs. Recently, intracellular recordings from hair cells in the turtle ear have shown that the outflow from the central nervous system produces large inhibitory potentials with a reversal

[51]McCloskey, D. I. 1978. *Physiol. Rev. 58*: 763–820.
[52]Matthews, P. B. C. 1982. *Annu. Rev. Neurosci. 5*: 189–218.

403
HOW SENSORY
SIGNALS ARISE
AND THEIR
CENTRIFUGAL
CONTROL

potential of about −80 mV.[53] These can produce a 25-fold reduction in the receptor potentials for tones presented at the hair cell's optimal frequency. In addition, the inhibition degrades the fine tuning properties of the cell so that it responds to sounds with less discrimination as well as less sensitivity.

A control system, with excitatory as well as inhibitory centrifugal fibers, has been demonstrated in the lateral line organs of fishes.[54] The comparative chemistry is unexpectedly and interestingly different in cells of the lateral line organs. Efferent nerves appear to liberate acetylcholine (ACh) as an inhibitory transmitter; γ-aminobutyric acid may be the excitatory transmitter for initiating sensory discharges (Chapter 12).[55]

Within the central nervous system itself, the distinction between afferent sensory fibers and efferent or motor fibers becomes at times difficult. The visual system offers a good example of progressive feedback control that occurs in successive stages as visual information ascends to higher centers. First, the amount of light allowed to fall on the receptors is regulated by the size of the pupil. The properties and functional significance of this feedback loop have been worked out in detail. Feedback also goes on within the retina: Receptors influence horizontal cells, and these cells then act back upon the receptors and modify their responses to light. In addition, as mentioned in Chapter 2, in birds there exists an efferent control from the brain to the eye.[56] The isthmooptic nucleus sends into the retina centrifugal fibers, ending on amacrine cells, that modulate the discharges set up by light stimulation. Stimulation of the axons of the isthmooptic tract acts indirectly by disinhibiting retinal ganglion cell activity. Thereby the higher centers are able to alter the responses from individual ganglion cells. In the cat, descending fibers run from the visual cortex to the superior colliculi, again modifying the responses of neurons to visual stimuli. A projection from the cortex to the geniculate nucleus is also known, but no role has yet been assigned to it. In the somatosensory system, centrifugal fibers descend from the cortex and end in the relay stations for touch and pressure neurons.[57] A corticofugal innervation to the olfactory bulb of the rat has been described.[58]

Centrifugal control has been mainly considered in peripheral sense organs because the principal mechanisms can be most easily demonstrated and understood in these examples. At the same time, it bears emphasizing that corticofugal pathways may have a much wider role

[53]Art, J. J., Crawford, A. C., Fettiplace, R. and Fuchs, P. A. 1982. *Proc. R. Soc. Lond. B 216*: 377–384.

[54]Flock, Å. and Russell, J. J. 1973. *J. Physiol. 235*: 591–605.

[55]Flock, Å. and Lam, D. M. K. 1974. *Nature 249*: 142–144.

[56]Pearlman, A. L. and Hughes, C. P. 1976a. *J. Comp. Neurol. 166*: 111–122.

[57]Towe, A. L. 1973. In A. Iggo (ed.). *Handbook of Sensory Physiology*, Vol. 2. Springer-Verlag, New York, pp. 700–718.

[58]Luskin, M. B. and Price, J. L. 1983. *J. Comp. Neurol. 216*: 264–291.

that is as yet insufficiently formulated and therefore remains obscure. The observations that ascending pathways can be suppressed at different levels suggest that efferent inhibition may be used to prevent unwelcome or distracting information from reaching certain parts of the brain, for example, during attention. Of particular interest is the possibility of suppressing or "gating" pathways that convey pain. Efferent control could also serve the opposite purpose—to enhance the flow of afferent information, thereby lowering the threshold for sensory stimuli. Many of these schemes can now be tested with available methods. An intriguing but fanciful idea is the use of efferent pathways to stir up subcortical centers for recall of stored information. Centrifugal control, therefore, seems to be part of the general organization of our higher centers that continuously edit and transform information at various stages of processing.

SUGGESTED READING

General reviews

Burgess, P. R. and Wei, J. Y. 1982. Signaling of kinesthetic information by peripheral sensory receptors. *Annu. Rev. Neurosci. 5*: 171–187.

Matthews, P. B. C. 1972. *Mammalian Muscle Receptors and Their Central Actions.* Edward Arnold, London.

Matthews, P. B. C. 1981. Evolving views on the internal operation and functional role of the muscle spindle. *J. Physiol. 320*: 1–30.

Matthews, P. B. C. 1982. Where does Sherrington's "muscular sense" originate? Muscles, joints, corollary discharges? *Annu. Rev. Neurosci. 5*: 189–218.

[These three reviews by Matthews provide a lucid and authoritative account of the history, structure, and experiments dealing with spindles.]

McCloskey, D. I. 1978. Kinesthetic sensibility. *Physiol. Rev. 58*: 763–820.

Vallbo, Å. B., Hagbarth, K.-E., Torebjörk, H. E. and Wallin, B. G. 1979. Somatosensory, proprioceptive and sympathetic activity in human peripheral nerves. *Physiol. Rev. 59*: 919–957.

Original papers

Arbuthnott, E. R., Ballard, K. J., Boyd, I. A., Gladden, M. H. and Sutherland, F. I. 1982. The ultrastructure of cat fusimotor endings and their relationship to foci of sarcomere convergence in intrafusal fibres. *J. Physiol. 331*: 285–309.

Art, J. J., Crawford, A. C., Fettiplace, R. and Fuchs, P. A. 1982. Efferent regulation of hair cells in the turtle cochlea. *Proc. R. Soc. Lond. B 216*: 377–384.

Banks, R. W., Barker, D., Bessou, P., Pages, B. and Stacey, M. J. 1978. Histological analysis of muscle spindles following direct observation of effects of stimulating dynamic and static motor axons. *J. Physiol. 283*: 605–619.

Crago, P. E., Houk, J. C. and Rymer, W. Z. 1982. Sampling of total muscle force by tendon organs. *J. Neurophysiol. 47*: 1069–1083.

Critchlow, V. and von Euler, C. 1963. Intercostal muscle spindle activity and its γ-motor control. *J. Physiol. 168*: 820–847.

405
HOW SENSORY
SIGNALS ARISE
AND THEIR
CENTRIFUGAL
CONTROL

Emonet-Dénand, F., Jami, L., Laporte, Y. and Tankov, N. 1980. Glycogen depletion of bag$_1$ fibres elicted by stimulation of static axons in cat peroneus brevis muscle spindles. *J. Physiol. 302*: 311–321.

Eyzaguirre, C. and Kuffler, S. W. 1955. Processes of excitation in the dendrites and in the soma of single isolated sensory nerve cells of the lobster and crayfish. *J. Gen. Physiol. 39*: 87–119.

Hunt, C. C., Wilkinson, R. S. and Fukami, Y. 1978. Ionic basis of the receptor potential in primary endings of mammalian muscle spindles. *J. Gen. Physiol. 71*: 683–698.

Jankowska, E. and McCrea, D. A. 1983. Shared reflex pathways from Ib tendon organ afferents and Ia muscle-spindle afferents in the cat. *J. Physiol. 338*: 99–112.

Kuffler, S. W., Hunt, C. C. and Quilliam, J. P. 1951. Function of medullated small-nerve fibers in mammalian ventral roots: Efferent muscle spindle innervation. *J. Neurophysiol. 14*: 29–54. (This paper definitively established the role of γ efferent fibers.)

Matthews, B. H. C. 1933. Nerve endings in a mammalian muscle. *J. Physiol. 78*: 1–53. (The classical paper describing the physiological properties of muscle spindles in the cat.)

McCloskey, D. I., Cross, M. J., Honnor, R. and Potter, E. K. 1983. Sensory effects of pulling or vibrating exposed tendons in man. *Brain 106*: 21–37.

Sears, T. A. 1964a. Efferent discharges in alpha and fusimotor fibers of intercostal nerves of the cat. *J. Physiol. 174*: 295–315.

TRANSFORMATION OF INFORMATION BY SYNAPTIC ACTION IN INDIVIDUAL NEURONS

CHAPTER SIXTEEN

Our higher centers continually receive and integrate information arising in a great variety of sources on the surface of the body and in the internal organs. A typical central neuron faces a task similar to that of the brain as a whole. It is the target of converging excitatory and inhibitory signals, whose information it synthesizes before taking action of its own in the form of generating new nerve impulses. This integrative activity of individual neurons—the combining of a variety of information—is centered on synaptic transmission. The universality of neural integration is illustrated by samples from three diverse cell systems—in crustaceans, fishes, and mammals.

A simple example is the opener muscle of the claw in lobsters and crayfish. All the muscle fibers are innervated at multiple sites by two nerve fibers only, one excitatory and the other inhibitory. With this innervation the muscle is able to execute finely graded movements. On an individual muscle fiber, the interaction of inhibitory and excitatory synaptic potentials determines the level of the summed postsynaptic potential and thereby the level of contraction. Besides its postsynaptic chemical action, the inhibitory axon also acts presynaptically and can reduce the amount of transmitter released by the excitatory terminal.

The Mauthner cell in the medulla of the goldfish is a large neuron with two dendrites, each several hundred micrometers long. Impulses arising in the neuron are controlled by five different types of synaptic actions: (1) excitatory chemical synapses, (2) inhibitory chemical synapses, (3) excitatory electrical synapses, (4) inhibitory electrical synapses, and (5) presynaptic chemical inhibition occurring on some of the excitatory nerve terminals. The large dendrites of the Mauthner neuron do not give regenerative impulses. The cell is, therefore, able to combine or integrate through its cable properties the diverse converging synaptic influences. All these are channeled to the crucial site, the axon hillock region, where conducted impulses are initiated.

The third example of integration, the spinal motoneuron, integrates the excitatory and inhibitory action of many thousands of chemically transmitting synapses. Here also presynaptic inhibition can selectively eliminate or reduce the action of certain pathways that reach the motoneuron. As in the Mauthner cell, the integrating action of the neuron is focused on the initial axon segment where impulses are initiated.

The synapse considered in greatest detail so far is the vertebrate myoneural junction; it differs in an important way from the majority of synapses within the central nervous system. Each skeletal muscle fiber receives its excitatory input from only one or sometimes two motor nerve fibers, and under normal conditions each impulse in the presynaptic nerve gives rise to an impulse in the muscle. The situation differs in neurons within the brain. They are, as a rule, supplied by many converging axons, some excitatory and others inhibitory. Each presynaptic impulse usually releases only a few quanta of transmitter, producing a small subthreshold effect. These synaptic potentials, like other graded localized potentials, can sum with each other either to reinforce or to inhibit. The main point of this chapter is to show that for a standard cell in the brain to discharge, the collective input from many other cells is required. The integration of synaptic events is crucial for information processing in general, since the rigidly determined all-or-none impulses cannot sum. It is at synapses that the flexibility of the nervous system resides.

The three preparations illustrated in Figure 1 have many features in common. Each is subjected to convergent excitatory and inhibitory influences, and each, directly or indirectly, gives rise to movement. The crustacean muscle contracts, the large Mauthner cell in the goldfish medulla activates motoneurons, and the motoneuron in the spinal cord innervates muscle fibers. However, whereas one crustacean muscle fiber receives just two or a few incoming fibers, the Mauthner cell and the motoneuron receive many hundreds or thousands. Furthermore, a new dimension is added through the placement of synaptic terminals on specific regions of the cell. Despite these differences, the underlying mechanisms of integration are highly similar in the three types of cells.

CRUSTACEAN MYONEURAL SYNAPSES

The nerve–muscle junctions in the lobster and the crayfish provide a preparation that is impressively simple yet shows the basic elements that nerve cells in vertebrate brains use for integration. The opener muscle of the claw is supplied by two axons only, each forming chemical synapses, one excitatory and the other inhibitory (Chapter 12). Yet the contraction of the entire muscle, composed of many individual muscle fibers, can be finely graded by these two axons. Both axons branch in unison (Figure 1) and distribute themselves over the surface of each muscle fiber, forming multiple synapses close to each other (Figure 2).[1-3] In this way, synaptic excitation and inhibition can be exerted at many spots almost simultaneously.

If such a muscle worked like human skeletal muscles, in which each muscle fiber gives a regenerative conducted response and a rapid

[1]Wiersma, C. A. G. and Ripley, S. H. 1952. *Physiol Comp. Oecol.* 2: 391–405.
[2]Atwood, H. L. and Morin, W. A. 1970. *J. Ultrastruct. Res.* 32: 351–369.
[3]Lang, F., Atwood, H. L. and Morin, W. A. 1972. *Z. Zellforsch.* 127: 189–200.

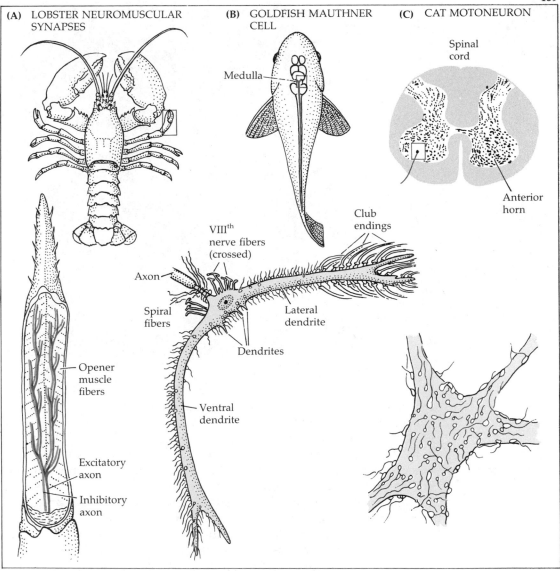

(A) LOBSTER NEUROMUSCULAR SYNAPSES

(B) GOLDFISH MAUTHNER CELL

(C) CAT MOTONEURON

Spinal cord

Medulla

Anterior horn

Club endings

VIIIth nerve fibers (crossed)

Axon

Spiral fibers

Lateral dendrite

Dendrites

Opener muscle fibers

Ventral dendrite

Excitatory axon

Inhibitory axon

1 **SYNAPSES IN LOBSTER, FISH AND CAT.** Synaptic arrangements in (A) a lobster (or crayfish) muscle, (B) a large neuron in the brain of a goldfish, and (C) a motoneuron in the spinal cord of a cat. The crustacean muscle is supplied by only one excitatory and one inhibitory axon, each of which forms scattered synapses on the muscle surface; the typical motoneuron receives thousands of synapses from hundreds of other neurons making excitatory and inhibitory connections; in the Mauthner cell some of the incoming nerve fibers with distinct function form synapses in specific regions of the cell. In the muscle and the motoneuron chemical synaptic excitation and chemical postsynaptic and presynaptic inhibition have been studied. The Mauthner cell receives, in addition, electrical excitatory and inhibitory synapses. (B after Bodian, 1942.)

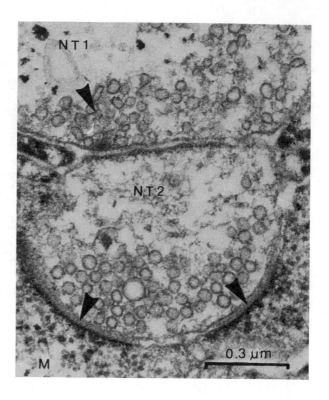

2

CHEMICAL SYNAPSES in crustaceans. A nerve terminal (NT2) forms chemical synapses (arrows), presumably excitatory, with a muscle fiber (M) in an opener muscle of the crayfish. The nerve terminal itself receives a chemical synaptic contact (NT1) which is, therefore, presynaptic and inhibitory. (From Lang, Atwood, and Morin, 1972.)

twitch, the entire muscle would contract maximally with each motor nerve impulse; it would behave as one motor unit. To achieve versatility, each muscle fiber in the opener of the claw gives graded local contractions, the amplitude being determined by the size of the synaptic potentials. In this way a crustacean muscle with only one excitor and one inhibitor axon is able to accomplish smoothly graded movements. The amplitudes of the synaptic potentials in turn depend on two processes: (1) TEMPORAL SUMMATION, which occurs when potentials arrive in quick enough succession so that each adds its effect to the preceding one, and (2) TEMPORAL FACILITATION, in which each successive synaptic potential becomes larger than the last.[4] A remarkable feature of the pattern of innervation of crustacean muscle is demonstrated by the characteristics of facilitation. The many muscle fibers, all multiply innervated by a single nerve axon, each show different facilitation characteristics; for example, in some fibers of a particular muscle the synaptic potentials occurring in a train become greatly increased in amplitude, in others less so. Yet on any one individual muscle fiber all of the endings of the axon facilitate to the same extent.[5] Somehow all the synapses on the fiber develop with like properties.

When the muscle surface is explored in detail by means of iono-

[4]Parnas, I., Parnas, H. and Dudel, J. 1982. *Pflügers. Arch. 393*: 232–236.
[5]Frank, E. 1973. *J. Physiol. 233*: 635–658.

411
TRANS-
FORMATION OF
INFORMATION
BY SYNAPTIC
ACTION IN
INDIVIDUAL
NEURONS

phoretic application of GABA and glutamate (the transmitter released by the excitatory axon),[6] the membrane is seen to be a mosaic of spots sensitive to one or the other of the substances in the region underlying the inhibitory or excitatory terminals. This experimental technique is used to determine the sites of inhibitory and excitatory synapses.[7] Through its cable properties, the postsynaptic cell shows SPATIAL SUM-MATION of all the discrete, distributed synaptic potentials so that in practice changes in potential tend to be distributed uniformly over the length of the fiber.

In the opener of the claw, then, two types of chemically transmitting synapses engage in opposing excitatory and inhibitory actions. These determine the state of the postsynaptic muscle membrane potential in a finely graded manner. The contraction, in turn, reflects the level of the membrane potential. An example of separate and joint synaptic excitation and inhibition is shown in Figure 3.

Integration at the crustacean neuromuscular junction displays two further types of synaptic interactions seen in the central nervous system of vertebrates. Presynaptic inhibition occurs through synapses formed by the inhibitory nerve on the excitatory terminals (Figure 2). The effect is to reduce the number of quanta liberated by excitatory terminals in response to each impulse. The timing is critical. The inhibitory trans-mitter—in this case, γ-aminobutyric acid (GABA)—must be released just before the action potential reaches the excitatory terminals.[8] Intra-cellular recordings from the excitatory axon close to its terminals have shown that impulses in the inhibitory axon give rise to presynaptic hyperpolarization.[9] Quite different, long-term alterations in release, which persist for minutes or hours, result from the actions of hormones or of circulating transmitters, such as biogenic amines (Chapter 12). For example, Kravitz[10] and his colleagues have shown that physiological concentrations of 5-hydroxytryptamine (5-HT) can increase the ampli-tude of the excitatory synaptic potentials in lobster muscles evoked by stimulation of the excitatory axon. Their results suggest that this mod-ulating effect is caused by a metabolic change induced in the presynaptic terminals by 5-HT.

The crustacean neuromuscular junction shows great versatility in its integrating action by making use of chemical synaptic excitation and inhibition, facilitation, temporal and spatial summation, presynaptic inhibition, and modulation. These mechanisms are also used by higher nervous systems for the same basic purpose. In addition, nonsynaptic mechanisms can influence the integration by muscle fibers. With repet-itive firing of the excitatory axon at naturally occurring frequencies, conduction of the impulse can fail. Parnas and his colleagues[11] have

[6]Kawagoe, R., Onodera, K. and Takeuchi, A. 1981. *J. Physiol. 312*: 225–236.
[7]Takeuchi, A. and Takeuchi, N. 1965. *J. Physiol. 177*: 225–238.
[8]Dudel, J. and Kuffler, S. W. 1961. *J. Physiol. 155*: 543–562.
[9]Fuchs, P. A. and Getting, P. A. 1980. *J. Neurophysiol. 43*: 1547–1557.
[10]Glusman, S. and Kravitz, E. A. 1982. *J. Physiol. 325*: 223–241
[11]Grossman, Y., Parnas, I. and Spira, M. E. 1979a. *J. Physiol. 295*: 283–305.

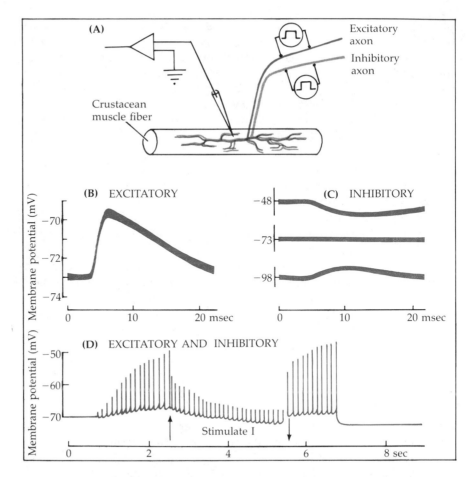

3 **CHEMICAL EXCITATION AND INHIBITION** in crustacean muscle.
**(A) Excitatory and inhibitory axons can be stimulated separately or
together. (B) An excitatory synaptic potential in the opener of the claw
at a membrane potential of −73 mV. (C) Stimulation of the inhibitory
axon causes no potential change at −73 mV, but hyperpolarizes at −48
mV and depolarizes at −98 mV. (D) Simultaneous excitatory and inhibitory stimulation at 10/sec (between arrows) greatly reduces the excitatory potentials. (C from Fatt and Katz, 1953; D from Atwood and Bittner,
1971.)**

shown that conduction failure also occurs at branch points where the
caliber of the axon branches is unfavorable for propagation. (Impulses
tend to become blocked particularly at sites where a small axon feeds
into larger diameter branches.) Hence, action potentials arising in a
nerve cell may not succeed in invading all of its terminals. Again, such
phenomena have also been observed within the central nervous system
(where impulses in dendrites, for example, may not give rise to firing
of the soma[12]).

[12]Llinás, R. and Sugimori, M. 1980. *J. Physiol. 305*: 171–195, 197–213.

413
TRANS-
FORMATION OF
INFORMATION
BY SYNAPTIC
ACTION IN
INDIVIDUAL
NEURONS

SYNAPTIC INTERACTIONS AT THE MAUTHNER CELL

In the medulla of teleost fishes, on each side of the midline, there is an unusually large, peculiarly shaped neuron—the Mauthner cell (Figure 1).[13] It has attracted the attention of many investigators, not only because of its size, but also because of the structural diversity of its synaptic contacts. The Mauthner cells represent a miniature nervous system that combines many of the synaptic mechanisms known at present. In addition to chemically mediated excitation and inhibition of the type seen at crustacean myoneural junctions, the Mauthner cell is excited and inhibited by electrical synapses.

Another attractive feature of Mauthner cells is the well-defined geometry of the synaptic inputs. In the crustacean muscle, knowledge of the particular distribution of synapses is not crucial for understanding integration. In contrast, in the Mauthner cell the various types of synapses occupy characteristic positions on specific parts of the neuronal surface and have different degrees of effectiveness. The regularity of the anatomical pattern in cell after cell in animals of the same species has greatly aided the search for typical structures for a detailed morphological analysis and for a correlation with physiological performance.

The distribution of synapses on the goldfish Mauthner cell is shown in Figures 1 and 4. The neuron lies about 1 mm beneath the surface of the medulla and cannot be seen in live preparations. In a comprehensive, detailed series of experiments, Furshpan and Furukawa[14] mapped the synaptic connections by using a coordinate system that enabled them to place electrodes within different regions of the Mauthner cell and accurately onto specific parts of its surface. They made extracellular and intracellular recordings, studied the membrane properties of the cells, and determined which structures gave rise to the various synaptic potentials.

The massive lateral and ventral dendrites can be several hundred micrometers long. The initial portion of the axon, close to the cell body, is the AXON HILLOCK, and again, as in other cells, it is here that impulses are initiated. This region plays a critical role in the integrative mechanisms of the cell because the dendrites and cell body do not generate action potentials. The segment of nerve between the axon hillock near the cell body and the starting point of the myelinated axon is surrounded by a wrapping of glial cells into which a number of fine axons penetrate to form a spiraling network. Some of the axons emerge again and farther on make synapses on the nearby cell body. This structure is the site of electrical inhibition and is called the AXON CAP.

The large axon (up to 50 μm in diameter) of each Mauthner neuron crosses to the opposite side, runs down along the spinal cord, and makes synapses with numerous motoneurons. Each time the Mauthner

Structure and function of Mauthner cells

[13]Bodian, D. 1937. *J. Comp. Neurol. 68*: 117–159.
[14]Furshpan, E. J. and Furukawa, T. 1962. *J. Neurophysiol. 25*: 732–771.

Synaptic inputs to
Mauthner cells

neuron discharges a single impulse, it causes a rapid, vigorous contraction of one side of the tail by synchronously activating motoneurons in the spinal cord. This escape reaction can be readily seen when one taps the side of a tank containing a number of fish; the fish execute sudden tail flips that propel them sideways.

Each neuron receives numerous fibers that come directly or indirectly from the eighth nerve on the two sides and from the opposite Mauthner cell. The excitatory inputs are derived from two groups of eighth nerve fibers (Figure 4). The larger fibers have diameters of 10 to 15 μm and are particularly conspicuous; they originate on the same side of the animal in a part of the labyrinth, a sensory structure in the inner ear concerned with proprioception. The sensory axons run directly to the distal portion of the lateral dendrite of the Mauthner cell, where they form the so-called club endings (Figure 5). Electron microscopy shows that gap junctions occur in the contact area of the club endings

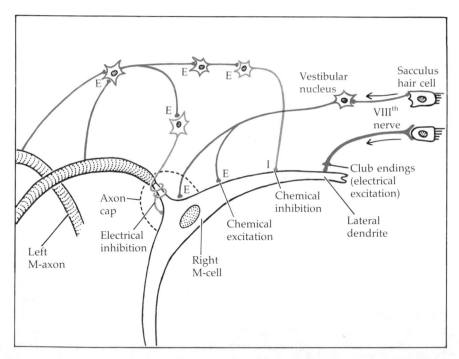

4 **SCHEME OF MAUTHNER CELL SYNAPSES.** The eighth nerve on the same side provides excitatory electrical synapses on the distal portion of the lateral dendrite and excitatory chemical synapses elsewhere on the cell. The inhibitory input comes mainly from the Mauthner cell and the eighth nerve on the other side and by way of axon collaterals from the same side. Probably these pathways involve interneurons. Electrical inhibition is confined to the axon cap region, and presynaptic chemical inhibition is seen in the large axons on the distal portion of the lateral dendrites. (After Furukawa, 1966.)

415
TRANS-
FORMATION OF
INFORMATION
BY SYNAPTIC
ACTION IN
INDIVIDUAL
NEURONS

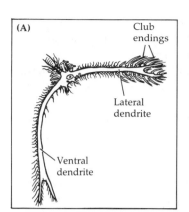

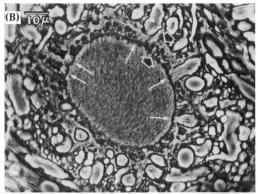

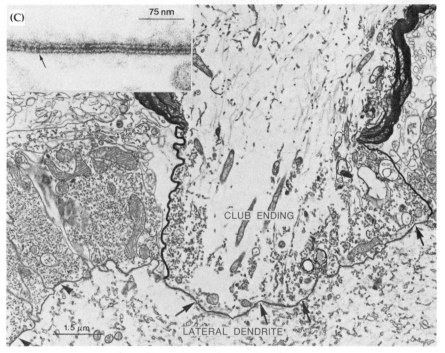

5 CHEMICAL AND ELECTRICAL SYNAPSES on the same neuron. (A) Scheme of the Mauthner cell. The electrically transmitting club endings lie on the distal portion of the lateral dendrite. (B) Histological view of several large club endings (thin arrows) in a transverse section through the distal portion of the lateral dendrite. Region of chemical synapses appears between thick arrows. (C) Electron micrograph of a club ending on the lateral dendrite. Arrows indicate special contact areas. One of these is shown at high magnification in the inset and is typical of electrical (gap) junctions, which are traversed by crossbridges (small arrow). The lateral boundaries of the unmyelinated portion of the club ending have been outlined in ink. To the left are smaller axons forming chemical synapses; two of these are marked by arrow heads. (A from Bodian, 1942; B from Robertson, Bodenheimer, and Stage, 1963; C courtesy of Y. Nakajima, see also Nakajima, 1974; other details on analogous chemical and electrical synapses in mammals are given in Figure 7, Chapter 9.)

and the dendrite;[15,16] these junctions, where the presynaptic and postsynaptic membranes are in close apposition, are the sites of electrical excitatory synapses. Current generated by impulses in the large axons spreads through the gap junctions into the lateral dendrites. The smaller eighth nerve fibers, which are responsible for chemical excitation, are believed to end more diffusely on the lateral dendrite.

The principal inhibitory inputs to the Mauthner cell come from the contralateral eighth nerve, from the contralateral Mauthner cell, and from the same Mauthner cell by way of recurrent collaterals. The interneurons involved in these pathways have been identified and injected with horseradish peroxidase so that their terminals on the Mauthner cell can be observed by electron microscopy. They give rise to three distinct forms of inhibition: (1) conventional chemical inhibitory potentials, similar to those in crustacean muscle, evoked by stimulation of the contralateral eighth nerve; (2) electrical inhibition produced by current flow in the region of the axon cap, evoked by stimulating the contralateral eighth nerve or either of the Mauthner cells; and (3) presynaptic inhibition occurring on the endings of eighth nerve fibers where they synapse upon the lateral dendrite of the Mauthner cell.[17]

Analysis of synaptic mechanisms: Excitatory electrical and chemical transmission

When the large axons in an ipsilateral eighth nerve are stimulated electrically and the responses are recorded with an intracellular electrode in the distal portion of the lateral dendrite, a depolarization of up to 50 mV appears with a latency of about 0.1 msec. This synaptic delay is so short that the transmission must be electrical (Figure 6B, potential 1). The size of the potential declines as the recording electrode is inserted at different distances away from the club endings toward the cell body. This potential, decreasing with distance, is therefore a nonregenerative, excitatory postsynaptic event. The spatial distribution of potentials also indicates that the site of electrical coupling is close to the end of the lateral dendrite.[18]

When the ipsilateral eighth nerve is stimulated more strongly, so that its smaller diameter axons are excited, an additional, delayed, postsynaptic potential can be recorded in the Mauthner cell after a latency of less than 1 msec (Figure 6B, potential 2). According to the usual criteria, this delayed excitatory synaptic potential is derived from chemically transmitting synapses, probably by way of interneurons. Either of these synaptic potentials can lead to conducted impulses in the Mauthner axon.

Chemical inhibitory synaptic transmission

Stimulation of the eighth nerve, which excites the Mauthner cell on the same side, inhibits the contralateral Mauthner cell. This, together with the electrical inhibitory circuit through axon collaterals of each Mauthner cell prevents simultaneous contraction in tail muscles on both sides of the animal. The sketch of connections in Figure 4 indicates the

[15]Robertson, J. D. 1963. *J. Cell Biol.* **19**: 201–221.
[16]Robertson, J. D., Bodenheimer, T. S. and Stage, D. E. 1963. *J. Cell Biol.* **19**: 159–199.
[17]Furukawa, T., Fukami, Y. and Asada, Y. 1965. *J. Neurophysiol.* **26**: 759–774.
[18]Furshpan, E. J. 1964. *Science* **144**: 878–880.

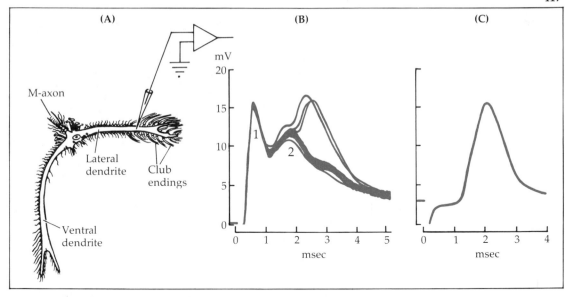

6 **EXCITATORY ELECTRICAL AND CHEMICAL TRANSMISSION in a Mauthner cell.** The eighth nerve is stimulated on the same side (Figure 4). (A) Recording from the distal portion of the lateral dendrite. (B) Ten to twelve superimposed records. The first response (1) appears with a negligible delay and is due to electrical synaptic transmission. The second, later synaptic response (2) is chemical; on three of the sweeps it gives rise to impulses in the axon hillock area at a distance of over 300 μm from the recording site. The impulses therefore appear small. (C) A single impulse generated by direct electrical stimulation of the Mauthner axon. It also appears relatively small because it spreads passively to the recording site in the lateral dendrite, which does not give regenerative responses. (From Diamond, 1968.)

consequences of ipsilateral and of contralateral stimulation of the Mauthner cells.

The chemical inhibition produced by these pathways is similar to other inhibitory processes already discussed in detail.[19] An example of the inhibitory potential and its reversal is shown in Figure 7. The chemical transmitter increases the conductance for chloride ions. Normally, the reversal potential for chloride is close to the resting potential. During the inhibitory transmitter action, the effect of all types of excitatory stimuli, chemical or electrical, is reduced. A convenient test consists of recording from a dendrite while an antidromic nerve impulse from the spinal cord is conducted into the axon hillock region. During an inhibitory potential, the height of the action potential recorded in the dendrite is reduced, and this decrease is a measure of the shunt, that is, of the inhibitory conductance in the cell body and dendrite region.

The fibers of the interneurons that give rise to electrical inhibition run into the axon cap region that is surrounded by a network of glial

Electrical inhibition at the axon cap

[19]Faber, D. S. and Korn, H. 1982. *J. Neurophysiol. 48*: 654–678.

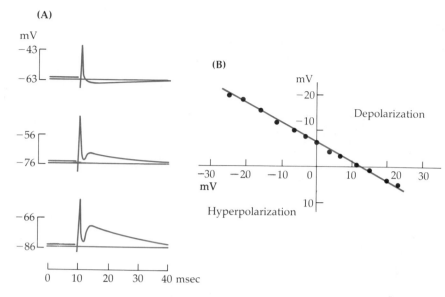

7 CHEMICAL INHIBITORY SYNAPTIC POTENTIALS in a Mauthner cell that was impaled with two microelectrodes, one for recording, the other for passing current through the cell, thereby displacing the membrane potential to various levels (as indicated). (A) Direct stimulation of the Mauthner axon produces first an impulse, followed by a slower inhibitory potential. It depolarizes at the resting potential of −76 mV, increases further at −86 mV, while at −63 mV the potential reverses its polarity. (B) The peak amplitudes of the inhibitory potential (ordinate) are plotted at different membrane levels (abscissa). The reversal potential occurs close to 10 mV depolarization. (After Furukawa and Furshpan, 1963.)

cells (Figure 1). The fibers can be activated by stimulating the eighth nerve or either Mauthner cell (Figure 8). After a brief delay, an extracellular electrode placed accurately in the cap region registers a potential that lasts for about 1 msec.[20] The potential is highly localized, and its sign indicates an increase in external positivity of about 15 mV. This restricted focal hyperpolarization of the axon membrane is generated by the current flow in fibers that penetrate the axon cap; it can be mimicked faithfully by current passed directly into the axon cap through an extracellular microelectrode. The injected currents and the potentials they create are confined to the axon cap region; therefore, they are not recorded as a potential difference between an intracellular electrode placed inside the cell body and a second, indifferent electrode at a distance from the cap region.

The hyperpolarization at the axon cap occurs at a strategically important site because this is the region where impulses are initiated. While it persists, the threshold for excitation is increased in the initial axon segment and conduction can be blocked. The nature of this effect

[20]Furukawa, T. Y. and Furshpan, E. J. 1963. *J. Neurophysiol.* 26: 140–176.

419
TRANS-
FORMATION OF
INFORMATION
BY SYNAPTIC
ACTION IN
INDIVIDUAL
NEURONS

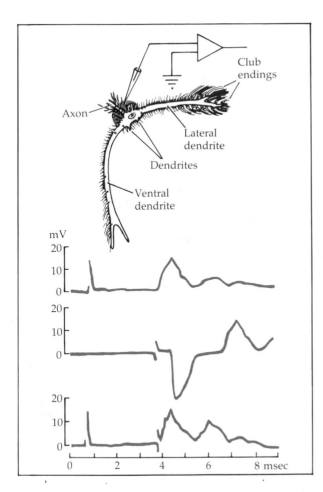

mV

8

ELECTRICAL INHIBITION caused by a hyper-polarization recorded with an electrode within the axon cap region just outside the axon. In record 1 the opposite Mauthner cell is stimulated; it causes hyperpolarization of almost 15 mV (up-ward deflection). In record 2 the Mauthner cell axon is stimulated directly, the first downward deflection being the axonal impulse conducting into the axon cap segment. If both Mauthner axons are stimulated (record 3), the axonal impulse is blocked by the hyperpolarization. (From Furukawa, 1966.)

is interesting because it is one of the few known physiological examples of inhibition produced by extrinsic electrical currents.[21] We do not know how the geometry of the glial packing is arranged to bring about a distribution of current flow through the extracellular spaces around the initial segment in such a way that the underlying membrane becomes hyperpolarized; nor do we know the physiological properties of the glial cells in the axon cap. In this context it is useful to recall another instance in which glial cells confine current flow to a restricted region: Oligodendrocytes make the myelin that channels current flow in nerve fibers through the nodes of Ranvier, thereby ensuring saltatory conduction.

There is experimental evidence for a third type of inhibition at the Mauthner cell. The reduction in the size of excitatory potentials observed during inhibition is too great to be accounted for by increases in conductance of the Mauthner cell. This may be explained by the observation that the action potentials in eighth nerve fibers are reduced in

Presynaptic inhibition at Mauthner cell synapses

[21]Korn, H., Triller, A. and Faber, D. S. 1978. *Proc. R. Soc. Lond.* B 202: 533–539.

amplitude during inhibition.[17] The decrease apparently is brought about by an inhibitory action on the eighth nerve axons themselves, making them less effective in evoking excitatory synaptic potentials in the Mauthner cell. The significance of this presynaptic inhibition is that one input to the postsynaptic cell can be selectively depressed without reducing the excitability of the cell as a whole.

INTEGRATION BY THE SPINAL MOTONEURON

The spinal motoneuron is the most studied mammalian nerve cell, and it has a pivotal role in all movements we perform.[22] Sherrington called the motoneuron the FINAL COMMON PATH because all the neural influences that have to do with movement or posture converge on it. Each motoneuron innervates a group of muscle fibers and together with them forms a functional unit called the MOTOR UNIT. When a motor nerve fiber discharges, all its muscle fibers contract. The motor unit is, therefore, the basic component of normal movement. This is so because individual mammalian skeletal muscle fibers give maximal, nongraded contractions in response to motor nerve impulses. (There is evidence for an exception to this rule in intrafusal muscle fibers and in some of the fibers in muscles that control eye movement.) By keeping track of motor unit discharges in muscles, one obtains a picture of the activity pattern of motoneurons in the spinal cord, even without recording from them. This technique, called electromyography, is a useful tool in clinical and physiological work. In muscles that produce fine movements, such as those of the fingers, each motoneuron innervates fewer fibers than in a larger muscle used to maintain posture. The smoothness and precision of our movements are brought about by varying the number and timing of motor units brought into play.[23] The rapid action of each motor unit is not apparent when the whole muscle contracts, because the individual contributions are asynchronous and are smoothed out by the elastic properties of muscles. For example, the 25,000 muscle fibers in the cat soleus are supplied by 100 α-motoneurons. Contractions of the whole muscle can therefore be graded in 100 steps, depending on the number of motor units, and can be controlled even more finely by the rates at which they fire.

The motor units of a muscle are not homogeneous: Some are faster in their contractions than others and fatigue more rapidly. In a detailed series of experiments in which the properties of motoneurons and muscles were compared, Burke[24] and his colleagues showed that there were four main classes, ranging from "fast twitch, fast fatiguing" to "slow twitch, fatigue resistant." Interestingly, all the muscle fibers of a motor unit had similar properties and histochemical characteristics. Moreover,

[22]Eccles, J. C. 1964. *The Physiology of Synapses.* Springer-Verlag, Berlin.
[23]Adrian, E. D. 1959. *The Mechanism of Nervous Action.* University of Pennsylvania Press, Philadelphia.
[24]Burke, R. E. 1978. *Am. Zool. 18:* 127–134.

421
TRANS-
FORMATION OF
INFORMATION
BY SYNAPTIC
ACTION IN
INDIVIDUAL
NEURONS

the type of motoneuron and its properties (see later) were appropriately matched to the muscle fibers.

Henneman and his colleagues[25] have demonstrated another important mechanism for grading the force with which a movement is executed. This depends on the differences in size of the various motor units that make up the total outflow to a muscle. Some motoneurons supply many muscle fibers, so that the impulses in a single motor axon give rise to a large increment of tension. Other motoneurons innervating relatively few muscle fibers generate less tension. There is evidence that the size of the motoneuron cell body and of its axon is directly related to the number of muscle fibers in the motor unit. This makes it possible to study how the population of motoneurons to a muscle is recruited one by one, by recording the action potentials in ventral roots or electromyograms. Thus, when records are made from extracellular electrodes placed on ventral roots, impulses from large axons appear larger than those from small axons. Motor units of different sizes can therefore be distinguished by their relative spike amplitudes.

When a reflex contraction occurs, small motor units fire first, producing a small increment in tension. As the strength of the reflex is increased, so larger and larger units are recruited, each contributing progressively more tension. A nice control is thereby achieved, enabling fine or coarse movements to be delicately graded. It is clearly efficacious for smaller units to fire at the beginning of a contraction rather than at the end when the percentage tension increment they produce would be far less. Thus, in the soleus muscle of the cat, the firing of a small motoneuron may give rise to an increase in tension of 4 g, whereas a larger unit causes 40 g; and the maximum contraction brought about by all the motor units firing is about 2200 g. Plainly the small motor unit would be ineffective if brought in late during the contraction, and the large unit firing at the beginning would impair the fine grading of the contraction.

This aspect of motor control is reminiscent of principles used for sensory signaling: As the intensity of a stimulus becomes greater, so the central nervous system becomes less able to detect small differences. For example, a change in weight from 2 to 3 g is readily appreciated, whereas a change from 2002 to 2003 g is not. Much of our perception of the world is logarithmic and we act accordingly. You would not mind much paying $2003 for an item costing $2002 but would be outraged at paying $3 for a $2 item.

The recruitment and fine control of motoneurons to produce coordinated movements requires that multiple influences should play upon them in the appropriate sequence and with appropriate balance. It is therefore not surprising that the average motoneuron has many connections on it, up to about 10,000 synapses by some estimates, connections providing information derived from all over the body and from higher centers.

The size principle

Synaptic inputs to motoneurons

[25]Henneman, E., Somjen, G. and Carpenter, D. O. 1965. *J. Neurophysiol.* 28: 560–580.

At the cellular level, the fine structure of the synapses and the mechanisms acting on motoneurons resemble those found in crustaceans and in Mauthner cells.[26] Once again, chemical postsynaptic excitation, inhibition, and presynaptic inhibition interact in determining the end result of converging signals. As in the Mauthner cell, there is good evidence that the impulse originates in one particular region of the cell, the axon hillock. Figure 9 shows a stained section through a motoneuron with numerous synaptic contacts in the relatively thin section that goes through the cell body. The real density of synaptic boutons covering the cell surface is more apparent in the drawing based on electron microscopic studies (Figure 9B).

Through extensive work, pioneered by Lloyd[27] and by Eccles[22] and his colleagues, much is now known about the mechanisms of synaptic transmission at this cell and the interaction of excitatory and inhibitory synapses (Figure 10). One important excitatory input arises from the muscle spindles: The group Ia and group II afferent fibers make monosynaptic excitatory connections. By painstakingly recording from the motor neurons supplying a muscle, it has been shown by Mendell and Henneman[28] that each Ia afferent fiber sends an input to as many as 300 motoneurons, virtually the entire population. Conversely, the motoneuron is the site of convergence of monosynaptic inputs arising from all of the spindles of the same muscle.

As with the Mauthner cell, there is order and precision in the anatomy of these connections. Individual Ia afferent fibers injected with horseradish peroxidase have been observed to terminate on the soma and the proximal regions of dendrites of the motoneuron.[29,30] Typically, Brown and Fyffe (Figure 11) found that a single Ia collateral may make two to five contacts in the form of presynaptic boutons on the dendritic tree and about two on the soma. Somehow during development the territories have been allocated in such a way that a single collateral of a Ia axon generally provides all of these boutons, while other branches of the same axon may pass close to that motoneuron to supply others, each one doing so on its own (see also Chapter 19). Moreover, all the contacts made by the single collateral on the various dendrites are

[26]Berthold, C.-H., Kellerth, J.-O. and Conradi, S. 1979. *J. Comp. Neurol. 184*: 709–740.
[27]Lloyd, D. P. C. 1943. *J. Neurophysiol. 6*: 317–326.
[28]Mendell, L. M. and Henneman, E. 1971. *J. Neurophysiol. 34*: 171–187.
[29]Brown, A. G. and Fyffe, R. E. W. 1981. *J. Physiol. 313*: 121–140.
[30]Burke, R. E., Walmsley, B. and Hodgson, J. A. 1979. *Brain Res. 160*: 347–352.

9 **SYNAPSES ON MAMMALIAN MOTONEURONS. (A)** Silver-stained section of a spinal motoneuron. Axons and their terminal synaptic boutons converging on the dendrites and cell body are stained black. Their diameters are several micrometers. **(B)** Drawing based on an electron-microscopic cell study of a motoneuron cell body. **(C)** Several nerve terminals are apposed to two dendrites, D, of a motoneuron. Three chemical synapses are marked by arrows. (A, unpublished photograph by F. DeCastro; B, from Poritsky, 1969; C, from Peters, Palay, and Webster, 1976.)

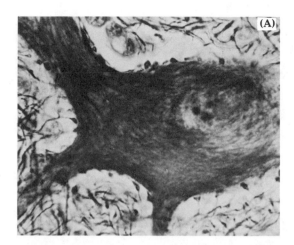

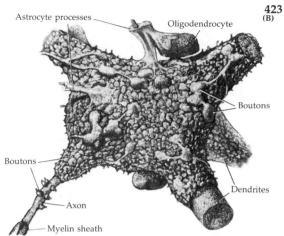

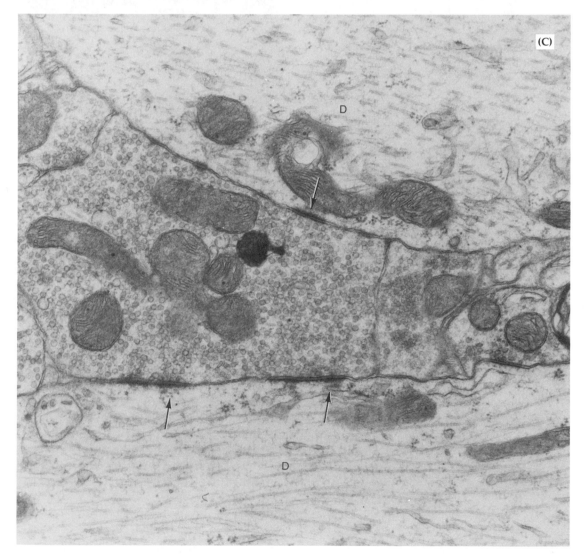

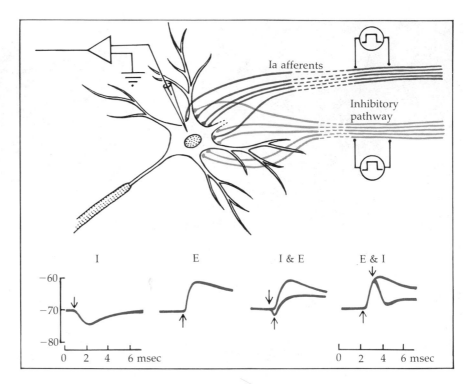

Ia afferents

Inhibitory
pathway

I   E   I & E   E & I

−60

−70

−80

0   2   4   6 msec   0   2   4   6 msec

10   INTERACTION between excitatory (E) and inhibitory (I) synaptic potentials in a motoneuron of the cat. The sum of the two effects determines whether an impulse is initiated at the initial segment of the axon.
(After Curtis and Eccles, 1959.)

situated at about the same distance from the soma. It has been suggested that, as in crustacean and cockroach axons, conduction failure
might perhaps occur at the branch points of Ia afferents.[31]

In numerous studies, the spatial aspects of integration have been
correlated with the structure and membrane properties of motoneurons.
Motoneurons have been filled with horseradish peroxidase or dyes and
their dimensions reconstructed. These findings, combined with electrical recordings from different sites in the cell, allow reasonable estimates
of the decline of a synaptic potential as it spreads passively from dendrite to cell body.

Unitary synaptic
potentials   As one might expect from the anatomy, the impulse in a single Ia
afferent fiber gives rise to only a very small monosynaptic excitatory
potential in the motoneuron, probably corresponding to only one or a
few quanta. One technique for demonstrating this is to dissect a muscle
nerve until only a single Ia afferent fiber remains and to examine its
effect.[32] A new, sensitive way of assessing small effects within the
central nervous system is to average the potentials evoked by a single

[31]Jack, J. J. B., Redman, S. J. and Wong, K. 1981. *J. Physiol. 321*: 65–96.
[32]Kuno, M. 1971. *Physiol. Rev. 51*: 647–678.

425
TRANS-
FORMATION OF
INFORMATION
BY SYNAPTIC
ACTION IN
INDIVIDUAL
NEURONS

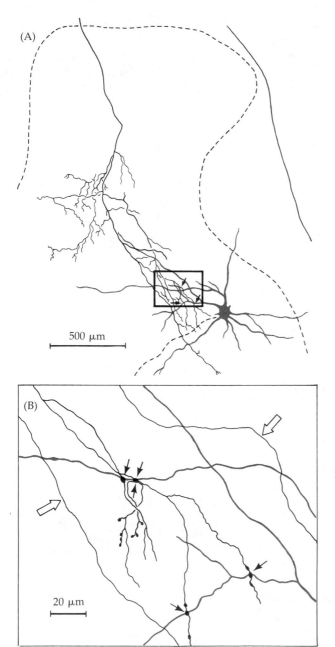

11

TERMINALS OF Ia AFFERENT COLLATERAL on a spinal motoneuron (cat). The Ia afferent fiber and the motoneuron were identified by intracellular microelectrodes and injected with horseradish peroxidase. (A) Reconstruction of the afferent fiber and the cell, made from serial sections. (B) Enlarged region showing contacts made on the dendrites by the Ia afferent (marked by dark arrows). These are the presumed sites of synapses. All the contacts are made by one collateral of the Ia afferent; the other collaterals of the same afferent (white arrows) pass by the dendrites without establishing contacts. Those fibers presumably end on other motoneurons of the pool. (Modified from Brown and Fyffe, 1981.)

input.[33,34] The scheme is illustrated in Figure 12. With time-locked events, such as the action potential of a Ia afferent fiber and the excitatory potential evoked in a motoneuron, repeated impulses are used to trigger an averager. This sums the voltages recorded from the moto-

[33]Kirkwood, P. A. and Sears, T. A. 1982. *J. Physiol. 322*: 287–314.
[34]Honig, M. C., Collins, W. F. and Mendell, L. 1983. *J. Neurophysiol. 49*: 886–901.

neuron in successive traces. Random deflections upward or downward cancel each other and become lost. Consistent signals arising at a particular latency, even if only a very few microvolts in amplitude, become progressively reinforced compared to the baseline. Since spindle afferents can fire typically at 50–400 per second, the average consequences

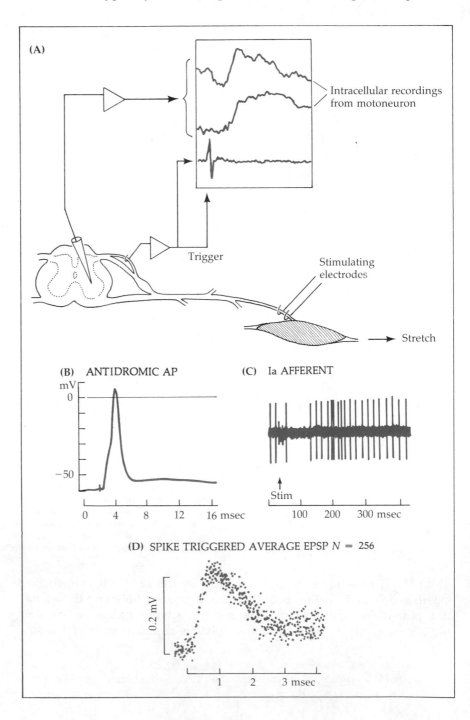

(A)

Intracellular recordings from motoneuron

Trigger

Stimulating electrodes

Stretch

(B) ANTIDROMIC AP

mV

0

−50

0   4   8   12   16 msec

(C) Ia AFFERENT

Stim

100   200   300 msec

(D) SPIKE TRIGGERED AVERAGE EPSP $N = 256$

0.2 mV

1   2   3 msec

427
TRANS-
FORMATION OF
INFORMATION
BY SYNAPTIC
ACTION IN
INDIVIDUAL
NEURONS

of several hundred or thousand potentials can be readily estimated (Figure 12). In this way it has been shown that the unitary potential evoked by a Ia afferent is approximately 150 μV.[33] A related and valuable technique for discerning the subtle influences of a presynaptic cell is to determine how it influences the frequency of firing of the postsynaptic cell, again by averaging the effects that a single Ia or II afferent has on the pattern of discharge of a motoneuron in a larger number of trials.[35] This obviates the necessity for recording intracellularly from the postsynaptic neuron. A small depolarizing potential can have a pronounced effect on the probability of impulses arising in the motoneuron. From knowledge of the relationship between frequency and membrane potential one can work backward and estimate from the alteration in the firing pattern the magnitude of a synaptic potential. In sum, the excitation of the motoneuron through the stretch reflex or by descending commands from higher centers depends upon roughly simultaneous activation of many converging fibers.

As in the crayfish and Mauthner cell synapses, inhibition is brought about by an increase in chloride conductance[22] and by presynaptic inhibition upon the endings of group Ia afferent fibers originating in muscle spindles. In response to synchronous activation of group Ib fibers from tendon organs, a large depolarization of muscle afferents, particularly the Ia axons, can be recorded with intracellular microelectrodes. The mechanism by which depolarization is produced and its precise role in modulating transmitter release for presynaptic inhibition are still not certain.[36] In addition, recurrent inhibition feeds back onto the cell itself as a result of activity through the Renshaw cell.[22]

Inhibition

COMPARISON OF INTEGRATION IN THE THREE CELL TYPES

The following paragraphs compare the various factors that contribute to the summation of different synaptic inputs on the three cell types:

[35]Kirkwood, P. A. 1979. *J. Neurosci. Methods* 1: 107–132.
[36]Rudómin, P., Engberg, I. and Jiménez, I. 1981. *J. Neurophysiol.* 46: 532–548.

12 SPIKE TRIGGERED AVERAGING OF UNITARY EPSPs. (A) Scheme for averaging: impulses from a single stretch receptor recorded in a dorsal root filament are used to trigger sweeps of the oscilloscope. An intracellular microelectrode records the membrane potential of a spinal motoneuron. The oscilloscope traces show the incoming sensory impulse (below) and two excitatory potentials (above). In practice a single epsp or ipsp is too small to measure reliably. (B) Stimulation of the muscle nerve evokes an impulse in the intracellular record, establishing its identity as a motoneuron. (C) Extracellular recording of Ia afferent discharge in dorsal root filament. Stimulation of the muscle (arrow) causes a twitch and cessation of spindle discharge (Chapter 15). (D) Epsps evoked by a single Ia afferent spike are small, about 50–250 μV, and are often buried in background noise. Since each epsp arises at about the same latency after the incoming spike (which triggers the oscilloscope trace), when many traces are averaged all random, non-time-linked events cancel out. As a result, the magnitude and time course of the averaged epsp become apparent. (After Hilaire, Nicholls, and Sears, 1983.)

crustacean muscle, Mauthner cell, and spinal motoneuron. The Mauthner cell and the spinal motoneuron have in common a single restricted site for impulse initiation, the axon hillock. The membrane potential of this part of the cell is the result of all that goes on in the cell body and the dendrites—the conflict or interplay between excitation and inhibition. The feature of making the initial axon segment the focal point of integrative activity has important consequences and can be most easily discussed in detail for the Mauthner cell. A similar analysis in the motoneuron is more difficult; it has, however, become more practicable with recent information about the sites of synaptic inputs and with techniques for assessing the small effects of single axons.

The integrative activity on the cell body of the Mauthner cell and its dendrites can be summarized roughly as follows, with the help of the sketch in Figure 4. When the ipsilateral eighth nerve fibers are stimulated, excitatory synaptic potentials arise at the distal end of the lateral dendrite. From there they spread passively to the axon hillock up to 300 μm away, decreasing by more than 50 percent through the cable properties of the cell. If at this point the depolarization reaches threshold, an impulse and a tail flip follow. Chemical postsynaptic inhibition tends to prevent firing in two ways: (1) To the extent that hyperpolarization spreads to the axon hillock region, it drives the potential there away from threshold; (2) more important, the increase in conductance "shorts out" the excitatory depolarization and, by reducing the length constant of the dendrite, attenuates its spread. In other words, excitatory current spread is curtailed before it can reach the critical axon hillock region. Clearly, what matters is not simply the presence of a synapse: The SITES OF THE SYNAPSES ON A NEURON play a key role in establishing their effectiveness. For example, the inhibition would be less useful if it occurred at the tip of the dendrite or if the excitatory synapses were close to the axon hillock; similarly, electrical inhibition blocks impulse initiation most effectively if it occurs in the critical area where impulses start. Finally, through presynaptic inhibition, the input from the ipsilateral eighth nerve can be reduced without influencing the ability of the cell to respond to other excitatory synapses on its surface.

In contrast to the motoneuron and the Mauthner cell, spatial considerations of the distribution of inhibitory and excitatory synapses on muscle fibers in crustaceans are less important. Since the excitatory and inhibitory neurons in the extensor make synapses close together, graded doses of excitatory and inhibitory action can counteract each other rather uniformly over the whole surface of the muscle fiber. Thus, with synapses scattered over its surface, the muscle fiber behaves evenly along its length, because the electrical spread of current between adjacent synapses blurs the distinction between active synaptic spots and the intervening inactive extrasynaptic regions. Certain nonspiking neurons within the central nervous system behave in a similar manner.[37] For

[37]Shepherd, G. M. 1983. *Neurobiology.* Oxford University Press, New York, pp. 134–147.

429
TRANS-
FORMATION OF
INFORMATION
BY SYNAPTIC
ACTION IN
INDIVIDUAL
NEURONS

example, horizontal cells in the retina sum excitatory and inhibitory inputs, as a result of which they can modulate without impulses the release of transmitter onto postsynaptic cells (Chapter 2).

In neurons such as the Purkinje cells of the cerebellum and pyramidal cells in the hippocampus, another factor influencing integration is the presence of dendritic action potentials.[12] These cells, like the motoneuron and the Mauthner cell, receive excitatory and inhibitory inputs that sum and initiate impulses at the axon hillock. But, in addition, their dendrites when sufficiently depolarized by synaptic action or by impulses in the soma give overshooting action potentials. These often occur in bursts and have been shown to be caused by calcium rather than sodium entry (Chapter 6). In such a cell with long dendrites, a distant input can have a large effect. Moreover, the dendrite itself can make synapses on another neuron;[38] thus, the impulse in a dendrite also allows it to act as a presynaptic element.

An important principle that bears on problems of development emerges from the preceding discussion of integration. Clearly, for all the various types of synaptic interactions to work, neurons must make highly precise and specific connections with the appropriate postsynaptic targets. Furthermore, during development, synapses are formed in the appropriate regions of those cells on particular dendrites or regions of the cell body.

SUGGESTED READING

General reviews

Atwood, H. L. 1982. Synapses and neurotransmitters. In H. L. Atwood and D. C. Sandeman (eds.). *Biology of Crustacea*, Vol. 3. Academic, New York, pp. 105–150.

Burke, R. E. 1981. Motor units: Anatomy, physiology and functional organization. In V. Brooks (ed.). *Handbook of Physiology. Section I: The Nervous System*, Vol. 2, Part 1. American Physiological Society, Bethesda, pp. 345–422.

Faber, D. S. and Korn, H. (eds.). 1978. *Neurobiology of the Mauthner Cell.* Raven, New York.

Govind, C. K. and Atwood, H. L. 1982. Organization of neuromuscular systems. In H. L. Atwood and D. C. Sandeman (eds.). *Biology of Crustacea*, Vol. 3. Academic, New York, pp. 63–98.

Henneman, E. and Mendell, L. M. 1981. Functional organization of motoneuron pool and its inputs. In V. Brooks (ed.). *Handbook of Physiology. Section I: The Nervous System*, Vol. 1, Part 1. American Physiological Society, Bethesda, pp. 423–507.

Shepherd, G. M. 1983. *Neurobiology.* Oxford University Press, New York.

Sypert, G. W. and Munson, J. B. 1981. Basis of segmental motor control: Motoneuron size or motor unit type? *Neurosurgery 8:* 608–621.

Original papers

Barrett, J. N. and Crill, W. E. 1974a. Specific membrane properties of cat motoneurones. *J. Physiol. 239:* 301–324.

[38]Shepherd, G. M. 1978. *Sci. Am. 238:* 92–103.

Barrett, J. N. and Crill, W. E. 1974b. Influence of dendrite location and membrane properties on the effectiveness of synapses on cat motoneurones. *J. Physiol. 239*: 325–345.

Brown, A. G. and Fyffe, R. E. W. 1981. Direct observations on the contacts made between Ia afferent fibres and α-motoneurones in the cat's lumbosacral spinal cord. *J. Physiol. 313*: 121–140.

Furshpan, E. J. and Furukawa, T. 1962. Intracellular and extracellular responses of the several regions of the Mauthner cell of the goldfish. *J. Neurophysiol. 25*: 732–771.

Furukawa, T. Y. and Furshpan, E. J. 1963. Two inhibitory mechanisms in the Mauthner neurons of goldfish. *J. Physiol. 26*: 140–176.

Honig, M. G., Collins, W. F. and Mendell, L. M. 1983. α-Motoneuron EPSPs exhibit different frequency sensitivities to single Ia-afferent fiber stimulation. *J. Neurophysiol. 49*: 886–901.

Kirkwood, P. A. and Sears, T. A. 1982. Excitatory post-synaptic potentials from single muscle spindle afferents in external intercostal motoneurones of the cat. *J. Physiol. 322*: 287–314.

Onodera, K. and Takeuchi, A. 1980. Distribution and pharmacological properties of synaptic and extrasynaptic glutamate receptors on crayfish muscle. *J. Physiol. 306*: 233–249.

INTEGRATIVE MECHANISMS IN THE CENTRAL NERVOUS SYSTEM FOR THE CONTROL OF MOVEMENT

How are complex, coordinated movements of the body such as rhythmical respiration, walking and running, eye movements and singing produced by the central nervous system? Experiments made in cats, monkeys, other vertebrates, and invertebrates have shown that feedback from sensory receptors in the periphery is not essential for the initiation of the movements or for producing the sequence in which they are executed. Thus, inspiration and expiration continue to alternate in a regular rhythm in cats in which the dorsal roots conveying information from inspiratory and expiratory muscles and joints have been severed. The drive for the rhythm to be maintained arises from neurons that are situated in the midbrain and medulla and that produce alternating excitatory and inhibitory drives on the appropriate motor neurons. Similarly, in cats with deafferented limbs, walking and stepping movements occur when a specific circumscribed region of the midbrain is stimulated electrically by regular pulses without periodic fluctuation: Flexors and extensors of the appropriate limbs undergo alternating contractions in a regular rhythm.

Information from the periphery is, however, essential for controlling the extent and the rate of movement. For example, increased levels of CO_2 in the air or stretch of the muscle spindles in thoracic muscles influence the rate and depth of respiration. Similarly, the walking, trotting, or galloping responses of cats to altered speeds of a treadmill depend on continuous inflow of information from sensory receptors in the limbs.

In an intact animal that is walking or running, the position of the head, neck, and trunk needs to be constantly adjusted so as to maintain an upright posture. For this to be achieved, the central nervous system requires information from sensory receptors in the vestibular apparatus (responding to tilt or rotation of the head) and in the neck muscles, and also from receptors in joints, skin, and muscles throughout the body. As in the visual system, there is an orderly projection of the body through successive relays to the cortex, where the primary somatosensory area (analogous to striate cortex) lies immediately posterior to the motor area. In both somatosensory and motor cortex, the body shape is represented unequally, with more neurons devoted to structures such as hands, fingers, or lips than to trunk or thighs. Again, as in the visual cortex, cells having similar properties are arranged in columns running through the thickness of the gray matter.

Clues about higher aspects of sensory and motor integration required for complex tasks such as speech, writing, or putting on a shirt are provided in patients by clinical disorders following lesions to selected areas of the brain that result in characteristic defects in sensory perception and coordinated movements.

For a leech or a fish to swim to its food, for an owl or a cat to catch a mouse, for a child to ride a bicycle or Vladimir Horowitz to perform the Liszt B Minor Sonata, virtually all the muscles of the body must be brought into play in rapid succession, contracted or relaxed in harmony. Where in the brain of a higher animal the decisions are made or how voluntary actions begin are complex questions with no solution in sight. At the same time the importance of studying the mechanisms involved is evident. As Adrian wrote in his remarkable book *The Mechanism of Nervous Action*, "The chief function of the central nervous system is to send messages to the muscles which will make the body move effectively as a whole"[1]–a somewhat cynical, but unchallengeable, statement.

In recent years new light has been shed on the ways in which the nervous system achieves the coordination necessary for purposeful movements. A general principle to emerge from experiments on both invertebrates and vertebrates is that the central nervous system contains within itself circuitry that is able to develop a programmed sequence of muscular contraction that can unfold and continue in the absence of sensory feedback. Thus, after the central nervous system of a cockroach has been disconnected from all sensory input, appropriate motoneurons that would activate the muscles of the legs in the intact animal continue to fire impulses in a regular order.[2] Toward muscles that bend or straighten or lift a leg are sent signals that would normally result in walking, one leg then the other being lifted, advanced, and lowered. Similarly, the central nervous system of the leech, even after having been completely removed from the animal, will send appropriate commands sequentially from its motoneurons through nerves along its length for the nonexistent body to swim in a coordinated manner[3] (Chapter 18). Several lines of evidence show that comparable mechanisms also exist within the vertebrate central nervous system. At first this may seem surprising since, as Sherrington suggested,[4] it appeared

[1]Adrian, E. D. 1959. *The Mechanism of Nervous Action*. University of Pennsylvania Press, Philadelphia.

[2]Pearson, K. G. and Iles, J. F. 1970. *J. Exp. Biol.* 52: 139–165.

[3]Stent, G. S. and Kristan, W. B. 1981. In K. J. Muller, J. G. Nicholls and G. S. Stent (eds.). *Neurobiology of the Leech*. Cold Spring Harbor Laboratory, Cold Spring Harbor, NY, Chapter 7.

[4]Sherrington, C. S. 1906. *The Integrative Action of the Nervous System*. Yale University Press, New Haven, pp. 68, 213 (1961 edition).

433
INTEGRATIVE
MECHANISMS
IN THE CNS
FOR THE CONTROL
OF MOVEMENT

reasonable that sensory feedback from the periphery, rather than centrally located mechanisms, might impose the patterns of limb movements associated with, for example, walking. According to such a chain reflex hypothesis, active bending or "flexion" of one leg would stretch antagonistic muscles, causing a sensory discharge of their muscle spindles. This, in turn, could cause the next movement by excitatory connections to the extensor motoneurons of the same muscles and of synergists in the same and other limbs and by inhibitory connections to antagonists. Thus, the central effects of interacting reflex arcs would shape the movements, each contraction of a particular muscle proceeding until the state of the other muscles reflexly stopped it or started it again.

In practice, certain movements could not be achieved in this way. For example, when birds sing the movements of the muscles follow each other in a rapid, orderly sequence without sufficient time for the feedback loop to be completed: The next instructions are sent out from the central nervous system before the first movement or sound has been perceived or sensed by the central nervous system. Similarly, the eyes in some animals, cats, and dogs are moved rapidly and accurately without feedback from the eye muscles, which contain virtually no muscle spindles or tendon organs. (Spindles are, however, present in the extraocular muscles of the pig, goat, giraffe, white-tailed gnu, chimpanzee, and man.)[5] The importance of central control in eye movements is demonstrated by a simple experiment: If the gaze is suddenly shifted from one point in the external world to another, the world appears stationary; but if the eye is pushed gently by a finger instead of moving voluntarily, the world appears to rotate. During the voluntary movement of the eye, signals have been sent not only to the eye muscles but also to regions where visual information is integrated. This allows the expected position of the visual field to be taken account of. In contrast, no such signals are sent to visual information centers when the finger movement is programmed to press on the eyeball. Consequently, the visual field is not expected to change position.

Two examples will be used to illustrate how specific groups of neurons within the mammalian central nervous system produce coordinated movements—respiration and walking. The automaticity of movements in invertebrates is considered in Chapter 18.

NEURONAL CONTROL OF RESPIRATION

Two antagonistic sets of muscles are responsible for drawing air into the lungs and expelling it. During inspiration, the ribcage is raised by the external intercostal muscles, and the diaphragm contracts. As a result, the volume of the chest is increased, the lungs expand, and air enters. Expiration is not due simply to recoil but is an active process

---

[5]Matthews, P. B. C. 1972. *Mammalian Muscle Receptors and Their Central Actions*. Edward Arnold, London, p. 52.

accompanied by contraction of the internal intercostal muscles. Other muscles of the thorax and abdomen also contribute to a variable extent, depending on the posture of the animal and the rate and depth of respiration.[6] An example of the respiratory rhythm is shown in Figure 1. Activity of each muscle can be registered by strain gauges or by recording the electrical activity with wire electrodes embedded in the body of the muscle—electromyography (EMG). Figure 1 shows that as the inspiratory and expiratory muscles contract, bursts of action potentials corresponding to motor units are recruited and the frequency of firing increases; it is apparent that the two sets of muscles contract out of phase.

As with limb muscles, the stretch reflex contributes to the inspiratory and expiratory movements by maintaining the excitability of the motoneurons. During inspiration and expiration, the internal and external

[6]Da Silva, K. M. C., Sayers, B. McA., Sears, T. A. and Stagg, D. T. 1977. *J. Physiol.* 266: 499–521.

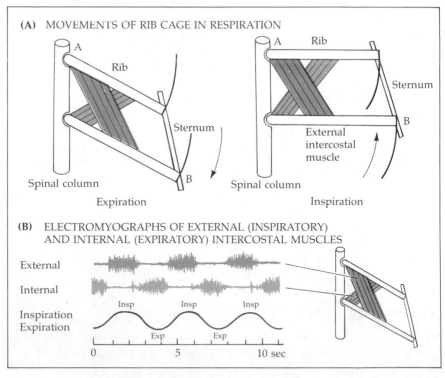

1   **MOVEMENTS OF RIB CAGE AND RESPIRATORY MUSCLES** during inspiration and expiration. **(A)** Actions of the external intercostal muscles (raise ribs in inspiration) and the internal intercostal muscles (depress ribs in expiration). **(B)** Activity of respiratory muscles in cat, recorded by needle electrodes. The activities of the external and internal intercostal muscles are out of phase.

435
INTEGRATIVE
MECHANISMS
IN THE CNS FOR
THE CONTROL
OF MOVEMENT

intercostal muscles are stretched alternately; throughout the cycle, as shown earlier, the afferent discharge from muscle spindles is maintained at a high frequency, even when the muscles are actively contracting, owing to the efferent output through the γ system[7,8] (Chapter 15, Figure 13). Each Ia afferent fiber firing at approximately 100/sec contributes excitatory synaptic potentials to the homonymous motor neurons (i.e., motor neurons supplying the same muscle) and, by an indirect pathway, inhibition to the motor neurons of antagonistic muscles. In the diaphragm of the cat, there are few if any muscle spindles and no true stretch reflex.

How is the respiratory rhythm generated? In cats, after the dorsal roots of the thoracic cord have been severed, a procedure depriving the central nervous system of all sensory feedback, respiration continues, inspiration being followed by expiration. Similarly, if the animal is curarized, a procedure paralyzing the respiratory muscles, a rhythmical motor outflow persists.[6,9] In contrast, section at the level of the medulla abolishes breathing and the animal dies. Within the midbrain and medulla are situated pools of neurons that fire during inspiration or expiration and that produce excitation and inhibition on the appropriate respiratory motor neurons. For example, motoneurons supplying an external intercostal muscle (inspiratory) show progressive depolarization during inspiration.[10] This is caused by a barrage of epsps arising from neurons in higher centers of the medulla or pons and leads to the production of action potentials. The inspiratory phase is terminated by a burst of inhibitory potentials to the inspiratory neurons from other central neurons in closely adjacent regions that are associated with expiration. Figure 2 shows the "central respiratory drive potentials." In Figure 2D, the potentials that terminate inspiration are reversed by hyperpolarization of the motor neuron.

At present the properties and interconnections of the neurons that generate the respiratory rhythm within the central nervous system are not fully understood.[11] In theory, one possibility is that central neurons possess inherent rhythmicity (as in heart muscle and certain neurons in invertebrates) and are therefore able to act as pacemakers. Alternatively, reciprocal inhibitory interactions could occur between pools of neurons that turn each other on and off sequentially. Although such interactions have actually been observed, the precise nature of the oscillatory mechanism has not yet been established. The situation is easier to analyze in invertebrates, such as the leech,[3] where both types of mechanism—endogenous pacemaker cells and rhythm generating circuits—have been observed.

The preceding considerations show that the role of feedback from the periphery is not to give rise to rhythmicity. Rather, stretch reflexes

[7]Sears, T. A. 1964a. *J. Physiol. 174*: 295–315.
[8]Critchlow, V. and von Euler, C. 1963. *J. Physiol. 168*: 820–847.
[9]Eldridge, F. L. 1977. *Fed. Proc. 36*: 2400–2404.
[10]Sears, T. A. 1964b. *J. Physiol. 175*: 404–424.
[11]Cohen, M. I. 1979. *Physiol. Rev. 59*: 1105–1173.

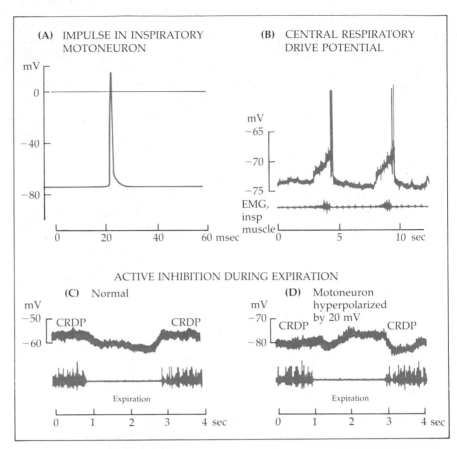

2 CENTRAL RESPIRATORY DRIVE POTENTIAL (CRDP) in an inspiratory motoneuron of cat. (A) Intracellular recording from a motoneuron supplying an inspiratory muscle, showing impulse at low gain and fast sweep speed. (B) At high gain and slower sweep speed, the bursts of excitatory potentials that depolarize the cell (CRDP) and give rise to impulses are apparent. The lower trace (EMG) shows the respiratory cycle and is the record from an inspiratory muscle. (C, D) Evidence that inhibition terminates the CRDP. (C) The depolarization and repolarization of the motoneuron at its normal "resting" potential. In (D) the motoneuron was hyperpolarized by 20 mV. Now during expiration, the membrane became depolarized as the inhibitory potentials that terminate inspiration were reversed. (Modified from Hilaire, Nicholls, and Sears, 1983.)

act to provide a tonic drive to the motor neurons by maintaining a steady depolarization and a raised level of excitability.[7] Although each individual Ia afferent fiber contributes only a small unitary synaptic potential approximately 200 μV in amplitude, the combined summated actions of a number of muscle spindles produce clear effects on the contractions of respiratory muscles. Figure 3 shows the effect of stretching or relaxing an inspiratory muscle (the levator costae) by pulling on its tendon. With each inspiration of the animal, the electromyogram of

437
INTEGRATIVE
MECHANISMS
IN THE CNS FOR
THE CONTROL
OF MOVEMENT

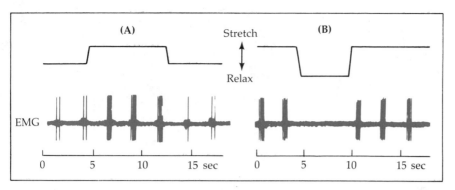

3 STRETCH REFLEX OF AN INSPIRATORY MUSCLE. During each inspiration a small muscle, the levator costae, contracts and the EMG shows bursts of spikes. (A) Stretch of the muscle by pulling on its tendon reflexly increases the activity and the force of contraction. (B) Relaxation of the muscle, which removes the reflex drive, gives rise to reduced activity and contraction. (From Hilaire, Nicholls, and Sears, 1983.)

the muscle shows a burst of spikes. The activity is reflexly enhanced by stretch (A) and inhibited by relaxation (B) of the muscle. In addition, stretch reflexes from lungs as well as muscle serve to modulate the rate and depth of breathing. For example, section of the vagus nerve carrying stretch receptor fibers from the lungs leads to prolongation of inspiration.[12]

Of key importance for rhythmicity is the level of CO_2 in the inspired air.[13] Under conditions of reduced inspiratory CO_2, respiration becomes weaker and can be suppressed; conversely the rate and depth of respiration are increased by raised levels of CO_2. This effect in part depends upon inputs from chemoreceptors in the periphery but mainly from neurons that are situated within the central nervous system in the medulla and are sensitive to the level of CO_2. The firing patterns of individual medullary neurons that control expiratory and inspiratory motoneurons have been shown to be influenced critically by CO_2: Changes in the steady level of CO_2 are translated into pronounced changes in the frequencies of firing of the interneurons and, therefore, of the motoneurons.

LOCOMOTION:
WALKING, TROTTING, AND GALLOPING IN CATS AND DOGS

A striking feature of locomotion is that in vertebrates there is a consistent, highly stereotyped pattern of limb movements (Figure 4).[14] In

---

[12]von Euler, C. 1977. *Fed. Proc. 36*: 2375–2380.
[13]Bainton, C. R., Kirkwood, P. A. and Sears, T. A. 1978. *J. Physiol. 280*: 249–272.
[14]Grillner, S. 1975. *Physiol. Rev. 55*: 274–304.

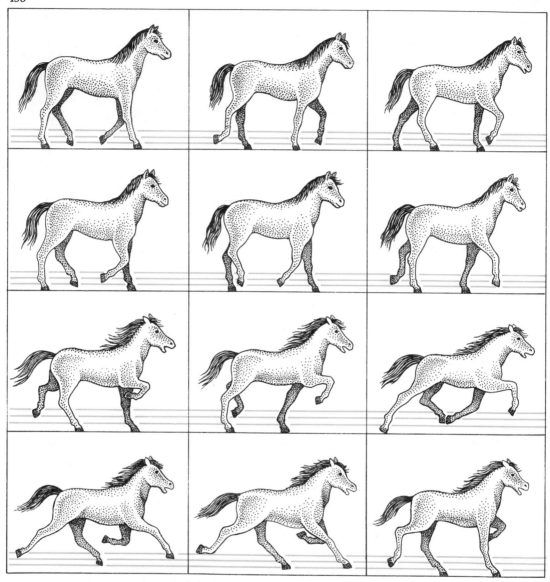

4 WALKING AND GALLOPING HORSE. While walking, only one foot at a time leaves the ground. During a gallop, the animal literally flies through the air. (Modified from Pearson, 1976.)

Figure 5, for a walking cat, the left hindlimb is lifted off the ground first, then the left forelimb, the right hindlimb, and the right forelimb. This sequence, which provides for stabilization by the forelimbs while the hindlimbs propel the animal, is common to frogs, crocodiles, rabbits, and cats (but not fishes): the tendency to turn (the turning couple), produced by the hindlimb, is counteracted by the forelimbs, thereby

439
INTEGRATIVE
MECHANISMS
IN THE CNS FOR
THE CONTROL
OF MOVEMENT

preventing rotation and enabling movement straight forward. Even in invertebrates with six legs, such as cockroaches, the same pattern is observed.[15] During locomotion each leg executes an elementary stepping movement that consists of two phases: (1) a "swing" phase during which the leg is flexed, raised off the ground, swung forward, and extended; (2) a "stance" phase during which the leg is in contact with the ground, moving backward in relation to the direction taken by the body (Figures 5 and 6).

That the gait of a cat undergoes striking changes as its speed increases is shown in Figure 5. While walking, only a single leg is raised off the ground at any one time. As the speed increases to a trot, two

[15]Pearson, K. 1976. *Sci. Am. 235*: 72–86.

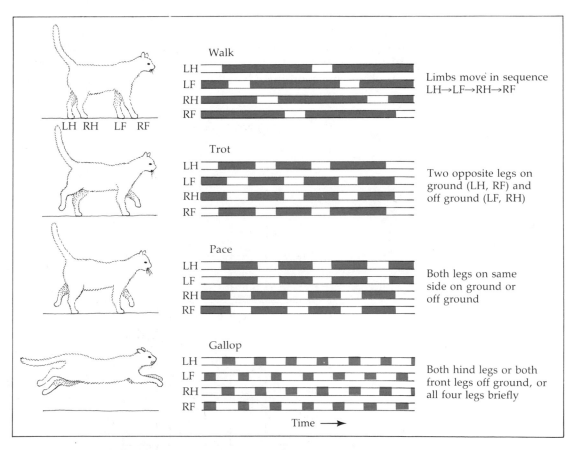

5 **STEPPING PATTERN OF A CAT** during walk, trot, pace, and gallop. The white bars show the time that a foot is off the ground, the blue bars the time that it is on the ground. While the animal walks, the legs are moved in sequence first on one side, then on the other. In a trot, two diagonally opposite limbs are raised at once. In a pace the rhythm changes again, both limbs on one side being raised at the same time. Faster still, hind limbs, then front limbs, leave the ground. (After Pearson, 1976.)

legs are raised off the ground at once—one front and one back on opposite sides of the animal. Still faster, at a gallop, the two front legs and then the two back legs alternate in leaving the ground. The increase in speed is accomplished by shortening the time that each leg stays on the ground—the stance phase. Thus, as the cat moves faster, each leg is extended for a briefer period before being bent, raised, and moved forward. At all speeds from a slow walk to a gallop, the time spent off the ground by each leg as it swings forward is little altered (Figure 6).

As early as 1911 Graham Brown[16] showed that elementary circuits required for walking movements in cats appeared to possess semiautomatic properties. The raising and placing of two hind feet in alternation could be achieved in a cat after its spinal cord had been transected. Moreover, the animal's hind legs have been shown to move appropriately after being placed on a treadmill when certain drugs are given, such as dopa (a precursor for biogenic amines; Chapter 12).[14]

Recent experiments made in the Soviet Union by Shik, Orlovsky, and Severin[17] and in Sweden by Lundberg, Grillner, and their colleagues[14] have provided new evidence for the role of central mechanisms in producing coordinated walking movements. When the upper

[16]Brown, T. G. 1911. *Proc. R. Soc. Lond. B 84*: 308–319.
[17]Shik, M. L. and Orlovsky, G. N. 1976. *Physiol. Rev. 56*: 465–501.

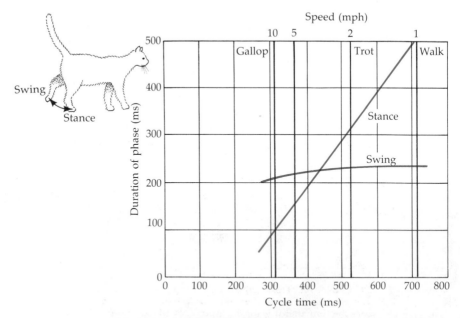

6 **CONSTANCY OF SWING PHASE during locomotion. As the animal runs faster and faster (abscissa), the time spent by each foot on the ground (stance phase) becomes shorter (ordinate). The time each foot spends in the air (swing phase) is almost the same in a walk and a gallop. (After Pearson, 1976.)**

441
INTEGRATIVE
MECHANISMS
IN THE CNS FOR
THE CONTROL
OF MOVEMENT

brain stem of a cat is transected, the animal can still stand but does not walk or run spontaneously. In the experiments made by Shik and his colleagues,[18] a cat with mesencephalic transection was held with its feet touching a treadmill. Continuous electrical stimulation at 30 to 60 per second was applied with electrodes placed in a specific region called the "mesencephalic locomotor region," which corresponds to a histologically identified structure—the cuneiform nucleus (Figure 7). The interesting result Shik and his colleagues obtained was that the cat walked, provided that the electrodes were in the correct area. Displacement by as little as 0.3 mm abolished the walking response to stimulation. The stance and swing of the forelegs and the electromyograms recorded during walking appeared normal. Stronger stimulation of the

[18]Shik, M. L., Severin, F. V. and Orlovsky, G. N. 1966. *Biophysics 11*: 756–765.

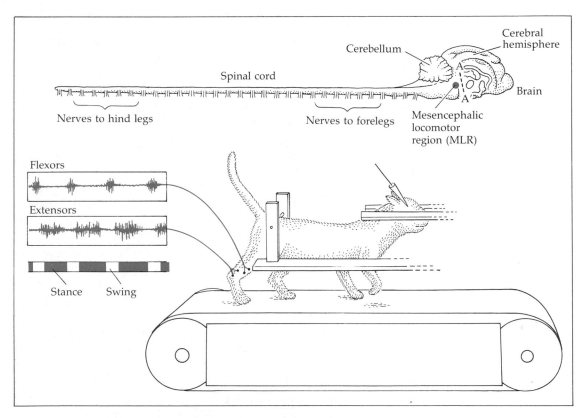

7 **LOCOMOTION BY CAT ON A TREADMILL** after a lesion in the brain stem (A——A′ in diagram). Such an animal does not walk spontaneously. Stimulation of a restricted area, the mesencephalic locomotor region, by electrical currents causes the animal to walk on the treadmill. The locomotor activity can be recorded by electrodes in the leg muscles. Speed of walking or galloping depends on the rate of the treadmill. The strength or frequency of stimulation by the electrode does not influence the speed of movement, but rather its strength (e.g., as though the animal were walking uphill). (After Pearson, 1976.)

mesencephalic locomotor region by larger currents at the same frequency caused the propulsive forces of the leg muscles to be increased. The increased effort achieved in this way was similar to that observed when a dog or cat was made to walk against a heavier load than normal, for example, uphill on a tilted treadmill or when being pulled backward. But while the strength of electrical stimulation could influence the force of the walking movements, the frequency of stepping was not altered, provided that the speed of the treadmill remained constant.

In response to acceleration of the treadmill with constant stimulation of the mesencephalic locomotor region, the animal changed from walking to trotting, and then to galloping. Sensory feedback plainly controls the rate of stepping with its shortened stance phase. Presumably, as the leg in contact with the treadmill becomes extended more rapidly, it reaches sooner the point at which afferent signals initiate the swing phase in which the leg is lifted and swung forward. How the organization within the central nervous system that controls the switching from walk to trot to gallop is switched is not yet understood. As expected, cutting dorsal roots abolishes the response to different treadmill speeds, but not the walking evoked by electrical stimulation.[19]

The general conclusion from these kinds of experiments is that within the central nervous system there exists a hierarchically ordered series of connections that can initiate and control a programmed series of movements. Within the spinal cord, interconnections between appropriate pools of motor neurons are acted upon by descending influences. In the absence of phasic sensory input from the limbs, each limb can be programmed by the central nervous system to be raised, swung forward, and lowered by contraction of the appropriate muscle groups acting on the joints. The role of feedback is to modulate these centrally initiated responses in accord with different needs or loads imposed by the external world.

From these considerations, certain striking parallels emerge in the automaticity of breathing and of walking. For both types of movement, a central program orders the contractions of appropriate groups of muscles in a predestined sequence. The drive for the alternation of leg movements or of inspiration and expiration depends on a descending stimulus—for respiration the influence of CO_2 on the medullary neurons and for walking the activity of neurons in the mesencephalic locomotor region (the connections of which are as yet unknown).

SENSORY–MOTOR INTEGRATION

A chicken with its head cut off can run across a yard and a cat with its spinal cord transected can still walk with its hindlimbs on a treadmill. But for purposeful movements—the avoidance of obstacles and the ability to hesitate or turn back—an animal needs more than motor

---

[19]Grillner, S. and Zangger, P. 1974. *Acta Physiol. Scand. 91*: 38A–39A.

443
INTEGRATIVE
MECHANISMS
IN THE CNS FOR
THE CONTROL
OF MOVEMENT

programs or simple reflex responses. Such decisions depend on the flow of sensory information to descending systems that coordinate muscular contractions throughout the body, maintain posture, and ensure that the head is kept upright. At present there is no clear way to follow the various signals arising from outside stimuli and from the muscles and joints as they are integrated within the central nervous system and result in the variation and shaping of the movements of the body as a whole. As a first step, however, it is possible to trace pathways by which sensory information ascends to reach the cortex and descends to control the motoneurons.

Figure 8 shows schematically the principal somatic sensory and motor pathways of the brain. (More detailed drawings are in Appendix B.) For information from skin, deep tissues, and joints, fibers enter the spinal cord, run in tracts situated dorsally (the dorsal columns), synapsing in the medulla at the gracile and cuneate nuclei. The axons of these cells cross and ascend through a bundle called the "medial lemniscus," to synapse on neurons situated in a nucleus of the thalamus (the ventroposterolateralis, or VPL); these then project to a region of the cortex in the postcentral gyrus—areas 3, 1, 2, known as the primary somatosensory area—an area that is analogous to the striate cortex.

At each successive level there is an orderly map of the body correlated with the modalities of touch, pressure, and joint position. Since the pathways are crossed, the left side of the body is represented on the right side of the brain. As in the visual system, where a small region of retina—the fovea—is represented most extensively within the lateral geniculate nucleus and visual cortex (Chapters 2 and 3), so there exists within the cortex comparable distortion of the representation of the body. In monkeys and human beings, areas within the central nervous system concerned with hands, fingers, and lips are larger than those concerned with trunk or legs. In different animals various other regions of the body predominate—the whiskers for a mouse (Figure 10) or the snout for a pig, which as Adrian[20] states "is its chief executive organ, spade as well as hand."

In the primary somatosensory cortex—corresponding to areas 3, 2 and 1 in Figure 12—the fine grain of the representation has been revealed by recording from individual neurons. Indeed, it was in this region of the cortex that Mountcastle, Powell, and their colleagues[21] first demonstrated columnar organization, preceding comparable work in the visual system. When a microelectrode penetration is made through the thickness of the somatosensory cortex, each neuron recorded from shares a number of receptive field properties with the other neurons encountered above and below. For example, as shown in Figure 9, all the cells encountered in one penetration were driven by light touch applied to one small spot on the finger of a monkey. Figure 9C shows that touching the skin around the central part of the receptive

Central nervous system organization for somatic sensation

[20]Adrian, E. D. 1946. *The Physical Background of Perception.* Clarendon Press, Oxford.
[21]Powell, T. P. S. and Mountcastle, V. B. 1959. *Bull. Johns Hopkins Hosp.* 105: 133–162.

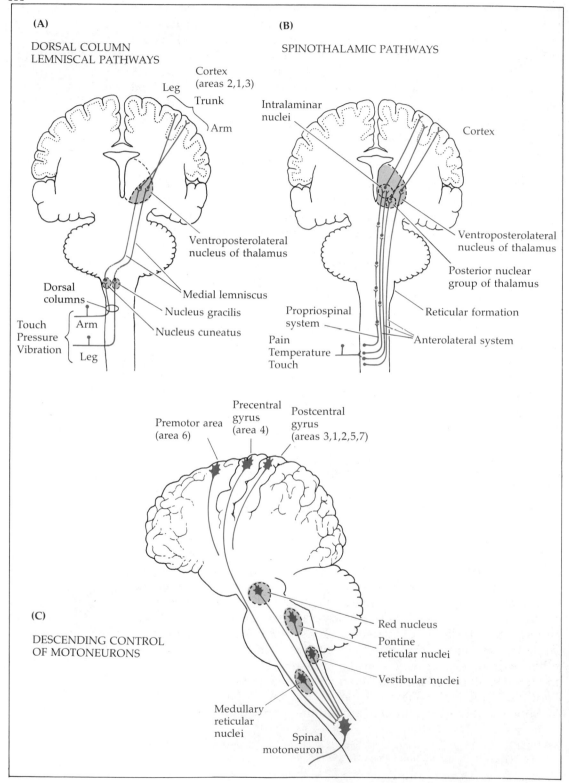

(A)

DORSAL COLUMN
LEMNISCAL PATHWAYS

Cortex
(areas 2,1,3)

Leg

Trunk

Arm

Ventroposterolateral
nucleus of thalamus

Dorsal
columns

Medial lemniscus

Nucleus gracilis

Nucleus cuneatus

Touch
Pressure
Vibration

Arm

Leg

(B)

SPINOTHALAMIC PATHWAYS

Intralaminar
nuclei

Cortex

Ventroposterolateral
nucleus of thalamus

Posterior nuclear
group of thalamus

Propriospinal
system

Reticular formation

Pain
Temperature
Touch

Anterolateral system

(C)

DESCENDING CONTROL
OF MOTONEURONS

Premotor area
(area 6)

Precentral
gyrus
(area 4)

Postcentral
gyrus
(areas 3,1,2,5,7)

Red nucleus

Pontine
reticular nuclei

Vestibular nuclei

Medullary
reticular
nuclei

Spinal
motoneuron

445
INTEGRATIVE
MECHANISMS
IN THE CNS FOR
THE CONTROL
OF MOVEMENT

field caused a decrease in the firing rate of such cells. Thus, there is an antagonistic inhibitory surround area similar to that found in the visual system. The presence of this surround makes possible greater discrimination between two closely spaced tactile stimuli applied to the skin. Penetrations through an area of cortex 1 mm or so away would result in a shift of the receptive field and perhaps of the modality, all the cells now responding to deep pressure or to displacement of the joint of the fingertip. Still farther away—with a shift of receptive field position to the forearm or elbow—the size of the receptive field of the cortical neuron would be considerably larger, measured in centimeters instead of millimeters.

The degree of precision with which parts of the body are represented in the somatosensory cortex is well illustrated by the projection of sensory innervation from whiskers of the mouse. Histological studies by Van der Loos, Woolsey, and their colleagues[22] have shown that the mouse somatosensory cortex contains characteristic groups of nerve cells clustered in the form of cylinders that stretch through layer IV of the cortex. These assemblies have been termed "barrels" from their shapes, which were determined by serial reconstructions. Each barrel is 100–400 μm in diameter and is composed of a ring of cells that surrounds a central "hollow" containing fewer cells. The center line is at right angles to the surface of the cortex. The array of barrels in the mouse is consistently organized (Figure 10C and D) with five rows. Van der Loos and Woolsey realized that this pattern corresponded exactly with the rows of whiskers or vibrissae on the mouse's face, one barrel for each whisker. The functional role of the barrels is confirmed by recording electrically from individual cells.[23,24] Each cell responds to movement only of the appropriate vibrissa, some firing when the whisker is moved in one direction, others when it is moved in the opposite

[22]Woolsey, T. A. and Van der Loos, H. 1970. *Brain Res. 17*: 205–242.
[23]Welker, C. 1976. *J. Comp. Neurol. 166*: 173–190.
[24]Simons, D. J. and Woolsey, T. A. 1979. *Brain Res. 165*: 327–332.

8 ASCENDING AND DESCENDING PATHWAYS. The schematic diagrams show only the principal relays and nuclei (see Appendix B for more details). (A) Sensory axons for touch, pressure, and vibration enter the dorsal roots and travel in dorsal columns of the spinal cord to synapse in the gracile and cuneate nuclei. Fibers ascend through the medial lemniscus and cross to the other side to end in the ventroposterolateral nucleus (VPL) of the thalamus; VPL cells project to the somatosensory cortex. (B) For pain and temperature, pathways are more diffuse. Incoming sensory axons synapse in the dorsal horn of the spinal cord. Fibers cross and ascend in the anterolateral spinothalamic tract, and others ascend ipsilaterally with numerous relays before terminating in both the posterior and VPL nuclei of the thalamus from which fibers project to widespread areas of cortex. (C) Diagram showing the principal areas involved in immediate control of motoneurons. (Thus, the cerebellum, which projects via vestibular nuclei, and the basal ganglia, which do not project directly to spinal motoneurons, are not shown.) (See also Appendix B.)

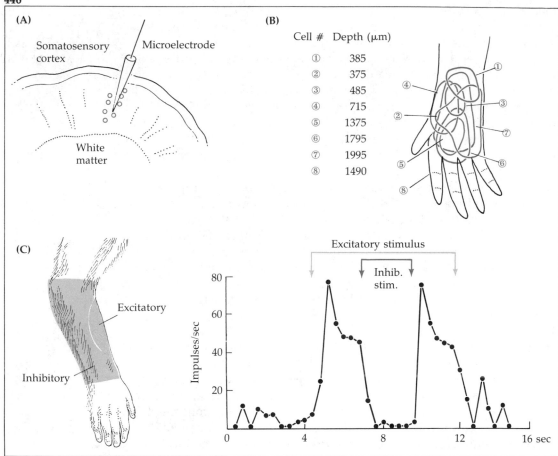

(A) Somatosensory cortex

Microelectrode

White matter

(B)

Cell #	Depth (μm)
①	385
②	375
③	485
④	715
⑤	1375
⑥	1795
⑦	1995
⑧	1490

(C) Excitatory

Inhibitory

Excitatory stimulus

Inhib. stim.

Impulses/sec

80

60

40

20

0   4   8   12   16 sec

9 RESPONSES AND RECEPTIVE FIELDS OF NEURONS in somatosensory cortex of monkey. **(A)** Diagram of electrode penetration through the somatosensory cortex at right angles to the surface. **(B)** Each cell encountered by the electrode responds to a touch of approximately the same area of skin of the hand. **(C)** Responses of a cortical cell to mechanical stimulation of the skin. Touch of one area (stippled) increases the firing rate, but touch within a large surrounding area inhibits firing (B after Powell and Mountcastle, 1959; C after Mountcastle and Powell, 1959.)

direction. The mouse uses the vibrissae as sensitive antennae, waving them backward and forward as it walks along to detect objects in its pathway on either side. The synaptic connections and overall design of circuitry within the barrels are not yet known.

In addition to the primary somatosensory cortex, multiple representations of the body occur in other secondary somatosensory areas and even within the primary area itself.[25,26] In area 5 of the parietal cortex,

[25]Merzenich, M. M., Kaas, J. H., Sur, M. and Lin, C.-S. 1978. *J. Comp. Neurol. 181*: 41–74.

[26]Kaas, J. H. 1983. *Physiol. Rev. 63*: 206–231.

447
INTEGRATIVE
MECHANISMS
IN THE CNS FOR
THE CONTROL
OF MOVEMENT

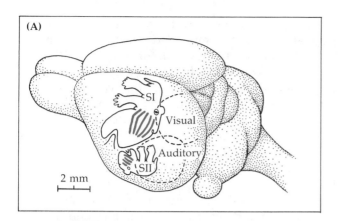

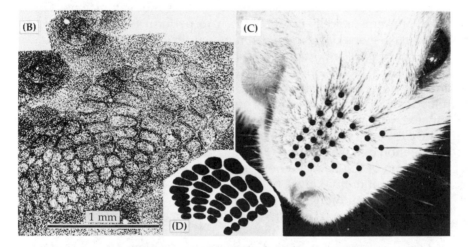

10 BARRELS IN MOUSE SOMATOSENSORY CORTEX corresponding to vibrissae. (A) Diagram of mouse brain to show representation of body on somatosensory cortex ("musculus" equivalent to "homunculus"), particularly face and vibrissae. SII is the second somatosensory area. (B) Horizontal section through cortex in area representing vibrissae, showing "barrels" in transverse section (like the hoops of a barrel). Scale, 1 mm. (C) Whiskers seen close up. (D) Schematic diagram of barrel arrangement from C, showing one barrel—one vibrissa. (After Woolsey and Van der Loos, 1970.)

neurons with more complex response properties have been found; these neurons are driven only by complex movements involving, not a single joint, but the entire limb in one direction only. Other columns of neurons that have been found respond to stroking of the skin in one direction but not the other.[27,28] What at present remains elusive is the way in which information from diverse areas is synthesized into a complete body image by the brain.

[27]Hyvärinen, J. and Poranen, A. 1978. *J. Physiol. 283*: 523–537.
[28]Costanzo, R. M. and Gardner, E. P. 1980. *J. Neurophysiol. 43*: 1319–1341.

The multiple descending motor systems shown in Figure 8 are able to respond to appropriate sensory stimuli and are highly complex. Moreover, it is considerably harder to study motor integration than to study the flow of sensory information. In sensory systems, a well-defined stimulus sets in motion neural responses that can in principle be followed step by step through clear-cut pathways; in contrast, during voluntary movement the best-defined event is the outcome—the muscular contraction produced by firing of the motor neurons. Upon those cells impinge not only reflexly mediated effects of groups Ia and II afferents from the same muscles, but also influences from other muscles and signals coming from the motor cortex, the basal ganglia, the cerebellum, the midbrain, and the reticular formation. These structures profoundly affect the initiation and control of movement, and in each there is an orderly map of sensory as well as motor functions. Often specific groups of cells can be shown to perform functions related to particular regions of the body from which they receive information.

Of particular importance for posture are the vestibular apparatus (which is a structure in the inner ear and is concerned with balance), and the reflexes originating in the neck muscles.[29] Both play essential roles in maintaining an upright posture. The vestibular apparatus is situated within the temporal bone of the skull and consists of two principal sensory organs: (1) otolith organs, which detect tilt and linear acceleration of the head, and (2) semicircular canals, which are activated by rotation of the head. In both end organs, displacement of hair cells by movement of fluid gives rise to electrical signals that are transmitted by the eighth (auditory) nerve to the vestibular nuclei. The output from these nuclei has marked effects on eye movements and on postural reflexes. These tend to keep the eyes stable in space, the head upright, and the legs appropriately bent or stretched. A falling cat will land with its forelegs extended and toes spread out. Similarly, if the body of a cat is suddenly tilted downward on a slope, the front legs and neck will extend while the hindlimbs flex, so as to keep the head horizontal. Afferent responses from neck muscle spindles initiated by twisting the head also play an essential role in such postural reflexes. It is only by afferents from the neck that the head knows how the rest of the body is oriented.

The complexity of descending integrative mechanisms is illustrated by the connections to interneurons situated within the spinal cord that give rise to inhibitory synaptic potentials in motor neurons (Figure 11). Such inhibitory interneurons have been studied in detail by Lundberg, Jankowska, and their colleagues,[30] using intracellular microelectrodes for recording and injection of horseradish peroxidase. They have shown that a single interneuron receives excitatory inputs from more than ten diverse regions of the central nervous system as well as inhibitory inputs.

[29]Shepherd, G. M. 1983. *Neurobiology*. Oxford University Press, New York, Chapter 15.

[30]Czarkowska, J., Jankowska, E. and Sybirska, E. 1981. *J. Physiol. 310*: 367–380.

449
INTEGRATIVE
MECHANISMS
IN THE CNS FOR
THE CONTROL
OF MOVEMENT

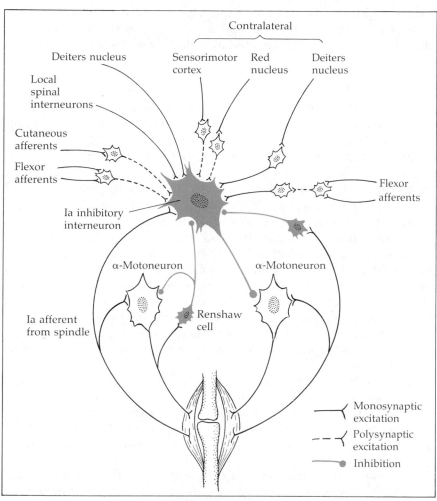

11   **MULTIPLE INPUTS TO INHIBITORY INTERNEURONS** in cat spinal cord. Jankowska, Lundberg, Lindström, and their colleagues have recorded from and stained interneurons mediating inhibition to motoneurons in the Ia pathway. The Ia afferent from a spindle gives rise to inhibition in antagonistic motoneurons (also in agonists, not shown). This single inhibitory interneuron receives excitatory inputs from higher centers, from skin, from other muscles, and from spinal interneurons. (Modified from Lindström, 1973.)

    These interneurons in turn constitute only a small fraction of the total input to the motor neurons, as described in Chapter 16. In monkeys, motor neurons receive a direct excitatory projection from the motor cortex[31] (Figure 12). In this motor region—area 4—cells are also arranged in columns.[32] Stimulation of each of the cells encountered in one penetration leads to a discrete movement—say, flexion of the con-

[31]Landgren, S., Philips, C. G. and Porter, R. 1962. *J. Physiol. 161*: 91–111.
[32]Asanuma, H. 1975. *Physiol. Rev. 55*: 143–156.

tralateral thumb; the same cortical cells are themselves activated by passive movement of the appropriate joint of that thumb in the same direction or else may be activated in anticipation of such movements.[33]

HIGHER FUNCTIONS

Like the mechanisms that produce the big picture of the visual field or of the body, the motor mechanisms responsible for producing integrated patterns of movement—say, putting on one's shirt, singing, or riding a bicycle—remain unknown at present. As for other unapproachable problems, clues about the structures involved can be provided in the first instance by neurologists studying clinical conditions in man.[34,35] Damage to selected regions of the brain by trauma, vascular disease, or tumors can often give rise to clear-cut, stereotyped disorders of higher functions, similar in patient after patient. The site of the lesion can be established at postmortem or sometimes by modern noninvasive techniques such as PET scanning[36] (see Chapter 20 and Glossary). Thus, disorders of motor performance after damage to area 4 appear as a loss of voluntary movement of a finger, hand, arm, leg, or one side of the body, depending on the position and extent of the lesion. Similarly, lesions in area 3 of somatosensory cortex may impair the ability to discriminate between a smooth surface and sandpaper with the fingertips.

Most of the cerebral cortex, however, unlike primary sensory and motor areas, is not simply concerned with receiving sensory information directly from the thalamus or sending out commands directly to motor neurons through the pyramidal tracts. Lesions of these "association" areas, some of which are shown in Figure 12, give rise to more subtle deficits.

Functional properties are not equally distributed between the two cerebral hemispheres. Most people are right-handed, with their left hemisphere possessing finer mechanisms for controlling movements on the right side of the body, particularly the hand. Language functions, which involve highly complex sensory and motor coordination, are even more strongly lateralized to the left hemisphere in right-handed people. Small lesions situated in the left cortex can selectively destroy the ability to speak or to understand written or spoken words. For example, with a lesion in the region of area 44 known as Broca's area (Figure 12), the patient may understand language and may be able to write but not to speak. Words are uttered only singly and with difficulty. Although there is no failure of comprehension, the words cannot be strung together. A lesion situated posterior to Broca's area in area 22 (Wernicke's area) results in a different deficit: failure to understand language. The patient

[33]Evarts, E. V. and Tanji, J. 1976. *J. Neurophysiol. 39*: 1069–1080.

[34]Kolb, B. and Whishaw, I. Q. 1980. *Fundamentals of Human Neuropsychology*. Freeman, San Francisco.

[35]Geschwind, N. 1979. *Sci. Am. 241*: 180–199.

[36]Crill, W. E. and Raichle, M. E. 1982. In J. G. Nicholls (ed.). *Repair and Regeneration of the Nervous System*. Springer-Verlag, New York, pp. 227–242.

451
INTEGRATIVE
MECHANISMS
IN THE CNS FOR
THE CONTROL
OF MOVEMENT

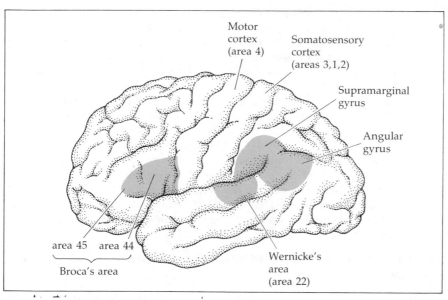

Motor
cortex
(area 4)

Somatosensory
cortex
(areas 3,1,2)

Supramarginal
gyrus

Angular
gyrus

area 45   area 44

Broca's area

Wernicke's
area
(area 22)

12 CORTICAL AREAS associated with disorders of speech and movement after injury.

can still speak, but words are run together and jumbled in an inappropriate manner. In contrast, for a right-handed person, more extensive damage to similar regions of the nondominant (right) hemisphere result in no impairment of language but might affect the ability to sing.

Interesting defects are produced by lesions of the supramarginal gyrus of the parietal lobe (Figure 12), particularly in the nondominant hemisphere. The patient can move normally and can perform simple actions such as picking up a pencil. But he is unable to carry through complex tasks that involve a series of movements such as putting on a shirt. The buttons may be done up first, then one sleeve put over the head. In such conditions, it is as though the elementary actions cannot be combined into a programmed sequence.

One can conjecture that a hierarchical organization exists within the brain for movement—from small components of movement (represented by the contractions of individual muscles) through movements of limbs, both of which are eventually combined by the appropriate groups of cells to include movement of the body as a whole. Automatic and voluntary functions work together in various combinations. Again, a lucid picture is given by Adrian[20] in *The Physical Background of Perception*. He uses the cerebellum to exemplify motor control:

> In spite of its resemblance to the cerebrum, the cerebellum has nothing to do with our mental activity. . . . The cerebellum has the more immediate and quite unconscious task of keeping the body balanced whatever the limbs are doing and of insuring that the limbs do what is required of them Its actions show what complex things can be done by the mechanism of the nervous system in carrying out the decisions of the mind. If I decide to raise my arm, a message is dispatched from the motor area of one cerebral

hemisphere to the spinal cord and a duplicate of the message goes to the cerebellum. There, as a result of interactions with other sensory impulses, supplementary orders are sent out to the spinal cord so that the right muscles come in at the exact moment when they are needed, both to raise the arm and to keep the body from falling over. The cerebellum has access to all the information from the muscle spindles and pressure organs and so can put in the staff work needed to prevent traffic jams and bad coordination. If it is injured the timing breaks down, muscles come in too early or too late and with the wrong force. The staff work needs to be elaborate, particularly when the body has to be balanced on two legs and uses its arms for all manner of movement, but it is done by the machinery of the nervous system after the mind has given its orders. The cerebellum has nothing to do with formulating the general plan of campaign. Its removal would not affect what we feel or think, apart from the fact that we should be aware that our limbs were not under full control and so should have to plan our activities accordingly.

Adrian also commented on the similarities between abstraction in sensory and motor systems.

The same principle seems to come in on the motor side as well as on the sensory. We may learn a skilled movement by employing certain muscles and therefore certain groups of nerve cells in the motor area of the brain, but when we have learnt it we can carry out the movement with an entirely different set of muscles and nerve cells—we can write our name with a pencil held between the toes when we have learnt to do it with the fingers. We can draw a triangle small or large when we have once learned its shape, just as we can recognize it small or large on the retina.

BOX 1 NOCICEPTIVE SYSTEMS AND PAIN

Information about noxious or painful stimuli is conveyed to higher centers by specific receptors and by pathways distinct from those used for proprioception, touch, or pressure. Nociceptive terminals appear as free nerve endings in the skin and the viscera; characteristically, they respond only to strong stimuli, such as pricking, excessive stretching, and extremes of temperatures, or to various chemicals, such as histamine and bradykinin.[37] The afferent axons have been shown to fall into two classes: (1) small myelinated axons 1 to 4 μm in diameter that conduct at 6 to 24 m/sec (corresponding to the Aδ class fibers; see Box 1, Chapter 15), and (2) unmyelinated C axons, 0.1–1 μm in diameter, that conduct more slowly at 0.5 to 2 m/sec.[38,39] A single afferent fiber, whether Aδ or C, responds to noxious stimuli applied to its receptive field in a characteristic manner: The rate of adaptation is slow and the discharge continues after removal of the stimulus. Nociceptive pathways within the

central nervous system are shown in Figure 8: Fibers enter the dorsal horn and ascend or descend through one or two segments to form synapses on cells in discrete layers—the marginal layer containing large neuronal cell bodies and the substantia gelatinosa.

Ascending fibers run through two major pathways: (1) the contralateral (and to a lesser extent ipsilateral) spinothalamic tract and (2) the medial spinoreticular thalamic pathways. The ascending fibers end in the ventroposterolateral and the intralaminar nuclei of the thalamus, respectively. Cells in these nuclei project to somatosensory cortex and other widespread areas of the nervous system. Unlike other interneurons in the somatosensory system, those in the thalamus and cortex receiving nociceptive inputs have large, ill-defined receptive fields, often covering wide areas of the body, contralateral as well as ipsilateral.

Although the afferent systems for nociception have their own through lines, it is clear that no noxious stimulus can fail to activate other receptors responding to touch, pressure, displacement, stretch, cooling, heating, and so on. In recent years numerous experiments have shown that the two systems interact. In particular, light touch or stroking of the skin can influence the receptive field properties of nociceptive neurons within the central nervous system. For example, Poggio and Mountcastle[40] showed that the activation (by stroking) of somatosensory afferents with nearby receptive fields had an inhibitory effect on the nociceptive discharges of cortical and thalamic neurons. The subject received considerable attention with the discovery of opiate receptors in the central nervous system—these opiate receptors are located in the dorsal horn and periaqueductal gray—and with the identification of naturally occurring opiate-like peptides in specific neurons concerned with nociceptive function[41] (see discussion of substance P and enkephalin in Chapter 12). As in other sensory systems, descending influences from higher centers can dramatically modify the flow of sensory information that reaches consciousness.[42]

Pain itself remains an elusive and difficult concept, beyond the scope of this book. In sharp contrast to the analysis of visual, auditory, or somatosensory systems, a discussion of "pain" with its high emotional content, of necessity deals with subjective matters, feelings akin to "anguish" and "suffering" that cannot at present be expressed in the language of neurobiology (just as "seeing a sunset" or "feeling warm" cannot be considered in those terms). Nevertheless, certain correlates between neural activity and pain are apparent. Thus, a noxious stimulus gives rise initially to a sharp well-localized pain, followed by a dull, gradually swelling, poorly localized pain; these two waves can be attributed to the activity of the Aδ and the C fibers conducting at different velocities and ascending through spinothalamic and spinoreticulothalamic pathways. Similarly, the analgesic effect of stroking the skin can be accounted for in part by synaptic interactions at the level of the spinal cord, the thalamus, and the cortex. Particularly appealing is the idea that the level of pain perceived may be reduced by descending influences that release morphinelike peptides, enabling the soldier charging into battle to sustain injuries without feeling them to the same extent as he would if they were to be inflicted while he sat in a chair.

Again, it is Adrian[20] who clearly enunciated a key problem for pain, namely, its function.

Why should we suffer pain, what is its purpose, and why should it be so unpleasant? Our ancestors may have believed the moralists (and the doctors) who told them that pain could be a valuable experience, not to be avoided by such unnatural means as anaesthetics for childbirth, but we have been made less hardy, and we need biological as well as moral grounds for the existence of such an evil. The chief biological excuse for pain is that it is a danger signal. The argument is convincing to a point. A successful type of animal, one which can look after itself, must have a sensory mechanism which will signal events likely to damage it and the signals must have priority over all others. . . . It must be a sensation to which we cannot manage to remain inattentive and one which we feel compelled to bring to an end as soon as possible. . . . The danger signal may get out of order, the warning sometimes sounds, although there is no chance of injury. . . . Nonetheless, the danger signal seems to wear thin when it comes to explaining the pain which you may have to suffer when the injury does not come from without so that there is some chance of avoiding it but from something happening inside the body, a slowly growing tumour or the movements of a renal calculus. Such pain may be a warning to call in the surgeon, but a few hundred years ago . . . calling in the surgeon would not have made much difference. . . . Medical science has already done so much to lessen pain that we need not be ashamed to confess that there is still a great deal more to do before we can fully understand how and why it arises.

[37]Ottoson, D. 1983. *Physiology of the Nervous System.* Oxford University Press, New York, Chapter 31.

[38]Kruger, L., Perl, E. R. and Sedivec, M. J. 1981. *J. Comp. Neurol.* 198: 137–154.

[39]Torebjörk, H. E. and Ochoa, J. 1980. *Acta Physiol. Scand.* 110: 445–447.

[40]Poggio, G. F. and Mountcastle, V. B. 1960. *Bull. Johns Hopkins Hosp.* 106: 266–316.

[41]Hughes, J. (ed.). 1983. *Br. Med. Bull. 39:* 1–106.

[42]Basbaum, A. I. and Fields, H. L. 1979. *J. Comp. Neurol.* 187: 513–532.

SUGGESTED READING

General reviews

Cohen, M. I. 1979. Neurogenesis of respiratory rhythm in the mammal. *Physiol. Rev. 59*: 1105–1173.

Grillner, S. 1975. Locomotion in vertebrates: Central mechanisms and reflex interaction. *Physiol. Rev. 55*: 247–304.

Kaas, J. H. 1983. What if anything is SI? Organization of first somatosensory area of cortex. *Physiol. Rev. 63*: 206–231.

Kosterlitz, H. W. and Terenius, L. Y. (eds.). 1980. Pain and society. *Life Sci. Res. Rep. 17*: Verlag Chemie, Weinheim.

Merrill, E. G. 1981. Where are the *real* respiratory neurons? *Fed. Proc. 40*: 2389–2394.

Mountcastle, V. B. 1975. The view from within: Pathways to the study of perception. *The Johns Hopkins Med. J. 136*: 109–131.

Pearson, K. G. 1976. The control of walking. *Sci. Am. 235*: 72–86.

Precht, W. 1979. Vestibular mechanisms. *Annu. Rev. Neurosci. 2*: 265–289.

Shik, M. L. and Orlovsky, G. N. 1976. Neurophysiology of locomotor automatism. *Physiol. Rev. 56*: 465–501.

Original papers

Akazawa, K., Aldridge, J. W., Steeves, J. D. and Stein, R. B. 1982. Modulation of stretch reflexes during locomotion in the mesencephalic cat. *J. Physiol. 329*: 553–567.

Bainton, C. R., Kirkwood, P. A. and Sears, T. A. 1978. On the transmission of the stimulating effects of carbon dioxide to the muscles of respiration. *J. Physiol. 280*: 249–272.

Christensen, B. N. and Perl, E. R. 1970. Spinal neurons specifically excited by noxious or thermal stimuli: Marginal zone of the dorsal horn. *J. Neurophysiol. 33*: 293–307.

Kirkwood, P. A. and Sears, T. A. 1982. Excitatory post-synaptic potentials from single muscle spindle afferents in external intercostal motoneurones of the cat. *J. Physiol. 322*: 287–314.

Mountcastle, V. B. and Powell, T. P. S. 1959. Neural mechanisms subserving cutaneous sensibility with special reference to the role of afferent inhibition in sensory perception and discrimination. *Bull. Johns Hopkins Hosp. 105*: 201–232.

Poggio, G. F. and Mountcastle, V. B. 1960. A study of the functional contributions of the lemniscal and spinothalamic systems to somatic sensibility. *Bull. Johns Hopkins Hosp. 106*: 266–316.

Shik, M. L., Severin, F. V. and Orlovsky, G. N. 1966. Control of walking and running by means of electrical stimulation of the midbrain. *Biofizika 11*: 659–666 (English trans. pp. 756–765).

Tanji, J. and Evarts, E. V. 1976. Anticipatory activity of motor cortex neurons in relation to the direction of an intended movement. *J. Neurophysiol. 39*: 1062–1068.

Woolsey, T. A. and Van der Loos, H. 1970. The structural organization of layer IV in the somatosensory region (S I) of mouse cerebral cortex: The description of a cortical field composed of discrete cytoarchitectonic units. *Brain Res. 17*: 205–242.

CHAPTER EIGHTEEN

SIMPLE NERVOUS SYSTEMS

Invertebrates perform various tasks that appear complex; yet they use only a relatively small number of nerve cells. This makes them useful for the study of some problems that at present seem too difficult to be approached in the mammalian brain. In particular, simple reflexes can be studied in terms of individual nerve cells that give rise to them and the various mechanisms of signaling. At a higher level of integration, it is possible to show how elementary units of behavior are combined into coordinated movements of the animal as a whole.

Two animals—the leech and the sea hare *Aplysia* (a mollusk)—are used to illustrate this approach. Their central nervous systems consist of discrete aggregates of neurons—the ganglia—which are linked to each other and to the periphery by bundles of axons. In *Aplysia* the ganglia contain several thousand nerve cells, whereas those in the leech are smaller and contain only about 400 nerve cells. In both animals the nerve cells can be seen clearly under the dissecting microscope and can be impaled with microelectrodes. Individual sensory and motor nerve cells (as well as interneurons) have been identified, their synaptic connections traced, and the fields they innervate in the periphery defined. This information is an essential requirement for tracing the normal wiring diagram of the nervous system and for studying how the properties of the neurons change with repeated use.

In ganglia of the leech and *Aplysia*, one can correlate reflex movements of the animal with the physiological characteristics of synapses between identified sensory and motor nerve cells. For example, when impulses are initiated at their normal frequencies in sensory neurons by natural stimulation of the skin, certain chemical synapses show sequentially both facilitation and depression. The reflexes mediated by cells connected through such synapses reflect this sequence to produce sequential motor acts that can be observed readily. In contrast, transmission mediated by cells with electrical connections shows little variability under similar conditions. In *Aplysia*, persistent changes in synaptic transmission that last for hours or even days are produced by regular stimuli applied to the skin. The depression of synaptic transmission can explain the way in which the animal becomes "habituated" to the stimulus.

The nervous systems of invertebrates also lend themselves well to following the formation of ganglia and specific connections during development and to the mechanisms used in repair of the nervous system during regeneration.

Throughout the previous chapters, discussion has been oriented toward problems of higher nervous systems. In the case of cellular mechanisms, it is quite satisfactory to take examples from a wide variety of species. The squid axon, for example, provides an excellent model of nearly universal validity for the conduction of impulses; cells of other invertebrates illustrate the mechanisms of permeability sequences found in synaptic inhibition and excitation in vertebrates. Similarly, neurochemical studies in ganglia of lobsters have direct relevance for neurons in the mammalian brain (Chapter 12).

The reasons for choosing invertebrate preparations are usually technical. Certain problems can be solved more easily in invertebrate nervous systems, which tend to survive in isolation; in addition, if their cells are large and readily accessible, they can be easily recognized and studied with electrical recording methods and biochemical techniques.

The relevance of the nervous systems of invertebrates, however, extends beyond basic mechanisms of conduction and synaptic transmission. These preparations also offer advantages for the exploration of some aspects of complex behavior that seem too difficult to handle in higher vertebrates. Along these lines, Kandel, Tauc, Strumwasser, and others have broken new ground by using cell groups in the sea snail *Aplysia* to analyze behavioral responses by tracing the neural circuits involved in their performance.[1] Some of these responses are performed by relatively few neurons, whereas analogous responses in mammals would require many thousand of neurons. Similarly, Wiersma,[2] Kennedy,[3] Wine, Krasne,[4] and their colleagues defined sensory and "command" interneurons (individual neurons that bring about coordinated movements) and traced the neural circuits for coordinated elementary units of behavior, such as postural reflexes and escape reactions and swimming, in crayfish. Central nervous systems of insects have also been used to study a variety of problems including flight, walking, and regeneration,[5] as well as communication by sound.[6] Each type of preparation has its own set of advantages for approaching specific problems at the cellular level.

We have singled out for fuller discussion the nervous systems of

[1]Kandel, E. R. 1976. *Cellular Basis of Behavior*. Freeman, San Fransicso.

[2]Wiersma, C. A. G. 1947. *J. Neurophysiol. 10*: 23–38.

[3]Kennedy, D. 1976. In J. C. Fentress (ed.). *Simpler Networks and Behavior*. Sinauer, Sunderland, MA, pp. 65–81.

[4]Wine, J. J. and Krasne, F. 1982. In D. C. Sandeman and H. L. Atwood (eds.). *Neural Integration and Behavior*, Vol. 4. *The Biology of Crustacea*. Academic, New York, pp. 241–292.

[5]Fentress, J. C. (ed.). 1976. *Simpler Networks and Behavior*. Sinauer, Sunderland, MA.

[6]Bentley, D. and Hoy, R. R. 1974. *Sci. Am. 231*: 34–44.

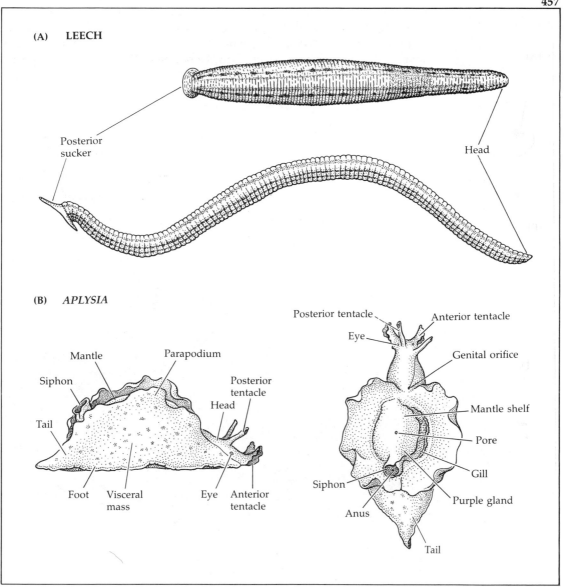

1 (A) THE LEECH, *Hirudo medicinalis,* has a segmented body with a sucker at each end. The animal can measure 5 inches in length after feeding. (B) *Aplysia californica,* the sea hare, viewed from the side and from above, with the parapodia retracted so as to show mantle, gill, and siphon (B modified from Kandel, 1976.)

two invertebrates: the leech, mentioned in Chapter 13 in relation to the properties of neuroglial cells, and *Aplysia.* These compact nervous systems contain relatively few cells arranged in an orderly manner; yet they provide useful models for the study of certain problems of neural organization found in the brains of higher animals. At the same time the ganglia use synaptic mechanisms similar to those that have been identified in vertebrates. Figures 1, 2, and 3 show the principal features of the animals and of their central nervous systems. In the leech, the central nervous system consists of a chain of stereotyped ganglia, whereas in *Aplysia* the ganglia are considerably larger and are special-

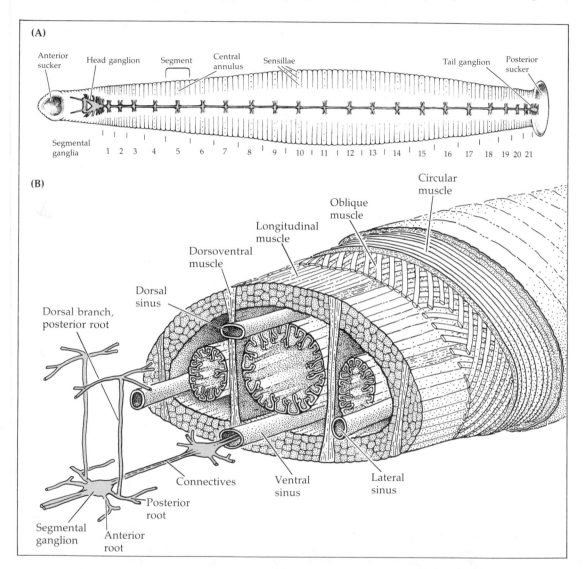

(A)

Anterior sucker   Head ganglion   Segment   Central annulus   Sensillae   Tail ganglion   Posterior sucker

Segmental ganglia   1 2 3 4 5 6 7 8 9 10 11 12 13 14 15 16 17 18 19 20 21

(B)

Circular muscle
Oblique muscle
Longitudinal muscle
Dorsoventral muscle
Dorsal sinus
Dorsal branch, posterior root
Connectives   Ventral sinus   Lateral sinus
Posterior root
Segmental ganglion   Anterior root

2   **CNS OF THE LEECH. (A) The central nervous system of the leech consists of a chain of 21 segmental ganglia, a head ganglion, and a tail ganglion. Over most of the body five circumferential annuli make up each segment; the central annulus is marked by sensory end organs responding to light (the sensillae). (B) The nerve cord lies in the ventral part of the body within a blood sinus. Ganglia, which are linked to each other by bundles of axons (the connectives), innervate the body wall by paired roots. The muscles are arranged in three principal layers: circular, oblique, and longitudinal. In addition, there are dorsoventral muscles that flatten the animal and fibers immediately under the skin that raise it into ridges.**

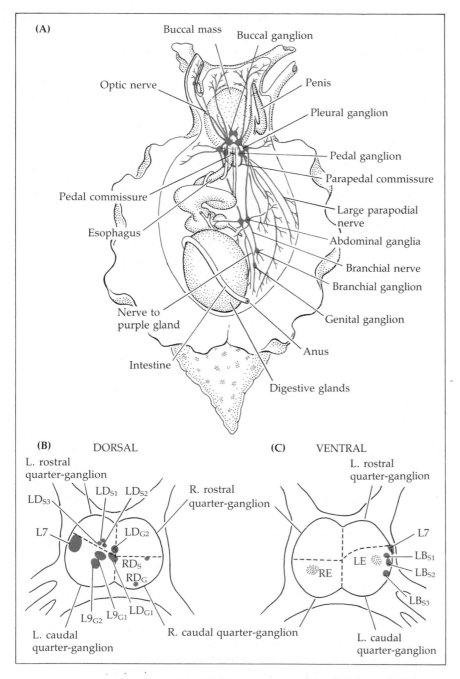

(A)

Buccal mass
Buccal ganglion
Optic nerve
Penis
Pleural ganglion
Pedal ganglion
Parapedal commissure
Large parapodial nerve
Abdominal ganglia
Branchial nerve
Branchial ganglion
Genital ganglion
Pedal commissure
Esophagus
Nerve to purple gland
Intestine
Anus
Digestive glands

(B) DORSAL

L. rostral quarter-ganglion

LD$_{S3}$
LD$_{S1}$ LD$_{S2}$
R. rostral quarter-ganglion
L7
LD$_{G2}$
RD$_S$
RD$_G$
L9$_{G2}$ L9$_{G1}$ LD$_{G1}$
L. caudal quarter-ganglion
R. caudal quarter-ganglion

(C) VENTRAL

L. rostral quarter-ganglion
L7
LB$_{S1}$
LE
LB$_{S2}$
RE
LB$_{S3}$
L. caudal quarter-ganglion

3   CNS OF *APLYSIA*. **(A) The various ganglia and their connections are shown in relation to the internal organs. Experiments described in this chapter were made largely on the abdominal ganglia. (B) Identified cells and cell clusters of abdominal ganglion involved in gill and siphon withdrawal reflexes. The groups of cells labeled RE and LE are sensory, the other cells are motoneurons. The connections of these cells are shown in Figure 9. (Modified from Kandel, 1976.)**

ized in different regions of the animal, thereby controlling different functions.

Working at the level of single cells, one can identify the individual neurons that mediate reflexes, trace their synaptic connections, and establish their characteristic properties—whether they are chemical or electrical, excitatory or inhibitory. This detailed knowledge of the wiring diagram sets the stage for studying how neural components act in concert to produce coordinated elements of behavior. As a next step, one can investigate how the properties of neurons and synapses change as a result of repeated natural sensory stimuli and how these changes are reflected in the performance of the animal. The advantage of using natural stimuli is that only in this way is one assured that impulses are arising in the appropriate fibers at frequencies similar to those in the intact animal.

Detailed knowledge of the circuitry in the adult also makes it possible to use the nervous systems of the leech and *Aplysia* to study at the cellular level problems concerned with development. Their embryos are transparent and are made up of large cells that develop rapidly—in days rather than weeks—into ganglia. In these preparations one can trace cell lineage, follow the growth of individual axons, measure changes in membrane properties, and assess the specificity and mechanisms involved in synapse formation. Moreover, the nervous system of the leech can repair itself after a lesion (see later).

The leech   Since the days of ancient Greece and Rome, leeches have been applied by physicians to patients suffering from diverse diseases such as epilepsy, angina, tuberculosis, meningitis, and hemorrhoids—an unpleasant treatment that almost certainly did more harm than good to the countless unfortunate victims.[7] By the nineteenth century, use of the medicinal leech was so prevalent that the leech became almost extinct in Western Europe, forcing Napoleon to import about six million leeches from Hungary in one year to treat his soldiers. This mania for leeching had at least one lasting benefit for contemporary biology—the medicinal application of leeches stimulated basic research on their reproduction, development, and anatomy. Thus, in the late nineteenth century, founders of experimental embryology, such as Whitman, chose the leech to follow the fates of early embryonic cells. Similarly, its nervous system was extensively studied by a roster of distinguished anatomists, including Sanchez, Ramón y Cajal, Gaskell, Del Rio Hortega, Odurih, and Retzius.[8] Interest in the leech thereafter declined, to be rekindled in 1960 when Stephen Kuffler and David Potter[9] at Harvard Medical School first applied modern neurophysiological techniques to its nervous system.

[7]Payton, W. B. 1981. In K. J. Muller, J. G. Nicholls and G. S. Stent (eds.). *Neurobiology of the Leech*. Cold Spring Harbor Laboratory, Cold Spring Harbor, NY, pp. 27–34.

[8]Muller, K. J., Nicholls, J. G. and Stent, G. S. (eds.). 1981. *Neurobiology of the Leech*. Cold Spring Harbor Laboratory, Cold Spring Harbor, NY.

[9]Kuffler, S. W. and Potter, D. D. 1964. *J. Neurophysiol.* 27: 290–320.

The principal advantages of the leech accrue from the simplicity of its body plan, which is reflected in the structure of its nervous system (Figures 2 and 4). Both body and nervous system are rigorously segmented and consist of a number of repeating units (SEGMENTS) that are similar throughout the length of the animal. Since the animal has no limbs, its behavior consists of a relatively simple repertoire of movements, such as swimming, walking, and shortening, performed by layered groups of muscles. Each segment is innervated by a stereotyped ganglion that is similar to the others within one animal and to those in other animals. Even the specialized head and tail "brains" consist of fused ganglia, in which many characteristic features of segmental ganglia are still recognized.

Each ganglion contains only about 400 nerve cells, which have distinctive shapes, sizes, positions, and branching patterns.[10,11] A ganglion innervates a well-defined territory of the body by way of paired axon bundles (ROOTS), and it communicates with neighboring and distant parts of the nervous system through another set of bundles (CONNECTIVES). Integration thus occurs in a succession of clear-cut steps: (1) Each segmental ganglion receives information from a circumscribed body segment, the performance of which it directly regulates; (2) Neighboring ganglia influence each other by direct interconnections; and (3) The coordinated operation of the whole nerve cord and the animal is governed by the brains at each end of the leech. The distinct segmental subdivisions can be studied on their own or together.

Leech ganglia: Semiautonomous units

Perhaps the main appeal of the leech is the beauty of the ganglion as it appears under the microscope, with its 400 or so neurons so recognizable and so familiar from segment to segment, from specimen to specimen, from species to species. As one looks at these limited aggregates of cells laid out in an orderly pattern, one cannot but marvel at how they, on their own, being the brain of the creature, are responsible for all its movements, hesitations, avoidance, mating, feeding, and sensations. In addition to the aesthetic pleasure provided by the preparation, there is the intellectual excitement in trying to solve the circuitry and logic of a finite, well-organized nervous system, one cell at a time.

Unlike the leech, the gastropod (i.e., belly-footed) mollusk *Aplysia* has a body with well-defined parts—a head, a foot, a shell, and a mantle that consists of a skirtlike flange of tissue that covers the mass of viscera.[1,12] Among specialized organs are tentacles, eyes, gills, and a gland that squirts jets of purple, inky fluid when the animal is disturbed. The *Aplysia* lives in seawater and, like the leech, its behavior is limited compared to that of other invertebrates such as bees or crickets. *Aplysia* walk, withdraw defensively in response to noxious stimuli, make respiratory movements, feed, squirt ink, and indulge in group sex. The

Aplysia

[10]Coggeshall, R. E. and Fawcett, D. W. 1964. *J. Neurophysiol.* 27: 229–289.
[11]Macagno, E. R. 1980. *J. Comp. Neurol.* 190: 283–302.
[12]Kandel, E. R. 1979. *Behavioral Biology of Aplysia.* Freeman, San Francisco.

name *Aplysia* comes from *Aplytos*, meaning "unwashed." Since the tentacles protruding from the body can make it look like a rabbit, Pliny called the animal *Lepus marinus* (sea hare). He considered it to be venomous, and Bacon in 1626 for some reason believed "that the sea hare hath an antipathy with the lungs (if it cometh near them) and erodeth them" (surely an unlikely event).

The account that follows of the behavior and neurobiology of *Aplysia* is based largely on the pioneering studies of Kandel, Castellucci, Carew, and their colleagues. The appeal and usefulness of the large nerve cells in *Aplysia* ganglia for neurobiology were first recognized by Arvanitaki[13] in 1941; she showed that they could be identified, and she analyzed their electrical activity and biochemical properties. Later Strumwasser[14] showed that it was possible to record continuously from an individual cell for days on end in the animal and in culture and thereby analyze the circadian rhythm exhibited by the bursts of action potentials. Moreover, in recent years Tauc,[15] Ascher,[16] Kehoe,[17] Strumwasser,[14] Levitan,[18] and others have made detailed studies of the circuitry, the mechanism of transmitter release, and the properties of postsynaptic chemoreceptors, as well as of the complex events under neurohumoral control, for example, egg-laying, which is regulated by specific secretory neurons (called "bag cells"). Thanks to the large amount of cytoplasm, biochemical analyses that would be quite impossible in leech neurons have been made in single cells of *Aplysia*.

Central nervous system of Aplysia

The buccal, cerebral, pleural, pedal, and abdominal ganglia of *Aplysia* each contain more than 1000 neurons. Some of the individual cells such as R2, R14, and R15 are more than 1 mm in diameter, bigger than an entire leech ganglion. Many of the identified neurons have been studied in great detail, particularly in the abdominal ganglion, which controls the mantle and the viscera. Some of the cells are distinctively colored—some yellow, others orange—and the ganglion as a whole has an orange hue. Neuronal circuits that are confined to the periphery and that give rise to reflex responses after the ganglia have been removed occur in *Aplysia* but not in the leech.

ANALYSIS OF REFLEXES MEDIATED BY INDIVIDUAL NEURONS

Sensory cells in leech and Aplysia ganglia

When one strokes, presses, or pinches the skin of a leech or an *Aplysia*, a sequence of movements follows. In the leech, one segment or more shortens abruptly, and the skin becomes raised into a series of distinct ridges. Subsequently, the animal swims away or executes writhing

[13]Arvanitaki, A. and Cardot, H. 1941. *C.R. Soc. Biol.* *135*: 1207–1211.
[14]Strumwasser, F. 1971. *J. Psychiat. Res. 8*: 237–257.
[15]Tauc, L. 1967. *Physiol. Rev. 47*: 521–593.
[16]Ascher, P., Marty, A. and Neild, O. 1978. *J. Physiol. 278*: 207–235.
[17]Kehoe, J. and Marty, A. 1980. *Annu. Rev. Biophys. Bioeng. 9*: 437–465.
[18]Adams, W. B. and Levitan, I. B. 1982. *Proc. Natl. Acad. Sci. USA 79*: 3877–3880.

movements. Similarly, in *Aplysia,* a touch or a jet of water applied to the siphon or mantle leads to a withdrawal reflex.

As a first step in defining the pathways, the synaptic mechanisms, and the changes that occur in time, the individual cells responsible for producing the reflexes must be identified. In living leech and *Aplysia* ganglia, one can reliably identify the sensory and motor cells according to their shapes, sizes, positions, and electrical characteristics.[1,19] Figures 3B, and 4 show the distribution of the identified sensory cells in the abdominal ganglion of *Aplysia* and in a leech ganglion. For example, the 14 leech neurons labeled T, P, and N in Figure 4 are all sensory and present three sensory modalities. Each cell responds selectively to touch, pressure, or noxious mechanical stimulation of the skin. Representative intracellular records are also shown in Figure 4. The impulses of T cells are always similar to, but smaller and briefer than, those in P and N cells. With practice, it is usually enough just to look at a single action potential to be certain which type of cell has been impaled. Figure 5 illustrates the responses of leech sensory cells to various forms of cutaneous stimuli. The T cells give transient responses to light touch of the skin surface or even to eddies in the solution bathing the skin. The sensory endings of touch cells consist of small dilatations situated between epithelial cells on the surface of the skin.[20] T cells adapt rapidly to a maintained step indentation and usually cease firing within a fraction of a second. The P cells respond only to a marked deformation of the skin and show a slowly adapting discharge. The frequency is graded with the extent of the indentation and light touch is ineffective in activating P cells. The N cells require still stronger mechanical stimuli, such as a radical deformation produced by pinching the skin with blunt forceps or scratching it with a pin. The N cells, like the P cells, are slowly adapting and often continue to fire after the stimulus has been removed. The specificity of these leech sensory cells is remarkable. For example, one of the two N cells has different chemosensitivity to transmitters.[21] Moreover, a monoclonal antibody can be made that selectively binds to molecules in one but not the other N cell, and that does not bind at all to T, P, Retzius, motor, or any other identified cells in the segmental ganglia.[22]

Sensory responses similar to those in the leech can be recorded from groups of identified neurons in *Aplysia* ganglia following comparable mechanical stimuli to the surface.[23] The number of cells, is, however, considerably larger. For example, in the abdominal ganglion, 25 distinctive cells—50 μm in diameter, orange colored, and with a dark rim (labeled LE in Figure 3)—send their axons out through the siphon nerve

[19]Blackshaw, S. 1981a. In K. J. Muller, J. G. Nicholls and G. S. Stent (eds.). *Neurobiology of the Leech.* Cold Spring Harbor Laboratory, Cold Spring Harbor, NY, pp. 51–78.

[20]Blackshaw, S. 1981b. *J. Physiol.* 320: 219–228.

[21]Sargent, P. B., Yau, K.-W. and Nicholls, J. G. 1977. *J. Neurophysiol. 40*: 453–460.

[22]Zipser, B. and McKay, R. 1981. *Nature 289*: 549–554.

[23]Byrne, J., Castellucci, V. F. and Kandel, E. R. 1974. *J. Neurophysiol. 37*: 1041–1064.

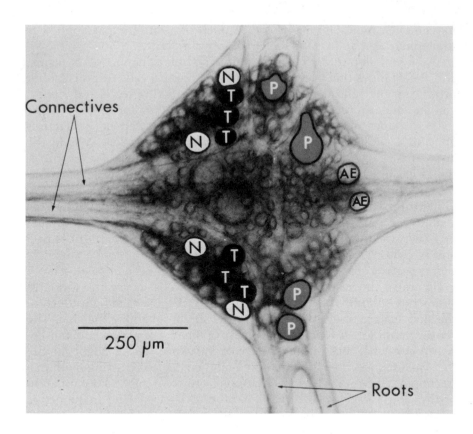

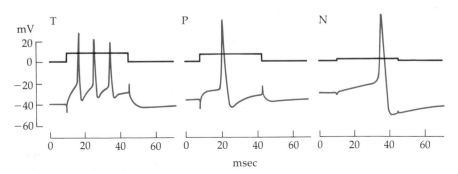

4 **VENTRAL VIEW OF LEECH SEGMENTAL GANGLION.** Individual cells are clearly recognized. The three sensory cells responding to touch (T) and the pairs of cell types responding to pressure (P) or noxious (N) mechanical stimulation of the skin are labeled. Each type of cell gives distinctive action potentials, as shown by the traces below. Impulses in T cells are briefer and smaller than those in P or N cells. Current injected into cells through the microelectrode is monitored on the upper traces. The cells outlined in the posterior part of the ganglion are the annulus erector (AE) motoneurons. (After Nicholls and Baylor, 1968.)

to innervate the skin of the siphon and the mantle shelf. They respond selectively to mechanical stimulation: some to touch, others to pressure. Unlike those in the leech, however, all these cells have similar electrical properties. A second group of 20 sensory cells, labeled RE in Figure 3B, innervate the mantle shelf and the purple gland. In general, the sensitivity and responses of *Aplysia* mechanoreceptor cells are similar to those of P cells in the leech (Figure 5B), except for one cell (L18), which responds only to severe mechanical deformation of the skin and resembles the N sensory neurons in the leech.

The modalities and responses of sensory neurons in leech and *Aplysia* resemble those of mechanoreceptors in the human skin, which also distinguish between touch, pressure, and noxious or painful stimuli. In the invertebrate, however, a single nerve cell does the job of many neurons supplying our own skin in a densely innervated region such as the fingertip. Examples of receptive fields for *Aplysia* and leech mechanoreceptor neurons are shown in Figure 6. The fields in *Aplysia* vary considerably in size from cell to cell and show considerable overlap.[23]

Receptive fields

In the leech each sensory cell innervates a defined territory of the periphery and responds only to stimuli applied within that area. The territory that a particular cell supplies is mapped by recording from it while applying mechanical stimuli to the skin. The boundaries can be conveniently identified by landmarks, such as segmentation or the coloring of skin, so that one can predict reliably which cell will fire when a particular area is touched, pressed, or pinched. Thus, one of the touch cells innervates dorsal skin, another ventral skin, and a third laterally situated skin. Similarly, the two P sensory cells divide the skin into roughly equal dorsal and ventral areas, whereas the two N cells respond to noxious stimulation over roughly equivalent overlapping areas.[19] In a system that has such clear-cut boundaries in the periphery, it is possible to study the dynamics of the formation and maintenance of receptive fields. How does the receptive field become established? What factors determine exactly where the endings of a particular cell can and cannot be formed? How is the degree of overlap by endings from the same cell and from different cells regulated? It has been shown that during development the axons of an individual sensory cell can be followed as they grow out to innervate skin for the first time (see later). They head for the correct area to form an appropriately shaped and located field of innervation of the type seen in the adult.[24] The dimensions and position of the receptive field are not, however, immutably fixed throughout the life of the leech. Thus, removal of the innervation from one area of skin causes an uninjured cell to expand its receptive field to occupy the vacant territory.[25] Interestingly, such sprouting is modality-specific: Selective killing of N cells in the leech causes uninjured N cells in the same ganglion (but not P or T sensory cells) to

[24]Kuwada, J. Y. and Kramer, A. P. 1983. *J. Neurosci. 3*: 2018–2111.
[25]Blackshaw, S., Nicholls, J. G. and Parnas. I. 1982. *J. Physiol. 326*: 261–268.

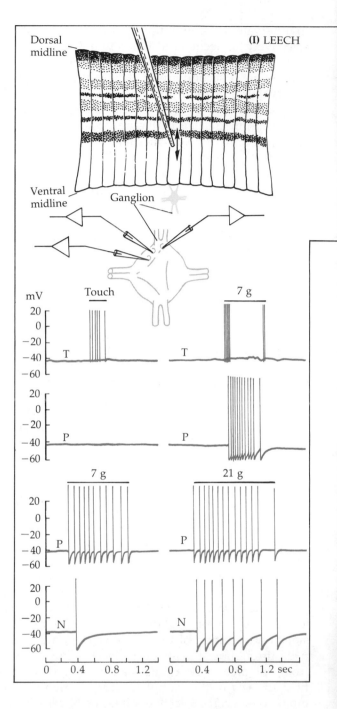

5

RESPONSES TO SKIN STIMULATION, by neurons in leech and *Aplysia*. I. Leech: intracellular records of T, P, and N sensory cells. The preparation consists of a piece of skin and the ganglion that innervates it. Cells are activated by touching or pressing their receptive fields in the skin. (A) A T cell responds to light touching that is not strong enough to stimulate the P cell. (B) Stronger, maintained pressure evokes a prolonged discharge from the P cell and a rapidly adapting "on" and "off" response from the T cell. (C, D) Still stronger pressure is needed to activate the N cell. (After Nicholls and Baylor, 1968.) II. Intracellular records from an *Aplysia* sensory neuron responding to pressure applied to the siphon. In their properties, the *Aplysia* sensory neurons in general resemble P cells. (Modified from Kandel, 1976.)

expand into the denervated territory. Figure 7 shows the receptive field of N cells: in a normal animal (A) and in a leech ganglion in which the other three N cells had been killed three months beforehand (B). The one N cell left expanded its territory to occupy skin denervated of N cells. P and T cells did not sprout. Similarly, selectively killing T cells

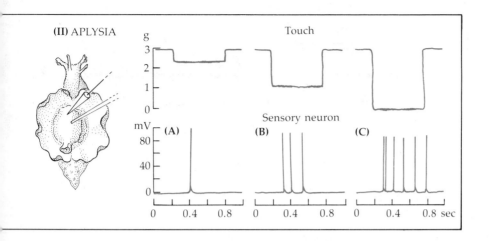

(II) APLYSIA

Touch

Sensory neuron

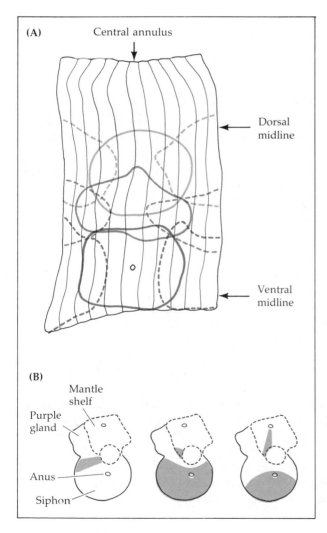

(A) Central annulus

Dorsal midline

Ventral midline

(B) Mantle shelf

Purple gland

Anus

Siphon

6

RECEPTIVE FIELDS OF SENSORY NEURONS
in leech and *Aplysia*. (A) Receptive fields of T
cells drawn on a tracing of the skin of a leech.
Solid lines show the boundaries of the receptive
fields of the three T cells; dashed lines show the
fields of cells in adjacent ganglia. Fields were
mapped successively for the nine cells using light
touch and marking the positions from which responses could be obtained on an enlarged photograph of the skin. An annulus width is approximately 1 mm. (B) Sensory neurons in
Aplysia have overlapping receptive fields. At
least 24 neurons cover the tip of the siphon skin.
The field of any one cell may be small, large, or
fractionated. (A modified from Nicholls and Baylor, 1968; B modified from Byrne, Castellucci, and
Kandel, 1974.)

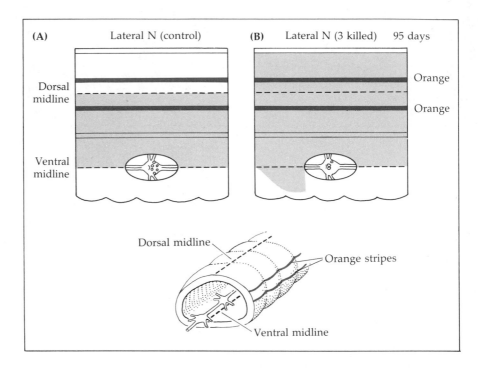

7 NORMAL AND EXPANDED RECEPTIVE FIELDS in leech skin sensory cells. (A) Schematic representation showing that each of the two N cells has a receptive field that extends from the dorsal midline to the ventral midline in normal animals. (B) Expanded receptive field of a lateral N cell in a ganglion in which the three other N cells had been killed 95 days beforehand. The cell innervated almost the entire territory, including contralateral skin up to the edge, an area it does not normally supply. Thresholds for electrical and mechanical activation were similar in the normal and expanded fields. The fields of T and P cells in this preparation did not cross the dorsal midline. (After Blackshaw, Nicholls, and Parnas, 1982.)

causes the remaining uninjured T cell in the ganglion to sprout without influencing N or P cells. These results suggest the presence of a dynamic relationship between a neuron and the target that it innervates—a problem considered again in this and subsequent chapters.

Motor cells   Individual motor cells in *Aplysia* and leech ganglia are shown in Figures 3 and 4. The criteria for showing that a cell is indeed motor is that each impulse in the cell gives rise to a conducted action potential in its axon leading to the muscle and then to a synaptic potential in the muscle fiber. In addition, deletion of a single cell can give rise to an obvious deficit in behavior. For example, one neuron called the annulus erector cell, labeled AE in Figure 4, causes the skin of the leech to be raised into ridges like a concertina (Figure 11). There are only two such cells in each ganglion—one on either side. If an AE cell is killed by injecting it with a mixture of proteolytic enzymes in an otherwise intact animal, which is then allowed to recover, the area of skin it innervates on its own fails to become erect in response to appropriate sensory stimuli.[26] In the leech, the individual motor cells supplying the various muscles that erect annuli, flatten, lengthen, shorten, and bend the body, as well as the motor cell that controls the heart, have all been identified in the segmental ganglia. Muscles can receive inhibitory inputs in addition to the excitatory inputs that are mediated by acetylcholine. So far more than 20 pairs of neurons have been unambiguously identified as directly supplying the muscles.[8] In *Aplysia*, larger numbers of motor neurons are involved in the production of reflex movements such as gill and siphon withdrawal.[1] Thirteen identified central motor neurons and about 30 peripheral motor neurons innervate muscles moving the siphon, the mantle shelf, and the gill. If these neurons are prevented from firing by hyperpolarizing them, reflex contractions of the muscles are abolished. Some of the motoneurons have a restricted action, causing only small movements, whereas others have a broad action, sometimes contracting completely separate effector organs, such as the gill and mantle shelf.

Here, then, are systems in which one can infer which sensory cells are activated when a mechanical stimulus is applied to a particular area of skin, and which motor cells are firing when the animal performs a movement.

In the nervous systems of invertebrates, the synapses between neurons are situated not on the cell bodies but on fine processes within a central region of the ganglion (the neuropil). The neuropil is highly complex and resembles that in the vertebrate brain. The synaptic potentials generated in the neuropil cannot be recorded directly at the sites of origin by microelectrodes because the processes are too small to impale. They do, however, spread into the nearby cell body where excitatory and inhibitory potentials several millivolts in amplitude can be recorded. Similarly, currents injected into the cell body can influence synaptic potentials and the release of transmitter.

Despite its apparent complexity, the neuropil is organized in an orderly manner: This fact is known from the constancy of connections between identified cells in leech and *Aplysia*. Another clear sign of orderliness is the characteristic configuration of the branching patterns of sensory and motor cells. For each type of cell, the pattern or the fingerprint is similar from ganglion to ganglion. Examples of the typical ramifications of identified neurons in the leech and in *Aplysia* are shown in Figure 8. These neurons were injected with horseradish peroxidase or cobalt, which penetrate into all the processes.[27] The enzyme can also be detected in electron micrographs, as in Figure 8C in which characteristic chemical synapses are marked by arrows. In this manner, then, one can identify the synaptic input to at least some of the synapses in the neuropil.[28]

Orderly distribution of synapses

[26]Bowling, D., Nicholls, J. G. and Parnas, I. 1978. *J. Physiol. 282*: 169–180.
[27]Muller, K. J. and McMahan, U. J. 1976. *Proc. R. Soc. Lond. B 194*: 481–499.
[28]Bailey, C. H., Kandel, P., and Chen, M. 1981. *J. Neurophysiol. 46*: 356–368.

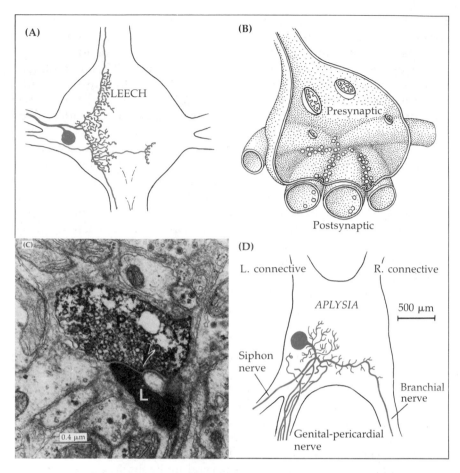

8 SHAPES AND SYNAPSES OF LEECH AND *APLYSIA* NEURONS. (A) Arborization of a pressure cell in a leech ganglion after injection with horseradish peroxidase. The cell body sends a single process to the neuropil where all the synaptic connections of the ganglion are made. Axons run to neighboring ganglia through connectives and to the body wall through roots. Small processes within the neuropil form synaptic contacts. (B) Reconstruction made from serial sections of synaptic specializations seen in leech neuropil. Typically, a single presynaptic process containing numerous vesicles makes synapses on two or more postsynaptic elements. This arrangement is seen also in the CNS of *Aplysia*, with similar thickenings and widened intercellular clefts. (C) Synapse made by P cell on L motoneuron in the neuropil of a leech ganglion. Both cells were injected with horseradish peroxidase for identification. (D) Cell L7 in an *Aplysia* abdominal ganglion injected with cobalt. This is a motor cell supplying the mantle shelf. (A, B from Muller, 1981; C by E. R. Macagno and K. J. Muller; D after Kandel, 1976.)

Electrical synapses in leech and *Aplysia* ganglia appear as contacts between neurons with a narrowed space 4–6 nm separating the two membranes;[29] as at other gap junctions, bridges span the cleft in these regions. Fluorescent dyes such as Lucifer Yellow injected into one cell

[29]Muller, K. J. 1981. In K. J. Muller, J. G. Nicholls and G. S. Stent (eds.). *Neurobiology of the Leech.* Cold Spring Harbor Laboratory, Cold Spring Harbor, NY, pp. 79–111.

will usually, but not always, cross such electrical junctions and spread into other coupled cells. The pattern of connections revealed in this way can be more complex than one might guess from physiological evidence. For example, intracellular recordings reveal that the six T cells in a leech ganglion are all weakly coupled to each other. This coupling, however, is not direct but is mediated by electrical synapses through two specific coupling interneurons. If these two cells are killed by injection of protease, coupling between the T cells is abolished.[30]

In *Aplysia* and the leech, the sensory cells responding to mechanical stimuli make excitatory connections on the motoneurons used for gill and siphon withdrawal or for shortening of the body. Several lines of evidence indicate that the connections are direct (i.e., that there are no unknown intermediary cells).[1,8] This fact is important, because only if each constituent and its properties are known can one pinpoint the sites at which any interesting modifications in signaling take place. In *Aplysia*, the sensory cells that make excitatory, chemically mediated connections on a population of motoneurons causing gill and siphon withdrawal also make excitatory synapses on other cells (interneurons) that in turn synapse on the motoneurons. A diagram of the connections that mediate gill withdrawal is shown in Figure 9.

Connections between sensory and motor cells

In the leech, the T, P, and N sensory cells all converge on one motor cell, the L motoneuron, which innervates longitudinal muscles and produces shortening. A remarkable feature, however, is that the mechanism of transmission onto the L motor cell is characteristically and consistently different for each of the sensory cells.[31] The N cells act through chemical synapses, the T cells through rectifying electrical synapses, and the P cells by a combination of both mechanisms. These simple pathways, therefore, exemplify the different forms of signaling encountered in the nervous systems of higher animals. The arrangement also offers an opportunity to dissect functionally the contribution that may be characteristic for a particular mode of synaptic transmission.

The modes of synaptic transmission between sensory cells and motoneurons of the leech and *Aplysia* have been distinguished from one another by (1) observing the latency of the synaptic potentials in the motoneuron; (2) bathing the ganglion in high concentrations of magnesium ions, which block chemical synaptic transmission; (3) changing the membrane potential of the presynaptic and postsynaptic neurons; and (4) observing the synapses in electron micrographs after injecting one or both of the cells with a marker such as horseradish peroxidase.

For simplicity, the initial analysis of signaling mechanisms is usually made by observing the effects of single impulses. However, while an animal is swimming in water or moving on a surface under normal conditions, stimulation of the skin causes trains of impulses. As a second step in analysis and to differentiate between the effects of chemical and electrical synapses on reflexes, the performance of the moto-

Performance of chemical and electrical synapses

[30]Muller, K. J. and Scott, S. A. 1981. *J. Physiol. 311*: 565–583.
[31]Nicholls, J. G. and Purves, D. 1972. *J. Physiol. 225*: 637–656.

472

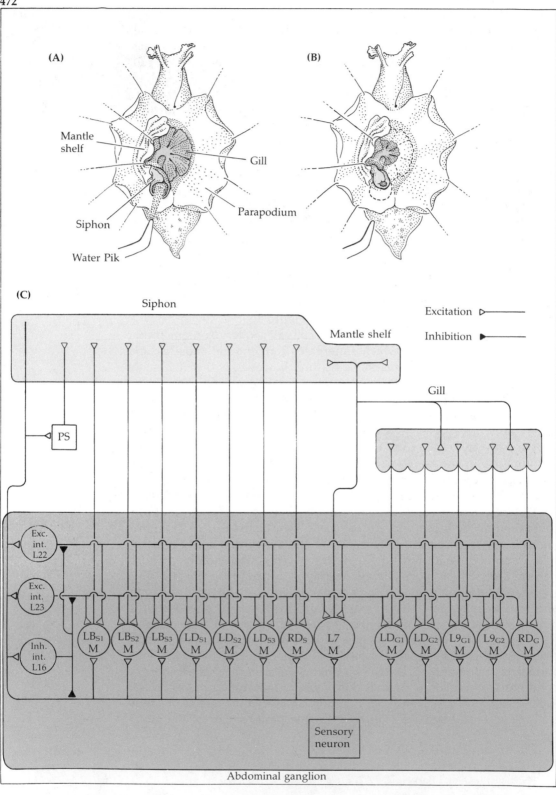

(A)

Mantle
shelf

Gill

Siphon

Parapodium

Water Pik

(B)

(C)

Siphon

Mantle shelf

Excitation ▷——

Inhibition ▶——

Gill

PS

Abdominal ganglion

Exc.
int.
L22

Exc.
int.
L23

Inh.
int.
L16

LB$_{S1}$
M

LB$_{S2}$
M

LB$_{S3}$
M

LD$_{S1}$
M

LD$_{S2}$
M

LD$_{S3}$
M

RD$_S$
M

L7
M

LD$_{G1}$
M

LD$_{G2}$
M

L9$_{G1}$
M

L9$_{G2}$
M

RD$_G$
M

Sensory
neuron

9 WITHDRAWAL REFLEX OF GILL AND SIPHON in *Aplysia.* (A, B)
Water squirted onto the siphon causes it and the gill to withdraw. (C)
Sensory cells, interneurons, motoneurons, and the connections mediating the withdrawal reflex. A single sensory neuron out of the population
of 24 or more is shown. Sensory neurons make monosynaptic and also
indirect connections with the pool of motoneurons (M). Excitatory (L22,
L23) and inhibitory (L16) interneurons mediate those actions. Motor
cells situated in the periphery (PS) also contribute to the reflex. (After
Kandel, 1976.)

neurons is tested by sustained sensory stimulation. From the material
presented in Chapter 9 on the characteristics of transmission processes,
the general expectation is that electrical synapses will remain relatively
stable in their performance under a variety of conditions, whereas chemical synapses will be much more variable. Figure 10 shows the difference
between the two forms of transmission when leech nociceptive and
touch cells fire. For example, when an N cell fires in response to stimulation or to pinching of the skin, the chemically evoked synaptic potentials recorded in the L cell during a train first increase (facilitation)
and then decrease (depression, not shown). In sharp contrast is the
synaptic effect of the touch neuron (the T cell) that makes an electrical
synapse on the L cell. With repeated stimulation or stroking of the skin
under the same conditions, the synaptic potentials evoked in the L cell
remain unchanged.[31]

Such observations raise the question whether the variable effectiveness of different chemical synapses may not be part of a functional
design that permits sequential activation or inactivation of different

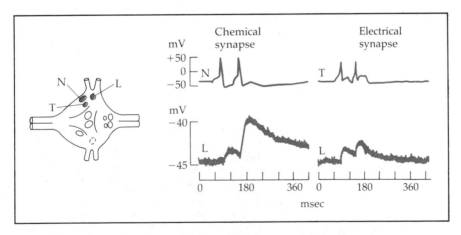

10 CHEMICAL AND ELECTRICAL TRANSMISSION in leech ganglion.
An N cell is stimulated twice in succession and at the same time its
impulses are recorded (upper records). At the chemical synapse between
N and L cells, facilitation occurs so that a second impulse leads to a
larger synaptic potential (bottom left). In contrast, two potentials evoked
by T-cell impulses in an L cell cause similar postsynaptic potentials
with double or multiple stimulation. This is typical of electrical synapses. (After Nicholls and Purves, 1972.)

postsynaptic structures. A single presynaptic neuron might, for example, activate first one and then another postsynaptic cell. This type of differential effect would adequately explain how pressing the skin of a leech leads first to a shortening reflex and then, after a delay, to erection of the annuli. In leech ganglia, the N and P sensory cells make chemical synapses not only on the L cell but also on the AE motor cell that raises the skin into ridges.[29] The synaptic potential recorded in the AE cell is considerably smaller than that in the L cell (Figure 11). With trains of impulses in a single sensory cell, the synaptic potentials in the AE and L motoneurons undergo phases of facilitation and depression; the facilitation, however, is characteristically greater and longer lasting at

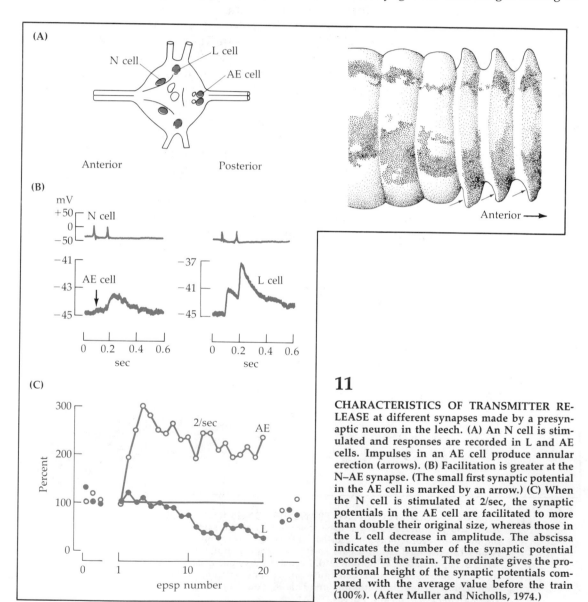

11

CHARACTERISTICS OF TRANSMITTER RE-LEASE at different synapses made by a presynaptic neuron in the leech. **(A)** An N cell is stimulated and responses are recorded in L and AE cells. Impulses in an AE cell produce annular erection (arrows). **(B)** Facilitation is greater at the N–AE synapse. (The small first synaptic potential in the AE cell is marked by an arrow.) **(C)** When the N cell is stimulated at 2/sec, the synaptic potentials in the AE cell are facilitated to more than double their original size, whereas those in the L cell decrease in amplitude. The abscissa indicates the number of the synaptic potential recorded in the train. The ordinate gives the proportional height of the synaptic potentials compared with the average value before the train (100%). (After Muller and Nicholls, 1974.)

synapses upon the AE motoneuron. A number of experiments suggest that the differences in synaptic transmission can be accounted for by variations in the amount of transmitter released at presynaptic N-cell terminals, rather than by differences in the postsynaptic cells. Similar observations have been made at crustacean neuromuscular synapses, where it was shown that not all synapses made by a single motor fiber behave in the same way.[32] Some of the synapses made on one muscle fiber cause pronounced facilitation while synapses on another muscle fiber by the same motoneuron show much less facilitation. The difference apparently lies in the presynaptic nerve terminals rather than in the muscle fibers, in that some terminals release more transmitter than others.

A satisfactory aspect of these studies is the general agreement between the properties of synaptic transmission and the behavioral reactions that occur in response to mechanical stimuli. In the intact leech, or in one of its isolated segments, the two reflexes follow different time courses: The shortening of the body wall occurs abruptly and is poorly maintained, whereas the annuli become erect more slowly and stay erect longer. This finding correlates well with the synaptic potentials that facilitate at different rates and trigger the two events in sequence. It will be of interest to learn how extensively other animals use differential effects in setting up a sequence of reactions to a maintained stimulus. So far there is little information about analogous situations in the central nervous system of vertebrates.

Short-term facilitation and depression of the sort seen at leech synapses or frog neuromuscular junctions are less evident at *Aplysia* synapses used for withdrawal reflexes, but there, as shown below, long-term changes persisting for hours, or even days, have been observed. Moreover, in *Aplysia*, it has been shown by Wachtel and Kandel[33] that a single interneuron can excite one postsynaptic target while it inhibits another. For example, in Figure 12, impulses in cell L10 evoke excitatory potentials in cell R15 and inhibitory potentials in cell L3, both mediated chemically by monosynaptic pathways. At both synapses the transmitter is acetylcholine, but the effect depends on the presence of different types of channels, opened by the acetylcholine receptors, in the two cells. The same cell—L10—also evokes either excitatory or inhibitory responses in one postsynaptic target (L7), the response depending on the frequency of firing. At certain synapses in *Aplysia*, the duration of excitatory or inhibitory potentials can be extremely long, persisting for minutes or even hours following one or a few impulses.[34] Acetylcholine, as well as dopamine and 5-hydroxytryptamine can produce both conductance decreases and increases, and these changes persist after the transmitter has been removed, like the slow potentials described in autonomic ganglia (Chapter 12).

Modulation of chemical synaptic transmission in Aplysia

[32]Frank, E. 1973. *J. Physiol. 233*: 635–638.
[33]Wachtel, H. and Kandel, E. R. 1971. *J. Neurophysiol. 34*: 56–68.
[34]Parnas, I. and Strumwasser, F. 1974. *J. Neurophysiol. 37*: 609–620.

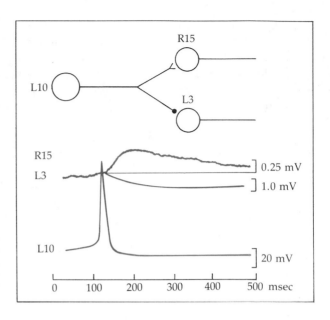

12

EXCITATION AND INHIBITION mediated by a single cell (L10) onto two postsynaptic neurons (L3 and R15) in *Aplysia*. At both synapses ACh is the transmitter, but the permeability changes produced are different. (After Wachtel and Kandel, 1971.)

Unusual synaptic mechanisms

In addition to conventional, frequency-dependent facilitation and depression of transmitter release, a variety of novel presynaptic mechanisms that influence the efficacy and duration of chemical synaptic transmission have been described in *Aplysia*, in the leech, and in other vertebrates. For example, at chemical synapses in the central nervous system of crustacea, insects, and mollusks, transmitter is released tonically by presynaptic neurons that do not give impulses.[35] Release occurs continuously at rest and is increased by graded depolarization of the presynaptic terminal (the depolarization resulting from excitatory synaptic inputs), cyclical alterations in membrane potential, or depolarization elicited elsewhere in the neuron and conducted electrotonically to the terminal. Conversely, hyperpolarization reduces the rate of release. Similar, "nonspiking" interneurons have been observed within the mammalian central nervous system, for example, the photoreceptor and bipolar cells in the retina. At such synapses, the tonic release of transmitter allows for finely graded, maintained action of one neuron on another.

Presynaptic depolarization and hyperpolarization have other subtle effects on chemical transmission at more conventional synapses at which the impulse is responsible for releasing transmitter. As a result, reflexes and behavior can be markedly altered. For example, at certain synapses in *Aplysia* and the leech, maintained depolarization of the presynaptic terminals enables an impulse to release more transmitter; conversely, if the membrane of the presynaptic terminal is hyperpolar-

[35]Siegler, M. V. S. and Burrows, M. 1979. *J. Comp. Neurol. 183*: 121–148.

ized, the impulse releases less transmitter.[36,37] A shift in resting potential of as little as 5 mV (from −40 to −35 mV) can cause a threefold increase in the amount of transmitter released by an impulse. The amplitude and the duration of the presynaptic impulse are not obviously altered by such small changes in resting potential. Similar mechanisms have now been shown to operate at the squid giant synapse.[38] The effects on reflexes are illustrated by a circuit in the leech, where the presynaptic terminals of certain interneurons controlling the heartbeat show naturally occurring hyperpolarization. This is produced cyclically by inhibitory inputs from other identified interneurons. As a result of the hyperpolarization of the terminal, the numbers of quanta liberated by each impulse invading it are reduced.[37] Here, then, is an instance of presynaptic inhibition that is mediated by hyperpolarization of the terminals. In the well-defined circuit of neurons controlling the heart of the leech, such presynaptic inhibition and cyclical modulation of the membrane potential of the presynaptic terminal have been shown to be essential for establishing the rhythmicity.[36] It is not yet known whether similar mechanisms may also be involved in the presynaptic inhibition observed in the mammalian spinal cord, where the role of presynaptic depolarization is not yet clear.

Failure of impulse conduction represents a further mechanism for altering the synaptic action of one cell upon its postsynaptic targets. As in crustacean motor axons (Chapter 16), in the central nervous system of the cockroach and the leech repeated trains of impulses occurring at natural frequencies can produce after effects that lead to conduction block at branch points. In crustacean axons,[39] the changes are associated with an increase in extracellular potassium concentration. In sensory neurons of the leech, the mechanism depends on the hyperpolarization caused by the electrogenic sodium pump and by long-lasting changes in a calcium-activated potassium conductance. Thus, stroking or pressing the skin of the leech repeatedly causes trains of impulses and a maintained hyperpolarization; as a result, propagation of impulses becomes blocked at certain branch points where the geometry for impulse conduction is unfavorable, with a small fiber feeding into a larger one.[40] Under these conditions, the terminals no longer invaded by impulses fail to release transmitter and fail to influence those postsynaptic cells while other branches of the same neuron continue to fire.[30] This, therefore, represents a nonsynaptic mechanism whereby with repeated use one set of postsynaptic targets becomes temporarily disconnected.

In a series of experiments, Kandel and his colleagues have analyzed the role played by various biophysical mechanisms on the production

Habituation and dishabituation in *Aplysia*

[36]Thompson, W. J. and Stent, G. S. 1976. *J. Comp. Physiol. 111*: 309–333.
[37]Nicholls, J. G. and Wallace, B. G. 1978. *J. Physiol. 281*: 157–170.
[38]Llinás, R. 1982. *Sci. Am. 247*: 56–65.
[39]Grossman, Y., Parnas, I. and Spira, M. E. 1979. *J. Physiol. 295*: 307–322.
[40]Yau, K.-W. 1976. *J. Physiol. 263*: 513–538.

of complex behavioral responses of *Aplysia*.[1,41] The gill withdrawal reflex (described earlier) becomes progressively weaker when water jets are applied to the mantle repeatedly at regular intervals (Figure 13). For example, if constant stimuli are used to elicit the reflex 15 times at intervals of 10 minutes, the strength of the gill withdrawal to the same stimulus becomes reduced to about 50 percent of its original value. It then recovers only slowly over a period of 30 minutes to several hours. The animal is thereby able to respond less vigorously to a stimulus that had previously been effective. The time course and the properties of this phenomenon resemble those of "habituation" in higher animals. For example, a sudden noise may elicit a startle reaction, but upon repetition will cease to do so. A characteristic of habituation is that, in addition to spontaneous recovery (if the stimulus is not presented for a period), immediate "dishabituation" occurs following a strong generalized stimulus of a different type. Dishabituation occurs in *Aplysia* if a strong electrical shock is delivered to the head: The animal will then once again withdraw its gill in response to a water jet.

To determine the mechanisms involved in habituation and dishabituation, recordings have been made from the sensory cells, the motor cells, and the interneurons in animals and in isolated preparations that show similar behavior to the whole animal. *Aplysia* are immobilized, their ganglia exposed, and stimuli delivered to the skin while intracellular recordings and tension measurements are made. The results of these experiments show that the reduced responsiveness occurs as the result of a long-lasting depression of transmitter release by the terminals of the sensory neurons upon the motor neurons.[42] The sensory response to the peripheral stimulus and the properties of the motor neurons show no comparable changes. However, intracellular recordings made from the sensory neuron indicate that the presynaptic action potential becomes shortened in duration. This gives rise to a smaller inward calcium current and therefore to a smaller amount of transmitter release. Since the presynaptic terminals are too small to be impaled by microelectrodes, inferences about their membrane properties are drawn entirely from records made in the soma.

How is dishabituation produced by strong electrical stimulation of the head? The suggestion is that the fibers mediating this effect release 5-HT, which can itself give rise to dishabituation when applied to the ganglion. The scheme proposed by Kandel and his colleagues[42] is shown in Figure 14. Appropriate nerve stimuli or the application of 5-HT lead to an increase in intracellular cyclic AMP in the presynaptic cell; this activates a specific protein, a cyclic AMP-dependent protein kinase. The action of the protein kinase is to decrease the potassium conductance of the membrane by a mechanism not yet understood, which thereby allows the presynaptic action potential to become restored to its original,

[41]Carew, T. J., Pinsker, H. M. and Kandel, E. R. 1972. *Science* 175: 451–454.

[42]Castellucci, V. F., Kandel, E. R., Schwartz, J. H., Wilson, F. D., Nairn, A. C. and Greengard, P. 1980. *Proc. Natl. Acad. Sci. USA* 77: 7492–7496.

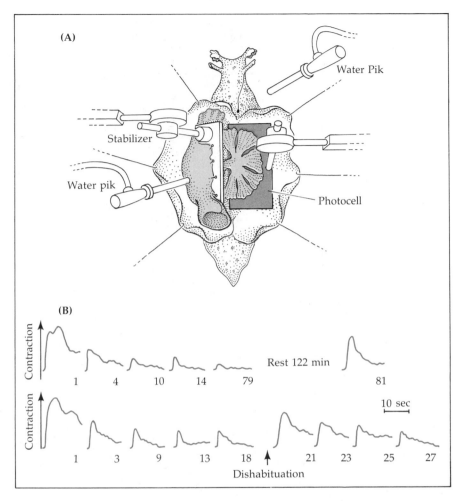

(A)

Water Pik

Stabilizer

Water pik

Photocell

(B)

Contraction

1   4   10   14   79   Rest 122 min   81

Contraction

10 sec

1   3   9   13   18 ↑ 21   23   25   27

Dishabituation

13 HABITUATION OF WITHDRAWAL REFLEX in *Aplysia*. (A) The animal is immobilized in circulating seawater with its mantle pinned. Contractions of the gill are monitored by a photocell. The stimulus is a jet of water. (B) Traces of gill contractions in response to standard stimuli. The depression is reversed after a rest of 122 min or if a strong mechanical stimulus is applied to the head (arrowhead). (After Kandel, 1976.)

more-prolonged duration. One likely suggestion is that the enzyme phosphorylates the potassium channel protein or a protein associated with it. In support, it has been found that substances that specifically block the protein kinase prevent this dishabituation. Conversely, injection of the active enzyme—the catalytic subunit of the protein kinase—into the appropriate sensory cell causes prolongation of the action potential and increased transmitter release as expected. Interestingly, 5-HT has the opposite effect on a different *Aplysia* neuron—R15. In R15, cyclic AMP also mediates the effect of 5-HT; but in this cell it causes an increase rather than a decrease of potassium conductance.[18]

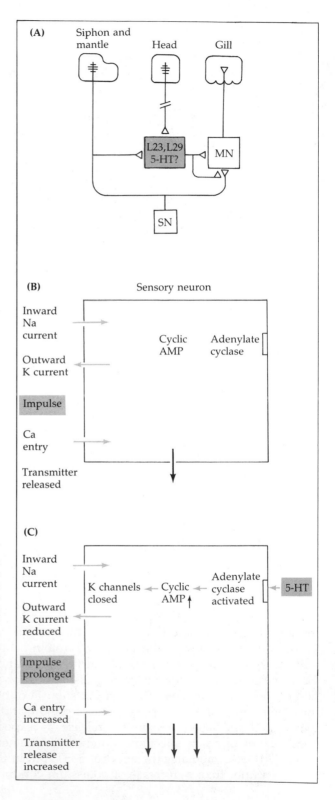

14

SCHEME TO EXPLAIN DISHABITUATION AND SENSITIZATION in *Aplysia*. **(A)** Simplified diagram of sensory and motor cells mediating withdrawal reflex. 5-HT application mimics stimuli that produce dishabituation. **(B,C)** Hypothesis for mode of action of 5-HT. The impulse in the sensory nerve terminals allows Ca to enter, the increased Ca concentration causing transmitter release (B). The effect of 5-HT is to prolong the duration of the impulse, thereby allowing more Ca to enter and more transmitter to be liberated. The steps involve 5-HT activation of adenylate cyclase and an increase in intracellular cyclic AMP; this increased concentration of cyclic AMP in turn, by decreasing g_K, would prolong the impulse (slower repolarization). Because the terminals are small and buried within the complex neuropil, recordings and injections are made at the cell body. (Modified from Klein, Shapiro, and Kandel, 1980.)

It is hard to see how such experiments could be made at the cellular level to analyze complex behavioral changes in the mammalian brain. They provide an exceptionally complete description of a process closely akin to learning. In related experiments, Kandel and his colleagues have studied associative conditioning in *Aplysia*:[43] A chemosensory conditioning stimulus (shrimp extract applied to the bath) that normally causes feeding is paired with noxious stimulation (the unconditioned stimulus, an electric shock applied to the animal). After a sufficient number of paired trials, the animal shows defensive responses instead of feeding when the chemosensory stimulus is applied. Similarly, weak touches (conditioned stimulus) to the siphon have been coupled with strong shocks to the tail (unconditioned stimulus). Interestingly, the area of skin touched can be discriminated so that responses from neighboring regions do not become conditioned. The cellular mechanisms for conditioning again are amenable to direct experimental analysis.

One aim of studies on an invertebrate like the leech is to analyze how complex behavioral acts are built up from simple, elementary reflexes. Smooth, coordinated movements of the animal as a whole are produced through interactions of individual ganglia with their neighbors, with distant ganglia, and with the head and tail brains. The leech, with its rigorously segmented ganglia and small number of neurons, has been particularly valuable for tracing the circuits and for identifying one by one the individual cells that act in concert to produce swimming. This complex movement has been approached by Stent, Kristan, Friesen, and their colleagues.[44] The individual muscle groups that participate and the motor cells that control them have now been identified and the central connections traced. The key problems are, first, to analyze the pattern of synaptic interconnections and the mechanisms that allow the waves of contraction to travel repeatedly from head to tail along the flattened body; and, second, to explain the source of the rhythm and the factors that modulate it. The head and tail brains are not essential for the swimming movements, which can occur in a few segments or even a single segment of the animal. As in other invertebrates (such as the cockroach, the locust, and the cricket) in which central motor programs involving a small number of individual cells have been shown to control complex patterns of movement, in the leech the basic rhythm is established by synaptic interactions within the ganglia; the role of peripheral receptors is to trigger, enhance, depress, or halt the swimming.

In order to swim, a leech first flattens and elongates its body. Next the animal bends up and down as one wave spreads from head to tail, followed by another bending wave. Cinephotography shows the leech's

Higher levels of integration

[43]Carew, T. J., Hawkins, R. D. and Kandel, E. R. 1983. *Science 219*: 397–400.
[44]Stent, G. S. and Kristan, W. B. 1981. In K. J. Muller, J. G. Nicholls and G. S. Stent (eds.). *Neurobiology of the Leech.* Cold Spring Harbor Laboratory, Cold Spring Harbor, NY, pp. 113–146.

body to bend in the form of a single sine wave as the crest and trough spread along its length. The wave is produced by alternate contraction and relaxation of ventral and dorsal muscles. Knowing the motor neurons that innervate those muscles, one can infer the pattern of activity that they generate, alternately firing and becoming quiescent. Thus, in a midbody ganglion, the motor neurons innervating ventral muscle fire, bending the tail downward, and then become inhibited as the dorsal motor neurons fire to bend the tail upward. In practice, it is not necessary to record intracellularly from motor neurons. Instead, it is simpler to monitor extracellularly the activity of the axons that run in specific peripheral nerve branches.[45] In a chain of ganglia removed from the body, sensory feedback is eliminated, yet the swimming rhythm remains intact, as evidenced by the firing pattern of the motor neurons. Even a single ganglion will, under certain conditions, "swim" rhythmically—that is to say, send appropriate commands down the appropriate axons to nonexistent muscles.

The basis for the swimming rhythm depends on a series of inhibitory connections made by a few interneurons on each other and on the motor neurons. The scheme outlined in Figure 15 can account quantitatively for the alternation of dorsal and ventral contractions spreading as a wave from head to tail. Certain specific unpaired neurons that have been identified initiate the swimming cycles. A second interesting modulatory role is played by 5-HT. Quiescent, sluggish (!), nonswimming leeches have lower blood levels of 5-HT than do active leeches.[46] Moreover, stimulation of cells known to secrete 5-HT promotes an increase in its concentration as well as swimming in the animal. When 5-HT is depleted in embryos (see later) by means of a specific chemical (5,6-dihydroxytryptamine) that selectively destroys the 5-HT neurons in the developing ganglia, the adult leeches do not swim spontaneously but will do so if immersed in a weak solution of 5-HT.[47]

For the swimming of the leech, the individual nerve cells and their connections are now largely known. The rhythm depends upon cyclical inhibition of one nerve by another in a recurrent chain. Facilitation and depression have not been shown to play an obvious role in this process. Other neurally mediated rhythms in invertebrates controlling the heartbeat or gastric motility are produced by different mechanisms. For example, the neurons controlling the heartbeat of the leech possess inherent rhythmicity—they fire spontaneously in bursts with an independently generated depolarization and repolarization that do not require synaptic inputs to turn the cells on or off.[36] Synaptic inputs may, however, modulate and coordinate the rhythm in several such cells.

Development and regeneration

Questions concerning the way in which vertebrate neurons develop, send out their axons, form specific synapses, and establish the wiring

[45]Stent, G. S., Kristan, W. B., Friesen, W. O., Ort, C. A., Poon, M. and Calabrese, R. L. 1978. *Science 200*: 1348–1357.

[46]Willard, A. 1981. *J. Neurosci. 1*: 936–944.

[47]Glover, J. C. and Kramer, A. P. 1982. *Science 216*: 317–319.

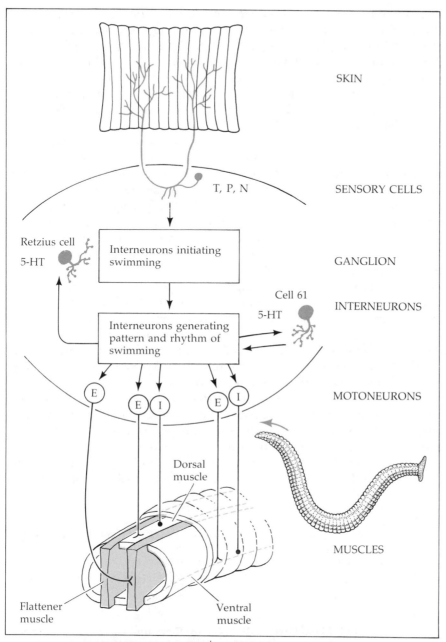

SKIN

T, P, N   SENSORY CELLS

Retzius cell

5-HT

Interneurons initiating
swimming

GANGLION

Cell 61

5-HT

INTERNEURONS

Interneurons generating
pattern and rhythm of
swimming

E   E   I   E   I   MOTONEURONS

Dorsal
muscle

MUSCLES

Flattener
muscle

Ventral
muscle

15 **SIMPLIFIED DIAGRAM OF NEURONAL CONNECTIONS** involved
in swimming by the leech. Interneurons within each ganglion that
generate the swimming rhythm have been identified (cells 27, 28, 33,
123, and 208); their connections, which are largely inhibitory, have been
traced within the ganglion and between ganglia. These cells act on the
motoneurons that control the muscles used for swimming (flatteners,
dorsal and ventral longitudinal muscles). Mechanosensory cells can
activate identified interneurons (cells 204, 205) that excite the pattern-
generating circuit. The system is also influenced by 5-HT either diffusely from the Retzius cells or at synapses made by cell 61. E, excitatory
motoneuron; I, inhibitory motoneuron. (After Kristan, 1983.)

diagram of the nervous system are considered in the following chapters. Invertebrates offer the opportunity for studying such problems directly at the cellular level, one cell at a time. Detailed and comprehensive developmental studies have been made in embryos of a tiny nematode worm, *Caenorhabditis elegans*, the adult nervous system of which contains only 302 neurons.[48] In leech and *Aplysia*, as well as in crickets and locusts, it is possible to follow cell division, migration, and development of ganglia from egg to adult.[49,50] In the leech, which undergoes no larval stage, the four individual cells that on their own directly give rise to the entire central nervous system of the animal, have been identified. Moreover, cell lineage can be followed visually by injecting one of the precursor cells with a label such as horseradish peroxidase that does not interfere with the process of development yet is transferred from cell to cell during division. Figure 16, for example, shows leech eggs, a developing leech embryo, and an embryo in which one cell was inoculated with horseradish peroxidase. As a result, the nervous system on that side contains the enzyme and stains black. It is also possible to produce precise deficits by killing a single precursor cell and thereby causing the animal to mature with its nervous system intact on one side but not the other. At later stages individual sensory cells responding to pressure can be detected in primordial ganglia before their axons have grown to the periphery. Injection of the soma with the fluorescent dye Lucifer Yellow has made it possible to follow the course of the axon as it establishes the receptive field in the skin.[24]

Neurons in the central nervous system of adult leeches retain their ability to sprout and can reform appropriate connections with their targets after an injury. Somehow the damaged cell is able to send its axons in the correct direction to the original region they innervated, there to form highly specific connections on some targets but not others.[51] Normal function can be restored in the animal after its nervous

[48]Horvitz, H. R. 1982. In J. G. Nicholls (ed.). *Repair and Regeneration of the Nervous System*. Springer-Verlag, New York, pp. 41–45.

[49]Weisblat, D. 1981. In K. J. Muller, J. G. Nicholls, and G. S. Stent (eds.). *Neurobiology of the Leech*. Cold Spring Harbor Laboratory, New York, pp. 173–195.

[50]Schacher, S., Kandel, E. R. and Wooley, R. 1979. *Devel. Biol. 71*: 163–175.

[51]Wallace, B. G., Adal, M. and Nicholls, J. G. 1977. *Proc. R. Soc. Lond. B 199*: 567–585.

16   **LEECH EGGS AND EMBRYOS. (A) Comparison of eggs from different species. The smallest eggs are from the medicinal leech (*Hirudo medicinalis*), the intermediate size eggs from *Helobdella triserialis* (which feeds on snails), and the large pair at the bottom from *Haementeria ghilianii* (which lives in Guyana, attains monstrous proportions, and feeds on crocodiles). Scale 0.5 mm. (B) *Haementeria* embryo about 2 mm long at 11 days, stained with hematoxylin. Ganglia are already apparent. (C) *Helobdella* embryo about 1.5 mm long in which one cell (known as the N teloblast), which gives rise to much of the CNS, had been injected with horseradish peroxidase 6 days beforehand. Note that the hemi-ganglia on one side that derive from this single precursor are stained. (Photos courtesy of Weisblat; from Weisblat, 1981.)**

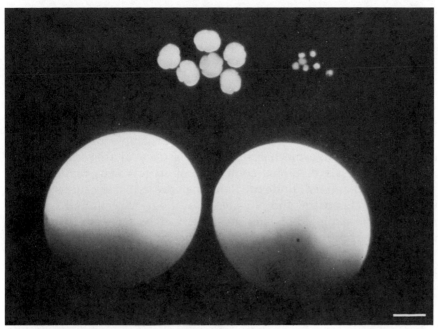

(A)

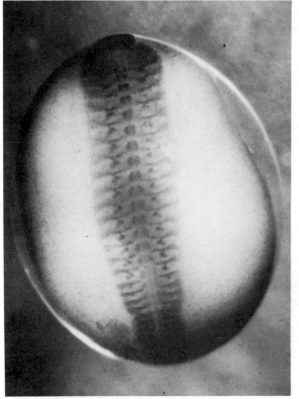

(B)

(C)

system has been transected (Chapter 19). Accurate regeneration of connections also occurs when the ganglia are maintained outside the animal in organ culture. A promising preparation for studying the mechanisms that play a part in cell–cell recognition of this type is provided by identified neurons that have been isolated from the ganglion and maintained in tissue culture. Such cells retain their membrane properties, sprout, and form specific chemical and electrical synapses in vitro that resemble in their properties those in the ganglion.[52]

This chapter illustrates the varied uses of simple preparations for exploring integrative mechanisms and plasticity at the cellular level. Although analysis of brains composed of so few cells may seem to present a well-defined, finite problem, many essential questions relating to anatomy, chemistry, and physiology of the nerve cells and synapses remain to be explored. Some questions can be answered simply by doing the appropriate experiments. Others must await technical advances. In addition, it seems promising that such nervous systems may provide clues to the mechanisms by which use, disuse, or abnormal use can alter the properties of synapses and thereby modify the performance of the animal's behavior. One hope for understanding molecular mechanisms involved in plasticity and development lies in the use of tissue culture and monoclonal antibodies, by which the neurons and their surface proteins can be studied in a controlled environment over long periods. Finally, as always, one expects a simple system to provide a stepping stone toward an understanding of complex nervous systems. We suspect that Nietzsche anticipated our problems when he wrote:[53]

> "Then you must be a scientist whose field is the leech," said Zarathustra, "and you must pursue the leech to its last rock-bottom, you conscientious man!"
>
> "Oh Zarathustra," answered the man, "that would be an enormity. How could I take up such a huge task! What I am the master and connoisseur of is the *brain* of the leech: that is *my* field! And it is a whole universe!"

SUGGESTED READING

General reviews

Bentley, D. and Konishi, M. 1978. Neural control of behavior. *Annu. Rev. Neurosci.* 1: 35–59.

Kandel, E. R. 1976. *Cellular Basis of Behavior.* Freeman, San Francisco.

Kandel, E. R. 1979. *Behavioral Biology of Aplysia.* Freeman, San Francisco.

Muller, K. J., Nicholls, J. G. and Stent, G. (eds.). 1981. *Neurobiology of the Leech.* Cold Spring Harbor Laboratory, Cold Spring Harbor, NY.

Shepherd, G. S. 1983. *Neurobiology.* Oxford University Press, New York.

[52]Fuchs, P. A., Henderson, L. P. and Nicholls, J. G. 1982. *J. Physiol. 323:* 195–210.
[53]Nietzsche, F. *Thus Spake Zarathustra.*

Original papers

Blackshaw, S. E., Nicholls, J. G. and Parnas, I. 1982. Expanded receptive fields of cutaneous mechanoreceptor cells after single neurone deletion in leech central nervous system. *J. Physiol. 326*: 261–268.

Carew, T. J., Castellucci, V. F. and Kandel, E. R. 1979. Sensitization in *Aplysia*: Rapid restoration of transmission in synapses inactivated by long-term habituation. *Science 205*: 417–419.

Carew, T. J., Hawkins, R. D. and Kandel, E. R. 1983. Differential classical conditioning of a defensive withdrawal reflex in *Aplysia*. *Science 219*: 397–400.

Castellucci, V. F., Kandel, E. R., Schwartz, J. H., Wilson, F. D., Nairn, A. C. and Greengard, P. 1980. Intracellular injection of the catalytic subunit of cyclic AMP-dependent protein kinase simulates facilitation of transmitter release underlying behavioral sensitization in *Aplysia*. *Proc. Natl. Acad. Sci. USA 77*: 7492–7496.

Klein, M., Shapiro, E. and Kandel, E. R. 1980. Synaptic plasticity and the modulation of the Ca^{++} current. *J. Exp. Biol. 89*: 117–157.

Kriegstein, A. R. 1977. Development of the nervous system of *Aplysia californica*. *Proc. Natl. Acad. Sci. USA 74*: 375–378.

Muller, K. J. and McMahan, U. J. 1976. The shapes of sensory and motor neurons and the distribution of their synapses in ganglia of the leech: A study using intracellular injection of horseradish peroxidase. *Proc. R. Soc. Lond. B 194*: 481–499.

Muller, K. J. and Scott, S. A. 1981. Transmission at a "direct" electrical connexion mediated by an interneuron in the leech. *J. Physiol. 311*: 565–583.

Nicholls, J. G. and Baylor, D. A. 1968. Specific modalities and receptive fields of sensory neurons in the CNS of the leech. *J. Neurophysiol. 31*: 740–756.

Stent, G. S. and Weisblat, D. 1982. The development of a simple nervous system. *Sci. Am. 246*: 136–146.

PART FIVE

NATURE AND NURTURE

IN SEARCH OF THE RULES OF FORMATION, MAINTENANCE, AND ALTERATIONS OF NEURAL CONNECTIONS

One of the most striking features of the nervous system is the high degree of precision with which nerve cells are connected to each other and to different tissues in the periphery, such as skeletal muscle and skin. This orderliness of the connections made during development is a necessary prerequisite for all the integrative mechanisms described so far.

The nervous system appears constructed as if each neuron had built into it an awareness of its proper place in the system. During development, the neuron grows toward its target, ignores some cells, selects others, and makes permanent contact not just anywhere on a cell but with a specified part of it. Conversely, neurons behave as if they were aware when they have received their proper connections. When they lose their synapses, they respond in various ways. For example, denervated neurons or muscle fibers develop supersensitivity to the chemical transmitter, owing to the appearance in their membranes of new receptor proteins. At times denervated cells provoke sprouting of processes from their neighbors. Even in the absence of denervation, simply as a result of changed patterns of activity, the effectiveness of synapses may change and new connections may be formed.

One of the tasks at present is to sharpen and widen our general recognition of these phenomena and, above all, to find out about the processes responsible for such behavior of cells. We therefore present selected examples that substantiate

these statements and point to instances where the rules are relaxed, as when cells become able to accept foreign nerves.

ORDERLINESS OF CONNECTIONS

Four examples illustrate the precise design of the architecture of the nervous system. First, Chapter 15 mentions the stretch reflex that arises as the result of impulses in the sensory nerve cell that innervates a muscle spindle. Its cell body, which lies in a dorsal root ganglion, sends some of its processes into the periphery to appropriate regions of the intrafusal muscle fibers. It also sends processes centrally to search out and make synapses exclusively on those motoneurons that innervate the same skeletal muscle in which the sensory neuron terminates. Other branches run in the dorsal columns to end in a localized region of the dorsal column nuclei, and still other branches end on additional interneurons. A second example (Chapter 2) is provided by individual neurons in the visual cortex that selectively recognize a vertically oriented light bar shone into their receptive fields. This is possible because the inputs are derived from selected lower order neurons, some of which are excitatory and others inhibitory. A third case of specificity is the Mauthner cell (Chapter 16), which initiates a rapid tail flip in the goldfish. Well-defined regions of its cell body and dendrites receive different inputs from the two auditory nerves and from the other Mauthner cell. One consistently finds certain recognizable synapses in the expected places, synapses that must be of the correct type—excitatory or inhibitory, electrical or chemical. Finally, a good example is the distribution of synapses made on Purkinje cells in the cerebellum, worked out by Ramón y Cajal, Eccles, Szentágothai, Ito, Llinás, Chan-Palay, and Palay. These large neurons (Chapter 1) provide the only known output from the cerebellum and are therefore the end stations for the integrative activity of all the other cells. They may receive more than 100,000 synapses, which end on appropriate parts of the neuron. Thus, the climbing fibers terminate on smooth dendrites and the basket cells on cell bodies and axons, whereas the granule cells make synapses with spiny processes of Purkinje neurons.

A number of questions naturally arise when one considers the preceding examples. What cellular mechanisms enable one neuron to select another out of myriad choices, to grow toward it, and to form synapses? Are both cells specified, or does the arrival of one determine the fate of the other? As for the precision of the wiring, how much variability is there in the connections between certain cells in different animals? What directs the systematic growth of nerve fibers along their well-defined paths? The answer to these questions influences thinking about the genetic blueprint for wiring up a brain containing 10^{10} to 10^{12} cells with a much smaller number of genes, 10^6 or fewer.

EXPERIMENTAL APPROACHES

One approach to these problems is to study the FORMATION OF CELLS DURING DEVELOPMENT. Using embryonic and newborn animals, Sidman and Rakič have followed cells as they migrate to their final destinations and form connections. This involves painstaking reconstructions of the positions, shapes, and distribution of various cell types at different stages of development. Layered structures with distinctive cells organized in a regular manner—the cerebellum, hippocampus, and cortex—lend themselves best to this sort of study. To follow the birthdays of cells, one

can label cell nuclei with tritiated thymidine. The label becomes incorporated into DNA only when a cell is in the process of dividing; in subsequent divisions the radioactivity is rapidly diluted out and lost. By labeling at different stages of development, one can establish by autoradiography the time when a cell underwent its last division (the birthday of a cell), where it arises, and into what type of cell it develops.

In addition, Sidman, Caviness, and Rakič have exploited genetic techniques and mutant animals.[1,2] If one class of cells fails to develop, one can ask what happens to the cells that would normally form synapses on them. Other promising approaches, which we merely mention, make use of the methods that have proved successful for molecular biology, the GENETIC DISSECTION of relatively simple nervous systems, including the brains of the fruit fly (*Drosophila*),[3] of a small crustacean (*Daphnia*),[4] and of various worms.[5] Because of the short life cycle of such animals, mutants can be selected and cloned with a view to determining how the presence or absence of a gene influences the behavior of an animal and the architecture of its nervous system.[6]

Lately, the formation of connections has been studied successfully in vitro, using TISSUE CULTURE techniques. Several types of nerve cells have been used, including neurons obtained from autonomic

[1]Rakič, P. and Sidman, R. L. 1973. *J. Comp. Neurol. 152*: 103–132, 133–162.

[2]Caviness, V. S. Jr. and Rakič, P. 1978. *Annu. Rev. Neurosci. 1*: 297–326.

[3]Benzer, S. 1971. *J.A.M.A. 218*: 1015–1022.

[4]Flaster, M. S., Macagno, E. R. and Schehr, R. S. 1982. In N. Spitzer (ed.). *Neuronal Development*. Plenum, New York, pp. 267–295.

[5]Horvitz, H. R., Sternberg, P. W., Greenwald, I. S., Fixsen, W. and Ellis, H. M. (eds.). 1984. *Cold Spring Harbor Symp. Quant. Biol. 48*: 453–464.

[6]Stent, G. S. 1981b. *Annu. Rev. Neurosci. 4*: 163–194.

and dorsal root ganglia and spinal cord.[7,8] When dissociated nerve cells are cultured together with cardiac or skeletal muscle fibers, functioning nerve–muscle junctions form. As the synapses develop, the chemistry of the transmitters and the associated enzymes undergoes changes. Dissociated embryonic cells can also form aggregates that resemble the tissue of origin. In cultures of cells from different regions of the brain one can observe that retinal, hippocampal, cerebellar, or cortical neurons form characteristic aggregates. Moreover, there is evidence for specific proteins that promote the aggregation.[9] The development of MONOCLONAL ANTIBODY techniques offers a new and powerful approach for identifying the molecules used by particular neurons;[10] by raising highly specific antibodies, it is possible to label individual neurons and to assess the functional roles of the particular molecule.

A major difficulty for culture studies on neurons compared with other tissues in the body is that neurons do not divide. A recent development is the introduction of NEURAL TUMOR CELLS, which are able to divide and can be cloned. Such neuroblastoma cultures have been shown to form chemical synapses in vitro.[11] In the long run, tissue culture methods promise to be most useful for determining the chemical requirements for neurons to form contacts and to manipulate experimentally the conditions under which contacts are made.

An alternative path to the study of the formation of connections is to make use of REGENERATION, the fact that the processes of mature nerve cells can grow back to form synapses after they have been severed by a lesion. For example, in lower vertebrates such as frogs or fishes, one can cut the optic nerve and allow the animal to recover. After a few weeks, the fibers grow back into the brain to reform connections that restore vision. In mammals, although no new neurons can be formed in the central nervous system and tracts do not regenerate, there is evidence that existing fibers do sprout. In any event, mammalian peripheral axons do grow back and innervate skeletal muscles and sensory end organs in the skin. This approach will gain in value if it turns out that the ground rules are similar during early development and regeneration or repair.

A different but related question concerns the STABILITY OF SYNAPSES once they have been formed. How is the effectiveness of transmission influenced by use, disuse, or inappropriate use? There are many examples to show that synapses can be changed in a number of ways, ranging from an increase in efficacy to complete block. On the short time scale, one impulse arriving at the chemical synapse can produce a residual change, so that the next impulse liberates more transmitter. This facilitation may last for seconds or minutes and has been studied in detail at many types of synapses. But there are also longer term changes that occur over days, weeks, or months. Among the most remarkable of these are the changes in the visual system produced by closing one eye or by producing a squint in a kitten. Subtle alterations of the sensory input disturb performance and disrupt pathways that had previously been effective.

Another important factor determining neural organization is CELL DEATH during development. In the embryo there is a considerable redundancy of nerve cells, such as those that supply limb muscles in amphibians or neuronal assemblies in the avian brain. Before connections become permanent, a sorting out process occurs

[7]Fischbach, G. D., and Dichter, M. A. 1974. *Dev. Biol.* 37: 100–116.

[8]Patrick, J., Heinemann, S. and Schubert, D. 1981. *Annu. Rev. Neurosci.* 1: 417–443.

[9]Rutishauser, U., Grumet, M. and Edelman, G. M. 1983. *J. Cell Biol.* 97: 145–152.

[10]McKay, R. D. G. 1983. *Annu. Rev. Neurosci.* 6: 527–546.

[11]Nirenberg, M. et al. 1984. *Cold Spring Harbor Symp. Quant. Biol.* 48: 707–716.

according to a predestined plan, and only a fraction survive.[12,13]

Nerve cells also die in the adult brain. This, however, is likely to represent attrition unlike the shaping process that occurs in development. Evidence for cell death in mature animals rests on counts of nerve cell nuclei, but hard numbers are difficult to come by in normal individuals.[14,15] Conversely, the relative number of glial cells increases. Such a constant attrition of neurons goes generally without special notice, perhaps because we regard the consequences as part of our normal aging process. One faces here the serious difficulty of diagnosing or defining "normal" and what the statement implies. For example (Chapter 15), about one-third of the spinal motoneuron pool is solely concerned with centrifugal control of sensory discharges rather than with direct execution of movement. How many cells, and which ones, would we miss if they gradually dropped out? Only after various injuries resulting from trauma, disease, or senility do clearcut symptoms appear of such magnitude that we can correlate them with confidence to specific lesions. Even with large cell deficits, as produced by some tumors in the frontal lobes, diagnosis may require considerable skill. To some extent, therefore, one can maintain that we possess a redundancy of nerve cells that automatically take care of continued neuron death. On the other hand, the brain in an adult has far less ability to compensate for injury than that in an immature animal or infant.

The scope of all the problems relating to development, synapse formation, neural specificity, and changes in efficiency is too great for a comprehensive discussion. Many aspects are covered in detail elsewhere.[16,17] The following two chapters deal with a few selected experimental approaches: (1) the use of skeletal muscles and autonomic nerve cells to study the effects of denervation and prolonged inactivity and the factors required for reinnervation; (2) the formation of nerve connections in the brain after lesions; (3) the properties and isolation of a nerve growth factor that influences the growth of sympathetic nerve fibers; and (4) the effects of sensory deprivation in the visual system of the cat and the monkey.

[12]Hamburger, V. 1975. *J. Comp. Neurol. 160*: 535–546.

[13]Spitzer, N. C. 1982. *J. Physiol. 330*: 145–162.

[14]Cowan, W. M. 1973. In M. Rockstein (ed.). *Development and Aging in the Nervous System*. Academic, New York.

[15]Brody, H. 1955. *J. Comp. Neurol. 102*: 511–556.

[16]Jacobson, M. 1978. *Developmental Neurobiology*, 2nd Ed. Plenum, New York.

[17]Patterson, P. H. and Purves, D. (eds.). 1982. *Readings in Developmental Neurobiology*. Cold Spring Harbor Laboratory, Cold Spring Harbor, NY.

SPECIFICITY OF NEURONAL CONNECTIONS

CHAPTER NINETEEN

How are the orderly and precise synaptic connections of the nervous system formed during development? Questions concerning neural specificity are discussed in the framework of denervation, synapse formation, and the growth of nerve fibers during development and regeneration.

Neurons in the heart of the frog or vertebrate skeletal muscle fibers, deprived of their synapses by denervation, develop new chemoreceptors and an increased sensitivity to the transmitter acetylcholine. Instead of being concentrated in the vicinity of the synapses, additional acetylcholine receptors appear over the entire neuronal and muscle surface, accounting for the supersensitivity. Direct electrical stimulation of supersensitive muscles causes the chemosensitivity to shrink back to the original end plate area. The distribution of chemoreceptors in the muscle membrane can be controlled by the level of muscle activity; additional unknown factors may also play a part. Supersensitivity goes hand in hand with the ability of muscle fibers to accept innervation, but no causal relation has been established. When muscles are denervated, even foreign nerves will form connections. Usually, however, the sites of the original synapses on the muscle fibers become reinnervated by the appropriate nerve. The basal lamina (an extracellular, protein-containing matrix that envelops the muscle membrane, nerve terminals, and Schwann cells) plays a key role in the differentiation of the nerve terminal and the postsynaptic membrane into a regenerated synapse.

A demonstration of more complex neural specificity can be obtained in amphibians and fish in which nerve fibers grow back to their specific targets; thus, regenerating optic nerve fibers can reform connections so that function is restored. Still more specificity is exemplified in the central nervous system of the leech, where an individual nerve cell can grow back and selectively restore functional connections with identified target cells. Contrary to previous ideas, it is now clear that neurons within the mammalian central nervous system are also able to send new, regenerating processes over long distances following injuries to the spinal cord, thalamus, cortex, and other areas if the right conditions are provided.

493

The closest approach to an identified substance that promotes the growth of certain nerve cells and their processes is nerve growth factor. This protein acts selectively on sympathetic neurons in the autonomic nervous system and dorsal root ganglion cells. Antibodies to nerve growth factor injected into immature mice can lead to failure of development of the sympathetic nervous system.

DENERVATION AND FORMATION OF CONNECTIONS

The denervated
muscle membrane

Neuromuscular synapses have provided a useful model for mechanisms of synaptic transmission between neurons in the higher centers. Similarly, changes occurring in denervated muscles are relevant for thinking about the formation and disappearance of neural connections in general.[1]

Some phenomena in skeletal muscles after severance of their nerve supply were originally described toward the end of the last century; they were usually noted in muscle fibers that could be readily seen, such as those in the tongue. After a variable period, individual muscle fibers start to exhibit spontaneous, asynchronous contractions called fibrillation. The onset of fibrillation may be as early as two to five days after denervation in rats, guinea pigs, or rabbits, or well over a week in monkeys and humans. Fibrillation occurs through changes in the muscle membrane and is not initiated by acetylcholine (ACh).[2]

Before or at the start of fibrillation, mammalian muscle fibers become supersensitive to a variety of chemicals. This means that the concentration of a substance required to produce depolarization, or shortening of a muscle, is reduced by a factor of several hundred to a thousand. For example, a denervated mammalian skeletal muscle is about 1000 times more sensitive to its transmitter—ACh—than is a normally innervated one if the ACh is either applied directly in the bathing fluid or injected into an artery supplying the muscle.[3] The increase in chemosensitivity is not restricted to the physiological transmitter—ACh— but occurs for a wide variety of chemical substances and even makes the muscle more sensitive to stretch or pressure.[4] Characteristic changes also occur in the passive electrical properties of denervated muscle fibers in frogs and mammals. The membrane resistance becomes approximately doubled owing to a decrease in the resting potassium conductance.[5] This effect, however, is far too small to account for the observed increases in sensitivity to ACh. The action potentials in denervated muscles can also change, becoming more resistant to tetrodotoxin, the puffer fish poison that blocks sodium channels (Chapter 6); thus, impulses can still be generated, with calcium substituting for sodium.[6]

[1]Cannon, W. B. and Rosenblueth, A. 1949. *The Supersensitivity of Denervated Structures: Law of Denervation.* Macmillan, New York.
[2]Purves, D. and Sakmann, B. 1974. *J. Physiol. 239:* 125–153.
[3]Brown, G. L. 1937. *J. Physiol. 89:* 438–461.
[4]Kuffler, S. W. 1943. *J. Neurophysiol. 6:* 99–110.
[5]Nicholls, J. G. 1956. *J. Physiol. 131:* 1–12.
[6]Harris, J. B. and Thesleff, S. 1971. *Acta Physiol. Scand. 83:* 382–388.

Other changes that occur in denervated muscle, such as the gradual atrophy or wasting and the many chemical changes that follow, are not discussed here.[7,8]

Formation of new
acetylcholine receptors
after denervation or
prolonged inactivity

Supersensitivity is explained by an altered distribution of ACh receptors in denervated muscles. This has been demonstrated by applying ACh locally to part of the muscle surface by ionophoretic release from an extracellular micropipette while recording the membrane potential with an intracellular microelectrode. As explained in Chapters 9 and 11, in normal frog, snake, and mammalian muscle the end plate region—where the nerve makes synapses—is sensitive to ACh, although the rest of the muscle membrane has a very low sensitivity.

When ACh is applied locally to denervated muscles, the results are very different from normal. The chemosensitivity increases day by day after a nerve to a mammalian muscle is cut, until by about seven days the surface of the muscle is almost uniformly sensitive to ACh (Figure 1).[9] In frog muscle, the changes are relatively small and take several weeks to develop.[10] The receptors in extrasynaptic areas do not appear as the result of spread away from the original end plate. In experiments by Katz and Miledi, frog muscles were cut in two[11] (Figure 2); nucleated fragments that had never been innervated and that were separated from the end plate survived and developed increased sensitivity to ACh. In this case the receptors could not have originated in the synaptic region, which had been physically separated.

The new receptors in denervated muscles are in many respects similar to those that are present normally. Acetylcholine still increases the permeability of the membrane to both potassium and sodium ions, but not to chloride,[12] and the action of acetylcholine is blocked by curare and the snake venom α-bungarotoxin (Chapter 9). There are, however, certain differences in the properties of the junctional and new extrajunctional receptors (see later).

How does section of a nerve lead to the appearance of new receptors? Is it simply inactivity of the muscle or is there some additional mechanism? Lømo and Rosenthal[13] investigated this problem by blocking conduction in rat nerves by application of a local anesthetic or of diptheria toxin. The substances were applied by means of a cuff to a short length of the nerve some distance from the muscle. With this technique, the muscles became completely inactive because motor impulses failed to conduct past the cuff. Occasional test stimulation of the nerve distal to the block produced a twitch of the muscle as usual, and miniature end plate potentials still occurred normally, a result showing that synaptic transmission was basically intact. And yet after seven days

[7]Gutmann, E. 1962. *Rev. Can. Biol. 21:* 353–365.

[8]Guth, L. 1968. *Physiol. Rev. 48:* 645–687.

[9]Axelsson, J. and Thesleff, S. 1959. *J. Physiol. 147:* 178–193.

[10]Miledi, R. 1960. *J. Physiol. 151:* 1–23.

[11]Katz, B. and Miledi, R. 1964. *J. Physiol. 170:* 389–396.

[12]Jenkinson, D. H. and Nicholls, J. G. 1961. *J. Physiol. 159:* 111–127.

[13]Lømo, T. and Rosenthal, J. 1972. *J. Physiol. 221:* 493–513.

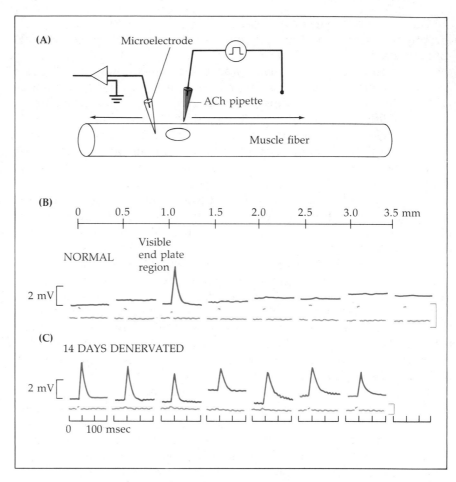

1 NEW ACh RECEPTORS appear in a muscle of the cat after denervation. (A) ACh-filled pipette is moved to different positions where it releases ACh onto the surface of a muscle fiber. (B) Pulses of ACh are applied to a muscle fiber with intact innervation. A response is seen only in the vicinity of the end plate. (C) After 14 days of denervation, muscle fibers respond to ACh along their entire length. (After Axelsson and Thesleff, 1959.)

of nerve block, the muscle had become supersensitive (Figure 3). Other experiments have shown that long-term application of curare or of α-bungarotoxin also leads to the appearance of new receptors. All these results show that "denervation" supersensitivity can be produced without interrupting the nerve; blockage of synaptic transmission is sufficient.[14,15]

The role of muscle activity itself as an important factor in controlling

[14]Berg, D. K. and Hall, Z. W. 1975. *J. Physiol. 244*: 659–676.
[15]Cohen, M. W. 1980. *J. Exp. Biol. 89*: 43–56.

supersensitivity was further shown in other experiments in which supersensitive denervated muscles in the rat were stimulated directly through electrodes permanently implanted around the muscle. Repetitive direct stimulation of muscles over several days caused the sensitive area to become restricted, so that once again only the synaptic region was sensitive to ACh (Figure 4).[13] The frequency of stimulation and the interval of quiescence were seen to be important variables in the development or reversal of supersensitivity. This explains the apparent inconsistency that denervated mammalian muscle fibers develop supersensitivity in spite of the ongoing contractions associated with fibrillation. Sampling the activity of individual fibers showed that fibrillation is cyclical, periods of activity alternating with inactivity. The level of spontaneous activity is, however, below that required to reverse the effects of denervation on the distribution of ACh receptors.[16,17]

There remains the question whether nerves provide the muscle with

[16]Purves, D. and Sakmann, B. 1974a. *J. Physiol. 237*: 157–182.
[17]Fambrough, D. M. 1979. *Physiol. Rev. 59*: 165–227.

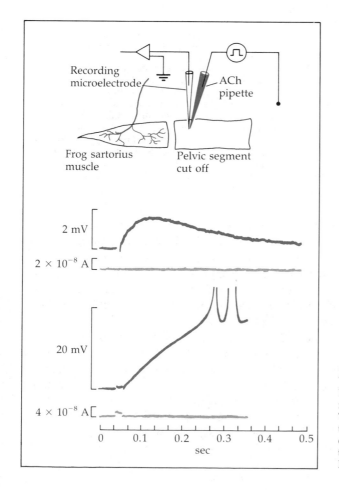

2

DEVELOPMENT OF NEW RECEPTORS in nerve-free isolated segment of a muscle in a frog. The pelvic muscle fibers had been severed from the rest of the sartorius muscle for 10 days. Ionophoretic pulses depolarize and (in the lower trace) initiate muscle action potentials. (After Katz and Miledi, 1964.)

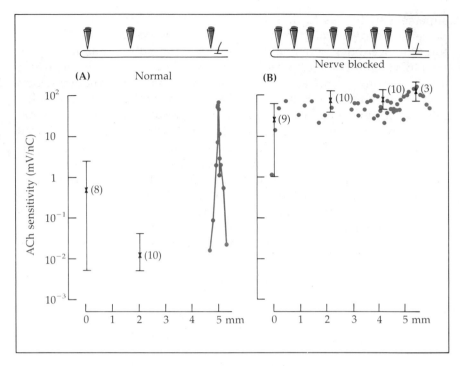

3 NEW ACh RECEPTORS in a muscle after block of nerve conduction. The nerve to the rat soleus muscle was blocked for 7 days by a local anesthetic. Neuromuscular transmission still functions if the nerve is stimulated below the block. **(A)** In the normal muscle the ACh sensitivity is restricted to the end plate region (near the 5 mm position). **(B)** In a muscle fiber whose nerve was blocked for 7 days, the ACh sensitivity is distributed over the entire muscle fiber surface. Sensitivity is expressed numerically in millivolts per nanocoulomb. The crosses and bars represent the mean and range of sensitivities in adjacent muscle fibers. (From Lømo and Rosenthal, 1972.)

some products other than transmitters that keep the muscles normal. For example, experiments in which slowly developing changes occur without activity per se playing an obvious role have been made on partially denervated frog and crustacean muscles. When fibers in the sartorius muscle are deprived of some of their multiple end plates, supersensitivity develops in the denervated portions of the muscle fibers; yet these fibers have kept contracting all along.[10] In principle, crustacean muscles that are supplied by an inhibitory as well as an excitatory axon can also be used to study the effects of partial denervation. Will selective removal of the inhibitory fiber result in changes in synaptic transmission in the excitatory fiber? In practice the experiments are difficult to perform since after section of a crustacean axon its distal segment can survive for weeks or months without degenerating. Accordingly, Parnas and his colleagues[18] destroyed the inhibitory axon by injecting proteolytic enzymes into it close to the terminals. This

[18]Parnas, I., Dudel, J. and Grossman, Y. 1982. *J. Neurophysiol. 47*: 1–10.

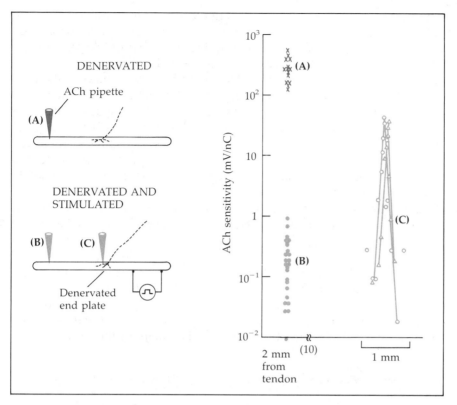

4   REVERSAL OF SUPERSENSITIVITY in denervated muscle of the rat by direct stimulation of the muscle fibers. (A) Increased ACh sensitivity in the nerve-free portion of a muscle fiber after 14 days of denervation. (B) Sensitivity in the nerve-free region of a muscle that had been denervated for 7 days without stimulation and then intermittently stimulated for another 7 days. This treatment reversed the denervation supersensitivity. (C) ACh sensitivity in two stimulated fibers of the same muscle near their denervated end plate regions. The high sensitivity is confined to this region in the stimulated muscle. (After Lømo and Rosenthal, 1972.)

procedure has been shown to affect only the injected axon and not its neighbors. Some days later clear augmentation was observed in the responses to stimulation of the excitatory nerve. Excitatory synaptic potentials became larger and more prolonged. This potentiation was shown to be caused by an increased sensitivity of the postsynaptic membrane to transmitter with a change in the open time of the excitatory channels. Presumably such effects cannot be simply explained as a result of inactivity of the muscle. This whole area of work is in an interesting state of flux and needs more direct experimental evidence that will allow assessment of the relative importance of activity and chemical factors for the control of receptor distribution.

Our knowledge of how nerve cells behave if some of their synaptic connections have degenerated while others are maintained is still restricted, largely because of technical difficulties. Loss of connections is

Development of extrasynaptic receptors in nerve cells

a common occurrence in the brain after injury or in disease, and there is no question that neurons in the central nervous system undergo changes when part of their synaptic input is destroyed. One of the consequences is an altered response to injected drugs.[1]

Recently, it has become possible to study certain nerve cells in as detailed a manner as that described earlier for muscles. The alterations of the neuronal surface membrane after loss of synapses have been studied in parasympathetic ganglion cells that innervate the heart of the frog. These nerve cells can be seen in the transparent interatrial septum and, like skeletal muscle fibers, are highly sensitive to the transmitter ACh only at selected spots on their surfaces, immediately under the presynaptic terminals (Figure 5; see also Chapter 10).

In studies on denervation, the two vagus nerves to the heart are cut and the frog is left to recover.[19] Synaptic transmission between vagal nerve terminals and ganglion cells fails rapidly, failure starting on the second day after denervation. At the same time, the chemosensitivity of the neuronal surface membrane starts to increase and is fully developed within four to five days. Thus, ACh applied ionophoretically causes a membrane depolarization wherever it is applied (Figure 5B). In this respect, the sensitivity of the neurons differs drastically from normal, although the membrane potentials remain in the normal range

[19]Kuffler, S. W., Dennis, M. J. and Harris, A. J. 1971. *Proc. R. Soc. Lond. B 177*: 555–563.

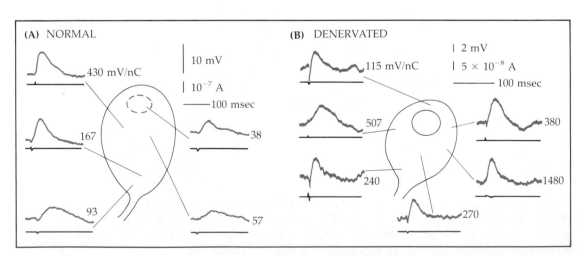

5   **DEVELOPMENT OF NEW RECEPTORS in parasympathetic nerve cells in the heart of the frog after denervation. (A)** In the normal neuron, the high ACh sensitivity is confined to synaptic regions. The large numbers indicate high sensitivity (expressed in millivolts per nanocoulomb). When ACh is applied to the extrasynaptic region, more must be released to have an effect. Such responses rise relatively slowly because ACh has to diffuse to a nearby sensitive synaptic spot. **(B)** After 21 days of denervation, the sensitivity of the neuronal surface is high wherever ACh is released. (After Kuffler, Dennis, and Harris, 1971.)

and the cells are able to fire impulses. Normal chemosensitivity re-appears if the original nerve is allowed to grow back into the heart. As in muscle (see later), the supersensitive area becomes restricted once more to the vicinity of the synapses.

Many questions arise about the ACh receptors that appear in supersensitive muscles, particularly in relation to their properties and the mechanisms for controlling the synthesis and turnover of the receptor proteins in the membrane. As mentioned, extrasynaptic receptors behave similarly to synaptic ones in some physiological tests but not in others. For example, the open times of channels activated by the ACh receptors of denervated muscle are longer than those at the motor end plate.[20] They differ also in their affinity to certain pharmacological agents, such as curare and α-bungarotoxin, and in their isoelectric points (the way in which they migrate in an electric field).[21]

A valuable technique for studying the turnover and development of new receptors is to label them with radioactive α-bungarotoxin, which binds strongly and with a high degree of specificity. The rat diaphragm is a convenient preparation because the muscle fibers run in parallel as a thin sheet and the end plates are restricted to a narrow zone that can be dissected out. The method used by several workers has been to bathe normal and denervated muscles in toxin and compare toxin binding at end plate and end plate-free areas. As expected, the amount and distribution of the labeled toxin were changed after denervation. Estimates of the binding sites at postsynaptic areas of the muscle are of the order of $10^4/\mu m^2$ compared with fewer than $10/\mu m^2$ at end plate-free areas. After denervation, however, receptor sites in the extrasynaptic regions increase to about $10^3/\mu m^2$.[22–24] The increase in numbers of receptors and development of supersensitivity have also been observed in muscles maintained outside the body in organ culture. In such isolated muscles, substances that block protein synthesis (for example, actinomycin and puromycin) prevent supersensitivity and the formation of new receptors.[17,25] The increase in receptors is attributable to increased synthesis and not to reduced degradation. This conclusion is supported by experiments in which receptor turnover has been measured by using labeled α-bungarotoxin: The average life of receptors in denervated rat diaphragm is about ten times shorter for extrasynaptic receptors than for those at the end plate.[17] In neonatal rat muscles (see later), the receptors at the end plate start off by resembling extrasynaptic receptors in denervated muscles and have a rapid turnover. Within a few days, the half-life of the new receptors at the end plate changes

Properties, synthesis, and turnover of new receptors

[20]Brenner, H. R. and Sakmann, B. 1983. *J. Physiol. 337*: 159–172.

[21]Brockes, J. P. and Hall, Z. W. 1975. *Biochemistry 14*: 2100–2106.

[22]Barnard, E. A., Wieckowski, J. and Chiu, T. H. 1971. *Nature 234*: 207–209.

[23]Hartzell, H. C. and Fambrough, D. M. 1972. *J. Gen. Physiol. 60*: 248–262.

[24]Berg, D. K., Kelly, R. B., Sargent, P. B., Williamson, P. and Hall, Z. W. 1972. *Proc. Natl. Acad. Sci. USA 69*: 147–151.

[25]Fambrough, D. M. 1970. *Science 168*: 372–373.

from one day to ten days and the channel open time decreases.[26] Hall and his colleagues have shown that concomitant changes also occur in the immunological properties of the receptors.[27] One of the antigenic sites changes from that characteristic of extrasynaptic receptors to that of end plate receptors in the mature animal.

Experiments made on a variety of preparations have opened up the possibility of studying in detail the biochemical steps involved in the synthesis of receptors, their regulations, and their insertion into membranes. Thus, studies made on the electric organs of torpedo and on embryonic, neonatal, and tissue-cultured muscles have provided much information about the steps involving (1) translation of the messenger RNAs for the subunits, (2) assembly of the subunits, and (3) insertion of receptor into the membrane,[17,28] Egg cells of the frog *Xenopus* provide a further, highly simplified preparation for studying receptor synthesis. Their membranes normally contain muscarinic ACh receptors that are blocked by atropine. Miledi and Sumikawa[29] showed that messenger RNA extracted from normal or denervated cat muscles injected into the oocyte induced the appearance of nicotinic ACh receptors on its surface. These new, alien ACh receptors were blocked by curare and α-bungarotoxin and when activated gave rise to inward currents that reversed at about −10 mV. Intracellular injection of ACh did not activate the receptors, a finding suggesting that they were inserted with the recognition site exposed to the external medium, as in muscle membrane.

SUPERSENSITIVITY, REINNERVATION OF MUSCLES, AND THE BASAL LAMINA

Some clues about the possible significance of supersensitivity come from studying the changes a muscle undergoes in the process of reinnervation. Normal skeletal muscle fibers in amphibians and mammals can be innervated by more than one nerve fiber.[30] A fully innervated muscle fiber will not, however, accept innervation by an additional accessory nerve. Thus, if a cut motor nerve is placed on an innervated muscle, it will not "take" to form additional new end plates on the muscle fibers. In contrast, nerve fibers do grow out and reinnervate a denervated or injured muscle.

What are the conditions that enable a denervated muscle to accept a nerve? Several tests have been made to see whether a correlation exists between increased chemosensitivity and synapse formation. In one such experiment Miledi[31] used the frog sartorius in which the

[26]Fischbach, G. D. and Schuetze, S. M. 1980. *J. Physiol. 303*: 125–137.
[27]Hall, Z. W., Roisin, M.-P., Gu, Y. and Gorin, P. D. 1984. *Cold Spring Harbor Symp. Quant. Biol., 48*: 101–108.
[28]Merlie, J. P., Sebbane, R., Gardner, S., Olson, E. and Lindström, J. 1984. *Cold Spring Harbor Symp. Quant. Biol., 48*: 135–146.
[29]Miledi, R. and Sumikawa, K. 1982. *Biomed. Res. 3*: 390–399.
[30]Hunt, C. C. and Kuffler, S. W. 1954. *J. Physiol. 126*: 293–303.
[31]Miledi, R. 1962. *Nature 193*: 281–282.

muscle fibers had been cut. As already shown by Katz and Miledi,[11] the pieces of muscle fibers that were separated from their end plate regions (Figure 2) became sensitive to ACh along their length. In addition, when a cut motor nerve was placed in apposition to these fragments, they became innervated. This is a remarkable result. It means that nerve fibers grow out to a muscle fiber and form synapses in a region that was never normally innervated. Perhaps even more remarkable is the observation that the nerve induces in the postsynaptic muscle fiber the formation of a subsynaptic specialization that this section of muscle fiber had never possessed. In particular, newly grown subsynaptic folds appear. In related experiments, muscles that had been cut into small fragments became reconstituted and reinnervated with normal end plates at which synaptic transmission occurred.[32]

Other investigations have shown that application of botulinum toxin produces neuromuscular block, not by destroying nerve terminals, but by preventing them from releasing transmitter (Chapter 10). Despite the presence of the intact-looking nerve terminals, the muscle membrane develops supersensitivity in previously extrasynaptic areas[33,34] and accepts additional new innervation. Similarly, after a muscle in the rat is made supersensitive as a result of blocking impulse transmission in the nerve, a foreign nerve is able to form additional synapses. An individual muscle fiber can then show synaptic potentials and contractions in response to simulation of each of the two nerves (Figure 6).[35] Conversely, when a denervated muscle is stimulated directly, the ability to accept extra innervation is lost together with the supersensitivity.

Is it a normal prerequisite that the muscle be supersensitive for innervation to occur? If this were so, one might expect muscles to be supersensitive when they become innervated for the first time in the fetus. This was shown in fetal and neonatal rat muscles, whose fibers are sensitive to ACh along their length. After innervation, the ACh-sensitive area shrinks over a period of about two weeks and becomes restricted to a region around the end plate.[36] A somewhat analogous situation exists in developing muscle fibers in tissue culture as they become innervated.[37]

Both initial innervation and reinnervation therefore occur when the muscle fibers are supersensitive. However, supersensitivity itself may be merely an outward expression of an unknown process that makes the muscle receptive for innervation. Several experiments indicate that synapse formation is not dependent upon the receptor itself, or at least not on that part of it to which α-bungarotoxin or curare binds. Thus,

[32]Bennett, M. R., Florin, T. and Woog, R. 1974. *J. Physiol. 238*: 79–92.

[33]Thesleff, S. 1960. *J. Physiol. 151*: 598–607.

[34]Fex, S., Sonessin, B., Thesleff, S. and Zelená, J. 1966. *J. Physiol. 184*: 872–882.

[35]Jansen, J. K. S., Lømo, T., Nicholaysen, K. and Westgaard, R. H. 1973. *Science 181*: 559–561.

[36]Diamond, J. and Miledi, R. 1962. *J. Physiol. 162*: 393–408.

[37]Frank, E. and Fischbach, G. D. 1979. *J. Cell Biol. 83*: 143–158.

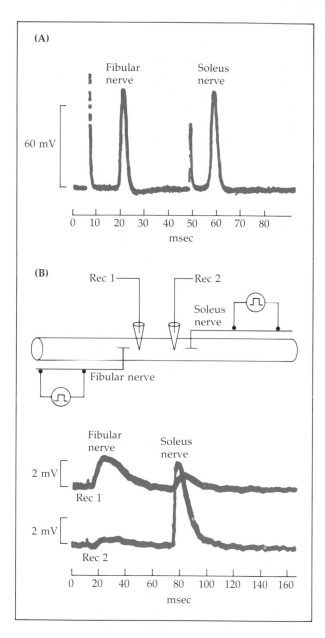

(A)

Fibular
nerve

Soleus
nerve

60 mV

0   10   20   30   40   50   60   70   80

msec

(B)

Rec 1 — Rec 2

Soleus
nerve

Fibular nerve

Fibular
nerve

Soleus
nerve

2 mV

Rec 1

2 mV

Rec 2

0   20   40   60   80   100   120   140   160

msec

6

DUAL INNERVATION of a single muscle fiber in the soleus of the rat by its original nerve and by a transplanted foreign nerve (the fibular nerve). (A) Both nerves initiate conducted action potentials in the same muscle fiber (recorded with an intracellular electrode). (B) Recordings with two microelectrodes (Rec 1, Rec 2) from one muscle fiber after transmission block by an increased magnesium concentration. Synaptic potentials only are caused by the original and foreign nerve stimulus. The synaptic potentials are relatively large near the point of innervation (see drawing) and become attenuated with distance. (After Jansen, Lømo, Nicolaysen, and Westgaard, 1973.)

reinnervation still occurs in denervated rat and toad muscles in the presence of α-bungarotoxin or curare.[38,39]

Role of basal lamina

An important factor that plays a key role in the regeneration of synapses between nerve and muscle is the extracellular material known as synaptic basal lamina. Lying between the nerve terminal and the

[38]Van Essen, D. and Jansen, J. K. 1974. *Acta Physiol. Scand. 91*: 571–573.
[39]Cohen, M. W. 1972. *Brain Res. 41*: 457–463.

muscle membrane, this material constitutes a densely staining extracellular matrix made up of polysaccharides and proteins, including collagen, cholinesterase, glycoproteins, and proteoglycans. As shown in Figure 7, the basal lamina surrounds the muscle, the nerve terminals, and the Schwann cell and dips into the folds in the postsynaptic membrane. McMahan and his colleagues have made a series of systematic and elegant studies on the physiological and structural effects that molecules in this noncellular material have on differentiation of nerve

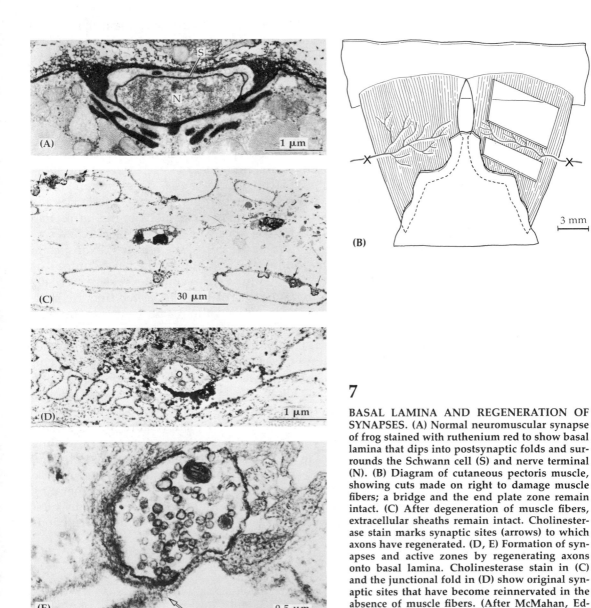

7

BASAL LAMINA AND REGENERATION OF SYNAPSES. (A) Normal neuromuscular synapse of frog stained with ruthenium red to show basal lamina that dips into postsynaptic folds and surrounds the Schwann cell (S) and nerve terminal (N). (B) Diagram of cutaneous pectoris muscle, showing cuts made on right to damage muscle fibers; a bridge and the end plate zone remain intact. (C) After degeneration of muscle fibers, extracellular sheaths remain intact. Cholinesterase stain marks synaptic sites (arrows) to which axons have regenerated. (D, E) Formation of synapses and active zones by regenerating axons onto basal lamina. Cholinesterase stain in (C) and the junctional fold in (D) show original synaptic sites that have become reinnervated in the absence of muscle fibers. (After McMahan, Edgington, and Kuffler, 1980.)

and muscle.[40-42] The key to their analysis was to use a thin transparent frog muscle—the cutaneous pectoris—in which regeneration occurs rapidly and where the end plates have a highly ordered arrangement. As a first step, regeneration by motor axons was carefully measured at denervated end plates. After the nerve had been crushed, axons grew back to the original end plates, filling the gutters and reforming active zones for transmitter release at precisely the original sites. How could this be demonstrated? Interestingly, denervated end plate zones in the muscle retained certain features making them recognizable: (1) Cholinesterase remained concentrated in the basal lamina of the gutters and folds for periods of weeks, thereby providing a convenient marker; (2) the gutters and folds within the postsynaptic membranes themselves remained intact in a constant pattern. Thus, it was possible to observe the regeneration of an axon by light and electron microscopy, to assess quantitatively that proportion of the original cholinesterase-stained end plate it filled, and to determine the position of the active zones in relation to postsynaptic folds.

To demonstrate the role of basal lamina in synapse formation, regeneration was studied in the absence of muscle fibers.[41] Slabs of the cutaneous pectoris muscle were removed from each side of the junctional region (Figure 7). This procedure causes degeneration of the remaining cut muscle fiber segments but, surprisingly, still leaves the basal lamina, and its associated cholinesterase, intact in the form of an exoskeleton or ghost of the muscle fiber. Muscle fiber regeneration was prevented by X-irradiation. When the nerve was crushed, axons again regenerated, grew to the former synaptic sites on the basal lamina—as marked by cholinesterase—and formed active zones for release precisely opposite portions of the basal lamina that had projected into the junctional folds (Figure 7C and D)—all this without a postsynaptic "target," the muscle fibers having been destroyed.

In addition, the basal lamina has been shown to have yet another effect—on the regenerating muscle fiber. After the muscle has been cut, myoblasts and eventually new muscle fibers grow to restore continuity. These new muscle fibers fill the basal lamina sheath. Acetylcholine receptors become clustered in high concentration on the muscle fibers, once again in the region of the old end plate (Figure 8). This same clustering occurs even when the nerve is prevented from regenerating by removing a section of it. That it is basal lamina and not the Schwann cell that is important was shown by killing Schwann cells and muscle fibers, leaving only basal lamina to provide cues for the regenerating nerve or muscle fibers.[43] Somehow, the information provided by molecules of the basal lamina at the end plate region provides instructions

[40]Letinsky, M. S., Fischbeck, K. H. and McMahan, U. J. 1976. *J. Neurocytol.* 5: 691–718.
[41]Sanes, J. R., Marshall, L. M. and McMahan, U. J. 1978. *J. Cell Biol.* 78: 176–198.
[42]Nitkin, R. M., Wallace, B. G., Spira, M. E., Godfrey, E. W. and McMahan, U. J. 1984. *Cold Spring Harbor Symp. Quant. Biol.*, 48: 653–666.
[43]McMahan, U. J., Edgington, D. R. and Kuffler, D. P. 1980. *J. Exp. Biol.* 89: 31–42.

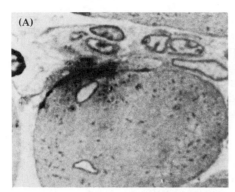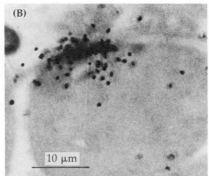

8 ACCUMULATION OF ACETYLCHOLINE RECEPTORS at original end plate sites of regenerated muscle fibers in the absence of nerve. The muscle was cut as in Figure 7A, but the nerve was prevented from regenerating. New muscle fibers formed within basal lamina sheaths. (A) Regenerated muscle stained for cholinesterase at the original end plate. (B) Same preparation labeled with radioactive α-bungarotoxin. Black dots represent silver grains and indicate presence of radioactivity and receptors at the original synaptic site. (After McMahan, Edgington, and Kuffler, 1980.)

for the nerve to form a synaptic terminal and for the muscle membrane to concentrate acteylcholine receptors. The nature of these molecular clues and the cells that synthesize them and insert them into basal lamina are as yet unknown, but studies with antibodies show that regional differences occur in the composition of the extracellular matrix at end plate and non-end plate regions.[42,44] Molecules that cause similar clustering of ACh receptors have also been extracted from other tissues such as brain, spinal cord, and the electric organ of torpedo.[45]

Under special circumstances, foreign or anomalous extra nerves can be made to form functional synapses with denervated skeletal muscles. Such experiments raise a number of questions regarding the specificity of synapse formation and the way in which nerve and muscle cells influence each other. What properties must the alien nerve have to be accepted? Does it alter the property of the muscle? Is the nerve itself altered as a result of innervating the wrong muscle?

Reinnervation of skeletal muscle by foreign nerves

Some observations on these questions date back to 1904, when Langley and Anderson showed that muscles of the cat could become innervated by cholinergic preganglionic sympathetic fibers,[46] which normally make synapses in ganglia. Similarly, formation of synaptic connections between vagal nerve fibers and the superior cervical ganglion has also been reported.[47]

The properties of synapses formed by vagus nerves on the frog

[44]Sanes, J. R. and Hall, Z. W. 1979. *J. Cell Biol. 83*: 357–370.
[45]Connolly, J. A., St. John, P. A. and Fischbach, G. D. 1982. *J. Neurosci. 2*: 1207–1213.
[46]Langley, J. N. and Anderson, H. K. 1904. *J. Physiol. 31*: 365–391.
[47]Vera, C. L., Vial, J. D. and Luco, J. V. 1957. *J. Neurophysiol. 20*: 365–373.

sartorius muscle have been studied by Landmesser.[48,49] The procedure was to transplant the denervated muscle to the thoracic region of the frog, where the cut gastric vagus nerve was then sutured to it. Within 50 days, stimulation of the vagal nerve produced synaptic potentials and visible contractions of the muscle. The synaptic potentials were different from those evoked by stimulation of a normally innervated muscle. They resembled those observed in the multiply innervated skeletal "slow" muscle fibers of the frog whose terminals make synapses at many widely distributed spots on the surface rather than at one or two discrete motor end plates. In addition, the individual synaptic potentials were small and facilitated with repetitive firing. There was no evidence to suggest that the properties of the nerve were changed or that the abnormal type of innervation had altered the electrical characteristics of the muscle.

The properties of certain other muscles in frogs do become markedly changed with foreign innervation. For example, the slow muscle fibers in the frog are quite distinctive. Apart from their diffuse innervation, they differ in their fine structure and cannot give regenerative impulses or twitches.[50] After denervation, the slow fibers can become reinnervated by nerves that normally innervate the twitch muscles at discrete end plates. Under these conditions, the slow fibers change and give conducted action potentials and twitches.[51]

The finding that muscles can accept additional, foreign innervation has led to speculation about the possible preference for one type of nerve over another. One idea is that when a muscle is reinnervated by its normal nerve after accepting innervation by a foreign one, the inappropriate additional synapses cease to function. This has been shown to be so in the salamander: The foreign synapses are eliminated after the normal nerve establishes its reconnection.[52] The salamander may be an exception since in both mammalian and fish muscles transplanted nerves can be as effective as the original ones in innervating fibers. Further, in muscle fibers with dual innervation—one original, the other foreign—both can function simultaneously (Figure 6).[35,53–55] There is evidence to show that a foreign nerve can even displace the normal appropriate motor axons in intact adult rat muscles, a finding suggesting a dynamic relationship required for maintenance of synaptic structure.[56]

Cross-innervation of mammalian muscles

Eccles, Buller, Close, and their colleagues have cut and interchanged

[48]Landmesser, L. 1971. *J. Physiol. 213*: 707–725.

[49]Landmesser, L. 1972. *J. Physiol. 220*: 243–256.

[50]Kuffler, S. W. and Vaughan Williams, E. M. 1953. *J. Physiol. 121*: 289–317.

[51]Miledi, R., Stefani, E. and Steinbach, A. B. 1971. *J. Physiol. 217*: 737–754.

[52]Dennis, M. J. and Yip, J. W. 1978. *J. Physiol. 274*: 299–310.

[53]Frank, E., Jansen, J. K. S., Lømo, T., and Westgaard, R. H. 1975. *J. Physiol. 247*: 725–743.

[54]Scott, S. A. 1975. *Science 189*: 644–646.

[55]Kuffler, D. P., Thompson, W. and Jansen, J. K. S. 1980. *Proc. R. Soc. Lond. B 208*: 189–222.

[56]Bixby, J. L. and Van Essen, D. C. 1979. *Nature 282*: 726–728.

the nerves to rapidly and more slowly contracting muscles in kittens and rats. After they have become reinnervated by the inappropriate nerves, the slow muscles become faster and the fast ones slower.[57-59] (It should be emphasized that both types of mammalian muscle fibers give conducted action potentials and therefore differ from the slow fibers in the frogs.) These results show, further, that the types of nerves that innervate a muscle can influence its mechanical properties. One cannot say as yet how this action of the nerve is brought about. It seems that a major factor is the pattern of impulses and contraction, since the motoneurons innervating slow and fast muscle fibers tend to fire at different frequencies. The type of use to which a muscle is subjected apparently influences its physiological performance and its structure.

NERVE GROWTH FACTOR: WHAT INDUCES SPROUTING, GROWTH, AND SURVIVAL?

There are no answers yet to the question of what makes a nerve select a certain muscle and what in the muscle attracts a nerve. There has, however, been support for many years for the idea that certain substances in tissues can attract neurons. For example, numerous experiments have shown that transplanting an extra leg onto the back of a tadpole, a lizard, or a newt causes the outgrowth of nerve fibers from the central nervous system.[60] This idea has received a considerable boost from the work of Levi-Montalcini and her colleagues,[61] who found a factor that selectively influences the growth of sympathetic and sensory neurons. These studies have helped in approaching the problems raised in this chapter, and the course of the investigations also illustrates the manner in which research progresses in the hands of perceptive investigators. The search for the growth factor is a remarkable sequence of coincidences—false but profitable leads, extraordinary and apparently fortunate choices, all leading to an important development in the area of the study of nerve growth.

To follow up the idea that there must be substances in transplanted limbs capable of attracting nerve fibers, it was reasonable to test the effect of rapidly growing tissues on the growth of neurons. The initial experiments were made by implanting onto chick embryos a connective tissue tumor (sarcoma) obtained from mice. On the side where the sarcoma had been implanted there was a profuse outgrowth of sensory and sympathetic nerve fibers from the embryo into the tumor. To show that the effect was caused by a humoral factor, sarcomas were grafted onto the chorioallantoic membrane, a tissue that surrounds the embryo. The only communication between the embryo and the tumor was in-

[57]Buller, A. J. 1970. *Endeavour 29*: 107–111.
[58]Buller, A. J., Eccles, J. C. and Eccles, R. M. 1960. *J. Physiol. 150*: 399–416.
[59]Close, R. I. 1972. *Physiol. Rev. 52*: 129–197.
[60]Hamburger, V. 1939. *Physiol. Zool. 12*: 268–284.
[61]Levi-Montalcini, R. 1982. *Annu. Rev. Neurosci. 5*: 341–362.

direct, but once again the dorsal root ganglia and sympathetic neurons on the side of the implant grew profusely.[62]

Next, it was shown that a similar dramatic effect is produced by sarcoma cells on tissue-cultured chick ganglia. The active factor in the sarcoma initially appeared to be a nucleoprotein. To see if nucleic acids were essential components of the growth-promoting factor, tumors were incubated with a snake venom whose action would hydrolyze the nucleic acids and thereby render the tumor fraction inactive. With venom present, however, the growth, far from being inactivated, was further increased. In fact, the control experiment of adding snake venom without the sarcoma extract revealed surprisingly that the venom itself was a far richer source of growth factor than was the sarcoma.[63] This, in turn, gave rise to the speculation that, since venom is secreted by the salivary gland, possibly salivary glands from other animals might also contain a similar factor.

The animal selected was the mouse.[63,64] It was fortunate that adult male mice were chosen because the salivary glands of female or immature mice contain far less growth factor than do those of other animals that have since been tried. Extracts of salivary glands of adult male mice are potent in causing the growth of cultured sympathetic ganglion cells (Figure 9). In this regard, the functional role of salivary glands in the animal is still not clear, especially since removal of these glands from young animals has minor effects on nerve growth.

The substance extracted from snake venom and salivary glands of mice has been called NERVE GROWTH FACTOR (NGF). In salivary glands, the protein is present as a complex made up of three types of subunits. The amino acid sequence has been worked out in the β-subunit, which is the biologically active part;[65,66] it consists of two identical peptide chains each containing 118 amino acids and 3 disulfide bridges. These chains show some resemblance to insulin (which acts on its targets in a somewhat similar manner).

Nerve growth factor causes an increase in the growth of sympathetic and dorsal root ganglion cells and decreases cell death. Nevertheless, it might be possible to argue that everything described so far is a purely pharmacological phenomenon—the effect of an agent that, although extractable from salivary glands, plays no part in the normal maturation or growth of neurons. This hypothesis was put to a test by producing a specific antibody to nerve growth factor by injecting it into rabbits.[67] When the serum was injected into newborn mice, they failed to develop

[62]See Levi-Montalcini, R. and Angeletti, P. U. 1968. *Physiol. Rev. 48*: 534–569, for references to earlier work.

[63]Cohen, S. 1959. *J. Biol. Chem. 234*: 1129–1137.

[64]Cohen, S. 1960. *Proc. Natl. Acad. Sci. USA 46*: 302–311.

[65]Angeletti, R. H., Mercanti, D. and Bradshaw, R. A. 1973. *Biochemistry 12*: 90–100, 100–115.

[66]Greene, L. A., Varon, S., Piltch, A. and Shooter, E. M. 1971. *Neurobiology 1*: 37–48.

[67]Levi-Montalcini, R. and Cohen, S. 1960. *Ann. N.Y. Acad. Sci. 85*: 324–341.

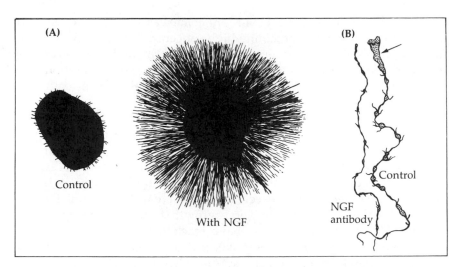

9 EFFECT OF NERVE GROWTH FACTOR on neuron in 7-day-old sensory ganglion from a chick embryo kept in a culture medium for 24 hours. (A) Ganglion in control medium is on the left. To the right, a ganglion maintained in a medium supplemented with nerve growth factor from the salivary gland of a male mouse shows prolific growth. (B) Thoracic sympathetic chain of ganglia from a control animal (mouse) is on the right. Arrow points to the stellate ganglion. To the left is a ganglion chain, much smaller in size, from a mouse injected 5 days after birth with the antiserum to nerve growth factor. (A after Levi-Montalcini, 1964; B after Levi-Montalcini and Cohen, 1960.)

normal sympathetic nervous systems (Figure 9). The parasympathetic nervous system was apparently not affected, and the dorsal root ganglia were only slightly smaller than normal. The antibodies act by neutralizing NGF circulating in the newborn animals. The animals lived normally but responded poorly to stress conditions. In adults, the antibody was much less effective. Neurons in explant cultures and dissociated cells from chick dorsal root ganglia will survive in culture only if NGF is provided in the medium.

The protein therefore seems to play a part in normal maturation and growth of certain cell types. Although other factors influencing the growth and survival of different types of neurons have been described, none is as yet as well characterized as nerve growth factor (see later).

The observations described in the preceding paragraphs have set the stage for an analysis of the mechanism of action of the nerve growth factor by a number of groups, including those of Levi-Montalcini,[62] Shooter,[68] and Thoenen.[69] Their studies have explored such questions as the fraction of the protein that is most effective in inducing growth, the receptors on the membrane that interact with nerve growth factor, and the metabolic events that subsequently occur.

[68]Yankner, B. A. and Shooter, E. M. 1982. *Annu. Rev. Biochem. 51*: 845–868.
[69]Thoenen, H. and Barde, Y. A. 1980. *Physiol. Rev. 60*: 1284–1335.

An early observation by Levi-Montalcini was that neurites tended to grow toward regions containing high concentrations of nerve growth factor. A dramatic, more recent example was provided by experiments in which nerve growth factor was injected into the neonatal rat brain.[70] Axons arising from sympathetic ganglia entered the spinal cord and ascended to the site of injection in the midbrain.

A clue to the mode of action is provided by the effects of nerve growth factor on cells in culture. When nerve growth factor is slowly released from a pipette close to one side of a sprout, the sprout will bend toward the region of higher concentration.[71] Similarly, experiments have been made in which sprouts are allowed to grow into a chamber separated from the cell body by a partition.[72] Initially, the central chamber must contain nerve growth factor for the neuron to survive. However, after sprouts have reached the side chamber, the neuron and its processes will survive, provided nerve growth factor is present in the chamber with the sprouts. Even though nerve growth factor is removed from the chamber containing the cell body, survival and growth continue. These results suggest that nerve growth factor can act via nerve terminals or processes.

Studies made with labeled nerve growth factor have shown that it is indeed taken up into nerve terminals and actively transported back to the soma.[73] As a first step, it binds to two distinct receptors on the surface, one type with a high affinity and the other with a lower affinity. The high-affinity receptors are found only on neurons and on pheochromocytoma cells (modified adrenal medullary cells), whereas the lower affinity receptors are also found on other types of nonneuronal cells.[74] In sympathetic neurons, the internalized nerve growth factor has been shown to regulate the synthesis of norepinephrine, the transmitter, by inducing two enzymes required for its synthesis: tyrosine hydroxylase and dopa β-hydroxylase. When NGF transport is impaired, the levels of these enzymes fall.[75]

The discovery of nerve growth factor offers hope for determining its site of action, the nature and distribution of its receptors, and the mechanisms by which a nerve cell is induced to grow. Since the amino acid sequence is known, it is in principle possible to isolate the gene itself and to follow the processing of the messenger RNA and precursor polypeptides to NGF as well as to determine the sites of synthesis within the cell and its regulation during development and regeneration.[68] However, as Levi-Montalcini has stated, "The NGF has still not

[70]Menesini-Chen, M. G., Chen, J. S. and Levi-Montalcini, R. 1978. *Arch. Natl. Biol.* *116*: 53–84.

[71]Gundersen, R. W. and Barrett, J. N. 1980. *J. Cell Biol. 87*: 546–554.

[72]Campenot, R. B. 1977. *Proc. Natl. Acad. Sci. USA 74*: 4516–4519.

[73]Hendry, I. A., Stöckel, K., Thoenen, H. and Iversen, L. L. 1974. *Brain Res. 68*: 103–

[74]Sutter, A., Riopelle, R. J., Harris-Warrick, R. M. and Shooter, E. M. 1979. *J. Biol. Chem. 254*: 5972–5982.

[75]Black, I. B. 1978. *Annu. Rev. Neurosci. 1*: 183–214.

found its place in the everyday broadening panorama of neuroscience. . . . It has disclosed only a few of its traits and keeps us wondering where it is heading for."[76]

Growth factors other than nerve growth factor

A growth-promoting protein with properties similar to those of nerve growth factor has been found for epithelia (epithelial growth factor).[77] Like nerve growth factor, it is present in high concentrations in the salivary gland of mice; the suggestion has been made that it may play a part in wound healing as the animal licks itself. Other molecules extracted from brain or peripheral nerves that influence the growth and survival of various types of neurons have been described.[78]

Numerous experimental observations suggest that specific neurons can be induced to grow selectively under the influence of molecules as yet unidentified. For example, regenerating motor axons heading toward denervated muscles will grow in a directed manner toward fragments of degenerating nerve, Schwann cells, or brain extracts. Undamaged axons can also be induced to grow under certain circumstances. Thus, in the leech (described in Chapter 18) and in salamanders, undamaged sensory axons will sprout to innervate a region of skin that has been deprived of its nerve supply.[79] Either the targets might, after denervation, secrete compounds that induce growth or, while innervated, secrete materials that prevent further growth of the sensory axons. The nature of such putative molecules is, however, not known. Nor is it clear how damage induces sprouting. A promising approach has been to examine proteins synthesized by neurons after injury.[80] It has been shown in frogs and mammals that new proteins are synthesized by the nerve cell body and transported to the growing terminals. Interestingly, proteins with similar molecular weights and charge are normally present in developing neurons.

A different type of action on growth is produced by a specific molecule that causes neurites to become aggregated and that is known as cellular adhesion molecule (CAM).[81] This substance is a plasma membrane glycoprotein that is extracted from chicken brain and retina and tends to promote not only aggregation of neurons but also the binding together of growing axons into fascicles. In culture, antibodies prepared against the molecule prevent the formation of nerve bundles by autonomic ganglia: The axons grow out normally but fail to form bundles.

Together these studies suggest that the induction and ordering of growth in regeneration and development may be mediated by a spectrum of molecules associated with the neurons, their targets, and extra-

[76]Levi-Montalcini, R. 1975. In F. G. Worden, J. P. Swaze and G. Adelman (eds.). *The Neurosciences: Paths of Discovery.* MIT Press, Cambridge, pp. 245–265.

[77]Carpenter, G. and Cohen, S. 1979. *Annu. Rev. Biochem. 48*: 193–216.

[78]Thoenen, H., Barde, Y.-A. and Edgar, D. 1982. In J. G. Nicholls (ed.), *Repair and Regeneration of the Nervous System.* Springer-Verlag, New York, pp. 173–185.

[79]Blackshaw, S. E., Nicholls, J. G. and Parnas, I. 1982. *J. Physiol. 326*: 261–268.

[80]Skene, J. H. P. and Willard, M. 1981. *J. Neurosci. 1*: 419–426.

[81]Rutishauser, U., Hoffman, S. and Edelman, G. M. 1982. *Proc. Natl. Acad. Sci. USA 79*: 685–689.

Retrograde effects on
neurons and synapses
following axotomy

cellular fluids, molecules performing roles similar to those of NGF for specific types of nerve cells.

Axotomy, which disconnects a nerve cell from its target organ, can lead to marked changes not only in the neuron itself but also in terminals synapsing upon it. Synaptic transmission becomes impaired in part because of morphological retraction of presynaptic terminals from the axotomized cell. The changes have been studied in detail in the autonomic ganglia of chicks and guinea pigs. In addition to nerve terminal retraction, the autonomic ganglion cells become less sensitive to acetylcholine, shrink in size, and may eventually die.[82,83] Moreover, the remaining presynaptic terminals release fewer quanta of transmitter. Thus, damage to the postsynaptic cell not only alters its ability to maintain its properties and "hold on" to the presynaptic terminals but also has a transynaptic retrograde effect on the terminals themselves. In the guinea pig, the effects of axotomy are mimicked by injecting the antibody to nerve growth factor and are largely prevented by application of nerve growth factor to the ganglion.[84] In the cat, axotomized spinal motor neurons also show synapse retraction (as well as histological changes and alteration in their membrane properties).[85] Recently, Rothshenker[86] has shown a novel transsynaptic effect on spinal motoneurons in the frog. When the motor axons to the cutaneous pectoris muscle of the frog are cut on one side of the animal, the intact, undamaged motor neurons to the same muscle on the other side of the animal sprout after a delay of a few weeks. The evidence suggests that a signal spreads from the cut neurons, crosses the spinal cord, and influences the motor neurons on the other side of the animal that have not been damaged. Motoneurons innervating other muscles are not affected. When dealing with the formation and maintenance of synapses, one must consider not only the effects of neurons upon their target cells but also the retrograde effects of the target cells on presynaptic terminals.

REINNERVATION IN THE CENTRAL NERVOUS SYSTEM

How accurately do
nerve fibers find their
targets?

There is evidence that when a motor nerve grows back into a muscle, the axons generally find the sites of former end plates. Similarly, in denervated mammalian skin, sensory axons grow back to reinnervate the original end organs that remain.[87] The studies with clearest results on the formation of connections during reinnervation in the central nervous system have been done on lower vertebrates and invertebrates. Even after extensive lesions to the brain, functions can be restored

[82]Brenner, H. R. and Johnson, E. W. 1976. *J. Physiol. 260*: 143–158.
[83]Brenner, H. R. and Martin, A. R. 1976. *J. Physiol. 260*: 159–175.
[84]Njå, A. and Purves, D. 1978. *J. Physiol. 277*: 53–75.
[85]Kuno, M. and Llinás, R. 1970. *J. Physiol. 210*: 823–838.
[86]Rotshenker, S. 1979. *J. Physiol. 292*: 535–547.
[87]Burgess, P. R., English, K. B., Horch, K. W. and Stensaas, L. J. 1974. *J. Physiol. 236*: 57–82.

through the regrowth of axons. For example, it was shown by Stone,[88] Sperry,[89] and their colleagues that if the optic nerve is cut in a frog or a salamander, fibers grow back to the appropriate region of the brain (the tectum). There they form synapses, and eventually the animal is able to see once again. Anatomical evidence in the goldfish has provided confirmation that optic nerve fibers can grow back in an orderly manner to their original destinations. In normal fishes, each point on the retina projects onto a particular part of the tectum, producing there an orderly and stereotyped map. During regeneration of the optic nerve, groups of fibers can be observed to course to the appropriate region of the tectum, following a similar route to that taken during development.[90]

In other experiments Sperry tested the ability of displaced regenerating fibers to find their targets.[91] The optic nerve of a frog was cut and the eye was rotated through 180°. About three weeks later, the nerve had regenerated and the frog could see again, but it behaved as though its vision were inverted. Thus, all its movements directed toward objects—as when the frog struck at a fly—were 180° out of phase. The simplest interpretation of this and other similar behavioral experiments is that fibers had grown back from the inverted retina to their original destinations in the tectum. This was confirmed by Gaze[92] and Jacobson,[93] using physiological techniques. A further important observation is that the animals never learned to correct their mistakes: Frogs with rotated eyes continued to strike downward at a fly held up in the air for as long as they lived after the operation. Evidently synaptic connections could not be reorganized through the stimulus of making behavioral errors. In this respect the behavior of frogs is different from that of higher animals; a man or a monkey can soon compensate for the effects of an inverting prism placed on the eyes (this does not imply that higher centers have reorganized their connections).

On the other hand, there is evidence that the projection of retinal fibers to tectal cells is not completely inflexible. For example, in the frog an extra eye can be implanted into the forebrain region of the embyro, causing two retinal projections to converge on a single tectal lobe.[94] Interestingly, the regions supplied by the two eyes are arranged in interleaved stripes. Similarly, in goldfish, a tectal cell is not irreversibly committed to receive its input from only one particular area of retina. For example, if half the tectum in a goldfish is removed, the entire retina can eventually project in an orderly fashion onto the remaining portion of the tectum so that the topography is preserved.[95] Conversely, after one-half the retina is removed, the remaining optic nerve fibers

[88]Stone, L. S. and Zaur, I. S. 1940. *J. Exp. Zool. 85*: 243–269.
[89]Sperry, R. W. 1944. *J. Neurophysiol. 7*: 57–69.
[90]Attardi, D. G. and Sperry, R. W. 1963. *Exp. Neurol. 7*: 46–64.
[91]Sperry, R. W. 1945. *J. Neurophysiol. 8*: 15–28.
[92]Gaze, R. M. 1970. *The Formation of Nerve Connections.* Academic, New York.
[93]Jacobson, M. 1978. *Developmental Neurobiology,* 2nd Ed. Plenum, New York.
[94]Law, M. I. and Constantine-Paton, M. 1981. *J. Neurosci. 1*: 741–749.
[95]Yoon, M. 1972. *Exp. Neurol. 37*: 451–462.

eventually expand to fill the entire tectum, again preserving normal topography.[96] Moreover, in the goldfish the system is in a dynamic state throughout its life: In the adult the retina is continually growing and the new optic nerve fibers form appropriate connections in the tectum[97] (a result similar to that found in the olfactory system of mammals; see later). From these results one can conclude that the connections in the central nervous system of the frog and the goldfish have some ability to become reorganized.

Accuracy of regeneration of individual neurons

The results obtained in vertebrates suggest that regenerating cells grow to predestined targets whenever possible. The degree of selectivity is not yet established, and it would be of interest to know how precisely cells reconnect in a system where individual neurons rather than whole populations can be examined. A convenient preparation for a study of the precision of regenerating nerve fibers is the leech central nervous system (Chapter 18), where individual cells can be recognized without ambiguity and their connections in normal animals traced.

To observe regeneration in the leech, the procedure is to sever the axons that link two ganglia and test whether the connections become reestablished. For example, an individual sensory cell in one ganglion is known to initiate a synaptic potential in a specific motor cell in the next ganglion. Such a connection between identified neurons can, in fact, be successfully reestablished, showing that an individual cell can discriminate among many targets in the complex neuropil so as to interact once more with one particular neuron in preference to others.[98] In one instance there is anatomical evidence for accurate regeneration.[99] In each ganglion a single neuron—called the S cell—sends an axon about 7 μm in diameter through one connective toward the neighboring ganglion on either side. This axon can be clearly recognized in cross sections since it has by far the largest diameter. It forms an electrical connection exclusively with the axon of its homologue in the midregion of the connective. When the nerve cord is cut or crushed, the regenerating portion of the axon searches for the cut end and reforms electrical synapses, so that transmission becomes reestablished. Interestingly, the initial contact is often made with the axon's own surviving distal stump and only later with the other S cell. The main point is the precision with which the target is recognized. No extraneous connections are made. An axon can also become reconnected to its old distal stump in the peripheral nervous system of crustaceans.[100] There is no evidence that repair by such mechanisms can occur in vertebrates.

Other invertebrates, such as the cricket, provide additional examples

[96]Schmidt, J. T., Cicerone, C. M. and Easter, S. S. 1978. *J. Comp. Neurol. 177*: 257–278.

[97]Easter, S. S. Jr., Rusoff, A. C. and Kish, P. E. 1981. *J. Neurosci. 1*: 793–811.

[98]Muller, K. J. and Nicholls, J. G. 1981. In K. J. Muller, J. G. Nicholls and G. S. Stent (eds.). *Neurobiology of the Leech.* Cold Spring Harbor Press, New York, pp. 197–226.

[99]Scott, S. A. and Muller, K. J. 1980. *Devel. Biol. 80*: 345–363.

[100]Hoy, R., Bittner, G. D. and Kennedy, D. 1967. *Science 156*: 251–252.

of regeneration with a high degree of precision.[101,102] In these creatures, it has been shown anatomically and physiologically that mechanoreceptor fibers regenerate to form functional connections on the appropriate neurons within the central nervous system. These experiments also provide evidence for reorganization of the central connections when regeneration is not allowed to occur after removal of parts of the sensory apparatus. The neurons within the centers that had lost their normal inputs now receive innervation from elsewhere. Similarly, supernumerary sensory fibers following transplantation regenerate in an orderly manner to appropriate targets in the central nervous system.

It has been generally believed that sprouting in the adult mammalian central nervous system is quite restricted, largely because transection of tracts is not followed by regeneration and restitution of function. In recent years, however, it has become apparent through the work of Aguayo and his colleagues that axons in the central nervous system can grow for distances of several centimeters under suitable circumstances.[103–105] Of prime importance is the immediate environment, which consists of the Schwann cells in the periphery and the glial cells in the central nervous system (Chapter 13). Clues to the role of satellite cells are provided by several types of experiment. First, it is well known that motor neurons, having their cell bodies situated within the spinal cord, can regenerate; whereas those of their axons in the periphery grow for long distances to reach the muscles after a lesion, those within the central nervous system sprout little if at all. Similarly, sensory axons regrow to their targets in the periphery; and if a sensory root running toward the spinal cord is cut close to the dorsal root ganglion, axons start to grow from the cell bodies to approach the cord. But they stop growing once they reach the astrocytic processes that delimit the surface of the central nervous system. Central nervous system glial cells can also prevent growth of peripheral axons.[106] Thus, although regenerating axons of the mouse sciatic nerve grow back to the periphery through a chain of remaining Schwann cells, in general they fail to enter chains of astrocytes and oligodendrocytes transplanted by placing a segment of optic nerve in the path of the regenerating nerve. Together these findings suggested either (1) that glial cells actively inhibit growth or (2) that Schwann cells provide factors that could stimulate the growth of injured neurons.

Whatever the mechanism, Schwann cells implanted into the central

Sprouting and repair of mammalian central nervous system

[101]Palka, J. and Edwards, J. S. 1974. *Proc. R. Soc. Lond. B 185*: 105–121.

[102]Murphey, R. K., Johnson, S. E. and Sakaguchi, D. S. 1983. *J. Neurosci. 3*: 312–325.

[103]David, S. and Aguayo, A. J. 1981. *Science 214*: 931–933.

[104]Benfey, M. and Aguayo, A. J. 1982. *Nature 296*: 150–152.

[105]Aguayo, A. J., Richardson, P. M., David, S. and Benfey, M. 1982. In J. G. Nicholls (ed.). *Repair and Regeneration of the Nervous System*. Springer-Verlag, New York, pp. 91–105.

[106]Aguayo, A. J., Dickson, R., Trecarten, J., Attiwell, M., Bray, G. M. and Richardson, P. 1978. *Neurosci. Lett. 9*: 97–104.

nervous system might therefore enable growth to occur. In a series of experiments, this has been shown to be the case. For example, when a segment of sciatic nerve is grafted between the cut ends of the spinal cord in a mouse, fibers grow across and fill the gap.[107] (The graft is composed of Schwann cells and connective tissue, the axons having degenerated.) Similarly, cultures of Schwann cells implanted into the spinal cord promote growth. A dramatic effect is observed by the use of "bridges" of the type shown in Figure 10. One end of a segment of sciatic nerve is implanted into the spinal cord, the other into a higher region of the nervous system: upper spinal cord, medulla, or thalamus. Bridges have even been made from cortex to another part of the CNS or to muscle. After several weeks or months, this graft resembles a normal nerve trunk filled with myelinated and unmyelinated axons (Figure 10). These neurons fire impulses and are electrically excited or

[107]Richardson, P. M., McGuiness, U. M. and Aguayo, A. J. 1980. *Nature 284*: 264–265.

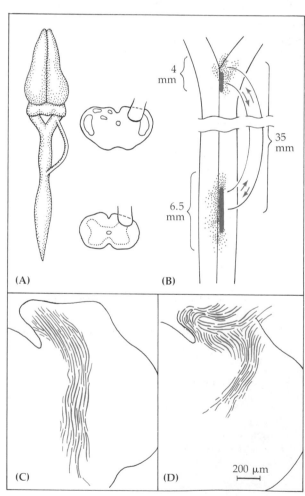

10

BRIDGES BETWEEN MEDULLA AND SPINAL CORD enabling neurons within the central nervous system to grow for prolonged distances. The graft consists of a segment of adult rat sciatic nerve in which axons have degenerated, leaving Schwann cells. These act as a conduit for central axons to grow along. (A) Sites of insertion of the graft. Neurons are labeled by cutting the graft and applying horseradish peroxidase to the tips. (B) Positions of 1472 neuronal cell bodies labeled by retrograde transport in seven grafted rats. Most of the cells sending axons into the graft are situated close to its insertion. (C, D) Cross sections of the medulla oblongata and spinal cord showing the course of axons in animals 26 and 30 weeks after grafting. (After David and Aguayo, 1981.)

inhibited by stimuli applied above or below the sites of implantation. By cutting the bridge and dipping the ends into horseradish peroxidase or other markers, the cells of origin become labeled and their distribution can be mapped. Examples such as those in Figure 10 show that the axons that have grown over distances of several centimeters arise from neurons whose cell bodies lie within the central nervous system. Such axons would never normally encounter a Schwann cell. Usually only those neurons with somata not more than a few millimeters from the bridge send axons into it. Similarly, axons leaving the bridge to enter the central nervous system grow only a short distance before terminating.

These results clearly hold out a promise for regeneration of connections by central nervous system axons. There is as yet, however, no evidence that the growing axons form synapses at their terminals. Even if they did, for function to be restored such connections would presumably have to be made only with appropriate targets.

A different approach used by Björklund and his colleagues to study regeneration in the adult mammalian central nervous system has been to transplant embryonic nerve cells.[108,109] Unlike neurons from adult central nervous system, which will not survive, explants of brain or dissociated cells taken from fetal or neonatal animals can be inserted into the central nervous system (Figure 11). There they not only sprout and differentiate but also release transmitters. An example of this is provided by experiments in which embryonic dopaminergic neurons taken from the substantia nigra (a region in the midbrain; Chapter 12 and Appendix B) were transplanted into the basal ganglia of rats in which the dopaminergic pathways had been previously destroyed. In these animals the behavioral deficits produced by the lesion were alleviated after transplantation, presumably because transmitter (dopamine) was released by the grafted cells. For example, if a lesion of the dopamine pathway is made on one side of a rat, a disorder of movement results. The animal turns toward the side of the lesion spontaneously or in response to stress. This asymmetry of movement can be counteracted by grafts of dopamine-containing neurons from the substantia nigra of immature animals placed in appropriate sites on the correct side. Fibers grow in to the host tissue from the graft and the animals then cease to rotate in response to the stressful stimuli.[110] Ultrastructural studies showed that axons originating in neuronal grafts could form synapses with host neurons. Presumably the transmitter dopamine released by the donor cells can act in a modulatory manner at synapses of the host; for such a "garden-sprinkler" diffuse type of action, specific one-to-one cell contacts are probably not required (Chapter 12).

[108]Björklund, A. and Stenevi, U. 1979. *Physiol. Rev. 59*: 62–100.

[109]Schmidt, R. H., Björklund, A. and Stenevi, U. 1981. *Brain Res. 218*: 347–356.

[110]Björklund, A., Dunnett, S. B., Stenevi, U., Lewis, N. E. and Iversen, S. D. 1980. *Brain Res. 199*: 307–333.

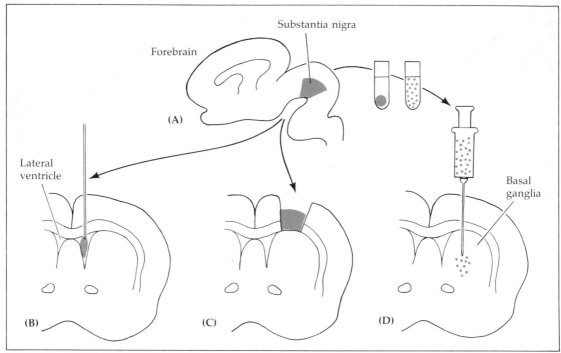

11 EXPERIMENTAL PROCEDURES FOR TRANSPLANTING EMBRY-ONIC TISSUE into adult rat brain. Tissue rich in cells containing do-pamine is dissected (A) from the substantia nigra (Chapter 12) and is injected into the lateral ventricle (B) or grafted into a cavity in the cortex overlying the basal ganglia (C). Alternatively, a suspension of disso-ciated substantia nigra cells can be injected directly into the basal gan-glia (D). Such embryonic cells survive, sprout, and secrete transmitter. (After Dunnett, Björklund, and Stenevi, 1983.)

As in invertebrates and lower vertebrates, uninjured neurons within the mammalian brain can form new processes following injuries to other cells that innervate the same target area. Experiments have been de-signed to examine well-defined structures in the brain such as the hippocampus, the colliculi, the visual cortex, or the red nucleus.[111–114] A lesion is made in a bundle of axons known to end in a circumscribed area of the tissue. After the terminals have degenerated, tests are made to see whether new growth has occurred into the denervated area from nearby regions. Lesions made in the hippocampus or colliculi resulted in orderly reinnervation by neurons whose field of innervation had expanded. In the red nucleus of the cat, Tsukahara has shown that

[111]Chow, K. L., Mathers, L. H. and Spear, P. D. 1973. *J. Comp. Neurol. 151*: 307–322.
[112]Lund, R. D. and Lund, J. S. 1973. *Exp. Neurol. 40*: 377–390.
[113]Raisman, G. and Field, P. 1973. *Brain Res. 50*: 241–264.
[114]Steward, O., Cotman, C. W. and Lynch, G. S. 1973. *Exp. Brain Res. 18*: 396–414.

there is an orderly projection of sensory and motor cortical neurons.[115] The topography is such that cortical neurons subserving forelimb or hindlimb make synapses specifically on the neurons of the red nucleus that are concerned with similar areas. Connections of cortical neurons onto these cells can be altered following lesions of the peripheral or the central nervous system.[116] Thus, when the ipsilateral projection from the cortex is cut, the fibers from the contralateral cortex sprout to innervate the denervated cells in the red nucleus. They do so—not randomly, but in the correct topological pattern, via the appropriate interneurons—and end on the correct dendrites. Sprouting and rein-nervation of denervated targets is a normal function of the olfactory system in mammals.[117] Throughout the life of mice and other verte-brates, olfactory epithelial cells and axons die. New receptors are con-tinuously generated and their axons grow to form connections on the appropriate cells in the olfactory bulb.

Chapter 20 describes how in the newborn monkey closure of one eye during a critical period causes shrinkage of the cortical area supplied by that eye. This change is accompanied by a corresponding increase in the area innervated and functionally activated by the normal eye.

A variety of mechanisms are involved in bringing about the forma-tion of specific connections between neurons and their targets during development. These include the lineage of neurons with specific prop-erties from common precursors, the guidance mechanisms directing growth of processes toward their correct destinations at the correct time, cell death, and the competition or sorting out of synapses on the post-synaptic cells. *Specificity in development*

Cell lineage is most readily followed in simple invertebrates like the leech,[118] the grasshopper,[119] or the tiny nematode *Caenorhabditis elegans*, which contains only about 300 neurons.[120] In these preparations, know-ing the final outcome one can follow the development cell by cell and can examine the expression of characteristics such as membrane prop-erties, transmitters, growth of axons, and branching patterns. The lin-eage history of a cell limits its developmental potential. Some cells have a unique fate (for example, the sensory cells in leech ganglia), whereas others seem to have multiple potential fates depending on temporal or positional cues.

How rigidly fixed are the properties of neurons that descend from a particular ancestor in vertebrate embryos? Are all the progeny des-

---

[115]Tsukahara, N. 1981. *Annu. Rev. Neurosci.* 4: 351–379.

[116]Fujito, Y., Tsukahara, N., Oda, Y. and Yoshida, M. 1982. *Exp. Brain Res.* 45: 1–12, 13–18.

[117]Graziadei, G. A. M. and Graziadei, P. P. C. 1978. In M. Jacobson (ed.). *Handbook of Sensory Physiology*, Vol. 9. Springer-Verlag, New York, pp. 55–85.

[118]Stent, G. S. and Weisblat, D. 1982. *Sci. Am.* 246: 136–146.

[119]Goodman, C. S. and Spitzer, N. C. 1979. *Nature 280*: 208–214.

[120]Horvitz, H. R. 1982. In J. G. Nicholls (ed.). *Repair and Regeneration of the Nervous System*. Springer-Verlag, New York, pp. 41–55.

tined to form cells with specified properties, such as being autonomic rather than sensory cells, or being cells using ACh as the transmitter rather than norepinephrine? Or can the transmitter made by a cell change under different influences, as has been shown for neonatal sympathetic neurons in culture (Chapters 12 and 13). These questions have been studied in chick and quail nervous systems by Le Douarin.[121] She has shown that cells from one region of the neural crest in the embryo of a quail can be transplanted to the same or a different region of a host chick embryo. In normal embryos these cells eventually give rise to dorsal root and autonomic ganglia, different regions of the crest being responsible for different ganglia and cells with characteristic transmitters. Cytological characteristics make the identification from chick or quail unambiguous. After transplantation, the quail cells were able to innervate structures they would never normally supply. For example, cells removed from a region that would normally become the adrenal gland could instead innervate gut.[122] Provided that the cells are transplanted at a sufficiently early stage, they can make a new transmitter—acetylcholine instead of norepinephrine. At later stages, the cells have become committed and lose this ability to differentiate in a manner determined by their new environment. The precise role played by the target tissues in determining the fates of these cells is not known.

In contrast to observations on neural crest cells, chick motor neurons seem specified to grow to the correct muscles. Even after reversal of segments of the spinal cord at early stages, motor neurons grow out to synapse upon the muscles they would innervate in a normal animal.[123] Moreover, in the chick the correct motor neurons receive synaptic inputs from sensory fibers innervating appropriate muscles.[124] Similarly, autonomic ganglia of the guinea pig receive their presynaptic inputs in an orderly pattern from fibers arising at appropriate segmental levels in the central nervous system. Purves and his colleagues have shown that ganglia of the sympathetic chain situated rostrally are supplied by fibers from three or four of the rostral segments in the spinal cord, those situated caudally by caudal segments.[125] If a ganglion is transplanted, presynaptic axons will grow back to form connections with the autonomic ganglion nerve cells. Again the reinnervation is specific, thoracic ganglia displaced to the cervical region becoming reinnervated preferentially by neurons originating in the thoracic segments of the spinal cord.[126]

Competition and formation of contacts with a target are factors that play a part in ensuring the survival of embryonic neurons. As mentioned earlier (page 491) large numbers of neurons die during devel-

[121]Le Douarin, N. 1982. *The Neural Crest.* Cambridge University Press, Cambridge, England.

[122]Le Douarin, N. 1980. *Nature 286*: 663–669.

[123]Lance-Jones, C. and Landmesser, L. 1980. *J. Physiol. 302*: 581–602.

[124]Eide, A.-L., Jansen, J. K. D. and Ribchester, R. R. 1982. *J. Physiol. 324*: 453–478.

[125]Purves, D. and Lichtman, J. W. 1983. *Annu. Rev. Physiol. 45*: 553–565.

[126]Purves, D., Thompson, W. and Yip, J. W. 1981. *J. Physiol. 313*: 49–63.

opment. In the ciliary ganglion of the chick, it has been clearly demonstrated by Landmesser, Pilar, and their colleagues that competition occurs between neurons, with considerable cell death.[127] Removal of the targets before death—at the time peripheral connections are being made—leads to the loss of virtually all neurons. The death of motor neurons has been studied extensively in the spinal cord of the chick by Hamburger and his colleagues.[128] Removal of limb buds has been shown to augment the process of cell death (from about 40 to 90 percent of the motor neuron population); conversely, an increase in the size of the target produced by grafting on a supernumerary limb reduced cell death (from 50 to 20 percent). Clearly the target plays an essential role in the survival of these cells, but the nature of the competition is not yet understood.[129,130] The results suggest that those cells that succeed in forming connections with the target myoepithelial cells may be less likely to die than those that fail.

At later stages of development, synapse elimination (without involving cell death) can further sharpen the precision with which targets are innervated (see also Chapter 20). An example that has been studied extensively is the neuromuscular junction in newly born rats. In such animals, an individual muscle fiber is supplied by a number of axons, each forming an effective synapse. Over the first two weeks or so after birth, axons lose their connections until the picture resembles that in the adult, with each end plate being supplied by only one axon.[131,132] Thus, the average size of the motor units becomes progressively smaller. These experiments raise a number of interesting problems: One concerns the mechanism by which the muscle rejects axons while allowing just one axon to remain. Although synapses are eliminated, no muscle fiber loses all its axons, and each motor neuron continues to innervate a number of muscle fibers.[132] Similar retraction of multiple inputs occurs in the autonomic ganglia of neonatal rats and guinea pigs.[125] Each ganglion cell is initially supplied by multiple inputs—about five—but by about five weeks after birth, only one usually remains.

At present it is not clear what mechanisms in the nerve and its target determine whether a particular terminal is maintained or retracts. If a skeletal muscle is partially denervated in early life, the remaining fibers can still maintain large territories; the motor units fail to retreat to the same extent.[133] Activity also plays a part. Thus, paralysis with tetrodotoxin delays the process of retraction, and stimulation accelerates it.

Synapse elimination during development

---

[127]Pilar, G., Landmesser, L. and Burstein, L. 1980. *J. Neurophysiol.* 41: 233–254.

[128]Hamburger, V., Brunso-Bechtold, J. K. and Yip, J. W. 1981. *J. Neurosci.* 1: 60–71.

[129]Hollyday, M., Hamburger, V. and Farris, J. 1977. *Proc. Natl. Acad. Sci. USA* 74: 3582–3586.

[130]Oppenheim, R. W. 1981. In W. M. Cowan (ed.). *Studies in Developmental Neurobiology.* Oxford University Press, New York, pp. 74–133.

[131]Redfern, P. A. 1970. *J. Physiol.* 209: 701–709.

[132]Brown, M. C., Jansen, J. K. S. and Van Essen, D. 1976. *J. Physiol.* 261: 387–422.

[133]Van Essen, D. C. 1982. In N. C. Spitzer (ed.). *Neuromuscular Synapse Elimination: Neuronal Development.* Plenum, London, pp. 333–376.

What seems important is not simply the amount of impulse traffic but the frequency characteristics. Stimulation by intermittent, high-frequency bursts is more effective in speeding up retraction than low-frequency stimuli delivered regularly—a patterning effect similar to that required for influencing extrajunctional acetylcholine receptors.[134,135]

Together these results suggest that possibly an "innervation factor" somewhat analogous to nerve growth factor is produced by the muscle fibers. The quantity would vary with the state of development and impulse traffic. the maintenance of the terminals would depend upon a steady supply involving some form of competition.

GENERAL CONSIDERATIONS OF NEURAL SPECIFICITY

The question of neural specificity—how nerve cells find their appropriate targets—is just one facet of a general property of the regulation of cell growth. Other more mundane cells in the body also appear to know

[134]Thompson, W. 1983. *Nature 302*: 614–616.
[135]Harris, W. A. 1981. *Annu. Rev. Physiol. 43*: 689–710.

BOX 1   MONOCLONAL ANTIBODIES

Distinctive molecular components of various types of neurons and glial cells can now be identified and characterized by the use of monoclonal antibody techniques. Surface or intracellular components of specific cells can be marked and tests made for the functions of those antigens. The principles involved in the production of mixed antibodies and monoclonal antibodies are shown in the figure. When an immunizing agent is injected into the mouse, its response is to make numerous diverse antibodies against antigenic molecules present on the injected material, as well as different antibodies against single antigenic determinants (i.e., subgroups of molecules). Hence, the production of a mixture of antibodies follows the injection of an antigen such as a polypeptide or a single protein. B lymphocytes, found in the spleen, are responsible for manufacturing these antibodies. Each B lymphocyte, however, synthesizes only a single type of antibody. Unfortunately, these lymphocytes cannot normally divide and produce a colony of cells in culture.

The key to the monoclonal technique is to fuse lymphocytes with cells that do divide and grow in culture, namely, malignant tumor cells obtained from a myeloma. The fused "hybridoma" cells retain (1) the ability of the tumor cell to divide and (2) the ability of the lymphocyte to manufacture the specific antibody against one particular antigenic determinant (for example, against a decapeptide). An individual hybrid cell cloned in culture therefore represents a potentially immortal source for obtaining large quantities of highly specific antibody. Once an appropriate clone has been identified, it can be frozen, stored, and then thawed, to reproduce again.

The advantages of this technique are that a highly complex mixture of antigenic determinants can be used to produce a broad spectrum of antibodies expressed one at a time, each in a different dish. Thus, it is no longer necessary to purify to homogeneity a molecular structure for which an antibody probe is desired. For example, one can use the entire central nervous system of the leech as an antigen and find with surprisingly little difficulty those clones that produce antibody binding to a single nerve cell body—say, the nociceptive cell (Chapter 18). In practice, the antibody molecule itself is combined with a marker, like horseradish peroxidase or a fluorescent dye, so as to label the molecular com-

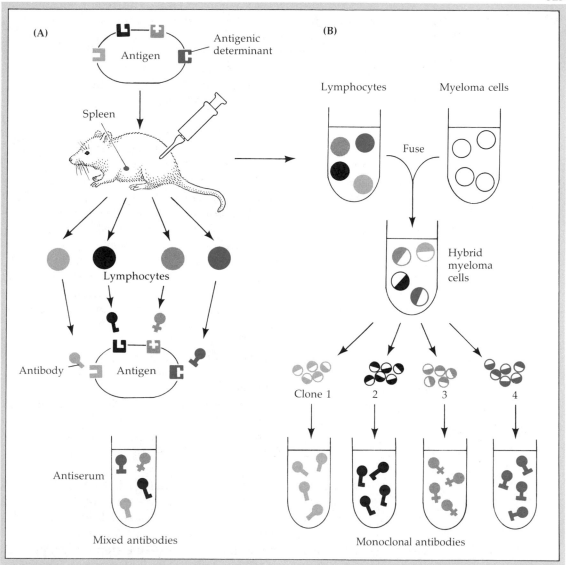

(A)

Antigen

Antigenic determinant

Spleen

Lymphocytes

Antibody   Antigen

Antiserum

Mixed antibodies

(B)

Lymphocytes   Myeloma cells

Fuse

Hybrid myeloma cells

Clone 1   2   3   4

Monoclonal antibodies

After Milstein 1980

ponents to which it binds. As a caution, it should be noted that since monoclonal antibodies recognize only a single antigenic determinant, which may be a short amino acid sequence, the presence of similar labeling in different types of cells does not unequivocally prove that they contain the same molecules, though they do each possess a common antigenic determinant.

The principal difficulties arise (1) in screening the large numbers of hybrid clones for the antibodies of interest and (2) in determining the pos-

sible functional roles played by the antigenic determinants.

Monoclonal antibodies produced so far have been used to label different types of neurons and glial cells, transmitters such as substance P and 5-hydroxytryptamine and surface and intracellular components, and extracellular materials such as synaptic basal lamina or cellular adhesion molecules. Monoclonal antibody techniques can also be used to assess at what stages of development particular antigens are expressed.

in what direction to grow and when to stop. The restitution of skin tissue after partial removal, the appropriate closing of a wound, and regrowth of an injured organ such as the liver to its proper size are related phenomena. Just as a denervated muscle seems to attract nerve fibers, so does a transplanted muscle attract new capillaries that grow into it by splitting off from an adjoining vascular bed. But how can we find out the precise mechanisms whereby nerve cells become connected and aggregated into tissues?

Perhaps a rather commonplace analogy may be encouraging. Let us assume that we are ignorant about the workings and design of the postal system. A chapter from a book on the nervous system, without its illustrations, is posted in Boston (or perhaps Denver?) and addressed to Tokyo, where it arrives a few days later. How does it get there? The writer knows only the closest letterbox and is unaware even of the post office in his district. The postal worker who empties the letterbox knows the post office; there the clerk who handles the mail may not know where Tokyo is but does know how to direct the package to the airport, and so on, to the right country, city, street, building, and eventually the correct person. If this were not enough, the illustrations that complement the chapter are posted by separate mail from Stanford (or Basel?) to the same destination, where they arrive almost simultaneously with the chapter from Boston.

Some aspects of neural specificity may not be too different. Suppose that a nerve cell in the left retina finds its specified target in the lateral geniculate nucleus and that a geniculate neuron finds the proper address in the visual cortex among the millions of other inhabitants of that particular visual area. And another nerve cell in the right eye also reaches its appropriate destination, so that eventually connections are made in the visual cortex at about the same time in order to complement the information sent from the left eye.

The comforting feature of this lengthy analogy is that the problem seems altogether baffling at first sight. Yet, one can solve the postal puzzle by following the mail step by step to its destination. This would reveal some of the logic and design of postal organization (albeit without disclosing the identity of the designer). At any one step, only a limited number of instructions are followed and a limited number of mechanisms operate. It would not necessarily be profitable to ask how the letterwriter knows the way to the right room and person in Tokyo, or what attracts the letter to the postal employee, the right franking machine, or airplane; equally, it may not be useful at this stage to ask how the neuron in the right eye finds its way to deliver its message to the correct column in the cortex to arrive in time for the rendezvous with the signals originating in the neuron from the left eye.

It seems that the problem of neural specificity can be broken down into analyzable parts. Perhaps the best tools at our disposal at present, in addition to those discussed in previous chapters, are a combination of genetic, tissue culture, molecular, and developmental approaches at the cellular level.

SUGGESTED READING

Papers marked with an asterisk (*) are reprinted in Patterson, P. H. and Purves, D. (eds.). 1982. *Readings in Developmental Neurobiology.* Cold Spring Harbor Laboratory, Cold Spring Harbor, NY.

General reviews

Barde, Y.-A., Edgar, D. and Thoenen, H. 1983. New neurotrophic factors. *Annu. Rev. Physiol. 45*: 601–612.

Björklund, A. and Stenevi, U. 1979. Regeneration of monoaminergic and cholinergic neurons in the mammalian central nervous system. *Physiol. Rev. 59*: 62–100.

Black, I. B. 1979. Neuronal responses to extracellular signals. In J. G. Nicholls (ed.), *The Role of Intercellular Signals: Navigation, Encounter, Outcome.* Dahlem Konferenzen, Berlin.

Graziadei, G. A. M. and Graziadei, P. P. C. Continuous nerve cell renewal in the olfactory system. In M. Jacobson (ed.). *Handbook of Sensory Physiology,* Vol. 9. Springer-Verlag, New York, pp. 55–85.

Harris, W. A. 1981. Neural activity and development. *Annu. Rev. Physiol. 43*: 689–710.

Landmesser, L. T. 1980. The generation of neuromuscular specificity. *Annu. Rev. Neurosci. 3*: 279–302.

Le Douarin, N. M., Smith, J. and Le Lievre, C. S. 1982. From the neural crest to the ganglia of the peripheral nervous system. *Annu. Rev. Physiol. 43*: 653–671.

Levi-Montalcini, R. 1975. NGF: An uncharted route. In F. G. Worden, J. P. Swaze and G. Adelman (eds.). *The Neurosciences: Paths of Discovery.* MIT Press, Cambridge, pp. 245–265.

McKay, R. D. G. 1983. Molecular approaches to the nervous system. *Annu. Rev. Neurosci. 6*: 527–546.

Molecular Neurobiology. 1984. *Cold Spring Harbor Symp. Quant. Biol.,* Vol. 48.

Nicholls, J. G. (ed.). *Repair and Regeneration of the Nervous System.* Springer-Verlag, New York.

Pumplin, D. W. and Fambrough, D. M. 1982. Turnover of acetylcholine receptors in skeletal muscle. *Annu. Rev. Physiol. 44*: 319–335.

Purves, D. and Lichtman, J. W. 1983. Specific connections between nerve cells. *Annu. Rev. Physiol. 45*: 553–565.

Sanes, J. R. 1983. Roles of extracellular matrix in neural development. *Annu. Rev. Physiol. 45*: 581–600.

Tsukahara, N. 1981. Synaptic plasticity in the mammalian central nervous system. *Annu. Rev. Neurosci. 4*: 351–379.

Van Essen, D. C. 1982. Neuromuscular synapse elimination. In N. C. Spitzer (ed.), *Neuronal Development.* Plenum Press, London, pp. 333–376.

Yankner, B. A. and Shooter, E. M. 1982. The biology and mechanism of action of nerve growth factor. *Annu. Rev. Biochem. 51*: 845–868.

Original papers

DENERVATION AND ACh RECEPTORS

Axelsson, J. and Thesleff, S. 1959. A study of supersensitivity in denervated mammalian skeletal muscle. *J. Physiol. 147*: 178–193.

Brenner, H. R. and Sakmann, B. 1983. Neurotrophic control of channel properties at neuromuscular synapses of rat muscle. *J. Physiol. 337*: 159–172.

Cohen, M. W. 1980. Development of an amphibian neuromuscular junction *in vivo* and in culture. *J. Exp. Biol. 89*: 43–56.

Hall, Z. W., Roisin, M.-P., Gu, Y. and Gorin, P. D. 1984. A developmental change in the immunological properties of acetylcholine receptors at the rat neuromuscular junction. *Cold Spring Harbor Symp. Quant. Biol., 48*: 101–108.

Kuffler, S. W., Dennis, M. J. and Harris, A. J. 1971. The development of chemosensitivity in extrasynaptic areas of the neuronal surface after denervation of parasympathetic ganglion cells in the heart of the frog. *Proc. R. Soc. Lond. B 167*: 555–563.

Lømo, T. and Rosenthal, J. 1972. Control of ACh sensitivity by muscle activity in the rat. *J. Physiol. 221*: 493–513.

REGENERATION, BASAL AND LAMINA, AND MOLECULAR MECHANISMS

Burden, S. J., Sargent, P. B. and McMahan, U. J. 1979. Acetylcholine receptors in regenerating muscle accumulate at original synaptic sites in the absence of the nerve. *J. Cell Biol. 82*: 412–425.

*Gunderson, R. W. and Barrett, J. N. 1979. Neuronal chemotaxis: Chick dorsal-root axons turn toward high concentrations of nerve growth factor. *Science 206*: 1079–1080.

Jessell, J. M., Siegel, R. E. and Fischbach, G. D. 1979. Induction of acetylcholine receptors on cultured skeletal muscle by a factor extracted from brain and spinal cord. *Proc. Natl. Acad. Sci. USA 76*: 5397–5401.

Miledi, R. and Sumikawa, K. 1982. Synthesis of cat muscle acetylcholine receptors by *Xenopus* oocytes. *Biomed. Res. 3*: 390–399.

Rutishauser, U., Grumet, M. and Edelman, G. M. 1983. Neural cell adhesion molecule mediates initial interactions between spinal cord neurons and muscle cells in culture. *J. Cell Biol. 97*: 145–152.

*Sanes, J. R., Marshall, L. M. and McMahan, U. J. 1978. Reinnervation of muscle fiber basal lamina after removal of muscle fibers. *J. Cell Biol. 78*: 176–198.

Skene, J. H. P. and Willard, M. 1981. Characteristics of growth-associated polypeptides in regenerating toad retinal ganglion cell axons. *J. Neurosci. 1*: 419–426.

CELL DEATH AND POLYNEURONAL INNERVATION

Betz, W. J., Caldwell, J. H. and Ribchester, R. R. 1980. The effects of partial denervation at birth on the development of muscle fibres and motor units in rat lumbrical muscle. *J. Physiol. 303*: 265–279.

*Brown, M. C., Jansen, J. K. S. and Van Essen, D. 1976. Polyneuronal innervation of skeletal muscle in new-born rats and its elimination during maturation. *J. Physiol. 261*: 387–422.

Hollyday, M. and Hamburger, V. 1976. Reduction of the naturally occurring motor neuron loss by enlargement of the periphery. *J. Comp. Neurol. 170*: 311–320.

*Pilar, G., Landmesser, L. and Burstein, L. 1980. Competition for survival among developing ciliary ganglion cells. *J. Neurophysiol. 41*: 233–254.

Thompson, W. 1983. Synapse elimination in neonatal rat muscle is sensitive to pattern of muscle use. *Nature 302*: 614–616.

DEVELOPMENT, SPROUTING AND SYNAPSE FORMATION

Benfey, M. and Aguayo, A. J. 1982. Extensive elongation of axons from rat brain into peripheral nerve grafts. *Nature 296*: 150–152.

David, S. and Aguayo, A. J. 1981. Axonal elongation into peripheral nervous system "Bridges" after central nervous system injury in adult rats. *Science. 214*: 931–933.

Lance-Jones, C. and Landmesser, L. 1981. Pathway selection by embryonic chick motoneurons in an experimentally altered environment. *Proc. R. Soc. Lond. B 214*: 19–52.

*Le Douarin, N. M. 1980. The ontogeny of the neural crest in avian embryo chimaeras. *Nature 286*: 663–669.

Rotshenker, S. 1979. Synapse formation in intact innervated cutaneous-pectoris muscles of the frog following denervation of the opposite muscle. *J. Physiol. 292*: 535–547.

Scott, S. A. and Muller, K. J. 1980. Synapse regeneration and signals for directed growth in the central nervous system of the leech. *Dev. Biol. 80*: 345–363.

GENETIC AND ENVIRONMENTAL INFLUENCES IN THE MAMMALIAN VISUAL SYSTEM

CHAPTER TWENTY

An approach has been made in the mammalian visual system to the questions of the relative importance of genetic factors and of the environment for the establishment and proper performance of synaptic interactions. In newborn, visually naive kittens and monkeys, many features of the neuronal organization are already present. Cells in the retina and the lateral geniculate nucleus respond to stimuli in much the same way as those in adult animals. In the visual cortex, the neurons have characteristic receptive fields and require oriented bars or edges. Certain differences are, however, apparent, particularly in layer IV, where geniculate fibers end. At the time of birth in kittens and monkeys, the ocular dominance columns are not fully formed: Cells in layer IV are driven by both eyes. The adult pattern—in which cells in layer IV are supplied by one or the other, but not both, eyes—is established in the first six weeks after birth. During this time geniculate fibers retract to form columns with clear boundaries.

In early life the connections of neurons in the visual cortex are susceptible to change and can be irreversibly affected by inappropriate use. For example, closure of the lids of one eye during the first three months of life leads to blindness in that eye. The abnormality occurs chiefly at the level of the cortex; although geniculate cells are still driven by the eye that had been closed, the great majority of cortical cells are not. The other eye develops and functions normally. The period of greatest susceptibility in cats and monkeys occurs during the first six weeks of life. Lid closure in adult animals has no effect.

Deprivation during the first six weeks leads to a shrinkage of the cortical dominance columns supplied by that eye. A corresponding increase is seen in the width of columns supplied by the normal eye. Thus, when retraction of geniculate inputs to layer IV occurs, columns that are normally of equal width become unequal after deprivation: Fibers supplied by the deprived eye retract more than usual, whereas those supplied by the normal eye hardly retract at all. During the critical period these effects can be reversed by opening the sutured eye and closing the undeprived eye. That competition between the two eyes for cortical territory occurs is further suggested

531

by the effects of bilateral deprivation. In monkeys with both eyes closed or enucleated, neither eye has an advantage and the columns develop normally. Each cell in the cortex is driven by only one or the other eye and not both.

Abnormal sensory input also leads to similar effects attributable to competition. When a squint, or strabismus, is produced by cutting extraocular muscles, each eye is exposed to the normal amount of visual input and only the fixation of the two eyes upon objects is altered. Yet the way cortical cells are driven by the two eyes is changed in kittens or immature monkeys that have been made to squint. The cells have normal receptive field properties, but only a few are driven by both eyes: Instead, one eye or the other eye is effective on its own. Since there is no disuse, it again appears that impulse traffic in convergent pathways must continue in an appropriately balanced manner for the normal functional organization to be maintained. This idea is reinforced by experiments in which all impulses, including spontaneous activity, were blocked in kittens by the use of tetrodotoxin injected into both eyes. Columns then failed to develop altogether, the areas in layer IV supplied by each eye remaining coextensive as at birth.

These results have a wide significance for considering a variety of aspects of development in the central nervous system. Other sensory systems and higher functions may also have critical periods during which their performance can be sharpened by appropriate use or irreversibly damaged by disuse or inappropriate use. Genetically determined errors of connections also occur. In the Siamese cat, some optic nerve fibers take the wrong pathway during development and become connected to inappropriate cells in the lateral geniculate nucleus and cortex.

We have emphasized repeatedly the constancy of the wiring that is necessary for the nervous system to function properly. It is also clear that development continues after birth for various periods in different animals; at the cellular level, we do not know whether synaptic connections are immutably fixed, even in the adult. For example, kittens are born with their eyes closed. If the lids are opened and light is shone into an eye, the pupil constricts, although the animal had not previously been exposed to light and appears to be completely blind. By ten days, the kitten shows evidence of vision and thereafter begins to recognize objects and patterns. When kittens are brought up in darkness instead of in their normal environment, they become blind, but the pupillary reflex continues to function.[1] It is as though there were a hierarchy of susceptibility, with "hard" and "soft" wiring in different parts of the brain.

Behavioral experiments support the view that simpler, more basic reflexes resist change and are altered only by drastic experimental pro-

[1]Riesen, A. H. and Aarons, L. 1959. *J. Comp. Physiol. Psychol. 52*: 142–149.

533
GENETIC AND
ENVIRONMENTAL
INFLUENCES IN
THE MAMMALIAN
VISUAL SYSTEM

cedures such as lesions; in contrast, higher functions may fail to develop, are far more susceptible, and can be modified by subtle changes in the environment.[2]

Changes in the performance of the nervous system raise a number of questions. What are the relative contributions of genetic factors and experience (summed up by the phrase "nature and nurture")? To what extent are the neuronal circuits required for vision already present and ready to work at birth? What effect has light falling into the eyes on their development? Does a kitten or a monkey brought up in darkness become blind because the connections fail to develop, or because some of the connections that had originally been there have withered away? The visual system offers great advantages for approaching directly questions such as these because the relay stations are accessible and the background of light and natural stimulation can readily be altered.

As in previous parts of this book, we have chosen to describe one system without presenting a full review of the field of sensory deprivation. And within the visual system we again emphasize the pathways from retina to cortex in cat and monkey. For our purposes it is convenient to focus largely on work that follows logically from the material presented in Part One.

THE VISUAL SYSTEM IN NEWBORN KITTENS AND MONKEYS

A good deal is known about the organization of the visual connections that underlie perception in the adult cat and monkey (Part One). Thus, a simple cell in the cat or monkey cortex selectively "recognizes" one well-defined type of visual stimulus, such as a narrow bar of light, oriented vertically, in a particular region of the visual field of either eye. It is natural to wonder whether cells of this type are already present in the newborn animal or whether visual experience or learning is required so that a random set of preexisting connections is reformed or at least modified for such a specific task. For technical reasons it is difficult to record from cells in newborn animals. Most experiments in kittens were made during the first three weeks after birth. To prevent form vision, the lids were sutured or the cornea was covered by a translucent occluder.[3] Similarly, visually naive monkeys were produced by suturing the animals' lids immediately or several days after birth; in some instances, monkeys were delivered by Caesarean section for later examination, care being taken to avoid exposure to light.[4]

A newborn monkey appears visually alert and is able to fixate. In contrast, a newborn kitten whose lids have been opened by surgery is behaviorally blind. Nevertheless, many of the features seen in adults are already present in the performance of cortical neurons in both animals. For example, recordings made from individual cells in area 17

---

[2]Held, R. and Bauer, J. A. 1974. *Brain Res. 71*: 265–271.
[3]Hubel, D. H. and Wiesel, T. N. 1963. *J. Neurophysiol. 26*: 994–1002.
[4]Wiesel, T. N. and Hubel, D. H. 1974. *J. Comp. Neurol. 158*: 307–318.

of the visual cortex show that the cells are not driven by diffuse illumination of the eyes. As in a mature animal, they fire best when light or dark bars with a particular orientation are shone onto a particular region of the retina of either eye. The responses of the receptive fields are also organized into antagonistic "on" and "off" areas that are similar in the two eyes. Moreover, with oblique penetrations the preferred orientation changes in a regular sequence as the electrode moves through the cortex (Figure 1).[4] In many instances the "tuning" curve or range of orientations in animals lacking prior visual experience cannot be distinguished from that in adults.[5] In newborn animals, however, the discharges of cortical cells tend to be weaker than those of the adult and some cells are unresponsive.[6]

A major difference, which has turned out to be of key importance in assessing the changes that occur during normal and abnormal development, is apparent in layer IV of the visual cortex. Unlike the pattern in the adult, where the ocular dominance columns corresponding to the inputs from the two eyes can be clearly distinguished, in the newly born kitten or monkey, extensive overlap occurs in the arborizations of geniculate fibers ending in layer IV.[7,8] This is shown in Figure

[5]Sherk, H. and Stryker, M. P. 1976. *J. Neurophysiol. 39*: 63–70.
[6]Blakemore, C. 1974. *Br. Med. Bull. 30*: 152–157.
[7]LeVay, S., Wiesel, T. N. and Hubel, D. H. 1980. *J. Comp. Neurol. 191*: 1–51.
[8]LeVay, S., Stryker, M. P. and Shatz, C. J. 1978. *J. Comp. Neurol. 179*: 223–244.

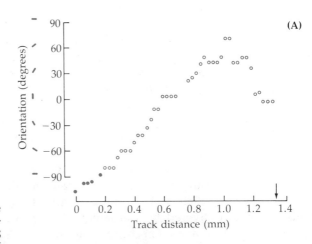

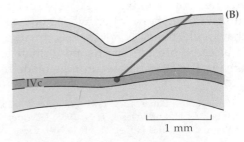

1

AXIS ORIENTATION COLUMNS in the absence of visual experience. **(A)** Axis orientation of receptive fields encountered by an electrode during an oblique penetration through the right cortex of a 17-day-old baby monkey whose eyes had been sutured closed on the second day after birth. **(B)** The circle at L marks a lesion made at the end of the electrode track in layer IV. The orientation of receptive fields in (A) changes progressively as orientation columns are traversed, indicating that normally organized orientation columns are present in the visually naive animal. Closed circles are from the ipsilateral eye; open circles are from the other eye. (B from Wiesel and Hubel, 1974.)

2A and schematically in Figure 3. Only a hint of ocular dominance can be resolved; the reason is that the individual fibers of the geniculate nucleus spread over a wide area in layer IV. During the first six weeks of the animal's life, axons retract, establishing more restricted and separate domains that become supplied exclusively by one eye or the other. In parallel, physiological changes develop: Initially, simple cortical cells of layer IV are driven by both eyes. By six weeks, as in the adult, each cell can be driven by only one eye. The postnatal development of ocular columns proceeds normally in animals reared in total darkness.[7]

That the sorting out of columns actually begins before birth and without visual experience was demonstrated by Rakič.[9] He injected the eyes of monkeys in utero with radioactive amino acids at different stages; this procedure made it possible to observe the outgrowth of fibers and their distribution in the lateral geniculate nucleus and layer IV of the cortex. At early stages the overlap of the territories supplied by the two eyes is virtually complete; by a few days before birth, hints of columnar organization are discernible. Similarly, in fetal cats it has been shown by Shatz[10] that the optic nerve fiber terminals overlap as

[9]Rakič, P. 1977b. *Phil. Trans. R. Soc. Lond. B 278*: 245–260.
[10]Shatz, C. J. 1983. *J. Neurosci. 3*: 482–499.

535
GENETIC AND
ENVIRONMENTAL
INFLUENCES IN
THE MAMMALIAN
VISUAL SYSTEM

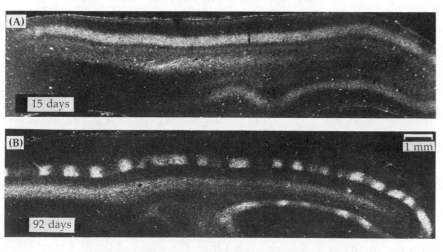

2 DEVELOPMENT OF OCULAR DOMINANCE COLUMNS in layer IV of cat cortex. Autoradiographs of horizontal sections of the ipsilateral and contralateral visual cortex of animals in which one eye had been previously injected with tritium-labeled proline. Photos taken in dark-field illumination show silver grains as white dots. (A) Brain of 15-day-old kitten. Layer IV on both sides is labeled continuously and uniformly with no evidence of columns. At this stage and for the next few weeks, fibers ending in the cortex from the geniculate overlap in layer IV. (B) Similar section through the visual cortex of an adult cat (92 days). Note the patchy distribution of label in layer IV, corresponding to ocular dominance columns for the injected eye. (After LeVay, Stryker, and Shatz, 1978.)

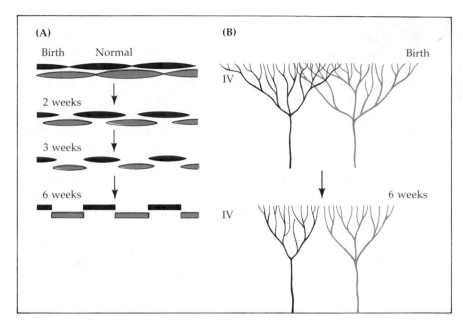

3 RETRACTION OF LATERAL GENICULATE NUCLEUS AXONS ending in layer IV of cortex during first 6 weeks of life. Diagrams on left and right show schematically the overlap present at birth and the subsequent segregation into separate clusters corresponding to ocular dominance columns. (Modified from Hubel and Wiesel, 1977.)

they reach the lateral geniculate nucleus and that the sorting out into separate layers (A, A₁, and C) begins during the last third of gestation and forms the full adult pattern about two weeks after birth. Ocular dominance columns only become recognizable at about 30 days after the birth of the cat.[8] The kitten is therefore born at a more immature stage of development than the monkey.

In newly born kittens and monkeys, as in adults, cortical cells outside layer IV tend to be driven by both eyes, some better by one eye, some by the other, and some equally well by both. About 20 percent of all the 1116 cells in Figure 4 from visually normal monkeys are driven solely by one eye and about the same percentage by the other. The degree of dominance can be conveniently expressed in a histogram by grouping neurons into seven categories according to the discharge frequency with which they respond to stimulation of one or the other eye. In the visually inexperienced kitten and the two-day-old monkey, the histogram appears rather normal, the majority of cells responding to appropriate illumination of either eye.

These findings in immature animals come as no great surprise. Although the development of the cortex is susceptible to environmental changes, one would be rather surprised if the basic outline of neural organization with its orderly and intricate visual connections were shaped by the vagaries of visual environment. The principal point to

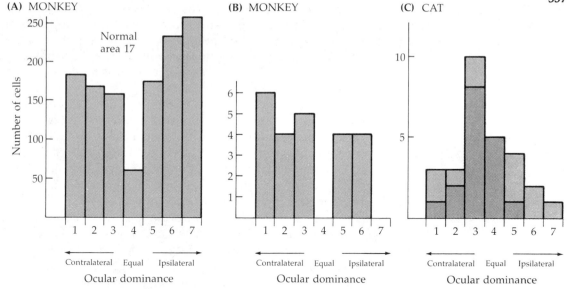

4 OCULAR DOMINANCE DISTRIBUTION in the visual cortex of normal monkeys. (A) Cells in groups 1 and 7 of the histogram are driven by one eye only (ipsilateral or contralateral). All other cells have input from both eyes. In groups 2,3 and 5,6 one eye predominates; in group 4 both have a roughly equal influence. (B) A similar ocular dominance distribution in a 2-day-old monkey. (C) Histogram of ocular dominance distribution in a 20-day-old kitten (light color) with normal visual experience and two kittens (8 and 16 days old) without previous visual exposure (dark color). (A, B from Wiesel and Hubel, 1974; C from Hubel and Wiesel, 1963b.)

be made here is that certain features of the basic wiring are already established at birth, whereas others become fully developed only in the first few weeks of life. This is reminiscent of the events occurring during the development of nerve–muscle synapses in neonatal rats: At birth, each motor end plate is supplied by numerous motoneurons, but in a few weeks, most retract, leaving each muscle fiber supplied by just one axon (Chapter 19).

In the remainder of this chapter we shall give an account of the ways in which abnormal sensory experience in early life can drastically affect the anatomy and physiology of the brain. However, abnormalities also occur as a result of genetic defects. Color blindness is one familiar example and others are provided by mutations occurring in Siamese cats and in mice, which we shall describe before dealing with sensory deprivation.

In Siamese cats, certain optic nerve fibers fail to grow along their usual pathways during development.[11] An outward sign of this disorder is that the animals are frequently cross-eyed. The defect arises from

Abnormal connections in the visual system of the Siamese cat and the "reeler" mouse

[11]Shatz, C. J. and Kliot, M. J. 1982. *Nature 300*: 525–529.

abnormal crossing by optic nerve fibers at the chiasm.[12] As a result, the lateral geniculate nucleus receives a disproportionately large input from the contralateral eye and a correspondingly diminished input from the ipsilateral eye. Interestingly, the optic nerve fibers that have taken a wrong course terminate in an unusual region of the lateral geniculate nucleus, the cells there receiving information from unaccustomed regions of the visual field and from the wrong eye. Farther on in the pathways, specific regions of the visual cortex receive an orderly representation of the abnormal portion of the visual field (by way of the incorrectly crossed optic nerve fibers relaying through the geniculate nucleus).[13,14] Complete scrambling of connections does not occur because the ground rules for normal cell connections are still followed— as a result the abnormal input is segregated from the normal.

The genetic aspects of the defect in development, which is shared by albinos of many species, are of considerable interest. They have been explored in detailed systematic studies by Guillery and his colleagues, who examined a large variety of animals, including rats, minks, mice, guinea pigs, Himalayan rabbits, ferrets, monkeys, and even one white tiger.[15] In all these species, the neurological defect goes hand in hand with a pigment deficit. A direct connection can be made between the defect and the albino gene. The absence of pigment in the eye is somehow associated with the abnormal course taken by the optic nerve fibers of certain ganglion cells. In view of this, it has been suggested that melanin itself or another related gene product of the pigment epithelial cell may influence the fate of retinal ganglion cell axons as they cross at the chiasm. Here, then, is an example of an intriguing genetic defect involving in its most obvious form the color of an animal but also causing errors in connections and a modification of specificity.

Another mutation of the visual system that has been studied in detail is the cortex of a mutant mouse known as "reeler." This provides an example of a genetic abnormality in which specific cells occupy aberrant positions in the brain yet appear to receive their normal connections. As its name suggests, the animal staggers and reels like a drunken sailor. The mutation involves laminated structures such as the cortex, the cerebellum, and the hippocampus. In the cortex, the normal pattern of layers is disrupted.[16] All the major classes of cells are still present, but the borders between layers are no longer distinct. Characteristic cell types are situated at the wrong depth. For example, the small pyramidal cells normally found in layers II and III are scattered and occur predominantly in deeper layers of the reeler cortex (Chapter

[12]Guillery, R. W. and Kaas, J. H. 1971. *J. Comp. Neurol. 143*: 73–100.
[13]Hubel, D. H. and Wiesel, T. N. 1971. *J. Physiol. 218*: 33–62.
[14]Shatz, C. J. 1979. *Soc. Neurosci. Symp. 4*: 121–141.
[15]Guillery, R. W. 1974. *Sci. Am. 230*: 44–54.
[16]Caviness, V. S. 1976. *J. Comp. Neurol. 170*: 435–448.

539
GENETIC AND
ENVIRONMENTAL
INFLUENCES IN
THE MAMMALIAN
VISUAL SYSTEM

3). Dräger,[17] Pearlman,[18,19] and their colleagues have recorded from various identified cortical neurons in normal and mutant mice. The identity of a cell type can be established by knowing its projection. (In Chapter 3, we saw that complex cells in layer V project to the colliculi; those in layers II and III to other cortical areas.) Surprisingly, perhaps, although cells projecting to specific areas are displaced, their connections are similar to those in the normal animal. With horseradish peroxidase labeling, the projections to the lateral geniculate nucleus, to the colliculus, and to other cortical areas appear normal. Moreover, the receptive field organization of simple and complex cells shows no major changes. Thus, in this animal, despite the genetic anomaly giving rise to incorrect positions of neurons during development, functionally appropriate connections can still be made in the cortex.

EFFECTS OF ABNORMAL EXPERIENCE

This section describes three types of experiments, mostly by Hubel and Wiesel, in which animals were deprived of normal visual stimuli. They studied the effects on the physiological responses of nerve cells in the visual system after (1) closing the lids of one or both eyes; (2) preventing form vision, but not access of light to the eye; and (3) leaving light and form vision intact, but producing an artificial strabismus (squint) in one eye. These procedures cause abnormalities in function; in some instances, the underlying anatomical changes have been demonstrated.

When the lids of one eye were sutured during the first two weeks of life, kittens and monkeys still developed normally and used their unoperated eye. However, at the end of one to three months, when the operated eye was opened and the normal one closed, it was clear that the animals were practically blind in the operated eye. For example, kittens would bump into objects and fall off tables.[20] There was no gross evidence of a defect within such eyes; pupillary reflexes appeared normal and so did the electroretinogram, which serves as an index of the average electrical activity of the eye (Chapter 13). Records made from retinal ganglion cells in deprived animals showed no obvious changes in their responses,[21] and their receptive fields appeared normal. Responses of cells in the lateral geniculate appeared relatively unchanged.[22] There were, however, striking changes in the responses of cortical cells.

Cortical cells after monocular deprivation

[17]Dräger, U. C. 1981. *J. Comp. Neurol. 201*: 555–570.
[18]Simmons, P. A., Lemmon, V. and Pearlman, A. L. 1982. *J. Comp. Neurol. 211*: 295–308.
[19]Lemmon, V. and Pearlman, A. L. 1981. *J. Neurosci. 1*: 83–93.
[20]Wiesel, T. N. and Hubel, D. H. 1963b. *J. Neurophysiol. 26*: 1003–1017.
[21]Sherman, S. M. and Stone, J. 1973. *Brain Res. 60*: 224–230.
[22]Sherman, S. M. and Spear, P. D. 1982. *Physiol. Rev. 62*: 738–855.

Relative importance of
diffuse light and form
for maintaining
normal responses

When electrical recordings were made in the visual cortex, only a few of the cells could be driven by the eye that had been closed. Figure 5 shows the ocular dominance histograms obtained from the cells examined in kittens and monkeys raised with closure of one eye during the first weeks of life. Only a few cells responded to stimulation of the deprived eye and many of those had abnormal fields.[7,20,23]

The results described so far indicate that if one eye is not used normally in the first weeks of life, its power wanes and it ceases to be effective. These far-reaching changes are produced by the relatively minor procedure of sewing the lids, without cutting any nerves. What is the important condition for maintaining and developing proper visual responses? Is diffuse light adequate? Lid closure reduces light that reaches the retina but does not exclude it. One would suspect, therefore,

[23]Wiesel, T. N. and Hubel, D. H. 1965a. *J. Neurophysiol. 28*: 1029–1040.

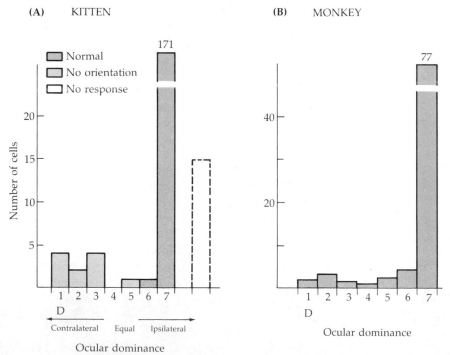

(A)   KITTEN   (B)   MONKEY

5   **DAMAGE PRODUCED BY CLOSURE OF ONE EYE. Ocular dominance distribution in kittens and in a monkey. (A) In 8- to 14-week-old kittens deprived of vision in the right eye, only 13 out of 199 cells responded to stimulation of the deprived eye (D). All but one of those cells had abnormal receptive fields. The column with interrupted lines represents cells that did not respond to either eye. (B) Ocular dominance distribution in a monkey whose right eye was closed from 21 to 30 days of age. In spite of a subsequent 4 years of binocular vision, most cortical neurons were unresponsive to stimulation of the right eye. (A from Hubel and Wiesel, 1965a; B from LeVay, Hubel, and Wiesel, 1980.)**

541
GENETIC AND
ENVIRONMENTAL
INFLUENCES IN
THE MAMMALIAN
VISUAL SYSTEM

that diffuse light alone would not keep an eye functioning normally. To test this idea, a series of experiments were made in which a plastic occluder (like frosted glass or a ping-pong ball) was placed over the cornea of a newborn kitten instead of closing eyelids; the occluder prevented form vision but did not exclude light by more than two log units.[20] In one experiment, a thin sheet of opaque tissue—the nictitating membrane—was sewn across the eye, a procedure causing blurred vision with even less attenuation of the light. All these cats were still functionally blind in the deprived eye. Furthermore, cortical cells were no longer driven by the deprived eye. Neither retinal nor geniculate responses were noticeably changed under such conditions. Thus, *form vision*, rather than the presence of light, is an important stimulus required to prevent abnormal development of cortical connections. Surprisingly, however, even form vision alone may not be adequate to maintain complete normality (see later).

Cells in the lateral geniculate nucleus of cat and of monkey are arranged in layers, each predominantly supplied by one or the other eye (Chapter 2). In the same animals that showed marked abnormalities in the cortex after lid closure, the geniculate cells seemed to behave normally. The cells in the appropriate layers responded with "on" or "off" discharges to small spots of light shone into the deprived or the normal eye and no clear-cut differences from normal firing patterns could be observed. Nevertheless, it was shown that marked changes in morphology occurred after lid closure of an eye: The cells were noticeably smaller in the layers supplied by the deprived eye.[24] The cell bodies were only about one-half as large as those in the normal layers, the reduction in size depending on the duration of lid closure. It seems surprising that cells in the lateral geniculate showed relatively obvious morphological changes but little significant physiological deficit. This may be due in part to a selective disappearance of one particular cell type, namely, the larger Y or fast-adapting neurons[22,25] (Chapter 2). A key factor controlling the size of cells may be the extent of their arborization in the cortex.[26]

Morphological changes in the lateral geniculate nucleus after visual deprivation

The morphological consequences of eye closure are particularly conspicuous in layer IV of the striate cortex where geniculate fibers terminate.[7,27,28] In monkeys, striking changes in ocular dominance columns develop after removal of one eye at birth or after closure of one eye. These have been revealed by the technique of autoradiography, which enables one to follow the axonal flow along visual pathways of radioactive materials injected into the eye (Chapter 3). After deprivation there occurs a marked reduction in width of the ocular dominance columns

Morphological changes in the cortex after visual deprivation

[24]Wiesel, T. N. and Hubel, D. H. 1963a. *J. Neurophysiol. 26*: 978–993.

[25]LeVay, S. and Ferster, D. 1977. *J. Comp. Neurol. 172*: 563–584.

[26]Guillery, R. W. and Stelzner, D. J. 1970. *J. Comp. Neurol. 139*: 413–422.

[27]Hubel, D. H., Wiesel, T. N. and LeVay, S. 1977. *Phil. Trans. R. Soc. Lond. B 278*: 377–409.

[28]Shatz, C. J. and Stryker, M. P. 1978. *J. Physiol. 281*: 267–283.

that receive their input from the eye that had been either removed or occluded. At the same time, the columns with input from the normal eye show a corresponding increase in width compared to that seen in normal adult monkeys. The expansion and shrinkage of occular dominance columns is evident in Figure 6 in which the normal columns can be compared with columns in animals in which one eye had been closed at 2 weeks for 18 months. The changes indicate that geniculate axons activated by the normal eye retained the territory in the cortex lost by their weaker, visually deprived neighbors. These results have also been confirmed physiologically by recording from layer IV where the geniculate fibers terminate. Almost all cells were driven by the eye that had not been deprived.

When the lids of one eye are closed in an adult cat or monkey, none of the abnormal consequences are seen. For example, in an adult animal, even if an eye is closed for over a year the cells in the cortex continue to be driven normally by both eyes and display the normal ocular dominance histogram. Moreover, even if one eye is completely enucleated in an adult monkey, the structure of layer IV remains normal when observed with autoradiography or the reduced silver staining

Critical period of susceptibility to lid closure

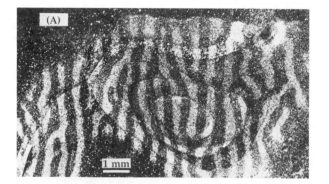

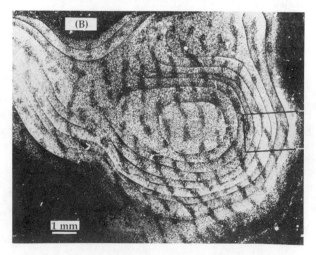

6

OCULAR DOMINANCE COLUMNS AFTER CLOSURE OF ONE EYE. (A) Normal control, adult rhesus monkey. The right eye had been injected with a radioactive proline-fucose mixture 10 days before a section was made tangential to the exposed dome-shaped primary visual cortex of the right hemisphere. The section passes through layer V, which is seen as a dark oval area near the center of the figure; just outside this is layer IV, seen as a ring of fingerlike alternating dark and light processes. With dark-field illumination, the radioactivity in the geniculate axon terminals in layer IV appears as fine white granules, forming the light stripes (columns) that correspond to the injected eye. The dark intervening bands correspond to the other eye. Roughly nine sets of stripes are shown. (B) A similar section in an 18-month-old monkey whose right eye had been closed at the age of 2 weeks. Proline-fucose injected into normal left eye. Here the plane of section grazes layer IV, which is seen as an oval. The white label again demonstrates the columns whose input is derived from the intact eye. The columns, however, are larger than normal and alternate with narrowed columns, seen as dark gaps, supplied by the eye whose lids had been closed. (From Hubel, Wiesel, and LeVay, 1966.)

543
GENETIC AND
ENVIRONMENTAL
INFLUENCES IN
THE MAMMALIAN
VISUAL SYSTEM

method.[7] This finding indicates a remarkable resistance to change of layer IV in the adult animal when compared to the changes seen in the immature animal, since enucleation leads to considerable degeneration in the lateral geniculate nucleus in adults.

The period during which susceptibility in kittens is highest has been narrowed down to the fourth and fifth weeks after birth. During the first three weeks or so of life, eye closure has little effect. This is not surprising since the kittens' eyes are normally closed at first. But, abruptly during weeks 4 and 5, sensitivity increases. Closure at that age for as little as three to four days leads to a sharp decline in the number of cells that can be driven by the deprived eye.[29] An experiment in which littermates are compared is shown in Figure 7. In this example, 6- and 8-day closures starting at the age of 23 and 30 days (Figure 7A, B) caused about as great an effect as three months of monocular deprivation from birth. The susceptibility to lid closure declines after the critical period has passed and eventually disappears by about three months of age (Figure 7C and D). The critical period can, however, be prolonged to more than six months by rearing kittens in the dark. In the absence of visual experience, the susceptibility to monocular closure

[29]Hubel, D. H. and Wiesel, T. N. 1970. *J. Physiol. 206*: 419–436.

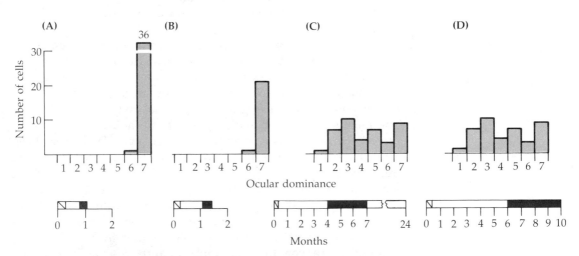

7 CRITICAL PERIOD in kittens. Histograms of ocular dominance distribution in the visual cortex in kittens that are littermates. (A) The right eye was sutured shut for 6 days (age 23 to 29 days). (B) The right eye was closed from age 30 to 39 days. In each animal only one cell was weakly influenced by the temporarily deprived eye (∗). The damage was about as great as eye closure for 3 months or longer. (C) The right eye was open for the first 4 months, then sutured for 3 months, then opened again. Recordings were made at the age of 2 years. (D) The eye was open for the first 6 months, then closed for 4 months. Ocular dominance was determined at 10 months of age. Ocular dominance distribution appeared normal for both eyes. The black segment below the abscissa indicates the closure period. (From Hubel and Wiesel, 1970.)

can still be demonstrated at these late times.[30] There is evidence that even a brief exposure of the kitten to light for a few hours may be sufficient to prevent such extension of the critical period.[31]

In monkeys, the greatest sensitivity to lid closure was during the first six weeks.[7,27] Before six weeks, substantial changes in eye preference and columnar architecture developed if one eye was closed for a few days. During the subsequent months (up to about 12 to 18 months), several weeks of closure were required to produce obvious changes in ocular dominance histograms or the width of columns in layer IV. At later times, no changes could be produced even by enucleation of one eye.

The susceptibility of kittens and monkeys during early life is reminiscent of some clinical observations made in man. It has long been known that removal of a clouded or opaque lens (cataract) can lead to a restoration of vision, even though the patient has been blind for many years. In contrast, a cataract that develops in a baby can lead to blindness without the possibility of recovery unless the operation is performed very early in the critical period. A familiar clinical procedure used in the past for the treatment of children with strabismus (or squint) was to patch the good eye for prolonged periods in order to encourage the weaker eye to be used. There is evidence that damage in acuity may result, depending on the child's age at the time and the duration of the patching. Clinical observations suggest that the greatest sensitivity occurs in babies during the first year but that the critical period may persist for several years.[32]

Recovery

To what extent is recovery possible after lid closure? Even if the deprived eye in cat or monkey is subsequently opened for months or years, the damage remains permanent, with little or no recovery: The animal continues to be blind in that eye, with shrunken columns and skewed ocular dominance histograms. In animals with monocular closure, experiments have been made in which the lids were opened in the deprived eye and closed over the normal eye. This procedure, termed "reverse suture" leads to a dramatic recovery of vision provided it is carried out during the critical period. Kittens[33] and monkeys[7] not only begin to see again with the initial deprived eye, but they become blind in the other eye. Accompanying these changes, the ocular dominance histograms switch, so that the newly opened eye drives most cells, while the eye that had been opened for the first weeks (now closed) cannot. Moreover, the anatomical pattern in layer IV revealed by autoradiography shows a similar switch in preference: The shrunken regions supplied by the initially closed eye expand at the expense of the other eye. Figure 8 shows recordings and autoradiographs of the cortex in a monkey in which the right eye had been closed at two days

[30]Cynader, M. and Mitchell, D. E. 1980. *J. Neurophysiol.* 43: 1026–1040.
[31]Mower, G. D., Christen, W. G. and Caplan, C. J. 1983. *Science* 221: 178–180.
[32]Jacobson, S. G., Mohindra, J. and Held, R. 1981. *Br. J. Ophthalmol.* 65: 727–735.
[33]Blakemore, C. and Van Sluyters, R. C. 1974. *J. Physiol.* 237: 195–216.

545
GENETIC AND
ENVIRONMENTAL
INFLUENCES IN
THE MAMMALIAN
VISUAL SYSTEM

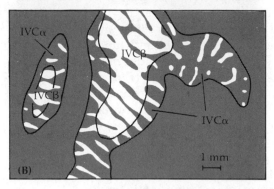

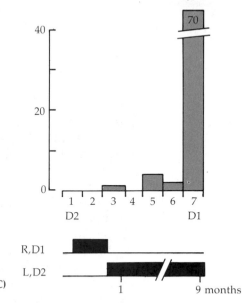

8   EFFECTS OF REVERSE SUTURE ON OCULAR DOMINANCE in monkey. At 2 days of age, the right eye was closed. At 3 weeks (19 days
later), the right eye was opened and the left closed. The left eye was
then kept closed for 9 months, at which time the right (initially deprived) eye was injected with tritium-labeled proline. (A) Tangential
section of cortex through layers IVCα and IVCβ (layer IVCα is superficial to IVCβ). In dark-field illumination, silver grains appear white.
The bands labeled by the right eye are expanded in layer IVCβ even
though it had been deprived of light for 19 days. (B) Diagram to show
expansion of territory supplied by the right, initially deprived eye in
layer IVCβ. In IVCα, the distribution is different: The territory supplied by the left eye is expanded and that by the right eye is reduced.
Note that the bands in IVCα and IVCβ are still in precise register as in
normal monkeys. (C) Ocular dominance histogram from the same monkey. Almost all cells are driven exclusively by the right eye (D1), virtually none by the left (D2). Had both eyes been open at 3 weeks, the
histogram and the arrangement in layer IVCβ would be reversed; almost
none of the cells would then be driven by the right eye. Accordingly,
fibers driven by the right eye have sprouted to recapture territory they
had previously lost. (After LeVay, Hubel, and Wiesel, 1980.)

for three weeks and then opened with the left eye closed for the next
eight months. Nearly all neurons responded only to the initially deprived right eye, and the areas of cortex supplied by it in layer IVCβ
had expanded.

The remarkable conclusions from these experiments are that (1)
during the critical period in a normal animal, geniculate fibers supplying
layer IV of the cortex retract so that each eye supplies areas of compar-

able extent; (2) lid closure of one eye during the critical period leads to unequal retraction; (3) reverse suture during the critical period produces *sprouting* of the geniculate axons so that an eye can recapture the cells it had lost. In adult animals, reverse suture is without effect. For example, in a monkey in which the reverse suture was performed at one year of age, the labeled columns for the initially deprived eye remained shrunken. Thus, though it is possible to cause change by lid closure at one year, connections that have already been changed once are not readily regained.

The concept of a well-defined, hard and fast critical period may be an oversimplification. Experiments on reverse suture in monkeys suggest that different layers of the striate cortex may develop at different rates, with the critical period over for one layer while an adjacent layer is still capable of being modified in structure and function. The biochemical mechanisms involved in maintaining and terminating the critical period are as yet unknown. There are, however, intriguing suggestions that the level of catecholamines, particularly norepinephrine, may be important. The effects of lid closure during the critical period can be prevented by injection of a neurotoxin—6-hydroxydopamine—into the ventricles or directly into the cortex: Ocular dominance columns mapped physiologically no longer show the shift from the deprived eye.[34,35] The prime action of 6-hydroxydopamine is on catecholamine neurons, leading to depletion of norepinephrine (Chapter 12). If norepinephrine is replaced locally after the 6-hydroxydopamine treatment, sensitivity to lid closure returns and cortical cells once again fail to be driven by the deprived eye. These results provide a new approach for studying chemical mechanisms that influence plasticity during development. At this time it is not yet clear what role the catecholamines play in the normal process of retraction and the establishment of fixed connections.

REQUIREMENT FOR MAINTENANCE OF FUNCTIONING CONNECTIONS IN THE VISUAL SYSTEM: THE ROLE OF COMPETITION

At this stage one might be tempted to conclude that loss of activity in the visual pathways is the main factor that tends to disrupt normal responses of cortical neurons. After all, cortical cells are driven not by diffuse illumination but by shapes and forms. The following discussion shows that some of the causes must be far more subtle. To maintain normal responses, even form vision is not enough. There must occur in addition a special interaction between the two eyes, an interaction we cannot yet explain.

[34]Kasamatsu, T., Pettigrew, J. D. and Ary, M. 1981. *J. Neurophysiol. 45*: 254–266.
[35]Daw, N. W., Rader, R. K., Robertson, T. W. and Ariel, M. 1983. *J. Neurosci. 3*: 907–914.

547
GENETIC AND
ENVIRONMENTAL
INFLUENCES IN
THE MAMMALIAN
VISUAL SYSTEM

Binocular lid closure

The first clue that loss of visually evoked activity cannot on its own account for the changed performance of neurons is shown by the following experiments. Both eyes were closed in monkeys, newborn or delivered by Caesarean section.[4,7] From the preceding discussion one would expect that cells in the cortex would subsequently be driven by neither eye. Surprisingly, however, after binocular closure for 17 days or longer, most cortical cells could still be driven by appropriate illumination; the receptive fields of simple and complex cells appeared largely normal. The columnar organization for orientation was similar to that in controls (Figure 1). The principal abnormality was that a substantial fraction of the cells could not be driven binocularly (Figure 9). In addition, some spontaneously active cells could not be driven at all, and others did not require specifically oriented stimuli. However, the areas of cortex supplied by each eye were equal and the pattern resembled that seen in normal or adult monkeys: In layer IV, cells were driven by one eye only and the columns were well defined when marked by autoradiography. Binocular closure in kittens leads to similar effects except that more cortical cells continue to be binocularly driven.[23] The conclusion one can draw from these experiments is that some, but not all, of the ill effects expected from closing one eye are reduced or averted by closing both eyes. Once more one might theorize that the two different pathways from the two eyes are somehow competing for representation in cortical cells, and with one eye closed the contest becomes unequal.

When the lids are closed or an animal is brought up in complete darkness, impulse traffic in the visual pathways does not stop entirely.

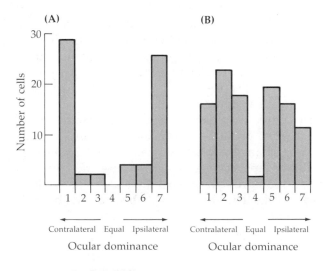

9

OCULAR DOMINANCE HISTOGRAMS after closure of both eyes at birth. (A) A monkey was delivered by Caesarean section and recordings were made at 30 days of age. In contrast to the effects of monocular deprivation, each of the two deprived eyes could drive cells in the visual cortex. The receptive fields appeared normal, except that relatively few cells were driven by both eyes. Black area at the bottom of the histogram indicates closure time. (B) Ocular dominance histogram from a normal 21-day-old monkey. (From Wiesel and Hubel, 1974.)

Neurons continue to fire spontaneously. Stryker[36] has shown that this presumably equal, low level of activity from the two eyes is important for normal development. Tetrodotoxin, which blocks sodium channels and prevents firing (Chapter 5), was injected into both eyes of newly born kittens. After removal of the toxin, the visual pathways from retina through the geniculate to the cortex conducted once again. The interesting result was that cells in layer IV were still driven by both eyes as in the newly born animal. In addition, the ocular dominance columns revealed by autoradiography resembled the neonatal pattern, with extensive overlap and no clear boundaries. Thus, in the absence of *all* firing, geniculate fibers failed to retract normally.

Effects of artificial squint

The abnormal effects described in the preceding discussion were produced by suturing eyelids or by using translucent diffusers, implicating loss of form vision. Following the clue that in children squint (strabismus) can produce severe loss of vision or blindness, Hubel and Wiesel produced artificial squint in cats and monkeys by cutting an eye muscle.[4,37,38] The optical axis of the eye is thereby deflected from normal. Under such conditions, illumination and pattern stimulation for each eye remain unchanged. The experiment at first seemed disappointing because after several months vision in both eyes of the operated kittens appeared normal; and Hubel and Wiesel were about to abandon a laborious set of experiments (personal communication). Nevertheless, they recorded from cortical cells and obtained the following surprising results. Individual cortical cells had normal receptive fields and responded briskly to precisely oriented stimuli. But almost every cell responded only to one eye; some were driven exclusively by the ipsilateral eye and others by the contralateral, but very few were driven equally well by both. The cells were, as usual, grouped in columns with respect to eye preference and field axis orientation. As expected, no atrophy occurred in the lateral geniculate body. Similar results were obtained in monkeys in which strabismus had been induced during the critical period. The almost complete lack of binocular representation on cortical cells is shown in a histogram (Figure 10) from a monkey with artificial squint. The critical period for squint to produce changes is comparable to that for monocular deprivation.[39]

Squint provides an example in which all the usual parameters of light are normal—the amount of illumination and form and pattern stimuli. The only apparent change consists in a failure of the images on the two retinas to fall on appropriate areas. Because the cortex in such an animal is rich in responsive cells and columns, it seems unlikely that the large percentage of cells that had originally been driven by both eyes could have dropped out.

[36]Stryker, M. P. 1981. *Soc. Neurosci. Abstr.* 7: 842.
[37]Hubel, D. H. and Wiesel, T. N. 1965b. *J. Neurophysiol.* 28: 1041–1059.
[38]Wiesel, T. N. 1982. *Nature 299*: 583–591.
[39]Van Sluyters, R. C. and Levitt, F. B. 1980. *J. Neurophysiol.* 43: 686–699.

549
GENETIC AND
ENVIRONMENTAL
INFLUENCES IN
THE MAMMALIAN
VISUAL SYSTEM

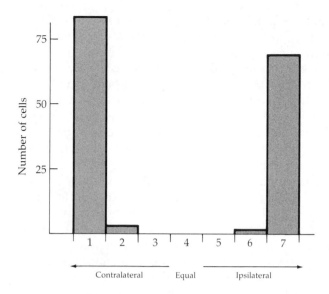

Number of cells

Contralateral   Equal   Ipsilateral

10

EFFECT OF SQUINT ON MONKEY ocular dominance. Histogram shows eye preference of cells in a 3-year-old-monkey in which strabismus had been produced by cutting one eye muscle at 3 weeks of age. Cells are driven by one eye or the other eye, but not both. These cells were grouped in a columnar fashion. (After Hubel and Wiesel, unpublished, in Wiesel, 1982.)

We have no detailed structural analysis of the possible mechanisms that may explain the loss of binocular connections with squint. The factor that seems important for maintaining the normal effectiveness of connections from the lateral geniculate body to the cortex is some form of congruity of input from the eyes. It is as though the homologous receptive fields in both eyes must be in register with, and superimposable on, each other, and excitation must be simultaneous. The following experiments further support this idea. During the first three months or longer, eyes of a kitten were occluded with one plastic occluder that was switched on alternate days from one eye to the other, so that the two eyes received the same total experience, but at different times.[37] Once again the result was the same as in the squint experiment: Cells were driven predominantly by either one eye or the other, but not by both. The maintenance of normal binocularity, therefore, depends not only on the amount of impulse traffic but also on the appropriate spatial and temporal overlap of activity in the different incoming fibers.

A logical extension is to ask whether the orientation preference of cortical cells can be changed by experience. Initial tests for this were made by raising kittens in an environment in which they saw only one orientation, but the results were somewhat difficult to interpret. However, clear changes in orientation preference have been found in kittens in which one eye was presented with bars or stripes of only one orientation while the other eye saw a normal spectrum of orientation.[40,41] Using a slightly different experimental approach Wiesel, Carlson, and Hubel sutured the lids of one eye in a newly born monkey (quoted in

Orientation preferences of cortical cells

[40]Cynader, M. and Mitchell, D. E. 1977. *Nature 270*: 177–178.
[41]Rauschecker, J. P. and Singer, W. 1980. *Nature 280*: 58–60.

reference 38). The animal was kept in darkness except when it placed its head in a holder. Then, with the head held vertically, it would see vertical stripes with the unsutured eye. Since the monkey received orange juice each time it pressed on the headholder correctly, it performed this maneuver frequently. Thus, during the critical period, one eye received no visual input while the other saw only vertical stripes. After 57 hours of experience between 12 and 54 days after birth, normal levels of cortical activity were found, with cells of all orientations arranged as unusual in columns. As expected, the open eye tended to dominate. When tests were made for orientation preference, the results shown in Figure 11 were obtained. Both eyes could drive cells equally well when horizontal lines were the stimulus. However, the right eye (the open eye) was considerably more effective for vertical stripes. The probable explanation for this result is that neither eye saw horizontal bars or edges during the critical period. Hence, the situation for horizontality is analogous to binocular closure with equal competition. For the vertical input, however, the open eye had an enriched experience and "captured" cells in vertical orientation columns that had previously been supplied by the deprived eye.

SENSORY DEPRIVATION IN EARLY LIFE

The effects produced by altered sensory inputs in kittens and immature monkeys have a number of important implications for our understanding of the nervous system. At the level of synaptic mechanisms it is not at all clear how use, disuse, and inappropriate use of the visual pathways can alter branching patterns and connections of neurons in such a drastic and permanent manner, with expansion and contraction of cortical dominance columns (Figure 12). Critical periods, retraction, and susceptibility to damage during the development of the nervous system have been observed in other animals and other sensory systems,

11 ORIENTATION PREFERENCES OF CORTICAL CELLS in a monkey with altered visual experience. The monkey was kept in a darkened room. At 12 days the right eye was closed. Whenever the monkey placed its head in a holder, it received orange juice. At that time it also saw vertical stripes with its left eye. (The head holder ensured that the head was not tilted.) Thus, for a total of 57 hours of self-exposure, one eye saw vertical lines, the other nothing. (A, B) Ocular dominance histograms. When horizontally oriented light stimuli were shone onto the screen, cortical cells responded equally well to the left eye and the right eye. Thus, no deprivation is apparent for such orientation. With vertically oriented stimuli, the left (open) eye was much more effective in driving cortical cells. The histogram resembles that seen after monocular deprivation. The results suggest competition that was equal for horizontal (which neither eye had ever seen) and unequal for vertical (favored by the left, open eye). (After Wiesel, Carlson, and Hubel, quoted in Wiesel, 1982).

551
GENETIC AND
ENVIRONMENTAL
INFLUENCES IN
THE MAMMALIAN
VISUAL SYSTEM

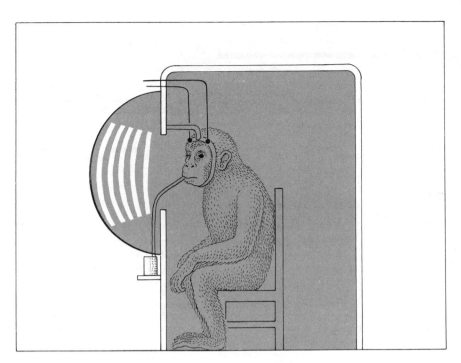

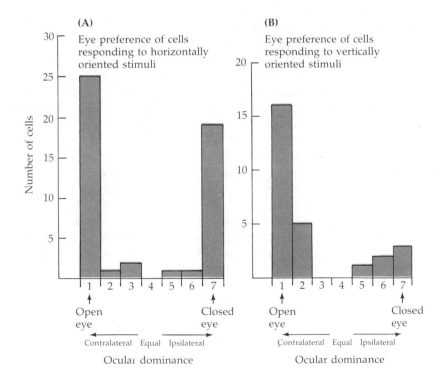

(A)

Eye preference of cells responding to horizontally oriented stimuli

(B)

Eye preference of cells responding to vertically oriented stimuli

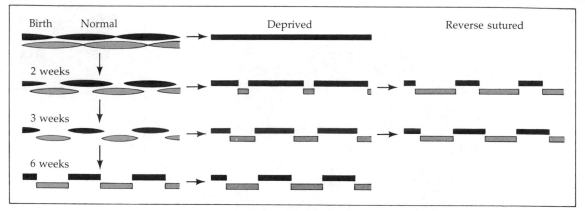

Birth   Normal   Deprived   Reverse sutured

2 weeks

3 weeks

6 weeks

12   SCHEME ILLUSTRATING EFFECTS OF EYE CLOSURE. As in Figure
3, in the normal cat or monkey by 6 weeks ocular dominance columns
have become well defined in layer IV of the cortex. Lid closure causes
excessive retraction of fibers supplied by the deprived eye. Those supplied by the open eye retract less than usual and in the adult supply
larger areas of cortex than in normal animals where the competition is
more equal. After reverse suture during the critical period, the initially
deprived eye can recapture the territory it had lost. (After Hubel and
Wiesel, 1977.)

as for example in the projections of the corpus callosum,[42] in the somatosensory cortex of mammals,[43] especially in barrel-fields of mice[44]
(Chapter 3), and in the auditory pathways of owls.[45]

At the level of behavior, the demonstration of a critical period of
vulnerability to deprivation or to abnormal experience is not new. In
the recent experiments, however, abnormalities in signaling have been
pinned down to relays in the cortex and not significantly to lower levels.
There exists a wealth of literature reporting other complex behavior
processes in a variety of animals that show periods of susceptibility.
Imprinting is one example.

Lorenz has shown that birds will follow any moving object presented
during the first day after hatching, as if it were their mother.[46] In higher
animals—for example, in dogs—behavioral studies indicate that if they
are handled by humans during a critical period of four to eight weeks
after birth, they are far more tractable and tame than animals that have
been isolated from human contact.[47] The critical period in an animal's
development may possibly represent a time during which a significant
sharpening of senses or faculties occurs.

[42]Innocenti, G. M. 1981. *Science 212*: 824–827.

[43]Kaas, J. H., Merzenich, M. M. and Killackey, H. P. 1983. *Annu. Rev. Neurosci. 6*:
325–356.

[44]Van Der Loos, H. and Woolsey, T. A. 1973. *Science 179*: 395–398.

[45]Knudsen, E. I., Knudsen, P. F. and Esterly, S. D. 1982. *Nature 295*: 238–240.

[46]Lorenz, K. 1970. *Studies in Animal and Human Behavior*, Vols. I and II. Harvard
University Press, Cambridge. Mass.

[47]Fuller, J. L. 1967. *Science 158*: 1645–1652.

553
GENETIC AND
ENVIRONMENTAL
INFLUENCES IN
THE MAMMALIAN
VISUAL SYSTEM

It is tempting to speculate about the effects of deprivation on higher functions in man. One can imagine, as Hubel has said,[48]

> Perhaps the most exciting possibility for the future is the extension of this type of work to other systems besides sensory. Experimental psychologists and psychiatrists both emphasize the importance of early experience on subsequent behavior patterns—could it be that deprivation of social contacts or the existence of other abnormal emotional situations early in life may lead to a deterioration or distortion of connections in some yet unexplored parts of the brain?

To find a physiological basis for such behavioral problems seems a distant, but not impossible, goal.

CONCLUDING REMARKS

At the beginning of this book we asked how far the cellular approach can take us in understanding the nervous system.

From our knowledge of signaling and the physical properties of neurons, we can form a picture of how groups of nerve cells in the brain integrate diverse signals and put together information. Many areas of the brain seem accessible to these approaches, but we still do not understand the basic mechanisms that underlie changes in membrane permeability and active transport. For the solution of these problems, it now appears that an important step will be a recognition and characterization of the membrane components that regulate transport and permeabilities, again at the cellular and subcellular levels.

We have witnessed in the past two decades promising steps that lead to a correlation of known groups of cells and specific signals with certain experiences during perception. The analysis of what happens to the information that is originally impressed upon the surface of the retina and then ascends stepwise from one group of cells to the next will undoubtedly be extended. So far, in following and deciphering visual clues we have progressed into the brain through only seven synaptic relays.

For further unraveling of neural organization, a more thorough recognition of the chemical architecture of cell assemblies—discussed in Chapter 12—may prove most valuable. Nevertheless, we must ask how useful the cellular approach will continue to be as we go still further, and at what stage the analysis will have to be enlarged. To interpret successfully the activity of large populations of cells, new methods are needed.

Although the field is by no means neglected, we have no clear-cut approach as yet for investigating the cellular basis of memory, learning, consciousness, or even how a simple act of movement is initiated in a higher animal. But perhaps we should feel encouraged by the vivid awareness of our many glaring deficiencies and by our ability to define

[48]Hubel, D. H. 1967. *Physiologist 10*: 17–45.

many areas of ignorance. Consider the question of how continued activity of nerve cells, particularly during growth and development, modifies their ability to influence other nerve cells. We can now at least point to specific processes in the visual cortex where certain groups of cells lose their synaptic drive from one eye that has been inappropriately used during a critical early period of life. In other words, some cortical neurons will be drastically changed for the rest of their lives because they have been used in the "wrong" way. But we do not know how normal impulse traffic acts to maintain and further develop anatomical connections between cells that are present at birth.

One could maintain that what is most needed is an understanding of the rules that govern the assembly of nerve cells in the first place. What is the ground plan according to which the instructions contained in the genes are translated into normal wiring? With such knowledge one might distinguish between genetically determined deviations and changes or abnormalities that are superimposed by the environment in the course of later experience.

Nonetheless, despite the enormous gaps in our knowledge, one can take some satisfaction by looking back a little more than two decades and recalling how impossible it seemed then to think of mechanisms by which nerve cells could recognize even a simple shape such as a corner or the letter L.

SUGGESTED READING

General reviews

Movshon, J. A. and Van Sluyters, R. C. 1981. Visual neural development. *Annu. Rev. Psychol. 32*: 477–522.

Sherman, S. M. and Spear, P. D. 1982. Organization of visual pathways in normal and visually deprived cats. *Physiol. Rev. 62*: 738–855.

Wiesel, T. N. 1982. The postnatal development of the visual cortex and the influence of environment. *Nature 299*: 583–592.

Original papers

Guillery, R. W. and Kaas, J. H. 1974. The effects of monocular lid suture upon the development of the visual cortex in squirrels (*Sciureus carolinensis*). *J. Comp. Neurol. 154*: 443–452.

Hubel, D. H. Wiesel, T. N. 1963b. Receptive fields of cells in striate cortex of very young, visually inexperienced kittens. *J. Neurophysiol. 26*: 994–1002.

Hubel, D. H. and Wiesel, T. N. 1965b. Binocular interaction in striate cortex of kittens reared with artificial squint. *J. Neurophysiol. 28*: 1041–1059.

Hubel, D. H. and Wiesel, T. N. 1970. The period of susceptibility to the physiological effects of unilateral eye closure in kittens. *J. Physiol. 206*: 419–436.

Hubel, D. H., Wiesel, T. N. and LeVay, S. 1977. Plasticity of ocular dominance columns in monkey striate cortex. *Phil. Trans. R. Soc. Lond. B 278*: 377–409.

Innocenti, G. M. 1981. Growth and reshaping of axons in the establishment of visual callosal connections. *Science 212*: 824–827.

555
GENETIC AND
ENVIRONMENTAL
INFLUENCES IN
THE MAMMALIAN
VISUAL SYSTEM

LeVay, S., Wiesel, T. N. and Hubel, D. H. 1980. The development of ocular dominance columns in normal and visually deprived monkeys. *J. Comp. Neurol. 191*: 1–51.

LeVay, S., Stryker, M. P. and Shatz, C. J. 1978. Ocular dominance columns and their development in layer IV of the cat's visual cortex: A quantitative study. *J. Comp. Neurol. 179*: 223–244.

Rakič, P. 1977b. Prenatal development of the visual system in rhesus monkey. *Phil. Trans. R. Soc. Lond. B 278*: 245–260.

Wiesel, T. N. and Hubel, D. H. 1963b. Single-cell responses in striate cortex of kitten deprived of vision in one eye. *J. Neurophysiol. 26*: 1003–1017.

Wiesel, T. N. and Hubel, D. H. 1965a. Comparison of the effects of unilateral and bilateral eye closure on cortical unit responses in kittens. *J. Neurophysiol. 28*: 1029–1040.

Wiesel, T. N. and Hubel, D. H. 1965b. Extent of recovery from the effects of visual deprivation in kittens. *J. Neurophysiol. 28*: 1060–1072.

APPENDIX A

CURRENT FLOW IN ELECTRICAL CIRCUITS

A few basic concepts are required to understand the electrical circuits used in this presentation. An especially clear and lively treatment is found in Rogers' book.[1] For our purposes it is sufficient to describe the properties of a few circuit elements and explain how they work when connected together in ways that correspond to the circuits described for nerves. The difficulties sometimes encountered on first reading accounts of electrical circuits often stem from the apparently abstract nature of the forces and movements involved. It is reassuring, therefore, to realize that many of the original pioneers in the field must have been faced with similar problems, since the terms devised in the last century are mainly related to the movement of fluids. Thus, the words "current," "flow," "potential," "resistance," and "capacitance" apply equally well to both electricity and hydraulics. The analogy between the two systems is illustrated by the fact that complex problems in hydraulics may be solved by using solutions to equivalent electrical circuits.

The analogy between a simple electrical circuit and its hydraulic equivalent is illustrated in Figure 1. The first point to be made is that a source of energy is required to keep the current flowing. In the hydraulic circuit, it is a pump; in the electrical circuit, a battery. The second point is that neither water nor electrical charge is created or lost within such a system. Thus, the flow rate of water is the same at points a, b, and c in the hydraulic circuit, since no water is added or removed between them. Similarly the electrical current in the equivalent circuit is the same at the three corresponding points. In both circuits, there are a number of *resistances* to current flow. In the hydraulic circuit, such resistance is offered by narrow tubes; similarly thinner wires offer greater resistance to electrical current flow.

The units used to express rate of flow is to some extent a matter of choice; one can measure flow of water through a pipe in cubic feet per minute, for example, although in some other situation milliliters per hour might be more suitable. Electrical current flow is conventionally measured in COULOMBS/SEC or AMPERES (abbreviated A). One coulomb is equal to the charge carried by 6.24×10^{18} electrons. In electrical

Terms and units describing electric currents

[1]Rogers, E. M. 1960. *Physics for the Enquiring Mind.* Princeton University Press, Princeton.

557

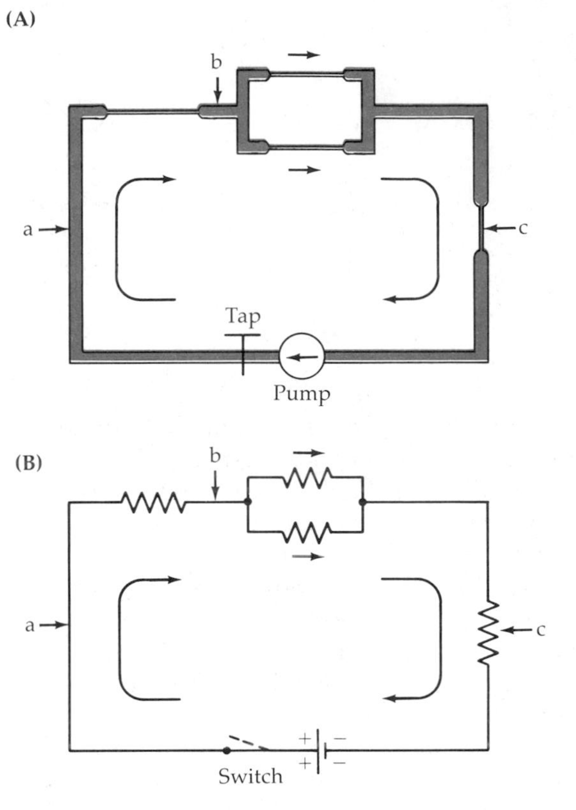

(A)

b

a

c

Tap

Pump

(B)

b

a

c

Switch

1

HYDRAULIC AND ELECTRICAL CIRCUITS.
(A), (B) Analogous circuits for the flow of water
and of electrical current. A battery is analogous
to a pump which operates at constant pressure,
the switch to a tap in the hydraulic line, and
resistors to constrictions in the tubes.

circuits and equations, current is usually designated by I or i. As with
flow of water, flow of current is a vector quantity, which is just a way
of saying that it has a specified direction. The direction of flow is often
indicated by arrows, as in Figure 1, current always being assumed to
flow from the positive to the negative pole of a battery.

What do POSITIVE and NEGATIVE mean with regard to current flow?
Here the hydraulic analogy does not help. It is useful instead to consider
the effects of passing current through a chemical solution. For example,
suppose two copper wires are dipped into a solution of copper sulfate
and connected to the positive and negative poles of a battery. Copper
ions in solution are repelled from the positive wire, move through the
solution, and are deposited from the solution onto the negative wire.
In short, positive ions move in the direction conventionally designated
for current: from positive to negative in the circuit. At the same time,
sulfate ions move in the opposite direction and are deposited onto the
positive wire. The direction specified for current, then, is the direction
in which positive charges move in the circuit; negative charges move
in the reverse direction. Students are usually told that the direction
specified for current in a wire is opposite to the direction in which the
electrons move. Such a statement is true, but irrelevant, and should be
ignored.

To explain the energy source for current flow and the meaning of electrical POTENTIAL, the hydraulic analogy is again useful. The flow of fluid depicted in Figure 1 depends on a pressure difference, the direction of flow being from high to low. No net movement occurs between two parts of the circuit at the same pressure. The overall pressure in the circuit is supplied by expenditure of energy in driving the pump. In the electrical circuit shown here, the electrical "pressure" or POTENTIAL is provided by a BATTERY, in which chemical energy is stored. Hydraulic pressure is measured in dynes/cm^2; electrical potential is measured in VOLTS.

Symbols used in electrical circuit diagrams and arrangements of circuit elements in series and in parallel are illustrated in Figure 2. As the names imply, a VOLTMETER measures electrical potential and is equivalent to a pressure gauge in hydraulics; an AMMETER measures current flowing in a circuit and is equivalent to a flowmeter.

In hydraulic systems, at least under ideal circumstances, the amount of current flowing through the system increases with pressure. The factor which determines the relation between pressure and flow rate is an inherent characteristic of the pipes, their RESISTANCE. Small-diameter, long pipes have greater resistances than large-diameter, short ones. Similarly, current flow in electrical circuits depends on the resistance in the circuit. Again, small, long wires have larger resistances than large, short ones. If current is being passed through an ionic solution, the resistance of the solution will increase as the solution is made more dilute. This is because there are fewer ions available to carry the current. In conductors such as wires, the relation between current and potential difference is described by Ohm's law, formulated by Ohm in the 1820s. The law says that the amount of current (I) flowing in a conductor is related to the potential difference (V) applied to it, $I = V/R$. The constant

Ohm's law and
electrical resistance

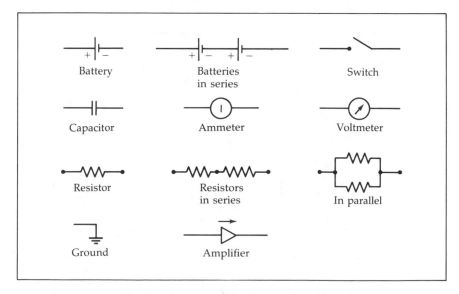

2   SYMBOLS IN ELECTRIC CIRCUIT DIAGRAMS.

R is the resistance of the wire. If I is in amperes and V is in volts, then R is in units of OHMS (Ω). The reciprocal of resistance is a measure of the ease with which current flows through a conductor and is called CONDUCTANCE and is indicated by $g = 1/R$; units of conductance are SIEMENS (S). Thus, Ohm's law may also be written $I = gV$.

Ohm's law holds whenever the graph of current against potential is a straight line. In any circuit or part of a circuit for which this is true, any one of the three variables in the equation may be calculated if the other two are known. For example,

1. We can pass a known current through a nerve membrane, measure the change in potential, and then calculate the membrane resistance ($R = V/I$).
2. If we measure the potential difference produced by the applied current and know the membrane resistance, we can calculate the applied current ($I = V/R$).
3. If we pass a known current through the membrane and know its resistance, then we can calculate the change in potential ($V = IR$).

Two additional simple, but important, rules (Kirchoff's laws) should be mentioned:

1. The algebraic sum of all the currents flowing toward any junction is zero. For example, at point a in Figure 4, $I_{total} + I_{R1} + I_{R3} = 0$, which means that I_{total} (arriving) $= -I_{R1} - I_{R3}$ (leaving) (this is merely a statement that charge is neither created nor destroyed anywhere in the circuit).
2. The algebraic sum of all the battery voltages is equal to the algebraic sum of all the IR voltage drops in a loop. An example of this is shown in Figure 3B: $V = IR_1 + IR_2$ (this is a statement of the conservation of energy).

We can now examine in more detail the circuits of Figures 3 and 4, which are needed to construct a model of the membrane. Figure 3A

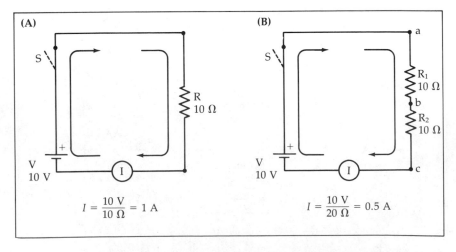

3 OHMS'S LAW applied to simple circuits. (A) Current $I = 10\ V/10\ \Omega = 1\ A$. (B) Current $= 10\ V/20\ \Omega = 0.5\ A$, and the voltage across each resistor is 5 V.

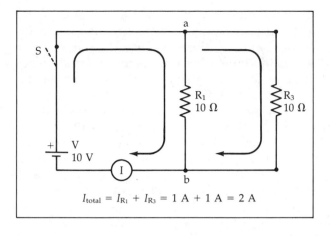

$I_{\text{total}} = I_{R_1} + I_{R_3} = 1 \text{ A} + 1 \text{ A} = 2 \text{ A}$

4

PARALLEL RESISTORS. When R_1 and R_3 are in parallel, the voltage drop across each resistor is 10 V and the total current is 2 A.

shows a battery (V) of 10 V connected to a resistance (R) of 10 Ω. The switch S can be opened or closed, thereby interrupting or establishing current flow. The voltage applied to R is 10 V; therefore the current measured by the ammeter, I, is, by Ohm's law, 1.0 A. In Figure 3B, the resistor is replaced by two resistors, R_1 and R_2, IN SERIES. By the first of Kirchoff's laws, the current flowing into point b must be equal to that leaving. Therefore, the same current, I, must flow through both the resistors. By the second of Kirchoff's laws, then, $IR_1 + IR_2 = V$ (10 V). It follows that the current, $I = V/(R_1 + R_2) = 0.5$ A. The voltage at b, then, is 5 V positive to that at c and a is 5 V positive to b. Note that because there is only one path for the current, the total resistance, R_{total}, seen by the battery is simply the sum of the two resistors; that is, $R_{\text{total}} = R_1 + R_2$.

What happens if, as shown in Figure 4, we add a second resistor, also of 10 Ω, IN PARALLEL, rather than in series? In the circuit, the two resistors R_1 and R_3 provide two separate pathways for current. Both have a voltage V (10 V) across them, so the respective currents will be

$$I_{R_1} = V/R_1 = 1 \text{ A}$$

$$I_{R_3} = V/R_3 = 1 \text{ A}$$

Therefore, to satisfy the first of Kirchoff's laws, there must be 2 A arriving at point a and 2 A leaving point b. The ammeter, then, will read 2 A. Now the combined resistance of R_1 and R_3 is $R_{\text{total}} = V/I = 10 \text{ V}/2 \text{ A} = 5$ Ω, or half that of the individual resistors. This makes sense if one thinks of the hydraulic analogy: Two pipes in parallel will offer less resistance to flow than one pipe alone. In the electrical circuit, the *conductances* add: $g_{\text{total}} = g_1 + g_3$, or $1/R_{\text{total}} = 1/R_1 + 1/R_3$.

If we now generalize to any number (n) of resistors, resistances in series add simply:

$$R_{\text{total}} = R_1 + R_2 + R_3 + \ldots + R_n$$

and in parallel:

$$1/R_{\text{total}} = 1/R_1 + 1/R_2 + 1/R_3 + \ldots + 1/R_n$$

Applying circuit
analysis to the
membrane model

Figure 5A shows a circuit similar to that used to represent nerve membranes. Notice that the two batteries drive current around the circuit in the same direction and that the resistances R_1 and R_2 are in series. What is the potential difference between points b and d (which represent the outside and inside of the membrane)? The total potential across the two resistors between a and c is 150 mV, a being positive to c. Therefore, the current flowing between a and c through the resistors is 150 mV/100,000 Ω = 1.5 μA. When 1.5 μA flows across 10,000 Ω, as between a and b, a potential drop of 15 mV is produced, a being positive with respect to b. The potential difference between outside and inside is therefore 100mV − 15 mV = 85 mV. We can obtain the same result by considering the voltage drop across R_2 (1.5 μA × 90,000 Ω = 135 mV) and adding it to V_2 (135 mV − 50 mV = 85 mV). This *must* be so, as the potential between b and d must have a unique value.

In Figure 5B, R_1 and R_2 have been exchanged. As the total resistance in the circuit is the same, the current must be the same, 1.5 μA. Now the potential drop across R_2, between a and b, is 90,000 Ω × 1.5 μA = 135 mV, a being positive. Now the potential across the membrane is 100 mV − 135 mV = −35 mV, OUTSIDE NEGATIVE; the same result can, of course, be obtained from the current through R_1. This simple circuit illustrates an important point about membrane physiology: THE POTENTIAL ACROSS A MEMBRANE CAN CHANGE AS A RESULT OF RESISTANCE CHANGES WHILE THE BATTERIES REMAIN UNCHANGED. A general expression for the membrane potential in the circuit shown in Figure 5A can

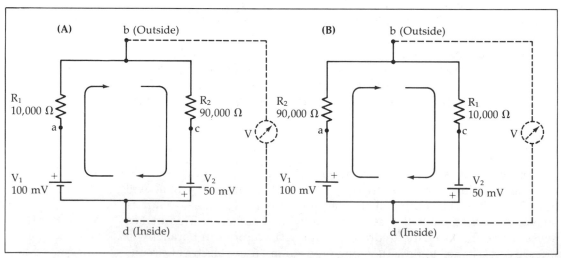

5 ANALOGUE CIRCUITS FOR NERVE MEMBRANES. In (A) and (B) the resistors R_1 and R_2 are reversed; otherwise the circuits are the same. The batteries V_1 and V_2 are in series. In (A), b (the "outside" of the membrane) is positive with respect to d (the "inside") by 85 mV; in (B) it is negative by 35 mV (see text). These circuits illustrate how changes in resistance can give rise to membrane potential changes even though the batteries (which represent ionic equilibrium potentials) remain constant.

be derived simply, as follows:

$$V_m = V_1 - IR_1$$

As $I = (V_1 + V_2)/(R_1 + R_2)$:

$$V_m = V_1 - \frac{(V_1 + V_2)R_1}{R_1 + R_2}$$

On rearranging:

$$V_m = \frac{V_1R_2/R_1 - V_2}{1 + R_2/R_1}$$

In the circuits described in Figures 3 and 4, closing or opening the switch produces instantaneous and simultaneous changes in current and potential. Capacitors introduce a time element into the consideration of current flow. These accumulate and store electrical charge and, when they are present in a circuit, current and voltage changes are no longer simultaneous. A capacitor consists of two conducting plates (usually of metal) separated by an insulator (air, mica, oil, or plastic). When voltage is applied between the plates (Figure 6A), there is an instantaneous displacement of charge from one plate to the other. Once the capacitor is fully charged, however, there is no further current, as none can flow across the insulator. the CAPACITANCE (C) of a capacitor is defined by how much charge (q) it can store for each volt applied to it:

$$C = q/V$$

The units of capacitance are coulombs/volt or FARADS (F). The larger the plates of a capacitor and the closer together they are, the greater its capacitance. A one-farad capacitor is very large; capacitances in common use are in the range of microfarads (μF) or smaller.

When the switch in Figure 6A is closed, then, there is an instantaneous charge separation at the plates. The amount of charge stored in the capacitor is proportional to its capacitance and to the magnitude of the applied voltage (V_C). When the switch is opened, as in B, the charge on the capacitor remains, as does the voltage (V) between the plates. (One can sometimes get a surprising shock from electronic apparatus after it has been turned off because some of the capacitors in the circuits may remain charged.) The capacitor can be discharged by shorting it with a second switch, as in Figure 6C. Again, the current flow is instantaneous, returning the charge and the voltage on the capacitor to zero. If, instead, the capacitor is discharged through a resistor (R; Figure 6D), the discharge is no longer instantaneous. This is because the resistor limits the current flow. If the voltage on the capacitor is V, then by Ohm's law the maximum current is $I = V/R$. With no resistor in the circuit, the current becomes infinitely large and the capacitor is discharged in an infinitesimal time period; if R is very large, the discharge process takes a very long time. The rate of discharge at any given time, dq/dt, is simply equal in magnitude to the current. In other words,

Electrical capacitance and time constant

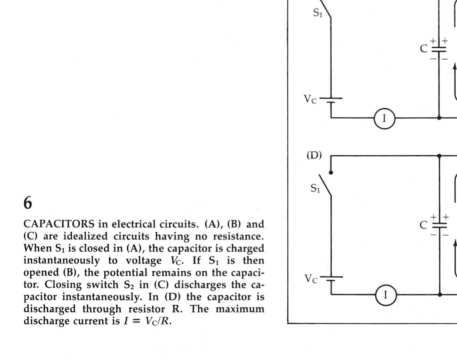

6

CAPACITORS in electrical circuits. (A), (B) and (C) are idealized circuits having no resistance. When S_1 is closed in (A), the capacitor is charged instantaneously to voltage V_C. If S_1 is then opened (B), the potential remains on the capacitor. Closing switch S_2 in (C) discharges the capacitor instantaneously. In (D) the capacitor is discharged through resistor R. The maximum discharge current is $I = V_C/R$.

$dq/dt = -V/R$ (negative because the charge is decreasing with time), where V initially is equal to the battery voltage and decreases as the capacitor is discharged. As $q = CV$, $dq/dt = CdV/dt$, and we can then write $CdV/dt = -V/R$, or

$$dV/dt = -V/RC$$

The equation says that the rate of loss of voltage from the capacitor is proportional to the voltage remaining. Thus, as the voltage decreases, the rate of discharge decreases. The constant of proportionality, $1/RC$, is the *rate constant* for the process: RC is its *time constant*. This kind of process arises over and over again in nature. For example, the rate at which water drains from a bathtub decreases as the depth, and hence the pressure at the drain, decreases. In this kind of situation, the discharge process is described by an exponential function:

$$V = V_0 \, e^{-t/\tau}$$

where V_0 is the initial charge on the capacitor and the time constant $\tau = RC$. Similarly, when the capacitor is charged through a resistor, as in Figure 7, the charging process takes a finite time. The voltage between

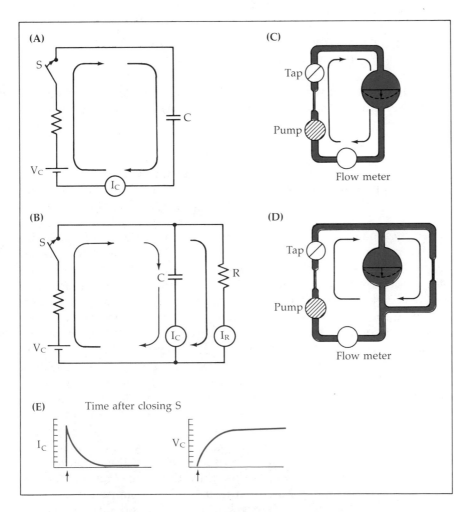

7 CHARGING OF A CAPACITOR. In (A) the capacitor is charged at a rate limited by the resistor, the initial rate being $I = V_C/R$. In (B) the charging rate depends on both resistors in the circuit. The capacitative current and the voltage across the capacitor are shown as functions of time in E. The voltage reaches its final value only when the capacitor is fully charged ($I_C = 0$). (C) and (D) are hydraulic analogues of the circuits in (A) and (B) (see text).

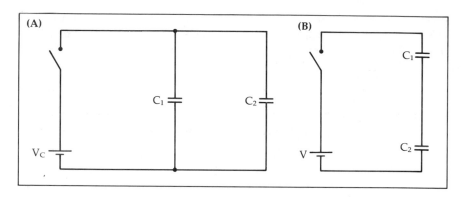

8   CAPACITORS IN PARALLEL (A) AND SERIES (B).

the plates increases with time until the battery voltage is reached and no further current flows. The charging process is now a rising exponential, with a time constant $\tau = RC$:

$$V = V_C(1 - e^{-t/\tau})$$

These examples illustrate another property of a capacitor. Current flows into and out of the capacitor only when the potential is changing:

$$I_C = dQ/dt = CdV/dt$$

When the voltage across the capacitor is steady ($dV/dt = 0$), the capacitive current, I_C, is zero. In other words, the capacitance has an "infinite resistance" for a steady potential difference and a "low resistance" for a rapidly changing potential. Figure 7B shows a circuit in which current flows through a resistor and capacitor in parallel and Figure 7E the time courses of the capacitative current and voltage.

The properties of a capacitor in a circuit can be illustrated by the slightly more elaborate hydraulic analogy shown in Figure 7C. The capacitor is represented by an elastic diaphragm that forms a partition in a fluid-filled chamber. When the tap is opened, the pressure generated by the pump causes fluid flow into the chamber, bulging the diaphragm until, because of its elasticity, it provides an equal and opposite pressure; then there is no more fluid flow and the chamber is fully charged. If a tube is placed alongside, as in Figure 7D, some fluid flows through the tube and some is used to expand the diaphragm. If the tube is of high resistance, the pressure difference between its two ends is larger for a given flow than for a lower resistance tube. In this case, the distention of the diaphragm is greater and takes longer to achieve. Similarly, if the capacity of the chamber is larger, more fluid is diverted during the filling (or "charging") process and a longer time is required to reach a steady state. Thus, the characteristic time constant of the system is determined by the product of resistance and capacitance.

When capacitors are arranged in parallel, as in Figure 8A, the total capacitance is increased. The total charge stored is the sum of that stored in each: $q_1 + q_2 = C_1 V_C + C_2 V_C$, or $q_{total} = C_{total} V_C$, where $C_{total} = C_1 + C_2$. In contrast, capacitance *decreases* when capacitors are arranged in series (Figure 8B). It turns out that the relation is the same as for resistors in parallel: their reciprocals sum. In summary, then, for a number (N) of capacitors in parallel,

$$C_{total} = C_1 + C_2 + C_3 + \ldots + C_n$$

and in series,

$$1/C_{total} = 1/C_1 + 1/C_3 + \ldots + 1/C_n$$

NUCLEAR MAGNETIC RESONANCE IMAGE OF A
SAGITTAL SECTION OF A LIVING HUMAN BRAIN

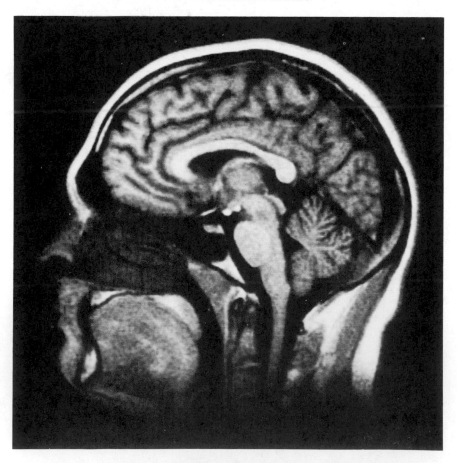

STRUCTURES
AND PATHWAYS
OF THE BRAIN

The following drawings show the brain viewed from different aspects and cut in different planes of section. Since the aim is to provide the visual equivalent of a glossary rather than a full atlas, only the key landmarks and the principal structures referred to in the text are labeled. Further information is provided by several comprehensive books:

Carpenter, M. B. 1978. *Core Text of Neuroanatomy*, 2nd Ed., Williams & Wilkins, Baltimore.

Gluhbegovic, N. and Williams, T. H. 1980. *The Human Brain*. Harper & Row, New York.

Nolte, J. 1981. *The Human Brain*. Mosby, St. Louis.

Williams, P. L. and Warwick, R. 1975. *Functional Neuroanatomy of Man*. Saunders, Philadelphia.

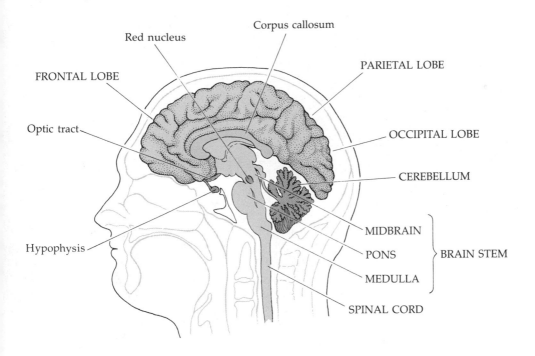

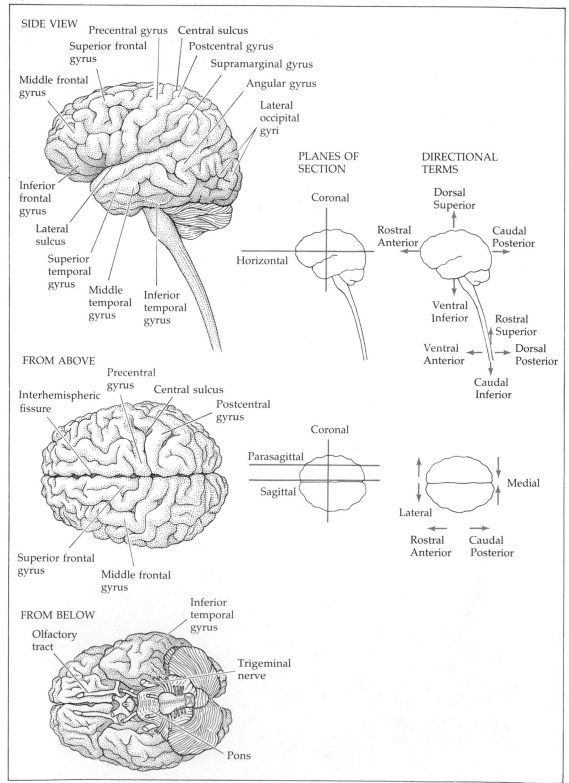

SIDE VIEW

Precentral gyrus   Central sulcus
Superior frontal gyrus   Postcentral gyrus
Supramarginal gyrus
Middle frontal gyrus   Angular gyrus
Lateral occipital gyri
Inferior frontal gyrus
Lateral sulcus
Superior temporal gyrus
Middle temporal gyrus   Inferior temporal gyrus

PLANES OF SECTION

Coronal

Horizontal

DIRECTIONAL TERMS

Dorsal Superior
Rostral Anterior
Caudal Posterior
Ventral Inferior
Rostral Superior
Ventral Anterior
Dorsal Posterior
Caudal Inferior

FROM ABOVE

Precentral gyrus
Interhemispheric fissure
Central sulcus
Postcentral gyrus

Coronal
Parasagittal
Sagittal

Lateral
Medial
Rostral Anterior
Caudal Posterior

Superior frontal gyrus
Middle frontal gyrus

FROM BELOW

Inferior temporal gyrus
Olfactory tract
Trigeminal nerve
Pons

SAGITTAL SECTIONS THROUGH BRAIN

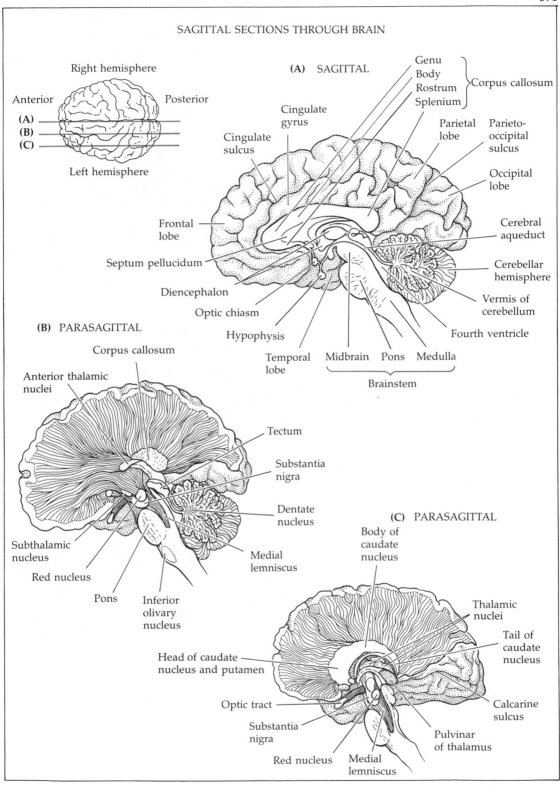

Right hemisphere

Anterior   Posterior

(A)
(B)
(C)

Left hemisphere

(A)   SAGITTAL

Genu
Body
Rostrum
Splenium
} Corpus callosum

Cingulate
gyrus

Cingulate
sulcus

Parietal
lobe

Parieto-
occipital
sulcus

Occipital
lobe

Cerebral
aqueduct

Frontal
lobe

Cerebellar
hemisphere

Septum pellucidum

Vermis of
cerebellum

Diencephalon

Fourth ventricle

Optic chiasm

Hypophysis

Temporal
lobe

Midbrain   Pons   Medulla

Brainstem

(B)   PARASAGITTAL

Corpus callosum

Anterior thalamic
nuclei

Tectum

Substantia
nigra

Dentate
nucleus

(C)   PARASAGITTAL

Body of
caudate
nucleus

Subthalamic
nucleus

Medial
lemniscus

Thalamic
nuclei

Red nucleus

Tail of
caudate
nucleus

Pons   Inferior
olivary
nucleus

Head of caudate
nucleus and putamen

Calcarine
sulcus

Optic tract

Substantia
nigra

Pulvinar
of thalamus

Red nucleus   Medial
lemniscus

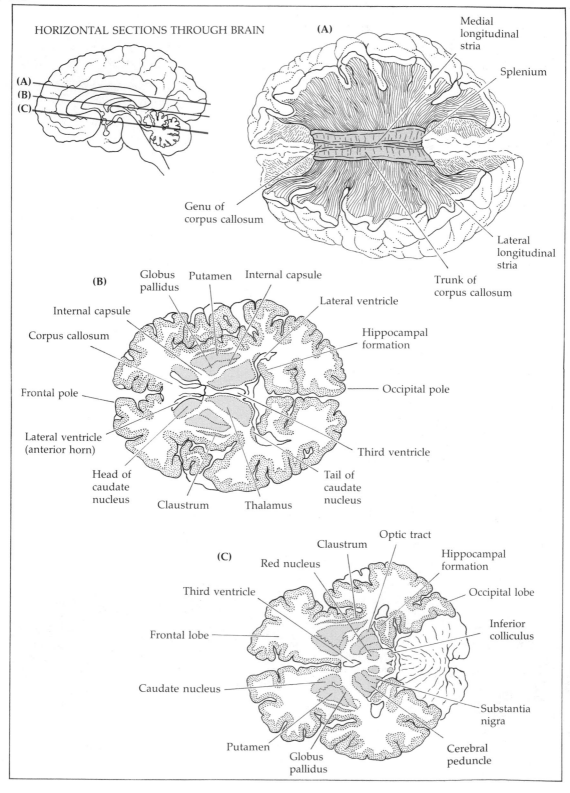

HORIZONTAL SECTIONS THROUGH BRAIN

(A)

Medial
longitudinal
stria

Splenium

Genu of
corpus callosum

Lateral
longitudinal
stria

Trunk of
corpus callosum

(B)

Globus
pallidus

Putamen

Internal capsule

Lateral ventricle

Internal capsule

Hippocampal
formation

Corpus callosum

Frontal pole

Occipital pole

Lateral ventricle
(anterior horn)

Third ventricle

Head of
caudate
nucleus

Claustrum

Thalamus

Tail of
caudate
nucleus

(C)

Optic tract

Claustrum

Hippocampal
formation

Red nucleus

Occipital lobe

Third ventricle

Inferior
colliculus

Frontal lobe

Caudate nucleus

Substantia
nigra

Putamen

Globus
pallidus

Cerebral
peduncle

CORONAL SECTION THROUGH BRAIN

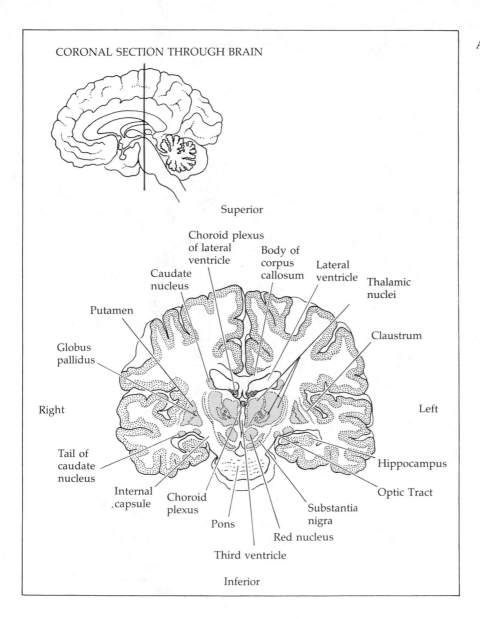

Superior

Choroid plexus
of lateral
ventricle

Body of
corpus
callosum

Lateral
ventricle

Thalamic
nuclei

Caudate
nucleus

Putamen

Claustrum

Globus
pallidus

Right

Left

Tail of
caudate
nucleus

Hippocampus

Optic Tract

Internal
capsule

Choroid
plexus

Substantia
nigra

Pons

Red nucleus

Third ventricle

Inferior

ASCENDING SENSORY PATHWAYS

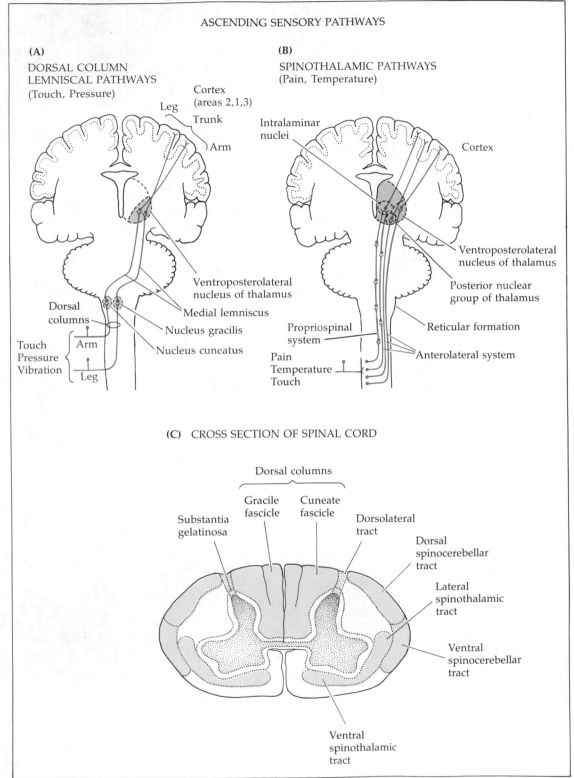

(A)
DORSAL COLUMN
LEMNISCAL PATHWAYS
(Touch, Pressure)

Leg

Cortex
(areas 2,1,3)

Trunk

Arm

Ventroposterolateral
nucleus of thalamus

Dorsal
columns

Medial lemniscus

Nucleus gracilis

Nucleus cuneatus

Touch
Pressure
Vibration

Arm

Leg

(B)
SPINOTHALAMIC PATHWAYS
(Pain, Temperature)

Intralaminar
nuclei

Cortex

Ventroposterolateral
nucleus of thalamus

Posterior nuclear
group of thalamus

Propriospinal
system

Reticular formation

Anterolateral system

Pain
Temperature
Touch

(C)   CROSS SECTION OF SPINAL CORD

Dorsal columns

Gracile
fascicle

Cuneate
fascicle

Substantia
gelatinosa

Dorsolateral
tract

Dorsal
spinocerebellar
tract

Lateral
spinothalamic
tract

Ventral
spinocerebellar
tract

Ventral
spinothalamic
tract

(A)   DESCENDING MOTOR PATHWAYS

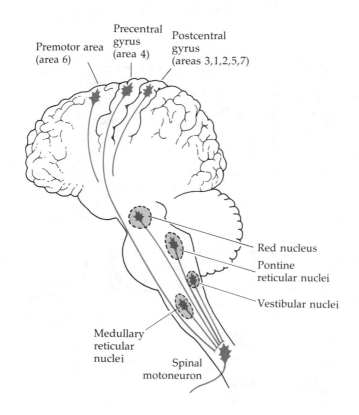

Premotor area
(area 6)

Precentral
gyrus
(area 4)

Postcentral
gyrus
(areas 3,1,2,5,7)

Red nucleus

Pontine
reticular nuclei

Vestibular nuclei

Medullary
reticular
nuclei

Spinal
motoneuron

(B) CROSS SECTION OF SPINAL CORD

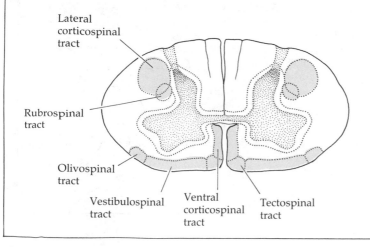

Lateral
corticospinal
tract

Rubrospinal
tract

Olivospinal
tract

Vestibulospinal
tract

Ventral
corticospinal
tract

Tectospinal
tract

CENTRAL PATHWAYS FOR NOREPINEPHRINE,
DOPAMINE, AND 5-HYDROXYTRYPTAMINE

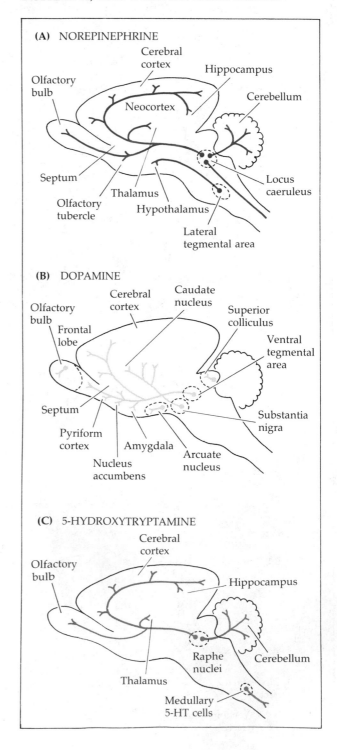

(A) NOREPINEPHRINE

Cerebral cortex
Hippocampus
Olfactory bulb
Cerebellum
Neocortex
Septum
Locus caeruleus
Olfactory tubercle
Thalamus
Hypothalamus
Lateral tegmental area

(B) DOPAMINE

Cerebral cortex
Caudate nucleus
Olfactory bulb
Superior colliculus
Frontal lobe
Ventral tegmental area
Septum
Substantia nigra
Pyriform cortex
Amygdala
Arcuate nucleus
Nucleus accumbens

(C) 5-HYDROXYTRYPTAMINE

Cerebral cortex
Olfactory bulb
Hippocampus
Raphe nuclei
Cerebellum
Thalamus
Medullary 5-HT cells

GLOSSARY

The definitions below apply to the terms used in the context of this book. Thus, **excitation, adaptation,** and **inhibition** all have additional meanings that are not included.

acetylcholine (ACh) CH_3—CO—O—CH_2—CH_2—N—$(CH_3)_3$. Transmitter liberated by vertebrate motoneurons, preganglionic sympathetic and parasympathetic neurons; hydrolyzed by cholinesterase.

action potential Brief regenerative, all-or-nothing electrical potential that propagates along an axon or muscle fiber.

active transport Movement of ions or molecules against an electrochemical gradient by utilization of metabolic energy.

adaptation Decline in response of a sensory neuron to a maintained stimulus.

afferent Axons conducting impulses toward the central nervous system.

anticholinesterase Cholinesterase inhibitor (e.g., neostigmine, eserine); such agents prevent the hydrolysis of ACh and thereby allow its action to be prolonged.

area centralis In the cat, the area of retina with highest discrimination, containing cones.

areas 17, 18, 19 Neurosensory areas of the cerebral cortex subserving vision, defined by histological and physiological criteria.

autonomic nervous system Part of the nervous system supplying viscera, skin, smooth muscle, glands, and heart and consisting of two distinct divisions, parasympathetic and sympathetic.

axon The process or processes of a neuron conducting impulses, usually over long distances.

axon cap Specialized region of Mauthner cell axon surrounded by glial cells at which electrical inhibition occurs.

axon hillock Region of the cell body at which the axon originates; often the site of impulse initiation.

axoplasm Intracellular fluid within an axon.

bipolar cell Neuron with two major processes arising from the cell body; in the vertebrate retina, interposed between receptors and ganglion cells.

blood-brain barrier Term denoting restricted access of substances to neurons and glial cells within the brain.

bouton Small terminal expansion of the presynaptic nerve fiber at a synapse; site of transmitter release.

α-bungarotoxin Toxin from venom of the snake *Bungarus multicinctus*; binds to ACh receptor with high affinity.

capacitance of the membrane (C_m) Property of the cell membrane enabling electrical charge to be stored and separated and introducing distortion in the time course of passively conducted signals; measured in farads (F).

caudal Posterior (in four-legged animals).

centrifugal control Regulation of performance of peripheral sense organs by axons coming from the central nervous system.

cerebrospinal fluid (CSF)   Clear liquid filling the ventricles of the brain and spaces between meninges, arachnoid, and pia. See **subarachnoid space, ventricles.**

channel   An opening or "pore" in a membrane through which ions or molecules can move.

cholinergic   Neurons releasing ACh as the transmitter.

choroid plexus   Folded processes that are rich in blood vessels and project into ventricles of the brain and secrete cerebrospinal fluid.

conductance (*g*)   Reciprocal of electrical resistance and thus a measure of the ability of a circuit to conduct electricity; in excitable cells, a useful measure of permeability for an ion or ions.

contralateral   Relating to the opposite side of the body.

convergence   Coming together and making synapses by a group of presynaptic neurons on one postsynaptic neuron.

coronal   Vertical section through the skull at right angles to the front-back (sagittal) axis.

cortical column   Aggregate of cortical neurons sharing common properties (e.g., sensory modality, receptive field position, eye dominance, orientation, movement sensitivity).

coulomb   Unit of electrical charge.

coupling potential   Synaptic potential seen as a result of current spread through an electrical synapse.

dendrite   Process of a neuron specialized to act as a receptor; postsynaptic region of a neuron.

depolarization   Reduction of membrane potential from the resting value toward zero.

divergence   Branching of a neuron to form synapses with several other neurons.

efferent   Axons conducting impulses outward from the central nervous system.

electroencephalogram (EEG)   Record taken of the electrical activity of the brain by external electrodes on the scalp.

electrogenic pump   Active transport of ions across a cell membrane resulting in a change in membrane potential.

electromyogram (EMG)   Record of muscular activity by external electrodes inserted into a muscle.

electroretinogram (ERG)   Potential change in response to light measured by external electrodes on the eye.

electrotonic potentials   Localized, graded potentials produced by subthreshold currents, determined by passive electrical properties of cells.

endothelial cells   Layer of cells lining blood vessels.

end plate   Postsynaptic area of vertebrate skeletal muscle fiber.

ependyma   Layer of cells lining cerebral ventricles and central canal of spinal cord.

epinephrine (adrenaline)   Hormone secreted by the adrenal medulla; certain of its actions resemble those of sympathetic nerves.

epp   End plate potential; synaptic potential in a skeletal muscle fiber produced by ACh liberated from presynaptic terminals.

epsp   Excitatory postsynaptic potential in a neuron.

equilibrium potential   Membrane potential at which there is no net passive movement of an ion species into or out of a cell.

eserine   Anticholinesterase (see above); also known as physostigmine.

excitation   Process tending to produce action potentials.

exocytosis   Process whereby synaptic vesicles fuse with presynaptic terminal membrane and empty transmitter molecules into the synaptic cleft.

extrafusal   Muscle fibers making up the mass of a skeletal muscle (i.e., not within the sensory muscle spindles).

facilitation   Greater effectiveness of synaptic transmission by successive presynaptic impulses usually due to increased transmitter release.

farad (F)   Unit of capacitance; more commonly used is **microfarad** ($\mu F = 10^{-6}F$).

faraday (F)   The number of coulombs carried by 1 mole of univalent ion (96,500).

field axis orientation   For simple and complex cortical neurons, the angle of the long axis of the receptive field (e.g., horizontal, vertical, oblique).

fovea   Central depression in the retina composed of slender cones; area of greatest visual resolution.

fusimotor   Motoneurons supplying muscle fibers in a muscle spindle.

g   Symbol for conductance.

γ-aminobutyric acid (GABA)   Inhibitory transmitter at crustacean neuromuscular synapses and a candidate in the central nervous system of vertebrates.

γ-efferent fiber   Small myelinated motor axon supplying intrafusal muscle fiber. See **fusimotor.**

ganglion   Discrete collection of nerve cells.

gap junction   Region of contact between cells at which intercellular space between adjacent membranes is reduced to about 2 nm; site of electrical coupling.

gate   Mechanism whereby a channel is opened and closed.

glia   See **neuroglia.**

gray matter   Part of the central nervous system composed predominantly of the cell bodies of neurons and fine terminals, as opposed to major axon tracts (white matter).

horseradish peroxidase   Enzyme used as a histochemical marker for tracing processes of neurons or spaces between cells.

hyperpolarization   Increase in membrane potential from the resting value, tending to reduce excitability.

impulse   See **action potential.**

inactivation   Gradual reduction in sodium conductance produced by depolarization.

inhibition   Effect of one neuron upon another tending to prevent it from initiating impulses.

> **Postsynaptic inhibition** is mediated through a permeability change in the postsynaptic cell, holding the membrane potential away from threshold.
> **Presynaptic inhibition** is mediated by an inhibitory fiber upon an excitatory terminal, reducing the release of transmitter.
> **Electrical inhibition** is mediated by currents in presynaptic fibers that hyperpolarize the postsynaptic cell and do not involve the secretion of a chemical transmitter.

initial segment   Region of an axon close to the cell body; often the site of impulse initiation.

integration   Process whereby a neuron sums the various excitatory and inhibitory influences converging upon it and synthesizes a new output signal.

intercellular clefts   Narrow fluid-filled spaces between membranes of adjacent cells; usually about 20 nm wide.

interneuron   Neuron that is neither purely sensory nor motor but connects other neurons.

internode Myelinated portion of a nerve axon lying between two nodes of Ranvier (see below).

intrafusal fiber Muscle fiber within a muscle spindle; its contraction initiates or modulates sensory discharge.

ionophoresis Transfer of ions by passing current through a micropipette; used for applying charged molecules with a high degree of temporal and spatial resolution.

ipsilateral On the same side of the body.

ipsp Inhibitory postsynaptic potential.

lateral geniculate nucleus Small knee-shaped nucleus; part of posteroinferior aspect of thalamus acting as a relay in the visual pathway.

length constant ($\lambda = \sqrt{r_m/r_i}$) Distance (usually in millimeters) over which a localized graded potential decreases to 1/e of its original size in an axon or a muscle fiber.

Mauthner cell Large nerve cell in the mesencephalon of fishes and amphibians, up to 1 mm in length.

mepp Miniature end plate potential; small depolarization at neuromuscular synapse caused by spontaneous release of a single quantum of transmitter from the presynaptic terminal.

modality Class of sensation (e.g., touch, vision, olfactory).

motoneuron (motor neuron) A neuron that innervates muscle fibers.

motor unit A single motoneuron and the muscle fibers it innervates.

muscle spindle Fusiform end organ in skeletal muscle in which afferent sensory fibers and a few motoneurons terminate.

myelin Fused membranes of Schwann cells or glial cells forming a high-resistance sheath surrounding an axon.

naloxone Specific antagonist against opiates.

neostigmine An anticholinesterase; also known as prostigmine.

neuroglia Non-neuron satellite cells associated with neurons. In the mammalian central nervous system the main groupings are astrocytes and oligodendrocytes; in peripheral nerves the satellite cells are called Schwann cells.

neuropil Network of axons, dendrites, and synapses.

node of Ranvier Localized area devoid of myelin occurring at intervals along a myelinated axon.

noise Fluctuations in membrane potential or current due to random opening and closing of channels.

norepinephrine (noradrenaline) Transmitter liberated by most sympathetic nerve terminals.

ocular dominance Greater effectiveness of one eye over the other for driving simple or complex cells in the visual cortex.

Ohm's law Relates current (I) to voltage (V) and resistance (R); $I = V/R$.

opiate Term denoting products derived from juice of opium poppy.

opioid Any directly acting compound whose actions are similar to those of opiates and are specifically antagonized by naloxone.

optic chiasm The point of crossing or decussation of the optic nerves. In cats and primates fibers arising from the medial part of the retina cross to supply the lateral geniculate nucleus on the other side of the animal.

ouabain G-strophanthidin, a glycoside that specifically blocks the sodium-potassium coupled pump.

overshoot Reversal of the membrane potential during the peak of the action potential.

patch clamp   A technique whereby a small patch of membrane is sealed to the tip of a micropipette, enabling currents through single membrane channels to be recorded.

permeability   Property of the membrane allowing substances to pass into or out of the cell.

pump   Active transport mechanism.

quantal release   Secretion of multimolecular packets (quanta) of transmitter by the presynaptic nerve terminal.

quantal size   Number of molecules of neurotransmitter in a quantum.

quantum content   Number of quanta in a synaptic response.

receptive field   Denotes the area of the periphery whose stimulation influences the firing of a neuron. For cells in the visual pathway, the receptive field refers to an area on the retina whose illumination influences the activity of a neuron.

receptor   1. Sensory nerve terminal. 2. A molecule in the cell membrane that combines with a specific chemical substance.

receptor potential   Graded, localized potential change in a sensory receptor initiated by the appropriate stimulus; the electrical sign of the transduction process.

reciprocal innervation   Interconnections of neurons arranged so that pathways exciting one group of muscles inhibit the antagonistic motoneurons.

reflex   Involuntary movement or other response elicited by a stimulus applied to the periphery, transmitted to the central nervous system, and reflected back out to the periphery.

refractory period   The time following an impulse during which a stimulus cannot elicit a second impulse.

resistance of the membrane (R_m)   Property of the cell membrane reflecting the difficulty ions encounter in moving across it. Inverse of conductance.

resting potential   The steady electrical potential across the membrane in the quiescent state.

reversal potential   The value of the membrane potential at which a chemical transmitter produces no change in potential.

Ringer's fluid   A saline solution containing sodium chloride, potassium chloride, and calcium chloride; named after Sydney Ringer.

sagittal   Section in the anteroposterior direction.

saltatory conduction   Conduction along a myelinated axon whereby the impulse leaps from node to node.

Schwann cell   Satellite cell in the peripheral nervous system, responsible for making the myelin sheath.

serotonin   Also known as 5-hydroxytryptamine or 5-HT; transmitter in molluscan nervous system and a candidate in the vertebrate central nervous system.

siemens (S)   Units of conductance; reciprocal of ohms.

soma   Cell body

striate cortex   Also known as area 17 or visual I; primary visual region of occipital lobe marked by striation of Gennari, visible with the naked eye.

subarachnoid space   Space filled by cerebrospinal fluid between two layers of connective tissue, the meninges, surrounding the brain.

supersensitivity   Increase in sensitivity to chemical transmitters in neurons, gland, or muscle cells following denervation.

synapse   Site at which neurons make functional contact; a term coined by Sherrington.

synaptic cleft   The space between the membranes of the pre- and post-synaptic cells at a chemical synapse across which transmitter must diffuse.

synaptic vesicles   Small membrane-bounded sacs contained in presynaptic nerve terminals. Those with dense cores contain catecholamines and serotonin; clear vesicles are presumed to be the storage sites for other transmitters.

tetraethylammonium (TEA)   Quaternary ammonium compound that selectively blocks potassium conductance channels in neurons and muscle fibers.

tetrodotoxin (TTX)   Toxin from puffer fish that selectively blocks the regenerative sodium conductance channel in neurons and muscle fibers.

threshold   1. Critical value of membrane potential or depolarization at which an impulse is initiated. 2. Minimal stimulus required for a sensation.

tight junction   Site at which fusion occurs between the outer leaflets of membranes of adjacent cells, resulting in a five-layered junction. It is called a **macula occludens** if the area is a spot and a **zonula occludens** if the junction is a circumferential ring. Such complete junctions prevent the movement of substances through the extracellular space between the cells.

time contant τ   A measure of the rate of buildup or decay of a localized graded potential, depending on the resistance and the capacitance of the membrane.

transducer   Device for converting one form of energy into another (e.g., a microphone, photoelectric cell, loudspeaker).

transmitter   Chemical substance liberated by a presynaptic nerve terminal causing an effect on the membrane of the postsynaptic cell, usually an increase in permeability to one or more ions.

undershoot   Transient hyperpolarization following an action potential; caused by increased potassium conductance.

ventricles   Cavities within the brain containing cerebrospinal fluid and lined by ependymal cells.

voltage clamp   Technique for displacing membrane potential abruptly to a desired value and keeping the potential constant while measuring currents across the cell membrane; devised by Cole and Marmont.

white matter   Part of the central nervous system appearing white; consisting of myelinated fiber tracts.

BIBLIOGRAPHY

THE NUMBERS THAT FOLLOW EACH ENTRY IDENTIFY THE CHAPTER(S) IN WHICH THE REFERENCE IS CITED.

Adams, P. R., Brown, D. A. and Constanti, A. 1982a. Pharmacological inhibition of the M-current. *J. Physiol. 332*: 223–262. (12)

Adams, P. R., Brown, D. A. and Constanti, A. 1982b. M-currents and other potassium currents in bullfrog sympathetic neurones. *J. Physiol. 330*: 537–572. (6, 9)

Adams, W. B. and Levitan, I. B. 1982. Intracellular injection of protein kinase inhibitor blocks the serotonic-induced increase in K^+ conductance in *Aplysia* neuron R 15. *Proc. Natl. Acad. Sci. USA 79*: 3877–3880. (9, 18)

Adrian, E. D. 1946. *The Physical Background of Perception*. Clarendon Press, Oxford. (1, 17)

Adrian, E. D. 1959. *The Mechanism of Nervous Action*. University of Pennsylvania Press, Philadelphia. (16, 17)

Agnew, W. S., Levinson, S. R., Brabson, J. S. and Raftery, M. A. 1978. Purification of the tetrodotoxin binding component associated with the voltage-sensitive sodium channel of *Electrophorus electricus* electroplax membrane. *Proc. Natl. Acad. Sci. USA 75*: 2606–2610. (6)

Aguayo, A. J., Bray, G. M. and Perkins, S. C. 1979. Axon–Schwann cell relationships in neuropathies of mutant mice. *Ann. N. Y. Acad. Sci. 317*: 512–531. (13)

Aguayo, A. J. and Bray, G. M. 1982. Developmental disorders of myelination in mouse mutants. In T. A. Sears (ed.). *Neuronal-Glial Cell Interrelationships*. Springer-Verlag, Berlin, pp. 57–75. (13)

Aguayo, A. J., Charron, L. and Bray, G. M. 1976. Potential of Schwann cells from unmyelinated nerves to produce myelin: A quantitative ultrastructural and radiographic study. *J. Neurocytol. 5*: 565–573. (13)

Aguayo, A. J., Dickson, R., Trecarten, J., Attiwell, M., Bray, G. M. and Richardson, P. 1978. Ensheathment and myelination of regenerating PNS fibres by transplanted optic nerve glia. *Neurosci. Let. 9*: 97–104. (19)

Aguayo, A. J., Kasarjian, J., Skamene, E., Kongshavn, P. and Bray, G. M. 1977. Myelination of mouse axons by Schwann cells transplanted from normal and abnormal human nerves. *Nature 268*: 753–755. (13)

Aguayo, A. J., Richardson, P. M., David, S. and Benfey, M. 1982. Transplantation of neurons and sheath cells: A tool for the study of regeneration. In J. G. Nicholls (ed.). *Repair and Regeneration of the Nervous System*. Springer-Verlag, New York, pp. 91–105. (19)

Akazawa, K., Aldridge, J. W., Steeves, J. D. and Stein, R. B. 1982. Modulation of stretch reflexes during locomotion in the mesencephalic cat. *J. Physiol. 329*: 553–567. (17)

Alexandrowicz, J. S. 1951. Muscle receptor organs in the abdomen of *Homarus vulgaris* and *Parlinurus vulgaris*. *Q. J. Microsc. Sci. 92*: 163–199. (15)

Almers, W. and Levinson, S. R. 1975. Tetrodotoxin binding to normal and depolarized frog muscle and the conductance of a single sodium channel. *J. Physiol. 247*: 483–509. (6)

Almers, W., Stanfield, P. and Stuhmer, W. 1983. Lateral distribution of sodium and potassium channels in frog skeletal muscle: Measurements with a patch-clamp technique. *J. Physiol. 336*: 261–284. (6)

Ames, A., Higashi, K. and Nesbett, F. B. 1965. Relation of potassium concentration in choroid-plexus fluid to that in plasma. *J. Physiol. 181*: 506–515. (14)

Andersen, P., Silfrenius, H., Sundong, S. H. and Sveen, O. 1980. A comparison of distal and proximal dendrite synapses on CA1 pyramids in guinea pig hippocampal slices *in vitro*. *J. Physiol. 307*: 273–299. (1)

Anderson, C. R. and Stevens, C. F. 1973. Voltage clamp analysis of acetylcholine produced end-plate current fluctuations at frog neuromuscular junction. *J. Physiol. 235*: 655–691. (11)

Angeletti, R. H., Hermodson, M. A. and Bradshaw, R. A. 1973. Amino acid sequences of mouse 2.5 S nerve growth factor. II. Isolation and characterization of the thermolytic and peptic peptides and the complete covalent structure. *Biochemistry 12*: 100–115. (19)

Angeletti, R. H., Mercanti, D. and Bradshaw, R. A. 1973. Amino acid sequences of mouse 2.5 S nerve growth factor. I. Isolation and characterization of the soluble tryptic and chymotryptic peptides. *Biochemistry 12*: 90–100, (19)

Arbuthnott, E. R., Ballard, K. J., Boyd, I. A., Gladden, M. H. and Sutherland, F. I. 1982. The ultrastructure of cat fusimotor endings and their relationship to foci of sarcomere convergence in intrafusal fibres. *J. Physiol. 331*: 285–309. (15)

Arbuthnott, E. R., Boyd, I. A. and Kalu, K. U. 1980. Ultrastructural dimensions of myelinated peripheral nerve fibres in the cat and their relation to conduction velocity. *J. Physiol. 308*: 125–127. (7)

Armstrong, C. M. 1981. Sodium channels and gating currents. *Physiol. Rev. 61*: 644–683. (6)

Armstrong, C. M., Bezanilla, R. and Rojas, E. 1973. Destruction of sodium conductance inactivation in squid axons perfused with pronase. *J. Gen. Physiol. 62*: 375–391. (6)

Armstrong, C. M. and Bezanilla, F. 1974. Charge movement associated with the opening and closing of the activation gates of Na channels. *J. Gen. Physiol. 63*: 533–552. (6)

Armstrong, C. M. and Hille, B. 1972. The inner quaternary ammonium ion receptor in potassium channels of the node of Ranvier. *J. Gen. Physiol. 59*: 388–400. (6)

Art, J. J., Crawford, A. C., Fettiplace, R. and Fuchs, P. A. 1982. Efferent regulation of hair cells in the turtle cochlea. *Proc. R. Soc. Lond. B 216*: 377–384. (15)

Arvanitaki, A. and Cardot, H. 1941. Les caractéristiques de l'activité rythmique ganglionnaire "spontanée" chez l'Aplysie. *C. R. Soc. Biol. 135*: 1207–1211. (18)

Asanuma, H. 1975. Recent developments in the study of the columnar arrangement of neurons within the motor cortex. *Physiol. Rev. 55*: 143–156. (3, 17)

Ascher, P. 1972. Inhibitory and excitatory effects of dopamine on *Aplysia* neurones. *J. Physiol. 225*: 173–209. (9)

Ascher, P., Marty, A. and Neild, O. 1978. The mode of action of antagonists of the excitatory response to acetylcholine in *Aplysia* neurones. *J. Physiol. 278*: 207–235. (18)

Attardi, D. G. and Sperry, R. W. 1963. Preferential selection of central pathways by regenerating optic fibers. *Exp. Neurol. 7*: 46–64. (19)

Atwood, H. L. 1982. Synapses and Neurotransmitters. In H. L. Atwood and D. C. Sandeman (ed.). *Biology of Crustacea*, Vol. 3, Academic, New York, pp. 105–150. (16)

Atwood, H. L. and Bittner, G. D. 1971. Matching of excitatory and inhibitory inputs to crustacean muscle fibers. *J. Neurophysiol. 34*: 157–170. (16)

Atwood, H. L., Lang, F. and Morin, W. A. 1972. Synaptic vesicles: Selective depletion of crayfish excitatory and inhibitory axons. *Science 176*: 1353–1355. (11)

Atwood, H. L. and Morin, W. A. 1970. Neuromuscular and axoaxonal synapses of the crayfish opener muscle. *J. Ultrastruct. Res. 32*: 351–369. (9, 16)

Axelrod, J. 1971. Noradrenaline: Fate and control of its biosynthesis. *Science 173*: 598–606. (12)

Axelsson, J. and Thesleff, S. 1959. A study of supersensitivity in denervated mammalian skeletal muscle. *J. Physiol. 147*: 178–193. (19)

Bainton, C. R., Kirkwood, P. A. and Sears, T. A. 1978. On the transmission of the stimulating effects of carbon dioxide to the muscles of respiration. *J. Physiol. 280*: 249–272. (17)

Baker, P. F., Blaustein, M. P., Hodgkin, A. L. and Steinhardt, R. A. 1969. The influence of calcium on sodium efflux in squid axons. *J. Physiol. 200*: 431–458. (8)

Baker, P. F., Blaustein, M. P., Keynes, R. D., Manil, J., Shaw, T. I. and Steinhardt, R. A. 1969. The ouabain-sensitive fluxes of sodium and potassium in squid giant axons. *J. Physiol. 200*: 459–496. (8)

Baker, P. F., Foster, R. F., Gilbert, D. S. and Shaw, T. I. 1971. Sodium transport by perfused giant axons of *Loligo*. *J. Physiol. 219*: 487–506. (8)

Baker, P. F., Hodgkin, A. L. and Ridgway, E. B. 1971. Depolarization and calcium entry in squid giant axons. *J. Physiol. 218*: 709–755. (5, 6, 8)

Baker, P. F., Hodgkin, A. L. and Shaw, T. I. 1962a. Replacement of the axoplasm of giant nerve fibres with artificial solutions. *J. Physiol. 164*: 330–354. (5)

Baker, P. F., Hodgkin, A. L. and Shaw, T. I. 1962b. The effects of changes in internal ionic concentrations on the electrical properties of perfused giant axons. *J. Physiol. 164*: 355–374. (5)

Baldessarini, R. J. and Tarsy, D. 1980. Dopamine and the pathophysiology of dyskinesias induced by antipsychotic drugs. *Annu. Rev. Neurosci. 3*: 23–41. (12)

Ballivet, M., Patrick, J., Lee, J. and Heinemann, S. 1982. Molecular cloning of cDNA coding for the gamma subunit of *Torpedo* acetylcholine receptor. *Proc. Natl. Acad. Sci. USA 79*: 4466–4470. (11)

Banks, R. W., Barker, D., Bessou, P., Pages, B. and Stacey, M. J. 1978. Histological analysis of muscle spindles following direct observation of effects of stimulating dynamic and static motor axons. *J. Physiol. 283*: 605–619. (15)

Barde, Y.-A., Edgar, D. and Thoenen, H. 1983. New neurotrophic factors. *Annu. Rev. Physiol. 45*: 601–612. (19)

Barker, D., Emonet-Dénand, F., Laporte, Y. and Stacey, M. J. 1980. Identification of the intrafusal endings of skeletofusimotor axons in the cat. *Brain Res. 185*: 227–237. (15)

Barker, D., Stacey, M. J. and Adal, M. N. 1970. Fusimotor innervation in the cat. *Phil. Trans. R. Soc. Lond. B 258*: 315–346. (15)

Barker, J. L., McBurney, R. N. and McDonald, J. F. 1982. Fluctuation analysis of neutral amino acid responses in cultured mouse spinal neurones. *J. Physiol. 322*: 365–387 (11)

Barlow, H. B. 1953. Summation and inhibition in the frog's retina. *J. Physiol. 119*: 69–88. (2)

Barlow, H. B., Hill, R. M. and Levick, W. R. 1964. Retinal ganglion cells responding selectively to direction and speed of image motion in the rabbit. *J. Physiol. 173*: 377–407. (2)

Barlow, H. B. and Levick, W. R. 1965. The mechanism of directionally selective units in rabbit's retina. *J. Physiol. 178*: 477–504. (2)

Barnard, E. A., Wieckowski, J. and Chiu, T. H. 1971. Cholinergic receptor molecules and cholinesterase molecules at mouse skeletal muscle junction. *Nature 234*: 207–209. (19)

Barrett, J. N. and Crill, W. E. 1974a. Specific membrane properties of cat motoneurones. *J. Physiol. 239*: 301–324. (16)

Barrett, J. N. and Crill, W. E. 1974b. Influence of dendrite location and membrane properties on the effectiveness of synapses on cat motoneurones. *J. Physiol. 239*: 325–345. (16)

Basbaum, A. I. and Fields, H. L. 1979. The origin of descending pathways in the dorsolateral funiculus of the spinal cord of the cat and rat: Further studies on the anatomy of pain modulation. *J. Comp. Neurol. 187*: 513–532. (17)

Bayliss, W. M. and Starling, E. H. 1902. The mechanism of pancreatic secretion. *J. Physiol. 28*: 325–353. (12)

Baylor, D. A. and Fettiplace, R. 1977a. Transmission from photoreceptors to ganglion cells in turtle retina. *J. Physiol. 271*: 391–424. (2)

Baylor, D. A. and Fettiplace, R. 1977b. Kinetics of synaptic transfer from receptors to ganglion cells in turtle retina. *J. Physiol. 271*: 425–448. (2)

Baylor, D. A., Fuortes, M. G. F. and O'Bryan, P. M. 1971. Receptive fields of cones in the retina of the turtle. *J. Physiol. 214*: 265–294. (2)

Baylor, D. A. and Hodgkin, A. L. 1973. Detection and resolution of visual stimuli by turtle photoreceptors. *J. Physiol. 234*: 163–198. (2)

Baylor, D. A., Lamb, T. D. and Yau, K.-W. 1979. The membrane current of single rod outer segments. *J. Physiol. 288*: 589–611. (2)

Baylor, D. A. and Nicholls, J. G. 1969. Changes in extracellular potassium concentration produced by neuronal activity in the central nervous system of the leech. *J. Physiol. 203*: 555–569. (13)

Benfey, M. and Aguayo, A. J. 1982. Extensive elongation of axons from rat brain into peripheral nerve grafts. *Nature 296*: 150–152. (19)

Bennett, M. R., Florin, T. and Woog, R. 1974. The formation of synapses in regenerating mammalian striated muscle. *J. Physiol. 238*: 79–92. (19)

Bennett, M. V. L. 1973. Function of electrotonic junctions in embryonic and adult tissues. *Fed. Proc. 32*: 65–75. (9)

Bennett, M. V. L. (ed.). 1974. *Synaptic Transmission and Neuronal Interaction.* Raven Press, New York, pp. 153–178. (9)

Bentley, D. and Hoy, R. R. 1974. The neurobiology of cricket song. *Sci. Am. 231*: 34–44. (18)

Bentley, D. and Konishi, M. 1978. Neural control of behavior. *Annu. Rev. Neurosci. 1*: 35–59. (18)

Benzer, S. 1971. From the gene to behavior. *J.A.M.A. 218*: 1015–1022. (Part 5)

Berg, D. K. and Hall, Z. W. 1975. Increased extrajunctional acetylcholine sensitivity produced by chronic postsynaptic neuromuscular blockade. *J. Physiol. 244*: 659–676. (19)

Berg, D. K., Kelly, R. B., Sargent, P. B., Williamson, P. and Hall, Z. W. 1972. Binding of α-bungarotoxin to acetylcholine receptors in mammalian muscle. *Proc. Natl. Acad. Sci. USA 69*: 147–151. (19)

Bernstein, J. 1902. Untersuchungen zur Thermodynamik der bioelektrischen Ströme. *Pflügers Arch. 92*: 521–562 (5)

Berlucchi, G. and Rizzolatti, G. 1968. Binocularly driven neurons in visual cortex of split-chiasm cats. *Science 159*: 308–310. (2)

Berthold, C.-H., Kellerth, J.-O. and Conradi, S. 1979. Electron microscopic studies of serially sectioned cat spinal α-motoneurons. *J. Comp. Neurol. 184*: 709–740. (16)

Betz, W. J., Caldwell, J. H. and Ribchester, R. R. 1980. The effects of partial denervation at birth on the development of muscle fibres and motor units in rat lumbrical muscle. *J. Physiol. 303*: 265–279. (19)

Betz, W. J. and Sakmann, B. 1973. Effects of proteolytic enzyme on function and structure of frog neuromuscular junctions. *J. Physiol. 230*: 673–688. (11)

Bignami, A. and Dahl, D. 1974. Astrocyte-specific protein and neuroglial differentiation: An immunofluorescence study with antibodies to the glial fibrillary acidic protein. *J. Comp. Neurol. 153*: 27–38. (13)

Birks, R., Katz, B. and Miledi, R. 1960. Physiological and structural changes at the amphibian myoneural junction in the course of nerve degeneration. *J. Physiol. 150*: 145–168. (10)

Birks, R. I. 1974. The relationship of transmitter release and storage to fine structure in a sympathetic ganglion. *J. Neurocytol. 3*: 133–160. (11)

Bishop, P. O., Coombs, J. S. and Henry, G. H. 1971. Responses to visual contours: Spatio-temporal aspects of excitation in the receptive fields of simple striate neurones. *J. Physiol. 219*: 625–657. (2)

Bittner, G. D. 1981. Trophic interactions of CNS giant axons in crayfish. *Comp. Biochem. Physiol. 68A*: 299–306. (13)

Bixby, J. L. and Van Essen, D. C. 1979. Competition between foreign and original nerves in adult mammalian skeletal muscle. *Nature 282*: 726–728. (19)

Björklund, A., Dunnett, S. B., Stenevi, U., Lewis, N. E. and Iversen, S. D. 1980. Reinnervation of the denervated striatum by substantia nigra transplants: Functional consequences as revealed by pharmacological and sensorimotor testing. *Brain Res. 199*: 307–333. (19)

Björklund, A. and Stenevi, U. 1979. Regeneration of monoaminergic and cholinergic neurons in the mammalian central nervous system. *Physiol. Rev. 59*: 62–100. (19)

Black, I. B. 1978. Regulation of autonomic development. *Annu. Rev. Neurosci. 1*: 183–214. (19)

Black, I. B. 1979. Neuronal responses to extracellular signals. In J. G. Nicholls, (ed.). *The Role of Intercellular Signals: Navigation, Encounter, Outcome.* Dahlem Konferenzen, Berlin. (19)

Blackman, J. G., Ginsborg, B. L. and Ray, C. 1963. Synaptic transmission in the sympathetic ganglion of the frog. *J. Physiol. 167*: 355–373. (9)

Blackman, J. G. and Purves, R. D. 1969. Intracellular recordings from ganglia of the thoracic sympathetic chain of the guinea-pig. *J. Physiol. 203*: 173–198. (10)

Blackshaw, S. 1981a. In K. J. Muller et al. (eds.). *Neurobiology of the Leech.* Cold Spring Harbor Laboratory, Cold Spring Harbor, NY, pp. 51–78. (18)

Blackshaw, S. 1981b. Morphology and distribution of touch cell terminals in the skin of the leech. *J. Physiol. 320*: 219–228. (18)

Blackshaw, S. E., Nicholls, J. G. and Parnas, I. 1982. Expanded receptive fields of cutaneous mechanoreceptor cells after single neurone deletion in leech central nervous system. *J. Physiol. 326*: 261–268. (18, 19)

Blakemore, C. 1974. Development of functional connections in the mammalian visual system. *Br. Med. Bull. 30*: 152–157. (20)

Blakemore, C. and Van Sluyters, R. C. 1974. Reversal of the physiological effects of monocular deprivation in kittens: Further evidence for a sensitive period. *J. Physiol. 237*: 195–216. (20)

Blasdel, G. G. and Lund, J. S. 1983. Termination of afferent axons in macaque striate cortex. *J. Neurosci. 3*: 1389–1413. (3)

Blaustein, M. P. 1977. Effects of internal and external cations and of ATP on sodium–calcium and calcium–calcium exchange in squid axons. *Biophys. J. 20*: 79–111. (8)

Blaustein, M. P. and Hodgkin, A. L. 1969. The effect of cyanide on the efflux of calcium from squid axons. *J. Physiol. 200*: 497–528. (8)

Bloom, F. E. 1972. Electron microscopy of catecholamine-containing structures. *Handbook Exp. Pharmacol. 33*: 46–78. (11)

Bodenheimer, T. S. and Brightman, M. W. 1968. A blood-brain barrier to peroxidase in capillaries surrounded by perivascular spaces. *Am. J. Anat. 122*: 249–268. (14)

Bodian, D. 1937. The structure of the vertebrate synapse: A study of the axon endings on Mauthner's cell and neighboring centers in the goldfish. *J. Comp. Neurol. 68*: 117–159. (16)

Bodian, D. 1942. Cytological aspects of synaptic function. *Physiol. Rev. 22*: 146–169. (16)

Bodian, D. 1966. Development of the fine structure of spinal cord in monkey fetuses. I. The motoneuron neuropil at the time of onset of reflex activity. *Bull. Johns Hopkins Hosp. 119*: 129–149. (13)

Boistel, J. and Fatt, P. 1958. Membrane permeability change during inhibitory transmitter action in crustacean muscle. *J. Physiol. 144*: 176–191. (12)

Bostock, H. and Sears, T. A. 1978. The internodal axon membrane: Electrical excitability and continuous conduction in segmental demyelination. *J. Physiol. 280*: 273–301. (7, 13)

Bostock, H., Sears, T. A. and Sherratt, R. M. 1981. The effects of 4-aminopyridine and tetraethylammonium ions on normal and demyelinated mammalian nerve fibres. *J. Physiol. 313*: 301–315. (7)

Bowery, N. G. and Brown, D. A. 1972. γ-Aminobutyric acid uptake by sympathetic ganglia. *Nature New Biol. 238*: 89–91. (13)

Bowery, N. G., Brown, D. A. and Marsh, S. 1979. γ-Aminobutyric acid efflux from sympathetic glial cells: Effect of 'depolarizing' agents. With an Appendix by R. R. Adams & D. A. Brown. *J. Physiol. 293*: 75–101. (13)

Bowery, N. G., Brown, D. A., White, R. D. and Yamini, G. 1979. (^{3}H)-γ-Aminobutyric acid uptake into neuroglial cells of rat superior cervical sympathetic ganglia. *J. Physiol. 293*: 51–74. (13)

Bowling, D. B. and Michael, C. R. 1980. Projection patterns of single physiologically characterized optic tract fibres in cat. *Nature 286*: 899–902. (2)

Bowling, D., Nicholls, J. G. and Parnas, I. 1978. Destruction of a single cell in the central nervous system of the leech as a means of analyzing its connexions and functional role. *J. Physiol. 282*: 169–180. (18)

Boycott, B. B. and Dowling, J. E. 1969. Organization of primate retina: Light microscopy. *Phil. Trans. R. Soc. Lond. B 255*: 109–184. (1)

Boyd, I. A. 1962. The structure and innervation of the nuclear bag muscle fibre system and the nuclear chain muscle fibre system in mammalian muscle spindles. *Phil. Trans. R. Soc. Lond. B 245*: 81–136. (15)

Boyd, I. A. and Martin, A. R. 1956. The end-plate potential in mammalian muscle. *J. Physiol. 132*: 74–91. (10)

Boyle, P. J. and Conway, E. J. 1941. Potassium accumulation in muscle and associated changes. *J. Physiol. 100*: 1–63. (5)

Bradbury, M. 1979. *The Concept of a Blood-Brain Barrier.* John Wiley & Sons, Chichester. (14)

Braitenberg, V. and Atwood, R. P. 1958. Morphological observations on the cerebellar cortex. *J. Comp. Neurol. 109*: 1–33. (1)

Bray, G. M., Rasminsky, M. and Aguayo, A. J. 1981. Interactions between axons and their sheath cells. *Annu. Rev. Neurosci. 4*: 127–162. (13)

Brenner, H. R. and Johnson, E. W. 1976. Physiological and morphological effects of post-ganglionic axotomy on presynaptic nerve terminals. *J. Physiol. 260*: 143–158. (19)

Brenner, H. R. and Martin, A. R. 1976. Reduction in acetylcholine sensitivity of axotomized ciliary ganglion cells. *J. Physiol. 260*: 159–175. (19)

Brenner, H. R. and Sakmann, B. 1983. Neurotrophic control of channel properties at neuromuscular synapses of rat muscle. *J. Physiol. 337*: 159–172. (19)

Brigant, J. L. and Mallart, A. 1982. Presynaptic currents in mouse motor endings. *J. Physiol. 333*: 619–636. (10)

Brightman, M. W. 1965. The distribution within the brain of ferritin injected into cerebrospinal fluid compartment. II. Parenchymal distribution. *Am. J. Anat. 117*: 193–219. (14)

Brightman, M. W., Klatzo, I., Olsson, Y. and Reese, T. S. 1970 The blood-brain barrier to proteins under normal and pathological conditions. *J. Neurol. Sci. 10*: 215–239. (14)

Brightman, M. W. and Reese, T. S. 1969. Junctions between intimately apposed cell membranes in the vertebrate brain. *J. Cell Biol. 40*: 668–677. (13, 14)

Brightman, M. W., Reese, T. S. and Feder, N. 1970. Assessment with the electronmicroscope of the permeability to peroxidase of cerebral endothelium in mice and sharks. In E. H. Thaysen (ed.). *Capillary Permeability*, Alfred Benzon Symposium II. Munskgaard, Copenhagen. (14)

Brinley, F. J. 1980. Regulation of intracellular calcium in squid axons. *Fed. Proc. 39*: 2778–2782. (8)

Brockes, J. P., Fryxell, K. J. and Lemke, G. E. 1981. Studies on cultured Schwann cells: The induction of myelin synthesis, and the control of their proliferation by a new growth factor. *J. Exp. Biol. 95*: 215–230. (13)

Brockes, J. P. and Hall, Z. W. 1975. Acetylcholine receptors in normal and denervated rat diaphragm muscle. II. Comparison of junctional and extrajunctional receptors. *Biochemistry 14*: 2100–2106. (19)

Brody, H. 1955. Organization of the cerebral cortex. IV. A study of aging in the human cerebral cortex. *J. Comp. Neurol. 102*: 511–556. (Part 5)

Brooks, V. B. 1956. An intracellular study of the action of the repetitive nerve volleys and of botulinum toxin on miniature end-plate potentials. *J. Physiol. 134*: 264–277. (10)

Brown, A. G. and Fyffe, R. E. W. 1981. Direct observations on the contacts made between Ia afferent fibres and α-motoneurones in the cat's lumbosacral spinal cord. *J. Physiol. 313*: 121–140. (16)

Brown, D. A., Caulfield, M. P. and Kirby, P. J. 1979. Relation between catecholamine-induced cyclic AMP changes and hyperpolarization in isolated rat sympathetic ganglia. *J. Physiol. 290*: 441–451. (9)

Brown, G. L. 1937. The actions of acetylcholine on denervated mammalian and frog's muscle. *J. Physiol. 89*: 438–461. (19)

Brown, H. M., Ottoson, D. and Rydqvist, B. 1978. Crayfish stretch receptor: An investigation with voltage-clamp and ion-sensitive electrodes. *J. Physiol. 284*: 155–179. (15)

590
BIBLIOGRAPHY

Brown, M. C., Jansen, J. K. S. and Van Essen, D. 1976. Polyneuronal inner-vation of skeletal muscle in new-born rats and its elimination during maturation. *J. Physiol. 261*: 387–422. (19)

Brown, T. G. 1911. The intrinsic factor in the act of progression in the mammal. *Proc. R. Soc. Lond. B 84*: 308–319. (17)

Buller, A. J. 1970. The neural control of the contractile mechanisms in skeletal muscle. *Endeavour 29*: 107–111. (19)

Buller, A. J., Eccles, J. C. and Eccles, R. M. 1960. Differentiation of fast and slow muscles in the cat hind limb. *J. Physiol. 150*: 399–416. (19)

Bullock, T. H. and Hagiwara, S. 1957. Intracellular recording from the giant synapse of the squid. *J. Gen. Physiol. 40*: 565–577. (10)

Bundgaard, M. and Cserr, H. F. 1981. A glial blood-brain barrier in elasmobranches. *Brain Res. 226*: 61–73. (14)

Bunge, R. P. 1968. Glial cells and the central myelin sheath. *Physiol. Rev. 48*: 197–251. (13)

Burden, S., Hartzell, H. C. and Yoshikami, D. 1975. Acetylcholine receptors at neuromuscular synapses: Phylogenetic differences detected by snake alpha-neurotoxins. *Proc. Natl. Acad. Sci. USA 72*: 3245–3249. (11)

Burden, S. J., Sargent, P. B. and McMahan, U. J. 1979. Acetylcholine receptors in regenerating muscle accumulate at original synaptic sites in the absence of the nerve. *J. Cell Biol. 82*: 412–425. (19)

Burgess, P. R., English, K. B., Horch, K. W. and Stensaas, L. J. 1974. Patterning in the regeneration of type I cutaneous receptors. *J. Physiol. 236*: 57–82. (19)

Burgess, P. R. and Wei, J. Y. 1982. Signaling of kinesthetic information by peripheral sensory receptors. *Annu. Rev. Neurosci. 5*: 171–187. (15)

Burke, R. E. 1978. Motor units: Physiological histochemical profiles, neural connectivity and functional specialization. *Am. Zool. 18*: 127–134. (16)

Burke, R. E. 1981. Motor units: Anatomy, physiology and functional organization. In V. Brooks (ed.). *Handbook of Physiology, Section I: The Nervous System, Volume 2, part 1*. Amer. Physiol. Soc., Bethesda, pp. 345–422. (16)

Burke, R. E., Walmsley, B. and Hodgson, J. A. 1979. HRP anatomy of group Ia afferent contacts on alpha motoneurones. *Brain Res. 160*: 347–352. (16)

Byrne, J., Castellucci, V. F. and Kandel, E. R. 1974. Receptive fields and response properties of mechanoreceptor neurons innervating siphon skin and mantle shelf in *Aplysia*. *J. Neurophysiol. 37*: 1041–1064. (18)

Caldwell, J. H. and Daw, N. W. 1978. Effects of picrotoxin and strychnine on rabbit retinal ganglion cells: Changes in centre surround receptive fields. *J. Physiol. 276*: 299–310. (12)

Caldwell, P. C., Hodgkin, A. L., Keynes, R. D. and Shaw, T. I. 1960. The effects of injecting 'energy-rich' phosphate compounds on the active transport of ions in the giant axons of *Loligo*. *J. Physiol. 152*: 561–590. (8)

Campenot, R. B. 1977. Local control of neurite development by nerve growth factor. *Proc. Natl. Acad. Sci. USA 74*: 4516–4519. (19)

Cannon, W. B. and Rosenblueth, A. 1949. *The Supersensitivity of Denervated Structures: Law of Denervation*. Macmillan, New York. (19)

Cantino, D. and Mugnaini, E. 1975. The structural basis for electrotonic coupling in the avian ciliary ganglion: A study with thin sectioning and freeze-fracturing. *J. Neurocytol. 4*: 505–536. (9)

Carew, T. J., Castellucci, V. F. and Kandel, E. R. 1979. Sensitization in *Aplysia*: Rapid restoration of transmission in synapses inactivated by long-term habituation. *Science 205*: 417–419. (18)

Carew, T. J., Hawkins, R. D. and Kandel, E. R. 1983. Differential classical conditioning of a defensive withdrawal reflex in *Aplysia. Science 219*: 397–400. (18)

Carew, T. J., Pinsker, H. M. and Kandel, E. R. 1972. Long-term habituation of a defensive withdrawal reflex in *Aplysia. Science 175*: 451–454. (18)

Carpenter, G. and Cohen, S. 1979. Epidermal growth factor. *Annu. Rev. Biochem. 48*: 193–216. (19)

Carpenter, M. B. 1978. *Core Text of Neuroanatomy*, 2nd Ed., Williams & Wilkins, Baltimore. (1, Appendix B)

Caspar, D. L. D., Goodenough, D. A., Makowski, L. and Phillips, W. C. 1977. Gap junction structures. I. Correlated electron microscopy and x-ray diffraction. *J. Cell Biol. 74*: 605–628. (9)

Castellucci, V. F., Kandel, E. R., Schwartz, J. H., Wilson, F. D., Nairn, A. C. and Greengard, P. 1980. Intracellular injection of the catalytic subunit of cyclic AMP-dependent protein kinase simulates facilitation of transmitter release underlying behavioral sensitization in *Aplysia. Proc. Natl. Acad. Sci. USA 77*: 7492–7496. (18)

Catterall, W. A. 1980. Neurotoxins that act on voltage-sensitive sodium channels in excitable membranes. *Annu. Rev. Pharmacol. Toxicol. 20*: 15–43. (6)

Caviness, V. S. Jr. 1976. Patterns of cell and fiber distribution in the neocortex of the reeler mutant mouse. *J. Comp. Neurol. 170*: 435–448. (20)

Caviness, V. S. Jr. and Rakič, P. 1978. Mechanisms of cortical developments: A view from mutations in mice. *Annu. Rev. Neurosci. 1*: 297–326. (Part 5)

Ceccarelli, B., Hurlburt, W. P. and Mauro, A. 1973. Turnover of transmitter and synaptic vesicles at the frog neuromuscular junction. *J. Cell Biol. 57*: 499–524. (11)

Chiu, S. Y., Ritchie, J. M., Rogart, R. B. and Stagg, D. A. 1979. A quantitative description of membrane currents in rabbit myelinated nerve. *J. Physiol. 292*: 149–166. (6)

Chiu, S. Y. and Ritchie, J. M. 1981. Evidence for the presence of potassium channels in the paranodal region of acutely demyelinated mammalian nerve fibres. *J. Physiol. 313*: 415–437. (7)

Chow, K. L., Mathers, L. H. and Spear, P. D. 1973. Spreading of uncrossed retinal projection in superior colliculus of neonatally enucleated rabbits. *J. Comp. Neurol. 151*: 307–322. (19)

Christensen, B. N. and Perl, E. R. 1970. Spinal neurons specifically excited by noxious or thermal stimuli: Marginal zone of the dorsal horn. *J. Neurophysiol. 33*: 293–307. (17)

Cleland, B. G., Dubin, M. W. and Levick, W. R. 1971. Sustained and transient projection in superior colliculus of neonatally enucleated rabbits. *J. Physiol. 217*: 473–496. (2)

Cleland, B. G., Levick, W. R. and Wässle, H. 1975. Physiological identification of a morphological class of cat retinal ganglion cells. *J. Physiol. 248*: 151–171. (2)

Clementi, F. and Palade, G. E. 1969. Intestinal capillaries. I. Permeability to peroxidase and ferritin. *J. Cell Biol. 41*: 33–58. (14)

Close, R. I. 1972. Dynamic properties of mammalian skeletal muscles. *Physiol. Rev. 52*: 129–197. (19)

Coggeshall, R. E. and Fawcett, D. W. 1964. The fine structure of the central nervous system of the leech, *Hirudo medicinalis. J. Neurophysiol. 27*: 229–289. (13, 18)

Cohen, L. B., Salzberg, B. B. and Grinvald, A. 1978. Optical methods for monitoring neuron activity. *Annu. Rev. Neurosci. 1*: 171–182. (7)

Cohen, M. I. 1979. Neurogenesis of respiratory rhythm in the mammal. *Physiol. Rev. 59*: 1105–1173. (17)

Cohen, M. W. 1970. The contribution by glial cells to surface recordings from the optic nerve of an amphibian. *J. Physiol. 210*: 565–580. (13)

Cohen, M. W. 1972. The development of neuromuscular connexions in the presence of D-tubocurarine. *Brain Res. 41*: 457–463. (19)

Cohen, M. W. 1980. Development of an amphibian neuromuscular junction *in vivo* and in culture. *J. Exp. Biol. 89*: 43–56. (19)

Cohen, M. W., Gerschenfeld, H. M. and Kuffler, S. W. 1968. Ionic environment of neurones and glial cells in the brain of an amphibian. *J. Physiol. 197*: 363–380. (14)

Cohen, S. 1959. Purification and metabolic effects of a nerve growth-promoting protein from snake venom. *J. Biol. Chem. 234*: 1129–1137. (19)

Cohen, S. 1960. Purification of a nerve-growth promoting protein from the mouse salivary gland and its neuro-cytotoxic antiserum. *Proc. Natl. Acad. Sci. USA 46*: 302–311. (19)

Cole, K. S. 1968. *Membranes, Ions and Impulses.* University of California Press, Berkeley. (6)

Coles, J. A. and Tsacopoulos, M. 1981. Ionic and possible metabolic interactions between sensory neurones and glial cells in the retina of the honeybee drone. *J. Exp. Biol. 95*: 75–92. (13)

Collier, B. and MacIntosh, F. C. 1969. The source of choline for acetylcholine synthesis in a sympathetic ganglion. *Can. J. Physiol. Pharmacol. 47*: 127–135. (12)

Colquhoun, D. and Sakmann, B. 1981. Fluctuations in the microsecond time range of the current through single acetylcholine receptor ion channels. *Nature 294*: 464–466. (11)

Conner, J. A. and Stevens, C. F. 1971. Voltage clamp studies of a transient outward membrane current in gastropod neural somata. *J. Physiol. 213*: 21–30. (6)

Connolly, J. A., St. John, P. A. and Fischbach, G. D. 1982. Extracts of electric lobe and electric organ from *Torpedo californica* increase the total number as well as the number of aggregates of chick myotube acetylcholine receptors. *J. Neurosci. 2*: 1207–1213. (19)

Conti, F., De Felice, L. J. and Wanke, E. 1975. Potassium and sodium ion current noise in the membrane of the squid giant axon. *J. Physiol. 248*: 45–82. (6)

Conti, F., Hille, B., Neumcke, B., Nonner, W. and Stämpfli, R. 1976. Measurement of the conductance of the sodium channel from current fluctuations at the node of Ranvier (frog). *J. Physiol. 262*: 729–742. (6)

Conti, F. and Neher, E. 1980. Single channel recordings of K^+ currents in squid axons. *Nature 285*: 140–143. (6)

Conti-Tronconi, B. M. and Raftery, M. A. 1982. The nicotinic cholinergic receptor: Correlation of molecular structure with functional properties. *Annu. Rev. Biochem. 51*: 491–530. (11)

Cooke, J. D. and Quastel, D. M. J. 1973. The specific effect of potassium on transmitter release by motor nerve terminals and its inhibition by calcium. *J. Physiol. 228*: 435–458. (13)

Coombs, J. S., Eccles, J. C. and Fatt, P. 1955. The specific ion conductances and the ionic movements across the motoneuronal membrane that produce the inhibitory post-synaptic potential. *J. Physiol. 130*: 326–373. (9)

Cooper, J. R., Bloom, F. E. and Roth, R. H., 1982. *The Biochemical Basis of Pharmacology,* 4th Ed. Oxford University Press, New York. (12)

Costanzo, R. M. and Gardner, E. P. 1980. A quantitative analysis of responses of direction-sensitive neurons in somatosensory cortex of awake monkeys. *J. Neurophysiol. 43*: 1319–1341. (17)

Couteaux, R. and Pécot-Dechavassine, M. 1970. Vésicules synaptiques et poches au niveau des zones actives de la jonction neuromusculaire. *C.R. Acad. Sci. (Paris) 271*: 2346–2349. (11)

Cowan, W. M. 1973. Neuronal death as a regulation mechanism in the control of cell number in the nervous system. In M. Rockstein (ed.). *Development and Aging in the Nervous System.* Academic, New York. (Part 5)

Cowan, W. M. and Powell, T. P. S. 1963. Centrifugal fibres in the avian visual system. *Proc. R. Soc. Lond. B 158*: 232–252. (2)

Crago, P. E., Houk, J. C. and Rymer, W. Z. 1982. Sampling of total muscle force by tendon organs. *J. Neurophysiol. 47*: 1069–1083. (15)

Craik, K. 1943. *The Nature of Explanation.* Cambridge University Press, London. (2)

Crill, W. E. and Raichle, M. E. 1982. In J. G. Nicholls (ed.). *Repair and Regeneration of the Nervous System.* Springer-Verlag, New York, pp. 227–242. (17)

Critchlow, V. and von Euler, C. 1963. Intercostal muscle spindle activity and its γ-motor control. *J. Physiol. 168*: 820–847. (15, 17)

Crowe, A. and Matthews, P. B. C. 1964. The effects of stimulation of static and dynamic fusimotor fibres on the response to stretching of the primary endings of muscle spindles. *J. Physiol. 174*: 109–131. (15)

Cserr, H. 1971. Physiology of the choroid plexus. *Physiol. Rev. 51*: 273–311. (14)

Cserr, H. and Rall, D. P. 1967. Regulation of cerebrospinal fluid (K^+) in the spiny dogfish, *Squalus acanthias. Comp. Biochem. Physiol. 21*: 431–434. (14)

Cserr, H. F. and Bundgaard, M. 1984. Blood-brain interfaces in vertebrates: A comparative approach. *Am. J. Physiol.* (in press). (14)

Cull-Candy, S. G., Miledi, R. and Parker, I. 1980. Single glutamate-activated channels recorded from locust muscle fibres with perfused patch-clamp electrodes. *J. Physiol. 321*: 195–210. (11)

Cull-Candy, S. G. and Mildei, R. 1981. Junctional and extra-junctional membrane channels activated by GABA in locust muscle fibres. *Proc. R. Soc. Lond. B 211*: 527–535. (11, 12)

Currie, D. N. and Kelly, J. S. 1981. Glial versus neuronal uptake of glutamate. *J. Exp. Biol. 95*: 181–193. (13)

Curtis, D. R. and Eccles, J. C. 1959. Repetitive synaptic activation. *J. Physiol. 149*: 43P–44P. (16)

Curtis, D. R. and Johnston, G. A. R. 1974. Amino acid transmitters in the mammalian central nervous system. *Ergeb. Physiol. 69*: 97–188. (12)

Curtis, H. J. and Cole, K. S. 1940. Membrane action potentials from the squid giant axon. *J. Cell. Comp. Physiol. 15*: 147–157. (5)

Cynader, M. and Mitchell, D. E. 1977. Monocular astigmatism effects on kitten visual cortex development. *Nature 270*: 177–178. (20)

Cynader, M. and Mitchell, D. E. 1980. Prolonged sensitivity to monocular deprivation in dark-reared cats. *J. Neurophysiol. 43*: 1026–1040. (20)

Czarkowska, J., Jankowska, E. and Sybirska, E. 1981. Common interneurones in reflex pathways from group Ia and Ib afferents of knee flexors and extensors in the cat. *J. Physiol. 310*: 367–380. (15, 17)

Dahlström, A. 1971. Axoplasmic transport (with particular respect to adrenergic neurones). *Phil. Trans. R. Soc. Lond. B 261*: 325–358. (12)

Dahlström, A. and Fuxe, K. 1964. Evidence for the existence of monoamine containing neurons in the central nervous system. I. Demonstration of monoamines in the cell bodies of brain stem neurons. *Acta. Physiol. Scand.* 62 (Suppl. 232): 1–55. (12)

Dale, H. H. 1953. *Adventures in Physiology.* Pergamon Press, London. (9)

Dale, H. H., Feldberg, W. and Vogt, M. 1936. Release of acetylcholine at voluntary motor nerve endings. *J. Physiol. 86*: 353–380. (9)

Daniel, P. M. and Whitteridge, D. 1961. The representation of the visual field on the cerebral cortex in monkeys. *J. Physiol. 159*: 203–221. (2)

Da Silva, K. M. C., Sayers, B. McA., Sears, T. A. and Stagg, D. T. 1977. The changes in configuration of the rib cage and abdomen during breathing in the anaesthetized cat. *J. Physiol. 266*: 499–521. (17)

David, S. and Aguayo, A. J. 1981. Axonal elongation into peripheral nervous system "Bridges" after central nervous system injury in adult rats. *Science 214*: 931–933. (19)

Daw, N. W., Rader, R. K., Robertson, T. W. and Ariel, M. 1983. Effects of 6-hydroxydopamine on visual deprivation in the kitten striate cortex. *J. Neurosci. 3*: 907–914. (20)

De Groat, W. C. 1972. GABA-depolarization of a sensory ganglion: Antagonism by picrotoxin and bicuculline. *Brain Res. 38*: 71–88. (12)

del Castillo, J. and Katz, B. 1954a. Quantal components of the end-plate potential. *J. Physiol. 124*: 560–573. (10)

del Castillo, J. and Katz, B. 1954b. Statistical factors involved in neuromuscular facilitation and depression. *J. Physiol. 124*: 574–585. (10)

del Castillo, J. and Katz, B. 1954c. Changes in the end-plate activity produced by presynaptic polarization. *J. Physiol. 124*: 586–604. (10)

del Castillo, J. and Katz, B. 1955. On the localization of acetylcholine receptors. *J. Physiol. 128*: 157–181. (9)

del Castillo, J. and Katz, B. 1956. Biophysical aspects of neuro-muscular transmission. *Prog. Biophys. 6*: 121–170. (9)

del Castillo, J. and Stark, L. 1952. The effect of calcium ions on the motor end-plate potentials. *J. Physiol. 116*: 507–515. (10)

Dennis, M. J., Harris, A. J. and Kuffler, S. W. 1971. Synaptic transmission and its duplication by focally applied acetylcholine in parasympathetic neurones in the heart of the frog. *Proc. R. Soc. Lond. B 177*: 509–539. (9, 11)

Dennis, M. J. and Miledi, R. 1974a. Electrically induced release of acetylcholine from denervated Schwann cells. *J. Physiol. 237*: 431–452. (13)

Dennis, M. J. and Miledi, R. 1974b. Characteristics of transmitter release at regenerating frog neuromuscular junctions. *J. Physiol. 239*: 571–594. (10)

Dennis, M. J. and Yip, J. W. 1978. Formation and elimination of foreign synapses on adult salamander muscle. *J. Physiol. 274*: 299–310. (19)

DeRobertis, E. 1967. Ultrastructure and cytochemistry of the synaptic region. *Science 156*: 907–914. (11)

Diamond, J. 1968. The activation and distribution of GABA and L-glutamate receptors on goldfish Mauthner neurones: An analysis of dendritic remote inhibition. *J. Physiol. 194*: 669–723. (16)

Diamond, J. and Miledi, R. 1962. A study of foetal and new-born muscle fibres. *J. Physiol. 162*: 393–408. (19)

Dickenson-Nelson, A. and Reese, T. S. 1983. Structural changes during transmitter release at synapses in the frog sympathetic ganglion. *J. Neurosci. 3*: 42–52. (11)

Dipolo, R., Requena, J., Brinley, F. J., Mullins, L. J., Scarpa, A. and Tiffert, T.

1976. Ionized calcium concentrations in squid axons. *J. Gen. Physiol. 67*: 433–467. (8)

Dodd, J. and Horn, J. P. 1983. Muscarinic inhibition of sympathetic C neurones in the bullfrog. *J. Physiol. 334*: 271–291. (9)

Dodge, F. A. and Rahamimoff, R. 1967. Cooperative action of calcium ions in transmitter release at the neuromuscular junction. *J. Physiol. 193*: 419–432. (10)

Douglas, W. W. 1978. Stimulus-secretion coupling: Variations on the theme of calcium-activated exocytosis involving cellular and extracellular sources of calcium. *Ciba Fdn. Symp. 54*: 61–90. (10)

Dowdall, M. J., Boyne, A. F. and Whittaker, V. P. 1974. Adenosine triphosphate. A constituent of cholinergic synaptic vesicles. *Biochem. J. 140*: 1–12. (11)

Dowling, J. E. and Boycott, B. B. 1966. Organization of the primate retina: Electron microscopy. *Proc. R. Soc. Lond. B 166*: 80–111. (2)

Dowling, J. E., Lasater, E. M., Van Buskirk, R. and Watling, K. J. 1983. Pharmacological properties of isolated fish horizontal cells. *Vision Res. 23*: 421–432. (2)

Dowling, J. E. and Werblin, F. S. 1971. Synaptic organization of the vertebrate retina. *Vision Res. 3*: 1–15. (2)

Dräger, U. C. 1981. Observations on the organization of the visual cortex in the reeler mouse. *J. Comp. Neurol. 201*: 555–570. (20)

Droz, B., Di Giamberardino, L., Koenig, N. J., Boyenval, J. and Hassig, R. 1978. Axon-myelin transfer of phospholipid components in the course of their axonal transport as visualized by radio-autography. *Brain Res. 155*: 347–353. (13)

Drujan, B. D. and Laufer, M. (eds.). 1982. *The S-Potential*. Vol. 13 of *Prog. Clin. Biol. Res.* Alan R. Liss, New York. (2)

Drujan, B. D. and Svaetichin, G. 1972. Characterization of different classes of isolated retinal cells. *Vision Res. 12*: 1777–1784. (2)

Du Bois-Reymond, E. 1848. *Untersuchungen über thierische Electricität* (Erster Band). Reimer, Berlin. (9)

Dubin, M. W. and Cleland, B. G. 1977. Organization of visual inputs to interneurons of the lateral geniculate nucleus of the cat. *J. Neurophysiol. 40*: 410–427. (2)

Dudel, J., Finger, W. and Stettmeier, H. 1980. ATPase activity in rapidly activated skinned muscle fibres. *Pflügers Arch. 387*: 167–174. (11)

Dudel, J. and Kuffler, S. W. 1961. Presynaptic inhibition at the crayfish neuromuscular junction. *J. Physiol. 155*: 543–562. (9, 10, 12, 16)

Dunnett, S. B., Björklund, A. and Stenevi, U. 1983. Dopamine-rich transplants in experimental parkinsonism. *T.I.N.S. 6*: 266–270. (19)

Dwyer, T. M., Adams, D. J. and Hille, B. J. 1980. The permeability of the endplate channel to organic cations in frog muscle. *J. Gen. Physiol. 75*: 469–472. (11)

Easter, S. S. Jr., Rusoff, A. C. and Kish, P. E. 1981. The growth and organization of the optic nerve and tract in juvenile and adult goldfish. *J. Neurosci. 1*: 793–811. (19)

Eccles, J. C. 1964. *The Physiology of Synapses*. Springer-Verlag, Berlin. (9, 16)

Eccles, J. C., Eccles, R. M. and Magni, F. 1961. Central inhibitory action attributable to presynaptic depolarization produced by muscle afferent volleys. *J. Physiol. 159*: 147–166. (9)

Eccles, J. C., Ito, M., and Szentágothai, J. 1967. *The Cerebellum as a Neuronal Machine*. Springer-Verlag, Berlin. (1)

Eccles, J. C., Katz, B. and Kuffler, S. W. 1942. Effect of eserine on neuromuscular transmission. *J. Neurophysiol. 5*: 211–230. (9)

Eccles, J. C. and O'Connor, W. J. 1939. Responses which nerve impulses evoke in mammalian striated muscles. *J. Physiol. 97*: 44–102. (9)

Eccles, J. C. and Sherrington, C. S. 1930. Numbers and contraction-values of individual motor-units examined in some muscles of the limb. *Proc. R. Soc. Lond. B 106*: 326–357. (15)

Edwards, C. 1982. The selectivity of ion channels in nerve and muscles. *Neuroscience 7*: 1335–1366. (9)

Edwards, C. and Ottoson, D. 1958. The site of impulse initiation in a nerve cell of a crustacean stretch receptor. *J. Physiol. 143*: 138–148. (5)

Ehrlich, B. E. and Diamond, J. M. 1980. Lithium, membranes, and manic-depressive illness. *J. Memb. Biol. 52*: 187–200. (8)

Eide, A.-L. Jansen, J. K. D. and Ribchester, R. R. 1982. The effect of lesions in the neural crest on the formation of synaptic connexions in the embryonic chick spinal cord. *J. Physiol. 324*: 453–478. (19)

Eldridge, F. L. 1977. Maintenance of respiration by central neural feedback mechanisms. *Fed. Proc. 36*: 2400–2404. (17)

Elliot, T. R. 1904. On the action of adrenalin. *J. Physiol. 31*: 20–26. (9)

Elliott, E. J. and Muller, K. J. 1983. Sprouting and regeneration of sensory axons after destruction of ensheathing glial cells in the leech central nervous system. *J. Neurosci. 3*: 1994–2006. (13)

Ellisman, M. H., Agnew, W. S., Miller, J. A. and Levinson, S. R. 1982. Electron microscope visualization of the tetrodotoxin binding protein from *Electrophorus electricus. Proc. Natl. Acad. Sci. USA 79*: 4461–4465. (6)

Emonet-Dénand, F., Jami, L. and Laporte, Y. 1980. Histophysiological observations on the skeleto-fusimotor innervation of mammalian spindles. *Progr. Clin. Neurophysiol. 8*: 1–11. (15)

Emonet-Dénand, F., Jami, L., Laporte, Y. and Tankov, N. 1980. Glycogen depletion of bag_1 fibres elicited by stimulation of static axons in cat peroneus brevis muscle spindles. *J. Physiol. 302*: 311–321. (15)

Enroth-Cugell, C. and Robson, J. G. 1966. The contrast sensitivity of retinal ganglion cells of the cat. *J. Physiol. 187*: 517–552. (2)

Erulkar, S. D. and Weight, F. F. 1977. Extracellular potassium and transmitter release at the giant synapse of squid. *J. Physiol. 266*: 209–218. (13)

Evarts, E. V. and Tanji, J. 1976. Reflex and intended responses in motor cortex pyramidal tract neurons of monkeys. *J. Neurophysiol. 39*: 1069–1080. (17)

Eyzaguirre, C. and Kuffler, S. W. 1955. Processes of excitation in the dendrites and in the soma of single isolated sensory nerve cells of the lobster and crayfish. *J. Gen. Physiol. 39*: 87–119. (15)

Faber, D. S. and Korn, H. 1978. In D. S. Faber and H. Korn (eds.). *Neurobiology of the Mauthner Cell*. Raven, New York. (16)

Faber, D. S. and Korn, H. 1982. Transmission at a central inhibitory synapse. I. Magnitude of unitary postsynaptic conductance change and kinetics of channel activation. *J. Neurophysiol. 48*: 654–678. (16)

Fahrenkrug, J. and Emson, P. C. 1982. Vasoactive intestinal polypeptide: Functional aspects. *Br. Med. Bull. 38*: 265–270. (12)

Falck, B., Hillarp, N.-Å., Thieme, G., and Thorp, A. 1962. Fluorescence of catecholamines and related compounds condensed with formaldehyde. *J. Histochem. Cytochem. 10*: 348–354. (12)

Fambrough, D. M. 1970. Acetylcholine sensitivity of muscle fiber membranes: Mechanism of regulation by motoneurons. *Science 168*: 372–373. (19)

Fambrough, D. M. 1974. Acetylcholine receptors: Revised estimates of extrajunctional receptor density in denervated rat diaphragm. *J. Gen. Physiol. 64*: 468–572. (11)

Fambrough, D. M. 1979. Control of acetylcholine receptors in skeletal muscle. *Physiol. Rev. 59*: 165–227. (19)

Famiglietti, E. V. Jr., Kaneko, A. and Tachibana, M. 1977. Neuronal architecture of on and off pathways to ganglion cells in carp retina. *Science 198*: 1267–1269. (2)

Fatt, P. and Katz, B. 1951. An analysis of the end-plate potential recorded with an intracellular electrode. *J. Physiol. 115*: 320–370. (9)

Fatt, P. and Katz, B. 1952. Spontaneous subthreshold activity at motor nerve endings. *J. Physiol. 117*: 109–128. (10)

Fatt, P. and Katz, B. 1953. The effect of inhibitory nerve impulses on a crustacean muscle fibre. *J. Physiol. 121*: 374–389. (9, 16)

Feldberg, W. 1945. Present views on the mode of action of acetylcholine in the central nervous system. *Physiol. Rev. 25*: 596–642. (9)

Fentress, J. C. (ed.). 1976. *Simpler Networks and Behavior.* Sinauer, Sunderland, MA. (18)

Fenwick, E. M., Marty, A. and Neher, E. 1982. Sodium and calcium channels in bovine chromaffin cells. *J. Physiol. 331*: 599–635. (6)

Ferster, D. 1981. A comparison of binocular depth mechanisms in areas 17 and 18 of the cat visual cortex. *J. Physiol. 311*: 623–655. (2)

Ferster, D. and LeVay, S. 1978. The axonal arborizations of lateral geniculate neurons in the striate cortex of the cat. *J. Comp. Neurol. 182*: 923–944. (3)

Ferster, D. and Lindström, S. 1983. An intracellular analysis of geniculocortical connectivity in area 17 of the cat. *J. Physiol. 342*: 181–215. (3)

Fertuck, H. C. and Salpeter, M. M. 1974. Localization of acetylcholine receptor by ^{125}I-labeled alpha-bungarotoxin binding at mouse motor endplates. *Proc. Natl. Acad. Sci. USA 71*: 1376–1378. (11)

Fex, S., Sonessin, B., Thesleff, S. and Zelená, J. 1966. Nerve implants in botchulinum poisoned mammalian muscle. *J. Physiol. 184*: 872–882. (19)

Fields, H. L. and Basbaum, A. I. 1978. Brainstem control of spinal pain-transmission neurons. *Annu. Rev. Physiol. 40*: 217–248. (12)

Fields, H. L., Evoy, W. H. and Kennedy, D. 1967. Reflex role played by efferent control of an invertebrate stretch receptor. *J. Neurophysiol. 30*: 859–874. (15)

Fischbach, G. D. and Dichter, M. A. 1974. Electrophysiologic and morphologic properties of neurons in dissociated chick spinal cord cell cultures. *Dev. Biol. 37*: 100–116. (13, Part 5)

Fischbach, G. D. and Schuetze, S. M. 1980. A post-natal decrease in acetylcholine channel open time at rat end-plates. *J. Physiol. 303*: 125–137. (19)

Fischer, B. and Poggio, G. F. 1979. Depth sensitivity of binocular cortical neurones of behaving monkeys. *Proc. R. Soc. Lond. B 204*: 409–414. (2)

Flaster, M. S., Macagno, E. R., and Schehr, R. S. Mechanisms for the formation of synaptic connections in isogenic nervous system of *Daphnia magna*. 1982. In N. Spitzer (ed.). *Neuronal Development.* Plenum Press, New York, pp. 267–295. (Part 5)

Fleischhauer, K. 1972. Ependyma and subependymal layer. In G. H. Bourne (ed.). *The Structure and Function of Nervous Tissue,* Vol IV. Academic, New York, pp. 1–46. (14)

Flock, Å. and Lam, D. M. K. Neurotransmitter synthesis in inner ear and lateral line organ. *Nature 249*: 142–144. (15)

Flock, Å. and Russell, J. J. 1973. The post-synaptic action of efferent fibres in the lateral line organ of the burbot *Lota lota. J. Physiol. 235*: 591–605. (15)

Florey, E. 1961. Comparative physiology: Transmitter substances. *Annu. Rev. Physiol. 23*: 501–528. (12)

Foote, S. L., Bloom, F. E. and Aston-Jones, G. 1983. Nucleus locus ceruleus: New evidence of anatomical and physiological specificity. *Physiol. Rev. 63*: 844–914. (12)

Frank, E. 1973. Matching of facilitation at the neuromuscular junction of the lobster: A possible case for influence of muscle on nerve. *J. Physiol. 233*: 635–658. (16, 18)

Frank, E. and Fischbach, G. D. 1979. Early events in neuromuscular junction formation *in vitro*: Induction of acetylcholine receptor clusters in the post-synaptic membrane and morphology of newly formed synapses. *J. Cell Biol. 83*: 143–158. (19)

Frank, E., Jansen, J. K. S., Lømo, T., and Westgaard, R. H. 1975. The interaction between foreign and original nerves innervating the soleus muscle of rats. *J. Physiol. 247*: 725–743. (19)

Frank, K. and Fuortes, M. G. F. 1957. Presynaptic and postsynaptic inhibition of monosynaptic reflexes. *Fed. Proc. 16*: 39–40. (9)

Frankenhaeuser, B. and Hodgkin, A. L. 1956. The after-effects of impulses in the giant nerve fibres of *Loligo*. *J. Physiol. 131*: 341–376. (13)

Frankenhaeuser, B. and Hodgkin, A. L. 1957. The action of calcium on the electrical properties of squid axons. *J. Physiol. 137*: 218–244. (6)

Friedlander, M. J., Lin, C.-S., Stanford, L. R. and Sherman, S. M. 1981. Morphology of functionally identified neurons in lateral geniculate nucleus of the cat. *J. Neurophysiol. 46*: 80–129. (2)

Fuchs, P. A. and Getting, P. A. 1980. Ionic basis of presynaptic inhibitory potentials at crayfish claw opener. *J. Neurophysiol. 43*: 1547–1557. (16)

Fuchs, P. A., Henderson, L. P. and Nicholls, J. G. 1982. Chemical transmission between individual Retzius and sensory neurones of the leech in culture. *J. Physiol. 323*: 195–210. (18)

Fujito, Y., Tsukahara, N., Oda, Y. and Maeda, J. 1982. Formation of functional synapses in the adult cat red nucleus from the cerebrum following cross-innervation of forelimb flexor and extensor nerves. I. Appearance of new synaptic potentials. *Exp. Brain Res. 45*: 1–12. (19)

Fujito, Y., Tsukahara, N., Oda, Y. and Yoshida, M. 1982. Formation of functional synapses in the adult cat red nucleus from the cerebrum following cross-innervation of forelimb flexor and extensor nerves. II. Analysis of newly appeared synaptic potentials. *Exp. Brain Res. 45*: 13–18. (19)

Fukami, Y. 1982. Further morphological and electrophysiological studies on snake muscle spindles. *J. Neurophysiol. 47*: 810–826. (15)

Fukami, Y. and Hunt, C. C. 1977. Structures in sensory region of snake spindles and their displacement during stretch. *J. Neurophysiol. 40*: 1121–1131. (15)

Fuller, J. L. 1967. Experimental deprivation and later behavior. *Science 158*: 1645–1652. (20)

Fuortes, M. G. F. and Poggio, G. F. 1963. Transient responses to sudden illumination in cells of the eye of *Limulus*. *J. Gen. Physiol. 46*: 435–452. (2)

Furshpan, E. J. 1964. "Electrical transmission" at an excitatory synapse in a vertebrate brain. *Science 144*: 878–880. (16)

Furshpan, E. J. and Furukawa, T. Y. 1962. Intracellular and extracellular responses of the several regions of the Mauthner cell of the goldfish. *J. Neurophysiol. 25*: 732–771. (16)

Furshpan, E. J. and Potter, D. D. 1959. Transmission at the giant motor synapses of the crayfish. *J. Physiol. 145*: 289–325. (9)

Furukawa, T. 1966. Synaptic interaction at the Mauthner cell of goldfish. *Prog. Brain Res. 21A*: 44–70. (16)

Furukawa, T., Fukami, Y. and Asada, Y. 1965. A third type of inhibition in the Mauthner cell of goldfish. *J. Neurophysiol. 26*: 759–774. (16)

Furukawa, T. Y. and Furshpan, E. J. 1963. Two inhibitory mechanisms in the Mauthner neurons of the goldfish. *J. Neurophysiol. 26*: 140–176. (9, 16)

Gainer, H., Tasaki, I. and Lasek, R. J. 1977. Evidence for the glia-neuron protein transfer hypothesis from intracellular perfusion studies of squid giant axons. *J. Cell Biol. 74*: 524–530. (13)

Garrahan, P. J. and Glynn, I. M. 1967. The incorporation of inorganic phosphate into adenosine triphosphate by reversal of the sodium pump. *J. Physiol. 192*: 237–256. (8)

Gaze, R. M. 1970. *The Formation of Nerve Connections.* Academic, New York. (19)

Gazzaniga, M. S. 1967. The split brain in man. *Sci. Am. 217*: 24–29. (2)

Gazzaniga, M. S. 1970. *The Bisected Brain.* Appleton, New York. (2)

Geffen, L. B. and Livett, B. G. 1971. Synaptic vesicles in sympathetic neurons. *Physiol. Rev. 51*: 98–157. (12)

Gerschenfeld, H. M. and Paupardin-Tritsch, D. 1974. Ionic mechanisms and receptor properties underlying the responses of molluscan neurones to 5-hydroxytryptamine. *J. Physiol. 243*: 427–456. (9)

Geschwind, N. 1979. Specializations of the human brain. *Sci. Am. 241*: 180–199. (17)

Gilbert, C. D. 1977. Laminar differences in receptive field properties of cells in cat primary visual cortex. *J. Physiol. 268*: 391–421. (2, 3)

Gilbert, C. D. 1983. Microcircuitry of the visual cortex. *Annu. Rev. Neurosci. 6*: 217–247. (3, 12)

Gilbert, C. D. and Wiesel, T. N. 1979. Morphology and intracortical projections of functionally characterised neurones in the cat visual cortex. *Nature 280*: 120–125. (2, 3)

Gilbert, C. D. and Wiesel, T. N. 1981. Laminar specialization and intracortical connections in cat primary visual cortex. In F. O. Schmitt, F. G. Worden, and F. Dennis (eds.). *The Organization of the Cerebral Cortex.* MIT Press, Cambridge, pp. 163–198. (3)

Gilbert, C. D. and Wiesel, T. N. 1983. Clustered intrinsic connections in cat visual cortex. *J. Neurosci. 3*: 1116–1133. (2, 3)

Glia-Neurone Interactions. 1981. J. E. Treherne (ed.). *J. Exp. Biol.* Vol. 95. (13)

Glover, J. C. and Kramer, A. P. 1982. Serotonin analog selectively ablates identified neurons in the leech embryo. *Science 216*: 317–319. (18)

Gluhbegovic, N. and Williams, T. H. 1980. *The Human Brain.* Harper & Row, New York. (Appendix B)

Glusman, S. and Kravitz, E. A. 1982. The action of serotonin on excitatory nerve terminals in lobster nerve-muscle preparations. *J. Physiol. 325*: 223–241. (16)

Glynn, I. M. 1968. Membrane adenosine triphosphatase and cation transport. *Br. Med. Bull. 24*: 165–169. (8)

Gold, M. R. 1982. The effects of vasoactive intestinal peptide on neuromuscular transmission in the frog. *J. Physiol. 327*: 325–335. (12)

Gold, M. R. and Martin, A. R. 1983a. Characteristics of inhibitory postsynaptic currents in brain-stem neurones of the lamprey. *J. Physiol. 342*: 85–98. (11)

Gold, M. R. and Martin, A. R. 1983b. Analysis of glycine-activated inhibitory post-synaptic channels in brain-stem neurones of the lamprey. *J. Physiol. 342*: 99–117. (11)

Goldman, D. E. 1943. Potential, impedance and rectification in membranes. *J. Gen. Physiol. 27*: 37–60. (5)

Golgi, C. 1903. *Opera Omnia,* Vols. I, II. U. Hoepli, Milan. (13)

Goodman, C. S. and Spitzer, N. C. 1979. Embryonic development of identified neurones: Differentiation from neuroblast to neurone. *Nature 280*: 208–214. (19)

Göpfert, H. and Schaefer, H. 1938. Uber den direkt und indirekt erregten Aktionsstrom und die Funktion der motorischen Endplate. *Pflügers Arch. 239*: 597–619. (9)

Govind, C. K. and Atwood, H. L. 1982. Organization of neuromuscular systems. In H. L. Atwood and D. C. Sandeman (eds.). *Biology of Crustacea,* Vol. 3. Academic, New York, pp. 63–98. (16)

Grafstein, B. and Forman, D. S. 1980. Intracellular transport in neurons. *Physiol. Rev. 60*: 1167–1283. (12)

Granit, R. 1947. *Sensory Mechanisms of the Retina.* Oxford University Press, London. (2)

Gray, J. A. B. 1959. Initiation of impulses at receptors. In J. Field (ed.). *Handbook of Physiology,* Vol. I. American Physiological Society, Bethesda, pp. 123–145. (15)

Graziadei, G. A. M. and Graziadei, P. P. C. 1978. Continuous nerve cell renewal in the olfactory system. In M. Jacobson (ed.). *Handbook of Sensory Physiology,* Vol. 9. Springer-Verlag, New York, pp. 55–85. (19)

Greene, L. A., Varon, S., Piltch, A. and Shooter, E. M. 1971. Substructure of the β subunit of mouse 7S nerve growth factor. *Neurobiology 1*: 37–48. (19)

Greengard, P. 1976. Possible role for cyclic nucleotides and phosphorylated membrane proteins in postsynaptic actions of neurotransmitters. *Nature 260*: 101–108. (9)

Gregory, R. A. (ed.). 1982. Regulatory peptides of gut and brain. *Br. Med. Bull. 38*: 219–318. (12)

Grillner, S. 1975. Locomotion in vertebrates: Central mechanisms and reflex interaction. *Physiol. Rev. 55*: 247–304. (17)

Grillner, S. and Zangger, P. 1974. Locomotor movements generated by the deafferented spinal cord. *Acta. Physiol. Scand. 91*: 38A-39A. (17)

Grinnell, A. D. 1970. Electrical interaction between antidromically stimulated frog motoneurones and dorsal root afferents: Enhancement by gallamine and TEA. *J. Physiol. 210*: 17–43. (9)

Grinvald, A., Cohen, L. B., Lesher, S. and Boyle, M. B. 1981. Simultaneous optical monitoring of activity of many neurons in invertebrate ganglia using a 124-element photodiode array. *J. Neurophysiol. 45*: 829–840. (7)

Grinvald, A. and Farber, I. C. 1981. Optical recording of calcium action potentials from growth cones of cultured neurons with a laser microbeam. *Science 212*: 1164–1167. (7)

Grinvald, A., Hildesheim, R., Farber, I. C. and Anglister, L. 1982. Improved fluorescent probes for the measurement of rapid changes in membrane potential. *Biophys. J. 39*: 301–308. (7)

Grinvald, A., Manker, A. and Segal, M. 1982. Visualization of the spread of electrical activity in rat hippocampal slices by voltage-sensitive optical probes. *J. Physiol. 333*: 269–291. (7)

Grossman, Y., Parnas, I. and Spira, M. E. 1979a. Differential conduction block in branches of a bifurcating axon. *J. Physiol. 295*: 283–305. (16)

Grossman, Y., Parnas, I. and Spira, M. E. 1979b. Ionic mechanisms involved in differential conduction of action potentials at high frequency in a branching axon. *J. Physiol. 295*: 307–322. (13, 18)

Guillery, R. W. 1970. The laminar distribution of retinal fibers in the dorsal lateral geniculate nucleus of the cat: A new interpretation. *J. Comp. Neurol.* *138*: 339–368. (2)

Guillery, R. W. 1974. Visual pathways in albinos. *Sci. Am. 230*: 44–54. (20)

Guillery, R. W. and Kaas, J. H. 1971. A study of normal and congenitally abnormal retinogeniculate projections in cats. *J. Comp. Neurol. 143*: 73–100. (20)

Guillery, R. W. and Kaas, J. H. 1974. The effects of monocular lid suture upon the development of the visual cortex in squirrels (*Sciureus carolinensis*). *J. Comp. Neurol. 154*: 443–452. (20)

Guillery, R. W. and Stelzner, D. J. 1970. The differential effects of unilateral lid closure upon the monocular and binocular segments of the dorsal lateral geniculate nucleus in the cat. *J. Comp. Neurol. 139*: 413–422. (20)

Gundersen, R. W. and Barrett, J. N. 1979. Neuronal chemotaxis: Chick dorsal root axons turn toward high concentrations of nerve growth factor. *Science 206*: 1079–1080. (19)

Gundersen, R. W. and Barrett, J. N. 1980. Characterization of the turning response of dorsal root neurites toward nerve growth factor. *J. Cell Biol. 87*: 546–554. (19)

Guth, L. 1968. "Trophic" influences of nerve. *Physiol. Rev. 48*: 645–687. (19)

Gutmann, E. 1962. Denervation and disuse atrophy in cross-striated muscle. *Rev. Can. Biol. 21*: 353–365. (19)

Gutnick, M. J., Connors, B. W. and Ransom, B. R. 1981. Dye-coupling between glial cells in the guinea pig neocortical slice. *Brain Res. 213*: 486–492. (13)

Hagins, W. A., Penn, R. D. and Yoshikami, S. 1970. Dark current and photo-current in retinal rods. *Biophys. J. 10*: 380–412. (2)

Hagiwara, S. and Byerly, L. 1981. Calcium channel. *Annu. Rev. Neurosci. 4*: 69–125. (6)

Hagiwara, S. and Harunon, H. 1983. Studies of single calcium channel currents in rat clonal pituitary cells. *J. Physiol. 336*: 649–661. (6)

Hagiwara, S., Kusano, K. and Saito, S. 1960. Membrane changes in crayfish stretch receptor neuron during synaptic inhibition and under action of gamma-aminobutyric acid. *J. Neurophysiol. 23*: 505–515. (15)

Hagiwara, S. and Tasaki, I. 1958. A study on the mechanism of impulse transmission across the giant synapse of the squid. *J. Physiol. 143*: 114–137. (10)

Hall, Z. W., Bownds, M. D. and Kravitz, E. A. 1970. The metabolism of gamma aminobutyric acid in the lobster nervous system. *J. Cell Biol. 46*: 290–299. (12)

Hall, Z. W., Hildebrand, J. G. and Kravitz, E. A. 1974. *Chemistry of Synaptic Transmission*. Chiron Press, Newton, MA. (9)

Hall, Z. W., Roisin, M.-P. Gu, Y. and Gorin, P. D. 1984. A developmental change in the immunological properties of acetylcholine receptors at the rat neuromuscular junction. *Cold Spring Harbor Symp. Quant. Biol. 48*: 101–108. (19)

Hamburger, V. 1939. Motor and sensory hyperplasia following limb-bud transplantations in chick embryos. *Physiol. Zool. 12*: 268–284. (19)

Hamburger, V. 1975. Cell death in the development of the lateral motor column of the chick embryo. *J. Comp. Neurol. 160*: 535–546. (Part 5)

Hamburger, V., Brunso-Bechtold, J. K. and Yip, J. W. 1981. Neuronal death in the spinal ganglia of the chick embryo and its reduction by nerve growth factor. *J. Neurosci. 1*: 60–71. (19)

Hamill, O. P., Marty, A., Neher, F., Sakmann, B. and Sigworth, F. J. 1981. Improved patch-clamp techniques for high-resolution current recording from cells and cell-free membrane patches. *Pflügers Arch. 391*: 85–100. (6, 11)

Hamill, O. P. and Sakmann, B. 1981. Multiple conductance states of single acetylcholine receptor channels in embryonic muscle cells. *Nature 294*: 462–464. (11)

Harris, A. J., Kuffler, S. W. and Dennis, M. J. 1971. Differential chemosensitivity of synaptic and extrasynaptic areas on the neuronal surface membrane in parasympathetic neurons of the frog, tested by microapplication of acetylcholine. *Proc. R. Soc. Lond. B 177*: 541–553. (11)

Harris, E. J. and Hutter, O. F. 1956. The action of acetylcholine on the movements of potassium ions in the sinus venosus of the heart. *J. Physiol. 133*: 58P–59P. (9)

Harris, G. W., Reed, M. and Fawcett, C. P. 1966. Hypothalamic releasing factors and the control of anterior pituitary function. *Br. Med. Bull. 22*: 266–272. (12)

Harris, J. B. and Thesleff, S. 1971. Studies on tetrodotoxin resistant action potentials in denervated skeletal muscle. *Acta Physiol. Scand. 83*: 382–388. (19)

Harris, W. A. 1981. Neural activity and development. *Annu. Rev. Physiol. 43*: 689–710. (19)

Hartline, H. K. 1940a. The receptive fields of optic nerve fibers. *Am. J. Physiol. 130*: 690–699. (2)

Hartline, H. K. 1940b. The nerve messages in the fibers of the visual pathway. *J. Opt. Soc. Am. 30*: 239–247. (2)

Hartzell, H. C. and Fambrough, D. M. 1972. Acetylcholine receptors: Distribution and extrajunctional density in rat diaphragm after denervation correlated with acetylcholine sensitivity. *J. Gen. Physiol. 60*: 248–262. (19)

Hartzell, H. C., Kuffler, S. W. and Yoshikami, D. 1975. Postsynaptic potentiation: Interaction between quanta of acetylcholine at the skeletal neuromuscular synapse. *J. Physiol. 251*: 427–463. (11)

Hartzell, H. C., Kuffler, S. W., Stickgold, R. and Yoshikami, D. 1977. Synaptic excitation and inhibition resulting from direct action of acetylcholine on two types of chemoreceptors on individual amphibian parasympathetic neurones. *J. Physiol. 271*: 817–846. (9)

Hecht, S., Shlaer, S. and Pirenne, M. H. 1942. Energy, quanta and vision. *J. Gen. Physiol. 25*: 819–840. (2)

Heidmann, T. and Changeux, J.-P. 1978. Structural and functional properties of the acetylcholine receptor protein in its purified and membrane-bound state. *Annu. Rev. Biochem. 47*: 317–357. (11)

Heineman, U. and Lux, H. D. 1977. Ceiling of stimulus induced rises in extracellular potassium concentration in the cerebral cortex of cats. *Brain Res. 120*: 231–249. (13)

Held, R. and Bauer, J. A. 1974. Development of sensorially guided reaching in infant monkeys. *Brain Res. 71*: 265–271. (20)

Helmholtz, H. 1889. *Popular Scientific Lectures.* Longmans, London. (1)

Hendrickson, A. E., Ogren, M. P., Vaughn, J. E., Barber, R. P., and Wu, J.-Y. 1983. Light and electron microscope immunocytochemical localization of glutamic acid decarboxylase in monkey geniculate complex: Evidence for GABAergic neurons and synapses. *J. Neurosci. 3*: 1245–1262. (12)

Hendrickson, A. E., Wilson, J. R. and Ogren, M. P. 1978. The neuroanatomical organizations of pathways between dorsal lateral geniculate nucleus and visual cortex in old and new world primates. *J. Comp. Neurol. 182*: 123–136. (3)

Hendry, I. A., Stöckel, K., Thoenen, H. and Iversen, L. L. 1974. The retrograde axonal transport of nerve growth factor. *Brain Res. 68*: 103–121. (19)

Henneman, E. and Mendell, L. M. 1981. Functional organization of motoneuron pool and its inputs. In V. Brooks (ed.). *The Nervous System, Volume 1, part 1*. Amer. Physiol. Soc., Bethesda, pp. 423–507. (16)

Henneman, E., Somjen, G. and Carpenter, D. O. 1965. Functional significance of cell size in spinal motoneurons. *J. Neurophysiol. 28*: 560–580. (16)

Henrickson, C. K. and Vaughn, J. E. 1974. Fine structural relationships between neurites and radial glial processes in developing mouse spinal cord. *J. Neurocytol. 3*: 659–679. (13)

Heuser, J. E. and Reese, T. S. 1973. Evidence for recycling of synaptic vesicle membrane during transmitter release at the frog neuromuscular junction. *J. Cell Biol. 57*: 315–344. (11)

Heuser, J. E., Reese, T. S. and Landis, D. M. D. 1974. Functional changes in frog neuromuscular junctions studied with freeze-fracture. *J. Neurocytol. 3*: 109–131. (10, 11)

Heuser, J. E., Reese, T. S., Dennis, M. J., Jan, Y., Jan, L. and Evans, L. 1979. Synaptic vesicle exocytosis captured by quick freezing and correlated with quantal transmitter release. *J. Cell Biol. 81*: 275–300. (11)

Heuser, J. E. and Reese, T. S. 1981. Structural changes after transmitter release at the frog neuromuscular junction. *J. Cell Biol. 88*: 564–580. (11)

Hilaire, G. G., Nicholls, J. G. and Sears, T. A. 1983. Central and proprioceptive influences on the activity of levator costae motoneurones in the cat. *J. Physiol. 342*: 527–548. (16, 17)

Hildebrand, J. G., Barker, D. L., Herbert, E. and Kravitz, E. A. 1971. Screening for neurotransmitters: A rapid radiochemical procedure. *J. Neurobiol. 2*: 231–246. (12)

Hille, B. 1970. Ionic channels in nerve membranes, *Prog. Biophys. Mol. Biol. 21*: 1–32. (6)

Hille, B. 1971. The permeability of the sodium channel to organic cations in myelinated nerve. *J. Gen. Physiol. 58*: 599–619. (6)

Hille, B. 1976. Ionic basis of resting and action potentials. In E. Kandel (ed.). *Handbook of the Nervous System*, Vol. I. American Physiological Society, Bethesda. (6)

Hille, B. 1984. *Ionic Channels of Excitable Membranes*. Sinauer, Sunderland, MA. (6)

Hille, B. and Campbell, D. T. 1976. An improved Vaseline gap voltage clamp for skeletal muscle fibers. *J. Gen. Physiol. 67*: 265–293. (6)

Hitchcock, D. I. 1945. In R. Höber (ed.). *Physical Chemistry of Cells and Tissues*. Blakiston, Philadelphia. (14)

Hobbs, A. S. and Albers, R. W. 1980. The structure of proteins involved in active membrane transport. *Annu. Rev. Biophys. Bioeng. 9*: 259–291. (8)

Hodgkin, A. L. 1937. Evidence for electrical transmission in nerve. I, II. *J. Physiol. 90*: 183–210, 211–232. (7)

Hodgkin, A. L. 1939. The relation between conduction velocity and the electrical resistance outside a nerve fibre. *J. Physiol. 94*: 560–570. (7)

Hodgkin, A. L. 1951. The ionic basis of electrical activity in nerve and muscle. *Biol. Rev. 26*: 339–409. (5)

Hodgkin, A. L. 1954. A note on conduction velocity. *J. Physiol. 125*: 221–224. (7)

Hodgkin, A. L. 1964. *The Conduction of the Nervous Impulse*. Liverpool University Press, Liverpool. (4, 5, 6)

Hodgkin, A. L. 1973. Presidential address. *Proc. R. Soc. Lond. B 183*: 1–19. (5)

Hodgkin, A. L. 1977. Obituary: Lord Adrian, 1889–1977. *Nature 269*: 543–544. (1)

Hodgkin, A. L. and Horowicz, P. 1959. The influence of potassium and chloride ions on the membrane potential of single muscle fibres. *J. Physiol. 148*: 127–160. (5)

Hodgkin, A. L. and Huxley, A. F. 1939. Action potentials recorded from inside a nerve fibre. *Nature 144*: 710–711. (5)

Hodgkin, A. L. and Huxley, A. F. 1952a. Currents carried by sodium and potassium ions through the membrane of the giant axon of *Loligo*. *J. Physiol. 116*: 449–472. (6)

Hodgkin, A. L. and Huxley, A. F. 1952b. The components of the membrane conductance in the giant axon of *Loligo*. *J. Physiol. 116*: 473–496. (6)

Hodgkin, A. L. and Huxley, A. F. 1952c. The dual effect of membrane potential on sodium conductance in the giant axon of *Loligo*. *J. Physiol. 116*: 497–506. (6)

Hodgkin, A. L. and Huxley, A. F. 1952d. A quantitative description of membrane current and its application to conduction and excitation in nerve. *J. Physiol. 117*: 500–544. (6)

Hodgkin, A. L., Huxley, A. F. and Katz, B. 1952. Measurement of current-voltage relations in the membrane of the giant axon of *Loligo*. *J. Physiol. 116*: 424–448. (6)

Hodgkin, A. L. and Katz, B. 1949. The effect of sodium ions on the electrical activity of the giant axon of the squid. *J. Physiol. 108*: 37–77. (5)

Hodgkin, A. L. and Keynes, R. D. 1955. Active transport of cations in giant axons from *Sepia* and *Loligo*. *J. Physiol. 128*: 28–60. (5, 8)

Hodgkin, A. L. and Keynes, R. D. 1956. Experiments on the injection of substances into squid giant axons by means of a microsyringe. *J. Physiol. 131*: 592–617. (5)

Hodgkin, A. L. and Rushton, W. A. H. 1946. The electrical constants of a crustacean nerve fibre. *Proc. R. Soc. Lond. B 133*: 444–479. (7)

Hollyday, M. and Hamburger, V. 1976. Reduction of the naturally occurring motor neuron loss by enlargement of the periphery. *J. Comp. Neurol. 170*: 311–320. (19)

Hollyday, M., Hamburger, V. and Farris, J. 1977. Localization of motor neuron pools supplying identified muscles in normal and supernumerary legs of chick embryo. *Proc. Natl. Acad. Sci. USA 74*: 3582–3586. (19)

Honig, M. C., Collins, W. F. and Mendell, L. M. 1983. α-Motoneuron EPSPs exhibit different frequency sensitivities to single Ia-afferent fiber stimulation. *J. Neurophysiol. 49*: 886–901. (16)

Horvitz, H. R. 1982. Factors that influence neural development in nematodes. In J. G. Nicholls (ed.). *Repair and Regeneration of the Nervous System*. Springer-Verlag, New York, pp. 41–55. (18, 19)

Horvitz, H. R., Sternberg, P. W., Greenwald, I. S., Fixsen, W. and Ellis, H. M. 1984. Mutations that affect neural cell lineages and cell fates during the development of the nematode *Caenorhabditis Elegans*. *Cold Spring Harbor Symp. Quant. Biol. 48*: 453–464. (Part 5)

Hoy, R., Bittner, G. D. and Kennedy, D. 1967. Regeneration in crustacean motoneurons: Evidence for axonal fusion. *Science 156*: 251–252. (19)

Hubbell, W. L. and Bownds, M. D. 1979. Visual transduction in vertebrate photoreceptors. *Annu. Rev. Neurosci. 2*: 17–34. (2)

Hubel, D. H. 1967. Effects of distortion of sensory input on the visual system of kittens. *Physiologist 10*: 17–45. (20)

Hubel, D. H. 1982a. Cortical neurobiology: A slanted historical perspective. *Annu. Rev. Neurosci. 5*: 363–370. (2)

Hubel, D. H. 1982b. Exploration of the primary visual cortex. *Nature 299*: 515–524. (2, 3)

Hubel, D. H. and Wiesel, T. N. 1959. Receptive fields of single neurones in the cat's striate cortex. *J. Physiol. 148*: 574–591. (2)

Hubel, D. H. and Wiesel, T. N. 1961. Integrative action in the cat's lateral geniculate body. *J. Physiol. 155*: 385–398. (2)

Hubel, D. H. and Wiesel, T. N. 1962. Receptive fields, binocular interaction and functional architecture in the cat's visual cortex. *J. Physiol. 160*: 106–154. (2, 3)

Hubel, D. H. and Wiesel, T. N. 1963a. Shape and arrangement of columns in cat's striate cortex. *J. Physiol. 165*: 559–568. (3)

Hubel, D. H. and Wiesel, T. N. 1963b. Receptive fields of cells in striate cortex of very young, visually inexperienced kittens. *J. Neurophysiol. 26*: 994–1002. (20)

Hubel, D. H. and Wiesel, T. N. 1965a. Receptive fields and functional architecture in two non-striate visual areas (18 and 19) of the cat. *J. Neurophysiol. 28*: 229–289. (2, 20)

Hubel, D. H. and Wiesel, T. N. 1965b. Binocular interaction in striate cortex of kittens reared with artificial squint. *J. Neurophysiol. 28*: 1041–1059. (20)

Hubel, D. H. and Wiesel, T. N. 1967. Cortical and callosal connections concerned with the vertical meridian of visual field in the cat. *J. Neurophysiol. 30*: 1561–1573. (2)

Hubel, D. H. and Wiesel, T. N. 1968. Receptive fields and functional architecture of monkey striate cortex. *J. Physiol. 195*: 215–243. (2, 3)

Hubel, D. H. and Wiesel, T. N. 1970. The period of susceptibility to the physiological effects of unilateral eye closure in kittens. *J. Physiol. 206*: 419–436. (20)

Hubel, D. H. and Wiesel, T. N. 1971. Aberrant visual projections in the Siamese cat. *J. Physiol. 218*: 33–62. (20)

Hubel, D. H. and Wiesel, T. N. 1972. Laminar and columnar distribution of geniculo-cortical fibers in the macaque monkey. *J. Comp. Neurol. 146*: 421–450. (2, 3)

Hubel, D. H. and Wiesel, T. N. 1974. Sequence regularity and geometry of orientation columns in the monkey striate cortex. *J. Comp. Neurol. 158*: 267–294. (3)

Hubel, D. H. and Wiesel, T. N. 1977. Ferrier Lecture. Functional architecture of macaque monkey visual cortex. *Proc. R. Soc. Lond. B 198*: 1–59. (2, 3, 20)

Hubel, D. H. and Wiesel, T. N. 1979. Brain mechanisms of vision. *Sci. Am. 241*: 150–162. (2)

Hubel, D. H., Wiesel, T. N. and LeVay, S. 1977. Plasticity of ocular dominance columns in monkey striate cortex. *Phil. Trans. R. Soc. Lond. B 278*: 377–409. (20)

Hubel, D. H., Wiesel, T. N. and Stryker, M. P. 1978. Anatomical demonstration of orientation columns in Macaque monkey. *J. Comp. Neurol. 177*: 361–380. (3)

Hudspeth, A. J., Poo, M. M. and Stuart, A. E. 1977. Passive signal propagation and membrane properties in median photoreceptors of the giant barnacle. *J. Physiol.* 272: 25–43. (15)

Hughes, J. (ed.). 1983. Opioid peptides. *Br. Med. Bull.* 39: 1–106. (12, 17)

Hughes, J., Beaumont, A., Fuentes, J. A., Malfroy, B. and Unsworth, C. 1981. Opioid peptides: Aspects of their origin, release and metabolism. *J. Exp. Biol.* 89: 239–255. (12)

Hughes, J., Smith, T. W., Kosterlitz, H. W., Fothergill, L. A., Morgan, B. A. and Morris, H. R. 1975. Identification of two related pentapeptides from the brain with potent opiate agonist activity. *Nature (Lond.)* 258: 577–579. (12)

Humphrey, A. L. and Hendrickson, A. E. 1983. Background and stimulus-induced patterns of high metabolic activity in the visual cortex (area 17) of the squirrel and macaque monkey. *J. Neurosci.* 3: 345–358. (3)

Humphrey, A. L., Skeen, L. C. and Norton, T. T. 1980. Topographic organization of the orientation column system in the striate cortex of the tree shrew (*Tupaia glis*). II. Deoxyglucose mapping. *J. Comp. Neurol.* 192: 549–566. (3)

Hunt, C. C. and Kuffler, S. W. 1951. Further study of efferent small nerve fibres to mammalian muscle spindles: Multiple spindle innervation and activity during contraction. *J. Physiol.* 113: 283–297. (15)

Hunt, C. C. and Kuffler, S. W. 1954. Motor innervation of skeletal muscle: Multiple innervation of individual muscle fibres and motor unit function. *J. Physiol.* 126: 293–303. (19)

Hunt, C. C. and Nelson, P. 1965. Structural and functional changes in the frog sympathetic ganglion following cutting of the presynaptic nerve fibre. *J. Physiol.* 177: 1–20. (13)

Hunt, C. C., Wilkinson, R. S. and Fukami, Y. 1978. Ionic basis of the receptor potential in primary endings of mammalian muscle spindles. *J. Gen. Physiol.* 71: 683–698. (15)

Huxley, A. F. and Stämpfli, R. 1949. Evidence for saltatory conduction in peripheral myelinated nerve fibres. *J. Physiol.* 108: 315–339. (7)

Hyvärinen, J. and Poranen, A. 1978. Movement-sensitive and direction and orientation-selective cutaneous receptive fields in the hand area of the postcentral gyrus in monkeys. *J. Physiol.* 283: 523–537. (2, 17)

Innocenti, G. M. 1981. Growth and reshaping of axons in the establishment of visual callosal connections. *Science* 212: 824–827. (20)

Iversen, L. L., Lee, C. M., Gilbert, R. F., Hunt, S. and Emson, P. C. 1980. Regulation of neuropeptide release. *Proc. R. Soc. Lond. B* 210: 91–111. (12)

Jack, J. J. B., Redman, S. J. and Wong, K. 1981. The components of synaptic potentials evoked in cat spinal motoneurones by impulses in single group Ia afferents. *J. Physiol.* 321: 65–96. (16)

Jackson, M. B., Lecar, H., Mathers, D. A. and Barker, J. L. 1982. Single channel currents activated by γ-aminobutyric acid, muscimol, and (−)-pentobarbital in cultured mouse spinal neurons. *J. Neurosci.* 2: 889–894. (11)

Jacobson, M. 1978. *Developmental Neurobiology*. 2nd Ed. Plenum, New York. (Part 5, 19)

Jacobson, S. G., Mohindra, J. and Held, R. 1981. Development of visual acuity in infants with congenital cataracts. *Br. J. Opthalmol.* 65: 727–735. (20)

Jan, Y. N., Jan, L. Y. and Kuffler, S. W. 1979. A peptide as a possible transmitter in sympathetic ganglia of the frog. *Proc. Natl. Acad. Sci. USA* 76: 1501–1505. (9)

Jan, Y. N., Jan, L. Y. and Kuffler, S. W. 1980. Further evidence for peptidergic transmission in sympathetic ganglia. *Proc. Natl. Acad. Sci. USA 77*: 5008–5012. (12)

Jankowska, E. and McCrea, D. A. 1983. Shared reflex pathways from Ib tendon organ afferents and Ia muscle spindle afferents in the cat. *J. Physiol. 338*: 99–112. (15)

Jansen, J. K. S., Lømo, T., Nicholaysen, K. and Westgaard, R. H. 1973. Hyper-innervation of skeletal muscle fibers: Dependence on muscle activity. *Science 181*: 559–561. (19)

Jansen, J. K. S. and Matthews, P. B. C. 1962. The central control of the dynamic response of muscle spindle receptors. *J. Physiol. 161*: 357–378. (15)

Jansen, J. K. S., Njå, A., Ormstad, K. and Walloe, L. 1971. On the innervation of the slowly adapting stretch receptor of the crayfish abdomen: An electrophysiological approach. *Acta Physiol. Scand. 81*: 273–285. (15)

Jasper, H. and Koyama, I. 1969. Rate of release of amino acids from the cerebral cortex in the cat as affected by brainstem and thalamic stimulation. *Can. J. Physiol. Pharmacol. 47*: 889–905. (12)

Jenkinson, D. H. and Nicholls, J. G. 1961. Contractures and permeability changes produced by acetylcholine in depolarized denervated muscle. *J. Physiol. 159*: 111–127. (9, 19)

Jessell, T. M. and Iversen, L. L. 1977. Opiate analgesics inhibit substance P release from rat nucleus. *Nature 268*: 549–551. (12)

Jessell, T. M., Siegel, R. E. and Fischbach, G. D. 1979. Induction of acetylcholine receptors on cultured skeletal muscle by a factor extracted from brain and spinal cord. *Proc. Natl. Acad. Sci. USA 76*: 5397–5401. (19)

Johnson, E. W. and Wernig, A. 1971. The binomial nature of transmitter release at the crayfish neuromuscular junction. *J. Physiol. 218*: 757–767. (10)

Kaas, J. H. 1983. What if anything is SI? Organization of first somatosensory area of cortex. *Physiol. Rev. 63*: 206–231. (3, 17)

Kaas, J. H., Merzenich, M. M. and Killackey, H. P. 1983. The reorganization of somatosensory cortex following peripheral nerve damage in adult and developing mammals. *Annu. Rev. Neurosci. 6*: 325–356. (20)

Kandel, E. R. 1976. *Cellular Basis of Behavior*. W. H. Freeman, San Francisco. (18)

Kandel, E. R. 1979. *Behavioral Biology of Aplysia*. W. H. Freeman, San Francisco. (18)

Kaneko, A. 1970. Physiological and morphological identification of horizontal, bipolar and amacrine cells in goldfish. *J. Physiol. 207*: 623–633. (2)

Kaneko, A. 1971. Electrical connexions between horizontal cells in the dogfish retina. *J. Physiol. 213*: 95–105. (2)

Kaneko, A. 1979. Physiology of the retina. *Annu. Rev. Neurosci. 2*: 169–191. (2)

Kaneko, A. and Tachibana, M. 1983. Double color-opponent receptive fields of carp bipolar cells. *Vision Res. 23*: 381–388. (2)

Kao, C. T. 1966. Tetrodotoxin, saxitoxin and their significance in the study of excitation phenomena. *Pharmacol. Rev. 18*: 977–1049. (6)

Kaplan, E. and Shapley, R. M. 1982. X and Y cells in the lateral geniculate nucleus of macaque monkeys. *J. Physiol. 330*: 125–143. (2)

Karahashi, Y. and Goldring, S. 1966. Intracellular potentials from "idle" cells in cerebral cortex of cat. *Electroencephalogr. Clin. Neurophysiol. 20*: 600–607. (13)

Karlin, A. 1980. Molecular properties of nicotinic acetylcholine receptors. In C. W. Cotman, G. Poste, and G. I. Nicholson (eds.). *Cell Surface and Neuronal Function*. Elsevier–North Holland, New York, pp. 191–260. (11)

Karnovsky, M. J. 1967. The ultrastructural basis of capillary permeability studied with peroxidase as a tracer. *J. Cell Biol. 35*: 213–236. (14)

Kasamatsu, T., Pettigrew, J. D. and Ary, M. 1981. Cortical recovery from effects of monocular deprivation: Acceleration with norepinephrine and suppression with 6-hydroxydopamine. *J. Neurophysiol. 45*: 254–266. (12, 20)

Katz, B. 1949. The efferent regulation of the muscle spindle in the frog. *J. Exp. Biol. 26*: 201–217. (15)

Katz, B. 1950. Depolarization of sensory terminals and the initiation of impulses in the muscle spindle. *J. Physiol. 111*: 261–282. (15)

Katz, B. 1966. *Nerve, Muscle and Synapse.* McGraw-Hill, New York. (5, 6, 9)

Katz, B. 1969. *The Release of Neural Transmitter Substances.* Liverpool University Press, Liverpool. (10)

Katz, B. and Miledi, R. 1964. The development of acetylcholine sensitivity in nerve-free segments of skeletal muscle. *J. Physiol. 170*: 389–396. (19)

Katz, B. and Miledi, R. 1965. The measurement of synaptic delay, and the time course of acetylcholine release at the neuromuscular junction. *Proc. R. Soc. Lond. B 161*: 483–495. (10)

Katz, B. and Miledi, R. 1967a. The release of acetylcholine from nerve endings by graded electrical pulses. *Proc. R. Soc. Lond. B 167*: 23–38. (10)

Katz, B. and Miledi, R. 1967b. The timing of calcium action during neuromuscular transmission. *J. Physiol. 189*: 535–544. (10)

Katz, B. and Miledi, R. 1967c. A study of synaptic transmission in the absence of nerve impulses. *J. Physiol. 192*: 407–436. (10, 13)

Katz, B. and Miledi, R. 1968a. The role of calcium in neuromuscular facilitation. *J. Physiol. 195*: 481–492. (10)

Katz, B. and Miledi, R. 1968b. The effect of local blockage of motor nerve terminals. *J. Physiol. 199*: 729–741. (10)

Katz, B. and Miledi, R. 1972. The statistical nature of the acetylcholine potential and its molecular components. *J. Physiol. 224*: 665–699. (11)

Katz, B. and Miledi, R. 1973. The effect of atropine on acetylcholine action at the neuromuscular junction. *Proc. R. Soc. Lond. B 184*: 221–226. (11)

Katz, B. and Miledi, R. 1977. Transmitter leakage from motor nerve endings. *Proc. R. Soc. Lond. B 196*: 59–72. (10)

Katz, B. and Miledi, R. 1982. An endplate potential due to potassium released by the motor nerve impulse. *Proc. R. Soc. Lond. B 216*: 497–507. (13)

Kawagoe, R., Onodera, K. and Takeuchi, A. 1981. Release of glutamate from the crayfish neuromuscular junction. *J. Physiol. 312*: 225–236. (16)

Kehoe, J. 1972a. Ionic mechanisms of a two-component cholinergic inhibition in *Aplysia* neurones. *J. Physiol. 225*: 85–114. (9)

Kehoe, J. 1972b. Three acetylcholine receptors in *Aplysia* neurones. *J. Physiol. 225*: 115–146. (9)

Kehoe, J. 1972c. The physiological role of three acetylcholine receptors in synaptic transmission in *Aplysia*. *J. Physiol. 225*: 147–172. (9)

Kehoe, J. and Marty, A. 1980. Certain slow synaptic responses: Their properties and possible underlying mechanisms. *Annu. Rev. Biophys. Bioeng. 9*: 437–465. (9, 12, 18)

Kelly, J. P. and Van Essen, D. C. 1974. Cell structure and function in the visual cortex of the cat. *J. Physiol. 238*: 515–547. (3, 13)

Kennedy, D. 1976. Neural elements in relation to network function. In J. C. Fentress (ed.). *Simpler Networks and Behavior.* Sinauer, Sunderland, MA, pp. 65–81. (18)

Kirby, A. W. and Enroth-Cugell, C. 1976. The involvement of gamma-amino-butyric acid in the organization of cat retinal ganglion cell receptive fields. A study with picrotoxin and bicuculline. *J. Gen. Physiol. 68*: 465–484. (12)

Kirby, A. W. and Schweitzer-Tong, D. E. 1981. GABA-antagonists alter spatial summation in receptive field centres of rod- but not cone-driven cat retinal ganglion cells. *J. Physiol. 320*: 303–308. (12)

Kirkwood, P. A. 1979. On the use and interpretation of cross-correlation measurements in the mammalian central nervous system. *J. Neurosci. Methods 1*: 107–132. (16)

Kirkwood, P. A. and Sears, T. A. 1974. Monosynaptic excitation of motoneurones from secondary endings of muscle spindles. *Nature 252*: 243–244. (15)

Kirkwood, P. A., Sears, T. A. and Westgaard, R. H. 1981. Recurrent inhibition of intercostal motoneurones in the cat. *J. Physiol. 319*: 111–130. (15)

Kirkwood, P. A. and Sears, T. A. 1982. Excitatory post-synaptic potentials from single muscle spindle afferents in external intercostal motoneurones of the cat. *J. Physiol. 322*: 287–314. (16, 17)

Kistler, J., Stroud, R. M., Klymkowsky, M. W., Lalancett, R. A. and Fairclough, R. H. 1980. Structure and function of an acetylcholine receptor. *Biophys. J. 37*: 371–383. (11)

Klein, M., Shapiro, E. and Kandel, E. R. 1980. Synaptic plasticity and the modulation of the Ca^{++} current. *J. Exp. Biol. 89*: 117–157. (18)

Knowles, A. 1982. Biochemical aspects of vision. In H. B. Barlow and J. D. Mollon (eds.). *The Senses.* Cambridge University Press, New York, pp. 82–100. (2)

Knudsen, E. I., Knudsen, P. F. and Esterly, S. D. 1982. Early auditory experience modifies sound localization in barn owls. *Nature 295*: 238–240. (20)

Kocsis, J. D., Malenka, R. C. and Waxman, S. 1983. Effects of extracellular potassium concentration on the excitability of the parallel fibres of the rat cerebellum. *J. Physiol. 334*: 225–244. (13)

Koike, H., Kandel, E. R. and Schwartz, J. H. 1974. Synaptic release of radio-activity after intrasomatic injection of choline-^{3}H into an identified cholinergic interneuron in abdominal ganglion of *Aplysia californica. J. Neurophysiol. 37*: 815–827. (12)

Kolb, B. and Whishaw, I. Q. 1980. *Fundamentals of Human Neuropsychology,* W. H. Freeman, San Francisco. (17)

Korn, H., Triller, A. and Faber, D. S. 1978. Structural correlates of recurrent collateral interneurons producing both electrical and chemical inhibitions of the Mauthner cells. *Proc. R. Soc. Lond. B 202*: 533–539. (16)

Kosterlitz, H. W. and Terenius, L. Y. (eds.). 1980. *Pain and Society. Life Sci. Res. Rep. 17.* Verlag Chemnie, Weinheim. (17)

Kravitz, E. A., Kuffler, S. W., Potter, D. D. and van Gelder, N. M. 1963. Gamma-aminobutyric acid and other blocking compounds in crustacea. II. Peripheral nervous system. *J. Neurophysiol. 26*: 729–738. (12)

Kravitz, E. A., Kuffler, S. W. and Potter, D. D. 1963. Gamma-aminobutyric acid and other blocking compounds in crustacea. III. Their relative concentrations in separated motor and inhibitory axons. *J. Neurophysiol. 26*: 739–751. (12)

Kriebel, M. E. and Gross, C. E. 1974. Multimodal distribution of frog miniature end-plate potentials in adult, denervated and tadpole leg muscle. *J. Gen. Physiol. 64*: 85–103. (10)

Kriegstein, A. R. 1977. Development of the nervous system of *Aplysia californica. Proc. Natl. Acad. Sci. USA 74*: 375–378. (18)

Kristan, W. B. 1983. The neurobiology of swimming in the leech. *T.I.N.S. 6*: 84–88. (18)

Krnjević, K. 1974. Chemical nature of synaptic transmission in vertebrates. *Physiol. Rev. 54*: 418–540. (12)

Kruger, L., Perl, E. R. and Sedivec, M. J. 1981. Fine structure of myelinated mechanical nociceptor endings in cat hairy skin. *J. Comp. Neurol. 198*: 137–154. (17)

Kuffler, D. P., Thompson, W. and Jansen, J. K. S. 1980. The fate of foreign endplates in cross-innervated rat soleus muscle. *Proc. R. Soc. Lond. B 208*: 189–222. (19)

Kuffler, S. W. 1942. Electrical potential changes at an isolated nerve-muscle junction. *J. Neurophysiol. 5*: 18–26. (9)

Kuffler, S. W. 1943. Specific excitability of the endplate region in normal and denervated muscle. *J. Neurophysiol. 6*: 99–110. (19)

Kuffler, S. W. 1948. Physiology of neuromuscular junctions: Electrical aspects. *Fed. Proc. 7*: 437–446. (9)

Kuffler, S. W. 1953. Discharge patterns and functional organization of the mammalian retina. *J. Neurophysiol. 16*: 37–68. (2)

Kuffler, S. W. 1954. Mechanisms of activation and motor control of stretch receptors in lobster and crayfish. *J. Neurophysiol. 17*: 558–574. (15)

Kuffler, S. W. 1967. Neuroglial cells: Physiological properties and a potassium mediated effect of neuronal activity on the glial membrane potential. *Proc. R. Soc. Lond. B 168*: 1–21. (13)

Kuffler, S. W. 1973. The single-cell approach in the visual system and the study of receptive fields. *Invest. Ophthalmol. 12*: 794–813. (2)

Kuffler, S. W. 1980. Slow synaptic responses in autonomic ganglia and the pursuit of a peptidergic transmitter. *J. Exp. Biol. 89*: 257–286. (9, 12)

Kuffler, S. W., Dennis, M. J. and Harris, A. J. 1971. The development of chemosensitivity in extrasynaptic areas of the neuronal surface after denervation of parasympathetic ganglion cells in the heart of the frog. *Proc. R. Soc. Lond. B 177*: 555–563. (19)

Kuffler, S. W. and Edwards, C. 1958. Mechanism of gamma aminobutyric acid (GABA) action and its relation to synaptic inhibition. *J. Neurophysiol. 21*: 589–610. (12)

Kuffler, S. W. and Eyzaguirre, C. 1955. Synaptic inhibition in an isolated nerve cell. *J. Gen. Physiol. 39*: 155–184. (9, 15)

Kuffler, S. W., Hunt, C. C. and Quilliam J. P. 1951. Function of medullated small-nerve fibers in mammalian ventral roots: Efferent muscle spindle innervation. *J. Neurophysiol. 14*: 29–54. (15)

Kuffler, S. W. and Nicholls, J. G. 1966. The physiology of neuroglial cells. *Ergeb. Physiol. 57*: 1–90. (13, 14)

Kuffler, S. W., Nicholls, J. G. and Orkand, R. K. 1966. Physiological properties of glial cells in the central nervous system of amphibia. *J. Neurophysiol. 29*: 768–787. (13, 14)

Kuffler, S. W. and Potter, D. D. 1964. Glia in the leech central nervous system: Physiological properties and neuron-glia relationship. *J. Neurophysiol. 27*: 290–320. (13, 18)

Kuffler, S. W. and Sejnowski, T. J. 1983. Peptidergic and muscarinic excitation at amhibian sympathetic synapses. *J. Physiol. 341*: 257–278. (12)

Kuffler, S. W. and Vaughan Williams, E. M. 1953. Small-nerve junctional potentials: The distribution of small motor nerves to frog skeletal muscle, and the membrane characteristics of the fibres they innervate. *J. Physiol. 121*: 289–317. (19)

Kuffler, S. W. and Yoshikami, D. 1975a. The distribution of acetylcholine sensitivity at the post-synaptic membrane of vertebrate skeletal twitch muscles: Iontophoretic mapping in the micron range. *J. Physiol. 244*: 703–730. (11)

Kuffler, S. W. and Yoshikami, D. 1975b. The number of transmitter molecules in a quantum: An estimate from iontophoretic application of acetylcholine at the neuromuscular junction. *J. Physiol. 251*: 465–482. (11)

Kuno, M. 1964a. Quantal components of excitatory synaptic potentials in spinal motoneurones. *J. Physiol. 175*: 81–99. (10)

Kuno, M. 1964b. Mechanism of facilitation and depression of the excitatory synaptic potential in spinal motoneurones. *J. Physiol. 175*: 100–112. (9, 10)

Kuno, M. 1971. Quantum aspects of central and ganglionic synaptic transmission in vertebrates. *Physiol. Rev. 51*: 647–678. (16)

Kuno, M. and Llinás, R. 1970. Alterations of synaptic action in chromatolysed motoneurones of the cat. *J. Physiol. 210*: 823–838. (19)

Kuno, M. and Rudomín, P. 1966. The release of acetylcholine from the spinal cord of the cat by antidromic stimulation of motor nerves. *J. Physiol. 187*: 177–193. (12)

Kuno, M. and Weakly, J. N. 1972. Quantal components of the inhibitory synaptic potentials in spinal motoneurones of the cat. *J. Physiol. 224*: 287–303. (10)

Kuo, J. F. and Greengard, P. 1969. Cyclic nucleotide-dependent protein kinases, IV. Widespread occurrence of adenosine 3′, 5′-monophosphate-dependent protein kinase in various tissues and phyla of the animal kingdom. *Proc. Natl. Acad. Sci. USA 64*: 1349–1355. (9)

Kuwada, J. Y. and Kramer, A. P. 1983. Embryonic development of the leech nervous system: Primary axon outgrowth of identified neurons. *J. Neuroscience 3*: 2098–2111. (18)

Kuypers, H. G. J. M., Catsman-Berrevoets, C. E. and Padt, R. E. 1977. Retrograde axonal transport of fluorescent substances in the rat's forebrain. *Neurosci. Lett. 6*: 127–135. (3)

Lam, D. M.-K. and Ayoub, G. S. 1983. Biochemical and biophysical studies of isolated horizontal cells from the teleost retina. *Vision Res. 23*: 433–444. (2, 12)

Lance-Jones, C. and Landmesser, L. 1980. Motoneurone projection patterns in the chick hind limb following early partial reversals of the spinal cord. *J. Physiol. 302*: 581–602. (19)

Lance-Jones, C. and Landmesser, L. 1981. Pathway selection by embryonic chick motoneurones in an experimentally altered environment. *Proc. R. Soc. Lond. B 214*: 19–52. (19)

Landgren, S., Philips, C. G. and Porter, R. 1962. Minimal synaptic actions of pyramidal impulses on some alpha motoneurones of the baboon's hand and forearm. *J. Physiol. 161*: 91–111. (17)

Landis, D. M. and Reese, T. S. 1981. Membrane structure in mammalian astrocytes: A review of freeze-structure studies on adult, developing, reactive and cultured astrocytes. *J. Exp. Biol. 95*: 35–48. (13)

Landmesser, L. 1971. Contractile and electrical responses of vagus-innervated frog sartorius muscles. *J. Physiol. 213*: 707–725. (19)

Landmesser, L. 1972. Pharmacological properties, cholinesterase activity and anatomy of nerve-muscle junctions in vagus-innervated frog sartorius. *J. Physiol. 220*: 243–256. (19)

Landmesser, L. T. 1980. The generation of neuromuscular specificity. *Annu. Rev. Neurosci. 3*: 279–302. (19)

Landowne, D. and Ritchie, J. M. 1970. The binding of tritiated ouabain to mammalian non-myelinated nerve fibres. *J. Physiol. 207*: 529–537. (8)

Lane, N. J. 1981. Invertebrate neuroglia-junctional structure and development. *J. Exp. Biol. 95*: 7–33. (13)

Lang, F., Atwood, H. L. and Morin, W. A. 1972. Innervation and vascular supply of the crayfish opener muscle. *Z. Zellforsch. 127*: 189–200. (16)

Langley, J. N. 1907. On the contraction of muscle, chiefly in relation to the presence of "receptive" substances. *J. Physiol. 36*: 347–384. (11)

Langley, J. N. and Anderson, H. K. 1892. The action of nicotin on the ciliary ganglion and on the endings of the third cranial nerve. *J. Physiol. 13*: 460–468. (9)

Langley, J. N. and Anderson, H. K. 1904. The union of different kinds of nerve fibres. *J. Physiol. 31*: 365–391. (19)

Lasek, R. J., Gainer, H. and Barker, J. L. 1977. Cell-to-cell transfer of glial proteins to the squid giant axon. *J. Cell Biol. 74*: 501–523. (13)

Lasek, R. J. and Tytell, M. A. 1981. Macromolecular transfer from glia to the axon. *J. Exp. Biol. 95*: 153–165. (13)

Lashley, K. S. 1941. Pattern of cerebral integration indicated by the scotomas of migraine. *Arch. Neurol. Psychiat. 46*: 331–339. (2)

Lassignal, N. and Martin, A. R. 1977. Effect of acetylcholine on postjunctional membrane permeability in eel electroplaque. *J. Gen. Physiol. 70*: 23–36. (9)

Latorre, R. and Miller, C. 1983. Conduction and selectivity in potassium channels. *J. Memb. Biol. 71*: 11–30. (6)

La Vail, J. H. and La Vail, M. M. 1974. The retrograde intraaxonal transport of horseradish peroxidase in the chick visual system: A light and electron microscopic study. *J. Comp. Neurol. 157*: 303–358. (3, 12)

Law, M. I. and Constantine-Paton, M. 1981. Anatomy and physiology of experimentally produced striped tecta. *J. Neurosci. 1*: 741–749. (19)

Le Douarin, N. M. 1980. The ontogeny of the neural crest in avian embryo chimaeras. *Nature 286*: 663–669. Reprinted by Patterson, P. H. and Purves, D. 1982. *Readings in Developmental Neurobiology,* Cold Spring Harbor Laboratory, Cold Spring Harbor, NY. (19)

Le Douarin, N. M., 1982. *The Neural Crest.* Cambridge University Press, Cambridge, England. (19)

Le Douarin, N. M., Smith, J. and Le Lièvre, C. S. 1982. From the neural crest to the ganglia of the peripheral nervous system. *Annu. Rev. Physiol. 43*: 653–671. (19)

Lee, C. Y. 1972. Chemistry and pharmacology of polypeptide toxins in snake venoms. *Annu. Rev. Pharmacol. 12*: 265–286. (11)

Leksell, L. 1945. The action potential and excitatory effects of the small ventral root fibres to skeletal muscle. *Acta Physiol. Scand. 10* (Suppl 31): 1–84. (15)

Lemmon, V. and Pearlman, A. L. 1981. Does laminar position determine the receptive field properties of cortical neurons? A study of cortico-tectal cells in area 17 of the normal mouse and the reeler mutant. *J. Neurosci. 1*: 83–93. (20)

Letinsky, M. S, Fischbeck, K. H. and McMahan, U. J. 1976. Precision of reinnervation of original postsynaptic sites in frog muscle after a nerve crush. *J. Neurocytol. 5*: 691–718. (19)

Leusen, I. 1972. Regulation of cerebrospinal fluid composition with reference to breathing. *Physiol. Rev. 52*: 1–56. (14)

LeVay, S. and Ferster, D. 1977. Relay cell classes in the lateral geniculate nucleus of the cat and the effects of visual deprivation. *J. Comp. Neurol. 172*: 563–584. (20)

LeVay, S., Hubel, D. H. and Wiesel, T. N. 1975. The pattern of ocular dominance columns in macaque visual cortex revealed by a reduced silver stain. *J. Comp. Neurol. 159*: 559–576. (3)

LeVay, S. and McConnell, S. K. 1982. On and off layers in the lateral geniculate nucleus of the mink. *Nature 300*: 350–351. (2)

LeVay, S., Stryker, M. P. and Shatz, C. J. 1978. Ocular dominance columns and their development in layer IV of the cat's visual cortex: A quantitative study. *J. Comp. Neurol. 179*: 223–244. (20)

LeVay, S., Wiesel, T. N. and Hubel, D. H. 1980. The development of ocular dominance columns in normal and visually deprived monkeys. *J. Comp. Neurol. 191*: 1–51. (20)

Levey, A. I., Armstrong, D. M., Atweh, S. F., Terry, R. D. and Wainer, B. G. 1983. Monoclonal antibodies to choline acetyltransferase: Production, specificity, and immunohistochemistry. *J. Neurosci. 3*: 1–9. (12)

Levi-Montalcini, R. 1975. NGF: An uncharted route. In F. G. Worden, J. P. Swaze and G. Adelman (eds.). *The Neurosciences: Paths of Discovery.* MIT Press, Cambridge, pp. 245–265. (19)

Levi-Montalcini, R. 1982. Developmental neurobiology and the natural history of nerve growth factor. *Annu. Rev. Neurosci. 5*: 341–362. (19)

Levi-Montalcini, R. and Angeletti, P. U. 1968. Nerve growth factor. *Physiol. Rev. 48*: 534–569. (19)

Levi-Montalcini, R. and Cohen, S. 1960. Effects of the extract of the mouse submaxillary salivary glands on the sympathetic system of mammals. *Ann. N.Y. Acad. Sci. 85*: 324–341. (19)

Levinson, S. R. and Meves, H. 1975. The binding of tritiated tetrodotoxin to squid giant axon. *Phil. Trans. R. Soc. Lond. B 270*: 349–352. (6)

Levitt, P. and Rakič, P. 1980. Immunoperoxidase localization of glial fibrillary acidic protein in radial glial cells and astrocytes of the developing rhesus monkey brain. *J. Comp. Neurol. 193*: 815–840. (13)

Lev-Tov, A. and Rahamimoff, R. 1980. A study of tetanic and post-tetanic potentiation of miniature end-plate potentials at the frog neuromuscular junction. *J. Phyiol. 309*: 247–273. (10)

Linder, T. M. and Quastel, D. M. J. 1978. A voltage-clamp study of the permeability change induced by quanta of transmitter at the mouse endplate. *J. Physiol. 281*: 535–556. (9)

Ling, G. and Gerard, R. W. 1949. The normal membrane potential of frog sartorius fibers. *J. Cell. Comp. Physiol. 34*: 383–396. (9)

Livingston, R. B., Pfenninger, K., Moor, H. and Akert, K. 1973. Specialized paranodal and interparanodal glial-axonal junctions in the peripheral and central nervous system: A freeze-etching study. *Brain Res. 58*: 1–24. (13)

Livingstone, M. S. and Hubel, D. H. 1982. Thalamic inputs to cytochrome oxidase-rich regions in monkey visual cortex. *Proc. Natl. Acad. Sci. USA 79*: 6098–6101. (3)

Livingstone, M. S. and Hubel, D. H. 1983. Specificity of cortico-cortical connections in monkey visual system. *Nature 304*: 531–534. (3)

Ljungdahl, A. and Hökfelt, T. 1973. Autoradiographic uptake patterns of [^{3}H]-GABA and [^{3}H]-glycine in central nervous tissues with special reference to the cat spinal cord. *Brain Res. 62*: 587–595. (13)

Llinás, R. R. 1981. Chapter 17. (In V. Brooks ed.). *Handbook of Physiology.* Amer. Physiol. Soc., Bethesda, pp. 831–976. (1)

Llinás, R. 1982. Calcium in synaptic transmission. *Sci. Am. 247*: 56–65. (10, 18)

Llinás, R., Baker, R. and Sotelo, C. 1974. Electrotonic coupling between neurons on cat inferior olive. *J. Neurophysiol. 37*: 560–571. (9)

Llinás, R. and Simpson, J. I. 1981. Cerebellar control of movement. *Handbook of Behavioral Neurobiology 5*: 231–302. (1)

Llinás, R. and Sugimori, M. 1980a. Electrophysiological properties of *in vitro* Purkinje cell somata in mammalian cerebellar slices. *J. Physiol. 305*: 171–195. (1, 6, 16)

Llinás, R. and Sugimori, M. 1980b. Electrophysiological properties of *in vitro* Purkinje cell dendrites in mammalian cerebellar slices. *J. Physiol. 305*: 197–213. (1, 6, 16)

Lloyd, D. P. C. 1943. Conduction and synaptic transmission of the reflex response to stretch in spinal cats. *J. Neurophysiol. 6*: 317–326. (16)

Lloyd, D. P. C. and Chang, H. T. 1948. Afferent fibers in muscle nerve. *J. Neurophysiol. 11*: 199–207. (15)

Loewenstein, W. 1981. Junctional intercellular communication: The cell-to-cell membrane channel. *Physiol. Rev. 61*: 829–913. (9)

Loewenstein, W. R. and Mendelson, M. 1965. Components of receptor adaptation in a Pacinian corpuscle. *J. Physiol. 177*: 377–397. (15)

Loewi, O. 1921. Über humorale Übertragbarkeit der Herznervenwirkung. *Pflügers Arch. 189*: 239–242. (9)

Lømo, T. and Rosenthal, J. 1972. Control of ACh sensitivity by muscle activity in the rat. *J. Physiol. 221*: 493–513. (19)

Longo, A. M. and Penhoet, E. E. 1974. Nerve growth factor in rat glioma cells. *Proc. Natl. Acad. Sci. USA 71*: 2347–2349. (13)

Lorenz, K. 1970. *Studies in Animal and Human Behavior*, Vols. I and II. Harvard University Press, Cambridge. (20)

Lund, R. D. and Lund, J. S. 1973. Reorganization of the retinotectal pathway in rats after neonatal retinal lesions. *Exp. Neurol. 40*: 377–390. (19)

Lund, J. S. 1980. Intrinsic organization of the primate visual cortex, area 17, as seen in Golgi preparations. In F. O. Schmitt, F. G. Worden, G. Adelman and S. G. Dennis (eds.). *The Organization of the Cerebral Cortex*. MIT Press, Cambridge, pp. 105–124. (3)

Lundberg, A. 1979. Multisensory control of spinal reflex pathways. *Progr. Brain Res. 50*: 11–28. (15)

Luskin, M. B. and Price, J. L. 1982. The distribution of axon collaterals from the olfactory bulb and the nucleus of the horizontal limb of the diagonal band to the olfactory cortex, demonstrated by double retrograde labeling techniques. *J. Comp. Neurol. 209*: 249–263. (1)

Luskin, M. B. and Price, J. L. 1983. The topographic organization of associational fibers of the olfactory system in the rat, including centrifugal fibers to the olfactory bulb. *J. Comp. Neurol. 216*: 264–291. (15)

Lux, H. D. and Nagy, K. 1981. Single channel Ca^{2+} currents in *Helix pomatia* neurons. *Pflügers Arch. 291*: 252–254. (6)

Macagno, E. R. 1980. Number and distribution of neurons in the leech segmental ganglion. *J. Comp. Neurol. 190*: 283–302. (18)

Magleby, K. L. and Miller, D. C. 1981. Is the quantum of transmitter release composed of subunits? A critical analysis in the mouse and frog. *J. Physiol. 311*: 267–287. (10)

Magleby, K. L. and Stevens, C. F. 1972a. The effect of voltage on the time course of end-plate currents. *J. Physiol. 223*: 151–171. (9, 11)

Magleby, K. L. and Stevens, C. F. 1972b. A quantitative description of end-plate currents. *J. Physiol. 223*: 173–197. (11)

Magleby, K. L. and Zengel, J. E. 1982. A quantitative description of stimulation-induced changes in transmitter release at the frog neuromuscular junction. *J. Gen. Physiol. 80*: 613–638. (10)

Malpeli, J. G. 1983. Activity of cells in area 17 of the cat in absence of input from layer A of lateral geniculate nucleus. *J. Neurophysiol. 49*: 595–610. (3)

Malpeli, J. G., Schiller, P. H. and Colby, C. L. 1981. Response properties of single cells in monkey striate cortex during reversible inactivation of individual lateral geniculate laminae. *J. Neurophysiol. 46*: 1102–1119. (3)

Marchiafava, P. L. and Weiler, R. 1982. The photoresponses of structurally identified amacrine cells in the turtle retina. *Proc. R. Soc. Lond. B 214*: 403–415. (2)

Marmont, G. 1949. Studies on the axon membrane. *J. Cell. Comp. Physiol. 34*: 351–382. (6)

Marr, D. 1969. A theory of cerebellar cortex. *J. Physiol. 202*: 437–470. (1)

Marr, D. 1982. *Vision*. W. H. Freeman, San Francisco. (2)

Martin, A. R. 1977. Junctional transmission II. Presynaptic mechanisms. In E. Kandel (ed.). *Handbook of the Nervous System*, Vol. 1. American Physiological Society, Baltimore, pp. 329–355. (10)

Martin, A. R. and Pilar, G. 1963. Dual mode of synaptic transmission in the avian ciliary ganglion. *J. Physiol. 168*: 443–463. (9)

Martin, A. R. and Pilar, G. 1964. Quantal components of the synaptic potential in the ciliary ganglion of the chick. *J. Physiol. 175*: 1–16. (10)

Matsuda, T., Wu, J.-Y., and Roberts, E., 1973a. Immunochemical studies on glutamic acid decarboxylase (EC 4.1.1.15) from mouse brain. *J. Neurochem. 21*: 159–166. (12)

Matsuda, T., Wu, J.-Y., and Roberts, E. 1973b. Electrophoresis of glutamic acid decarboxylase (EC 4.1.1.15) from mouse brain in sodium dodecyl sulphate polyacrylamide gels. *J. Neurochem. 21*: 167–172. (12)

Matthews, B. H. C. 1931. The response of a single end organ. *J. Physiol. 71*: 64–110. (15)

Matthews, B. H. C. 1931. The response of a muscle spindle during active contraction of a muscle. *J. Physiol. 72*: 153–174. (15)

Matthews, B. H. C. 1933. Nerve endings in mammalian muscle. *J. Physiol. 78*: 1–53. (15)

Matthews, G. and Wickelgren, W. O. 1979. Glycine, GABA and synaptic inhibition of reticulospinal neurones of the lamprey. *J. Physiol. 293*: 393–415. (12)

Matthews, P. B. C. 1964. Muscle spindles and their motor control. *Physiol. Rev. 44*: 219–288. (15)

Matthews, P. B. C. 1972. *Mammalian Muscle Receptors and Their Central Action*. Edward Arnold, London. (15, 17)

Matthews, P. B. C. 1981. Evolving views on the internal operation and functional role of the muscle spindle. *J. Physiol. 320*: 1–30. (15)

Matthews, P. B. C. 1982. Where does Sherrington's "muscular sense" originate? Muscles, joints, corollary discharges? *Annu. Rev. Neurosci. 5*: 189–218. (15)

Maturana, H. R., Lettvin, J. Y., McCulloch, W. S. and Pitts, W. H. 1960. Anatomy and physiology of vision in the frog (*Rana pipiens*). *J. Gen. Physiol. 43*: 129–175. (2)

McAfee, D. A. and Greengard, P. 1972. Adenosine 3', 5'-monophosphate: Electrophysiological evidence for a role in synaptic transmission. *Science 178*: 310–312. (9)

McCloskey, D. I. 1978. Kinesthetic sensibility. *Physiol. Rev. 58*: 763–820. (15)

McCloskey, D. I., Cross, M. J., Honner, R. and Potter, E. K. 1983. Sensory effects of pulling or vibrating exposed tendons in man. *Brain 106*: 21–37. (15)

McKay, R. D. G. 1983. Molecular approaches to the nervous system. *Annu. Rev. Neurosci. 6*: 527–546. (Part 5, 19)

McLachlan, E. M. 1978. The statistics of transmitter release at chemical synapses. *International Review of Physiology, Neurophysiology III 17*: 49–117. (10)

McMahan, U. J., Edgington, D. R. and Kuffler, D. P. 1980. Factors that influence regeneration of the neuromuscular junction. *J. Exp. Biol. 89*: 31–42. (19)

McMahan, U. J. and Kuffler, S. W. 1971. Visual identification of synaptic boutons on living ganglion cells and of varicosities in postganglionic axons in the heart of the frog. *Proc. R. Soc. Lond. B 177*: 485–508. (9, 11).

McMahan, U. J., Spitzer, N. C. and Peper, K. 1972. Visual identification of nerve terminals in living isolated skeletal muscle. *Proc. R. Soc. Lond. B 181*: 421–430. (11)

Meech, R. W. 1974. The sensitivity of *Helix aspersa* neurones to injected calcium ions. *J. Physiol. 237*: 259–277. (6)

Mendell, L. M. and Henneman, E. 1971. Terminals of single Ia fibers: Location, density, and distribution within a pool of 300 homonymous motoneurons. *J. Neurophysiol. 34*: 171–187. (16)

Mensini-Chen, M. G., Chen, J. S. and Levi-Montalcini, R. 1978. Sympathetic nerve fibers in growth in the central nervous system of neonatal rodents upon intracerebral NGF injections. *Arch. Natl. Biol. 116*: 53–84. (19)

Merlie, J. P., Sebbane, R., Gardner, S., Olson, E. and Lindström, J. 1984. The regulation of acetylcholine receptor expression in mammalian muscle. *Cold Spring Harbor Symp. Quant. Biol. 48*: 135–146. (19)

Merrill, E. G. 1981. Where are the *real* respiratory neurons? *Fed. Proc. 40*: 2389–2394. (17)

Merzenich, M. M., Kaas, J. H., Sur, M. and Lin, C.-S. 1978. Double representation of the body surface within cytoarchitectonic areas 3b and 1 in "S-I" in the owl monkey (*Aotus trivigatus*). *J. Comp. Neurol. 181*: 41–74. (17)

Michael, C. R. 1973. Color Vision. *N. Engl. J. Med. 288*: 724–725. (2)

Michael, C. R. 1981. Columnar organization of color cells in monkey's striate cortex. *J. Neurophysiol. 46*: 587–604. (3)

Middlebrooks, J. C., Dykes, R. W. and Merzenich, M. M. 1980. Binaural response-specific bands in primary auditory cortex (A1) of the cat: Topographical organization orthogonal to isofrequency contours. *Brain Res. 181*: 31–48. (3)

Miledi, R. 1960a. The acetylcholine sensitivity of frog muscle fibres after complete or partial denervation. *J. Physiol. 151*: 1–23. (19)

Miledi, R. 1960b. Junctional and extra-junctional acetylcholine receptors in skeletal muscle fibres. *J. Physiol. 151*: 24–30. (9)

Miledi, R. 1962. Induced innervation of end-plate free muscle segments. *Nature 193*: 281–282. (19)

Miledi, R., Parker, I. and Sumikawa, K. 1982. Synthesis of chick brain GABA receptors by frog oocytes. *Proc. R. Soc. Lond. B 216*: 509–515. (12)

Miledi, R., Stefani, E. and Steinbach, A. B. 1971. Induction of the action potential mechanism in slow muscle fibres of the frog. *J. Physiol. 217*: 737–754. (19)

Miledi, R. and Sumikawa, K. 1982. Synthesis of cat muscle acetylcholine receptors by *Xenopus* oocytes. *Biomed. Res. 3*: 390–399. (19)

Miller, R. F. and Dacheux, R. F. 1983. Intracellular chloride in retinal neurons: Measurement and meaning. *Vision Res. 23*: 399–412. (2)

Miller, R. F. and Dowling, J. E. 1970. Intracellular responses of the Müller (glial) cells of mudpuppy retina: Their relation to b-wave of the electroretinogram.

J. Neurophysiol. 33: 323–341. (13)

Miller, R. F., Frumkes, T. E., Slaughter, M. and Dacheux, R. F. 1981. Physiological and pharmacological basis of GABA and glycine action on neurons of mudpuppy retina. II. Amacrine and ganglion cells. *J. Neurophysiol. 45*: 764–782. (2)

Milstein, C. 1980. Monoclonal antibodies. *Sci. Am. 243*: 66–74. (19)

Minchin, M. C. W. and Iversen, L. L. 1974. Release of [^{3}H]-gamma-aminobutyric acid from glial cells in rat dorsal root ganglia. *J. Neurochem. 23*: 533–540. (13)

Mirsky, R. 1982. The use of antibodies to define and study major cell types in the central and peripheral nervous system. In J. Brockes (ed.). *Neuroimmunology*. Plenum Press, New York, pp. 141–181. (13)

Molecular Neurobiology. 1984. *Cold Spring Harbor Symp. Quant. Biol. 48*. (19)

Mollon, J. D. 1982. Colour vision and colour blindness. In H. B. Barlow and J. D. Mollon (eds.). *The Senses*. Cambridge University Press, New York, pp. 165–191. (2)

Moody, W. J. 1981. The ionic mechanism of intracellular pH regulation in crayfish neurones. *J. Physiol. 316*: 293–308. (8)

Moore, J. W., Blaustein, M. P., Anderson, N. C. and Narahashi, T. 1967. Basis of tetrodotoxin's selectivity in blockage of squid axons. *J. Gen. Physiol. 50*: 1401–1411. (6)

Moore, R. Y. and Bloom, F. E. 1978. Central catecholamine neuron systems: Anatomy and physiology of the dopamine systems. *Annu. Rev. Neurosci. 1*: 129–169. (12)

Moore, R. Y. and Bloom, F. E. 1979. Central catecholamine neuron systems: Anatomy and physiology of the norepinephrine and epinephrine systems. *Annu. Rev. Neurosci. 2*: 113–168. (12)

Morell, P. and Norton, W. T. 1980. Myelin. *Sci. Am. 242*: 88–118. (13)

Mountcastle, V. B. 1957. Modality and topographic properties of single neurons of cat's somatic sensory cortex. *J. Neurophysiol. 20*: 408–434. (3)

Mountcastle, V. B. 1975. The view from within: Pathways to the study of perception. *The Johns Hopkins Med. J. 136*: 109–131. (17)

Mountcastle, V. B. and Powell, T. P. S. 1959. Neural mechanisms subserving cutaneous sensibility with special reference to the role of afferent inhibition in sensory perception and discrimination. *Bull. Johns Hopkins Hosp. 105*: 201–232. (17)

Movshon, J. A., Thompson, I. D. and Tolhurst, D. J. 1978a. Spatial summation in the receptive fields of simple cells in the cat's striate cortex. *J. Physiol. 283*: 53–77. (2)

Movshon, J. A., Thompson, I. D. and Tolhurst, D. J. 1978b. Spatial and temporal contrast sensitivity of neurones in areas 17 and 18 of the cat's visual cortex. *J. Physiol. 283*: 101–120. (3)

Movshon, J. A. and Van Sluyters, R. C. 1981. Visual neural development. *Annu. Rev. Psychol. 32*: 477–522. (20)

Mower, G. D., Christen, W. G. and Caplan, C. J. 1983. Very brief visual experience eliminates plasticity in the cat visual cortex. *Science 221*: 178–180. (20)

Mugnaini, E. 1982. Membrane specializations in neuroglial cells and at neuronglial contacts. In T. A. Sears (ed.). *Neuronal-Glial Cell Interrelationships*. Springer-Verlag, New York, pp. 39–56. (13)

Muller, K. J. 1981. Synapses and synaptic transmission. In K. J. Muller, J. G. Nicholls and G. S. Stent (eds.). *Neurobiology of the Leech*. Cold Spring Harbor Laboratory, Cold Spring Harbor, NY, pp. 79–111. (18)

Muller, K. J. and McMahan, U. J. 1976. The shapes of sensory and motor neurones and the distribution of their synapses in ganglia of the leech: A study using intracellular injection of horseradish peroxidase. *Proc. R. Soc. Lond. B 194*: 481–499. (1, 3, 18)

Muller, K. J. and Nicholls, J. G. 1974. Different properties of synapses between a single sensory neurone and two different motor cells in the leech CNS. *J. Physiol. 238*: 357–369. (18)

Muller, K. J. and Nicholls, J. G. 1981. Regeneration and plasticity. In K. J. Muller, J. G. Nicholls and G. S. Stent (eds.). *Neurobiology of the Leech*. Cold Spring Harbor Laboratory, Cold Spring Harbor, NY, pp. 197–226. (19)

Muller, K. J., Nicholls, J. G. and Stent, G. S. (eds.). 1981. *Neurobiology of the Leech*. Cold Spring Harbor Laboratory, Cold Spring Harbor, NY. (18)

Muller, K. J. and Scott, S. A. 1981. Transmission at a "direct" electrical connexion mediated by an interneurone in the leech. *J. Physiol. 311*: 565–583. (18)

Mullins, L. J. and Brinley, F. J. 1967. Some factors influencing sodium extrusion by internally dialyzed squid axons. *J. Gen. Physiol. 50*: 2333–2355. (8)

Mullins, L. J. and Noda, K. 1963. The influence of sodium-free solutions on the membrane potential of frog muscle fibers. *J. Gen. Physiol. 47*: 117–132. (8)

Murphey, R. K. Johnson, S. E. and Sakaguchi, D. S. 1983. Anatomy and physiology of supernumerary cercal afferents in crickets: Implication for pattern formation. *J. Neurosci. 3*: 312–325. (19)

Nageotte, J. 1910. Phénomènes de sécrétion dans le protoplasma des cellules névrogliques de la substance grise. *C. R. Soc. Biol. (Paris) 68*: 1068–1069. (13)

Nakajima, S. and Onodera, K. 1969a. Membrane properties of the stretch receptor neurones of crayfish with particular reference to mechanisms of sensory adaptation. *J. Physiol. 200*: 161–185. (8, 15)

Nakajima, S. and Onodera, K. 1969b. Adaptation of the generator potential in the crayfish stretch receptors under constant length and constant tension. *J. Physiol. 200*: 187–204. (15)

Nakajima, S. and Takahashi, K. 1966. Post-tetanic hyperpolarization and electrogenic Na pump in stretch receptor neurone of crayfish. *J. Physiol. 187*: 105–127. (15)

Nakajima, Y. and Reese, T. S. 1983. Inhibitory and excitatory synapses in crayfish stretch receptor organs studied with direct rapid-freezing and freeze-substitution. *J. Comp. Neurol. 213*: 66–73. (15)

Nakajima, Y., Tisdale, A. D. and Henkart, M. P. 1973. Presynaptic inhibiton at inhibitory nerve terminals: A new synapse in the crayfish stretch receptor. *Proc. Natl. Acad. Sci. USA 70*: 2462–2466. (9)

Nastuk, W. L. 1953. Membrane potential changes at a single muscle end-plate produced by transitory application of acetylcholine with an electrically controlled microjet. *Fed. Proc. 12*: 102. (9)

Nauta, W. J. H. and Gygax, P. A. 1954. Silver impregnation of degenerating axons in the central nervous system: A modified technic. *Stain Technol. 29*: 91–93. (3)

Neher, E. and Steinbach, J. H. 1978. Local anaesthetics transiently block currents through single acetycholine-receptor channels. *J. Physiol. 277*: 153–176. (11)

Nelson, R., Famiglietti, E. V. Jr. and Kolb, H. 1978. Intracellular staining reveals different levels of stratification for on- and off-center ganglion cells in cat retina. *J. Neurophysiol. 41*: 472–483. (2)

Nicholls, J. G. 1956. The electrical properties of denervated skeletal muscle. *J. Physiol. 131*: 1–12. (19)

Nicholls, J. G. (ed.). 1982. *Repair and Regeneration of the Nervous System*. Springer-Verlag, New York. (19)

Nicholls, J. G. and Baylor, D. A. 1968. Specific modalities and receptive fields of sensory neurons in the CNS of the leech. *J. Neurophysiol. 31*: 740–756. (13, 18)

Nicholls, J. G and Kuffler, S. W. 1964. Extracellular space as a pathway for exchange between blood and neurons in the central nervous system of the leech: Ionic composition of glial cells and neurons. *J. Neurophysiol. 27*: 645–671. (13, 14)

Nicholls, J. G. and Kuffler, S. W. 1965. Na and K content of glial cells and neurons determined by flame photometry in the central nervous system of the leech. *J. Neurophysiol. 28*: 519–525. (13)

Nicholls, J. G. and Purves, D. 1972. A comparison of chemical and electrical synaptic transmission between single sensory cells and a motoneurone in the central nervous system of the leech. *J. Physiol. 225*: 637–656. (18)

Nicholls, J. G. and Wallace, B. G. 1978. Modulation of transmission at an inhibitory synapse in the central nervous system of the leech. *J. Physiol. 281*: 157–170. (18)

Nicoll, R. A., Schenker, C. and Leeman, S. E. 1980. Substance P as a transmitter candidate. *Annu. Rev. Neurosci. 3*: 227–268. (12)

Nietzsche, F. *Thus Spake Zarathustra*. (18)

Nirenberg, M., Wilson, S., Higashida, H., Rotter, A., Krueger, K., Busis, N., Ray, R., Kenimer, K., Adler, M. and Fukui, H. 1984. Synapse formation by neuroblastoma hybrid cells. *Cold Spring Harbor Symp. Quant. Biol. 48*: 707–716. (Part 5)

Nishsi, S. and Koketsu, K. 1960. Electrical properties and activities of single sympathetic neurons in frogs. *J. Cell Comp. Physiol. 55*: 15–30. (9)

Nishi, S. and Koketsu, K. 1968. Early and late afterdischarges of amphibian sympathetic ganglion cells. *J. Neurophysiol. 31*: 109–118. (9)

Nitkin, R. M., Wallace, B. G., Spira, M. E., Godfrey, E. W. and McMahan, U. J. 1984. Molecular components of the synaptic basal lamina that direct differentiation of regenerating neuromuscular junctions. *Cold Spring Harbor Symp. Quant. Biol. 48*: 653–666. (19)

Njå, A. and Purves, D. 1978. The effects of nerve growth factor and its antiserum on synapses in the superior cervical ganglion of the guinea-pig. *J. Physiol. 277*: 53–75. (19)

Noda, M., Takahashi, H., Tanabe, T., Toyosato, M., Furutani, Y., Hirose, T., Asai, M., Inayam, S., Miyata, T. and Numa, S. 1982. Primary structure of alpha-subunit precursor of *Torpedo californica* acetylcholine receptor deduced from cDNA sequence. *Nature 299*: 793–797. (11)

Noda, M., Takahashi, H., Tanabe, T., Toyosato, M., Furutani, Y., Hirose, T., Asai, M., Takashima, H., Inayam, S., Miyata, T. and Numa, S. 1983. Primary structure of beta- and gamma-unit precursor of *Torpedo californica* acetylcholine receptor deduced from cDNA sequence. *Nature 301*: 251–255. (11)

Nolte, J. 1981. *The Human Brain*. Mosby, St. Louis. (Appendix B)

Nunn, B. J. and Baylor, D. A. 1982. Visual transduction in retinal rods of the monkey *Macaca fascicularis*. *Nature 299*: 726–728. (2)

Obaid, A. L., Socolar, S. J. and Rose, B. 1983. Cell-to-cell channels with two independently regulated gates in series: Analysis of junctional conductance modulation by membrane potential, calcium and pH. *J. Memb. Biol. 73*: 68–89. (9)

Obata, K. 1969. Gamma-aminobutyric acid in Purkinje cells and motoneurones. *Experientia 25*: 1283. (12)

Obata, K. 1974. Transmitter sensitivies of some nerve and muscle cells in culture. *Brain Res. 73*: 71–88. (12)

Obata, K. 1980. Biochemistry and physiology of amino acid transmitters. In *Handbook of Physiology. The Nervous System*, Vol. 1, Part 1. American Physiological Society, Bethesda, pp. 625–650. (12)

Obata, K., Takeda, K. and Shinozaki, H. 1970. Further study on pharmacological properties of the cerebellar-induced inhibition of Deiters neurones. *Exp. Brain Res. 11*: 327–342. (12)

Onodera, K. and Takeuchi, A. 1980. Distribution and pharmacological properties of synaptic and extrasynaptic glutamate receptors on crayfish muscle. *J. Physiol. 306*: 233–249. (16)

Oppenheim, R. W. 1981. Neuronal cell death and some related regressive phenomena during neurogenesis: A selective historical review and progress report. In W. M. Cowan (ed.). *Studies in Developmental Neurobiology.* Oxford University Press, New York, pp. 74–133. (19)

Optican, L. M. and Robinson, D. A. 1980. Cerebellar dependent adaptive control of primate saccadic system. *J. Neurophysiol. 44*: 1058–1076. (1)

Orkand, P. M., Bracho, H. and Orkand, R. K. 1973. Glial metabolism: Alteration by potassium levels comparable to those during neural activity. *Brain Res. 55*: 467–471. (13)

Orkand, P. M. and Kravitz, E. A. 1971. Localization of the sites of γ-aminobutyric acid (GABA) uptake in lobster nerve-muscle preparations. *J. Cell Biol. 49*: 75–89. (12, 13)

Orkand, R. K. 1982. Signalling between neuronal and glial cells. In T. A. Sears (ed.). *Neuronal-Glial Cell Interrelationships.* Springer-Verlag, New York, pp. 147–157. (13)

Orkand, R. K., Nicholls, J. G. and Kuffler, S. W. 1966. Effect of nerve impulses on the membrane potential of glial cells in the central nervous system of amphibia. *J. Neurophysiol. 29*: 788–806. (13)

Orkand, R. K., Orkand, P. M. and Tang, C.-M. 1981. Membrane properties of neuroglia in the optic nerve of *Necturus. J. Exp. Biol. 95*: 49–59. (13)

Otsuka, M., Iversen, L. L., Hall, Z. W. and Kravitz, E. A. 1966. Release of gamma-aminobutyric acid from inhibitory nerves of lobster. *Proc. Natl. Acad. Sci. USA 56*: 1110–1115. (9, 12)

Otsuka, M., Kravitz, E. A. and Potter, D. D. 1967. Physiological and chemical architecture of a lobster ganglion with particular reference to γ-aminobutyrate and glutamate. *J. Neurophysiol. 30*: 725–752. (12)

Otsuka, M., Obata, K., Miyata, Y. and Tanaka, Y. 1971. Measurement of γ-aminobutyric acid in isolated nerve cells of cat central nervous system. *J. Neurochem. 18*: 287–295. (12)

Ottoson, D. 1983. *Physiology of the Nervous System.* Oxford University Press, New York. (1, 17)

Overton, E. 1902. Beiträge zur allgemeinen Muskel- und Nervenphysiologie. II. Über die Unentbehrlichkeit von Natrium- (oder Lithium-) Ionen für den Kontraktionsakt des Muskels. *Pflügers Arch. 92*: 346–386. (5, 8)

Palay, S. L. and Chan-Play, V. 1974. *Cerebellar Cortex.* Springer-Verlag, New York. (1)

Palay, S. L. and Chan-Play, V. (eds.). 1982. *The Cerebellum—New Vistas.* Springer-Verlag, New York. (1)

Palay, S. L. and Palade, G. E. 1955. Fine structure of neurons. *J. Biophys. Biochem. Cytol. 1*: 69–88. (11)

Palka, J. and Edwards, J. S. 1974. The cerci and abdominal giant fibers of the house cricket *Acheta domesticus.* II. Regeneration and effects of chronic deprivation. *Proc. R. Soc. Lond. B 185*: 105–121. (19)

Palmer, L. A. and Rosenquist, A. C. 1974. Visual receptive fields of single striate cortical units projecting to the superior colliculus in the cat. *Brain Res. 67*: 27–42. (2)

Pappenheimer, J. R. 1953. Passage of molecules through capillary walls. *Physiol. Rev. 33*: 387–423. (14)

Pappenheimer, J. R. 1967. The ionic composition of cerebral extracellular fluid and its relation to control of breathing. *Harvey Lect. 61*: 71–94. (14)

Parnas, H., Dudel, J. and Parnas, I. 1982. Neurotransmitter release and its facilitation in crayfish. *Pflügers Arch. 393*: 1–14. (10)

Parnas, I., Dudel, J. and Grossman, Y. 1982. Chronic removal of inhibitory axon alters excitatory transmission in a crustacean muscle fiber. *J. Neurophysiol. 47*: 1–10. (19)

Parnas, I., Parnas, H. and Dudel, J. 1982. Neurotransmitter release and its facilitation in crayfish. II. Duration of facilitation and removal processes of calcium from the terminal. *Pflügers Arch. 393*: 232–236. (16)

Parnas, I. and Strumwasser, F. 1974. Mechanisms of long-lasting inhibition of a bursting pacemaker neuron. *J. Neurophysiol. 37*: 609–620. (18)

Patlak, J. B., Gration, K. A. F. and Usherwood, P. N. R. 1979. Single glutamate-activated channels in locust muscle. *Nature 278*: 643–645. (11)

Patrick, J., Heinemann, S. and Schubert, D. 1981. Biology of cultured nerve and muscle. *Annu. Rev. Neurosci. 1*: 417–443. (Part 5)

Patterson, P. 1978. Environmental determination of autonomic neurotransmitter functions. *Annu. Rev. Neurosci. 1*: 1–17. (13)

Patterson, P. H. and Chun, L. L. Y. 1977. Induction of acetylcholine synthesis in primary cultures of dissociated rat sympathetic neurons. I. Effects of conditioned medium. *Devel. Biol. 56*: 263–280. (12)

Patterson, P. H. and Purves, D. (eds.). 1982. *Readings in Developmental Neurobiology.* Cold Spring Harbor Laboratory, Cold Spring Harbor, NY. (Part 5, 19)

Payton, W. B. 1981. History of medicinal leeching and early medical references. In K. J. Muller, J. G. Nicholls and G. S. Stent (eds.) *Neurobiology of the Leech.* Cold Spring Harbor Laboratory, Cold Spring Harbor, NY, pp. 27–34.

Pearlman, A. L. and Hughes, C. P. 1976a. Functional role of efferents to the avian retina. I. Analysis of retinal ganglion cell receptive fields. *J. Comp. Neurol. 166*: 111–122. (15)

Pearlman, A. L. and Hughes, C. P. 1976b. Functional role of efferents to the avian retina. II. Effects of reversible cooling of the isthmo-optic nucleus. *J. Comp. Neurol. 166*: 123–132. (2)

Pearson, K. G. 1976. The control of walking. *Sci. Am. 235*: 72–86. (17)

Pearson, K. G. and Iles, J. F. 1970. Discharge patterns of coxal levator and depressor motorneurons of the cockroach *Periplaneta americana. J. Exp. Biol. 52*: 139–165. (17)

Penfield, W. 1932. *Cytology and Cellular Pathology of the Nervous System,* Vol. II., Hafner, New York. (13)

Pentreath, V. M. and Kai-Kai, M. A. 1982. Significance of the potassium signal from neurones to glial cells. *Nature 295*: 59–61. (13)

Peper, K., Bradley, R. J. and Dreyer, F. 1962. The acetylcholine receptor at the neuromuscular junction. *Physiol. Rev. 62*: 1271–1340. (11)

Peper, K., Dreyer, F., Sandri, C., Akert, K. and Moor, H. 1974. Structure and ultrastructure of the frog motor end-plate: A freeze-etching study. *Cell Tissue Res. 149*: 437–455. (11)

Peracchia, C. 1981. Direct communication between axons and glia cells in crayfish. *Nature 290*: 597–598. (13)

Pert, C. B. and Snyder, S. H. 1973. Opiate receptor: Demonstration in nervous tissue. *Science 179*: 1011–1014. (12)

Peters, A., Palay, S. L. and Webster, H. de F. 1976. *The Fine Structure of the Nervous System.* Saunders, Philadelphia. (1, 13)

Phelps, M. E., Mazziotta, J. C. and Hueng, S.-C. 1982. Study of cerebral function with positron computed tomography. *J. Cerebr. Blood Flow Metab. 2*: 113–162. (3)

Piccolino, M. and Gerschenfeld, H. M. 1980. Characteristics and ionic processes involved in feedback spikes of turtle cones. *Proc. R. Soc. Lond. B 206*: 439–463. (2)

Picker, S., Pieper, C. F. and Goldring, S. 1981. Glial membrane potentials and their relationship to $[K^+]_o$ in man and guinea pig. A comparative study of intracellularly marked normal, reactive, and neoplastic glia. *J. Neurosurg. 55*: 347–363. (13)

Pilar, G., Landmesser, L. and Burstein, L. 1980. Competition for survival among developing ciliary ganglion cells. *J. Neurophysiol. 41*: 233–254. (19)

Poggio, G. F. and Mountcastle, V. B. 1960. A study of the functional contributions of the lemniscal and spinothalamic systems to somatic sensibility. *Bull. Johns Hopkins Hosp. 106*: 266–316. (17)

Poritsky, R. 1969. Two and three dimensional ultrastructure of boutons and glial cells on the motoneuronal surface in the cat spinal cord. *J. Comp. Neurol. 135*: 423–452. (16)

Porter, C. W. and Barnard, E. A. 1975. The density of cholinergic receptors at the endplate postsynaptic membrane: Ultrastructural studies in two mammalian species. *J. Membr. Biol. 20*: 31–49. (11)

Post, R. L. 1981. The sodium and potassium ion pump. In T. P. Singer and R. N. Ondarza (eds.). *Molecular Basis of Drug Action.* Elsevier, New York, pp. 299–331. (8)

Powell, T. P. S. and Mountcastle, V. B. 1959. Some aspects of the functional organization of the cortex of the postcentral gyrus of the monkey: A correlation of findings obtained in a single unit analysis with cytoarchitecture. *Bull. Johns Hopkins Hosp. 105*: 133–162. (3, 17)

Precht, W. 1979. Vestibular mechanisms. *Annu. Rev. Neurosci. 2*: 265–289. (17)

Pumplin, D. W. and Fambrough, D. M. 1982. Turnover of acetylcholine receptors in skeletal muscle. *Annu. Rev. Physiol. 44*: 319–335. (19)

Purves, D. 1975. Functional and structural changes in mammalian sympathetic neurones following interruption of their axons. *J. Physiol. 252*: 429–463. (13)

Purves, D. and Lichtman, J. W. 1983. Specific connections between nerve cells. *Annu. Rev. Physiol. 45*: 553–565. (19)

Purves, D. and Sakmann, B. 1974a. The effect of contractile activity on fibrillation and extrajunctional acetylcholine sensitivity in rat muscle maintained in organ culture. *J. Physiol. 237*: 157–182. (19)

Purves, D. and Sakmann, B. 1974b. Membrane properties underlying spontaneous activity of denervated muscle fibres. *J. Physiol. 239*: 125–153. (19)

Purves, D., Thompson, W. and Yip, J. W. 1981. Re-innervation of ganglia transplanted to the neck from different levels of the guinea-pig sympathetic chain. *J. Physiol. 313*: 49–63. (19)

Quick, D. C., Kennedy, W. R. and Donaldson, L. 1979. Dimensions of myelinated nerve fibers near the motor and sensory terminals in cat tenuissimus muscles. *Neuroscience 4*: 1089–1096. (7)

Raichle, M. E. 1983. Positron emission tomography. *Annu. Rev. Neurosci. 6*: 249–267. (3)

Raisman, G. and Field, P. 1973. A quantitative investigation of the development of collateral innervation after partial deafferentation of the septal nuclei. *Brain Res. 50*: 241–264. (19)

Rakič, P. 1971. Neuron-glia relationship during granule cell migration in developing cerebellar cortex: *J. Comp. Neurol. 141*: 283–312. (13)

Rakič, P. 1977a. Genesis of the dorsal lateral geniculate nucleus in the rhesus monkey: Site and time of origin, kinetics of proliferation, routes of migration and pattern of distribution of neurons. *J. Comp. Neurol. 176*: 23–52. (20)

Rakič, P. 1977b. Prenatal development of the visual system in rhesus monkey. *Phil. Trans. R. Soc. Lond. B 278*: 245–260. (20)

Rakič, P. 1982. The role of neuronal-glial cell interaction during brain development. In T. A. Sears (ed.). *Neuronal-Glial Cell Interrelationships*. Springer-Verlag, New York, pp. 25–38. (13)

Rakič, P. and Sidman, R. L. 1973a. Sequence of developmental abnormalities leading to granule cell deficit in cerebellar cortex of weaver mutant mice. *J. Comp. Neurol. 152*: 103–132. (Part 5)

Rakič, P. and Sidman, R. L. 1973b. Organization of cerebellar cortex secondary to deficit of granule cells in weaver mutant mice. *J. Comp. Neurol. 152*: 133–162. (Part 5)

Ramón y Cajal, S. 1955. *Histologie du Système Nerveux*. II., C.S.I.C., Madrid. (1, 2)

Ransom, B. R. and Goldring, S. 1973. Slow depolarization in cells presumed to be glia in cerebral cortex of cat. *J. Neurophysiol. 36*: 869–878. (13)

Rasminsky, M. and Sears, T. A. 1972. Internodal conduction in undissected demyelinated nerve fibres. *J. Physiol. 227*: 323–350. (13)

Rauschecker, J. P. and Singer, W. 1980. Changes in the circuitry of the kitten visual cortex are gated by postsynaptic activity. *Nature 280*: 58–60. (20)

Rawlins, F. 1973. A time-sequence autoradiographic study of the in vivo incorporation of [1,2-^{3}H] cholesterol into peripheral nerve myelin. *J. Cell Biol. 58*: 42–53. (13)

Redfern, P. A. 1970. Neuromuscular transmission in new-born rats. *J. Physiol. 209*: 701–709. (19)

Reese, T. S. and Karnovsky, M. J. 1967. Fine structural localization of a blood-brain barrier to exogenous peroxidase. *J. Cell Biol. 34*: 207–217. (14)

Richardson, P. M., McGuiness, U. M. and Aguayo, A. J. 1980. Axons from CNS neurones regenerate into PNS grafts. *Nature 284*: 264–265. (19)

Riesen, A. H. and Aarons, L. 1959. Visual movement and intensity discrimination in cats after early deprivation of pattern vision. *J. Comp. Physiol. Psychol. 52*: 142–149. (20)

Ritchie, J. M. 1982. On the relation between fibre diameter and conduction velocity in myelinated nerve fibres. *Proc. R. Soc. Lond. B 217*: 29–35. (7)

Ritchie, J. M. and Rogart, R. B. 1977. The binding of saxitoxin and tetrodotoxin to excitable tissue. *Rev. Physiol. Biochem. Pharmacol. 79*: 1–50. (6)

Roberts, A. and Bush, B. M. H. 1971. Coxal muscle receptors in the crab: The receptor current and some properties of the receptor nerve fibres. *J. Exp. Biol. 54*: 515–524. (15)

Robertson, J. D. 1963. The occurrence of a subunit pattern in the unit membranes of club endings in Mauthner cell synapses in goldfish brains. *J. Cell Biol. 19*: 201–221. (16)

Robertson, J. D., Bodenheimer, T. S. and Stage, D. E. 1963. The ultrastructure of Mauthner cell synapses and nodes in goldfish brains. *J. Cell Biol. 19*: 159–199. (16)

Rodieck, R. W. 1973. *The Vertebrate Retina: Principles of Structure and Function.* W. H. Freeman, San Francisco. (2)

Rodieck, R. W. 1979. Visual pathways. *Annu. Rev. Neurosci. 2:* 193–225. (2)

Rogers, E. M. 1960. *Physics for the Enquiring Mind.* Princeton University Press, Princeton. (Appendix A)

Rotshenker, S. 1979. Synapse formation in intact innervated cutaneous-pectoris muscles of the frog following denervation of the opposite muscle. *J. Physiol. 292:* 535–547. (19)

Rovainen, C. M. 1967. Physiological and anatomical studies on large neurons of central nervous system of the sea lamprey (*Petromyzon marinus*). II. Dorsal cells and giant interneurons. *J. Neurophysiol. 30:* 1024–1042. (9)

Rudomín, P., Engberg, I. and Jiménez, I. 1981. Mechanisms involved in presynaptic depolarization of group I and rubrospinal fibers in cat spinal cord. *J. Neurophysiol. 46:* 532–548. (16)

Rushton, W. A. H. 1951. A theory of the effects of fibre size in medullated nerve. *J. Physiol. 115:* 101–122. (7)

Russell, J. M. 1983. Cation-coupled chloride influx in squid axon. Role of potassium and stoichiometry of the transport process. *J. Gen. Physiol. 81:* 909–925. (8)

Rutishauser, U., Grumet, M. and Edelman, G. M. 1983. Neural cell adhesion molecule mediates initial interactions between spinal cord neurons and muscle cells in culture. *J. Cell Biol. 97:* 145–152. (Part 5, 19)

Rutishauser, U., Hoffman, S. and Edelman, G. M. 1982. Binding properties of a cell adhesion molecule from neural tissue. *Proc. Natl. Acad. Sci. USA 79:* 685–689. (19)

Salem, R. D., Hammerschlag, R., Bracho, H. and Orkand, R. K. 1975. Influence of potassium ions on accumulation and metabolism of [^{14}C]-glucose by glial cells. *Brain Res. 86:* 499–503. (13)

Sanes, J. R. 1983. Roles of extracellular matrix in neural development. *Annu. Rev. Physiol. 45:* 581–600. (19)

Sanes, J. R. and Hall, Z. W. 1979. Antibodies that bind specifically to synaptic sites on muscle fiber basal lamina. *J. Cell Biol. 83:* 357–370. (19)

Sanes, J. R., Marshall, L. M. and McMahan, U. J. 1978. Reinnervation of muscle fiber basal lamina after removal of muscle fibers. *J. Cell Biol. 78:* 176–198. (19)

Sargent, P. B., Yau, K.-W. and Nicholls, J. G. 1977. Synthesis of acetylcholine by excitatory motoneurons in central nervous system of the leech. *J. Neurophysiol. 40:* 453–460. (18)

Schacher, S., Kandel, E. R. and Woolley, R. 1979. Development of neurons in the abdominal ganglion of *Aplysia californica.* I. Axosomatic synaptic contacts. *Devel. Biol. 71:* 163–175. (18)

Schachner, M. 1982. Cell type-specific antigens in the mammalian nervous system. *J. Neurochem. 39:* 1–8. (13)

Schachner, M., Sommer, I., Lagenaur, C. and Schnitzer, J. 1982. Developmental expression of antigenic markers in glial subclasses. In T. A. Sears (ed.). *Neuronal-Glial Cell Interrelationships.* Springer-Verlag, New York, pp. 321–336. (13)

Scharrer, E. 1944. The blood vessels of the nervous tissue. *Q. Rev. Biol. 19:* 308–318. (14)

Schiller, P. H. and Malpeli, J. G. 1977. Properties and tectal projections of monkey retinal ganglion cells. *J. Neurophysiol. 40:* 428–445. (2)

Schiller, P. H. and Malpeli, J. G. 1978. Functional specificity of lateral geniculate nucleus laminae of the rhesis monkey. *J. Neurophysiol. 41:* 788–797. (2)

Schmidt, J. T., Cicerone, C. M. and Easter, S. S. 1978. Expansion of the half retinal projection to the tectum in goldfish: An electrophysiological and anatomical study. *J. Comp. Neurol. 177*: 257–278. (19)

Schmidt, R. F. 1971. Presynaptic inhibition in the vertebrate central nervous system. *Ergeb. Physiol. 63*: 20–101. (9)

Schmidt, R. H., Björklund, A. and Stenevi, U. 1981. Intracerebral grafting of dissociated CNS tissue suspensions: A new approach for neuronal transplantation to deep brain sites. *Brain Res. 218*: 347–356. (19)

Schon, F. and Kelly, J. S. 1974. Autoradiographic localization of [^{3}H] GABA and [^{3}H] glutamate over satellite glial cells. *Brain Res. 66*: 275–288. (13)

Schultzberg, M., Hökfelt, T. and Lundberg, J. M. 1982. Coexistence of classical neurotransmitters and peptides in the central and peripheral nervous systems. *British Med. Bull. 38*: 309–313. (12)

Schuurmans Stekhoven, F. and Bonting, S. L. 1981. Transport adenosine triphosphatases: Properties and functions. *Physiol. Rev. 61*: 1–76. (8)

Schwartz, J. H. 1979. Axonal transport: Components, mechanisms, and specificity. *Annu. Rev. Neurosci. 2*: 467–504. (12)

Schwartz, J. H. 1982. Chemical basis of synaptic transmission (10). Biochemical control mechanisms in synaptic transmission (11). In E. R. Kandel and J. H. Schwartz (eds.). *Principles of Neuroscience*. Elsevier, New York, Chapters 10 and 11. (12)

Schwarz, W. and Passow, H. 1983. Ca^{2+}-activated K^+ channels in erythrocytes and excitable cells. *Annu. Rev. Physiol. 45*: 359–374. (6)

Scott, S. A. 1975. Persistence of foreign innervation on reinnervated goldfish extraocular muscles. *Science 189*: 644–646. (19)

Scott, S. A. and Muller, K. J. 1980. Synapse regeneration and signals for directed growth in the central nervous system of the leech. *Dev. Biol. 80*: 345–363. (19)

Sears, T. A. 1964a. Efferent discharges in alpha and fusimotor fibers of intercostal nerves of the cat. *J. Physiol. 174*: 295–315. (15, 17)

Sears, T. A. 1964b. The slow potentials of thoracic respiratory motoneurones and their relation to breathing. *J. Physiol. 175*: 404–424. (17)

Sears, T. A. 1982. *Neuronal-Glial Cell Interrelationships*. Springer-Verlag, New York. (13)

Shapovalov, A. I. and Shiriaev, B. I. 1980. Dual mode of junctional transmission at synapses between single primary afferent fibres and motoneurones in the amphibian. *J. Physiol. 306*: 1–15. (9)

Shatz, C. J. 1977. Anatomy of interhemispheric connections in the visual system of Boston Siamese and ordinary cats. *J. Comp. Neurol. 173*: 497–518. (2)

Shatz, C. J. 1979. Abnormal connections in the visual system of Siamese cats. *Soc. Neurosci. Symp. 4*: 121–141. (20)

Shatz, C. J. 1983. The prenatal development of the cat's retinogeniculate pathway. *J. Neurosci. 3*: 482–499. (20)

Shatz, C. J. and Kliot, M. J. 1982. Prenatal misrouting of the retinogeniculate pathway in Siamese cats. *Nature 300*: 525–529. (20)

Shatz, C. J., Lindström, S. and Wiesel, T. N. 1977. The distribution of afferents representing the right and left eyes in the cat's visual cortex. *Brain Res. 131*: 103–116. (3)

Shatz, C. J. and Stryker, M. P. 1978. Ocular dominance in layer IV of the cat's visual cortex and the effects of monocular deprivation. *J. Physiol. 281*: 267–283. (20)

Shepherd, G. M. 1978. Microcircuits in the nervous system. *Sci. Am. 238*: 92–103. (16)

Shepherd, G. M. 1983. *Neurobiology.* Oxford University Press, New York, pp. 134–147. (12, 16, 17, 18)

Sheridan, J. D. 1978. Junctional formation and experimental modification. In J. Feldman, N. B. Gilula and J. D. Pitts (eds.). *Intercellular Junctions and Synapses.* Chapman and Hall, London, pp. 39–59. (9)

Sherk, H. and Stryker, M. P. 1976. Quantitative study of cortical orientation selectivity in visually inexperienced kitten. *J. Neurophysiol. 39*: 63–70. (20)

Sherman, S. M. and Spear, P. D. 1982. Organization of visual pathways in normal and visually deprived cats. *Physiol. Rev. 62*: 738–855. (20)

Sherman, S. M. and Stone, J. 1973. Physiological normality of the retina in visually deprived cats. *Brain Res. 60*: 224–230. (20)

Sherrington, C. S. 1906. *The Integrative Action of the Nervous System.* Yale University Press, New Haven. (1961 Edn.) (2, 4, 17)

Sherrington, C. S. 1933. *The Brain and its Mechanism.* Cambridge University Press, London. (2, 4)

Sherrington, C. S. 1951. *Man on His Nature.* Cambridge University Press, Cambridge, England. (2)

Shik, M. L. and Orlovsky, G. N. 1976. Neurophysiology of locomotor automatism. *Physiol. Rev. 56*: 465–501. (17)

Shik, M. L., Severin, F. V. and Orlovsky, G. N. 1966. Control of walking and running by means of electrical stimulation of the mid-brain. *Biofizika 11*: 659–666 (English trans. pp. 756–765). (17)

Shkolnik, L. J. and Schwartz, J. H. 1980. Genesis and maturation of serotonergic vesicles in identified giant cerebral neuron of *Aplysia. J. Neurophysiol. 43*: 945–967. (12)

Sidman, R. L. and Rakič, P. 1973. Neuronal migration with special reference to developing human brain: A review. *Brain Res. 62*: 1–35. (13)

Siegler, M. V. S and Burrows, M. 1979. The morphology of local nonspiking interneurons in the metathoracic ganglia of the locust. *J. Comp. Neurol 183*: 121–148. (18)

Sigworth, F. J. and Neher, E. 1980. Single Na^+ channel currents observed in cultured rat muscle cells. *Nature 287*: 447–449. (6)

Sillito, A. M. 1979. Inhibitory mechanisms influencing complex cell orientation selectivity and their modification at high resting discharge levels. *J. Physiol. 289*: 33–53. (12)

Simmons, P. A., Lemmon, V. and Pearlman, A. L. 1982. Afferent and efferent connections of the striate and extrastriate visual cortex of the normal and reeler mouse. *J. Comp. Neurol. 211*: 295–308. (20)

Simons, D. J. and Woolsey, T. A. 1979. Functional organization in mouse barrel cortex. *Brain Res. 165*: 327–332. (3, 17)

Skene, J. H. P. and Shooter, E. M. 1983. Denervated sheath cells secrete a new protein after nerve injury. *Proc. Natl. Acad. Sci. USA 80*: 4169–4173. (13)

Skene, J. H. P. and Willard, M. 1981. Characteristics of growth-associated polypeptides in regenerating toad retinal ganglion cell axons. *J. Neurosci. 1*: 419–426. (19)

Skou, J. C. 1957. The influence of some cations on an adenosine triphosphatase from peripheral nerves. *Biochim. Biophys. Acta 23*: 394–401. (8)

Skou, J. C. 1964. Enzymatic aspects of active linked transport of Na^+ and K^+ through the cell membrane. *Prog. Biophys. Mol. Biol. 14*: 133–166. (8)

Sokoloff, L. 1977. Relation between physiological function and energy metabolism in the central nervous system. *J. Neurochem. 29*: 13–26. (3)

Sokolove, P. G. and Cooke, I. M. 1971. Inhibition of impulse activity in a sensory neuron by an electrogenic pump. *J. Gen. Physiol.* *57*: 125–163. (15)

Somjen, G. G. 1979. Extracellular potassium in the mammalian central nervous system. *Annu. Rev. Physiol.* *41*: 159–177. (13)

Sotelo, C., Llinás, R., and Baker, R. 1974. Structural study of inferior olivary nucleus of the cat: Morphological correlates of electrotonic coupling. *J. Neurophysiol.* *37*: 541–559. (9)

Sotelo, C. and Taxi, J. 1970. Ultrastructural aspects of electrotonic junctions in the spinal cord of the frog. *Brain Res.* *17*: 137–141. (9)

Specht, S. and Grafstein, B. 1973. Accumulation of radioactive protein in mouse cerebral cortex after injection of ^{3}H-fucose into the eye. *Exp. Neurol.* *41*: 705–722. (3)

Sperry, R. W. 1944. Optic nerve regeneration with return of vision in anurans. *J. Neurophysiol.* *7*: 57–69. (19)

Sperry, R. W. 1945. Restoration of vision after crossing of optic nerves and after contralateral transplantation of eye. *J. Neurophysiol.* *8*: 15–28. (19)

Sperry, R. W. 1970. Perception in the absence of neocortical commissures. *Proc. Res. Assoc. Nerv. Ment. Dis.* *48*: 123–138. (2)

Spitzer, N. C. 1982. Voltage- and stage-dependent uncoupling of Rohon-Beard neurones during embryonic development of *Xenopus* tadpoles. *J. Physiol.* *330*: 145–162. (9, Part 5)

Stent, G. S. 1980. Thinking about seeing: A new approach to visual perception. In *The Sciences* (May/June issue), The New York Academy of Sciences, pp. 6–11. (2)

Stent, G. S. 1981a. Cerebral hermeneutics. *J. Social Biol. Struc.* *4*: 107–124. (2)

Stent, G. S. 1981b. Strength and weakness of the genetic approach to the development of the nervous systems. *Annu. Rev. Neurosci.* *4*: 163–194. (Part 5)

Stent, G. S., Kristan, W. B., Friesen, W. O., Ort, C. A., Poon, M. and Calabrese, R. L. 1978. Neuronal generation of the leech swimming movement. *Science* *200*: 1348–1357. (18)

Stent, G. S. and Kristan, W. B. 1981. Neural circuits generating rhythmic movements. In K. J. Muller, J. G. Nicholls and G. S. Stent (eds.). *Neurobiology of the Leech*. Cold Spring Harbor Laboratory, Cold Spring Harbor, NY, pp. 113–146. (17, 18)

Stent, G. S. and Weisblat, D. 1982. The development of a simple nervous system. *Sci. Am.* *246*: 136–146. (18, 19)

Sterling, P. 1983. Microcircuitry of the cat retina. *Annu. Rev. Neurosci.* *6*: 149–185. (2, 12)

Steward, O., Cotman, C. W. and Lynch, G. S. 1973. Reestablishment of electrophysiologically functional entorhinal cortical input to the dentate gyrus deafferented by ipsilateral entorhinal lesions: Innervation by the contralateral entorhinal cortex. *Exp. Brain Res.* *18*: 396–414. (19)

Stone, L. S. and Zaur, I. S. 1940. Reimplantation and transplantation of adult eyes in the salamander (*Triturus viridescens*) with return of vision. *J. Exp. Zool.* *85*: 243–269. (19)

Strumwasser, F. 1971. The cellular basis of behavior in *Aplysia*. *J. Psychiat. Res.* *8*: 237–257. (18)

Stryker, M. P. 1981. Late segregation of geniculate afferents to the cat's visual cortex after recovery from binocular impulse blockade. *Soc. Neurosci. Abstr.* *7*: 842. (20)

Susz, J. P., Haber, B. and Roberts, E. 1966. Purification and some properties of mouse brain L-glutamic decarboxylase. *Biochemistry* *5*: 2870–2877. (12)

Sutter, A., Riopelle, R. J., Harris-Warrick, R. M. and Shooter, E. M. 1979. Nerve growth factor receptors characterization of two distinct classes of binding sites on chick embryo sensory ganglia cells. *J. Biol. Chem.* *254*: 5972–5982. (19)

Svaetichin, G. 1953. The cone action potential. *Acta Physiol. Scand.* *29*: 565–600. (2)

Syková, E. 1981. K$^+$ changes in the extracellular space of the spinal cord and their physiological role. *J. Exp. Biol.* *95*: 93–109. (13)

Syková, E., Shirayev, B., Kriz, N. and Vycklický L. 1976. Accumulation of extracellular potassium in the spinal cord of frog. *Brain Res.* *106*: 413–417. (13)

Sypert, G. W. and Munson, J. B. 1981. Basis of segmental motor control: Motoneuron size or motor unit type? *Neurosurgery 8*: 608–621. (16)

Szentágothai, J. 1973. Neuronal and synaptic architecture of the lateral geniculate nucleus. In H. H. Kornhuker (ed.). *Handbook of Sensory Physiology*, Vol. VI, *Central Visual Information.* Springer-Verlag, Berlin, pp. 141–176. (2)

Takato, M. and Goldring, S. 1979. Intracellular marking with Lucifer Yellow CH and horseradish peroxidase of cells electrophysiologically characterized as glia in the cerebral cortex of the cat. *J. Comp. Neurol.* *186*: 173–188. (13)

Takeuchi, A. 1977. Junctional transmission. I. Postsynaptic mechanisms. In E. Kandel (ed.). *Handbook of the Nervous System*, Vol. 1. American Physiological Society, Baltimore, pp. 295–327. (9)

Takeuchi, A. and Takeuchi, N. 1960. On the permeability of the end-plate membrane during the action of transmitter. *J. Physiol.* *154*: 52–67. (9)

Takeuchi, A. and Takeuchi, N. 1965. Localized action of gamma-aminobutyric acid on the crayfish muscle. *J. Physiol.* *177*: 225–238. (16)

Takeuchi, A. and Takeuchi, N. 1966. On the permeability of the presynaptic terminal of the crayfish neuromuscular junction during synaptic inhibition and the action of γ-aminobutyric acid. *J. Physiol.* *183*: 433–449. (9, 12)

Takeuchi, A. and Takeuchi, N. 1967. Anion permeability of the inhibitory postsynaptic membrane of the crayfish neuromuscular junction. *J. Physiol.* *191*: 575–590. (9, 12)

Takeuchi, A. and Takeuchi, N. 1969. A study of the action of picrotoxin on the inhibitory neuromuscular junction of the crayfish. *J. Physiol.* *205*: 377–391. (12)

Talbot, S. A. and Marshall, W. H. 1941. Physiological studies on neural mechanisms of visual localization and discrimination. *Am. J. Ophthalmol.* *24*: 1255–1264. (2)

Tang, C.-M., Cohen, M. W. and Orkand, R. K. 1980. Electrogenic pumps in axons and neuroglia and extracellular potassium homeostasis. *Brain Res.* *194*: 283–286. (13)

Tang, C.-M., Strichartz, G. R. and Orkand, R. K. 1979. Sodium channels in axons and glial cells of the optic nerve of *Necturus maculosa*. *J. Gen. Physiol.* *74*: 629–642. (13)

Tanji, J. and Evarts, E. V. 1976. Anticipatory activity of motor cortex neurons in relation to the direction of an intended movement. *J. Neurophysiol.* *39*: 1062–1068. (17)

Tao-Cheng, J.-H., Hirosawa, K. and Nakajima, Y. 1981. Ultrastructure of the crayfish stretch receptor in relation to its function. *J. Comp. Neurol.* *200*: 1–21. (15)

Tasaki, I. 1959. Conduction of the nerve impulse. In J. Field (ed.). *Handbook of Physiology*, Section 1, Vol. I., Chapter III. American Physiological Society, Bethesda, pp. 75–121. (7)

Tauc, L. 1967. Transmission in invertebrate and vertebrate ganglia. *Physiol. Rev.* 47: 521–593. (18)

Tauc, L. 1982. Nonvesicular release of neurotransmitter. *Physiol. Rev.* 62: 857–893. (11)

Taylor, A. and Prochazka, A. 1981. *Muscle Receptors and Movement.* Macmillan, London. (15)

Teschemacher, H., Ophein, K. E., Cox, B. M. and Goldstein, A. 1975. A peptide-like substance from pituitary that acts like morphine. *Life Sci.* 16: 1771–1776. (12)

Thesleff, S. 1960. Supersensitivity of skeletal muscle produced by botulinum toxin. *J. Physiol.* 151: 598–607. (19)

The Synapse. 1976. *Cold Spring Harbor Symp. Quant. Biol.,* Vol. 40. (9)

Thoenen, H. and Barde, Y.-A. 1980. Physiology of nerve growth factor. *Physiol. Rev.* 60: 1284–1335. (19)

Thoenen, H., Barde, Y.-A. and Edgar, D. 1982. In J. G. Nicholls (ed.). *Repair and Regeneration of the Nervous System.* Springer-Verlag, New York, pp. 173–185. (13, 19)

Thomas, R. C. 1969. Membrane current and intracellular sodium changes in a snail neurone during extrusion of injected sodium. *J. Physiol.* 201: 495–514. (8)

Thomas, R. C. 1972a. Electrogenic sodium pump in nerve and muscle cells. *Physiol. Rev.* 52: 563–594. (8)

Thomas, R. C. 1972b. Intracellular sodium activity and the sodium pump in snail neurones. *J. Physiol.* 220: 55–71. (8)

Thomas, R. C. 1974. Intracellular pH of snail neurones with a new pH-sensitive glass micro-electrode. *J. Physiol.* 238: 159–810. (8)

Thomas, R. C. 1977. The role of bicarbonate, chloride and sodium ions in the regulation of intracellular pH in snail neurones. *J. Physiol.* 273: 317–338. (8)

Thompson, W. J. and Stent, G. S. 1976. Neuronal control of heartbeat in the medicinal leech. I. Generation of the vascular constriction rhythm by heart motor neurons. *J. Comp. Physiol.* 111: 309–333. (18)

Thompson, W. 1983. Synapse elimination in neonatal rat muscle is sensitive to pattern of muscle use. *Nature* 302: 614–616. (19)

Tomita, T. 1965. Electrophysiological study of the mechanisms subserving color coding in the fish retina. *Cold Spring Harbor Symp. Quant. Biol.* 30: 559–566. (2)

Torebjörk, H. E. and Ochoa, J. 1980. Specific sensations evoked by activity in single identified sensory units in man. *Acta Physiol. Scand.* 110: 445–447. (17)

Towe, A. L. Somatosensory cortex: Descending influences on ascending systems. 1973. In A. Iggo (ed.). *Handbook of Sensory Physiology, Somatosensory System.* Vol. II. Springer-Verlag, New York, pp. 700–718. (15)

Tsien, R. W. 1983. Calcium channels in excitable cells. *Annu. Rev. Physiol.* 45: 341–358. (6)

Tsukahara, N. 1981. Synaptic plasticity in the mammalian central nervous system. *Annu. Rev. Neurosci.* 4: 351–379. (19)

Unwin, P. N. T. and Zampighi, G. 1980. Structure of the junction between communicating cells. *Nature* 283: 545–549. (9)

Vallbo, Å. B., Hagbarth, K.-E., Torebjörk, H. E. and Wallin, B. G. 1979. Somatosensory, proprioceptive and sympathetic activity in human peripheral nerves. *Physiol. Rev.* 59: 919–957. (15)

Van der Loos, H. and Woolsey, T. A. 1973. Somatosensory cortex: Structural alterations following early injury to sense organs. *Science* 179: 395–398. (20)

Van Deures, B. 1980. Structural aspects of brain barriers, with special reference to the permeability of the cerebral endothelium and choroidal epithelium. *Int. Rev. Cytol.* 65: 117–191. (14)

Van Essen, D. C. 1979. Visual areas of the mammalian cerebral cortex. *Annu. Rev. Neurosci.* 2: 227–263. (2, 3)

Van Essen, D. C. 1982. Neuromuscular synapse elimination. In N. C. Spitzer (ed.). *Neuronal Development.* Plenum, London, pp. 333–376. (19)

Van Essen, D. and Jansen, J. K. 1974. Reinnervation of rat diaphragm during perfusion with α-bungarotoxin. *Acta Physiol. Scand.* 91: 571–573. (19)

Van Essen, D. C. and Zeki, S. M. 1978. The topographic organization of rhesus monkey prestriate cortex. *J. Physiol.* 277: 193–226. (3)

Van Sluyters, R. C. and Levitt, F. B. 1980. Experimental strabismus in the kitten. *J. Neurophysiol.* 43: 686–699. (20)

Venosa, R. A. and Horowicz, P. 1981. Density and apparent location of the sodium pump in frog sartorius muscles. *J. Memb. Biol.* 59: 225–232. (8)

Vera, C. L., Vial, J. D. and Luco, J. V. 1957. Reinnervation of nictitating membrane of cat by cholinergic fibers. *J. Neurophysiol.* 20: 365–373. (19)

Verveen, A. A. and DeFelice, L. J. 1974. Membrane noise. *Prog. Biophys. Mol. Biol.* 28: 189–265. (11)

Vigh, B., Vigh-Teichmann, I., Koritsánszky, S. and Aros, B. 1970. Ultrastruktur der Liquorkontaktneurone des Ruckenmarkes von Reptilien. *Z. Zellforsch.* 109: 180–194. (14)

Villegas, J. 1981. Schwann-cell relationships in the giant nerve fibre of the squid. *J. Exp. Biol.* 95: 135–151. (13)

Virchow, R. 1859. *Cellularpathologie.* (F. Chance, trans.) Hirschwald, Berlin. Excerpts are from pp. 310, 315, and 317. (13)

Vizi, S. E. and Vyskočil, F. 1979. Changes in total and quantal release of acetylcholine in the mouse diaphragm during activation and inhibition of membrane ATPase. *J. Physiol.* 286: 1–14. (10)

von Euler, C. 1977. The functional organization of the respiratory phase-switching mechanisms. *Fed. Proc.* 36: 2375–2380. (17)

Wachtel, H. and Kandel, E. R. 1971. Conversion of synaptic excitation to inhibition at a dual chemical synapse. *J. Neurophysiol.* 34: 56–68. (9, 18)

Wallace, B. G., Adal, M. and Nicholls, J. G. 1977. Regeneration of synaptic connexions of sensory neurones in leech ganglia in culture. *Proc. R. Soc. Lond. B* 199: 567–585. (18)

Wässle, H., Peichl, L. and Boycott, B. B. 1981. Morphology and topography of on- and off-alpha cells in the cat retina. *Proc. R. Soc. Lond. B* 212: 157–175. (2)

Weight, F. and Votava, J. 1970. Slow synaptic excitation in sympathetic ganglion cells: Evidence for synaptic inactivation of potassium conductance. *Science* 170: 755–758. (9)

Weisblat, D. 1981. Development of the nervous system. In K. J. Muller, J. G. Nicholls and G. S. Stent (eds.). *Neurobiology of the Leech.* Cold Spring Harbor Laboratory, Cold Spring Harbor, NY, pp. 173–195. (18)

Welker, C. 1976. Microelectrode delineation of fine grain somatotopic organization of Sm1 cerebral neocortex in albino rat. *J. Comp. Neurol.* 166: 173–190. (17)

Wernig, A. 1972. Changes in statistical parameters during facilitation at the crayfish neuromuscular junction. *J. Physiol.* 226: 751–759. (10)

Westbury, D. R. 1982. A comparison of the structures of α- and γ-spinal motoneurones of the cat. *J. Physiol.* 325: 79–91. (15)

Whittaker, V. B., Essman, W. B. and Dowe, G. H. C. 1972. The isolation of pure cholinergic synaptic vesicles from the electric organs of elasmobranch fish of the family *Torpedinidae*. *Biochem. J.* 128: 833–846. (11)

Wiersma, C. A. G. 1947. Giant nerve fiber system of the crayfish. A contribution to comparative physiology of synapse. *J. Neurophysiol.* 10: 23–38. (18)

Wiersma, C. A. G. and Ripley, S. H. 1952. Innervation patterns of crustacean limbs. *Physiol. Comp. Oecol.* 2: 391–405. (12, 16)

Wiesel, T. N. 1982. The postnatal development of the visual cortex and the influence of environment. *Nature* 299: 583–591. (20)

Wiesel, T. N. and Hubel, D. H. 1963a. Effects of visual deprivation on morphology and physiology of cells in the cat's lateral geniculate body. *J. Neurophysiol.* 26: 978–993. (20)

Wiesel, T. N. and Hubel, D. H. 1963b. Single-cell responses in striate cortex of kittens deprived of vision in one eye. *J. Neurophysiol.* 26: 1003–1017. (20)

Wiesel, T. N. and Hubel, D. H. 1965a. Comparison of the effects of unilateral and bilateral eye closure on cortical unit responses in kittens. *J. Neurophysiol.* 28: 1029–1040. (20)

Wiesel, T. N. and Hubel, D. H. 1965b. Extent of recovery from the effects of visual deprivation in kittens. *J. Neurophysiol.* 28: 1060–1072. (20)

Wiesel, T. N. and Hubel, D. H. 1974. Ordered arrangement of orientation columns in monkeys lacking visual experience. *J. Comp. Neurol.* 158: 307–318. (3, 20)

Willard, A. 1981. Effects of serotonin on the generation of the motor program for swimming by the medicinal leech. *J. Neurosci.* 1: 936–944. (18)

Williams, P. L. and Warwick, R. 1975. *Functional Neuroanatomy of Man.* Saunders, Philadelphia. (1, Appendix B)

Wine, J. J. and Krasne, F. 1982. The cellular organization of crayfish escape behavior. In D. C. Sandeman and H. L. Atwood (eds.). *The Biology of Crustacea, Vol. 4. Neural Integration and Behavior.* Academic, New York, pp. 241–292. (18)

Wong-Riley, M. 1979. Changes in the visual system of monocularly sutured or enucleated cats demonstrable with cytochrome oxidase histochemistry. *Brain Res. 171*: 11–28. (3)

Woolsey, T. A. and Van der Loos, H. 1970. The structural organization of layer IV in the somatosensory region (S I) of mouse cerebral cortex: The description of a cortical field composed of discrete cytoarchitectonic units. *Brain Res.* 17: 205–242. (17)

Wright, E. M. 1978. Transport processes in the formation of the cerebrospinal fluid. *Rev. Physiol. Biochem. Exp. Pharmacol.* 83: 1–34. (14)

Yankner, B. A. and Shooter, E. M. 1982. The biology and mechanism of action of nerve growth factor. *Annu. Rev. Biochem.* 51: 845–868. (19)

Yau, K.-W. 1976. Receptive fields, geometry and conduction block of sensory neurones in the central nervous system of the leech. *J. Physiol.* 263: 513–538. (8, 18)

Yoon, M. 1972. Transposition of the visual projection from the nasal hemi-retina onto the foreign rostral zone of the optic tectum in goldfish. *Exp. Neurol.* 37: 451–462. (19)

Yoshikami, S., George, J. S. and Hagins, W. A. 1980. Light-induced calcium fluxes from outer segment layer of vertebrate retinas. *Nature* 286: 395–398. (2)

Young, J. Z. 1936. The giant nerve fibres and epistellar body of cephalopods. *Q. J. Microsc. Sci.* 78: 367–386. (5)

Zeki, S. M. 1978. The third visual complex of rhesus monkey prestriate cortex. *J. Physiol. 277*: 245–272. (2)

Zeki, S. M. 1980. The response properties of cells in the middle temporal area (area MT) of owl monkey visual cortex. *Proc. R. Soc. Lond. B 207*: 239–248. (3)

Zeuthen, T. and Wright, E. M. 1981. Epithelial potassium transport: Tracer and electrophysiological studies in choroid plexus. *J. Membr. Biol. 60*: 105–128. (14)

Zimmerman, H. 1979. Vesicle recycling and transmitter release. *Neuroscience 4*: 1773–1804. (11)

Zipser, B. and McKay, R. 1981. Monoclonal antibodies distinguish identifiable neurones in the leech. *Nature 289*: 549–554. (18)

Zucker, R. S. 1973. Changes in the statistics of transmitter release during facilitation. *J. Physiol. 229*: 787–810. (10)

INDEX

ABOUT THE BOOK

This book is set in Palatino, a face designed by the contemporary German typographer Hermann Zapf. Inspired by the typography of the Italian Renaissance, Palatino letters derive their elegance from the natural motion of the edged pen. Palatino is a highly readable face especially suited for extended use in books.

Drawings for the book were prepared by Laszlo Meszoly, who worked in close collaboration with the authors. Many of the charts and graphs were rendered by Vantage Art, Inc. and Fredric J. Schoenborn. Jodi Simpson was copyeditor.

Type was set by DEKR Corporation, and the book was manufactured at The Murray Printing Company.

Joseph Vesely supervised production.